CONTROLS ON CARBONATE PLATFORM AND BASIN DEVELOPMENT

Based on a Symposium

Sponsored by the Society of
Economic Paleontologists and Mineralogists

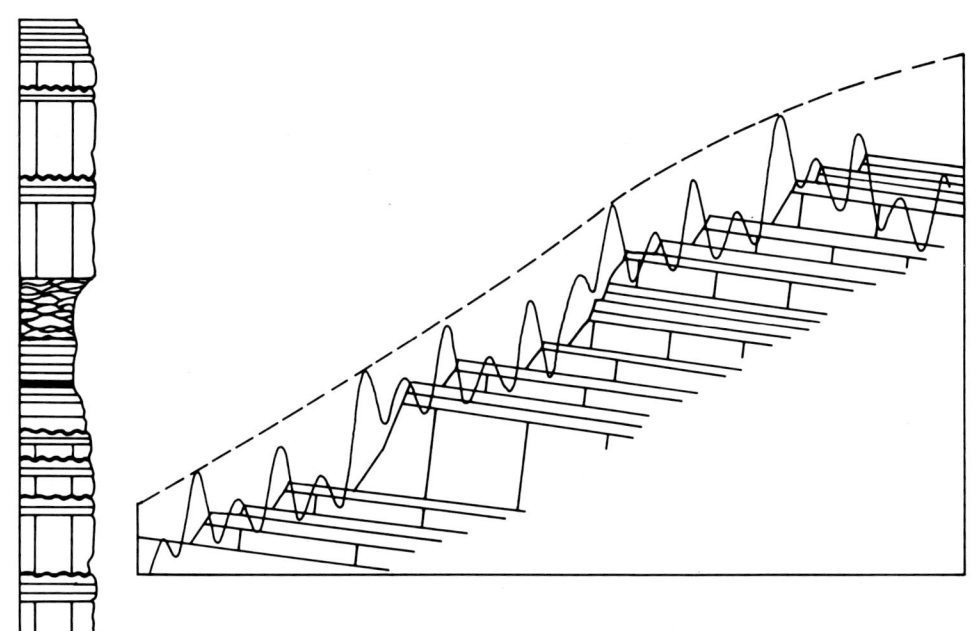

Edited by

Paul D. Crevello, Marathon Oil Company, Littleton, Colorado;
James L. Wilson, New Braunfels, Texas;
J. Frederick Sarg, Exxon Production Research, Houston, Texas;
and
J. Fred Read, Virginia Tech, Blacksburg, Virginia

Copyright © 1989 by
SOCIETY OF ECONOMIC PALEONTOLOGISTS AND MINERALOGISTS

Barbara H. Lidz, Editor of Special Publications
Special Publication No. 44

Tulsa, Oklahoma, U.S.A.

July 1989

A Publication of
The Society of Economic Paleontologists And Mineralogists

ISBN # 0-918985-79-X

© 1989 by
Society of Economic Paleontologists and Mineralogists
P.O. Box 4756
Tulsa, Oklahoma 74131

Printed in the United States of America

PREFACE

This volume, and a succeeding group of papers soon to be published by the International Association of Sedimentologists, are derived from the SEPM Symposium entitled *Controls on Carbonate Platform and Basin Development*, formulated by the co-editors for the Los Angeles meetings of AAPG and SEPM in 1987. The multinational Symposium was the outgrowth of an international Penrose Conference held on the Isle of Capri (Bay of Naples) in 1981 and a subsequent field conference held in the Guadalupe and Sacramento Mountains of New Mexico and West Texas in 1983. Initiation of the subject and the inspiration for the first Conference was from Professor Bruno D'Argenio of the Geological Institute of the University of Naples.

The present volume covers many subjects relative to geology of carbonate platforms and adjoining slopes and basins. A preliminary section, based on principles of deposition and computer modeling studies, is followed by a group of about a dozen papers devoted to examples of carbonate platforms on passive cratonal margins resulting from rifting. The volume also considers halos of carbonate developed as a fringe around the pericratonic Permian basin, as well as some examples of isolated offshore platforms. Some of the examples are from major platform systems along Paleozoic miogeoclines around North America and the Mesozoic of Tethys. Other studies are of local and individual platform-basin systems. Examples of both ramps and rimmed platforms are included. The case histories presented span the whole of geologic time from early in the Proterozoic to the Holocene. Four papers discuss Cenozoic platforms, and about eleven concern Mesozoic and Paleozoic strata. The Cenozoic and Holocene examples offer opportunities to understand the influence of modern oceanographic factors in shaping the platforms.

The papers furnish important data relative to carbonate facies and platform construction. Five papers deal extensively with seismic and sequence stratigraphy. The description and origin of sedimentary cycles at various scales are discussed in several of the papers. The common occurrence of shoaling-upward sequences in platform interiors is documented thoroughly. The papers also provide information on rates of subsidence. The difficult problem of estimating amount of sea-level drops is addressed by some authors.

The formulators of the 1987 Symposium, this compilation, and its successor volume are very grateful to the authors for their enterprise in the undertaking and their patience in reworking parts of some papers. We also appreciate the considerable interest by the SEPM-AAPG membership, whose large attendance at both the earlier conferences, and the 1987 Symposium and poster sessions, indicates the importance of the subject to petroleum exploration and development.

Paul D. Crevello
James L. Wilson
J. Frederick Sarg
J. Fred Read

CONTENTS

PREFACE

I. PRINCIPLES OF PLATFORM DEVELOPMENT AND COMPUTER MODELING OF PLATFORM EVOLUTION

1. DYNAMICS OF TETHYAN CARBONATE PLATFORMS . . . . *Alfonso Bosellini* 3
2. DROWNING UNCONFORMITIES ON CARBONATE PLATFORMS . . . . *Wolfgang Schlager* 15
3. QUANTITATIVE MODELING OF CARBONATE PLATFORMS: SOME EXAMPLES . . . . *Thomas Aigner, Mark Doyle, David Lawrence, Manfred Epting, and Arthur Van Vliet* 27
4. ROLE OF THERMAL SUBSIDENCE, FLEXURE, AND EUSTASY IN EVOLUTION OF EARLY PALEOZOIC PASSIVE-MARGIN CARBONATE PLATFORMS . . . . *Gerard Bond, Michelle A. Kominz, Michael S. Steckler and John P. Grotzinger* 39
5. JUDY CREEK: A CASE STUDY FOR A TWO-DIMENSIONAL SEDIMENT DEPOSITION SIMULATION . . . . *David M. Scaturo, John S. Strobel, Christopher G. St. C. Kendall, Jack C. Wendte, Gautam Biswas, James Bezdek, and Robert Cannon* 63

II. EXAMPLES OF CARBONATE RIMMED PLATFORMS AND RAMPS DEVELOPED ON PASSIVE (RIFTED) CRATONAL MARGINS

6. FACIES AND EVOLUTION OF PRECAMBRIAN CARBONATE DEPOSITIONAL SYSTEMS: EMERGENCE OF THE MODERN PLATFORM ARCHETYPE . . . . *John P. Grotzinger* 79
7. TECTONIC CONTROL ON THE FORMATION OF A CARBONATE PLATFORM: THE CAMBRIAN OF SOUTHWESTERN SARDINIA . . . . *Thilo Bechstädt and Maria Boni* 107
8. EVOLUTION OF A LOWER PALEOZOIC CONTINENTAL-MARGIN CARBONATE PLATFORM, NORTHERN CANADIAN APPALACHIANS . . . . *Noel P. James, Robert K. Stevens, Christopher R. Barnes, and Ian Knight* 123
9. CONTROLS ON EVOLUTION OF CAMBRIAN-ORDOVICIAN PASSIVE MARGIN, U.S. APPALACHIANS . . . . *J. F. Read* 147
10. ARCHITECTURE AND EVOLUTION OF A WHITEROCKIAN (EARLY MIDDLE ORDOVICIAN) CARBONATE PLATFORM, BASIN RANGES OF WESTERN U.S.A. . . . . *Reuben J. Ross, Jr., Noel P. James, Lehi F. Hintze, and Forrest G. Poole* 167
11. REEFAL PLATFORM DEVELOPMENT, DEVONIAN OF THE CANNING BASIN, WESTERN AUSTRALIA . . . . *Phillip E. Playford, Neil F. Hurley, Charles Kerans, and Michael F. Middleton* 187
12. SEDIMENTARY AND TECTONIC CONTROLS ON THE DEVELOPMENT OF AN EARLY MISSISSIPPIAN CARBONATE RAMP, SACRAMENTO MOUNTAINS AREA, NEW MEXICO . . . . *Wayne M. Ahr* 203
13. SILICICLASTIC INFLUENCE ON MESOZOIC PLATFORM DEVELOPMENT: BALTIMORE CANYON TROUGH, WESTERN ATLANTIC . . . . *Franz O. Meyer* 213
14. THE EVOLUTION OF THE CARBONATE PLATFORMS OF NORTHEAST AUSTRALIA . . . . *Peter J. Davies, Philip A. Symonds, David A. Feary, and Christopher J. Pigram* 233
15. RECENT CARBONATE SLOPE SEDIMENTS AND SEDIMENTARY PROCESSES BORDERING A NON-RIMMED PLATFORM: SOUTHWEST FLORIDA CONTINENTAL MARGIN . . . . *Gregg R. Brooks and Charles W. Holmes* 259

III. EXAMPLES FROM THE PERMIAN BASIN CARBONATE FRINGE AND SLOPE

16. SLOPE SEDIMENTATION ASSOCIATED WITH A VERTICALLY BUILDING SHELF, BONE SPRING FORMATION, MESCALERO ESCARPE FIELD, SOUTHEASTERN NEW MEXICO . . . . *Arthur H. Saller, Jane W. Barton, and Ricky E. Barton* 275
17. EVOLUTION AND DESTRUCTION OF A CARBONATE BANK AT THE SHELF MARGIN: GRAYBURG FORMATION (PERMIAN), WESTERN ESCARPMENT, GUADALUPE MOUNTAINS, TEXAS . . . . *Evan K. Franseen, Thomas E. Fekete, and Lloyd C. Pray* 289
18. LOWER PERMIAN PLATFORM AND BASIN DEPOSITIONAL SYSTEMS, NORTHERN MIDLAND BASIN, TEXAS . . . . *S. J. Mazzullo and A. M. Reid* 305

IV. PINNACLE REEFS–ISOLATED OFFSHORE BANKS ON PASSIVE MARGINS

19. EUSTATIC CONTROLS ON THE STRATIGRAPHY AND GEOMETRY OF THE LATEMAR BUILDUP (MIDDLE TRIASSIC), THE DOLOMITES OF NORTHERN ITALY . . . . *Robert K. Goldhammer and Mark T. Harris* 323
20. CENOZOIC PROGRADATION OF NORTHWESTERN GREAT BAHAMA BANK, A RECORD OF LATERAL PLATFORM GROWTH AND SEA-LEVEL FLUCTUATIONS . . . . *Gregor P. Eberli and Robert N. Ginsburg* 339

21. PLATFORM EVOLUTION AND SEQUENCE STRATIGRAPHY OF THE NATUNA PLATFORM, SOUTH CHINA SEA .... *Kurt W. Rudolph and Patrick J. Lehmann*   353

V. BANK DEVELOPMENT IN A FORELAND BASIN AND PELAGIC SEDIMENTATION IN AN ACTIVE-MARGIN BASIN

22. UPPER CRETACEOUS PLATFORM-TO-BASIN DEPOSITIONAL-SEQUENCE DEVELOPMENT, TREMP BASIN, SOUTH-CENTRAL PYRENEES, SPAIN .... *Antonio Simo*   365

23. SYNSEDIMENTARY TECTONICS IN THE LATE CRETACEOUS-EARLY TERTIARY PELAGIC BASIN OF THE NORTHERN APENNINES, ITALY .... *Alessandro Montanari, Lung S. Chan, and Walter Alvarez*   379

SUBJECT INDEX

PART I
PRINCIPLES OF PLATFORM DEVELOPMENT AND COMPUTER MODELING OF PLATFORM EVOLUTION

DYNAMICS OF TETHYAN CARBONATE PLATFORMS

ALFONSO BOSELLINI

Istituto di Geologia, Università di Ferrara, 44100 Ferrara, Italy

ABSTRACT: Tethyan carbonate platforms are relatively short-lived depositional systems that were born, developed, died, and were resurrected in a tectonically active area. In fact, they experienced the entire geodynamic spectrum (rifting, drifting, transtension, transpression, and collision) of the Wilson cycle. Many of their peculiar features, such as lack of ramps and common occurrence of faulted boundaries and megabreccia wedges, suggest tectonics as a primary control, with eustatic sea-level oscillations being only a secondary overprint.

During the rifting stages, many platforms drowned or underwent tectonic retreat, while others were able to survive the pulsating Liassic subsidence. Shedding of ooid sands during sea-level highstands, and subaerial exposure with karstification and bauxite formation during lowstands, are some typical events that occurred during the Middle Jurassic-Middle Cretaceous period of drifting and convergence. Beginning from Late Jurassic, the Tethyan platforms were progressively involved in the Alpine-Himalayan orogenic systems, and, after an early uplift, were generally buried under thick piles of siliciclastics. Some, however, were resurrected before the final collisional stages.

INTRODUCTION

The Tethyan Carbonate Platforms and their Plate-Tectonic Setting

The concept of Tethys, as understood by its originator Suess (1893), was of an east-west seaway lying north of Gondwanaland and across the area that now constitutes the Alpine-Mediterranean mountain chains and the Himalayas. In the 1970s, it was widely recognized that the present Alpine-Himalayan system of orogenic belts is the product of the closure of two separate Tethyan oceans, the *Paleotethys* and *Neotethys* (Stöcklin, 1974; Laubscher and Bernoulli, 1977; Jenkyns, 1980; Sengör and others, 1984).

The Tethyan seaway proper (Neotethys) was the smaller Jurassic ocean, or a complex of oceans, generated in the Alpine-Mediterranean region synchronously with the separation of Africa from North America, and between the laterally rifting continents of Africa and Europe. During the rifting and drifting stages, the different basins of the Tethyan ocean have been visualized as pertaining to a zone of small ocean basins of the size of the present Red Sea or Gulf of California (Bosellini and Winterer, 1975; Bosellini, 1981; Kelts, 1981; Weissert and Bernoulli, 1985).

The reorganization of plate boundaries and of the global spreading-ridge system in the Middle and Late Cretaceous, and the opening of the North and South Atlantic, initiated the elimination of the Tethyan ocean. Sinistral movements in the eastern Tethys were replaced by dextral ones and, combined with an increased north-south component of Africa with respect to Eurasia, led to the destruction of Tethys.

The geodynamic evolution of the Alpine-Mediterranean region, however, also implies important crustal deformation after the closure of the oceanic areas and after the major continental collisions (Tapponier, 1977). This deformation continued to affect the mobile areas within an ensialic regime and produced orogenic belts such as the Apennines, the Maghrebides of North Africa and Sicily, and the Hellenic Arc (Boccaletti and others, 1982).

The passive (rifted and sheared) continental margins of the Tethyan seaways, and especially the southern ones, were the sites of extensive colonization by carbonate platforms that developed and flourished from Late Triassic to Oligo-Miocene times. They therefore experienced the entire geodynamic spectrum of rifting, transtension, transpression, collision, and ensialic deformation; they were relatively short-lived depositional systems, which were born, lived, died, and were resurrected under predominantly tectonic control.

In contrast with pericratonic platforms (Cook and Taylor, 1977; Read, 1980, 1982; Jansa, 1981), which are areally widespread, show broad facies belts, very gentle slopes (ramps), and transitional boundaries, Tethyan carbonate platforms were smaller and had steep, often faulted boundaries. I interpret here that the overall control exerted on these two contrasting systems is the geodynamic milieu: relatively stable in the pericratonic case, where eustatic oscillations and long-term subsidence are probably the main controls, and quite dynamic in the Tethyan case.

The primary focus of this paper is to describe the fundamental steps of the contrasted and active life of these Tethyan platforms. This paper represents the writer's view of the Mesozoic-Cenozoic carbonate geology of the Tethyan realm and is founded on the following basic assumptions.

Some Thoughts and Assumptions on Carbonate Platform Evolution

The gross architecture, i.e., size and geometry, of carbonate platforms is mainly controlled by relative changes in sea level due primarily to tectonics and eustasy (Kendall and Schlager, 1981). On the contrary, their internal facies association is a function of a number of factors, including bathymetric profile, shelf width, leeward or windward position, currents, storms, and climate (Wilson, 1975; Read, 1985; Tucker, 1985). Also, the particular type of carbonate-producing benthos, which had changed considerably from the Mesozoic to Cenozoic, had an impact on slopes and margins of carbonate platforms.

Because of their peculiar origin, carbonate platforms may be considered ecosystems whose population has a biologic productivity that, during climax conditions, largely exceeds the receptivity of the environment. Recent shallow-water carbonates and reefs show an astounding capability to grow upward and outward and to keep pace with the rapid Holocene rise of sea level.

A platform with a sediment-covered flat top that simultaneously builds upward and progrades basinward must produce more sediment than it needs to match the relative rise of sea level. Consequently, its growth potential must be greater than its vertical growth rate. As prograding platforms are known practically throughout the entire Phanerozoic (for example, Newell and others, 1953; Playford, 1980; Purdy, 1980; Hurst and Surlyk, 1983; Bosellini, 1984; Mitchum and Uliana, 1985; Simò, 1986; Bosellini and others, 1987; Eberli and Ginsburg, 1987; but see also Schlager, 1981, fig. 4 and table 2), it seems reasonable to infer that growth potential of reefs and platforms was also more than sufficient to keep pace with rise of relative sea level in the geologic past.

The presumption is made that with relative sea level rising at a rate less than 400–500 m/ma[1] (but with favorable ecological conditions, i.e., warm and clear sea water), carbonate platforms should generally expand laterally while growing upward. An exception to this may occur at the beginning of a new cycle, when the rate of relative sea-level rise is very high, and the platform may aggrade without substantial lateral accretion.

From my experience with Tethyan platforms, I conclude that whenever the pattern of gradual aggradation and progradation is disturbed, an exceptional event, such as tectonic pulse, subaerial exposure, or environmental deterioration (water pollution, cooling), has to be invoked.

DYNAMICS OF TETHYAN CARBONATE PLATFORMS

In this section, growth and evolution of the Tethyan carbonate platforms are metaphorically discussed using biologic terms. In fact, a carbonate platform can be envisaged as a biologically controlled ecosystem depending largely on a healthy carbonate-producing benthos.

Platform Initiation (Birth)

At the end of Triassic time (Norian-Rhaetian), the margins of the western Tethys were dominated by few widespread facies belts (Bosellini and Hsü, 1973), including:

(a) a continental facies (red beds), representing semi-arid regions with extensive fans dissected by braided streams, alluvial plains with meandering rivers, and interior depressions in which clayey and silty sediments accumulated; prevalently denuded areas, covered with thin pediment deposits, occurred interspersed in this vast terrestrial domain (European Hercynian massifs, the Sahara and Arabian shields);

(b) a transitional facies (marls, shales, dolostones, gypsum, anhydrite, and halite), representing coastal plains, playa flats, coastal sabkhas, and restricted lagoons;

(c) a shallow-water carbonate facies (thick peritidal successions of dolostones and limestones), representing a vast shelf of tidal flats and lagoons, incised by intrashelf, elongated and restricted basins, where organic-rich anoxic carbonates and shales accumulated;

(d) a deeper water facies (cherty radiolarian mudstones, in places crowded with the thin-shelled pelecypod *Halobia*, and scattered occurrences of red nodular pelagic limestone), representing deposition in deep and narrow straits, sounds, and oceanic tongues, segmenting the outer parts of the paleo-Tethyan margins.

The Tethyan carbonate platforms were initiated by flooding of these Late Triassic pericratonic margins (Fig. 1) over any type of substratum (e.g., igneous or metamorphic basement, alluvial sediments, evaporites, or shallow-water carbonates). Paleomorphology dictated initial sedimentation: peritidal muds on lowlands (coastal plains, tidal flats, and lagoons), oolite sands on irregular topography (moderate paleorelief, inselbergs); but, in general, the Tethyan platforms evolved typically as peritidal cyclic successions until Sinemurian-Early Pliensbachian (Carixian) time. Afterward, they began to be segmented into smaller blocks, and shelf-edge facies (oolite and skeletal sands) became quite common.

The Liassic flooding of the Triassic pericratonic margins was the result of the rifting processes, which affected the entire western Tethys, from the Mediterranean region to Middle East, Oman, Somalia, and Madagascar.

The Atlantic continental margins (Scotian Shelf and Georges Banks), where the Paleozoic basement is overlain by Triassic red beds and evaporites, were also transgressed by an open sea in Sinemurian time, and shallow-water carbonates were deposited (Jansa, 1981).

Platform Development (Growth and Life)

During their evolution, which lasted, in some cases, until the Late Tertiary, the Tethyan platforms developed a number of characteristic features, including the following.

Cyclic stratigraphy.—

The bulk of the kilometers-thick Mesozoic platforms consists of a repetitive succession of shallowing-upward

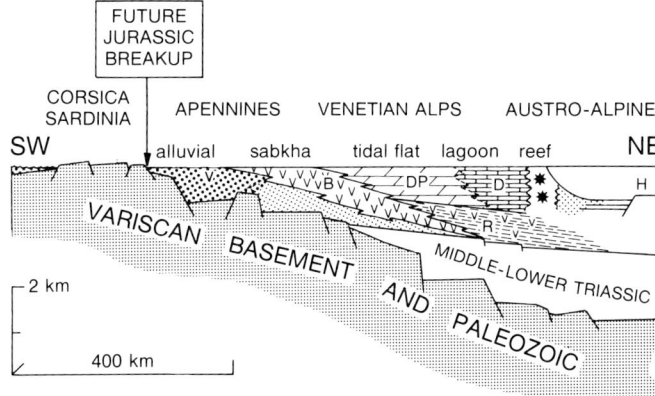

FIG. 1.—A restored Late Triassic pericratonic margin across the present Tyrrhenian-Alpine area; the paleo-Tethyan ocean was to the east. Flooding and segmentation of these margins created the bewildering number of Tethyan platforms. V (Verrucano), red beds; B (Burano Anhydrite); R (Raibl Formation), evaporite, lagoon, and shallow-water sediments; DP (Dolomia Principale), peritidal dolomite; D (Dachsteinkalk), peritidal limestone; H (Hallstatt Kalk), basinal and condensed limestone.

---

[1]Davies and Hopley (1983) report growth rates of 8–10 m/1,000 years for the Holocene reefs in the Great Barrier Reef, and the Triassic platforms of the Dolomites (Bosellini, 1984) were able to prograde while growing upward at a rate of 200–250 m/million years.

cycles, which, in the rugged peri-Mediterranean mountain chains, appear as spectacular layer-cake stacks of meter-scale strata (Fig. 2). This internal geometry, best developed in the Late Triassic and Cretaceous platform interiors, can be explained according to two types of models: (1) some form of autocyclicity involving progradation (lateral shifting of bars and tidal flats) and (2) some form of small-scale eustasy based on Milankovitch cycles (Hardie and others, 1986). The fact that this cyclicity does not appear to be well defined in the Jurassic successions could be related to an attenuation of the high-frequency sea-level oscillations or to dominantly tectonic control on sedimentation (the Jurassic was a time of major rifting and drowning for the Tethyan carbonate platforms).

Progradation.—

Carbonate platforms are not only large sediment accumulations but also prolific sediment factories that commonly produce more material than can be accommodated on their flat top (Droxler and Schlager, 1985). The excess sediment is exported to the adjacent basin areas, generating a halo of periplatform sediments and allowing the platform to expand laterally, i.e., to prograde.

When the exported material consists mainly of periplatform ooze, skeletal or oolitic sand, and minor clasts, clinoform declivity is usually less than 20°. Part of the sand may be redeposited as a turbidite wedge at the base of slope, in a basinal setting. Because much platform-derived sediment is characterized by an unstable mineralogy (aragonite, Mg-calcite), however, abundant and widespread cemented surfaces (Schlager and James, 1978; Hine and others, 1981) may act to stabilize the slope.

This kind of *physiologic progradation,* displayed by many Jurassic and Cretaceous Tethyan platforms, is the result of redeposition of mainly loose carbonate mud and sand previously accumulated at the platform edge, and it is typically associated with sea-level highstands (Schlager and Chermak, 1979; Hine and others, 1981; Droxler and Schlager, 1985).

Progradation may also result from repetitive pathologic events, i.e., erosion and partial dismantling of the platform margin. It is prevalent during sea-level lowstands. Actually, the prograding system, made largely of coarse debris (megabreccias), is an advancing submarine talus with an angle of repose as high as 30°–35°. The scanty turbidites deposited at the base of the talus are also made of clasts of already cemented carbonate sediment. To enable the system to prograde, a large number of repetitive lowstand events is necessary. Tectonic pulses are too rare, but high-frequency eustasy can account for the observed pattern. This *pathological progradation* is typical in the Triassic and, to a lesser extent, in the Cretaceous, times in which Milankovitch sea-level fluctuations have been shown to affect the sedimentary pattern considerably (Fischer, 1964, 1982; Hardie and others, 1986).

It has been observed that, at the beginning of a new cycle, platforms tend to grow mainly upward (fast rise of relative sea level); only later do they begin to prograde decisively, reaching the highest progradation rate at the end of the cycle (rate of rising relative sea level decreasing exponentially). The line of contact between the flat-lying beds of the platform interior and the clinoforms is a curve (*offlap curve*), which, in many cases, can be considered an inverted subsidence curve (Fig. 3).

The rate and style of progradation can also be a function of windward versus leeward position of the platform margin (Hine and Neumann, 1977; Hine and others, 1981; Hubbard and others, 1986). If this control is fairly well documented in the case of Recent sand and mud offbank transport, and possibly also in some Jurassic Tethyan platforms (Bosellini and others, 1981b), however, very little is known about coarse rubble and megabreccia deposits. The progradation patterns of the Triassic platforms of the Dolomites show clear asymmetry in most cases, but so far there is no independent control on whether the asymmetry is related to the windward versus leeward position.

Tectonic retreat and lateral grafting.—

As stated earlier, the Tethyan carbonate platforms also developed and grew with stationary margins. This pattern has been observed only where the platform margin coincided with a structural boundary, i.e., growth faults and tectonic scarps delimiting strongly subsiding adjacent areas. The writer has never observed retrogradation, i.e., a gradual and continuous "sedimentary" retreat of the platforms. *Tectonic retreat* has been observed and is a phenomenon well documented during the Jurassic rifting stages (Bernoulli, 1964; Castellarin, 1972; Bosellini, 1973; Bernoulli

FIG. 2.—Platform interiors are typically represented by thick successions of meter-scale cycles. The photograph shows the Upper Triassic-Lower Jurassic platform of the Tofane Group near Cortina, in the Dolomites. Thickness of the succession is about 900 m.

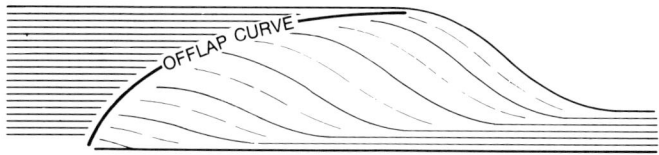

FIG. 3.—During the early stages of platform evolution, the line of separation between the flat-lying beds of the platform interior and the inclined slope deposits is steep; the same line flattens toward the top. The "angle of offlapping" appears to be a measure of the rate of relative sea-level rise and quite often is a measure of subsidence rate.

and Jenkyns, 1974; Colacicchi and others, 1975; D'Argenio, 1976; Castellarin and others, 1978; Catalano and D'Argenio, 1978, among others). The platforms retreated by successive tectonic collapses, becoming progressively narrower, while huge masses of megabreccias were being released at the foot of the tectonic scarps (Fig. 4). Where the platform sectors submerged only a little, say 200–400 m, skeletal and oolitic sands could reconquer the shallow-water realm by prograding and lateral filling of the subsided block. Shoaling-upward sequences, 100–300 m thick (Fig. 5), produced by these peripheral recoveries are common at the margins of many Tethyan platforms (Morocco, Alps, Greece; Barbujani and others, 1986; Masetti and Bianchin, 1987).

Catastrophic slope failures, producing large truncation surfaces, also forced the margins to retreat, and huge megabreccia debris flows are related to these scalloped margins (Castellarin and others, 1978). The Gargano Peninsula, which is part of the Apulia Platform (D'Argenio and others, 1973) of southern Italy, is a case in point. The stratigraphic framework of the eastern part of the peninsula is largely the result of catastrophic events, which produced the collapse of large portions of the platform margin in Early Turonian and Early Eocene times, and of successive lateral graftings and progradations of Late Cretaceous and Eocene platforms (Fig. 6). With regard to the size of slide scars and the unconformity geometries, a striking similarity exists with those recently recognized on the west Florida carbonate platform margin (see fig. 4 in Mullins and others, 1986).

The Turonian collapse was clearly related to a pronounced lowstand (Haq and others, 1987), whose effects are well documented all over the southern Apennines. A general emersion surface with associated bauxitic horizons (Crescenti and Vighi, 1970; Istituto di Geologia e Geofisica Università di Napoli, 1978) occurs in all the Cretaceous platforms. Clear evidence of an abrupt lowering of relative sea level is also documented in the Lower Eocene (Sgrosso, 1968; Selli, 1971; Istituto di Geologia e Geofisica Università di Napoli, 1978). Possible causes of these failures and of the pathologic retreat of the platform margins include tectonic collapses (Castellarin, 1972; Castellarin and others, 1978, among others), earthquake shocks (Cook and others, 1972; Mutti and others, 1984; Marjanac, 1985, among others) and local sediment overloading and gravitational instability (Crevello and Schlager, 1980; Mullins and others, 1986).

FIG. 5.—A shallowing- (and thickening) upward sequence produced by progradation over a collapsed, peripheral block of the platform (Upper Liassic, M. Baldo, northern Italy).

It is important to point out, however, that after each emersion, another platform grafted onto the older dead one, quickly healing the scar and expanding the area of shallow-water sedimentation. In conclusion, the carbonate succession of the Gargano Peninsula is the result of the superposition of at least three different platforms grafted on top of one another (Fig. 6).

The grafting of carbonate platforms onto older ones is a common phenomenon throughout the Phanerozoic, which has, so far, been underestimated in literature. It produces superimposed edifices or *palimpsest platforms,* which can be erroneously interpreted as continuous and uniform carbonate successions. Spectacular examples occur in the Triassic of the Dolomites, where the Carnian platforms mantle the underlying Ladinian platforms that had been exposed during the eustatic lowstand at the Ladinian-Carnian boundary (Biddle, 1984). Another case has been recently documented in the Oligo-Miocene succession of the Somali continental margin (Bosellini and others, 1987), and similar examples can probably be envisaged in well-known carbonate platforms, such as the Great Bahama Bank (Eberli and Ginsburg, 1987) and the Maldives (Purdy, 1980).

Oolite shedding.—

A distinctive event of the Tethyan paleoceanography is the export of oolitic sand into the adjacent basins during Mid-Jurassic time. Actually, the reposited sediments consists of peloidal-oolitic grainstone with minor amounts of skeletal debris. These gravity-displaced deposits generally form wedge-shaped slope and base-of-slope aprons and can reach thicknesses of as much as several hundred meters (800

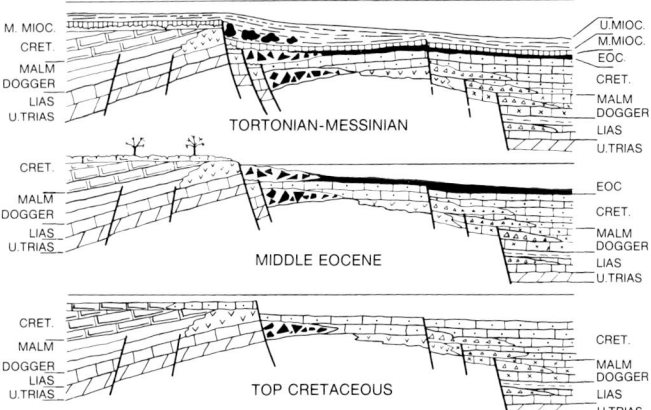

FIG. 4.—The Late Cretaceous to Late Miocene evolution of the Marsica Platform margin in the central Apennines (from Colacicchi and others, 1975): a classic scheme of platform retreat produced by successive tectonic collapses; megabreccias are indicated by large black spots.

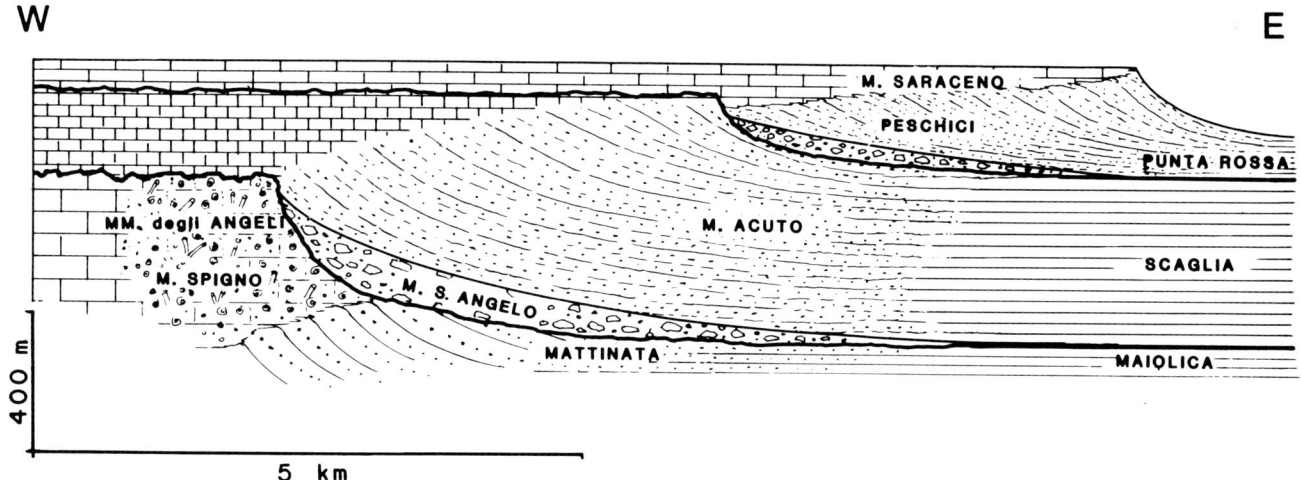

FIG. 6.—Stratigraphic section across the southern part of the Gargano Peninsula, southern Italy. Two catastrophic events, in Early Turonian and Early Eocene times, respectively, produced the collapse of large portions of the platform margin. As a result of successive lateral graftings and progradations, the Gargano platform appears now as a composite edifice, a palimpsest of three different platforms grafted and superimposed on top of one another.

m in the Belluno Trough of northern Italy; 500 m in the Budva Basin of Jugoslavia, near Titograd). They have been documented in many Tethyan continental margins, from the Atlantic (Nova Scotia, Portugal) through the Mediterranean (Morocco, southern Spain, Balearic Islands, Alps, Apennines, Jugoslavia, Greece), Oman, and Somalia (Bosellini and others, 1981a, 1981b; Ruiz-Ortiz, 1983; Searle and others, 1983; Barnolas-Cortinas, 1984; Wright and Wilson, 1984; Watts and Garrison, 1986; Bernoulli and Weissert, 1987). The Mid-Jurassic deep-water oolitic deposits, which typically overlie Toarcian black shales (Jenkins, 1985; Jenkyns and Clayton, 1986) and are sealed by pelagic shale and radiolarite blankets, appear to be attractive exploration targets all along the Tethyan passive margins.

The major problem is to understand why the Tethyan platforms exported oolitic sediments in Mid-Jurassic time. Clearly, the occurrence of redeposited oolites of the same age, from Nova Scotia to Oman (at least as known so far), rules out local factors and demands general, probably worldwide causes.

The Dogger was a time of prolific oolite formation on many carbonate platforms around the world, and it was the breakup time, in large part, of the Tethyan realm, an event probably preceded by accelerated subsidence in many carbonate platforms and associated with the Bajocian-Bathonian eustatic highstand (Haq and others, 1987). Off-platform transport of sediment was therefore strongly enhanced during this period.

Platform Termination (Death)

The various Tethyan carbonate platforms happened to be interrupted at different stages of their evolution, between the Early Jurassic and Mid-Tertiary. They "died" owing to a number of phenomena and circumstances, including drowning, subaerial exposure, terrigenous influx and burial, and hypothermia.

Drowning.—

Shallow-water carbonates are sensitive indicators of light and temperature. The light dependence of carbonate production limits the growth environment of platforms to the euphotic zone, approximately the uppermost 75–100 m of the water column. Submergence below this zone will end carbonate-secreting benthos production and will transform the living platform into a plateau or guyot.

In optimal conditions for carbonate growth, drowning requires such a rapid relative rise of sea level that few processes can be called upon; they include a tectonic collapse, i.e., a spasmodic pulse of subsidence (Bosellini, 1973), glacio-eustasy, and desiccation of an ocean basin (Schlager, 1981). If environmental conditions are in stress for carbonate production (e.g., pronounced anoxia, changes in water temperature or salinity, or terrigenous pollution), even a normal, steady subsidence can cause drowning (Grigg, 1982).

Two main types of drowning have been observed in the Tethyan platforms, namely a *starved drowning* and an *anoxic drowning*. The first type occurs when the submerged platform becomes a seamount, a guyot, or a deep-water plateau. The neritic deposits are abruptly and unconformably overlain by pelagic condensed sediments (Fig. 7). Hardgrounds with crusts of ferro-manganese oxide and red, nodular limestone rich in ammonites, the well-known Tethyan facies of the *Ammonitico rosso*, are typically developed (Jenkyns, 1971; Ogg, 1981).

The second type of drowning occurs when the subsiding platform becomes a topographic depression. This is accomplished through a relatively gradual increase in pelagic components, such as planktonic biota, clay, and chert, and contemporaneous decrease of bed thickness and texture. The overlying, thin-bedded, often laminated, pelagic carbonates are very rich in organic content and are dark gray or black in color.

The starved drowning was probably caused mainly by tectonic collapse plus minimal sediment input as a result of

FIG. 7.—Sharp, disconformable contact between Pliensbachian lagoonal limestone below, crowded with oyster-like *Lithiotis* lamellibranchs, and the overlying Bajocian red nodular limestone (Ammonitico rosso) in the central part of Trento Platform. The surface, indicated by the arrow, is a hardground with ferro-manganese encrustations and nodules, representing the physical expression of the drowning event; afterward, the platform became a pelagic plateau.

are interpreted to have caused a number of platforms to become subaerially exposed. These movements appear to be time-transgressive across an orogenic profile and occurred in different intervals of the Late Cretaceous-Miocene time span, according to the Tethyan sector involved. The case history of the southern Apennines will be briefly described.

The southern Apennines are the collided Mesozoic passive margin of the Adria Plate, where elongate carbonate platforms and intervening basins occurred grossly parallel to the margin (D'Argenio, 1976). Starting in the Early Miocene, these paleogeographic domains were successively involved, from west to east, in the orogenic "wave." First they were uplifted and the platforms subaerially exposed, then buried under huge thicknesses of turbidites and finally caught up into the overthrust pile. The easternmost carbonate platform, the Apulia Platform, which is now the relatively undeformed foreland, is completely emerged and separated from the Apennine orogen by a foreland basin (*Fossa Bradanica*), whose substratum is partly represented by the downfaulted western part of the Apulia Platform itself (Mostardini and Merlini, 1986).

It is here suggested that the geologic evolution described above is the isostatic effect produced by the advancing overthrust edifice (Fig. 9). As shown by Quinlan and Beaumont (1984), the response of the lithosphere to an applied load creates a downwarped flexural moat around the load, with this moat in turn surrounded by an upwarped peripheral bulge. As the overthrust edifice advances, both the bulge and the moat advance, and, like floating corks, the carbonate platforms undergo a "yo-yo" movement.

Hypothermia.—

The wandering lithospheric plates carry along the shallow-water carbonate platforms, which undergo profound effects during these movements. A particularly drastic environmental change occurs when platforms leave the low-latitude belt of coral reef growth (the *Darwin Point* of Grigg, 1982) and enter into temperate, subtropical waters. It does not seem, however, that the platforms with which we are dealing have been substantially affected by the relatively

dissolution and bypassing of sediments by currents (Jenkins, 1971; Winterer and Bosellini, 1981). Anoxic drowning is considered to be the result of the suppression of carbonate production by the unfavorable water conditions in the narrow and restricted Jurassic rift basins.

Tectonic "yo-yoing".—

Many Tethyan platforms (southern Apennines, eastern Alps, Jugoslavia, Greece, and so forth) were subaerially exposed and subsequently submerged and buried by deepwater clastics (flysch) in Tertiary time (Fig. 8). The emersion is documented by paleokarst, bauxite horizons, and pronounced unconformities (Aubouin, 1959; D'Argenio, 1970, 1976; D'Argenio and others, 1973; Blanchet, 1975; Instituto di Geologia e Geofisica Università di Napoli, 1978; Cousin, 1980, among others). Rapid tectonic movements

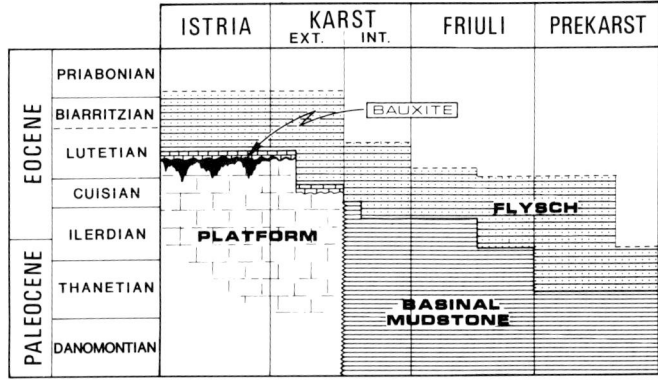

FIG. 8.—Stratigraphic relation in the Karst-Friuli region (Yugoslavia-Italy), according to Cousin (1980, simplified). The Karst platform appears subaerially exposed and karstified (bauxite) before its definitive burying below the advancing deep-water clastic wedge.

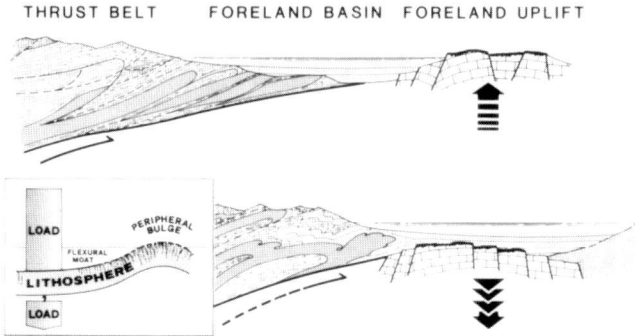

FIG. 9.—The isostatic effect produced by the progressive advance of the overthrust edifice forces foreland carbonate platforms to undergo an up-and-down movement. Insert from Quinlan and Beaumont (1984).

limited latitudinal plate movements in the Tethyan belt (except for India).

The carbonate platforms of the Mediterranean area suffered instead from the Tertiary cooling of the poles and the mid-latitudes that transformed the globally warm and equable Mesozoic climate into a colder regime with cyclic glaciations during the Pliocene to Holocene (Pantiĉ, 1986).

Evidence for these climatic effects is present in the Oligocene coral reefs of the Vicentin of northern Italy (Bosellini and Russo, 1988). Here, a quasi-monospecific community dominated by *Actinacis rollei,* an extinct poritid that probably occupied the same ecologic niche as present *Porites* or *Goniopora,* characterizes the reefs. The overlying marly and shaly Miocene sediments, although of shallow-water environment, document the cessation of carbonate platform sedimentation. The *Actinacis* "explosion" in the Early Oligocene has been interpreted (Bosellini and Russo, 1988) as an effect of the pronounced lowering of sea-water temperature documented at the Eocene-Oligocene boundary (Shackleton and Kennett, 1975; Keigwin, 1980; Vergnaud Grazzini and Oberhaensli, 1986).

In the central and southern Apennines, Lower Miocene organogenic sediments, known as "Bryozoa and Lithothamnium Limestones" (Barbera and others, 1978, 1980), uconformably overlie Late Cretaceous platform limestones and grade upward into pelagic lime mud and flysch. Lack of the most typical constituent of Recent tropical carbonates, such as aragonitic mud, peloids, ooids, green algae, and hermatypic corals, and the common occurrence of bryozoa and nodules of coralline algae (rhodolites) suggest sedimentation in a temperate carbonate shelf (Barbera and others, 1978, 1980; Carannante and others, 1981; Simone and Carannante, 1985). The Miocene carbonates are interpreted as deposits of open shelf circumlittoral areas, at depths exceeding 80 m, with water temperature similar to the present Atlantic and Mediterranean subtropical-temperate open-shelf areas (Foramol-type Association of Lees, 1975).

The terminal-Miocene (Tortonian-Messinian) low-diversity *Porites* reefs of the western Mediterranean also suggest the periodic invasion of supposedly colder Atlantic waters into a warmer and partly desiccated Mediterranean as a major factor controlling their distribution and development (Esteban, 1979). After the Burdigalian, the coral communities show a progressive reduction in diversity until their extinction in the Messinian.

In conclusion, the general cooling conditions, established through Tertiary time, seriously diminished carbonate productivity and the growth potential of the surviving Tethyan carbonate platforms: they died by hypothermia. Those of the Mediterranean region were particularly affected by the climatic change.

Re-initiation (Resurrection)

The Trento Platform of the southern Alps is a major structural and paleogeographic domain of the Adria Plate continental margin (Winterer and Bosellini, 1981). It was a region of shallow-water carbonate accumulation until Late Liassic time, when it was drowned to depths below the euphotic zone and became a pelagic plateau. The immediately overlying deposits are Bajocian to Kimmeridgian pelagic red nodular and cherty limestone (*Ammonitico rosso*; Fig. 7), which are, in turn, overlain by a succession of white (*Maiolica*) and reddish (*Scaglia rossa*) Early Cretaceous to Early Eocene deep-water sediments. The pelagic section capping the Trento Platform is 150–300 m thick and encompasses a time span of about 120 m.y.

The complex collision between Europe and the Adria Plate generated the Alpine Chain, and the resulting tectonic patterns were largely controlled by the inherited structural grain of the margins. The former Trento Platform reacted rigidly during the collision and was block-faulted. Deep-seated basic and ultrabasic volcanics poured out on its carapace, while transpressive wedges and flower structures developed along its western margin (Doglioni and Bosellini, 1987). The submerged Trento Platform was segmented into variously uplifted blocks and was punctuated by several volcanic piles. These "highs" (Figs. 10, 11) acted as centers of initiation of shallow-water carbonates (reefs, nummulitic banks; Luciani, 1988), which prograded and coalesced and gave rise to form the *Lessini Shelf,* the resurrected Trento Platform, scattered with reefs, lagoons, islands, and volcanoes (Fig. 12).

CONCLUSIONS: A GEODYNAMIC OVERVIEW

In relatively short time, the carbonate platforms of the Tethyan ocean passed through the entire geodynamic spectrum of the Wilson cycle: rifting, drifting, transtension, transpression, and collision. In contrast to pericratonic platforms, they are characterized by a lack of ramps, common occurrence of tectonic boundaries and megabreccia wedges, sudden drowning surfaces, and close relation with orogenic systems. Many of these peculiar features point to tectonics as a primary control on their evolution, with eustatic sea-level oscillations being only a secondary modulator.

Except for platforms located along sheared margins, whose evolution might have been erratic and without any polarity, a general evolutionary trend can be envisaged for the others (Fig. 13). The Tethyan platforms were born, mainly in Late Triassic to Early Liassic time, as widespread pericratonic shelves, after an important Middle Triassic tectonic and magmatic event, whose real geodynamic significance is still uncertain. The onset of rifting, linked to the incipient breakup

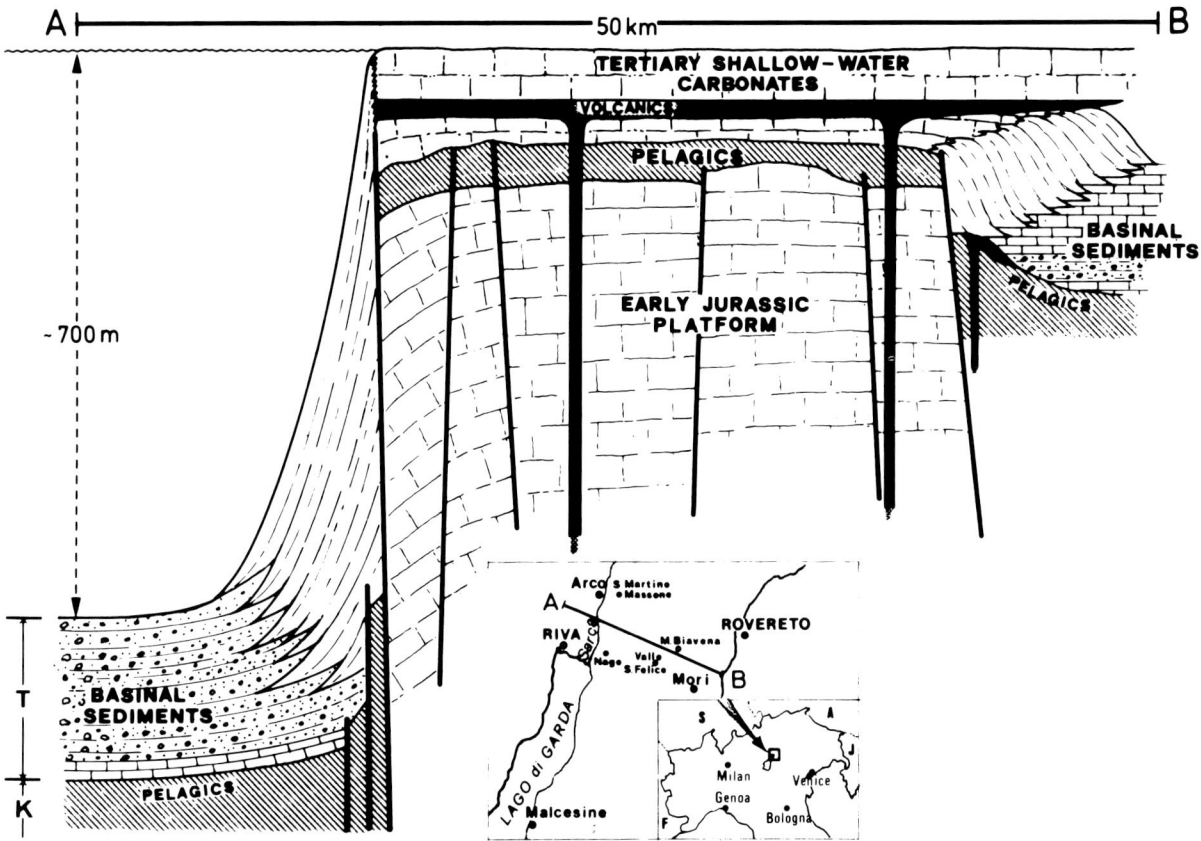

Fig. 10.—The block-faulted and uplifted horst of the Monte Baldo (a flower structure according to Doglioni and Bosellini, 1987), where Eocene shallow-water carbonate deposition initiated and, prograding westward, began to coalesce with other points of shallow-water carbonate radiation (from Luciani, 1988).

Fig. 11.—The Eocene shallow-water carbonates (SW) on top of the Late Jurassic-Cretaceous-Paleocene, pelagic succession (PS) (Doss D'Abramo, nea Rovereto, in the southern Alps). Thickness of the Eocene carbonates is about 100 m. The former drowned Liassic platform (LP) is also visible to the right.

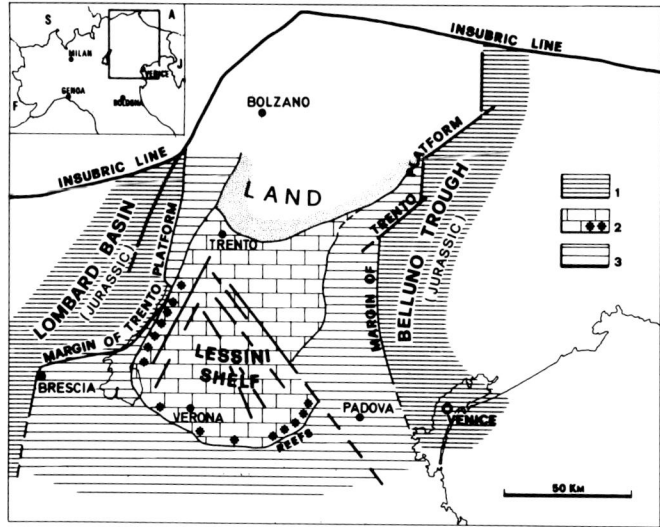

FIG. 12.—The Lessini Shelf, a fringing, protruding carbonate platform resurrected on the block-faulted and uplifted carapace of the Liassic Trento Platform, which was drowned in Dogger time. (1) Deep-water mudstone of the surrounding Jurassic basins (Lombard Basin and Belluno Trough); (2) Paleogene lagoon and barrier reefs of the Lessini Shelf; (3) Paleogene pelagic claystone and marlstone.

of Pangea (opening of the Central Atlantic, separation of Madagascar-India from Africa), segmented the Late Triassic shelves and gave rise to a large number of smaller platforms. During the rifting stage, which lasted until the Middle Jurassic (Dogger), many platforms experienced tectonic retreat or drowned, while others were able to survive the spasmodic Liassic subsidence and could continue to expand and grow upward. During sea-level highstands, some platforms could export sediment to the adjacent basins, while during lowstands, others were subaerially exposed and killed. Only by grafting onto the karstified surfaces were they able to continue building multistage edifices.

The first collisions (eastern Alps, Dinarids) started to occur toward the end of the Jurassic. The carbonate platforms were involved in orogenesis after a relatively quiet period of drifting apart and convergence: first they were uplifted in the foreland area and subsequently buried into the foredeep basins under huge piles of clastics. Some of these platforms were even resurrected for a while before they finally emerged from the Tethyan sea during the last collisional stages. Today, "the folded and crumpled" Tethyan platforms "stand forth to heaven in Thibet, Himalaya, and the Alps" (Suess, 1893).

ACKNOWLEDGMENTS

The concepts developed in this paper are the result of more than two decades of study of carbonate platform geology around the Mediterranean (Alps, Apennines, Jugoslavia, Greece, Morocco) and East Africa. Research was funded mainly by the Italian Corsiglio Nazionale delle Ricerche and Ministero della Pubblica Istruzione. I thank A. Bally, C. Doglioni, L. A. Hardie, D. Hubbard, J. Hurst, D. Masetti, C. Neri, M. Sarti, W. Schlager, and P. Vail for helpful discussions and suggestions. The manuscript has benefited greatly from critical reviews by D. Bernoulli, R. N. Ginsburg, H. Jenkyns, A. Montanari, R. Sarg, and W. Schlager. Technical assistance has been given by R. Brandoli and F. Nalin.

REFERENCES

AUBOUIN, J., 1959, Contribution à l'étude géologique de la Grèce septentrionale: Les confins de l'Epire et de la Thessalie: Annales Géologiques des Pays Helléniques, v. 10, 583 p.
BARBERA, C., CARANNANTE, G., D'ARGENIO, B., AND SIMONE, L., 1980, Il Miocene calcareo dell'Appennino meridionale: Contributo della paleoecologia alla costruzione di un modello ambientale: Annali Università di Ferrara, sez. IX, v. 6, p. 281–299.
―――, SIMONE, L., AND CARANNANTE, G., 1978, Depositi circalittorali di Piattaforma Aperta nel Miocene Campano. Analisi sedimentologica e paleoecologica: Bollettino Società Geologica Italiana, v. 97, p. 821–834.
BARBUJANI, C., BOSELLINI, A., AND SARTI, M., 1986, L'Oolite di S. Vigilio nel Monte Baldo (Giurassico, Prealpi Venete): Annali Università di Ferrara, sez. IX, v. 9, p. 19–47.
BARNOLAS-CORTINAS, A., ed., 1984, Sedimentologia del Jurasico de Mallorca: Excursion Guide Book, Palma de Mallorca, 263 p.
BERNOULLI, D., 1964, Zur Geologie des Monte Generoso (Lombardische Alpen): Beiträge zur Geologische Karte Schweiz, Neue Folge, Bern, 134 p.
―――, AND JENKYNS, H., 1974, Alpine, Mediterranean, and central Atlantic Mesozoic facies in relation to the early evolution of the Tethys, in Dott, R. H., and Shaver, R. H., eds., Modern and Ancient Geosynclinal Evolution: Society of Economic Paleontologists and Mineralogists Special Publication 19, p. 129–160.
―――, AND WEISSERT, H., 1987, The upper Hawasina nappes in the central Oman Mountains: Stratigraphy, palinspastics and sequence of nappe emplacement: Geodinamica Acta, v. 1, p. 47–58.
BIDDLE, K. T., 1984, Triassic sea level change and the Ladinian-Carnian stage boundary: Nature, v. 308, p. 631–633.
BLANCHET, R., 1975, De l'Adriatique au bassin pannonique; Essai d'un modèle de chaîne alpine: Mémoires Societé geologique de France, n. 120, 172 p.
BOCCALETTI, M., CONEDERA, L., AND DAINELLI, P., 1982, The Recent (Miocene-Quarternary) regmatic system of the Western Mediterranean Region: Journal of Petroleum Geology, v. 5, p. 31–49.
BOSELLINI, A., 1973, Modello geodinamico e paleotettonico delle Alpi Meridionali durante il Giurassico-Cretacico. Sue possibili applicazioni agli Appennini: Accademia Nazionale Lincei, Quaderno n. 183, p. 163–205.
―――, 1981, The Emilia Fault: A Jurassic fracture zone that evolved into a Cretaceous-Paleogene sinistral wrench fault: Bollettino Società Geologica Italiana, v. 100, p. 161–169.
―――, 1984, Progradation geometries of carbonate platforms: Examples from the Triassic of the Dolomites, northern Italy: Sedimentology, v. 31, p. 1–24.
―――, FAZZUOLI, M., MASETTI, D., MATTAVELLI, L., AND SARTI, M., 1981a, Le torbiditi oolitiche della Falda Toscana (Giurassico mediosuperiore): Provenienza e implicazioni tettoniche: Rivista Italiana Paleontologia e Stratigrafia, v. 87, p. 177–192.

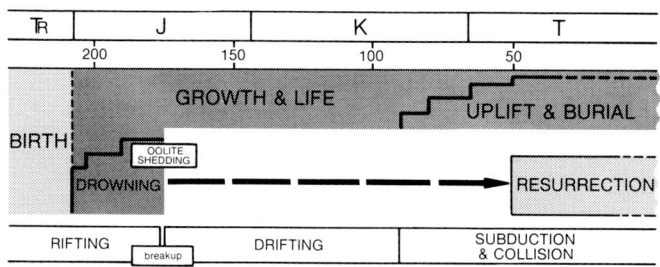

FIG. 13.—Evolutionary stages of the Tethyan platforms as related to the geodynamics of the Tethyan continental margins.

———, AND HSÜ, K. J., 1973, Mediterranean plate tectonics and Triassic paleogeography: Nature, v. 224, p. 144–146.

———, MASETTI, D., AND SARTI, M., 1981b, A Jurassic "Tongue of the Ocean" infilled with oolitic sands: The Belluno Trough, Venetian Alps, Italy: Marine Geology, v. 44, p. 59–95.

———, RUSSO, A., ARUSH, M. A., AND CABDULQADIR, M. M., 1987, The Oligo-Miocene of Eil (NE Somalia): A prograding coral-*Lepidocyclina* system: Journal of African Earth Sciences, v. 6, p. 583–593.

———, AND WINTERER, E. L., 1975, Pelagic limestone and radiolarite of the Tethyan Mesozoic: A genetic model: Geology, v. 3, p. 279–282.

BOSELLINI, F. R., AND RUSSO, A., 1988, The Oligocene *Actinacis* coral community of the southern Alps (Italy): Temperature vs. terrigenous control (abst.): Proceedings, Sixth International Coral Reef Symposium, Townsville, Australia.

CARANNANTE, G., SIMONE, L., AND BARBERA, C., 1981, "Calcari a Briozoi e Litotamni" of southern Apennines. Miocene analogs of Recent Mediterranean rhodolitic sediments (abst.): International Association of Sedimentologists, Second European Meeting, Bologna, p. 17–20.

CASTELLARIN, A., 1972, Evoluzione paleotettonica sinsedimentaria del limite tra piattaforma veneta e bacino lombardo a nord di Riva del Garda: Giornale di Geologia, v. 38, p. 11–212.

———, COLACICCHI, R., AND PRATURLON, A., 1978, Fasi distensive, trascorrenze e sovrascorrimenti lungo la Linea Ancona-Anzio, dal Lias medio al Pliocene: Geologica Romana, v. 17, p. 161–189.

CATALANO, R., AND D'ARGENIO, B., 1978, An essay of palinspastic restoration across the western Sicily: Geologica Romana, v. 17, p. 145–159.

COLACICCHI, R., PIALLI, G., AND PRATURLON, A., 1975, Megabreccias as a product of tectonic activity along a carbonate platform margin (abst.): Eleventh International Congress of Sedimentology, Nice, p. 61–70.

COOK, H. E., MCDANIEL, P. N., MOUNTJOY, E. W., AND PRAY, L. C., 1972, Allochtonous carbonate debris flows at Devonian bank ("reef") margins, Alberta, Canada: Bulletin of Canadian Petroleum Geologists, v. 20, p. 439–497.

———, AND TAYLOR, M. E., 1977, Comparison of continental slope and shelf environments in the Upper Cambrian and Lowest Ordovician of Nevada, *in* Cook, H. E., and Enos, Paul, eds., Deep-Water Carbonate Environments: Society of Economic Paleontologists and Mineralogists Special Publication 25, p. 51–82.

COUSIN, M., 1980, Les Rapports Alpes-Dinarides: Le confins de l'Italie et de la Yougoslavie: Unpublished Ph. D. Dissertation, Université de Pierre et Marie Curie, Paris, 517 p.

CRESCENTI, V., AND VIGHI, L., 1970, Risultati delle ricerche eseguite sulle formazioni bauxitiche cretaciche del Casertano e del Matese in Campania: Memorie Società Geologica Italiana, v. 9, p. 401–434.

CREVELLO, P. D., AND SCHLAGER, W., 1980, Carbonate debris sheets and turbidites, Exuma Sound, Bahamas: Journal of Sedimentary Petrology, v. 50, p. 1121–1148.

D'ARGENIO, B. 1970, Central and southern Italy Cretaceous bauxite stratigraphy and paleogeography: Annales of Hungarian Institute of Geology, Budapest, v. 54, p. 221–233.

———, 1976, Le piattaforme carbonatiche periadriatiche. Una rassegna di problemi nel quadro geodinamico mesozoico dell'area Mediterranea: Memorie Società Geologica Italiana, v. 13, p. 1–28.

———, PESCATORE, T., AND SCANDONE, P., 1973, Schema geologico dell'Appennino meridionale (Campania e Lucania): Accademia Nazionale Lincei, Quaderno n. 183, p. 49–72.

DAVIES, P. J., AND HOPLEY, D., 1983, Growth fabrics and growth rates of Holocene reefs in the Great Barrier Reef: Bureau of Mineral Resources Journal of Australian Geology and Geophysics, v. 8, p. 237–251.

DOGLIONI, C., AND BOSELLINI, A., 1987, Eoalpine and Mesoalpine tectonics in the southern Alps: Geologische Rundschau, v. 76, p. 735–754.

DROXLER, A. W., AND SCHLAGER, W., 1985, Glacial versus interglacial sedimentation rates and turbidite frequency in the Bahamas: Geology, v. 13, p. 799–802.

EBERLI, G. P., AND GINSBURG, R. N., 1987, Segmentation and coalescence of Cenozoic carbonate platforms, northwestern Great Bahama Bank: Geology, v. 15, p. 75–79.

ESTEBAN, M., 1979, Significance of the Upper Miocene coral reefs of the western Mediterranean: Palaeogeography, Palaeoclimatology, Palaeoecology, v. 29, p. 169–188.

FISCHER, A. G., 1964, The Lofer cyclothems of the Alpine Triassic: Bulletin of Kansas State Geological Survey, v. 169, p. 107–149.

———, 1982, Long-term climatic oscillations recorded in stratigraphy, *in* Climate in Earth History, Studies in Geophysics: National Academy Press, Washington, D.C., p. 97–104.

GRIGG, R. W., 1982, Darwin Point: A threshold for atoll formation: Coral Reefs, v. 1, p. 29–34.

HAQ, B. U., HARDENBOL, J., AND VAIL, P. R., 1987, Chronology of fluctuating sea levels since the Triassic: Science, v. 235, p. 1156–1167.

HARDIE, L. A., BOSELLINI, A., AND GOLDHAMMER, R. K., 1986, Repeated subaerial exposure of subtidal carbonate platforms, Triassic, northern Italy: Evidence for high frequency sea level oscillations on a 10^4 year scale: Paleoceanography, v. 1, p. 447–457.

HINE, A. C., AND NEUMANN, A. C., 1977, Shallow carbonate-bank-margin growth and structure, Little Bahama Bank, Bahamas: American Association of Petroleum Geologists Bulletin, v. 61, p. 376–406.

———, WILBER, R. J., BANE, J. M., NEUMANN, A. C., AND LORENSON, K. R., 1981, Offbank transport of carbonate sands along open, leeward bank margins: northern Bahamas: Marine Geology, v. 42, p. 327–348.

HUBBARD, D. K., BURKE, R. B., AND GILL, I. P., 1986, Styles of reef accretion along a steep, shelf-edge reef, St. Croix, U. S. Virgin Islands: Journal of Sedimentary Petrology, v. 56, p. 848–861.

HURST, J. M., AND SURLYK, F., 1983, Depositional environments along a carbonate ramp to slope transition in the Silurian of Washington Land, North Greenland: Canadian Journal of Earth Sciences, v. 20, p. 473–499.

ISTITUTO DI GEOLOGIA E GEOFISICA UNIVERSITÀ DI NAPOLI, 1978, Processi paleocarsici e neocarsici e loro importanza economica nell'Italia Meridionale: Guida alle escursioni, 210 p.

JANSA, L. F., 1981, Mesozoic carbonate platforms and banks of the eastern North American margin: Marine Geology, v. 44, p. 97–117.

JENKYNS, H., 1971, The genesis of condensed sequences in the Tethyan Jurassic: Lethaia, v. 4, p. 327–352.

———, 1980, Tethys: Past and present: Proceedings, Geological Association, v. 91, p. 107–118.

———, 1985, The Early Toarcian and Cenomanian-Turonian anoxic events in Europe: Comparisons and contrasts: Geologische Rundschau, v. 74, p. 505–518.

———, AND CLAYTON, C. J., 1986, Black shales and carbon isotopes in pelagic sediments from the Tethyan Lower Jurassic: Sedimentology, v. 33, p. 87–106.

KEIGWIN, L. D., 1980, Palaeo-oceanographic changes in the Pacific at the Eocene/Oligocene boundary: Nature, v. 287, p. 722–725.

KELTS, K., 1981, A comparison of some aspects of sedimentation and translational tectonics from the Gulf of California and the Mesozoic Tethys, northern Penninic margin: Eclogae Geologicae Helvetiae, v. 74, p. 317–338.

KENDALL, C. G., ST. C., AND SCHLAGER, W., 1981, Carbonates and relative changes in sea level: Marine Geology, v. 44, p. 181–212.

LAUBSCHER, H. P., AND BERNOULLI, D., 1977, Mediterranean and Tethys, *in* Nairn, A. E. M., Kanes, W. H., and Stehli, F. G. eds., The Ocean Basins and Margins, v. 4A, The Eastern Mediterranean: Plenum Press, New York, p. 1–22.

LEES, A., 1975, Possible influence of salinity and temperature on modern shelf carbonate sedimentation: Marine Geology, v. 19, p. 159–198.

LUCIANI, V., 1988, La dorsale paleogenica M. Baldo-M. Bondone (Trentino Meridionale): Significato paleogeografico e paleotettonico: Rivista Italiana di Paleontologia e Stratigrafia, v. 93, p. 507–520.

MARJANAC, T., 1985, Composition and origin of the megabed containing huge clasts: Flysch Formation, middle Dalmatia, Yugoslavia (abst.): International Association of Sedimentologists, Sixth European Regional Meeting, Lleida, Spain, p. 270–273.

MASETTI, D., AND BIANCHIN, G., 1987, Geologia del Gruppo dello Schiara (Dolomiti Bellunesi): Memorie di Scienze Geologiche, v. 34, p. 187–212.

MITCHUM, R. M., AND ULIANA, M. A., 1985, Seismic stratigraphy of carbonate depositional sequences, Upper Jurassic-Lower Cretaceous, Neuquén Basin, Argentina, *in* Berg, O. R., and Woolverton, D. G., eds., Seismic Stratigraphy II, An Integrated Approach to Hydrocarbon Exploration: American Association Petroleum Geologists Memoir 39, p. 255–274.

MOSTARDINI, F., AND MERLINI, S., 1986, Appennino Centro Meridionale. Sezioni geologiche e proposta di modello strutturale: 73 Congresso Società Geologica Italiana, Roma, 46 p.

MULLINS, H. T., GARDULSKI, A. F., AND HINE, A. C., 1986, Catastrophic collapse of the west Florida carbonate platform margin: Geology, v. 14, p. 167–170.

MUTTI, E., RICCI LUCCHI, F., SEGURET, M., AND ZANZUCCHI, G., 1984, Seismoturbidites: A new group of resedimented deposits: Marine Geology, v. 55, p. 103–116.

NEWELL, N. D., RIGBY, J. K., FISCHER, A. G., WHITEMAN, A. J., HICKOX, J. E., AND BRADLEY, J. S., 1953, The Permian Reef Complex of the Guadalupe Mountains Region, Texas and New Mexico: Freeman and Company, San Francisco, 236 p.

OGG, J. G., 1981, Sedimentology and Paleomagnetism of Jurassic Pelagic Limestones ("Ammonitico Rosso" Facies): Unpublished Ph. D. Dissertation, University of California, San Diego, 203 p.

PANTIĆ, N. K. 1986, Global Tertiary climatic changes, paleo-phytogeography and phytostratigraphy, in Walliser, O., ed., Global Bio-Events: Lecture Notes in Earth Sciences, v. 8, p. 419–427.

PLAYFORD, P. E., 1980, Devonian "Great Barrier Reef" of Canning Basin, Western Australia: American Association of Petroleum Geologists Bulletin, v. 64, p. 814–840.

PURDY, E. G., 1980, Evolution of the Maldive Atolls, Indian Ocean (abst.): Geological Society of America, Annual Meeting, Atlanta, p. 504.

QUINLAN, G. M., AND BEAUMONT, C., 1984, Appalachian thrusting, lithospheric flexure, and the Paleozoic stratigraphy of the Eastern Interior of North America: Canadian Journal of Earth Sciences, v. 21, p. 973–996.

READ, J. F., 1980, Carbonate ramp-to-basin transitions and foreland basin evolution, Middle Ordovician, Virginia Appalachians: American Association of Petroleum Geologists Bulletin, v. 64, p. 1575–1612.

———, 1982, Carbonate platforms of passive (extensional) continental margins-types, characteristics and evolution: Tectonophysics, v. 81, p. 195–212.

———, 1985, Carbonate platforms facies models: American Association of Petroleum Geologists Bulletin, v. 69, p. 1–21.

RUIZ-ORTIZ, P. A., 1983, A carbonate submarine fan in a fault controlled basin of the Upper Jurassic, Betic Cordillera, southern Spain: Sedimentology, v. 30, p. 33–48.

SHACKLETON, N. J., AND KENNETT, J. P., 1975, Paleotemperature history of the Cenozoic and the initiation of Antarctic glaciation: Oxygen and carbon isotope analyses in DSDP Sites 277, 279, 281, in Kennett, J. P., Houtz, R. E., and others, Initial Reports of the Deep Sea Drilling Project, v. 29, U.S. Government Printing Office, Washington, D.C., p. 743–756.

SCHLAGER, W., 1981, The paradox of drowned reefs and carbonate platforms: Geological Society of America Bulletin, v. 92, p. 197–211.

———, AND CHERMAK, A., 1979, Sediment facies of platform-basin transition, Tongue of the Ocean, Bahamas, in Doyle, L. J., and Pilkey, O. H., eds., Geology of Continental Slopes: Society of Economic Paleontologists and Mineralogists Special Publication 27, p. 193–208.

———, AND JAMES, N. P., 1978, Low-magnesian calcite limestones forming at the deep-sea floor, Tongue of the Ocean, Bahamas: Sedimentology, v. 25, p. 675–702.

SEARLE, M. P., JAMES, N. P., CALON, T. J., AND SMEWING, J. D., 1983, Sedimentological and structural evolution of the Arabian continental margin in the Musandam Mountains and Dibba Zone, United Arab Emirates: Geological Society of America Bulletin, v. 94, p. 1381–1400.

SELLI, R., 1971, Isole Tremiti e Pianosa, in Cremonini, G., Elmi, C., and Selli, R., Foglio 156, S. Marco in Lamis: Note Illustrative della Carta Geologica d'Italia, Servizio Geologico d'Italia, Roma, p. 49–59.

SENGÖR, A. M. C., YILMAZ, Y., AND SUNGURLU, O., 1984, Tectonics of the Mediterranean Cimmerides: Nature and evolution of the western termination of Palaeo-Tethys, in Dixon, J. E., and Robertson, A. H. F., eds., The Geological Evolution of the Eastern Mediterranean, The Geological Society of London, Blackwell, London, p. 77–112.

SGROSSO, I., 1968, Note biostratigrafiche sul M. Vesole (Cilento): Bollettino Società Naturalisti in Napoli, v. 77, p. 159–180.

SIMÒ, A., 1986, Carbonate platform depositional sequences, Upper Cretaceous, south-central Pyrenees (Spain): Tectonophysics, v. 129, p. 205–231.

SIMONE, L., AND CARANNANTE, G., 1985, Evolution of a Miocene carbonate open shelf from inception to drowning: The case of the southern Apennines: Società Nazionale di Scienze, Lettere e Arti in Napoli, serie IV, v. 52, p. 1–43.

STÖCKLIN, J., 1974, Possible ancient continental margins in Iran, in Burk, C. A., and Drake, C. L., eds., The Geology of Continental Margins: Springer-Verlag, Berlin, p. 873–887.

SUESS, E., 1893, Are great oceans depths permanent?: Natural Science, v. 2, p. 180–187.

TAPPONIER, P., 1977, Evolution tectònique du système alpin en Méditerranée: Poinçonnement et écrasement rigide-plastique: Bulletin de la Societé géologique de France, v. 19, p. 437–460.

TUCKER, M. E., 1985, Shallow-marine carbonate facies and facies models, in Brenchley, P. J., and Williams, B. P. J., eds., Recent Developments and Applied Aspects: Sedimentology, v. 18, p. 147–169.

VERGNAUD GRAZZINI, C., AND OBERHAENSLI, H., 1986, Isotopic events at the Eocene/Oligocene transition. A review, in Pomerol, Ch., and Premoli-Silva, I., eds., Terminal Eocene Events: Elsevier, Amsterdam, p. 311–329.

WATTS, K. F., AND GARRISON, R. E., 1986, Sumeini Group, Oman—Evolution of a Mesozoic carbonate slope on a south Tethyan continental margin: Sedimentary Geology, v. 48, p. 107–168.

WEISSERT, H. J., AND BERNOULLI, D., 1985, A transform margin in the Mesozoic Tethys: Evidence from the Swiss Alps: Geologische Rundschau, v. 74, p. 665–679.

WILSON, J. L., 1975, Carbonate Facies in Geologic History: Springer-Verlag, New York, 417 p.

WINTERER, E. L., AND BOSELLINI, A., 1981, Subsidence and sedimentation on Jurassic passive continental margin, southern Alps, Italy: American Association of Petroleum Geologists Bulletin, v. 65, p. 394–421.

WRIGHT, V. P., AND WILSON, R. C. L., 1984, A carbonate submarine-fan sequence from the Jurassic of Portugal: Journal of Sedimentary Petrology, v. 54, p. 394–412.

DROWNING UNCONFORMITIES ON CARBONATE PLATFORMS

WOLFGANG SCHLAGER

Free University, Institute for Earth Sciences, P.O. Box 7161, Amsterdam 1007 MC, The Netherlands

ABSTRACT: Flanks of carbonate platforms steepen as the platform rises higher above the basin floor. Furthermore, platform slopes are steeper on average than siliciclastic slopes. Termination of platform growth through rapid submergence or suffocation by siliciclastics produces an unconformity, because the clastics cannot assume the steep carbonate slope angle and because they are shed from different directions. This "drowning unconformity" resembles the unconformity produced by a lowstand of sea level, even though it is associated with a rise or a highstand of sea level.

Examples of drowning unconformities include the mid-Cretaceous unconformity in the Gulf of Mexico, the unconformities on the flanks of the Wilmington platform, the Lahave platform, and the platforms off Morocco, all drowned in the earliest Cretaceous, and the mid-Jurassic unconformities on certain platforms of the High Atlas.

Drowning unconformities are best developed on platforms that rise 800 m or more above the basin, have concave upper flanks of 6° or more, and commonly possess an elevated rim. Drowning and burial of smaller platforms with gentle flanks still produce unconformities, because the pattern of sediment input and dispersal is different for carbonates and siliciclastics.

INTRODUCTION

In the past decade, seismic interpreters have expanded their techniques to include mapping of seismic unconformities besides tracing high-amplitude reflectors. The success and popularity of this approach warrants a concomitant effort by sedimentologists to learn more about the changes in dip of bedding planes that form the basis for the seismic unconformities (and subsequently for definition of stratigraphic sequences).

Vail and others (1977) have proposed sea level as a prime cause of changes in bedding geometry, and this principle has been widely applied as a stratigraphic tool. This paper focuses on changes in lithology as a cause of seismic unconformities, specifically, on the change from shallow-water carbonate deposition to siliciclastic deposition.

The difference between slopes in siliciclastic and carbonate terrains had been noticed by Dailly (1982) and Cook and Mullins (1983). Schlager and Camber (1982, 1986) have quantified these differences in a survey of extant platforms and siliciclastic slopes. They pointed out that abrupt drowning of platforms and subsequent burial by siliciclastics produces an unconformity that resembles geometrically the unconformity resulting from a lowstand of sea level. Even where platform flanks are not steeper than siliciclastic slopes, an unconformity is likely to develop when a platform is drowned, simply because the pattern of sediment dispersal will differ for carbonate and siliciclastic material.

Schlager and Camber (1986) introduced the term "drowning unconformity" for an unconformity created by the drowning of a platform and the subsequent onlap of siliciclastics or other non-platform sediments. I propose to broaden this term to include not only platforms killed by submergence below the euphotic zone (drowning *sensu stricto*) but also platforms "drowned" by prograding marine siliciclastic sediments or by outpourings of marine volcanics or volcaniclastic sediments. The term excludes termination by significant falls of sea level with the formation of lowstand wedges in the sense of Vail and others (1977).

This report discusses drowning unconformities on carbonate platforms. It presents the principles that explain their development, describes several examples, and finally discusses the implications for seismic stratigraphy.

PRINCIPLES OF SLOPE SEDIMENTATION

Slope angles of carbonate and siliciclastic systems.—

Schlager and Camber (1986) measured about 500 slope profiles on carbonate platforms and siliciclastic continental margins. They found that angles increase with height (i.e., top-to-basin relief) of the slope for carbonate platforms. In siliciclastic terrains this trend holds only for small heights. Above a height of 500 m, declivity of siliciclastics is no longer coupled to slope height. On average, siliciclastic slopes are distinctly flatter than the flanks of carbonate platforms. The increase of declivity with height at the beginning of slope construction is a geometric necessity—for a slope of declivity zero, the height must also be zero. The carbonate and siliciclastic slope populations start to separate at heights of about 200 m; above 3,000 m, the separation is almost complete (Fig. 1).

A survey of seismic profiles confirms that the principle of carbonate oversteepening also holds for the more distant geologic past. Figure 2 shows the relation of carbonates and overlying clastics qualitatively in three categories. In over 90 percent of all cases, carbonates exceed siliciclastics in slope angle.

The reasons for the oversteepening of carbonate slopes are still a matter of debate, and several processes merit mention in this context. (1) Periplatform carbonates contain high quantities of the metastable minerals magnesian calcite and aragonite; consequently, they lithify faster than purely pelagic carbonate ooze or terrigenous muds (e.g., Austin and others, 1986). (2) Periplatform sediments lithify extremely fast where exposed on the sea floor (Schlager and James, 1978; Mullins and others, 1980). This process benefits from the shift from deposition to erosion associated with increasing slope angle. (3) Organic framebuilding stabilizes the sea floor in the euphotic zone to support near-vertical walls, and some encrusting organisms invade the dysphotic zone to several hundred meters depth (James and Ginsburg, 1979). (4) There is some indication that carbonate sediments have higher shear strength than siliciclastic muds, even in an unlithified state (J. Kenter, unpubl. report).

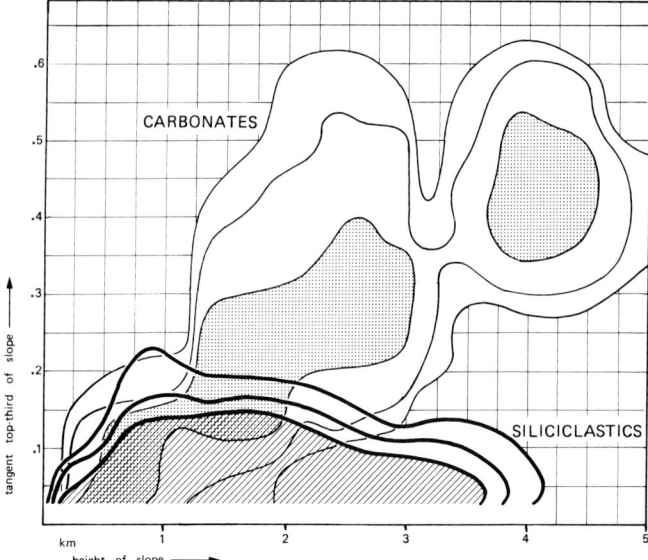

FIG. 1.—Angle of upper one-third of slope versus height of slope. Contours indicate concentrations of 0.5, 1, and 2 percent of total sample in unit area of 0.25 km × 0.05 tan S, measured as moving average of 9 unit cells. Carbonate sample includes Bahamas and central Pacific atolls ($N = 413$); siliciclastic sample is based on Atlantic continental slopes ($N = 72$). Data from Schlager and Camber (1986). Carbonate slopes steepen with height to heights of 5,000 m or more; siliciclastics follow this trend only to 500 m; over 500 m, slope height has no influence on declivity of siliciclastic slopes.

most common form of gravity flows, flow power is proportional to the 1.5 power of the slope (Allen, 1968). Seismic profiles and field observations in the Bahamas confirm the predictions from flow theory (Fig. 3; Schlager and Ginsburg, 1981; Austin and others, 1988). On the most gentle slopes, sediment is largely trapped on the slope, and sedimentation rates decrease with distance from the platform source. As the slope steepens, the depocenter migrates from the slope to the debris apron at the basin floor, and the regime on the slope changes from accretion to erosion. On steep accretionary slopes, bypassing becomes important. Dense turbidity currents no longer deposit sediment on the slope; rather, they bypass it and deposit their load as a series of graded beds on the debris apron at the toe of slope. The slope collects mud from pelagic rain and from dilute turbidity currents.

The change from accretion to erosion favors seafloor lithification by exposing sediment on the sea floor for extended periods. This sets off an important feedback loop. Winnowing and erosion by turbidity currents create hardgrounds on the slope; these bare surfaces prevent turbidity currents from becoming saturated with sediment, leading to more erosion. The results are extensive crusts on the steep upper slopes of carbonate platforms (James and Ginsburg, 1979; Land and Moore, 1977). These slopes are too hard and too steep to allow much sediment accumulation, but they are not easily eroded either.

Slope angle and depositional regime.—

On most submarine slopes, deposition is controlled by slumping and other sediment gravity transport. An increase of slope angle will shift the activity of sediment gravity flows from deposition to erosion. For turbidity currents, the

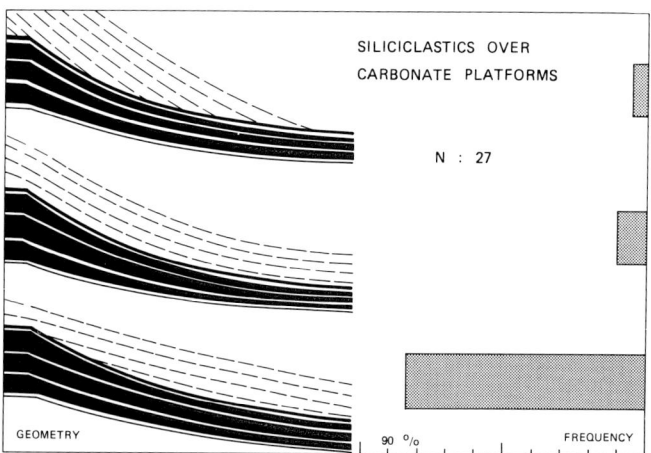

FIG. 2.—Slope angles of carbonates and siliciclastics in seismic dip lines, where carbonate platforms are overlain by siliciclastics. In most instances, platform flanks dip more steeply than overlying siliciclastics. Based on Vail and others (1977); U.S. Geological Survey Seismic lines from east coast of United States; Bally and others (1981); Jansa (1981); Grow and others, Crutcher and others, Petrobras and others, Veeken and others, all in Bally (1983); Hubbard and others (1985); and unpublished industry data.

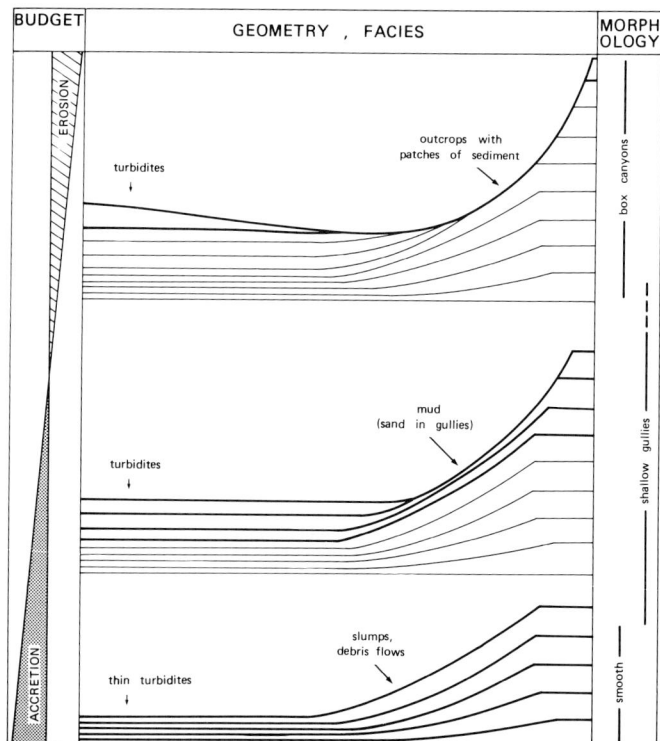

FIG. 3.—Changes in sequence geometry, sediment facies, and slope morphology related to change in sediment budget on a slope. As slope steepens, depositional regime on slope changes from accretion to erosion with concomitant changes in sediment cover and morphology of slope (after Schlager and Camber, 1986).

The drowning unconformity.—

The oversteepening of carbonate slopes has one important implication: the demise of high-rising, steep-sided carbonate platforms, and its burial by siliciclastics, produces an unconformity. The siliciclastic sediments, unable to assume the steep carbonate slope angle, will onlap the flank of the platform and bury it from the bottom upward (Fig. 4). Geometrically, such a relation resembles a lowstand unconformity in the sense of Vail and others (1977). Onlap of the debris apron onto steep accretionary platform slopes (Fig. 3, middle panel, and Fig. 4b) is a normal feature of platform growth and therefore is not considered in this discussion of drowning unconformities. It should be noted, however, that this phenomenon also generates marine onlap independent of sea level.

The demise of a platform by rapid submergence below the euphotic zone is one way to generate a drowning unconformity; suffocation and "drowning" of the platform under prograding siliciclastics is another. For the final geometry, it makes little difference whether a platform was killed by rapid submergence and later buried by siliciclastics, or whether burial by the siliciclastics caused the demise. The relation to sea level may be different though. Demise by submergence (i.e., drowning *sensu stricto*) requires a relative rise of sea level that exceeds the growth potential of the platform. Suffocation by marine siliciclastics requires a relative highstand of the sea to provide accommodation for the siliciclastics on top of the platform.

Drowning of free-standing platforms in the oceans and their subsequent covering by pelagic sediments also produces a drowning unconformity, albeit one that is not commonly encountered in classic sea-level studies.

In summary, a drowning unconformity is created by rapid termination of platform growth during a rise or a highstand of sea level. It forms without significant fall of sea level. One should note, however, that platforms grow so close to sea level that island formation and minor exposures are a normal ingredient of their growth record.

EXAMPLES

Mid-Cretaceous unconformity, Gulf of Mexico.—

Buffler and others (1980) discovered and mapped a basin-wide unconformity and sequence boundary in the Gulf of Mexico. Its stratigraphic expression varies considerably. On the margins, the unconformity occurs between carbonate platforms and overlying, largely terrigenous deep-water sediments; on the basin slopes, the unconformity separates periplatform debris from onlapping deep-water siliciclastics (Fig. 5); in the abyssal plain of the western Gulf, the sequence boundary is represented by a correlative conformity with high-amplitude reflections, indicating carbonate-rich deposits below, and a nearly transparent zone of shales above.

The age of this unconformity in the deep part of the Gulf of Mexico is not well constrained. Only DSDP Site 540 penetrated this horizon, unfortunately at a location where several unconformities are bundled and the mid-Cenomanian through Campanian interval is represented by a hiatus (Buffler and others, 1984). Correlations with stratigraphy on the shelves suggest a Cenomanian age for the Middle Cretaceous Unconformity (MCU; Addy and Buffler, 1984; Corso 1988). Ties to shelf stratigraphy are somewhat uncertain, however, because of jump correlations. It seems that the mid-Cretaceous age of this basin-wide marker is indeed very likely, but the exact position within this interval is not constrained by definitive stratigraphic evidence.

Ideas on the origin of the MCU have varied. Buffler and others (1980) interpreted the MCU as a lowstand unconformity and linked it to the Cenomanian lowstand of sea level postulated by Vail and others (1977). Schlager and others (1984) pointed out that the mid-Cretaceous was a time of major reorganization of sedimentation patterns in the Gulf of Mexico, where most of the carbonate platforms rimming the basin were drowned or reduced in size, and sedimentation began its stepwise change to siliciclastic input (Fig. 6). Winker and Buffler (1988, p. 335) argue along a similar line when they state: "This basinwide change in the nature of carbonate sedimentation is not directly attributable to terrigenous sediment supply, nor can it be attributed solely to a short-term perturbation such as a drop in sea level."

Both Winker and Buffler (1988) and Schlager and others (1984) relate the MCU to the termination of platform growth around the Gulf. I now consider it a prime example of a drowning unconformity.

In detail, this scenario invokes stepwise platform drowning with major events in the late Aptian, the late Albian, and the middle or late Cenomanian (Arthur and Schlanger, 1979; Corso, 1988; Winker and Buffler, 1988). Each time

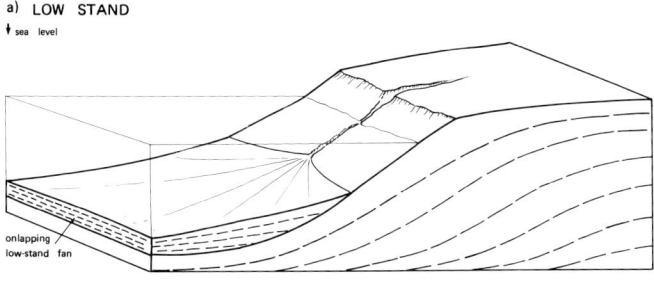

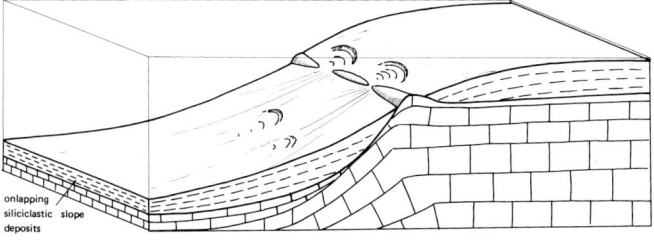

FIG. 4.—Lowstand unconformity and drowning unconformity, a comparison. (a) Lowstand of sea level exposes shelf, and efficient downslope transport leads to growth of deep-sea fan that onlaps deposits from previous highstand. Lower unit is carbonate or siliciclastics; upper unit is siliciclastics. After Vail and others, (1977, modified). (b) Stressed by rapid relative rise of sea level, a carbonate platform has built steep relief, including erosional upper slope, elevated rim, and deep lagoon ("empty bucket"). After termination of platform growth, siliciclastics bury platform with more gentle slope angles and create drowning unconformity that resembles lowstand unconformity.

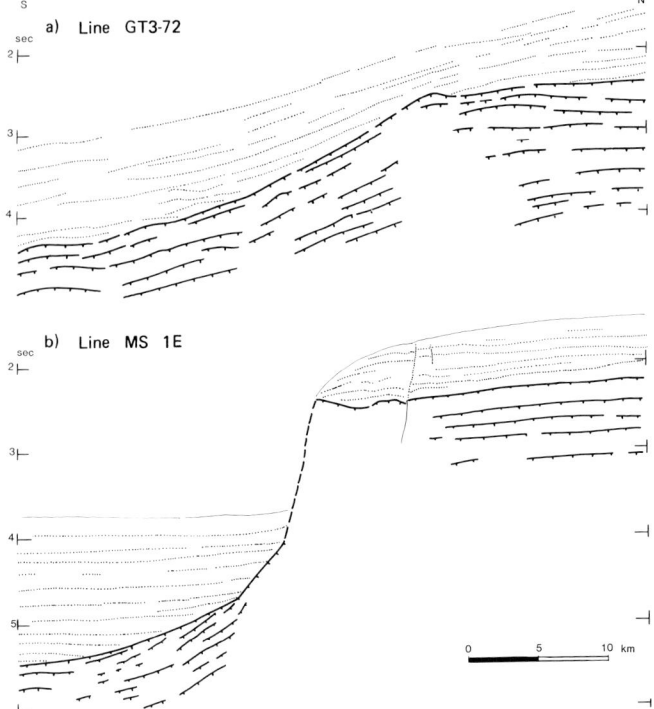

FIG. 5.—Mid-Cretaceous unconformity, Gulf of Mexico; line drawings of seismic profiles, demonstrating two different styles in the expression of the unconformity (hachured lines—carbonates, dotted lines—siliciclastics and pelagics). In (a) platform is about 1,000 m high and has a relatively gentle accretionary slope. Siliciclastics burying platform were shed down the same slope as carbonate debris during preceding phase of platform accretion. Siliciclastics simultaneously onlap platform flank and downlap on platform top. In (b) platform is about 2,500 m high and possesses erosional upper slope, indicated by truncation of reflectors and outcrops of platform interior facies on platform flank (Corso, 1988). After termination of growth, platform was buried by two different depositional systems. Platform top was covered by hemipelagics prograding from east to west (parallel to section) and gently downlapping on platform. Platform flank is being buried by sediments of Mississippi fan transported roughly perpendicular to plane of section. Seismic lines courtesy of R. T. Buffler, Texas Institute for Geophysics.

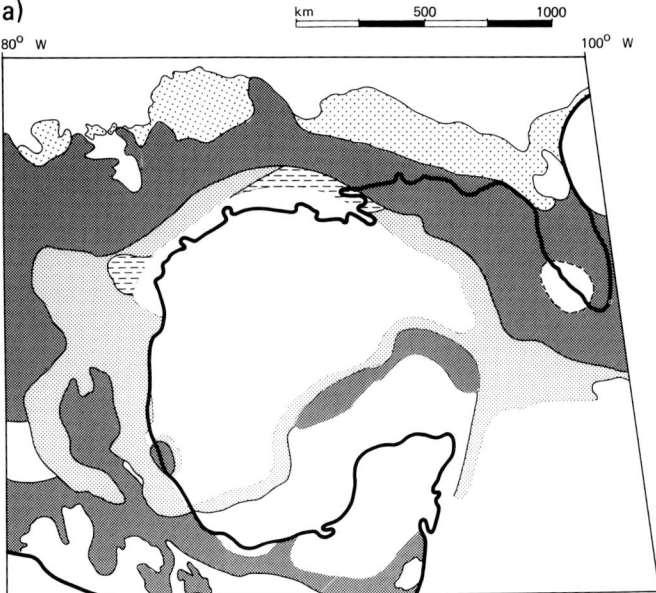

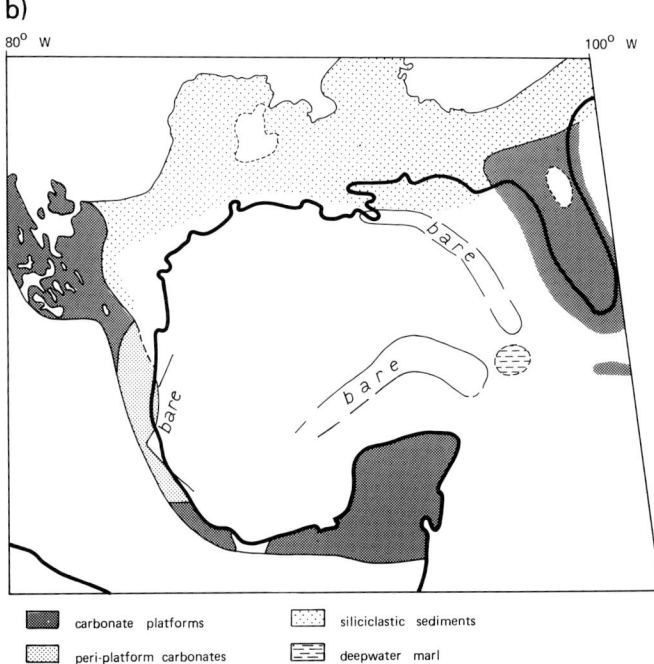

FIG. 6.—Mid-Cretaceous drowning of carbonate platforms around the Gulf of Mexico. (a) Paleogeography of mid-Aptian to mid-Cenomanian: Gulf almost completely surrounded by carbonate platforms. (b) Paleogeography of mid-Cenomanian to top Turonian: platforms greatly reduced, former platform margins appear as bare submarine outcrops, and siliciclastics dominate northern rim. After Cook and Bally (1975, modified).

the rim of platforms around the Gulf was reduced. Some platforms, such as the Golden Lane or the Jordan Knoll, were drowned completely; others were forced to step back and abandon territory. Between crises, the platforms grew and often prograded rapidly but were generally unable to reoccupy the margins of the previous cycle. The demise of the platforms was followed by widespread pelagic sedimentation in the deep part of the Gulf (Winker and Buffler, 1988). The record of the starved pelagic interval may not exceed 50–100 m (at sedimentation rates of 5 m/m.y.) and thus go unrecognized in most seismic profiles. What dominates the seismic images is the burial of the inactive debris aprons and steep flanks of the carbonate platforms (Addy and Buffler, 1984; Corso, 1988). In the southeastern Gulf, the platform flanks were buried by the flysch in front of the Cuban arc (Fig. 7); in other parts they stood bare until they were covered by the fringes of the Mississippi fan (Fig. 5b). Burial of the old platform system is incomplete to this day, however, particularly in the eastern and southern Gulf, where terrigenous input is small. The platform flanks are exposed in the Florida and Campeche Escarpments, as well as on the flanks of isolated guyots such as Jordan Knoll.

Paull and others (1987) suggested that dissolution through acid etching from oxidized sulphide in brine seeps may be an important erosional process on the Florida Escarpment. I agree, but consider it unlikely that this process did more than modify a primary, erosional slope. Similar escarp-

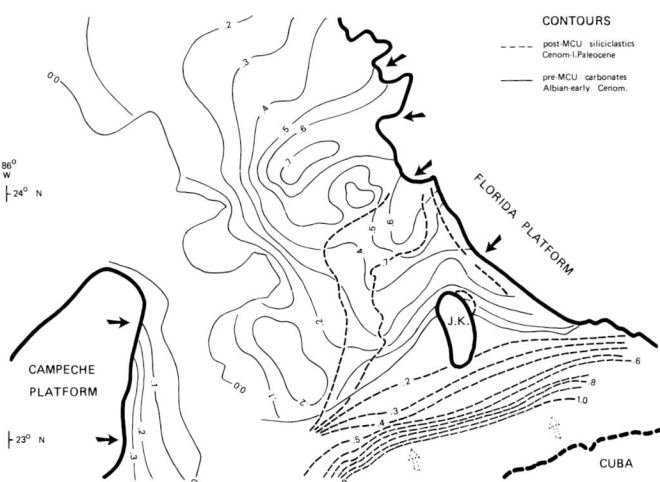

FIG. 7.—Thickness variations below and above MCU in southeastern Gulf of Mexico. Thickness based on interval travel times reported in Angstadt and others (1985) and Phair (1984). Note change in pattern of sediment input and dispersal. At pre-MCU time, carbonate platforms are active and dominate sediment input to basin; at post-MCU time, platforms are drowned and inactive; sedimentation after drowning is at first slow and pelagic; later, advancing Cuban island arc becomes principal sediment supplier to southeastern Gulf. Contour interval 0.1 sec two-way time.

ments surround Jordan Knoll and the eastern lobe of Campeche Bank. Both platforms are too small to contain significant volumes of evaporites and to support large-scale internal circulation. On seismic lines, the escarpment profile continues unbroken below the sea floor, indicating that erosion at the level of the brine seeps generally is below the resolution of the standard seismic tool. Another argument for the largely pristine character of the escarpment is that the slope follows the steepening-with-height rule, being gentle and accretionary in the northern part at a height of 1,000 m, and changing to a steep, erosional slope southward as height increases to over 2,000 m (Corso, 1988).

Early Cretaceous platform off Morocco.—

The Late Jurassic was a time of rapid subsidence and widespread platform growth on the margins of the North Atlantic. Growth was terminated in areas such as the Scotian shelf, offshore New Jersey, and offshore Morocco in the earliest Cretaceous (Valanginian; Vail and others, 1977; Eliuk, 1978; Karlo, 1986; Ringer and Patten, 1986; Meyer, this volume).

The Moroccan example is particularly well known through the repeated reference in Vail and others (1977). I will discuss it first. The platform rises 2,500–3,000 m above the abyssal sea floor. The uppermost contour shows an elevated rim and a steep, erosional upper slope with clastics onlapping the toe of slope (Fig. 8). The platform margin was penetrated by a borehole. Vail and others (1977) interpreted this configuration as the result of a lowstand of sea level, the boundary of cycles J3.2 and K1.1 at the Berriasian/Valanginian boundary. The arguments for platform demise by a significant fall of sea level are, besides the overall sequence geometry, the occurrence of a hiatus at the platform margin and karst caves in the upper part of the platform (P. Vail, oral commun.).

The features do not contradict a significant lowstand of sea level, but they do not prove it either. Geometry and top hiatus are equally well compatible with platform drowning without preceding fall of sea level. The karst caves do indicate exposure, but it is unclear whether they are indeed associated with the hiatus at the top. Caves in carbonate platforms originate very commonly at the water table by mixing-zone corrosion (Back and others, 1986). A sea-level fall of a few meters or even island formation during a highstand are sufficient to create this phenomenon. Minor falls and short-term exposures are common with mature platforms that are built almost to low-tide level. Alternatively, formation of the caves may postdate the demise of the platform. The seismic interpretation by Vail and others (1977) implies burial of the platform margin during supercycle K1.2, that is 3–5 m.y. after termination of growth and after deposition of what Vail and others (1977) interpret as the Valanginian lowstand wedge.

The arguments for a drowning unconformity, as proposed here, are the following.

1. The downward shift of marine onlap is disproportionately large, much larger than with other sea-level supercycles.

2. The platform is 2,500–3,000 m high. Our data from the Recent indicate that these very high platforms maintain steep upper slopes that almost always operate in an erosional regime. The oversteepening of the upper slope and truncation of platform reflectors are easily explained as typical features of platform growth at this advanced stage. There is no need to invoke a fall in sea level.

3. Turbidites in DSDP Site 416 indicate drowning of the carbonate platform in the Valanginian (Schlager, 1980). DSDP Site 416 drilled into what was then the continental rise of the Moroccan margin and penetrated a sequence of siliciclastic and carbonate turbidites (Lancelot and others 1980). The carbonate turbidites record the growth and demise of the Mazagan carbonate platform, i.e., the northern continuation of the platform shown by Vail and others (1977). The record of growth extends from the Tithonian through early Valanginian (Fig. 9). During the Valanginian, turbidite composition changes. Input of platform mud ceases and phosphatic grains (including phosphate ooids) and carbon-

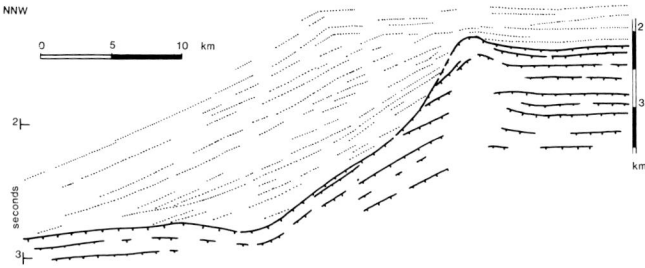

FIG. 8.—Valanginian unconformity off Morocco. Line drawing of seismic line in Vail and others (1977, p. 140 and 156). Hachured lines—carbonates; dotted lines—siliciclastics and pelagics. Note elevated rim and erosional upper slope of platform. Siliciclastics show shelf edge at about same elevation as preceding platform margin.

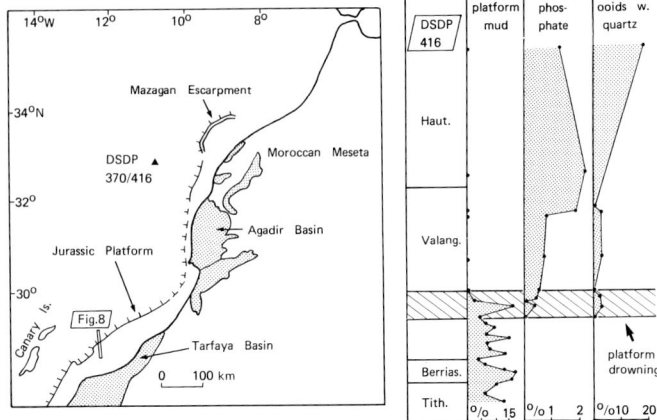

FIG. 9.—Record of turbidites in DSDP Site 416 indicates Valanginian drowning of Mazagan platform through disappearance of platform mud and appearance of phosphate and ooids with quartz nuclei. See text for details. After Schlager (1980).

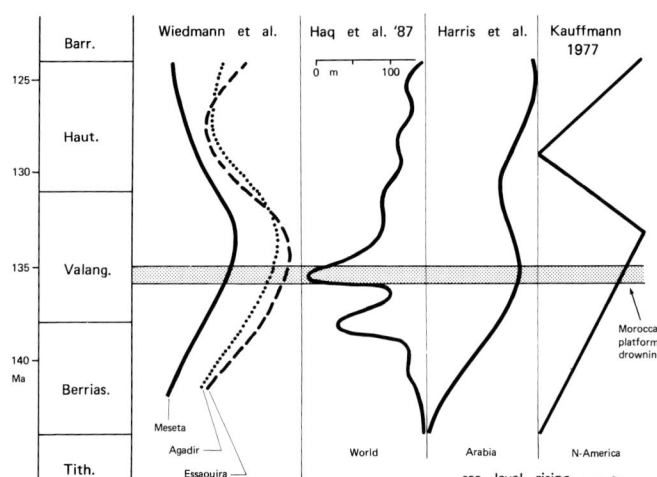

FIG. 10.—Sea-level curves and Valanginian drowning. Platform drowning suggested by turbidites agrees with sea-level curves derived from onshore basins in Morocco, Arabia, and North America, but is at variance with sea-level curve of Haq and others (1987).

ate ooids with quartz nuclei appear in the sand fraction. Calcium phosphate requires similar chemical conditions for precipitation as calcium carbonate; in the euphotic zone, it is outpaced by organically driven precipitation of the latter. Sediment-starved environments below the euphotic zone, such as the tops of drowned platforms, are a favored site of phosphate deposition (e.g., terraces off Florida; Mullins and Neumann, 1979). The appearance of carbonate ooids with quartz nuclei heralds encroaching siliciclastic deposition, while proving at the same time that the top of the platform was still flooded, allowing ooid production to continue in the most shallow parts. Formation of carbonate ooids requires current action on the flat platform top that is incompatible with complete exposure of the platform. I conclude from the turbidite record of DSDP 416 that platform growth in the Mazagan sector was terminated by drowning in the Valanginian and that exposure played no significant role in this process.

4. Submersible dives and deep-sea drilling on the Mazagan Escarpment led Von Rad and others (1985) to postulate termination by drowning in post-Berriasian and probably pre-Hauterivian time. In fact, the sequence of events deduced by Steiger and Cousin (1985), i.e., inter-Berriasian exposure and Valanginian drowning, closely matches the sequence described from the Wilmington platform by Meyer (this volume).

5. Sea-level curves gleaned from transgressions and regressions in the onshore basins of western Morocco corroborate the turbidite record and show continuous transgression in the Tithonian-Valanginian interval with a peak in the late Valanginian (Wiedmann and others, 1982). The sea-level curves from the Moroccan basins agree with those reported by Harris and others (1984) from eastern Arabia and by Kauffmann (1977) from the Western Interior basin of North America (Fig. 10).

Early Cretaceous Wilmington Platform.—

Almost exactly in conjugate position to the Moroccan platform lies the coeval carbonate belt of the Lahave and Wilmington platforms (Sheridan, 1974; Eliuk, 1978; Meyer, this volume). Several wells have recently been drilled on the Wilmington platform (Karlo, 1986; Ringer and Patten, 1986; Meyer, 1986 and this volume). The platform represents the leading edge of a continental shelf, passing landward into siliciclastics. On the seaward side, the platform drops off 3,000 m to the abyssal Atlantic. The topmost contour outlines a very concave, steep slope that culminates in an elevated platform rim. This marginal rim rose 180 m above the adjacent lagoon during the final stages of growth (Meyer, this volume). According to Karlo (1986), Ringer and Patten (1986), and Meyer (this volume), the platform aggraded nearly vertically in the Tithonian and early Berriasian, became exposed in the late Berriasian, resumed growth along its elevated rim, and was finally drowned at the Berriasian/Valanginian boundary. It was subsequently buried by siliciclastics that prograded across the top and onlapped the steep seaward flank (Fig. 11).

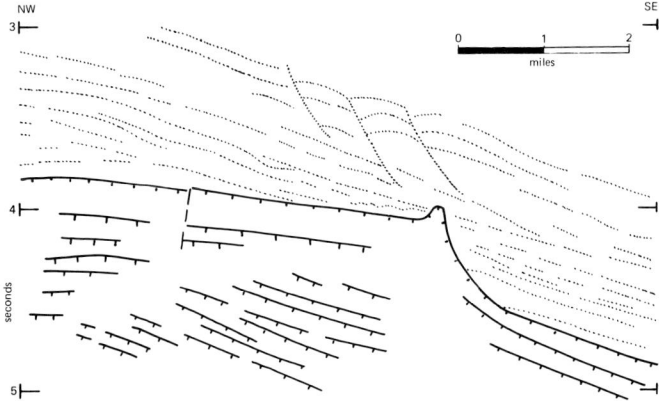

FIG. 11.—Drowning unconformity of Wilmington platform off New Jersey. Line drawing from seismic profile no. 19355. Hachured lines—carbonates; dotted lines—siliciclastics and pelagics. (Line courtesy of Shell Oil Company).

The top contour of the Wilmington platform is an excellent example of a drowning unconformity. Typical features include: high platform-to-basin relief, concave, steep upper slope and termination by drowning, as indicated by a deep lagoon and raised rim, lack of fresh-water dissolution in the youngest platform deposits, and by a cover of deep-water carbonates predating the burying clastics (Meyer, this volume). The sicliciclastics bury this huge platform with a greatly reduced slope angle that leads to a pronounced onlap on the ocean-facing slope.

Early Cretaceous Lahave (Abenaki) platform.—

The Abenaki Formation on the Scotian shelf was analysed in exemplary fashion by Eliuk (1978). The platform is not quite as high as the Wilmington platform, yet clearly displays an unconformity with siliciclastic onlap of its seaward flank (e.g., Fig. 12; Eliuk, 1978; figs. 11, 12; Jansa, 1981, fig. 4). Where not covered by an advancing delta, the demise of the platform is attributed to drowning, i.e., "relative sea-level rise . . . rapid and of sufficient magnitude to stop carbonate deposition" (Eliuk, 1978, p. 472). The facies succession in the drowning interval resembles that of the Wilmington platform (Ringer and Patten, 1986). The timing of drowning may be intra-Valanginian (Eliuk, 1978) rather than basal Valanginian, as at Wilmington. Intra-Valanginian drowning is also indicated for at least part of the Moroccan platform belt by the turbidite record at DSDP Site 416 (see earlier discussion).

Toarcian, Djebel Bou Dahar, Morocco.—

The Early Jurassic platforms in the South Atlas trough display what seem to be drowning unconformities in outcrop. One example is the Djebel Bou Dahar (Fig. 13). The platform grew in the Liassic on a basement high. It was drowned in the Toarcian and subsequently onlapped by deep-water shales (Agard and Du Dresnay, 1965, Crevello, 1987). The angle between carbonate slope and shales frequently exceeds 15°.

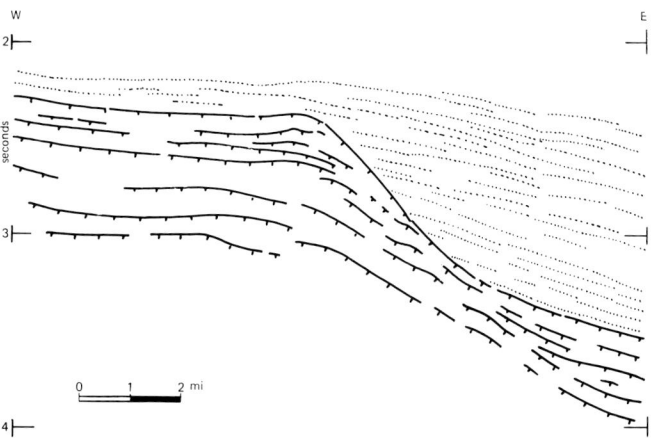

FIG. 12.—Drowning unconformity of Lahave platform off Nova Scotia. Line drawing after figure 4 in Eliuk (1978). Hachured lines—carbonates; dotted lines—siliciclastics and pelagics.

DISCUSSION

Drowning unconformities, sea-level fall and subaerial exposure.—

For lowstand unconformities in the sense of Vail and others (1977), a fall of sea level is obviously a basic requirement. This is not so for the geometrically similar drowning unconformities. Both the Lahave platform (Eliuk, 1978) and the Wilmington platform (Meyer, this volume) were repeatedly exposed during their growth period, but the last growth increment was immediately followed by drowning. There is no evidence that the drowning unconformity at the tops of these platforms is an exposure surface. Similar lack of exposure characterizes some of the mid-Cretaceous platforms around the Gulf of Mexico, e.g., parts of the Stuart City Reef (Bebout and others, 1977) and parts of the El Abra platform in Mexico (Enos and others, 1983). In these examples, the drowning unconformity is part of a deepening-upward sequence.

FIG. 13.—Drowning unconformity in outcrop, Djbel Bou-Dahar, Morocco. After Agard and du Dresnay (1965) and P. D. Crevello (pers. commun). Arrow marks boundary of Early Jurassic (Domerian) platform slope deposits to left and Middle Jurassic (Aalenian) onlapping flat-lying basinal limestones to right. Platform was drowned in late Early Jurassic (Toarcian). (Photo courtesy of P. D. Crevello.)

The frequent lack of subaerial exposure at drowning unconformities agrees with theoretical considerations. Termination by drowning requires a relative rise of sea level, and suffocation by clastics requires a relatively highstand of sea level to permit encroachment of clastics onto the platform top. Sea-level fall and exposure will also halt platform growth. Unless the old platform edifice is destroyed or covered during the time of exposure, however, the platform will resume growth when re-flooded. Exposure during a normal sea-level cycle will terminate platform growth only in conjunction with a subsequent rise. The discussion is reduced to the question whether exposure preceding a rise facilitates the demise of a platform. This is indeed the case, but the negative effect of exposure is a short one, as demonstrated by the Holocene transgression (Adey, 1978; Schlager, 1981). When platforms were flooded in the early Holocene, production commenced slowly, and reef growth and sediment accumulation fell behind sea level by tens of meters. This lag period lasted only 3,000–5,000 years. It was followed by extremely rapid growth, so that 10,000 years after flooding, the systems had again caught up with sea level. To damage a platform permanently through the initial lag effect, it must be submerged below the zone of optimum growth, while still operating in the lag phase. This requires very rapid rates of sea-level rise. If we assume a thickness of 20–40 m for the water layer of optimum carbonate growth, and a lag period of 5,000 years, we require a sea-level rise of 4,000–8,000 mm/ka. Rates of this magnitude were indeed common during the Holocene transgression (Bloom, 1977), but they are one or two orders of magnitude larger than those postulated for longer term sea-level changes in the geologic record (Hancock & Kauffman, 1979; Haq and others, 1987). Thus, I see little indication at present that exposure preceding a "Vail-type" sea-level rise will significantly enhance the chances of drowning carbonate platforms.

Platform-specific relief.—

We have seen that the flanks of carbonate platforms tend to steepen as the platform matures and rises higher above the basin floor. This oversteepened carbonate relief deviates progressively from the more gentle relief created by siliciclastic deposition. (It also differs significantly from the relief built by evaporites—but this point is beyond the scope of this paper).

Excellent examples for the repeated development of carbonate-specific relief are to be found in the Persian Gulf syncline (Murris, 1980) and in the Permian of the Delaware Basin (Meissner, 1972; Wilson, 1975). The tendency toward typical platform relief is also reflected in the frequently cited principle of margin evolution from ramp to rimmed margin (Read, 1982).

Demise of the platform will "freeze" the carbonate relief formed gradually during the preceding growth period. Burial of this platform relief, however, and re-configuration into siliciclastic morphology may take longer than its construction. Dailly (1982) called this process "slope readjustment." In the case of the mid-Cretaceous platforms of the Gulf of Mexico, it is still not completed—100 m.y. after termination of platform growth.

The boundary between syn-platform and post-platform deposits is commonly sharp, unconformable, and likely to involve a significant lithologic change. It meets the criteria for a seismostratigraphic sequence boundary in the sense of Vail and others (1977). Sequence geometry at this stratigraphic turning point depends largely on how far the platform-specific relief has evolved. On platforms with heights of 500–800 m or more, the effect of carbonate oversteepening will mask any sea-level effects.

Recognition of drowning unconformities.—

Drowning unconformities typically include a change in lithology from shallow-water carbonates to deeper water siliciclastics or pelagics of variable composition. The succession has the form of a deepening-upward sequence. Subaerial exposure may occur in the platform sequence but not higher in the section. The boundary between the drowned platform and its deeper water cover is commonly a condensed horizon or submarine hardground (Schlager, 1981), but it may also be a rapid transition in outcrop. The last point is important to note. During the drowning of a carbonate platform and the subsequent change to another mode of deposition, the two lithosomes may mix for a short time. The outcrop section will record a rapid transition of lithologies, whereas the seismic tool indicates an unconformity and sequence boundary (see Stuart City/Georgetown boundary, Bebout and others, 1977, p. 237 and Bebout and others, 1981, fig. 7).

Drowning unconformities in seismic profiles are best developed on high platforms of 800 m or more. These platforms tend to develop strongly concave flanks with upper slope 6° or steeper, causing a disproportionately large seaward shift in onlap upon drowning. On moderately high platforms, the effect of slope oversteepening is minor, and the unconformity is largely caused by a change in the pattern of sediment input into the basin. Stoakes and Wendte (1987) and Dravis (1987) present good examples of drowning unconformities in low-relief settings. When platform growth is terminated by submergence below the photic zone, rather than by invasion of siliciclastics, the final relief on the platform commonly displays a raised rim and deep lagoon (Figs. 8, 11; Meyer, this volume). This "empty-bucket" profile is characteristic of platforms that are forced to grow very near their maximum rate.

The effects of platform oversteepening will be felt most strongly on passive continental margins during periods of long, uninterrupted carbonate sedimentation. Such a setting provides deep basins floored with thin ocean crust and long time for the relief to evolve. Conversely, the effects will be minimal in small, shallow, intracratonic basins that fill rapidly and frequently interrupt carbonate deposition with siliciclastic episodes. The lithology-specific effects on relief will be small in this setting, and changes in input pattern or sea-level signals may dominate the record.

Figure 14 illustrates the last point by comparing the sequence geometry around the Jurassic/Cretaceous boundary in a cratonic basin (Mitchum and Uliana, 1985) and a passive margin (Vail and others, 1977). In the cratonic basin, the postulated base-Valanginian lowstand is an event much like others in this succession. The postulated coeval event

The top contour of the Wilmington platform is an excellent example of a drowning unconformity. Typical features include: high platform-to-basin relief, concave, steep upper slope and termination by drowning, as indicated by a deep lagoon and raised rim, lack of fresh-water dissolution in the youngest platform deposits, and by a cover of deep-water carbonates predating the burying clastics (Meyer, this volume). The sicliciclastics bury this huge platform with a greatly reduced slope angle that leads to a pronounced onlap on the ocean-facing slope.

Early Cretaceous Lahave (Abenaki) platform.—

The Abenaki Formation on the Scotian shelf was analysed in exemplary fashion by Eliuk (1978). The platform is not quite as high as the Wilmington platform, yet clearly displays an unconformity with siliciclastic onlap of its seaward flank (e.g., Fig. 12; Eliuk, 1978; figs. 11, 12; Jansa, 1981, fig. 4). Where not covered by an advancing delta, the demise of the platform is attributed to drowning, i.e., "relative sea-level rise . . . rapid and of sufficient magnitude to stop carbonate deposition" (Eliuk, 1978, p. 472). The facies succession in the drowning interval resembles that of the Wilmington platform (Ringer and Patten, 1986). The timing of drowning may be intra-Valanginian (Eliuk, 1978) rather than basal Valanginian, as at Wilmington. Intra-Valanginian drowning is also indicated for at least part of the Moroccan platform belt by the turbidite record at DSDP Site 416 (see earlier discussion).

Toarcian, Djebel Bou Dahar, Morocco.—

The Early Jurassic platforms in the South Atlas trough display what seem to be drowning unconformities in outcrop. One example is the Djebel Bou Dahar (Fig. 13). The platform grew in the Liassic on a basement high. It was drowned in the Toarcian and subsequently onlapped by deep-water shales (Agard and Du Dresnay, 1965, Crevello, 1987). The angle between carbonate slope and shales frequently exceeds 15°.

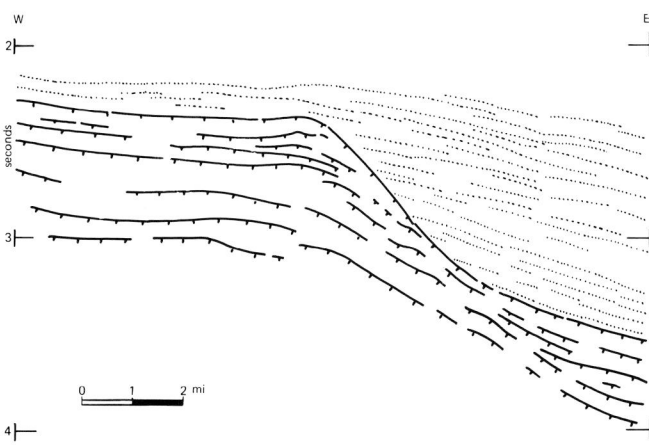

Fig. 12.—Drowning unconformity of Lahave platform off Nova Scotia. Line drawing after figure 4 in Eliuk (1978). Hachured lines—carbonates; dotted lines—siliciclastics and pelagics.

DISCUSSION

Drowning unconformities, sea-level fall and subaerial exposure.—

For lowstand unconformities in the sense of Vail and others (1977), a fall of sea level is obviously a basic requirement. This is not so for the geometrically similar drowning unconformities. Both the Lahave platform (Eliuk, 1978) and the Wilmington platform (Meyer, this volume) were repeatedly exposed during their growth period, but the last growth increment was immediately followed by drowning. There is no evidence that the drowning unconformity at the tops of these platforms is an exposure surface. Similar lack of exposure characterizes some of the mid-Cretaceous platforms around the Gulf of Mexico, e.g., parts of the Stuart City Reef (Bebout and others, 1977) and parts of the El Abra platform in Mexico (Enos and others, 1983). In these examples, the drowning unconformity is part of a deepening-upward sequence.

Fig. 13.—Drowning unconformity in outcrop, Djbel Bou-Dahar, Morocco. After Agard and du Dresnay (1965) and P. D. Crevello (pers. commun). Arrow marks boundary of Early Jurassic (Domerian) platform slope deposits to left and Middle Jurassic (Aalenian) onlapping flat-lying basinal limestones to right. Platform was drowned in late Early Jurassic (Toarcian). (Photo courtesy of P. D. Crevello.)

The frequent lack of subaerial exposure at drowning unconformities agrees with theoretical considerations. Termination by drowning requires a relative rise of sea level, and suffocation by clastics requires a relatively highstand of sea level to permit encroachment of clastics onto the platform top. Sea-level fall and exposure will also halt platform growth. Unless the old platform edifice is destroyed or covered during the time of exposure, however, the platform will resume growth when re-flooded. Exposure during a normal sea-level cycle will terminate platform growth only in conjunction with a subsequent rise. The discussion is reduced to the question whether exposure preceding a rise facilitates the demise of a platform. This is indeed the case, but the negative effect of exposure is a short one, as demonstrated by the Holocene transgression (Adey, 1978; Schlager, 1981). When platforms were flooded in the early Holocene, production commenced slowly, and reef growth and sediment accumulation fell behind sea level by tens of meters. This lag period lasted only 3,000–5,000 years. It was followed by extremely rapid growth, so that 10,000 years after flooding, the systems had again caught up with sea level. To damage a platform permanently through the initial lag effect, it must be submerged below the zone of optimum growth, while still operating in the lag phase. This requires very rapid rates of sea-level rise. If we assume a thickness of 20–40 m for the water layer of optimum carbonate growth, and a lag period of 5,000 years, we require a sea-level rise of 4,000–8,000 mm/ka. Rates of this magnitude were indeed common during the Holocene transgression (Bloom, 1977), but they are one or two orders of magnitude larger than those postulated for longer term sea-level changes in the geologic record (Hancock & Kauffman, 1979; Haq and others, 1987). Thus, I see little indication at present that exposure preceding a "Vail-type" sea-level rise will significantly enhance the chances of drowning carbonate platforms.

Platform-specific relief.—

We have seen that the flanks of carbonate platforms tend to steepen as the platform matures and rises higher above the basin floor. This oversteepened carbonate relief deviates progressively from the more gentle relief created by siliciclastic deposition. (It also differs significantly from the relief built by evaporites—but this point is beyond the scope of this paper).

Excellent examples for the repeated development of carbonate-specific relief are to be found in the Persian Gulf syncline (Murris, 1980) and in the Permian of the Delaware Basin (Meissner, 1972; Wilson, 1975). The tendency toward typical platform relief is also reflected in the frequently cited principle of margin evolution from ramp to rimmed margin (Read, 1982).

Demise of the platform will "freeze" the carbonate relief formed gradually during the preceding growth period. Burial of this platform relief, however, and re-configuration into siliciclastic morphology may take longer than its construction. Dailly (1982) called this process "slope readjustment." In the case of the mid-Cretaceous platforms of the Gulf of Mexico, it is still not completed—100 m.y. after termination of platform growth.

The boundary between syn-platform and post-platform deposits is commonly sharp, unconformable, and likely to involve a significant lithologic change. It meets the criteria for a seismostratigraphic sequence boundary in the sense of Vail and others (1977). Sequence geometry at this stratigraphic turning point depends largely on how far the platform-specific relief has evolved. On platforms with heights of 500–800 m or more, the effect of carbonate oversteepening will mask any sea-level effects.

Recognition of drowning unconformities.—

Drowning unconformities typically include a change in lithology from shallow-water carbonates to deeper water siliciclastics or pelagics of variable composition. The succession has the form of a deepening-upward sequence. Subaerial exposure may occur in the platform sequence but not higher in the section. The boundary between the drowned platform and its deeper water cover is commonly a condensed horizon or submarine hardground (Schlager, 1981), but it may also be a rapid transition in outcrop. The last point is important to note. During the drowning of a carbonate platform and the subsequent change to another mode of deposition, the two lithosomes may mix for a short time. The outcrop section will record a rapid transition of lithologies, whereas the seismic tool indicates an unconformity and sequence boundary (see Stuart City/Georgetown boundary, Bebout and others, 1977, p. 237 and Bebout and others, 1981, fig. 7).

Drowning unconformities in seismic profiles are best developed on high platforms of 800 m or more. These platforms tend to develop strongly concave flanks with upper slope 6° or steeper, causing a disproportionately large seaward shift in onlap upon drowning. On moderately high platforms, the effect of slope oversteepening is minor, and the unconformity is largely caused by a change in the pattern of sediment input into the basin. Stoakes and Wendte (1987) and Dravis (1987) present good examples of drowning unconformities in low-relief settings. When platform growth is terminated by submergence below the photic zone, rather than by invasion of siliciclastics, the final relief on the platform commonly displays a raised rim and deep lagoon (Figs. 8, 11; Meyer, this volume). This "empty-bucket" profile is characteristic of platforms that are forced to grow very near their maximum rate.

The effects of platform oversteepening will be felt most strongly on passive continental margins during periods of long, uninterrupted carbonate sedimentation. Such a setting provides deep basins floored with thin ocean crust and long time for the relief to evolve. Conversely, the effects will be minimal in small, shallow, intracratonic basins that fill rapidly and frequently interrupt carbonate deposition with siliciclastic episodes. The lithology-specific effects on relief will be small in this setting, and changes in input pattern or sea-level signals may dominate the record.

Figure 14 illustrates the last point by comparing the sequence geometry around the Jurassic/Cretaceous boundary in a cratonic basin (Mitchum and Uliana, 1985) and a passive margin (Vail and others, 1977). In the cratonic basin, the postulated base-Valanginian lowstand is an event much like others in this succession. The postulated coeval event on the Moroccan margin is disproportionately large. Again, I propose that the cause for this discrepancy lies in the fact that the sequence boundary off Morocco is really a drowning unconformity that preserved an extremely oversteepened platform relief. What sea level did in this time slice is of little importance compared to the degree of oversteepening that had developed at the margin.

The similarity between drowning unconformity and lowstand unconformity is not restricted to geometry. Sedimentologic features contribute to this parallel. The shift from deposition to erosion that accompanies an increase in slope angle leads to a situation where high-rising platforms are surrounded by erosional slopes even though their tops are still growing. On seismic profiles, this self erosion of a platform is difficult to distinguish from erosion by a lowstand of sea level. Evidence from the rocks may be required to make the distinction.

Sequence boundaries—changes in input and dispersal of sediment.—

The concept of depositional sequences as introduced by Sloss (1963) and Vail and others (1977, p. 53) is one of the most fundamental contributions to basin analysis. Vail and others (1977) defined the term as ". . . a stratigraphic unit composed of a relatively conformable succession of genetically related strata and bounded at its top and base by unconformities or their correlative conformities . . . it is determined by a single objective criterion—the physical relations of the strata themselves." In seismic stratigraphy, this means that subtle differences in the dip and configuration of bedding planes are used to set boundaries of depositional sequences.

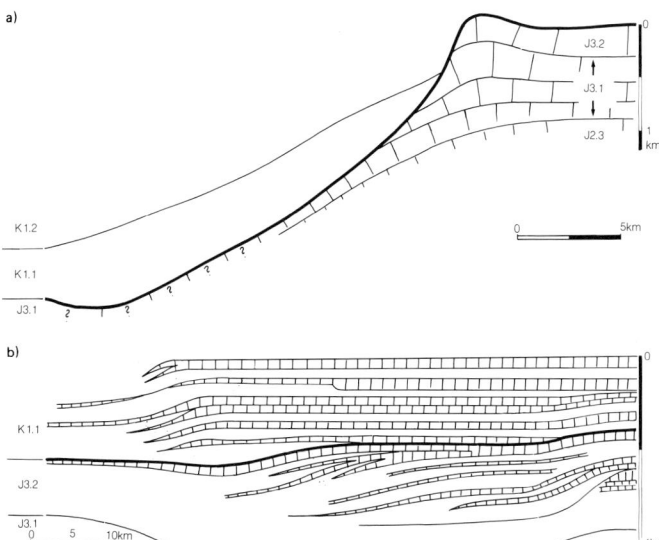

FIG. 14.—Sequence geometry of Late Jurassic and earliest Cretaceous sections in Neuquen basin, Argentina, and on African passive margin off Morocco. (a) Section off Morocco (after Vail and others, 1977). (b) Neuquen basin (after Mitchum and Uliana, 1985). Off Morocco, unconformity on top of platform indicates disproportionately large seaward shift in marine onlap; in shallow, intracratonic Neuquen basin, shifts in onlap are all of rather similar magnitude.

Vail and others (1977) proposed that sea level exerts the prime control on sequence geometry. It must be emphasized, however, that relative changes of sea level are not the only way to generate depositional sequences and sequence boundaries. The drowning unconformity is a sequence boundary formed by a change in sediment source. Changing the loci of sediment input in a basin may also create sequence boundaries.

In a very general way, we can say that sequence boundaries reflect changes in the pattern of sediment input and dispersal in a basin. Relative changes of sea level are a common but not the exclusive cause of such changes. The seismic stratigraphy of the deep Gulf of Mexico is a case in point (Addy and Buffler, 1984; Shaub and others, 1984). The units were explicitly defined by the criteria of Vail and others (1977). The units extend over several supercycles on the sea level chart (Haq and others, 1987), however, and their isopach maps clearly show different input patterns for each of these units (Shaub and others, 1984; figs. 7–12): input from carbonate platforms for the Challenger Unit, input from the west (Cordilleras) for the Campeche and Lower Mexican Ridges Unit, input from north and west for Upper Mexican Ridges Unit, and input mainly from north (Mississippi) for Cinco de Mayo and Sigsbee Units. The mid-Cretaceous unconformity stands out as the most dramatic stratigraphic turning point in this sequence. It signals both a change in sediment composition as well as changes in the points of entry of sediment into the basin (e.g., fig. 7). The other unit boundaries reflect changes in the points of entry only.

CONCLUSIONS

1. "Drowning" of platforms by submergence below the euphotic zone or by invasion of siliciclastics produces unconformities. These drowning unconformities are related to a rise or a highstand of sea level, yet they resemble the classic lowstand unconformity in geometry.

2. Two effects contribute to drowning unconformities: first, flanks of carbonate platforms are often steeper than the maximum angles tolerated by siliciclastic sediments, so that the clastics must onlap the platform flank and bury it from the bottom upward. Second, the pattern of sediment input and dispersal is different for shallow-water carbonates and siliciclastic sediments; rapid change from one to the other creates a seismic unconformity.

ACKNOWLEDGMENTS

I gratefully acknowledge the help of F. O. Meyer (Shell Oil Company) and R. T. Buffler and W. Corso (both University of Texas, Austin), who made important manuscripts available to me. Paul D. Crevello drew my attention to the drowning unconformities of Djebel Bou Dahar. Robert N. Ginsburg (University of Miami), J. Wendte (Esso Canada), and J. Dravis (Rice University, Houston) provided significant input at critical junctures of this study. Peter Vail (Rice University, Houston) and J. Kenter (Free University, Amsterdam) contributed important unpublished information, and Shell Oil Company and the Texas Institute for Geophysics (R. T. Buffler) made seismic lines available to me. To T.

Aigner (Royal Dutch Shell) and J. Dravis, I owe thoughtful reviews of this paper.

REFERENCES

ADDY, S. K., AND BUFFLER, R. T., 1984, Seismic stratigraphy of shelf and slope, northeastern Gulf of Mexico: American Association of Petroleum Geologists Bulletin, v. 68, p. 1782–1789.

ADEY, W. H., 1978, Coral reef morphogenesis: A multidimensional model: Science, v. 202, p. 831–837.

AGARD, J. AND DU DRESNAY, R., 1965, La region mineralisee du jbel Bou-Dahar, pres de Beni-Tajite (Haut-Atlas oriental): Etude geologique et metallogenique: Royaume du Maroc, Notes et Memoires du Service Geologique, v. 181, p. 135–152.

ALLEN, J. R. L., 1968, On criteria for the continuance of flute marks, and their implications: Geologie en Mijnbouw, v. 47, p. 3–16.

ANGSTADT, D. M., AUSTIN, J. A., JR., AND BUFFLER, R. T., 1985, Early Late Cretaceous to Holocene seismic stratigraphy and geologic history of southeastern Gulf of Mexico: American Association of Petroleum Geologists Bulletin, v. 69, p. 977–995.

AUSTIN, J. A., JR., SCHLAGER, W., PALMER, A., AND OTHERS, 1986, Proceedings of the Ocean Drilling Program; U.S. Government Printing Office, Washington D.C., v. 101A, p. 1–569.

———, ———, ———, ———, 1988, Leg 101—An overview: Proceedings of the Ocean Drilling Program U.S. Government Printing Office, Washington, D.C., v. 101B, 10 p.

ARTHUR, M. A., AND SCHLANGER, S. O., 1979, Cretaceous "Oceanic Anoxic Events" as causal factors in development of reef-reservoired giant oil fields: American Association of Petroleum Geologists Bulletin, v. 63, p. 870–885.

BACK, W., HANSHAW, B. W., HERMAN, J. S., AND VAN DRIEL, J. N., 1986, Differential dissolution of a Pleistocene reef in the groundwater mixing zone of coastal Yucatan, Mexico: Geology, v. 14, p. 137–140.

BALLY, A. W., ed., 1983, Seismic Expression of Structural Styles—A Picture and Work Atlas: American Association of Petroleum Geologists, Studies in Geology No. 15, V.I, 39 p., 28 pls; v. II, 56 pls; v. III, 53 pls.

———, WATTS, A. B., GROW, J. A., MANSPEIZER, W., BERNOULLI, D., SCHREIBER, C., AND HUNT, J. M., 1981, Geology of Passive Continental Margins: History, Structure and Sedimentologic Record: American Association of Petroleum Geologists, Education Course Note Series No. 19, p. 1–324.

BEBOUT, D. G., BUDD, D. A., AND SCHATZINGER, R. A. 1981, Depositional and diagenetic history of the Sligo and Hosston Formations (Lower Cretaceous) in South Texas: Bureau of Economic Geology, University of Texas at Austin, Report of Investigations, v. 109, p. 1–70.

———, SCHATZINGER R. A., AND LOUCKS, R. G., 1977, Porosity distribution in the Stuart City trend, Lower Cretaceous, South Texas, in Bebout, D. G., and Loucks, R. G., eds., Cretaceous Carbonates of Texas and Mexico: Bureau of Economic Geology, University of Texas at Austin, Report of Investigations No. 89, p. 4.

BLOOM, A. L., 1977, Atlas of sea-level curves: Cornell University, Ithaca, New York, p. I–VII, A1–14, B1–16, C1–40, D1–D27, E1–5, 1–12.

BUFFLER, R. T., SCHLAGER, W., AND LEG 77 SCIENTIFIC PARTY, 1984, Initial Reports of the Deep Sea Drilling Project: U.S. Government Printing Office, Washington D.C., v. 77, p. 1–747.

———, WATKINS, J. S., SHAUB, F. J., AND WORZEL, J. L., 1980, Structure and early geologic history of the deep central Gulf of Mexico, in Pilger, R. H., Jr., ed., The Origin of the Gulf of Mexico and the Central North Atlantic: Proceedings of a symposium at Louisiana State University, p. 3–16.

COOK, H. E., AND MULLINS, H. T., 1983, Basin margin environment, in Scholle, P. A., Bebout, D. G., and Moore, C. H., eds., Carbonate Depositional Environments: American Association of Petroleum Geologists Memoir 33, p. 540–617.

COOK, T. D., AND BALLY, A. W., eds., 1975, Stratigraphic Atlas of North America: Princeton University Press, Princeton, p. 1–272.

CORSO, W., 1988, Development of the Early Cretaceous northwest Florida carbonate platform: unpublished Ph.D. Dissertation, University of Texas at Austin, p. 1–180.

CREVELLO, P. D., WARME, J. E., SEPTFONTAINE, M., BURKE, R. B., 1987, Evolution of Jurassic carbonate platforms in an active transtensional rift: High Atlas of Morocco: American Association of Petroleum Geologists Bulletin, v. 71, p. 543.

DAILLY, G. C., 1982, Slope readjustment during sedimentation on continental margins, in Watkins, J. S., and Drake, C. L., eds., Studies in Continental Margin Geology: American Association of Petroleum Geologists Memoir 34, p. 593–608.

DRAVIS, J. J., 1987, Regional facies and porosity relationships in Jurassic Haynesville limestones of East Texas (abs.): Houston Geological Society Bulletin, p. 10.

ELIUK, L. S., 1978, The Abenaki Formation, Nova Scotia shelf, Canada—A depositional and diagenetic model for a Mesozoic carbonate platform: Bulletin of Canadian Petroleum Geology, v. 26, p. 424–514.

ENOS, P., MINERO, C. J., AND AGUAYO-C., J. E., 1983, Sedimentation and diagenesis of Mid-Cretaceous platform margin east-central Mexico: Dallas Geological Society Field Guide, Dallas, Texas, p. 1–168.

HANCOCK, J. M., AND KAUFFMAN, E. G., 1979, The great transgressions of the Late Cretaceous: Journal of the Geological Society of London, v. 136, p. 175–186.

HAQ, B. U., HARDENBOL, J., AND VAIL, P. R., 1987, Chronology of fluctuating sea levels since the Triassic (250 million years ago to present): Science, v. 235, p. 1156–1167.

HARRIS, P. M., FROST, S. H., SEIGLIE, G. A., AND SCHNEIDERMANN, N., 1984, Regional unconformities and depositional cycles, Cretaceous of the Arabian Peninsula, in Schlee, J. S., ed., Interregional Unconformities and Hydrocarbon Accumulation: American Association of Petroleum Geologists Memoir 36, p. 67–80.

HUBBARD, R. J., PAPE, J., AND ROBERTS, D. G., 1985, Depositional sequence mapping as a technique to establish tectonic and stratigraphic framework and evaluate hydrocarbon potential on a passive continental margin, in Berg, O. R., and Woolverton, D. G., eds., Seismic Stratigraphy II: American Association of Petroleum Geologists Memoir 39, p. 79–91.

JAMES, N. P., AND GINSBURG, R. N., 1979, The seaward margin of Belize barrier and atoll reefs: International Association of Sedimentologists Special Publication 3, p. 191.

JANSA, L. F., 1981, Mesozoic carbonate platforms and banks of the eastern North American margin: Marine Geology, v. 44, p. 97–117.

KARLO, J. F., 1986, Results of exploration in Mesozoic shelf-edge carbonates, Baltimore Canyon basin: American Association of Petroleum Geologists Bulletin, v. 70, p. 605–606.

KAUFFMANN, E. G., 1977, Geological and biological overview: Western Interior Cretaceous basin: The Mountain Geologist, v. 14, p. 75–79.

LANCELOT, Y., WINTERER, E. L., AND OTHERS, 1980, Initial Reports of the Deep Sea Drilling Project: U.S. Government Printing Office, Washington D.C., v. 50, p. 1–868.

LAND, L. S., AND MOORE, C. H., JR., 1977, Deep forereef and upper island slope, North Jamaica: American Association of Petroleum Geologists, Studies in Geology No. 4, p. 53–65.

MEISSNER, F. F., 1972, Cyclic sedimentation in Middle Permian strata of the Permian Basin, West Texas and New Mexico, in Elam, J. C., and Chuber, S., eds., Cyclic Sedimentation in the Permian Basin: West Texas Geological Society, Midland, Texas, p. 203–232.

MEYER, F. O., 1986, Facies specificity of megaporosity in Mesozoic shelf-edge carbonates, Baltimore Canyon Basin: American Association of Petroleum Geologists Bulletin, v. 70, p. 621.

MITCHUM, R. M., JR., AND ULIANA, M. A., 1985, Seismic stratigraphy of carbonate depositional sequences, Upper Jurassic-Lower Cretaceous, Neuquen Basin, Argentina, in Berg, O. R., and Woolverton, D. G., eds., Seismic Stratigraphy II: American Association of Petroleum Geologists Memoir 39, p. 255–274.

MULLINS, H. T., NEUMANN, A. C., WILBER, R. J., AND BOARDMAN, M. R., 1980, Nodular carbonate sediment on Bahamian slopes: Possible precursors to nodular limestones: Journal of Sedimentary Petrology, v. 50, p. 117–131.

———, AND ———, 1979, Geology of the Miami terrace and its paleoceanographic implications: Marine Geology, v. 30, p. 205–232.

MURRIS, R. J., 1980, Middle East: Stratigraphic evolution and oil habitat: American Association of Petroleum Geologists Bulletin, v. 64, p. 597–618.

PAULL, C. K., AND NEUMANN, A. C., 1987, Continental margin brine seeps: Their geological consequences: Geology, v. 15, p. 545–548.

PHAIR, R. L., 1984, Lower Cretaceous seismic stratigraphy of the southeastern Gulf of Mexico and southwestern Straits of Florida: Unpublished M. A. Thesis, University of Texas at Austin, p. 1–300.

READ, J. F., 1982, Carbonate platforms of passive (extensional) type continental margins: Types, characteristics and evolution: Tectonophysics, v. 81, p. 195–212.

RINGER, E. R., AND PATTEN, H. L., 1986, biostratigraphy and depositional environments of Late Jurassic and Early Cretaceous carbonate sediments in Baltimore Canyon Basin: American Association of Petroleum Geologists Bulletin, v. 70, p. 639–640.

SCHLAGER, W., 1980, Mesozoic calciturbidites in DSDP hole 416A—Petrographic recognition of a drowned carbonate platform, in Lancelot, Y., Winterer, E. L., and others, Initial Reports Deep Sea Drilling Project: U.S. Government Printing Office, Washington, D.C., v. 50, p. 733–749.

———, 1981, The paradox of drowned reefs and carbonate platforms: Geological Society of America Bulletin, v. 92, p. 197–211.

———, BUFFLER, R. T., ANGSTADT, D., AND PHAIR, R., 1984, Geologic history of the southeastern Gulf of Mexico, in Buffler, R. T., Schlager, W., and others, Initial Reports of the Deep Sea Drilling Project: U.S. Government Printing Office, Washington, D.C., v. 77, p. 715–738.

———, AND CAMBER, O., 1982, Depositional, erosional and bypass slopes on carbonate platforms (abst.): International Association of Sedimentologists, Eleventh Congress on Sedimentology, Hamilton, Ontario, Canada, p. 179.

———, AND ———, 1986, Submarine slope angles, drowning unconformities, and self-erosion of limestone escarpments: Geology, v. 14, p. 762–765.

———, AND GINSBURG, R. N., 1981, Bahama carbonate platforms—The deep and the past: Marine Geology, v. 44, p. 1–24.

———, AND JAMES, N. P., 1978, Low Mg-calcite limestones forming on the deep-sea floor (Tongue of the Ocean, Bahamas): Sedimentology, v. 25, p. 675–702.

SHAUB, E. J., BUFFLER, R. T., AND PARSONS, J. G., 1984, Seismic stratigraphic framework of deep central Gulf of Mexico Basin: American Association of Petroleum Geologists Bulletin, v. 68, p. 1790–1802.

SHERIDAN, R. E., 1974, Atlantic continental margin of North America, in Burk, C. A., and Drake, C. L., eds., The Geology of Continental margins: Springer-Verlag, New York, p. 391–407.

SLOSS, L. L., 1963, Sequences in the cratonic interior of North America: Geological Society of America Bulletin, v. 74, p. 93–114.

STEIGER, T., AND COUSIN, M., 1985, Microfacies of the Late Jurassic to Early Cretaceous carbonate platform at the Mazagan Escarpment (Morocco): Oceanologica Acta, Special v. 5, p. 111–126.

STOAKES, F. A. AND WENDTE, J. L., 1987, The Woodbend Group: Canadian Society of Petroleum Geology, Core Conference Manual at the Second International Symposium on the Devonian System, Calgary, Canada, p. 153–170.

VAIL, P. R., MITCHUM, R. M., TODD, R. G., WIDMIER, J. M., THOMPSON, S., SANGREE, J. B., BUBB, J. N., AND HATLELID, W. G., 1977, Seismic stratigraphy and global changes of sea level, in Payton, C. E., ed., Seismic Stratigraphy—Applications to Hydrocarbon Exploration: American Association of Petroleum Geologists Memoir 26, p. 49–212.

VON RAD, U., AUZENDE, J. M., RUELLAN, E., AND CYAMAZ GROUP, 1985, Stratigraphy, structure, paleoenvironment and subsidence history of the Mazagan Escarpment off central Morocco: A CYAMAZ synthesis: Oceanologica Acta, Special v. 5, p. 161–182.

WIEDMANN, J., BUTT, A., AND EINSELE, G., 1982, Cretaceous stratigraphy, environment, and subsidence history at the Moroccan continental margin, in Von Rad, U., Hinz, K., Sarnthein, M., and Seibold, E., eds., Geology of the Northwest African Continental Margin: Springer-Verlag, Berlin, p. 366–395.

WILSON, J. L., 1975, Carbonate Facies in Geologic History: Springer-Verlag, New York, p. 1–411.

WINKER, C. D., AND BUFFLER, R. T., 1988, Paleogeographic evolution of early deep-water Gulf of Mexico and margins, Jurassic to Middle Cretaceous (Comanchean): American Association of Petroleum Geologists Bulletin, v. 72, p. 318–346.

QUANTITATIVE MODELING OF CARBONATE PLATFORMS: SOME EXAMPLES

THOMAS AIGNER AND MARK DOYLE
Shell Research B.V., Volmerlaan 6, 2288 GD Rijswijk, The Netherlands
DAVID LAWRENCE
Shell Development Company, 3737 Bellaire Blvd., Houston, Texas 77025 USA
AND
MANFRED EPTING AND ARTHUR VAN VLIET
Shell Research B.V., Volmerlaan 6, 2288 GD Rijswijk, The Netherlands

ABSTRACT: A deterministic computer program has been developed to simulate the stratigraphic evolution of two-dimensional transects across sedimentary basins. Clastic, carbonate, and mixed clastic-carbonate systems can be simulated. The main application of the program is to simulate basin stratigraphy using all available data as constraints. In this paper we concentrate on the simulation of carbonate systems and illustrate the following two specific applications of the program.

(1) The history of sea-level fluctuations is reconstructed using the stratigraphy and geometry of carbonate systems as constraints. In a first example, the architecture of isolated Miocene carbonate buildups is simulated using sea-level curves similar to published eustatic charts. In a second example, the stratigraphic patterns at the margin of a Mesozoic cratonic carbonate basin are used to derive an estimate of the sea-level history.

(2) Our understanding of the controls on carbonate platform architecture is improved by isolating individual processes. In this respect, we investigate the possible significance of isostasy in controlling the formation of intra- and inter-platform basins and in the initiation of carbonate pinnacles.

INTRODUCTION

We present some examples of the numerical simulation of carbonate platform architecture using a computer program designed to model the stratigraphic development of two-dimensional transects through sedimentary basins. The program has been developed over a number of years at the Shell Research Laboratories in Houston (USA) and Rijswijk (The Netherlands). It is a forward-modeling program operating within the framework of generally accepted stratigraphic, sedimentologic, and geophysical principles. The program uses algorithms that describe the deposition and erosion of clastic and carbonate sediments, sediment compaction, and the isostatic response of the basement to surface loads and to horizontally directed tectonic forces (Lawrence and others, 1987). A principle limitation of the program is that it is only two-dimensional.

Detailed descriptions of the concepts employed and their numerical representation in the program code is beyond the intent and scope of this paper; however, the key input parameters into the model are: (1) tectonic subsidence history (from user-specified subsidence rates, or from lithosphere stretching models); (2) sea-level history (either freely specified by the user or from a library of digitized time/amplitude sequences from available sea-level curves); (3) clastic sediment input rate (specified by the user as the rate of sediment introduction, the rate of sediment dispersion, erosion, and slope stability); and (4) carbonate growth potential (following empirical relations between average carbonate production and water depth within the photic zone).

A simulation is performed on a spatial grid and in a sequence of small steps from a prescribed set of initial conditions. Clastic, carbonate, and mixed carbonate/clastic depositional systems can be handled. During program application, the user varies the input parameters within the limits of known geological constraints, until the model is consistent with all available data and observations. Model results are displayed in two-dimensional cross sections showing the geometric relations between stratigraphic units and the overall facies distribution (Fig. 1).

The program is mainly applied to simulate the distribution of depositional facies in a basin using all available data as constraints. A subsidiary application is in the development, evaluation, and demonstration of conceptual models. Here we illustrate the potential of numerical simulation of carbonate platforms as a means of constraining sea-level variations and improving our understanding of the controls on carbonate platform evolution. In the latter respect, we highlight the possible significance of isostasy on carbonate systems.

CONSTRAINTS ON SEA-LEVEL VARIATIONS

Sea-level changes are a fundamental control on the geometry and the internal architecture of carbonate systems. While periods of rapidly rising sea level often lead to partial drowning of carbonate platforms (Kendall and Schlager, 1981; Schlager, 1981), sea-level lowstands may cause subaerial exposure with leaching and other diagenetic modifications. Computer simulation of platform architecture can be used to constrain the amplitude and timing of the sea-level changes and hence either to evaluate independently derived eustatic curves, or to derive a local curve. In this section we demonstrate this application with two examples from the subsurface.

Simulation of Buildup Architecture

Carbonate buildups of central Luconia, offshore Sarawak (Malaysia) (Fig. 2), display various geometries during their evolution. Transgressive, buildup, buildout, buildin, and subaerial exposure phases can be distinguished (for details see Epting, 1988). A qualitative comparison of the architecture with Exxon's latest eustatic chart (Haq and others, 1987) suggests a relation to sea level, which is supported by the fact that the neighboring Miocene Terumbu carbonates also seem eustatically controlled (May and Eyles, 1985). This possibility can be tested by computer simulation of three of the buildups using a sea-level history optimized to reproduce the observed architecture individually (Fig. 3). This "optimization" is performed via a series of modeling

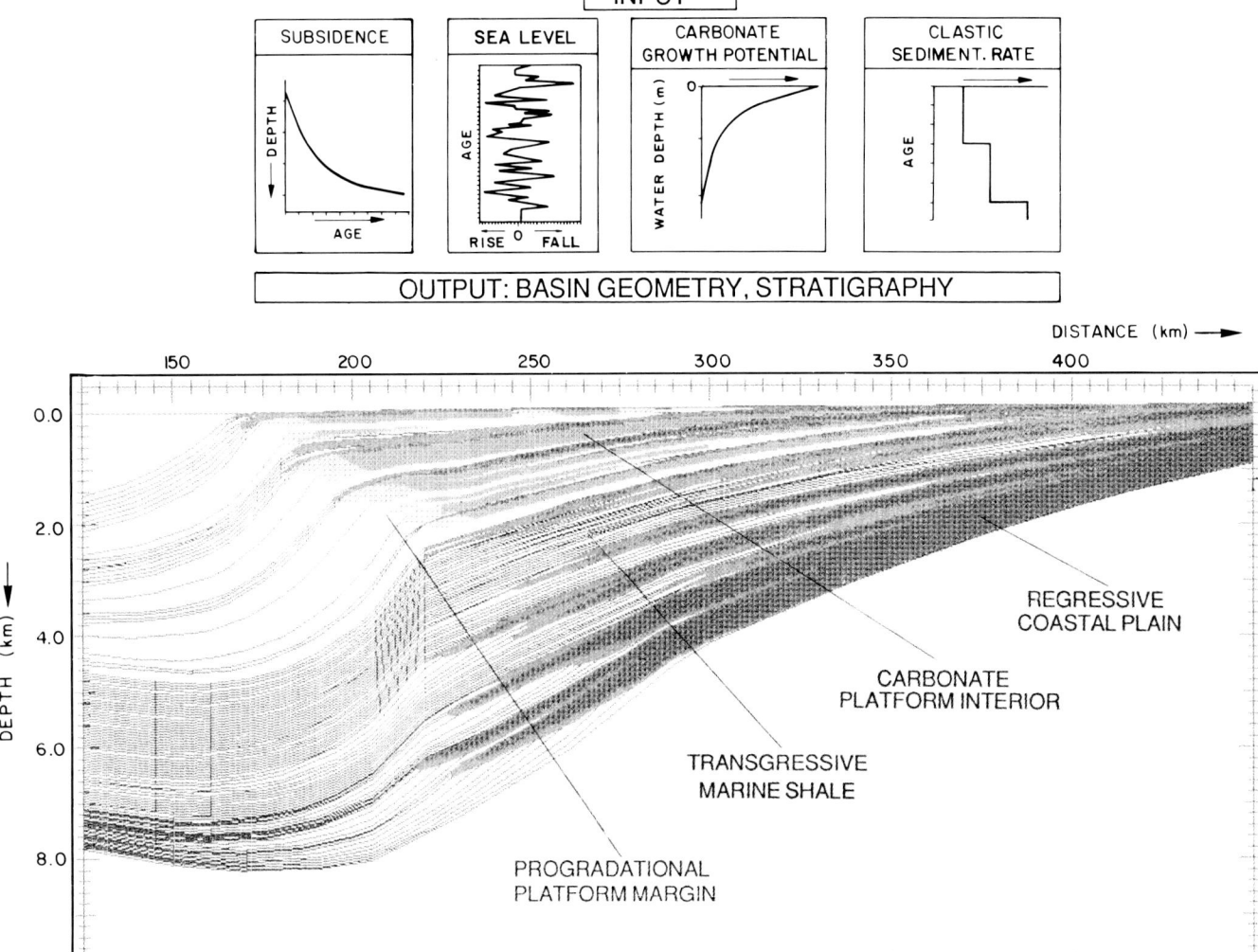

Fig. 1.—Illustration of the computer program that simulates basin stratigraphy. Principle input parameters are the sea-level and subsidence history, clastic sedimentation rate, and the carbonate growth potential. Typical model output is a cross section displaying basin geometry and stratigraphy. Black lines are time lines spaced at 1-m.y. intervals.

runs, changing the input parameters on a trial-and-error basis until a best fit is achieved.

Figure 3A shows a simulation of the E.11 buildup. Parameters used for modeling include uniform tectonic subsidence (30 m/m.y.) and a carbonate growth potential of 450 m/m.y. For the F.6 and G.10 buildups (Fig. 3B, C), situated more basinward and further away from the clastic source, higher tectonic subsidence rates (60 m/m.y. and 70 m/m.y., respectively) and higher carbonate growth potentials were used (580 m/m.y. and 650 m/m.y., respectively).

The three iteratively derived sea-level curves are all similar in shape and in absolute amplitudes (Fig. 4). The departure of the E.11 curve in the final drowning phase suggests an additional control that is not included in the simulation (such as clastic pollution from prograding deltas). Whereas the timing of falls and rises in the three derived curves is similar to the eustatic curve of Haq and others (1987), the absolute magnitudes of the events differ (Fig. 4). This difference probably reflects the effects of the global smoothing implicit in the derivation of the eustatic curve, or may again be due to local events that have not been included in the simulation.

Simulation of Basin Margin Stratigraphy

A cross section based on well logs from the Mesozoic of the Rub'al Khali basin margin of the Arabian Peninsula (Fig. 5A) displays an alternating sequence of sheetlike units of uniform thickness, and wedge-shaped units that thin toward the margin. The regional geology implies a simple structural and tectonic subsidence history. Slow rates of tectonic subsidence and the persistence of shallow-marine environments suggest that the stratigraphic geometries are mainly controlled by sea-level variations. A simple conceptual model involving the interaction between sea-level changes and rotational tectonic subsidence can explain the relations. This model predicts the following: during periods of slow rates of eustatic sea-level changes, space creation is dominated by the rotating substrate, and wedge-shaped units, thinning toward the basin margin, are generated. During pulses of rapid eustatic sea-level rises, however, the contemporaneous rotation of the substrate becomes negligible, and regional sheetlike units are deposited. Computer simulation allows us to demonstrate the concept easily and to constrain the likely magnitudes of rate changes that explain the observed stratigraphy.

The simulated stratigraphic cross section and the derived sea-level curve are shown in Figure 5C and D. This model explains with reasonable accuracy the gross basin margin stratigraphy, major sequence-bounding unconformities, relative thicknesses, and geometrical relations of the depositional units. Input parameters used to construct the model include rotational tectonic subsidence increasing to a maximum of 4.5 m/m.y. and a constant carbonate growth potential of 100 m/m.y. Again, the sea-level history was derived iteratively until the best fit was achieved.

Whereas computer simulation of basin stratigraphy can be used to constrain the magnitude and timing of sea-level variations, it is important to recognize that not all high-frequency signals should be attributed to eustasy. Curves derived from the analysis of individual systems only constrain the relative sea-level history and may contain signals from, for instance, the regional isostatic response to variations in plate boundary forces, and the local effects of fault reactivation and mobile substratum.

ISOSTATIC CONTROLS ON CARBONATE PLATFORMS

It is well known that, at the sites of continental collision, foredeep basins develop as a result of downflexing of the overridden plate under the load of the advancing thrust belt (Beaumont and others, 1982). While much smaller in magnitude, carbonate platforms also represent surface loads, and the lithosphere should flex accordingly. The isostatic response to this load, however, and its possible control on platform architecture have received little attention. Computer simulation can be used to investigate and isolate the role of isostasy in controlling carbonate platform architecture. In this section we highlight some features of carbonate platform evolution that could be attributed to isostasy.

Numerous studies (e.g., Watts and Cochran, 1974; Watts and others, 1982, among others) have suggested that a

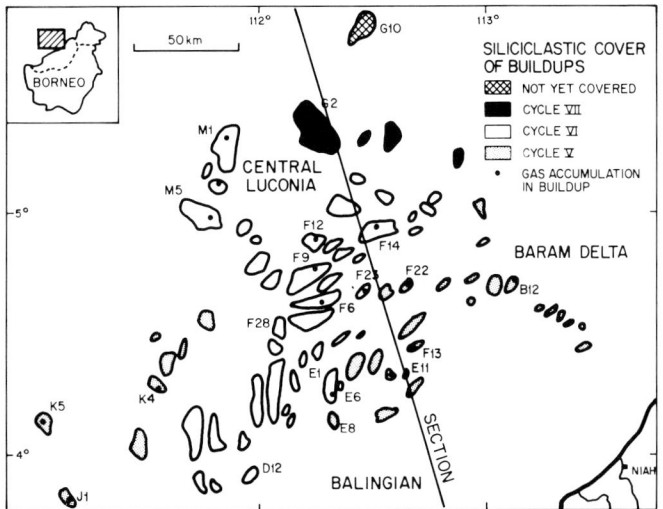

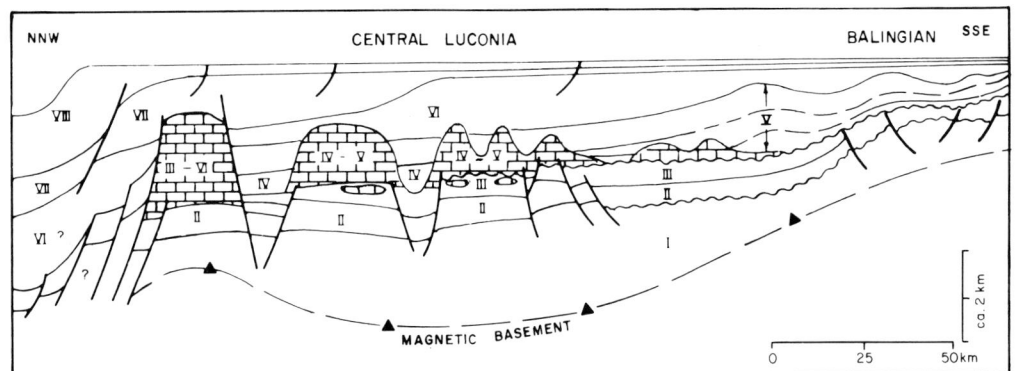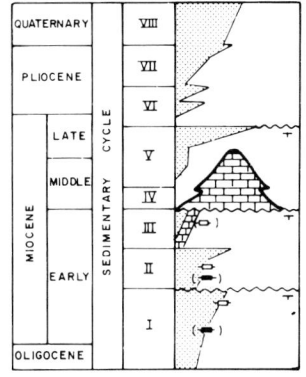

FIG. 2.—Miocene carbonate buildups in central Luconia, offshore Borneo (from Epting, 1988). (A) Areal distribution of major buildups. The buildups are covered by siliciclastic sediments of progressively younger sequences toward the north. (B) Stratigraphic relation between buildups and depositional sequences.

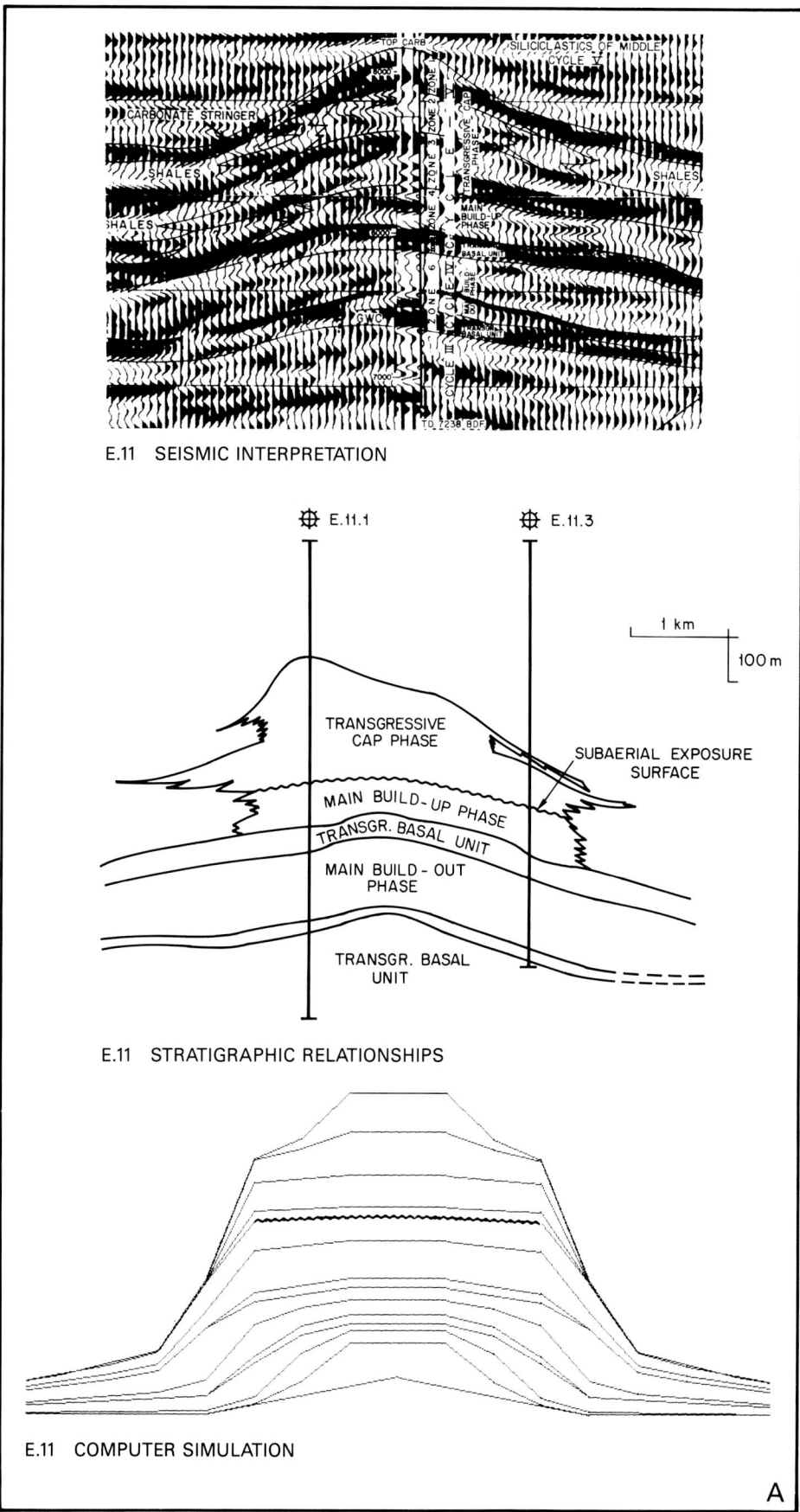

Fig. 3.—Comparisons of the observed and the simulated buildup architectures. (A) The E.11 buildup; (B) the F.6 buildup; (C) the G.10 buildup. Wavy lines in the simulations of (A) and (B) represent surfaces of subaerial exposure. Time lines in all simulations are spaced at 0.5-m.y. intervals.

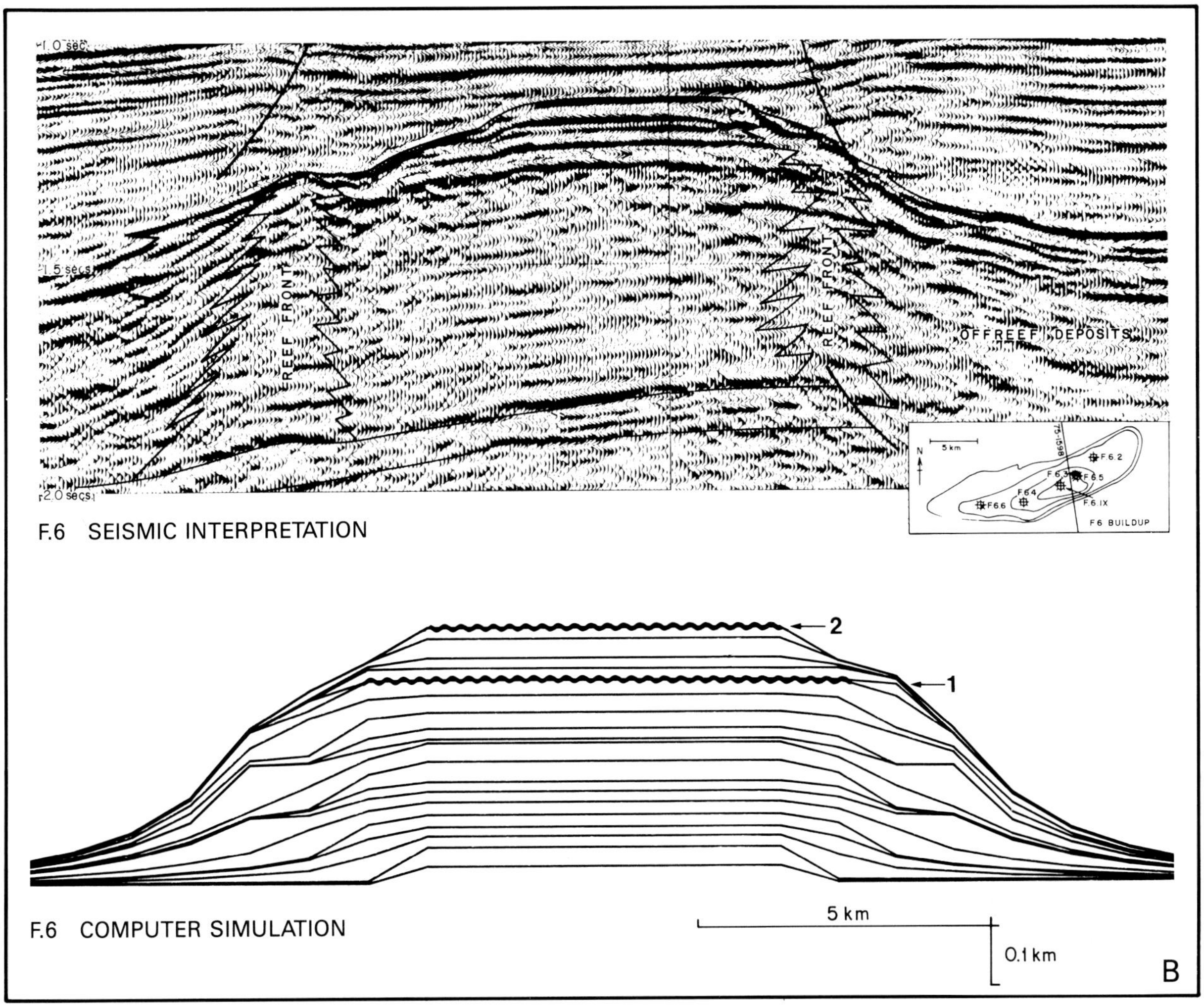

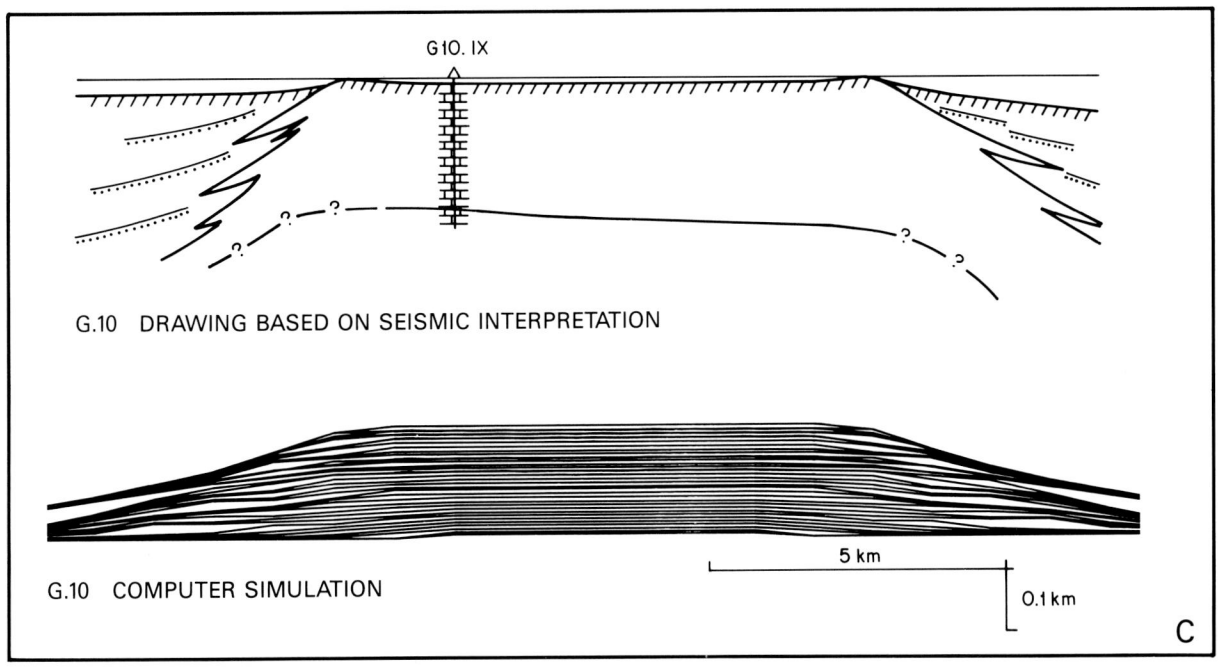

1.0 sec
1.5 secs
2.0 secs
REEF FRONT
REEF FRONT
OFFREEF DEPOSITS
F6 BUILDUP
F.6 SEISMIC INTERPRETATION
2
1
F.6 COMPUTER SIMULATION
5 km
0.1 km
B

G10. IX
G.10 DRAWING BASED ON SEISMIC INTERPRETATION
G.10 COMPUTER SIMULATION
5 km
0.1 km
C

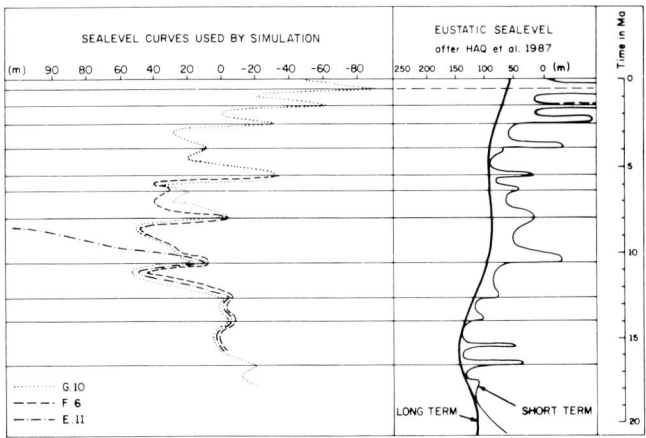

FIG. 4.—Comparison of eustatic curve of Haq and others (1987) with sea-level curves used to simulate three buildups in central Luconia. Note that the general shapes of the curves are remarkably similar, but absolute amplitudes differ. The sea-level rise required to drown the E.11 buildup is unrealistically high, suggesting that some other cause, e.g., clastic influx, contributed to buildup termination.

lithosphere plate responds rigidly in response to long-term surface loads. An isolated surface load will induce a deflection on a regional scale both beneath and surrounding the load. Beyond the edges of the load, a flexural bulge may develop (i.e., a localized region of relative basement uplift). The form of the induced deflections depends on the magnitude and geometry of the load and on the strongly temperature-dependent rheology of the lithosphere. These studies suggest that the isostatic response to long-term surface loads can be modeled in terms of the deflections of a thin elastic plate overlying a weak fluid. The appropriate equivalent elastic-plate thickness depends on the thermal age of the basement (defined by the time since the last major heating event), and appears to vary from 0–10 km near mid-ocean ridges to perhaps 100 km in continental interiors (Karner and others, 1983). Some examples of the modeled isostatic response to idealized surface loads are shown in Figure 6 for effective elastic thicknesses of 10, 25, and 50 km.

Isostatic Model for Intra- and Inter-Platform Basins

Intra-platform (or intra-shelf) basins.—

Intra-platform (or intra-shelf) basins are relatively shallow depressions in the interior of extensive carbonate platforms. Although they occur in carbonate platforms of different ages (see Murris, 1980; Markello and Read, 1982; Read, 1985), no generally applicable and satisfactory explanation for their formation has yet been proposed. Figure 7 shows such intra-shelf basins on the Arabian Peninsula, including the Hanifa basin which hosts prolific source rocks and was separated from the open ocean by a high-energy platform margin (Murris, 1980). Figure 8 shows a series of simulations of a platform where parameters were chosen to model the Arabian platform at the time just prior to the deposition of the Hanifa source rocks. An intra-platform basin forms during a major rise in sea level, which drowns the isostatically sagged platform interior. The sea-level history used corresponds to that derived by an earlier modeling exercise from the margin of same basin (cf. Fig. 5D). Since this cross section is situated more basinward, however, this simulation used a higher tectonic subsidence rate (6 m/m.y.) and a lower carbonate growth potential (55 m/m.y.) to account for the restricted circulation in this epeiric sea. Elastic thicknesses varied between 19 and 95 km.

Inter-platform basins.—

Carbonate complexes are often differentiated into an array of steep-sided platforms and deep inter-platform basins that may be underlain by an extensive sheetlike "megabank." The Bahamas are a modern example. In some cases there is clear evidence that these interplatform basins are fault-controlled tectonic grabens formed during rifting (e.g., Winterer and Bosselini, 1981). Figure 9 demonstrates, however, an atectonic model in which a major sea-level rise around 100 Ma. (taken from the curve of Haq and others, 1987) triggers the differentiation of the megabank into an array of platforms and inter-platform troughs simply in response to flexural loading (elastic thickness of 30 km), post-rift thermal subsidence (stretching factor beta = 2), and a constant carbonate growth potential (170 m/m.y.). During this sea-level rise, carbonate production keeps pace with rising sea level only along the margins of the isostatically sagged megabank (Fig. 9B). Subsequently, the isostatic response to the modified load distribution, combined with a fall in sea-level, brings two additional areas into the photic zone (Fig. 9C), resulting in four isolated platforms (Fig. 9D).

Discussion.—

The isostatic model for the generation of inter- and intra-platform basins is sensitive to the assumed value of carbonate growth potential. The mechanism only works for an effective, long-term growth potential of less than 200 m/m.y., which is consistent with compilations made by Schlager (1981) for the pre-Holocene. While there are many other controls on carbonate systems that are capable of producing the observed features, the isostatic control is present in all cases and under certain conditions can be expected to predominate. In general terms, in areas with high rates of tectonic subsidence (e.g., young passive margins), a sea-level rise may trigger the differentiaton of an extensive platform into an archipelago of isolated platforms and inter-platform basins. The platform-and-basin topography may persist for substantial periods of time since the rate of tectonic subsidence will be similar to the rate of sea-level change, at least during non-glacial intervals. In areas of lower rates of tectonic subsidence (e.g., stable cratonic basins), major sea-level rises can be expected to transform carbonate platform interiors into intra-platform basins. These basins will tend to be short-lived and may be filled during a subsequent sea-level cycle.

Isostatic Model for Carbonate Pinnacle Initiation

In a number of basins, isolated carbonate pinnacles are known to occur preferentially at short distances (10–50 km) basinward of carbonate platforms (see Wilson, 1975). There

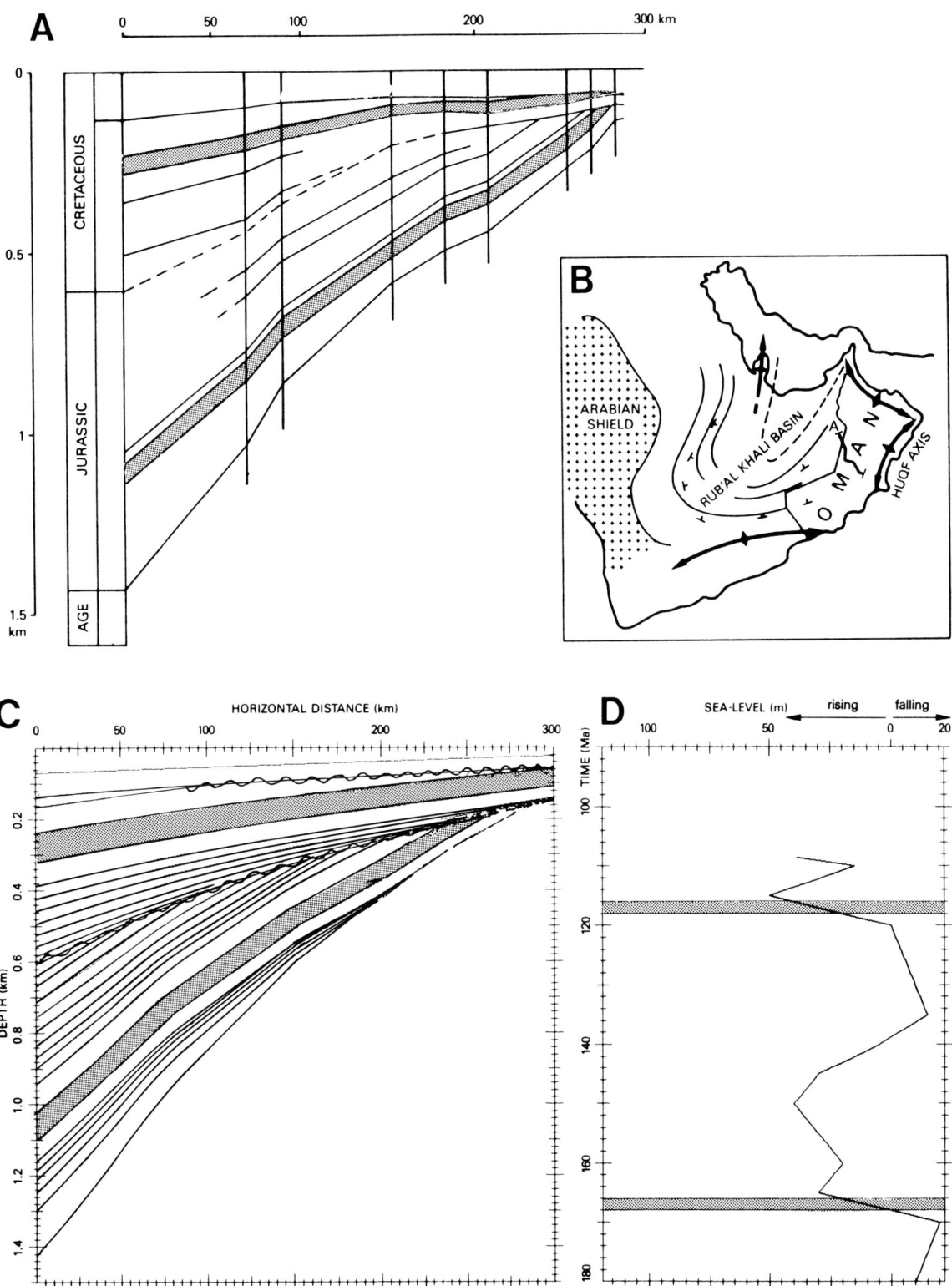

Fig. 5.—(A) Stratigraphic cross section through Mesozoic strata of Rub'al Khali basin margin based on log correlation. Note differences between wedgelike and sheetlike units (shaded). (B) Map of Arabian Peninsula. Cross section of (A) is located at the southeast margin of the Rub'al Khali basin. (C) Computer simulation of the stratigraphy shown in (A). Note again difference between wedge- and sheetlike units (shaded). Time lines are spaced at 2-m.y. intervals, wavy lines represent surfaces of subaerial exposure. (D) Sea-level curve used for the simulation. Note that sheetlike units are generated during rapid sea-level rises.

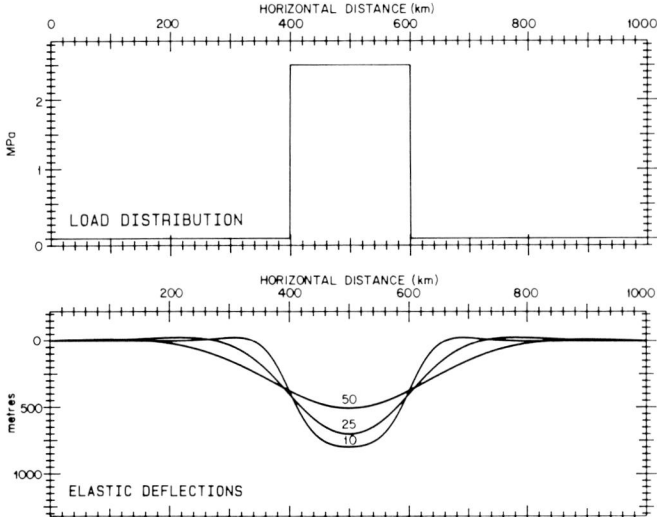

FIG. 6.—Illustration of the elastic-plate model for the isostatic response to surface loads. Computed basement deflection curves in response to a 200-km-wide rectangular load. The response has been computed for elastic-plate thicknesses of 10, 25, and 50 km. As the plate thickness decreases, the plate becomes weaker, and the deflection is more localized beneath the load. Note flexural bulges on either side of main depressions.

is probably a variety of factors controlling pinnacle initiation, and a number of models have been proposed. The observation that pinnacles often occur close to the platform edge suggests a causal relation and led us to examine if such pinnacles could, in principle, also be generated by a simple isostatic mechanism. This does not imply that we propose to explain all carbonate pinnacles with such a model.

The isostatic model of pinnacle initiation involves the interaction between a eustatic cycle and the isostatic response of the basement to the load of a carbonate platform. Figure 10 A-C illustrates this mechanism with two stationary carbonate platforms developing in a uniformly subsiding basin. The deflection induced by the load displays two small-amplitude flexural bulges basinward of each platform. During a sea-level fall, buildup of the platform ceases due to subaerial exposure, but growth is initiated in the basinal area (Fig. 10B). Since carbonate growth rates are a function of water depth (Schlager, 1981), the areas on the flexural bulges grow at a faster rate and small pinnacles begin to develop. During a sea-level rise and resumption of platform growth, only the pinnacle reefs in the basinal area have sufficient growth potential to keep pace with rising sea level (Fig. 10C). The position of the flexural bulge, and hence

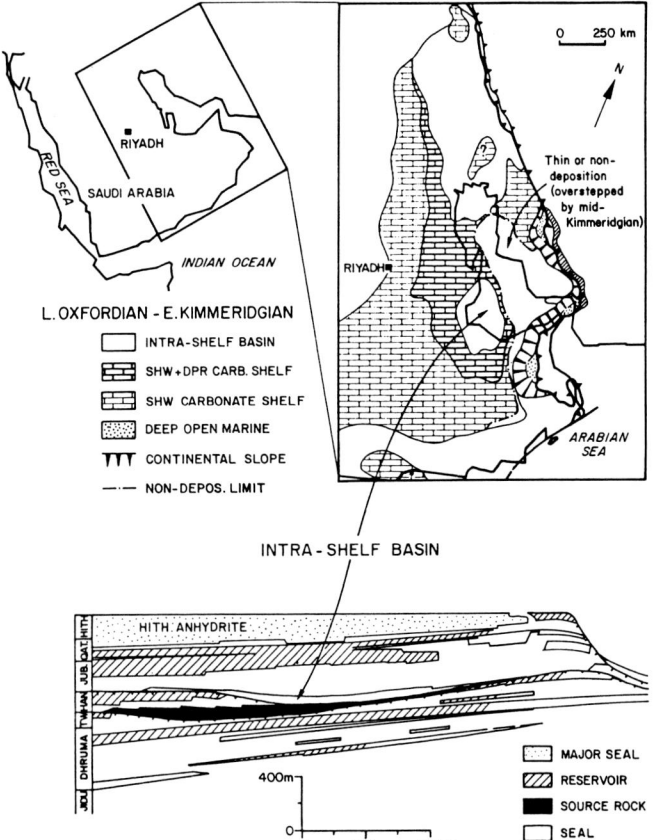

FIG. 7.—Paleogeographic map of Arabian Peninsula during the Mesozoic showing intra-platform basins and cross section illustrating source rock-bearing Hanifa basin and Jurassic hydrocarbon habitat (from Murris, 1980). The simulated section shown in Figure 8 refers to parts of this cross section (Hanifa intra-shelf basin).

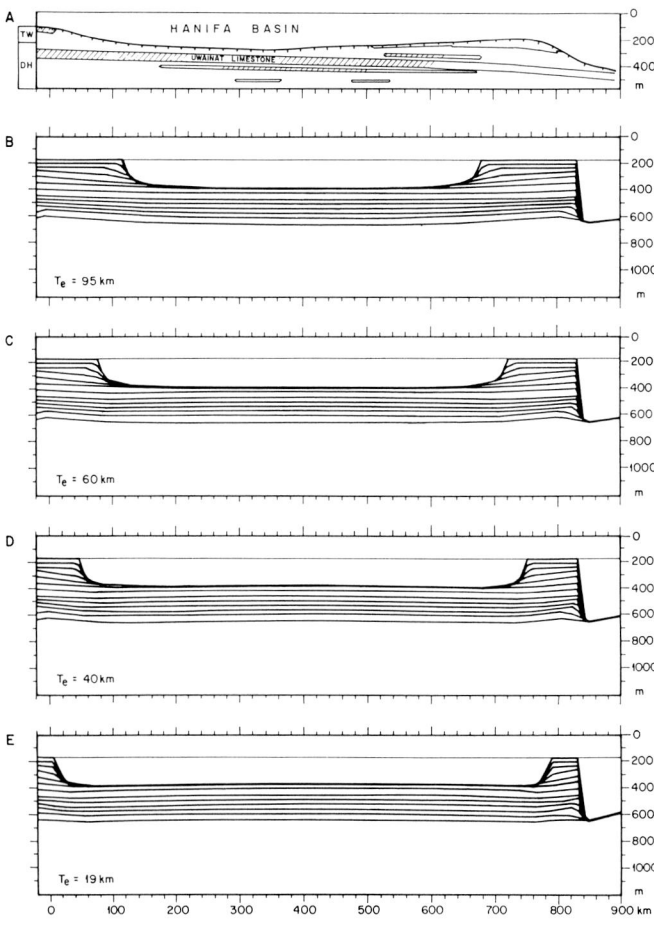

FIG. 8.—Isostatic model for initiation of intra-platform basins, as exemplified by the source rock-bearing Hanifa basin. A sea-level rise causes differential drowning of the interior of an extensive carbonate platform for a wide range of elastic thicknesses (T_e = 19–95 km).

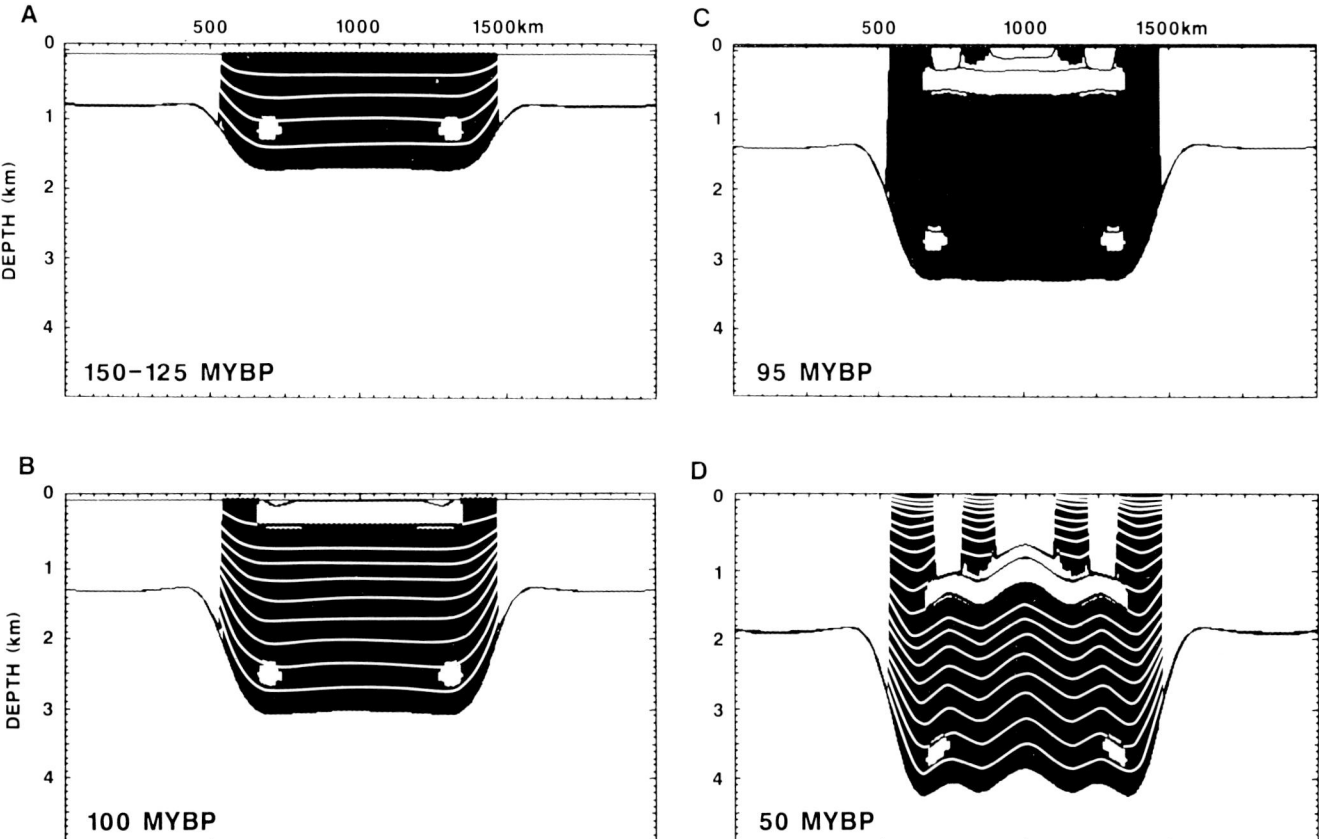

FIG. 9.—Isostatic model for the differentiation of an extensive carbonate "megabank" (A) into a series of isolated platforms and inter-platform basins. A major sea-level rise at 100 m.y. triggers the differentiation (B), and the subsequent pattern of development (C and D) is controlled by the isostatic response to the surface load. Time lines are spaced at 5-m.y. intervals. Black areas indicate carbonates built up to sea level.

the pinnacle reef relative to the platform, is dependent on the elastic-plate thickness. As shown in Figure 11 D and E, an increase in the elastic-plate thickness results in broader pinnacles forming at greater distances from the platforms and eventually leads to their amalgamation in the center of the basin.

The development of pinnacles by the proposed mechanism has two requirements. First, it requires a cycle of falling sea level followed by a sea-level rise. Second, it requires an isostatic response characterized by an elastic-plate thickness (< 10 km) that is normally associated with a hot lithosphere plate. Many studies, however (see compilation of Karner and others, 1983), suggest that the effective elastic thickness of stable continental interiors is on the order of 50 km, and elastic-plate thicknesses of less than 10 km may only be achieved at the culmination of major heating events during periods of continental rifting. The proposed mechanism thus has restricted applications; its greatest potential is in intervals associated with rifting and/or decreased elastic thickness of the lithosphere. We note, however, that the analysis of Barton and Wood (1984) of a gravity profile across the central North Sea indicated an elastic-plate thickness of less than 10 km. Therefore, the choice of appropriate rheological parameters to describe the continental lithosphere is difficult; recent work of Watts (1988) indicates a much weaker continental lithosphere than commonly assumed. It is therefore interesting to speculate if, in addition to constraining sea-level variations, studies of the development of carbonate platforms may also increase our understanding of the long-term thermo-mechanical properties of continental lithosphere.

CONCLUSIONS

Computer simulations employing algorithms that describe the complex interaction of geologic processes, can be used to improve our understanding of the controls on carbonate platform evolution. This approach is particularly powerful in evaluating and demonstrating conceptual models, since parameters can be easily varied and components isolated and investigated. In applying this approach to the simulation of observed carbonate systems, the input parameters must be as rigorously constrained as possible. The principal problems are the difficulty in separating the eustatic signal from local and/or regional tectonic processes, quantifying the spatial and temporal controls on carbonate growth potential, and in obtaining high-resolution age control on carbonate environments. Any step toward resolving these problems would greatly enhance the predictive power of computer simulations of carbonate systems.

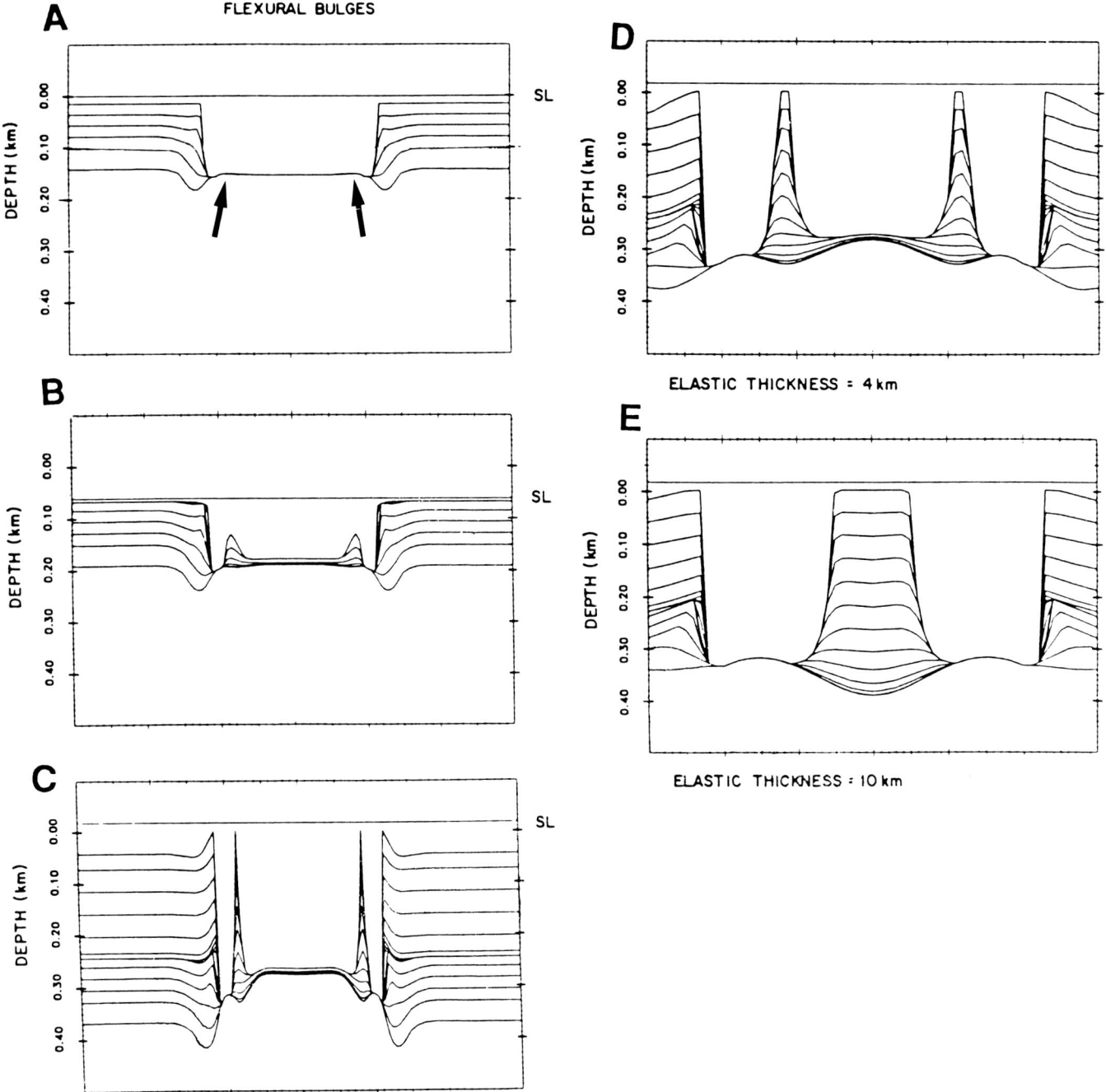

Fig. 10.—Isostatic model for the initiation of carbonate pinnacles. (A) Flexural bulges developing basinward of two stationary platforms. (B) During a sea-level lowstand, platform buildup ceases, but growth is initiated in the basinal areas. (C) A sea-level rise leads to renewed growth of the platforms and drowning in the basinal area, except on the relatively uplifted flexural bulges where carbonate pinnacles develop (elastic thickness 2 km). (D) and (E) show the same configuration as in (A) to (C) but with an elastic thickness of 4 km (D) and 10 km (E). A more rigid plate results in pinnacles being initiated at greater distances away from the platforms, leading eventually to their amalgamation in the center of the basinal areas. Time lines are spaced at 1-m.y. intervals.

ACKNOWLEDGMENTS

We thank our colleagues in the Geology Departments of the Shell Laboratories in Rijswijk and in Houston for discussions and for their general support. In addition, thanks are due to Shell Research and Shell Development for their permission to publish this article.

REFERENCES

BARTON, P., AND WOOD, R. 1984, Tectonic evolution of the North Sea basin: Crustal stretching and subsidence: Geophysical Journal of the Royal Astronomical Society, v. 79, p. 987–1022.

BEAUMONT, C., KEEN, C. E. AND BOUTILIER, R., 1982, A comparison of foreland and rift margin sedimentary basins: Philosophical Transactions of the Royal Society of London, v. A305, p. 295–317.

EPTING, M., 1988, The Miocene carbonate buildups of central Luconia, offshore Sarawak, in Bally, A. W. B., ed., Atlas of Seismic Stratigraphy: American Association of Petroleum Geologists, Studies in Geology, v. 27, p. 170–178.

HAQ, B. U., HARDENBOL, J., AND VAIL, P., 1987, Chronology of fluctuating sea levels since the Triassic (250 million years ago to present): Science, v. 235, p. 1156–1167.

KARNER, G., STECKLER, M., AND THORNE, J., 1983, Long-term thermomechanical properties of continental lithosphere: Nature, v. 304, p. 250–253.

KENDALL, C. G. St. C., AND SCHLAGER, W., 1981, Carbonates and relative changes in sea level: Marine Geology, v. 44, p. 181–212.

LAWRENCE, D., DOYLE, M., SNELSON, S. AND HORSFIELD, W., 1987, Stratigraphic modelling of sedimentary basins: Society of Exploration Geophysicists, 57th Annual International Meeting, Expanded Abstracts of Technical Program, p. 407–408.

MARKELLO, J. R., AND READ, J. F., 1982, Upper Cambrian intrashelf basin, Nolichucky Formation, Southwest Virginia Appalachians: American Association of Petroleum Geologists Bulletin, v. 66, p. 860–878.

MAY, J. A., AND EYLES, D. R., 1985, Well log and seismic character of Tertiary Terumbu carbonate, South China Sea, Indonesia: American Association of Petroleum Geologists Bulletin, v. 69, p. 1339–1358.

MURRIS, R. J., 1980, Middle East: stratigraphic evolution and oil habitat: American Association of Petroleum Geologists Bulletin, v. 64, p. 579–618.

READ, J. F., 1985, Carbonate platform facies models: American Association of Petroleum Geologists Bulletin, v. 69, p. 1–21.

SCHLAGER, W., 1981, The paradox of drowned reefs and carbonate platforms: Geological Society of America Bulletin, v. 92, p. 197–211.

WATTS, A., AND COCHRAN, J., 1974, Gravity anomalies and flexure of the lithosphere along the Hawaiian-Emperor seamount chain: Geophysical Journal of the Royal Astronomical Society, v. 38, p. 119–141.

———, KARNER, G., AND STECKLER, M., 1982, Lithosphere flexure and the evolution of sedimentary basins: Philosophical Transactions of the Royal Society of London, v. A305, p. 249–281.

WATTS, A. B., 1988, Basin subsidence and lithospheric flexure (abst.): 78th Annual Meeting of the Geological Society, Julich, West Germany, p. 18–19.

WILSON, J., 1975, Carbonate facies in geologic history: Springer-Verlag, New York, 471 p.

WINTERER, E. L., AND BOSELLINI, A., 1981, Subsidence and sedimentation on Jurassic passive continental margin, southern Alps, Italy: American Association of Petroleum Geologists Bulletin, v. 65, p. 394–421.

ROLE OF THERMAL SUBSIDENCE, FLEXURE, AND EUSTASY IN THE EVOLUTION OF EARLY PALEOZOIC PASSIVE-MARGIN CARBONATE PLATFORMS

GERARD C. BOND, MICHELLE A. KOMINZ AND MICHAEL S. STECKLER
Lamont-Doherty Geological Observatory of Columbia University, Palisades, New York 10964
AND
JOHN P. GROTZINGER
Department of Earth, Atmospheric and Planetary Science, Massachusetts Institute of Technology, Cambridge, Massachusetts 02139

ABSTRACT: Modeling of early Paleozoic passive margins in the Cordilleran and Appalachian orogens indicates that factors controlling growth of early Paleozoic passive-margin carbonate platforms were thermally controlled subsidence, time-dependent flexure of the lithosphere, and at least two orders of eustatic sea-level changes. Initiation of the carbonate platforms in Middle Cambrian time followed a marked reduction in supply of Lower Cambrian coarse siliciclastic material to the passive margins. Two-dimensional modeling of palinspastically restored cross sections implies that the reduction in relief of onshore sediment sources resulted mainly from increased time-dependent flexural rigidity and extension of the area of subsidence into the craton. Continued increase in rigidity and bending of the craton edge, combined with a long-term eustatic sea-level rise, further reduced the supply of siliciclastic material to the carbonate platforms, resulting in a progressive cratonward shift of the siliciclastic shoreline and cratonward expansion of the carbonate platforms.

Additional evidence of eustatic controls on growth of the platforms is obtained from one-dimensional analyses of post-rift subsidence of the platforms. The effects of sediment loading and lithification are removed from cumulative subsidence curves, producing reduced cumulative curves, designated R1 curves. The first-order form of the R1 curves is exponential, matching closely the form of theoretical curves calculated from cooling plate models for passive margins. After subtracting best-fit model cooling curves from the R1 curves, the residual curves, designated R2 curves, contain evidence of two orders of "events" superimposed on the thermally controlled subsidence of the margins. One event is the long-term rise and fall of sea level observed in the two-dimensional modeling. The long-term event coincides temporally with the Sauk transgression-regression on the craton. The other consists of repeating short-term sea-level changes with wave lengths of 2 to 6 Ma. The short-term sea-level events have similar timing in the southern Canadian Rockies, in the Great Basin, and in the Virginia-Tennessee Appalachians, suggesting a eustatic control.

These inferred eustatic events appear to have exerted a major influence on the lithologic framework of the carbonate platforms. The long-term eustatic fall in Late Cambrian and Ordovician time augmented the reduction in rate of net subsidence of the platforms resulting from decay of the thermal anomaly. The much slower subsidence probably was the principal cause of the marked expansion in Late Cambrian and Ordovician time of carbonate shoal facies within the platforms. The short-term eustatic events produced distinct cycles composed of fine-grained shaley material in their lower halves and coarser grained shoal facies in their upper halves. Apparently, each short-term sea-level rise reduced the rate of carbonate production sufficiently to allow widespread deposition of subtidal facies with large amounts of interbedded siliciclastic mud. During each short-term fall, rates of carbonate production increased and led to expansion of shoal facies across the platforms.

INTRODUCTION

A large number of the early Paleozoic carbonate platforms throughout the world occurs within passive continental margins that formed after a major episode of continental breakup in latest Proterozoic time (Heckel, 1974; Wilson, 1975; Read, 1982; Bond and others, 1983, 1984; Read, 1985). Although most of these platforms have been folded and thrust faulted, their stratigraphy and sedimentology are known reasonably well in many localities, especially in North America. Few attempts have been made, however, to interpret their evolution in terms of thermal and mechanical processes that operate in passive margins, especially in two dimensions. An effort to do so has become feasible owing to recent advances in thermo-mechanical modeling of modern passive margins and to accurate palinspastic reconstructions of some of the carbonate platforms, particularly in the southern Canadian Rockies.

In this paper, we apply one- and two-dimensional geophysical models of passive margins to selected portions of early Paleozoic carbonate platforms in the Appalachian and Cordilleran orogens of North America. We refer to these platforms as passive-margin carbonate platforms, following Read (1982), to distinguish them from platforms that formed in different geodynamic settings, such as in active rifts or foreland basins. We emphasize the importance of eustatic sea-level changes, thermally controlled subsidence, flexure, and rift-flank uplift in controlling the evolution of the platforms and their stratigraphic and sedimentologic framework.

We have modeled passive-margin carbonate platforms in the southern Canadian Rockies, in parts of the Great Basin in Utah and Nevada, and in the Virginia-Tennessee Appalachians (Fig. 1). The stratigraphic sections we have analyzed occur within intact and well-exposed intervals between major faults so that no section is a composite of strata from different thrust panels originally separated by large distances. In the Great Basin, we avoided areas that lie within a possible fault-controlled embayment of Middle Cambrian age (Rees, 1986). Finally, the sections contain reasonably well-constrained Middle Cambrian and younger stratigraphic boundaries to which numerical ages are assigned (Harland and others 1982).

GENERALIZED STRATIGRAPHIC AND SEDIMENTOLOGIC FRAMEWORK OF THE PASSIVE-MARGIN PLATFORMS IN THE CORDILLERAN AND APPALACHIAN MIOGEOCLINES

The carbonate platforms overlie an extensive wedge of mature, predominantly marine quartz sandstone that accumulated during the early post-rift stages of the passive margins (Bond and others, 1983, 1984). In the Great Basin and in the southern Canadian Rockies, the transition from these early post-rift siliciclastic materials to carbonate sedimentation occurs within an interval a few tens of meters thick broadly correlative with the *Bonnia-Olenellus* trilobite zone

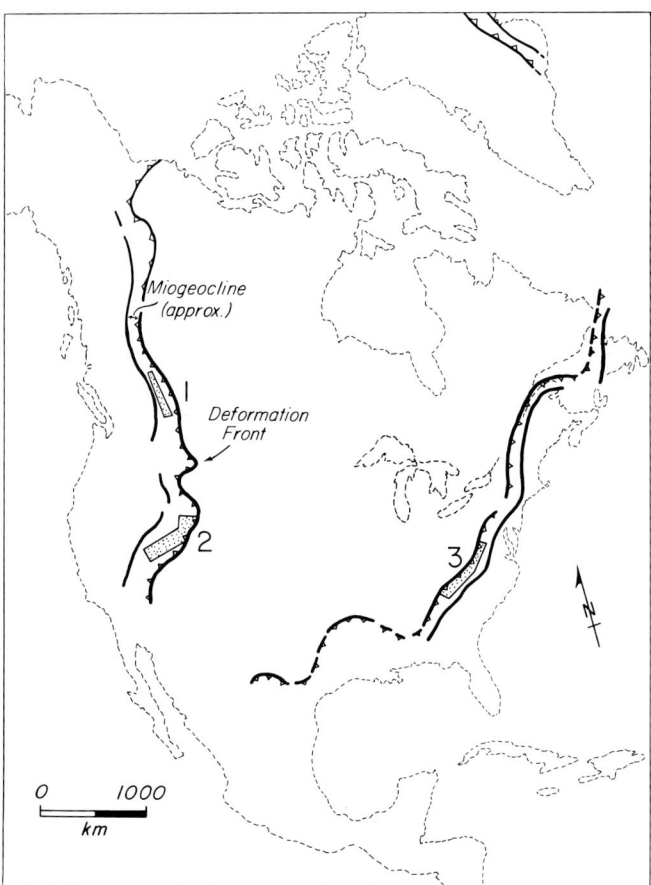

FIG. 1.—Location of areas of analysis of carbonate platforms in southern Canadian Rockies (1), Great Basin (2), and Virginia-Tennessee Appalachians (3).

(Aitken, 1966; Palmer 1971a). In the Virginia-Tennessee Appalachians, carbonate platform deposition began after a transitional interval of similar thickness and also in late Early Cambrian time, but the fossil control for its precise timing is poor (Palmer, 1971b; Read, pers. commun.).

Although carbonate platforms evolved through most of Paleozoic time, only their lower portions formed in passive margins. In the Appalachians, the passive margin was terminated in Middle Ordovician time by uplift associated with flexural bending of the lithosphere during emplacement of the Taconic allochthons (Jacobi, 1981; Shammugam and Lash, 1982; Mussman and Read, 1986). In the southern Canadian Rockies, the passive-margin stage ended between Early Silurian and Middle Devonian time with formation of an enigmatic regional unconformity (e.g., Porter and others, 1982). The longest record of passive-margin platform growth is in the Great Basin, where thermally controlled subsidence continued into Late Devonian time and was terminated by uplift associated with the initial phases of tectonic loading in the Antler Orogenic Belt (Poole, 1974; Armin and Meyer, 1983).

In all three areas, the passive-margin platforms subsided most rapidly in their outermost portions owing to the increase in lithospheric thinning with increasing distance from the craton edge, giving rise to their distinct wedge shape. Thicknesses of the platforms range from about 1 km in the innermost parts to 4 to 6 km along the outermost preserved segments (Aitken, 1966; Palmer, 1971a,b; Read, pers. commun.). The sediments within the platforms accumulated in subtidal to intertidal environments and typically consist of complexly interlayered siliciclastic shales and siltstones, wavy to nodular bedded limestones with shaley partings, bedded to thoroughly burrowed calcisiltites, cross-bedded grainstones, vertically stacked stromatolites, cryptalgalaminites, and thrombolites. These facies are commonly arranged in shoaling-upward cycles, ranging in thickness from a few meters (Aitken, 1966; Kepper, 1972; Lohmann, 1976; Bova and Read, 1986) to a few hundred meters (Aitken, 1966, 1978, 1981; Kepper, 1976; Palmer and Halley, 1979; Palmer, 1981a; Chow and James, 1987). The thickest cycles, termed Grand Cycles, have been thoroughly documented in the southern Canadian Rockies, where they consist of a shaley half-cycle overlain gradationally by a carbonate half-cycle (Aitken, 1966, 1978).

ONE-DIMENSIONAL MODELING OF SUBSIDENCE

Our one-dimensional modeling of subsidence in the lower Paleozoic passive margins involves a series of reductions of cumulative stratigraphic thickness curves that are constructed from measured sections. The methods we use are refinements of procedures that have been developed for analyses of subsidence in modern passive margins (Sleep, 1971; Steckler and Watts, 1978).

First reduction-construction of R1 curves.—

The first reduction removes the effects of sediment loading, compaction, and water depth changes from the cumulative thickness curves. This step is the basis for identifying the tectonic component of subsidence and comparing it with subsidence models for passive margins. The reduced subsidence for each stratigraphic unit is obtained from equation (1) below and is plotted against age using the time scale of Harland and others (1982) to produce reduced cumulative curves. We refer to the curves produced by this first reduction as R1 curves.

$$\text{R1 (1st reduction)} = Y + \Delta SL \frac{\rho_m}{\rho_m - \rho_w}$$
$$= S^* \frac{\rho_m - \rho_s}{\rho_m - \rho_w} + Wd \quad (1)$$

where Y = tectonic subsidence, S^* = sediment thickness corrected for lithification, ρ_s = mean bulk density of sediments, ρ_m = mean density of the mantle, ρ_w = mean density of sea water, ΔSL = eustatic change in sea level relative to its present elevation, and Wd = average depth of water in which the unit was deposited. Airy compensation is assumed for sediment and water loads, which is justified as long as only the form (and not the amplitude) of postrift subsidence is of interest (Bond and others, 1988).

Since the sections are from inner-shelf environments, the difficulties in estimating the effects of large changes in water depths (>100 m) on the forms of R1 curves are avoided (Lohmann, 1976; Aitken, 1978; Demicco, 1983; Walker and others, 1983). Smaller changes in water depths are important in evaluating small-scale events such as eustatic changes, however, and are considered in a following section. The values for the densities of water, mantle, and crust are taken from McKenzie (1978).

We calculate S^* from maximum and minimum delithification factors derived from empirical porosity-depth curves for each of six lithologies commonly found in the lower Paleozoic strata (see Bond and Kominz, 1984, for detailed discussion). The procedure also includes a choice of early cementation in the carbonate strata and in the siliciclastic sandstones and siltstones. Although our subsidence curves extend only to the base of the Middle Cambrian, the R1 analyses always include from 500 m to 1 km of the underlying lower Cambrian deposits in the margins in order to include the effects of lithification of these strata on the higher part of the sections. We have two options for calculating S^*. In the first, we assume maximum and minimum factors, with and without early cementation, that put probable upper and lower boundaries on the true delithified thicknesses (Fig. 2). We have used this option to place limits on the true value of the long-term component of subsidence. In the second option, we specify a delithification factor for each of the six lithologic categories by defining a specific empirical porosity-depth curve and choosing either purely mechanical lithification or early cementation for each lithology. The procedure generates a single R1 curve with the same overall form as the first option but also with small changes in slope that reflect vertical changes in lithologies and lithification processes (R1 Max. Diff., Fig. 2). This option is appropriate for assessing the effects of different lithification mechanisms and gradual changes in grain size (over tens of meters) on the forms of the subsidence curves.

In the next step the R1 subsidence curves are fit to theoretical curves for the post-rift subsidence of a passive margin (e.g., Fig. 2), using least-squares methods. The theoretical curves are exponential with a decay constant of 62.8 m.y. These curves have been calculated from lithospheric plate models, with a thermal equilibrium thickness of 125 km, that have been applied successfully to the post-rift evolution of modern passive margins (Watts, 1981; Keen, 1982). We ignore the non-exponential form of the model subsidence curves, which occurs in the first 16 m.y. after rifting (McKenzie, 1978), on the basis of geologic evidence that the carbonate platforms were initiated at least 15 to 20 m.y. after post-rift subsidence began (Bond and others, 1984; Simpson and Sundberg, 1987). As has been shown previously (Bond and others, 1983, 1984; Armin and Meyer, 1983; Bond and Kominz, 1984), there is a good fit between the R1 curves and the theoretical cooling curves, providing strong evidence that the tectonic subsidence was controlled mainly by cooling of a heated lithospheric plate.

Our previous analyses of the long-term R1 subsidence have been limited to strata no younger than middle Ordovician, and the terminal portion of the exponentially decaying subsidence of the carbonate platforms is poorly constrained. In the Great Basin the stratigraphic record is complete enough to analyze the mature stages of the evolution

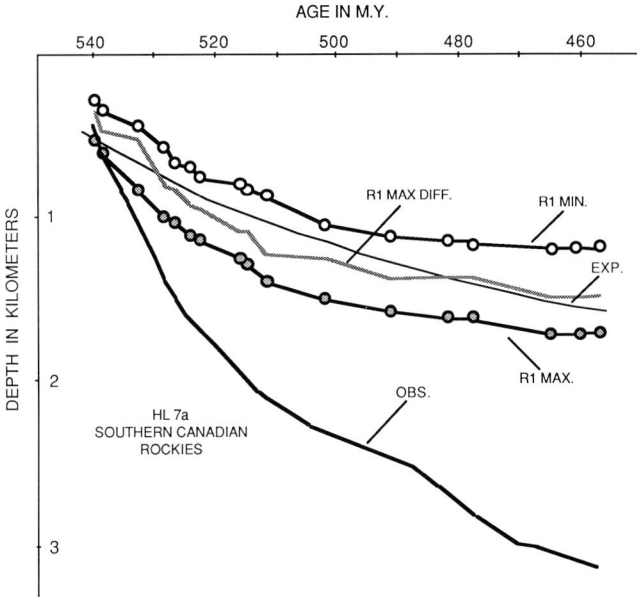

FIG. 2.—Example of an R1 curve from a section in the southern Canadian Rockies (HL7a, located in Fig. 6A) produced by different delithification options. R1 Max. and R1 Min. are from the maximum and minimum delithification factors, respectively, in our first delithification option, as described in the text. R1 Max. Diff. is from our second delithification option and maximizes the differences in delithification factors for fine- and coarse-grained, calcareous and non-calcareous sediment. Note that the curve for maximum difference does not span the range between the maximum and minimum limits. Exp. is the best-fit exponential with a decay constant of 62.8 m.y. Obs. is the cumulative stratigraphic thickness curve from the observed or present measured thicknesses.

of the passive margin extending to the base of Upper Devonian strata. The forms of extended R1 curves from this region document continued reduction in the rate of subsidence closely following the model exponential cooling curves for a 125-km plate (Fig. 3A, B).

Second reduction-construction of R2 curves.—

By subtracting the best-fit exponential curves from the corresponding R1 curves, a second set of curves is produced that gives an approximation of external events superimposed on the thermally controlled subsidence. These curves are a second reduction of the cumulative subsidence curves and are referred to as R2 subsidence curves. Ideally, by comparing the forms of the R2 curves from widely separated areas, local external events induced by local climate or tectonics can be distinguished from events that correlate well enough temporally and on large enough geographic scales to be consistent with a eustatic mechanism.

Bond and others (1983) and Bond and others (1988) documented a long-wavelength (~70 m.y.) event consisting of a sea-level rise in Middle and Late Cambrian time followed by a relative fall in Late Cambrian through early Ordovician time. The event has similar timing in the southern Canadian Rockies, in Utah, and in the Virginia-Tennessee Appalachians and corresponds closely to the Sauk sequence on the craton (Sloss, 1963), strongly suggesting that it is a

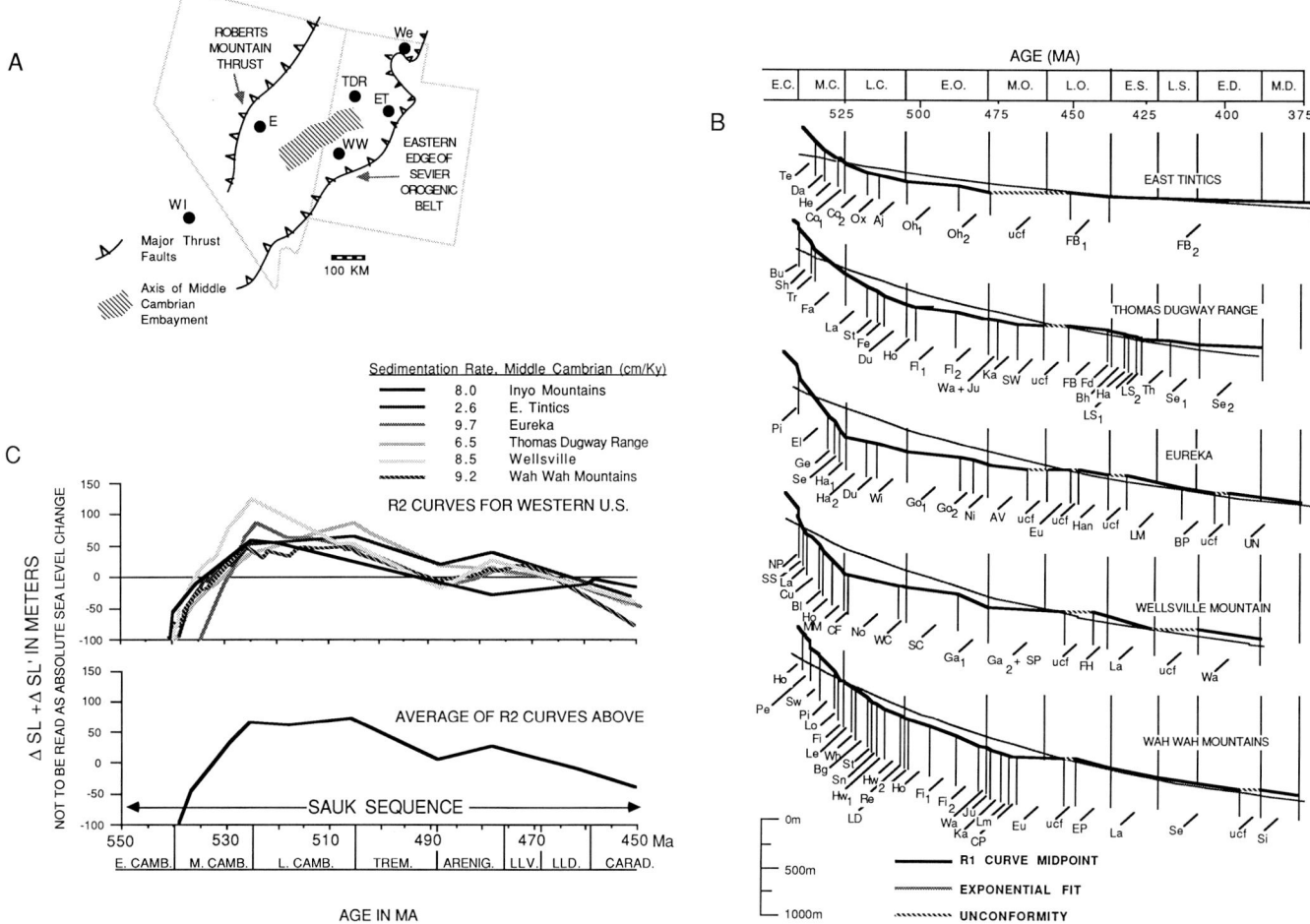

Fig. 3.—(A) Map showing locations of sections of early Paleozoic carbonate platforms in the Great Basin used to construct extended R1 for the mature stages of passive-margin subsidence. Location of the Middle Cambrian embayment is from Rees (1986). Sections as follows: We—Wellsville; TDR—Thomas Dugway Range; ET—East Tintic district; E—Eureka; WW—Wah Wah Range. (B) Extended R1 subsidence curves for sections in the Great Basin, located in Figure 3A, showing only the midpoint of range of delithification factors in our first delithification option, as described in the text. Long vertical lines mark numerically dated stratigraphic boundaries. The exponential curves are the best-fit exponential curves with a decay constant of 62.8 m.y. Formation names as follows:

East Tintics: Te—Teutonic Limestone; Da—Dagmar Dolomite; He—Herkimer Limestone and Bluebird Dolomite; Co—Cole Canyon Dolomite; Ox—Opex Formation; Aj—Ajax Dolomite; Oh—Opohonga Limestone; FB_1—Fish Haven Dolomite and Bluebell Dolomite (Ordovician); FB_2—Bluebell Dolomite (Silurian and Devonian).

Thomas Dugway Range: Bu—Busby Quartzite; Sh—Shadscale Formation; Tr—Trailer Limestone; Fa—Fandangle Limestone; La—Lamb Dolomite; St—Straight Canyon Formation; Fe—Fera Limestone; Du—Dugway Ridge Dolomite; Ho—House Limestone; Fl—Fillmore Limestone; Wa—Wah Wah Limestone; Ju—Jaub Limestone; Ka—Kanosh Shale; SW—Swan Peak Quartzite; FB—Fish Haven Dolomite; Fd—Floride Dolomite; Bh—Bell Hill Dolomite; Ha—Harrisite Dolomite; LS—Lost Sheep Dolomite; Th—Thursday Dolomite; Se—Sevy Dolomite.

Eureka: Pi—Pioche Shale; El—Eldorado; Ge—Geddes; Se—Secret Canyon; Ha—Hamburg; Du—Dunderberg; Wi—Windfall; Go—Goodwin; Ni—Ninemile; AV—Antelope Valley; Eu—Eureka Quartzite; Han—Hanson Creek; LM—Lone Mountain; BP—Beacon Peak; UN—Upper Nevada.

Wellsville Mountain: NP—Naomi Peak Member (Langston Formation); SS—Spence Shale Member (Langston Formation); La—Langston Formation (upper member); CU—Ute Formation; Bl—Blacksmith Formation; Ho—Hodges Shale; MM—middle member (Bloomington Formation); CF—Calls Fort Member (Bloomington Formation); No—Nounan Formation; WC—Worm Creek Quartzite; SC—Saint Charles Formation; Ga—Garden City Limestone; SP—Swan Peak Quartzite; FH—Fish Haven Dolomite; La—Laketown Dolomite; Wa—Water Canyon Formation.

Wah Wah Mountains: Ho—Howell Limestone and Chisholm Shale; Pe—Peasley Formation, Dome Limestone, and Whirlwind Formation; Sw—Swasey Limestone and Milford Limestone; Pi—Pierson Cove Formation; Lo—Lower Member Trippe Limestone; Fi—Fish Springs; Le—Ledgy Member (Wah Wah Summit Formation); Wh—White Marker Member (Wah Wah Summit Formation); Bg—Big Horse Canyon Member (Orr Formation); St—Steamboat Pass Shale Member (Orr Formation); Sn—Sneakover Pass Limestone Member (Orr Formation); Hw—Hellnmaria Canyon Member (Notch Peak Formation); Re—Red Tops Calcarenite Member (Notch Peak Formation); LD—Lava Dam Limestone Member (Notch Peak Formation); Ho—House Limestone; Fi—Fillmore Formation; Wa—Wah Wah Limestone; Ka—Kanosh Shale; Lm—Lehman Formation; CP—Crystal Peak Dolomite, Watson Ranch Quartzite, and Eureka Quartzite; Eu—Eureka Quartzite; EP—Ely Springs Dolomite; La—Laketown Dolomite; Se—Sevy Dolomite; Si—Simpson Dolomite.

Sources of data are: Sandberg and Mapel, 1967; Hintze, 1973; Poole and others, 1977; Ross, 1977; Sandberg and others, 1982; Bond and others, 1983. (C) The upper half is a composite of extended R2 curves from the Great Basin calculated from R1 curves in Figure 3B. The patterned lines are keyed to the sections in Figure 3B. The sedimentation rates are calculated from the Middle Cambrian segments in the R1 curves, Figure 3B. The lower half is the average of the curves above and is an approximation of the form (but not necessarily the amplitude) of the eustatic sea-level change from Middle Cambrian to the end of Ordovician time, corresponding to the Sauk sequence of Sloss (1963).

eustatic cycle with a long wavelength. We have further documented the timing and form of the long-wavelength eustatic cycle and its striking correlation with the Sauk sequence with the extended curves from the Great Basin (Fig. 3C). The average of the curves from the Great Basin (Fig. 3C) is probably the best approximation of the long-term eustatic sea-level change during early Paleozoic time that can be obtained from North America.

Evaluation of error in one-dimensional analyses.—

Bond and others (1988) evaluated error in construction of R1 and R2 curves and concluded that as long as only the form of subsidence is of interest, errors arising from time scales, assumptions in delithification procedures, and use of simplified one-dimensional models are of secondary importance and can be neglected in the context of our analyses of the controls on carbonate platform evolution. Using the extended R1 and R2 curves from the Great Basin, we have evaluated two additional sources of error, the insulating effect of sediment and uncertainties in the decay constants for the exponential cooling curves. The accumulation of cold sediment in a passive margin retards the cooling of the lithosphere relative to that for an empty basin, potentially causing the subsidence to mimic the deviation we observe between the R1 and best-fit exponential curves (e.g., Beaumont and others, 1982; Stephenson 1987). This effect is strongly dependent on the sedimentation rate, however (Stephenson and others, 1987), and its amplitude will be proportional to the thickness of strata within equivalent intervals of time in the margins. This is clearly not the case for sections with significantly different sedimentation rates in the Great Basin (Fig. 3C).

The decay constant of the best-fit exponential curves is 62.8 m.y., a value obtained from the average age-depth curve for modern ocean floor (Parsons and Sclater, 1977). In modern ocean basins, the decay constant actually varies considerably from one part of the ocean to another and presumably varied in the early Paleozoic as well. Although a decay constant as low as 40 m.y., a value that is lower than the lowest values observed in modern oceans (Cazenave and others, 1983) and in passive margins (Watts and Steckler, 1979), reduces the misfit with the R1 curves, the distinctive shape and timing of the R2 curves is retained (Fig. 4A, B). Although the long-term R2 curves could be constructed using exponential curves whose decay constants are determined with best-fit procedures, this is not justified because the amplitudes of the R1 curves are not well constrained and are model dependent. We emphasize that regardless of the procedure for constructing R2 curves, their amplitude is not necessarily a correct indication of the magnitude of the inferred early Paleozoic eustatic change. For example, Watts and Steckler (1979), using a similar procedure, may have underestimated the amplitudes of Mesozoic and Tertiary sea-level changes.

TWO-DIMENSIONAL MODELING OF SUBSIDENCE

Two-dimensional processes, such as flexure and lateral heat flow, have an important effect on subsidence in a pas-

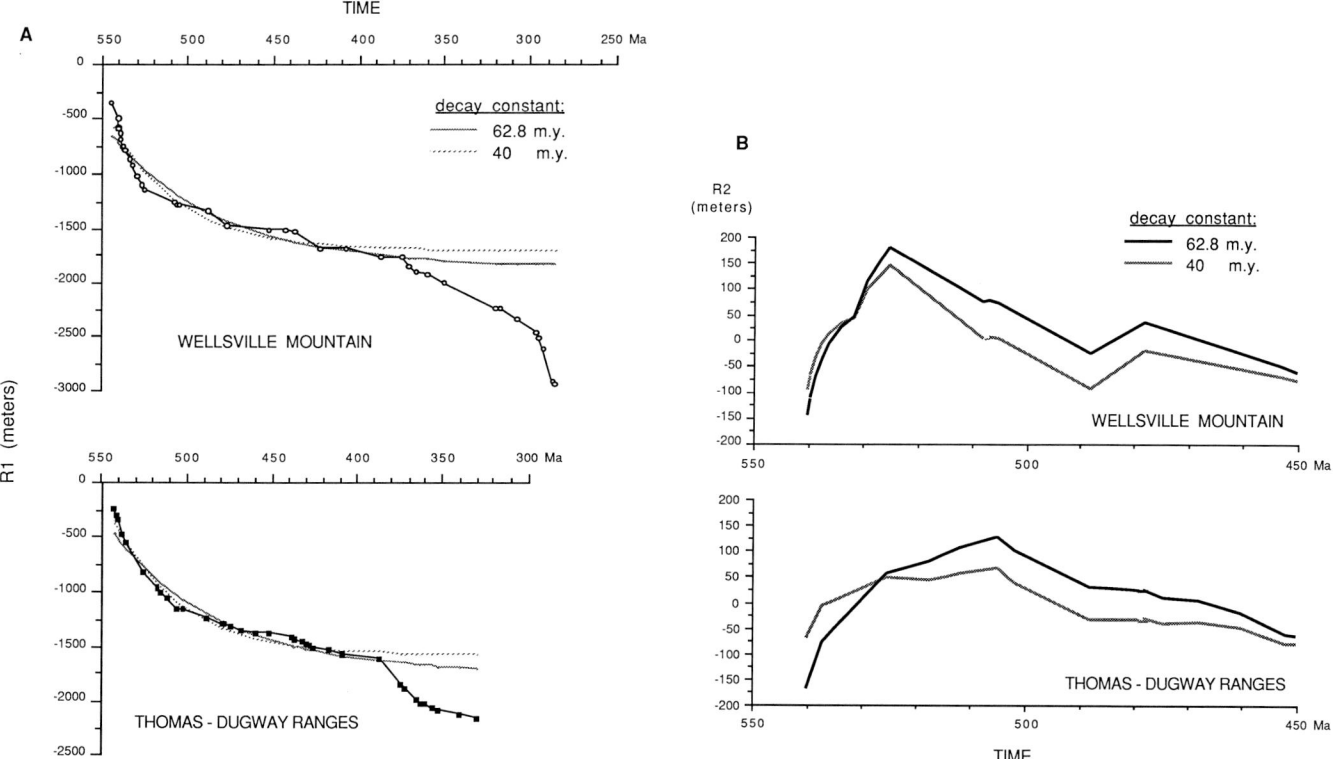

FIG. 4.—(A) R1 subsidence curves from two sections in the Great Basin in Figure 3B and best-fit exponentials with decay constants of 62.8 m.y. and 40 m.y. (B) R2 curves from the R1 curves in (A) using the two decay constants. While the amplitude of the R2 signal is reduced with a very low decay constant of 40 m.y., the form of the R2 signal is only slightly changed.

sive margin after rifting ends (Watts, 1981; Beaumont and others, 1982). During the early stages of margin evolution, heat introduced during rifting produces a thermal uplift in lithosphere adjacent to the rift. In addition, tectonic denudation (e.g., detachment) induces mechanical uplift of the rift shoulders, amplifying the thermally driven uplift. This rift-flank uplift is potentially a source of a significant amount of siliciclastic sediment during the rifting and early post-rift stages. As heated lithosphere cools and the flexural rigidity increases, the flanking uplifts are pulled down, reducing the basement relief adjacent to the rift and extending subsidence well into the adjacent craton. These processes affecting the vertical motion history of crust exert an important influence on the evolution of passive-margin carbonate platforms.

In order to analyze the effects of flexure and the formation of rift-flank uplifts, it is necessary to apply two-dimensional passive-margin models to palinspastically restored cross sections of the passive margins. We use two end-member models, the simplest of which is the one-layer or uniform model of McKenzie (1978). This model assumes that the lithosphere is stretched uniformly and instantaneously by a specified amount (β). During rifting, the surface subsides isostatically as light crust is replaced by denser mantle, and the lithosphere is heated by reduction of the distance between isotherms. At the onset of drift, the lithosphere cools and contracts, and the surface subsides further at a rate which is proportional to the amount of extension. In the limit, if the lithosphere were thinned to zero thickness, the subsidence and heat flow would be identical to those of oceanic crust. There is considerable evidence that the heat advected into the lithosphere during rifting is greater than the uniform model predicts (Royden and Keen, 1980; Steckler, 1985; Buck, 1986). We simulate this possibility and other asymmetric rift processes (Wernicke, 1985) numerically, using the depth-dependent stretching model developed by Royden and Keen (1980). This is accomplished with the depth-dependent model by thinning the upper part of the lithosphere, usually corresponding to the crust, by a different amount than the remainder of the lithosphere. With the depth-dependent model, it is possible to obtain relatively small amounts of synrift subsidence or even uplift during rifting, together with relatively high values of lithospheric heating and large amounts of post-rift subsidence.

A model passive margin illustrating structures that potentially influence platform growth was generated using a computer simulation from Steckler (1981) of the instantaneous uniform stretching mechanism (Fig. 5). The hinge zone is a fundamental structural boundary that separates relatively unstretched crust ($\beta \sim 1$) on the left of the model from highly thinned crust on the right. The post-rift strata thin abruptly across this structure, especially in the early stages of post-rift subsidence, and thicken markedly in the direction of increasing crustal thinning, producing the characteristic wedge shape of the passive-margin strata. After rifting, the lithosphere cools and becomes more rigid with time. As rigidity and the sediment infill increase, flexural bending of the craton edge shifts landward, leading to deposition of a thinner sedimentary prism, the inner flexural wedge.

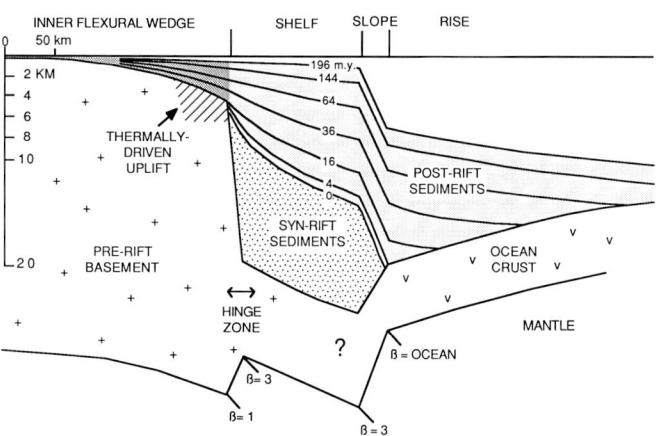

FIG. 5.—Model showing the major structural elements of a passive margin produced by a computer simulation of the uniform model for a lithospheric plate with an equilibrium thermal thickness of 125 km (from Steckler, 1981). β refers to the amount of crustal stretching. Mantle, crust, sediment, and water densities = 3.33, 2.80, 2.50, 1.03 g/cc, respectively. Other model parameters are given in Steckler (1981).

Palinspastic restorations, southern Canadian Rockies.—

Two-dimensional modeling of the early Paleozoic carbonate platforms requires reasonably accurate palinspastic restorations of the deformed strata. Suitable restorations are available only in the southern Canadian Rockies, where subsurface data and superb exposures provide excellent control, and errors probably are no larger than about 15 percent (Price, 1980). We have modeled restorations along AA' to EE' (Fig. 6A) but limit our discussion to the results for the representative section AA', the North Saskatchewan River section. Here, the carbonate platform thickens across the ancient margin in a seaward direction and thins by onlap onto the craton to the east (Fig. 6B). During Middle Cambrian and part of Late Cambrian time, the platform was rimmed by an algal reef complex, the Kicking Horse Rim (Aitken, 1971). To the west, beyond the restored platform, the carbonate strata pass abruptly into a thick shale facies. Farther west, the strata thin over an outer ridge, which may have developed above a block of North American continental basement that was not stretched as much as the basement on either side (Price, 1980). The transition to oceanic lithosphere is not preserved but presumably lay west of the outer ridge.

The strata in the shale basin and in the parts of the margin farther west cannot be restored palinspastically owing to a large amount of ductile deformation and the lack of adequate surface and subsurface control. We have assumed that the shale basin was shortened by the same amount as the carbonate bank and that the outer ridge lay immediately west of the restored shale basin (Fig. 7A, B). A series of model tests indicates, however, that except for the outermost model block in the carbonate platform (block A in Fig. 7A, B), the modeling is not significantly affected by assuming different widths and thicknesses of the shale basin and by the presence or absence of an outer ridge. The outermost model block required slightly different stretching

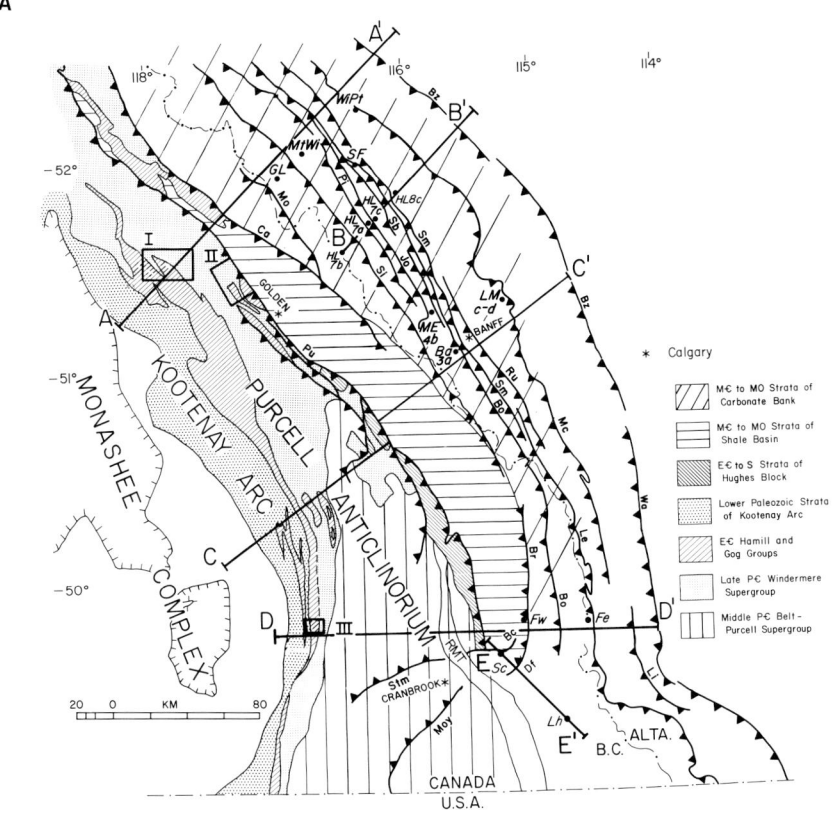

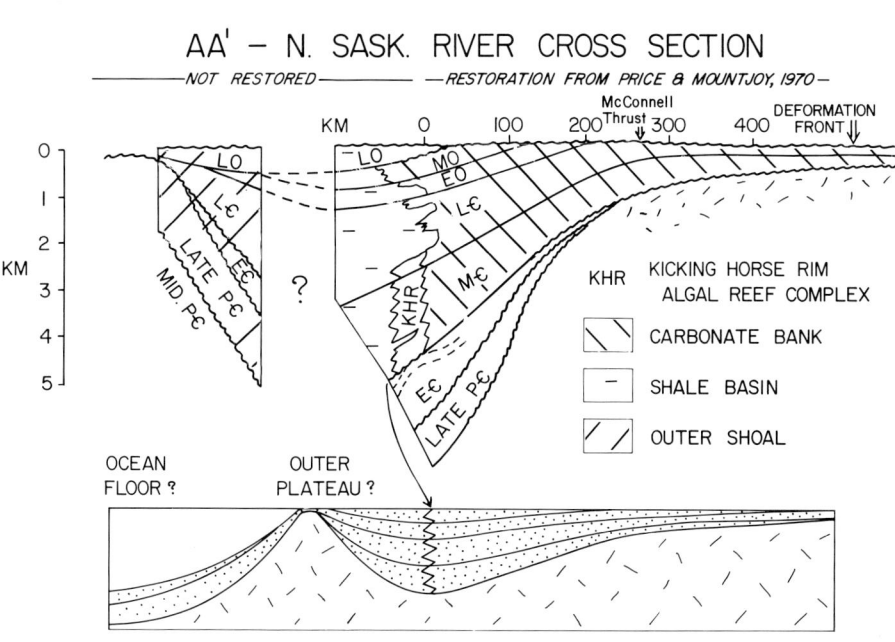

Fig. 6.—(A) Map of major late Mesozoic and early Tertiary structures and early Paleozoic depositional elements in the southern Canadian Rockies. Solid dots with slanted abbreviations (WiPt and so forth) are locations of columnar sections used to construct R1 curves in Bond and Kominz (1984). Thrusts are as follows: Bz, Brazeau; Mc, McConnell; Li, Livingstone; Ru, Rundle; Le, Lewis; Sm, Sulphur Mountain; Bo, Bourgeau; Sb, Sawback; Jo, Johnston Creek, Pi, Pipestone; Si, Simpson Pass; Br, Bull River; Mo, Mons; Ca, Chatter Creek; Pu, Purcell; Stm, St. Mary; Moy, Moyie. AA' to DD' are locations of palinspastically restored sections of the carbonate platform; AA' is the location of the North Saskatchewan River section that is modeled in Figure 7. Open rectangles are locations of igneous rocks in the Hamill Group formed during rifting of the margin. Modified from Bond and Kominz (1984). (B) North Saskatchewan River cross section in the southern Canadian Rockies, located as AA' in Figure 6A. Carbonate bank facies is equivalent to the carbonate platform in this region. See text for discussion.

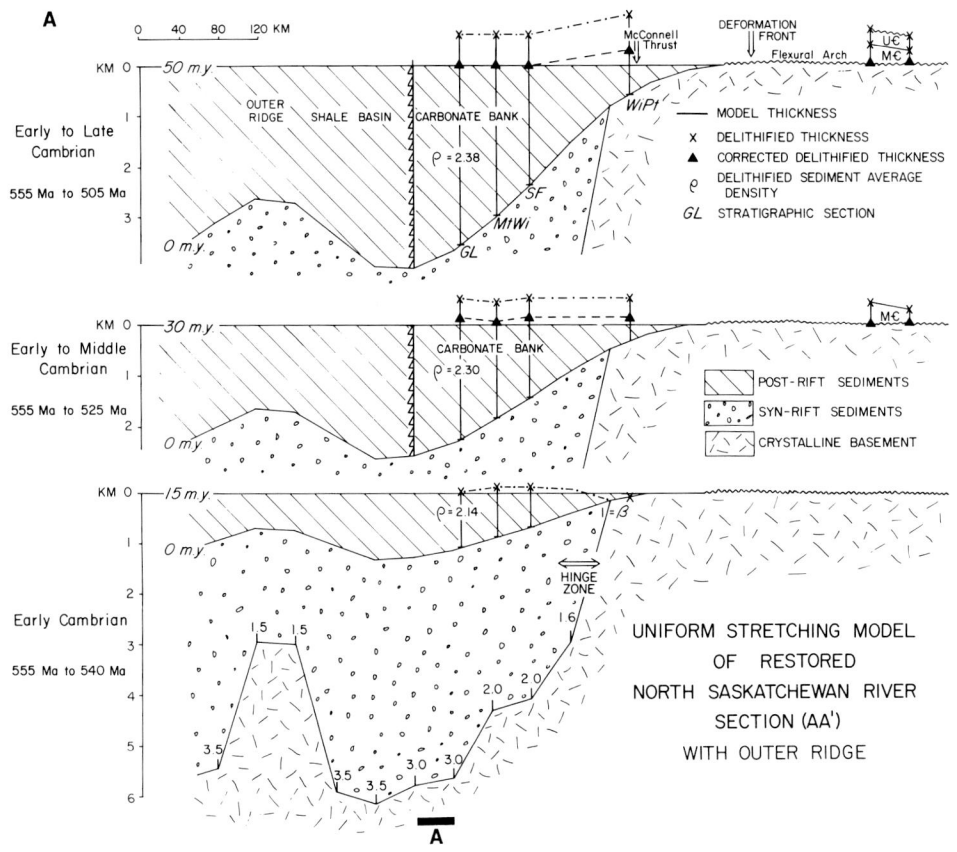
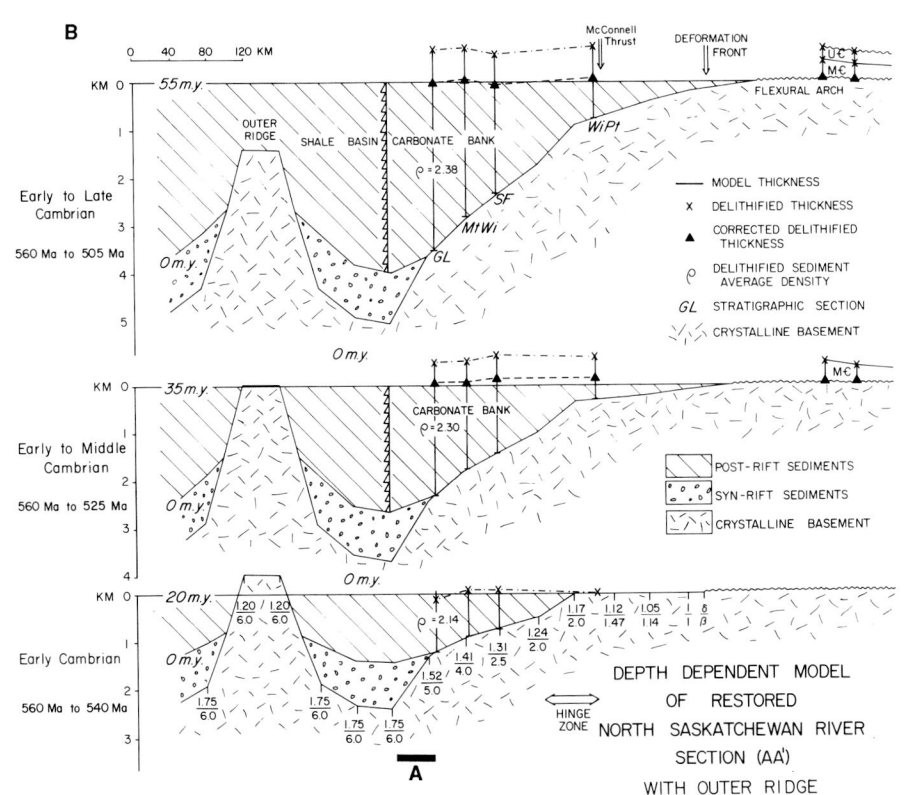

factors for the different assumptions in each test, varying by about 10 percent, in order to produce the required amount of post-rift subsidence.

Modeling procedure.—

The original thicknesses and densities of strata were obtained using the midpoint of the range of delithified thicknesses given by the first of our two delithification options. Using software from Steckler (1981), the delithified section was modeled in terms of the uniform and depth-dependent models by first finding a set of stretching factors, through trial and error, that reproduced the restored thickness of the first post-rift stratigraphic interval, the wedge of Lower Cambrian mature quartz sandstones of the Gog Group (Fig. 7A, B). The best age for the initiation of cooling was found by trial and error to be 555 Ma for the one-layer model and 560 Ma for the two-layer model. These earliest Cambrian ages agree well with ages of onset of post-rift cooling derived from other geological and geophysical analyses (Armin and Meyer, 1983; Bond and others, 1983, 1984; Bond and Kominz, 1984; Devlin and Bond, 1988). Next, using the stretching factors obtained from the first step, the model was allowed to run from the age of initiation of cooling to 525 Ma, the end of Middle Cambrian time, which gives a model prediction for the thickness of the Middle Cambrian stratigraphic interval. In the last step, the model was allowed to run from the age of initiation of cooling to 505 Ma, or to the top of the Upper Cambrian strata, again using the stretching factors obtained in the first step. The thicknesses predicted by the model were then compared with the delithified thicknesses at the corresponding stratigraphic levels in the restored section.

The parameters in both stretching models (Table 1) are from Steckler (1981), except that the initial crustal density of 2.80 g/cc and initial crustal thickness of 31.2 km were changed to 2.73 g/cc and 27.6 km, respectively, on the assumption that the pre-rift basement contains 6 km of sedimentary rocks of the Belt-Purcell Supergroup (Price and Mountjoy, 1970; Price, 1980) with a density of 2.60 g/cc. The crustal thickness was calculated assuming that the surface lay at sea level when rifting began, following Steckler (1981) and Cochran (1981). The density of the sediments infilling the margin is the average density for each of the three stratigraphic intervals that is modeled (Fig. 7A, B) as calculated from the delithification procedure in Bond and Kominz (1984). The flexural rigidity was calculated in both models assuming a simple elastic plate with thickness defined as the depth to the 350°C isotherm (Watts and others, 1980). In the two-layer model, the boundary between the upper and lower layers is assumed to be the base of the continental crust.

TABLE 1.—PARAMETERS USED FOR TWO DIMENSIONAL MODELING OF EARLY PALEOZOIC CARBONATE PLATFORMS

1.	Coefficient of thermal expansion	$3.40 \times 10^{-5}/°C$
2.	Temperature of asthenosphere	1333°C
3.	Time constant of thermal decay	62.8 m.y.
4.	Thermal conductivity of lithosphere	7.5×10^{-3} cal/cm/s/°C
5.	Thermal diffusivity of lithosphere	$8 \times 10^{-3}/cm^2/s$
6.	Equilibrium thermal thickness of lithosphere	125 km
7.	Density of mantle	3.33 g/cm^3
8.	Density of water	1.03 g/cm^3
9.	Young's modulus	6.5×10^{10} N$^-$m
10.	Poisson's ratio	0.25
11.	Density of sediment	ρ_s = varies with depth
12.	Average density of crust	2.73 g/cm^3
13.	Initial thickness of crust	27.6 km
14.	Isotherm defining effective elastic thickness	350°C

Results of two-dimensional modeling.—

A set of stretching factors can be found for both models that simulates the overall geometry of the post-rift restored section (Fig. 7A, B). The model predictions are significantly different for the synrift strata, however. The uniform model produces a thick synrift deposit (Fig. 7A), whereas the depth-dependent model generates a family of stretching factors, all of which simulate the post-rift stratigraphy but have different amounts of synrift sediments. The stretching factors for the depth-dependent model in Figure 7B generate synrift uplift and formation of an unconformity be-

FIG. 7.—(A) Results of applying the instantaneous uniform stretching model to the restored delithified cross section of the Cambrian post-rift strata along AA' as discussed in text. Misfit between model thickness predictions and delithified thicknesses in the carbonate platform are given by the difference between the dash-dot lines connecting "x"s and the solid horizontal line in each of the three model runs. The short sections labeled MC and UC, just to the right (east) of the flexural arch in each of the three cross sections, are the delithified thicknesses in the Alberta Trough for Middle Cambrian and Upper Cambrian strata, based on data from Pugh (1971). The dashed lines connecting solid triangles are the corrected delithified thicknesses of the platform. The corrected thicknesses are produced by subtracting the delithified thicknesses of the Middle Cambrian (MC) and Upper Cambrian (UC) strata in the Alberta Trough, which are assumed to have formed mainly in response to a long-term eustatic sea-level rise, from the correlative delithified strata in the carbonate platform (see text for discussion). ρ = average delithified density of strata for each time interval as given for the midpoint of our delithification procedure, using our first option. Location of McConnell Thrust is the restored position of that thrust and can be compared with its position in Figure 6A and B. Deformation front is the eastern limit of the early Cenozoic compressional deformation of the margin. Flexural arch is a basement uplift of 100 to 200 m produced by flexural bending of the craton edge as sediment loading and rigidity of the lithosphere increase with time after rifting. Barbed vertical line is the outer edge of the carbonate platform as described in the text. Solid bar labeled A locates outer model block of the carbonate platform, which is affected somewhat by the presence or absence of the outer ridge, as discussed in text. β = amount of lithospheric stretching for each model run. The vertical lines, labeled GL and so forth, are locations of stratigraphic sections that have been analyzed using one-dimensional procedures (Bond and Kominz, 1984) and are located in Figure 6A. (B) Results of applying the instantaneous depth-dependent stretching model to the restored delithified cross section AA' as discussed in text. δ = amount of stretching in upper layer, defined as crust in the modeling; β = amount of stretching in the lower layer, defined as the subcrustal lithosphere in the modeling. All other data and symbols as in Figure 7A.

tween the basement and post-rift sedimentary wedge. This is an upper limit on the range of depth-dependent stretching factors that would simulate the post-rift restoration. Geologic constraints for the rifting history and synrift sediments are not sufficient to distinguish between the two models, and the range in results must be regarded as possible simulations of the rifting history and synrift stratigraphy of the margin. We should emphasize, however, that with respect to the rifting phase, the stretching factors are strongly model dependent, particularly since we assumed instantaneous stretching, and they should be regarded only as two sets of stretching factors that can simulate the margin.

The two-dimensional modeling also provides additional evidence of the long-term sea-level rise in Middle and Late Cambrian time inferred from the one-dimensional analyses. The delithified thicknesses of the Middle and Upper Cambrian carbonate platform strata are consistently slightly larger than the thicknesses given by both models (Fig. 7A, B), implying that a systematic misfit exists between the model predictions and the restored section. The excess delithified thicknesses in the carbonate platform closely match the thicknesses of correlative Middle and Upper Cambrian strata in the subsurface of the Alberta Trough (Pugh, 1971) for both models (Fig. 7A, B). The strata in the Alberta Trough lie on the craton beyond the limit of flexural bending, and they were most likely deposited in response to the long-term sea-level rise inferred from the R2 curves. Consequently, they must be subtracted from the correlative delithified thicknesses in the restored section. The subtraction brings the restored section into remarkably close agreement with the thicknesses predicted by both models (Fig. 7A, B). Although a eustatic rise may have been underway in Early Cambrian time as well, its magnitude is unknown. The error resulting from an unrecognized sea-level change is systematic across the margin, however, and will mainly affect the magnitude of the stretching factors required to simulate the post-rift stratigraphy. Since the stretching factors for both models produced the Middle and Upper Cambrian thicknesses reasonably well after correcting for Middle and Late Cambrian eustasy, the magnitude of an Early Cambrian sea-level rise probably was relatively small.

NEW EVIDENCE FOR SHORT-TERM EUSTATIC SEA-LEVEL CHANGES IN MIDDLE AND LATE CAMBRIAN STRATA OF THE CARBONATE PLATFORMS

Bond and others (1988), using our second option for delithifying strata, found evidence in R2 curves of short-term (2 to 6 Ma) external events comparable to the third-order events of Vail and others (1977). These events appeared to correlate in time, at least approximately, in sections from the passive-margin carbonate platforms in the southern Canadian Rockies, Utah, and the Virginia-Tennessee Appalachians, implying a eustatic mechanism. The results were based on generalized data from published sections, however. We have documented these results further with new data from sections we have measured on meter scales, and we have examined the influence of these short-term events on the evolution of the passive-margin carbonate platforms.

Procedure for identifying short-term events.—

The R2 curves were constructed so that all shaley siliciclastic and calcareous rocks were delithified using values in the upper ranges of porosity-depth curves in Bond and Kominz (1984), whereas all sand to silt-size siliciclastic and calcareous rocks were delithified using the minimum porosity-depth curves and assuming porosity reduction by early, external cementation. Whether these assumptions are correct for all lithologies is uncertain, but early cementation is well documented in coarser grained sediments, and shales commonly undergo more compaction than sandstones and siltstones (Fuchtbauer, 1974; Chilingarian and Wolf, 1975, 1976; Magara, 1980). We have found that in the early Paleozoic strata, most oolitic rocks and calcarenites have open packing, suggesting early cementation, and fluid escape structures and severely flattened worm tubes occur in both siliciclastic and calcareous shales, indicating substantial amounts of compaction during burial.

All sections were measured on a meter scale to recover the precise form of the short-term stratigraphic cycles as accurately as possible. Each measured unit was assigned to a facies on the basis of criteria summarized in Table 2. In addition, we recorded the average grain size of each unit in one of three categories, arenite, siltite, or mud (for both siliciclastic and carbonate compositions), and we visually estimated the percentage of siliciclastic mud within parted limestones (fine-grained, thin bedded to nodular limestones with tan partings or laminae of dolomitic shale). The percentage of siliciclastic mud in parted limestones is especially important because its concentration is rarely quantified in published sections, yet even in moderate quantities (~ about 15 percent of the unit), its delithification has a significant effect on the shapes of the R2 curves. Where the grain sizes were mixed in interbedded lithologies, we

TABLE 2.—FACIES CATEGORIES IDENTIFIED IN THE EARLY PALEOZOIC CARBONATE PLATFORMS AND THE WATER DEPTH RANGES ASSIGNED TO EACH

Facies Category	Water Depths (m)
1) Cryptalgalaminites with prism cracks	0 to 1
2) Cross-bedded, bioclastic, ooid, and pellet grainstones; stromatolitic and thrombolitic boundstone.	1 to 10
3) Fine to coarse, winnowed bioclastic ribbon limestone and dolomite with burrows; fine to medium, thin-bedded siliciclastic sandstone with glauconite. Both facies contain current- and wave-generated ripples, and lack mudcracks.	5 to 20
4) Pervasively bioturbated calci/dolosiltite, with distinctive rod-shaped calcite-filled burrows, thin beds of intraclastic conglomerate (tempestites?), and locally abundant thrombolites; thinly interbedded siliciclastic siltstone and fine to medium sandstone, burrowed, cross-laminated, possible wave ripples.	15 to 30
5) Nodular, burrowed calci/dolosiltite with locally abundant *Girvenella* oncoliths, lacks rod-shaped calcite-filled burrows and contains uncommon whole-body fossils; burrowed, laminated siliciclastic siltstone and fine sandstone with plane lamination and possible hummocky cross-stratification.	20 to 40
6) Laminated to nodular limestone and minor dolomite with locally abundant spicules and whole-body fossils, minor burrowing; siliciclastic shale with minor limy layers and concretions.	30 to greater than 60

visually estimated the percentage of each size category. We assigned ages to each unit using the delithified thicknesses to interpolate linearly between stratigraphic boundaries with numerical ages from the time scale of Harland and others (1982; base of the Middle Cambrian, base of the Upper Cambrian, and the Cambro-Ordovician boundary).

Results from the southern Canadian Rockies.—

Two examples from the Middle and Upper Cambrian strata of the southern Canadian Rockies illustrate the typical short-term cyclic patterns recorded in R2 curves (Fig. 8). The unconformity near the top of the Lyell Formation is indicated by the absence of the *Dunderbergi* Zone, a gap in Upper Cambrian strata recognized over much of North America (Lochman-Balk, 1970; Palmer, 1981b). The time lost at the unconformity is unknown and was chosen arbitrarily in the curves. Formation names and section locations are in Table 3. The tops of the classic Grand Cycles of the southern Canadian Rockies (Aitken, 1966, 1978) are clearly defined by distinct upward deflections or positive changes in slope. These changes correspond to the fairly abrupt boundary between the coarse-grained upper halves of the Grand Cycles and the predominantly shaley material of the lower halves. The changes in slope reflect the larger amount of subsidence required by the delithified thicknesses of the finer grained material relative to that of the coarser grained material. A result that we did not anticipate, however, is the presence of distinct events within three of the Grand Cycles, producing a total of 10 cycles from the base of the Mt. Whyte Formation to the top of the Mistaya Formation (Fig. 8). Three smaller cycles occur in the Mt. Whyte-Cathedral (MW-CA) Grand Cycle, corresponding directly to the three subcycles that Aitken (1981) recognized within the same Grand Cycle. A twofold division of the Stephen-Eldon cycle results from the shaley "Black

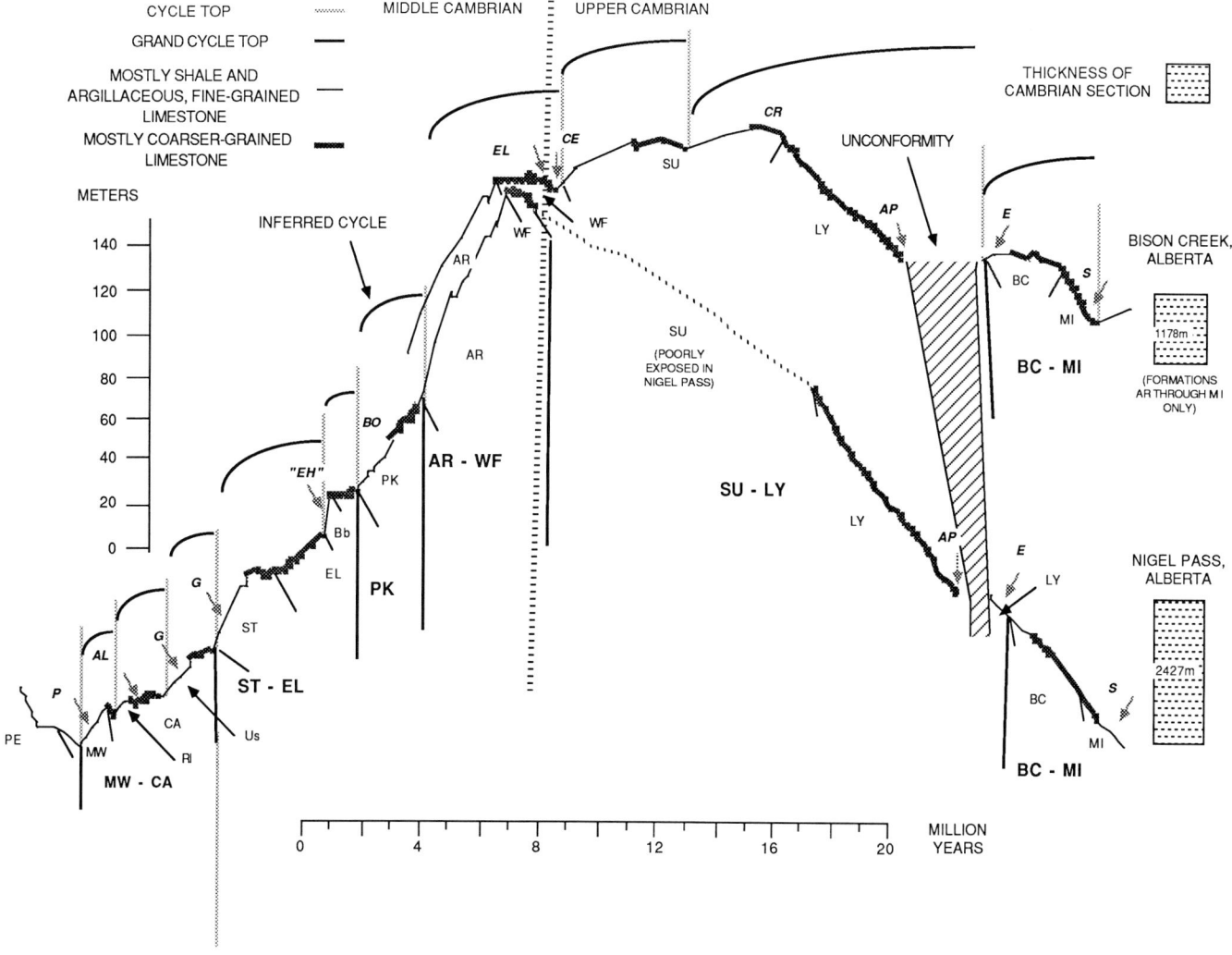

FIG. 8.—R2 curves from two sections in the southern Canadian Rocky Mountains showing Grand Cycles, as defined by Aitken (1966, 1978), and the cycles identified by the procedure for constructing R2 curves described in the text. Note the appearance of distinct subcycles within the Grand Cycles.

TABLE 3.—KEY TO LITHOLOGIC UNITS IN SECTIONS IN THE SOUTHERN CANADIAN ROCKIES, THE GREAT BASIN, AND TENNESSEE

TINTIC SECTION, UTAH
Begins as base of Ophir Formation at northeast end of Broad Canyon
(Formations from Morris and Lovering, 1961)

O1—Lower Shale Member of the Ophir Formation
O2—Middle Limestone Member of the Ophir Formation
O3—Upper Shale Member of the Ophir Formation
Te—Teutonic
Da—Dagmar
He—Herkimer
Bl—Bluebird
Cc—Cole Canyon
Op—Opex
Aj—Ajax

WAH WAH MOUNTAINS SECTION, UTAH
Section measured on west flank of Wah Wah Mountains, north of Utah Highway 21
(Formations from Hintze, 1974)

Ta—Tatow Formation
H1—Millard Member of Howell Limestone
H2—Upper Member of Howell Limestone
Ch—Chisholm Shale
Pe—Peasley Limestone
Do—Dome Limestone
Wh—Whirlwind Formation
Sw—Swasey Limestone
En—Limestone Member of Eye of Needle Formation
Pc—Pierson Cove Formation
T1—Lower Member of Trippe Limestone
T2—Shaley Limestone Member of Fish Springs of Trippe Limestone
Lm—Ledgy Member of Wah Wah Summit Formation
Wm—White Marker Member of Wah Wah Summit Formation
Bc—Limestone Member of Big Horse Canyon of Orr Formation
Sp—Shale Member of Sneakover Pass of Orr Formation
HC1—Lower part of Hellnmaria Canyon Member of Notch Peak Formation
HC2—Upper part of Hellnmaria Canyon Member of Notch Peak Formation
RT—Calcarenite Member of Red Tops of Notch Peak Formation
LD1—Lower part of Lava Dam Member of Notch Peak Formation

WELLSVILLE MOUNTAIN SECTION, UTAH
Measured on Wellsville Mountain north of Brigham City
(Formations from Doelling, 1980, and Hintze, 1973)

La—Shale + Sandstone at base of Langston Formation
Np—Naomi Peak Member of Langston Formation
S—Spence Shale Member of Langston Formation
U1—Upper Member of Langston Formation
Ut—Ute Formation
Bk—Blacksmith Formation
Ho—Hodges Shale Member of Bloomington Formation
Mn—Middle Member of Bloomington Formation
Cf—Calls Fort Member of Bloomington Formation
No—Nounan Formation
Wc—Worm Creek Quartzite
Sc—St. Charles Formation

NIGEL PASS SECTION, ALBERTA, CANADA
Measured in Nigel Pass and above Columbia Icefields chalet
(Formations from Aitken and Greggs, 1967)

Pe—Peyto Formation
Mw—Mt. Whyte Formation
Ca—Cathedral Formation
St—Stephen Formation
El—Eldon Formation (Bb is the Black Band Member)
Pk—Pika Formation
Ar—Arctomys Formation
Wf—Waterfowl Formation
Su—Sullivan Formation
Ll—Lyell Formation
Bc—Bison Creek Formation
Mi—Mistaya Formation

LEE VALLEY SECTION, TENNESSEE
(From section measured in Lee Valley, Tennessee, by Rodgers and Kent, 1948)

Pu—Pumpkin Valley Formation
Ru—Rutledge Formation
Ro—Rogersville Formation
My—Marysville Formation
N1—Lower Shale Member of Nolichucky Formation
N2—Limestone Member of Nolichucky Formation
N3—Upper Shale of Nolichucky Formation
Ma—Maynardville Formation
CR—Copper Ridge Formation

Band," a tongue of the outer detrital belt in carbonate rocks of the upper Eldon Formation (Aitken and others, 1972), and the division of the Sullivan-Lyell cycle reflects a relatively coarse limestone interval about 30 m thick in the middle of the shaley Sullivan Formation. Only the Pika cycle is poorly developed and is evident in a slight negative change in slope in the upper half of the formation.

Comparison with R2 curves from the Great Basin and Appalachians.—

Aitken (1978, 1981) has argued that the correlation of the Grand Cycles over a large portion the southern Canadian Rockies is evidence of a eustatic mechanism. Standard field procedures, however, have produced mixed evidence that temporal equivalents of these cycles are present in other regions such as the Great Basin and the Appalachians. A striking implication of our R2 analyses is that temporal equivalents of the Grand Cycles, and even the subdivisions of most of them, are regionally developed within the passive margins. This is clearly evident in a comparison of the forms of R2 curves, together with the facies that constitute each cycle, from the southern Canadian Rockies, the Great Basin, and the Appalachians (Fig. 9). All but one of the cycles found in Canada also occur in the Great Basin (Fig. 9). The MW-CA Grand Cycle is present only in the Wah Wah section, and evidence of its three subcycles is lacking. In the Lee Valley area of Tennessee, equivalents of the MW-CA, ST-EL, and AR-WF Grand Cycles, and the two subcycles in the SU-LY Grand Cycle are clearly present, further supporting the suggestion of Aitken (1981) that Grand Cycle equivalents exist in the Virginia-Tennessee Appalachians. Our data from Virginia-Tennessee are from a measured section in Rodgers and Kent (1948) and lack sufficient information on grain size and shale content to test for the presence of other short-term cycles and the generally poorly developed Pika cycle. Additional work on strata in this area is in progress. It should be emphasized that, although the cryptalgalaminites are very fine grained, the segments of the R2 curves in which they occur have negative changes in slopes because they are interbedded with coarser shoal facies containing calcarenites.

Tests of the cycle correlations and their compatibility with a eustatic forcing mechanism.—

There are two significant tests of our physical correlation of the short-term cycles in the R2 curves. First, the match of the R2 curves produces a minimum distortion or "bending" of the positions of assemblage zone fossils, to the extent that they are known, from one section to the next (Figs. 9, 10), indicating that the physical correlation of the cycles is compatible with the faunal correlations. Second and most important, beginning at least with the St-El cycle and its equivalents, the same *number* of cycles (eight) always occurs in the Middle and Upper Cambrian strata (Figs. 9, 10). This is significant because it is unlikely that a precise, one-to-one correlation in time will exist between a global external forcing event, such as eustasy, and its manifestation in the stratigraphic record. Over large areas, the timing of sedimentologic responses to external events varies with changes in depositional slopes, changes in rates of subsidence, and distances required for the lateral migration of facies (e.g., Pitman and Golovchenko, 1988). This could explain the apparent diachroneity in the rising segment of the first subcycle of the SU-LY Grand Cycle, which occurs in the *Eldoradia* Zone in the Great Basin and in the lower *Cedaria* Zone in Canada and in the Appalachians (Figs. 9, 10). (The magnitude of the diachroneity of the shale is not accurately given by the time-scale bar in the figures.) Given the relative low level of precision of many of the Middle and Upper Cambrian faunal correlations, other cycles in the R2 curves could have a similar degree of diachroniety. In addition, variable facies responses to sea-level changes produce different proportions of facies within the same cycle in different areas. This is probably the explanation for the different forms of R2 cycles in different sections, such as the thining of the Middle Cambrian shale segments in the Wah Wah section (Fig. 13A), which is located within a carbonate shoal (e.g., Brady and Koepnick, 1979; Rees, 1986). It is reasonable to expect, however, that in all but the most "unresponsive" environments, such as perhaps deep subshelf basins, each external event should produce a facies response of some type, particularly for carbonate environments, where the source of much of the sediment is organic, lies within the depositional environment, and is highly sensitive to changing environmental conditions. Thus, in

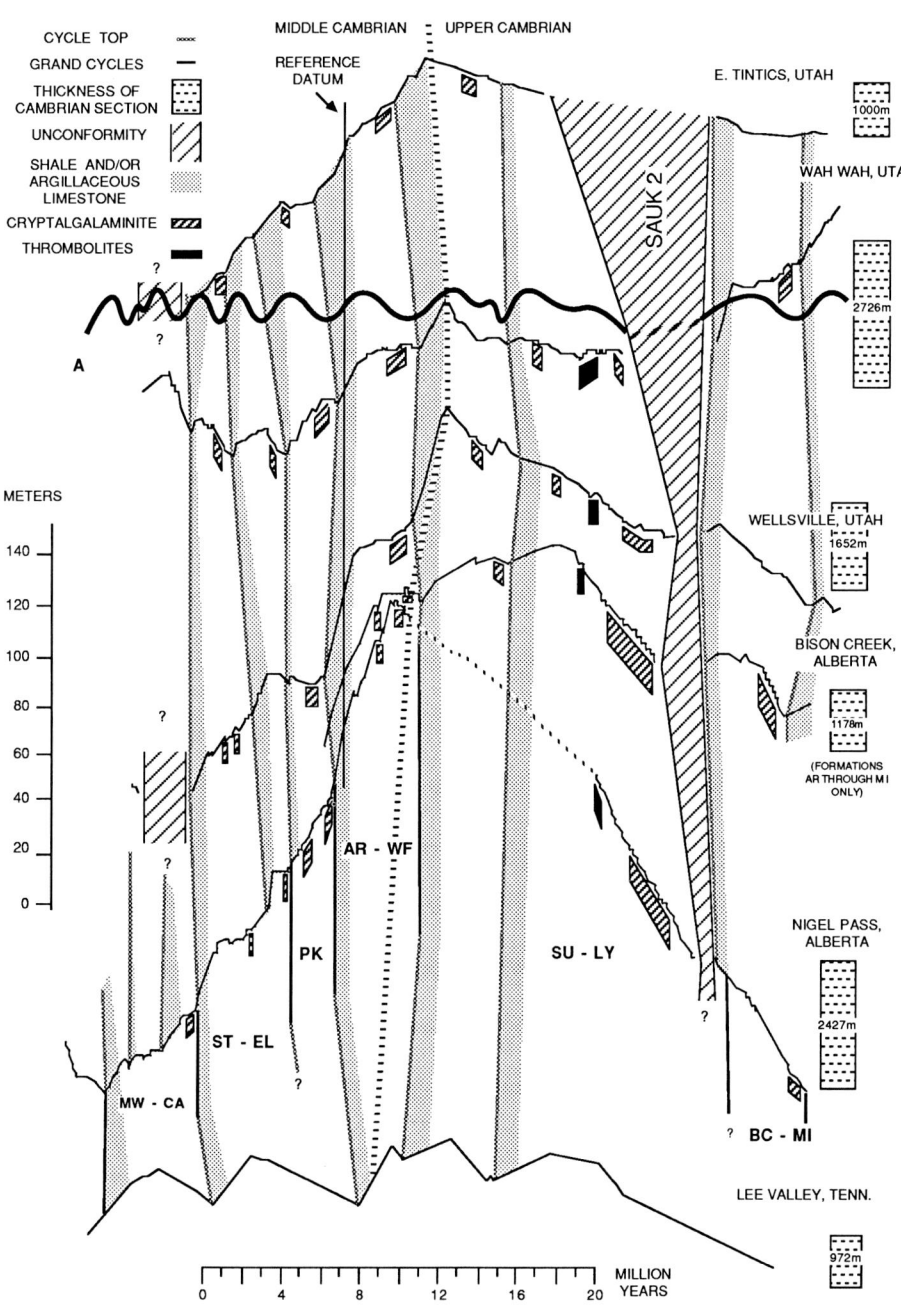

Fig. 9.—Diagram of our preferred correlation between R2 curves from the Great Basin, the southern Canadian Rockies, and the Central Appalachians. Cycle tops are identified by abrupt positive (upward deflection) changes in slope of the R2 curves that correspond to the base of the shaley and/or fine-grained portions of cycles. The solid curve "A" is an approximation of the average short-term eustatic curve derived by eye from the R2 curves. The curve is drawn assuming (1) that the rising segments of the R2 curves correspond to rising eustatic sea level and that the facies corresponding to the maximum flooding lies within this rising segment, close to the inflection point, and (2) that the segments with lower slopes or that are falling correspond to falling eustatic sea level and that the sequence boundary (either the unconformity or the correlative conformity) lies within this segment, close to the inflection point (see Fig. 14 also).

In constructing this figure, the R2 curves from Canada were matched with those in the Great Basin using a distinctive upward deflection corresponding to a well-developed shaley interval in the upper part of the Middle Cambrian strata as a datum (marked "reference datum" in the figure). The R2 curve from the Appalachians was matched with the western sections using the occurrences of *Albertella* and *Glossopleura* Zone fossils in the first cycle (PV-RU formations). In the Great Basin, the first chronostratigraphic interval above the widespread Upper Cambrian unconformity, which nearly always contains *Elvinia* Zone trilobites, is strongly distorted with respect to the trilobite zones below the unconformity. This distortion is a direct measure of the time lost at the unconformity relative to that assumed in the Canadian sections. Accordingly, we "pulled back" and shortened the Upper Cambrian segments of the R2 curves beginning with *Elvinia* so that this zone and the *Saukia* Zone were aligned above the unconformity. Unconformities may also be present in the lower Middle Cambrian strata in parts of the Great Basin, based on difficulty in finding equivalents of part or all of the Canadian MW-CA cycle in the East Tintic and Wellsville areas (see also Fig. 10).

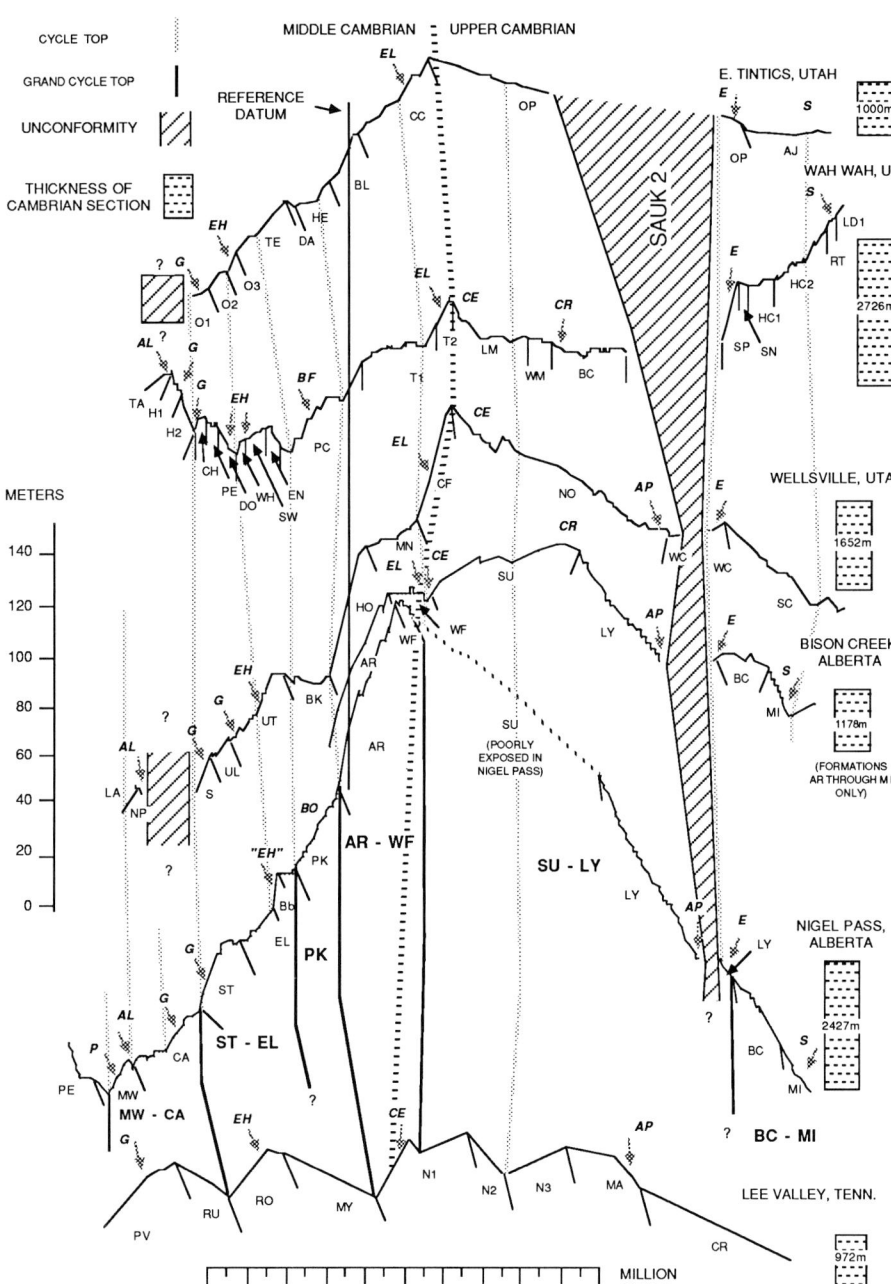

FIG. 10.—Diagram of our preferred correlation between R2 curves from the Great Basin, the southern Canadian Rockies, and the Central Appalachians. In this figure the boundaries of the formations are shown beneath each R2 curve and the biostratigraphic faunal control is given above each curve. Faunal zone abbreviations as follows: P: *Plagiura—Poliella;* AL: *Albertella;* G: *Glossopleura;* EH: *Ehmaniella;* BO: *Bolaspidella;* BF: (subdivision of *Bolaspidella* Zone); EL: *Eldoradia;* CE: *Cedaria;* CR: *Crepicephalus;* AP: *Aphelaspis;* E: *Elvinia;* S: *Saukia.* Note that the cycle correlations do not severely distort the vertical (temporal) alignment of the faunal zones. Faunal data from Rassetti (1965), Willoughby (1969), Palmer (1971a, 1971b, 1981a), Aitken and others (1972), Hintze (1973), and Hintze and Robison (1975).

such settings, the best evidence of eustatic or other global external forcing mechanisms is not the constancy of age and form but the constancy of cycle number within a chronostratigraphic interval, especially one that contains a large number of those cycles.

Analysis of error due to variations in water depths and sedimentation rates.—

There clearly are sufficient differences in facies in the Middle and Upper Cambrian strata to require corrections for water depths and sedimentation rates in the construction of the R2 curves. Neither is known well enough to assign specific values; instead, we assume a reasonable range of water depths for each facies and then examine the effect of that range on the aggradation rates (rate of vertical growth of the sediment surface), ages, and R1 values for each unit. We use the Wellsville section as an example. The water depth range for each facies category from Bond and others (1988) is given in Table 2. Although the specific value of water depth is open to dispute, the relative arrangement of water depths is probably generally correct. The limits on water depths were obtained from: (1) summaries of well-documented ancient platforms (Halley, 1983; Wilson and Jordon, 1983); (2) analysis of specific lithofacies and modern analogues (Aitken, 1978; Mullins and others, 1980; Pratt and James, 1982; Demicco, 1983) (3) studies of modern sediment/water depth relations (Neumann and Land, 1975;

Hine and Neumann, 1977; Mullins and others, 1980; Schlager and Ginsburg, 1981); (4) geometric analysis of facies/depositional slope/water depth relations of ancient platforms (Lohmann, 1976; Grotzinger, 1986); and (5) stratigraphically controlled forward modeling of water depths over ancient carbonate platforms (Grotzinger, 1986; Read and others, 1986). Overlap of water depth ranges is one measure of the uncertainty inherent in such assignments. In some instances, a specific facies may belong in different categories of water depth (e.g., thrombolites) and must be treated on an individual basis, where water depths are probably best derived from information contained in associated facies (Pratt and James, 1982). In other cases, the chosen range of water depths for a specific category (e.g., that for "ribbon rocks") is subject to arbitration between various sources of information and represents a reasonable compromise.

First, we consider whether it is possible to eliminate the zig-zag pattern in the R2 curves by changing the interpolated ages. We do this by calculating the R1 value for each unit with the minimum water depth included (Wd in equation 1). We then recalculate the ages of each unit so that the R1 curve is linear between the numerically dated stratigraphic boundaries; that is, we deliberately adjusted the ages so that the R1 curve has no deflections from the smooth thermally controlled subsidence of the margin (Fig. 11A, ii). This results in an impossible overlap in calculated ages (Fig. 11A, iii). We then calculated the aggradation rates for each unit using the delithified thicknesses and the ages determined in this step. This results in a large variation in aggradation rates, several of which have, in effect, negative values (Fig. 11B). A negative aggradation rate is impossible and indicates that the aggradation must be less than that which would produce a smooth R1 curve, implying that there must be sea-level falls. The highest concentration of negative aggradation rates by far occurs in the upper halves of the cycles, corresponding to the downward deflections or negative changes in slope of the R2 curves constructed without water depth corrections (Fig. 11B). Therefore, corrections for water depths and aggradation rates most likely cannot eliminate the changes in slopes that define the cycles in the R2 curves.

In Figure 11C, we compare the variation in ages of the fine-grained cycle halves in the Wellsville section produced by linear interpolation using (1) the observed unit thicknesses, (2) the delithified unit thicknesses, and (3) the R1 values that incorporate the minimum and maximum water depth estimates for each unit. In the case of the R1 values with minimum and maximum water depth estimates, however, we assign an age to each unit that just eliminates the negative aggradation rates (Fig. 11B). The R1 subsidence curve produced by this step is not smooth and can be regarded as the minimum sea-level change from the assumed smooth, thermally controlled subsidence of the margin that is required to just prevent negative aggradation rates. The maximum variation in ages of the fine-grained cycle halves for the different age assignment options in the Wellsville section is not much different from the variation in ages of the fine-grained cycle halves from one section to another (compare Fig. 11C with Fig. 9). The Wellsville section is broadly representative of the typical grain-size variations in the cycles in the thick sections we have analyzed, which includes all of the Canadian sections and all but the East Tintic section in Utah. It can be assumed, therefore, that the age differences resulting from the different options for liner interpolation in the Wellsville section are representative of those that could occur between the thick sections in Canada and Utah.

It could be argued that larger variations in ages of the finer grained cycle halves than those given in Figure 11C can be produced by making appropriate adjustments in aggradation rates. Such adjustments, however, would result in water depths for facies that fall outside of the ranges we have assigned to the facies, and it would be necessary to assume a deviation from smooth thermally controlled subsidence to maintain the water depth values within our assigned ranges. Assuming that thermally controlled subsidence and lithification are smooth processes on the time scales of the cycles in Figure 9, the most likely cause of such deviations from smooth subsidence will be sea-level changes. A sea-level change, however, will affect all sections equally. Consequently, it is unlikely that further changing aggradation rates will significantly increase the differences in ages of fine-grained cycle halves in the thicker sections beyond that which can be inferred from the Wellsville section (Fig. 11C) as a representative example. We have not attempted, however to compare the thicker sections with the much thinner sections in the East Tintics and in the Appalachians in terms of the variation in interpolated ages.

DISCUSSION

Growth of carbonate platforms on a cooling lithospheric plate.—

The results of the one- and two-dimensional modeling are strong evidence that the subsidence of the carbonate platforms can be simulated using the cooling lithospheric plate model with a thermal equilibrium thickness of 125 km that has been applied to modern passive margins. This is significant because the rate and form of decay of thermally controlled subsidence depend on plate thickness. The rate of decay of subsidence is faster for a plate thinner than 125 km and much slower for one thicker than 125 km (Fig. 12). Thus, all other factors being equal, the "lifetime" (potential maximum time for significant vertical growth) of a passive-margin carbonate platform is a function of the plate thickness. For a 125-km plate, the maximum lifetime cannot exceed about 200 Ma (Fig. 12), and the half-life of the total potential subsidence is about 44 Ma. Of course, the full-life expectancy is not achieved if initiation of the platform postdates onset of post-rift cooling and/or the passive margin is destroyed before the thermal anomaly has essentially decayed. Even so, the predictable subsidence history and limits on maximum time available for growth set passive-margin carbonate platforms apart from platforms in other tectonic settings. In foreland basins, for example, the form and duration of subsidence depends on the history of loading and is not predictable. Platform growth tends to accel-

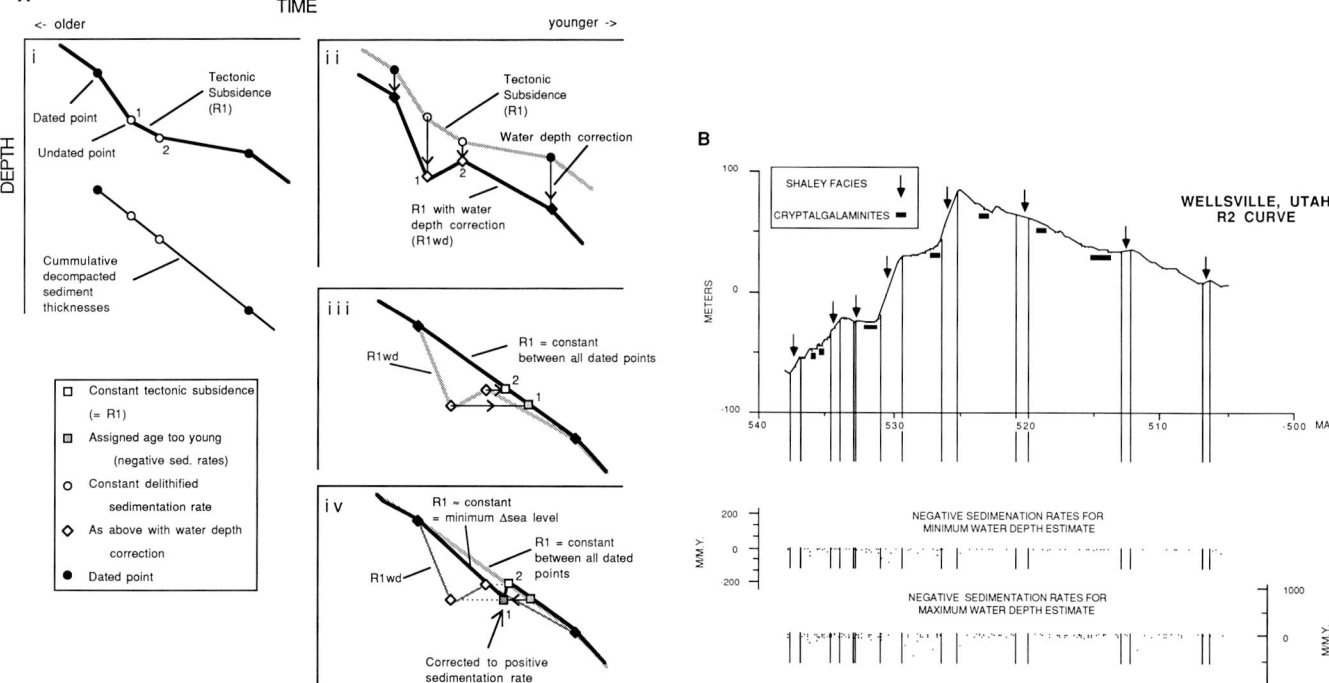

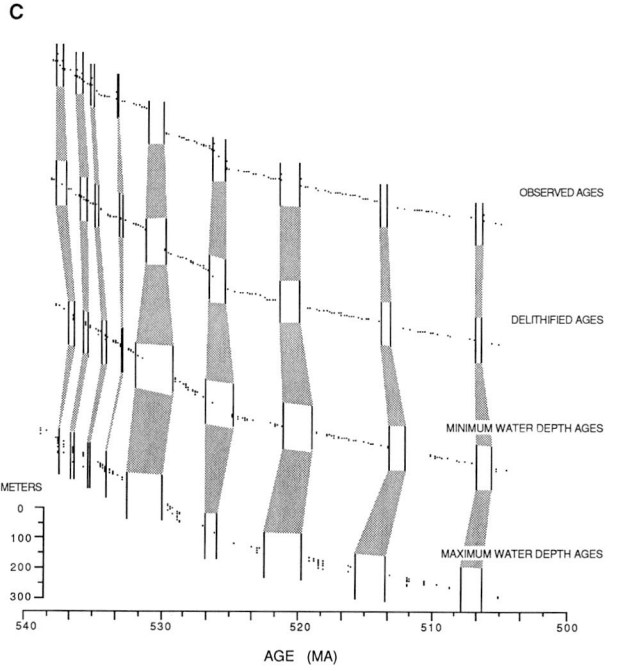

FIG. 11.—(A) Illustration depicting the method used for defining ages in order to obtain either: (1) constant tectonic subsidence (R1) between dated intervals or (2) minimum changes in sea level with R1 approximately constant between dated intervals. For simplicity, only four sediment units are shown, two dated (black) and two between (points 1 and 2) which have no age control. In graph i, an example is given of an R1 curve generated assuming constant aggradation rates. Shown at the base is the cumulative decompacted thicknesses of the units. A straight line between the two dated points on the cumulative delithified thickness curve constrains the horizontal locations (the ages) of the two undated points between; and the resulting R1 curve is also shown. In graph ii, water depth corrections from Table 2 have been added to the R1 curve, resulting in a different form of the R1 curve (R1wd). While the magnitude of R1 is defined for each point, the zig-zag form of the R1wd curve can be totally eliminated by changing the ages of points 1 and 2 so that they fall on a straight line between the two surrounding dated points (graph iii). Because the R1 value of point 1 is greater than that of point 2, however, it is given a younger age than point 2. This is physically impossible and results in the prediction of negative aggradation rates for the sediments deposited in unit 2. In graph iv, the negative sedimentation rate is eliminated by assigning an age to point 1 which is slightly older than that of unit 2 (0.001 m.y.). Thus, an R1 curve is calculated which shows minimum fluctuation in R1 values and does not generate negative sedimentation rates. (B) R2 curve from Wellsville showing a strong correspondence between the shoaling, or falling, portions of the short-term cycles and times of negative aggradation rates, indicated by small dots, for the extremes in the ranges of water depths from Table 2. This indicates that the assumption that R1 subsidence with water depth corrections is constant between dated intervals is not reasonable and strongly suggests that the falling segments of the short-term cycles require a sea-level fall. (C) R1 curves calculated for the Wellsville section assuming different methods for assigning ages. The R1 value for each measured unit is given by a single dot. In the top curve, the fully lithified units, as measured in the field, are assumed to have been deposited at a constant rate between dated intervals. In the second curve down, each unit is assumed to have accumulated at a constant rate between dated intervals at the time that it was deposited. In the bottom two curves, the R1 subsidence, with water depth corrections, is assumed to be constant between dated intervals with a correction to remove negative aggradation rates as described in the text and Figure 11B. The finer grained portion of each cycle, as depicted in Figure 9, is marked by the grey tone pattern. See text for discussion.

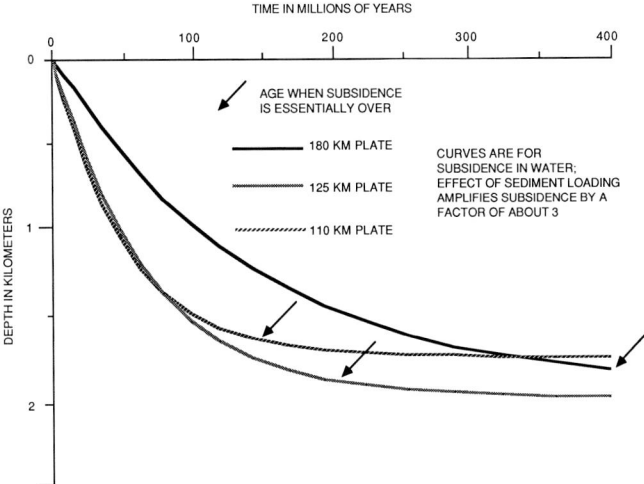

FIG. 12.—Subsidence curves in water for the cooling plate model of McKenzie (1978), assuming a thermal equilibrium plate thickness of 110, 125, and 180 km calibrated to an oceanic crustal thickness of 5 km, following Steckler (1981) and Cochran (1981). See text for discussion.

erate with time as tectonic loads increase or shift cratonward toward the platform, or the platform is repeatedly drowned and reestablished in response to episodic tectonic loading.

The close similarity in the subsidence of the early Paleozoic passive margins with that predicted for modern passive margins also is evidence that little change has occurred in the first-order thermal structure of the lithosphere during the Phanerozoic. Thus, the tectonic subsidence of all Phanerozoic passive-margin carbonate platforms must have essentially the same exponential form, and all Phanerozoic platforms have had a maximum life expectancy of about 200 Ma. The thermal structure of the lithosphere could have been significantly different in the earlier history of the earth, however, and the subsidence histories and maximum lifetimes of Proterozoic passive-margin carbonate platforms could have been quite different from those of the Phanerozoic.

The results of our one-dimensional modeling suggest that the growth of the early Paleozoic carbonate platforms can be divided into three stages, which are best be seen in the first derivative of R1 curves, especially in the outer parts of the margin such as in the Wellsville, Utah, area (Fig. 13). The earliest stage, in Middle Cambrian time, was characterized by high and rapidly falling rates of net subsidence, from over 95 m/m.y. to about 35 m/m.y., resulting from the high and rapidly falling rates of thermally controlled subsidence combined with the eustatic sea-level rise. The second stage, in Late Cambrian through earliest Ordovician (Tremodocian) time, was characterized by much lower and more slowly declining rates of net subsidence, from 35 m/m.y. to less than 10 m/m.y., reflecting the later stages of exponential decay of thermally controlled subsidence combined with the eustatic sea-level fall. In the final or mature stage of growth, from Middle Ordovician though Middle Devonian time, the rate of thermally controlled subsidence approaches zero, reflecting the essential removal of

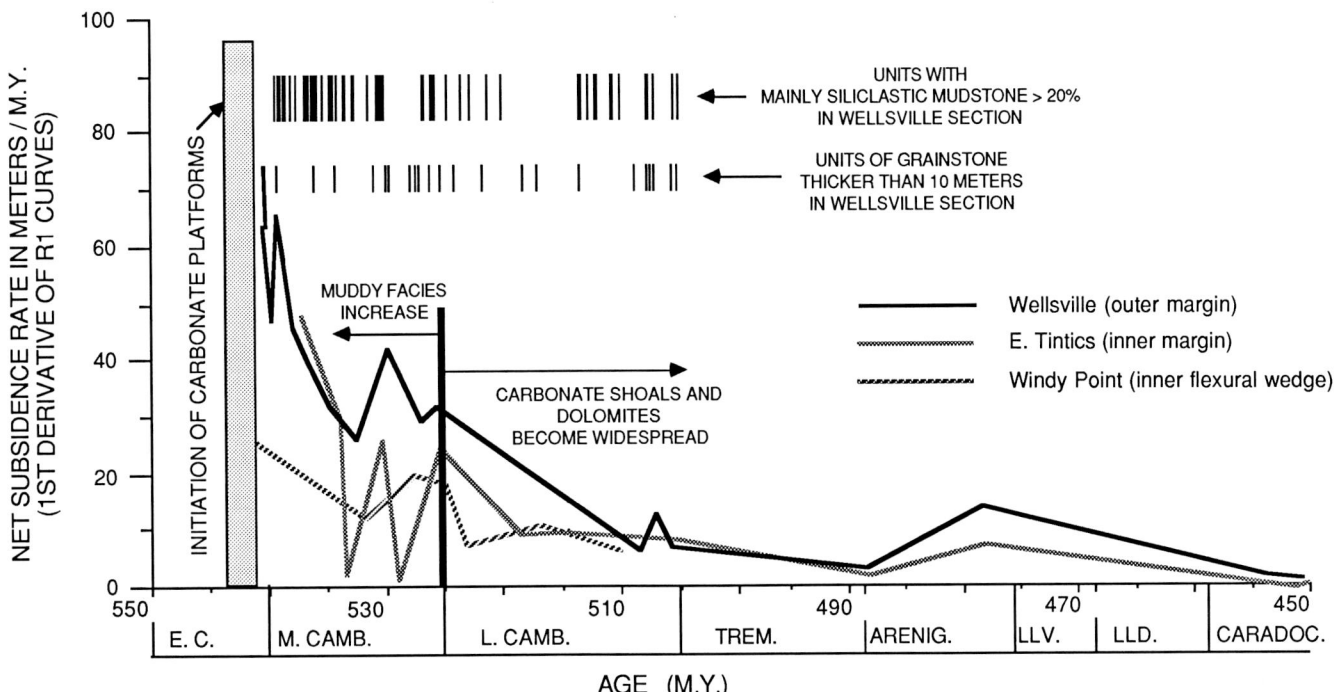

FIG. 13.—Relation between lithology in the carbonate platforms and net subsidence rates, as given by the first derivative of R1 curves for selected sections in the Cordillera. The Wellsville and E. Tintic sections and corresponding R1 curves are given in Figure 3A and B, respectively. The Windy Point section is in the southern Canadian Rockies, located in Figure 6A. The corresponding R1 curve is from Bond and Kominz (1984). See text for discussion.

the thermal anomaly in the subjacent lithosphere. In this portion of the platform stratigraphy, marked unconformities are present that appear to be correlative with unconformities on the craton (e.g., in Middle Ordovician, Lower Silurian, Upper Silurian, and lower Devonian strata, some of which are shown in Fig. 3B; Sloss, 1963; Hintze, 1973; Ross, 1977; Poole and others, 1977). Thermally controlled subsidence apparently was slow enough that the subsidence of the carbonate platforms approached that of the stable craton, and most of the limited growth of the platforms probably occurred in response to eustatic rises in sea level that are thought to account for the transgressions of the craton (Vail and others, 1977; Barnes, 1984; Fortey, 1984; Johnson and others, 1985).

Effect of thermally controlled subsidence and multiple orders of eustatic changes on lithologic composition.—

The vertical distribution of lithologies in the carbonate platforms corresponds closely to the stages of platform growth combined with the multiple orders of sea-level change implied by the R2 curves. The relation has been quantified for Middle and Upper Cambrian strata in the section from Wellsville, Utah (Fig. 13). Here, the highest concentration of units with a substantial portion of siliciclastic shale (>20 percent) occurs in Middle Cambrian strata (Bond and Grotzinger, unpubl. measured section), corresponding to the highest rates of net subsidence. Most of these fine-grained units are wavy-bedded to nodular limestones with argillaceous partings, and a few are true shales that tend to be concentrated in the rising segments of the R2 curves (Fig. 9). Less shaley carbonate strata, commonly consisting of calcisiltites and grainstones, tend to be concentrated in the flat or falling segments of the R2 curves (Fig. 9). In contrast, in Upper Cambrian strata of the Wellsville section, corresponding to the time of slowly changing and low long-term subsidence rates, there are fewer shaley units. In addition, the number of thick grainstone units tends to increase somewhat beginning in the upper third of the Middle Cambrian strata. Comparable reductions in shaley material from Middle Cambrian to Upper Cambrian strata are typical of the thick sections (>1.5 km) elsewhere in the Great Basin (Robison, 1964; Hintze, 1973; Brady and Rowell, 1976; Bond and Grotzinger, unpubl. field data). In the southern Canadian Rockies, siliciclastic muddy sediments are abundant in the shaley halves of the Grand Cycles in the both the Middle and Upper Cambrian strata of the carbonate platform. Aitken (1966, 1978), however, has shown that in the carbonate halves of the Grand Cycles, muddy, parted limestones and shales decrease, and oolitic grainstones and peritidal facies increase from Middle to Upper Cambrian strata, leading to the distinction between the overall finer grained Stephen-type cycles of Middle Cambrian age and overall coarser grained Sullivan-type cycles of Upper Cambrian age.

We suggest that the long-term and short-term vertical changes in lithologic compositions of the platforms are a direct consequence of the changes in the subsidence rates as given by the R1 and R2 curves. In Middle Cambrian time, when subsidence rates were high owing to the combination of rapid cooling and the rise in long-term eustatic sea level, rates of net subsidence during the short-term eustatic rises must have become large relative to the growth potential of the carbonate platform (as defined by Kendall and Schlager, 1981, and Schlager, 1981). This resulted in slower rates of carbonate sedimentation within the platform and widespread deposition of fine-grained subtidal facies. During the short-term eustatic falls, rates of carbonate production must have increased relative to the net subsidence rate, leading to an increase in carbonate production and widespread deposition of carbonate shoal facies. As the thermally controlled subsidence slowed and long-term sea level fell in Late Cambrian time, carbonate production kept pace more easily with subsidence. Consequently, the short-term eustatic rises were not as effective in reducing carbonate production as during Middle Cambrian time, and the proportion of shaley strata decreased from Middle to Late Cambrian time.

These superimposed eustatic and thermal subsidence mechanisms and the inferred lithologic responses are summarized diagramatically in Figure 14. The diagram can be viewed as representing the long-term evolution of an early Paleozoic carbonate platform at a single location within a passive margin, assuming that short-term eustatic cycles, such as those in Middle and Upper Cambrian time, continued to occur throughout the growth of the platforms. Stages 4 and 5 represent the conditions in Middle Cambrian time favoring a relatively high percentage of deeper, finer grained facies in each cycle. These stages will also tend to contain fewer and/or smaller unconformities at sequence boundaries (Vail and others, 1977) relative to the later stages (1–2, Fig. 14). As the rate of cooling slows and long-term eustatic sea level falls, as during Late Cambrian and Ordovician time, the percentage of deeper, finer grained facies decreases, and unconformities will become more common at sequence boundaries (2 and 3, Fig. 19). As the rate of cooling approaches zero, as in Silurian and Devonian time (stages 1 and 2, Fig. 19), the stratigraphy of the platforms will approach that of the stable craton. In these stages deeper water facies will tend to be limited, and unconformities with large time gaps will form during each sea-level fall.

The diagram in Figure 14 can also be viewed as representing the subsidence and facies changes from the inner (1) to outer (5) part of the platforms within a small increment of time, although the long-term eustatic change should be considered fixed. In this context the diagram implies that the finer grained deeper water facies and the coarser grained shallower water facies will always tend to be located close to specific segments of the short-term eustatic curves. The deepest facies, corresponding to condensed intervals, as defined by Vail and others (1977), will tend to be close to the inflection points of the rising segments of a short-term sea-level curve and the shallowest facies, corresponding to sequence boundaries, will tend to be close to the inflection points of the falling segments. This relation will hold even though the specific types of facies and their grain size vary from the inner to outer parts of margins as long as there are relative changes in grain size associated with each segment of the short-term curve. In other words, although the

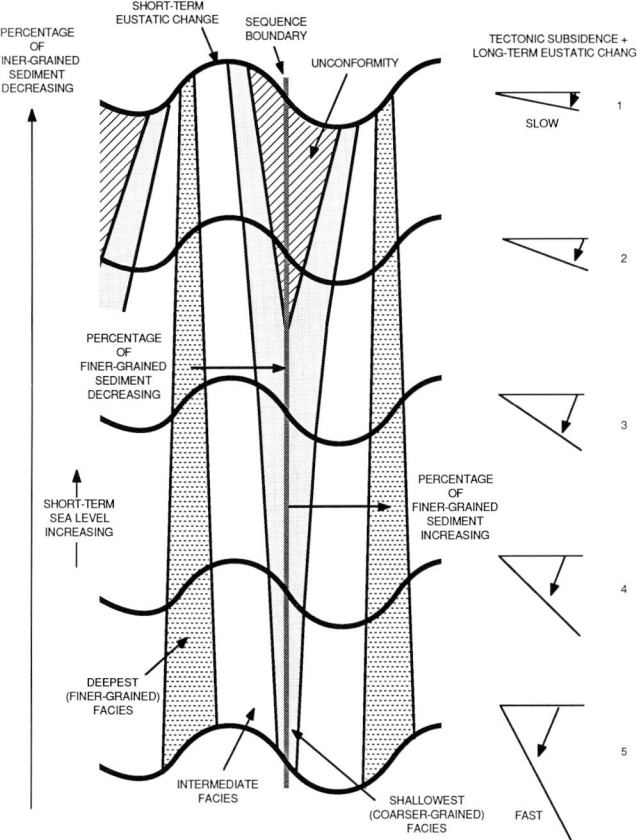

FIG. 14.—Lithostratigraphic model for early Paleozoic carbonate platforms based on the lithofacies response to the combined effects of long-term eustatic sea-level changes, thermally controlled subsidence, and short-term eustatic sea-level changes that are assumed to be sinusoidal in form. See text for discussion.

specific facies that occurs within the region labeled deepest (finer grained) facies or condensed interval (Fig. 14) may become somewhat coarser and shallower from curves 5 to 1, that facies will always tend to contain the finest grained material deposited during each cycle. It should also be clear that the cycle tops and the abrupt changes in slopes of the R2 curves do not correspond to sequence boundaries. The sequence boundaries in the carbonate platforms should occur within the flat or falling segments of the R2 curves in Figure 9, although they have not yet been identified in outcrop.

The diagram in Figure 14 also implies that for settings in which the facies response to eustasy is rapid, as appears to be the case in the early Paleozoic carbonate platforms, the deepest and generally finest grained facies, which forms during rising sea level, will be essentially coeval across the margin. This facies potentially is a somewhat better time-stratigraphic marker than the sequence boundary in outcrop studies, where it can be recognized, because it is least affected by erosion during oscillating sea level.

Advantages of using R2 curves to identify stratigraphic cycles.—

Our experience with the construction of R2 curves has clearly underscored several advantages of using this method in the early Paleozoic carbonate platforms. As was pointed out by Bond and others (1988), the equivalents of the Canadian cycles in the Great Basin tend to be manifested as rather subtle changes in grain size within predominantly carbonate intervals, instead of distinct shale-to-carbonate changes as in the southern Canadian Rockies that are easily seen even from great distances. These differences probably have been one of the principal sources of difficulty in recognizing equivalents of the Canadian Grand Cycles in the Great Basin. Where such differences occur between regions, a quantitative and systematic method such as ours can emphasize similarities in key facies indicators, such as trends in relative grain size and mud content, that may not be evident using more conventional, descriptive methods.

As with other procedures based on physical criteria for correlation, such as Fischer plots (Fischer, 1964) and sequence stratigraphy, construction of R2 curves has the advantage of providing relatively fine-scale correlation where fossil control is poor or lacking, providing that a few reliable tie points for correlation are available. Our procedure is significantly different from methods based on Fischer plots, because it is not dependent on an *a priori* recognition of cycles and an assumption that they have constant period. An advantage over sequence stratigraphic methods is that the multi-ordered components of a composite eustatic signal can be identified and compared in a single curve. Our procedure is subject to much of the same error in ages and correlation as these other methods, but it provides a basis for quantitatively assessing these errors.

Implications of two-dimensional modeling for the initiation of the carbonate platforms.—

The results of the two-dimensional modeling strongly suggest that the carbonate platforms in all three areas we have analyzed formed largely in response to the first-order thermo-mechanical mechanisms associated with the evolution of passive margins. Both the uniform and depth-dependent models predict that during, and in the first few millions of years after rifting, a basement uplift occurred along the edge of the craton adjacent to the hinge zone (Fig. 7A, B). In the uniform model, this uplift is small (<200 m) and is a result of thermal expansion of the lithosphere as heat flows laterally away from the rifted margin. The magnitude and duration of this uplift, however, are not sufficient to reproduce the abrupt termination of the Lower Cambrian quartz sandstones at the hinge zone, resulting in a slight misfit between the model and the restored Lower Cambrian strata at the inner part of the margin (Fig. 7A). In the depth-dependent model, a set of stretching factors for the inner part of the margin can produce the amount and duration of uplift required to fit the restored stratigraphy across the hinge zone (Fig. 7B). Possibly, the depth-dependent model is a better approximation of the subsidence of the margin, at least for the inner portion, as has been found in some modern margins (Royden and Keen, 1980; Celerier, 1986;

Steckler and others, 1988). The significance of the modeling, however, is that regardless of the stretching mechanism, a basement uplift occurred along the edge of the craton that could have been the source of the Lower Cambrian siliciclastic sediments. This interpretation is further supported by the generally westward-to-southwestward transport of the quartz sandstones indicated by paleocurrent data (Mountjoy and Aitken, 1963; Young, 1979). The modeling implies that, as the margin cooled and the flexural rigidity increased, the uplifted edge of the craton was deflected downward and sedimentation began to shift cratonward, forming a broad inner flexural wedge, so that by Middle Cambrian time, flexural bending had caused subsidence along a substantial portion of the craton edge (Fig. 7A, B). At the same time, the long-term sea-level rise shifted the shoreline well into the craton, at least as far as central Saskatchewan. Thus, the flexural bending of the craton edge and the sea-level rise together must have substantially reduced the siliciclastic sediment sources along the edge of the passive margin.

We suggest that the timing of initiation of the carbonate platforms was a direct consequence of the combined effects of eustasy, thermal cooling, erosion, and flexure after rifting ended. Even though stable, shallow-marine shelves in warm, low latitudes must have been established soon after rifting ended in Early Cambrian time, formation of a carbonate platform was inhibited by the high rates of deposition of siliciclastic material derived from basement uplifts adjacent to the margin. Both models predict that the sediment sources were reduced or eliminated by early Middle Cambrian time, coinciding closely with the time of initiation of the carbonate platforms. Thus, the platforms began to grow only after flexure, combined with the eustatic sea-level rise, thermal subsidence, and erosion, had essentially eliminated the basement highs and substantially reduced the supply of siliciclastic sediment to the margin.

Although this interpretation applies strictly to the North Saskatchewan River section, it probably has general application to all of the areas we have analyzed. We have argued previously that the early Paleozoic passive margins in North America, as well as in other parts of the world, probably formed within a relatively narrow interval of time during a major episode of continental breakup (Bond and others, 1984). The growth of the inner flexural wedge and burial of nearby basement uplifts will have been broadly correlative in these margins, since the increase of rigidity and flexural bending of lithosphere is, to first order, a function of the rate of decay of the thermal anomalies formed during rifting. We suggest that the predictable thermo-mechanical evolution of a passive margin after rifting, combined with the effects of the long-term eustatic sea-level rise, are a plausible explanation for the striking degree of synchroneity in the initiation of the carbonate platforms in all of the sections we have analyzed.

SUMMARY AND CONCLUSIONS

In summary, our one- and two-dimensional analyses of the subsidence of early Paleozoic carbonate platforms in North America have led to the following conclusions:

(1) One-dimensional R1 subsidence analyses indicate that the first-order control on subsidence and growth of the carbonate platforms in the southern Canadian Rockies, the Great Basin, and in the Appalachians was thermal contraction of heated lithosphere following the rifting of the margins. The subsidence of the platforms can be modeled in terms of a cooling lithospheric plate with a thermal equilibrium thickness of 125 km, implying that significant vertical growth of passive-margin carbonate platforms cannot exceed about 200 Ma after rifting.

(2) R2 curves in all three areas indicate that eustatic changes in sea level were the second-order control on subsidence and growth of the platforms. The eustatic component of subsidence is complex, consisting of a long-term rise and fall spanning about 100 Ma and repeating short-term cycles about 2 to 6 Ma in duration.

(3) The combined effects of the thermally controlled subsidence and multiple cycles of eustasy were a primary control on the lithostratigraphic framework of the carbonate platforms. During the rising segments of the shorter term eustatic cycles, net subsidence was rapid enough to reduce rates of carbonate production, leading to widespread deposition of fine-grained subtidal facies. During the falling segments of the shorter term cycles, rates of carbonate production increased, leading to widespread deposition of shallow subtidal and intertidal facies with a large component of coarser grained sediment. On a longer time scale of tens of millions of years, the proportion of finer grained sediment decreased in response to the reduction in subsidence rates caused by the exponential decay of thermally controlled subsidence and the long-term eustatic sea-level fall.

(4) Two-dimensional modeling implies that initiation of the carbonate platforms was a result of thermo-mechanical processes related to formation of the margins and the long-term eustatic sea-level rise. The modeling suggests that until the end of Early Cambrian time, uplifts of basement existed along the craton edge. These could have been the source of large amounts of Lower Cambrian quartz sandstones that suppressed carbonate production, except locally at the outermost parts of some margins. By the beginning of Middle Cambrian time, the modeling implies that flexural bending of the craton edge, together with the long-term sea-level rise, will have submerged and buried these basement sources, thereby reducing the supply of siliciclastic sediment sufficiently to initiate widespread production of carbonate sediment. The similarity in timing of initiation of the platforms in all three areas probably reflects the broad synchroneity in the timing of continental breakup and initiation of the passive margins around the edge of North America, together with the extensive and coeval submergence of the craton by the eustatic sea-level rise beginning in Middle Cambrian time.

(5) Integration of the results of one- and two-dimensional modeling suggests a distinct pattern of growth of the carbonate platforms that may be generally applicable to the evolution of platforms in passive margins. Carbonate sedimentation was initiated only sporadically during the latest part of the Early Cambrian as local buildups along the outermost, rapidly subsiding parts of the passive margins, where siliciclastic influx from cratonic sources was lowest. Dur-

ing Middle and Late Cambrian time, the effects of long-term eustasy plus increasing flexural bending of the cratonic edge resulted in retreat of the cratonic siliciclastic shoreline and progradation of carbonate in a cratonward direction. This phase marked the development of the "inner detrital belt" Palmer (1971a,b) and the "intrashelf basin" of Markello and Read (1982). Ultimately, from Ordovician time to the end of the passive-margin phase of growth, the effect of repeated sea-level rises, together with continued cratonward expansion of the flexural wedge, generated further backstepping and eventual near-elimination of the cratonic shoreline and influx of siliciclastic sediment. Toward the end of this phase, the carbonate platforms merged with probably isolated shoals of the craton interior, forming a laterally continuous veneer of carbonate deposits over North America. Thus, the overall expansion of the carbonate platforms generally proceeded from the outer edges of the passive margins toward the interior of the craton.

(6) In view of the fact that the vertical growth of the early Paleozoic platform in the Great Basin spanned the interval from about 540 Ma to 375 Ma, the platform in this region had achieved at least 80 percent of its maximum potential growth. Only the earliest stage, when thermally controlled subsidence rates were highest, is missing, as is the case in all of the areas we have analyzed. In contrast, in the southern Canadian Rockies and in the Virginia-Tennessee Appalachians, the passive-margin platforms had achieved only about 45 to 50 percent of their maximum potential growth before the thermally controlled phase of subsidence was terminated. The growth of the platform up to Late Devonian time in the Great Basin, therefore, serves as a good example of the complete, or nearly complete, evolution of a passive-margin carbonate platform on a 125-km lithospheric plate. This, together with the large amount of detailed data available in the region, makes the Great Basin an ideal setting in which to develop a comprehensive geodynamic model for the stratigraphic, sedimentologic, and biostratigraphic evolution of a Phanerozoic passive-margin platform. Such a model could be useful for comparison with the evolution of other Phanerozoic carbonate platforms and with Proterozoic passive-margin platforms that might have formed on lithospheric plates with thicknesses significantly different from 125 km.

ACKNOWLEDGMENTS

We thank Paul Olsen and Walter Pitman, III, for constructive reviews of the manuscript. The research was supported by National Science Foundation Grants EAR 83-06241, EAR 84-17439, EAR 88-17403, EAR 85-18644, and L-DGO contribution number 4411.

REFERENCES

AITKEN, J. D., 1966, Middle Cambrian to Middle Ordovician cyclic sedimentation, Southern Canadian Rocky Mountains of Alberta: Bulletin of Canadian Petroleum Geology, v. 14, p. 405–441.
———, 1971, Control of lower Paleozoic sedimentary facies by the Kicking Horse Rim, Southern Canadian Rocky Mountains, Canada: Bulletin of Canadian Petroleum Geology, v. 19, p. 557–569.
———, 1978, Revised models for depositional grand cycles, Cambrian of the Southern Rocky Mountains, Canada: Bulletin of Canadian Petroleum Geology, v. 26, p. 515–542.
———, 1981, Generalizations about grand cycles, in Taylor, M. E., ed., Short Papers for the Second International Symposium on the Cambrian System: U.S. Geological Survey Open-File Report 81-743, p. 8–14.
———, FRITZ, W. H., AND NORFORD, B. S., 1972, Cambrian and Ordovician biostratigraphy of the southern Canadian Rocky Mountains, in Glass, D. J., ed., Guidebook for Excursion A-19: XXIV International Geological Congress, Montreal, Quebec, 57 p.
———, AND GREGGS, R. G., 1967, Upper Cambrian formations, southern Canadian Rocky Mountains of Alberta, an interim report: Geological Society of Canada Paper 66-49, 91 p.
ARMIN, R. A., AND MAYER, L., 1983, Subsidence analysis of the Cordilleran miogeocline: Implications for timing of late Proterozoic rifting and amount of extension: Geology, v. 11, p. 702–706.
BARNES, C. R., 1984, Early Ordovician eustatic events in Canada, in Bruton, D. L., ed., Aspects of the Ordovician System: Palaeontological Contributions, University of Oslo, Norway, Universitetsforlaget, p. 51–63.
BEAUMONT, C. R., KEEN, C. E., AND BOUTILLIER, R., 1982, On the evolution of rifted continental margins: Comparisons of models and observations for the Nova Scoia margin: Geophysical Journal of the Royal Astronomical Society, v. 70, p. 667–715.
BOND, G. C., AND KOMINZ, M. A., 1984, Construction of tectonic subsidence curves for the early Paleozoic miogeocline, southern Canadian Rocky Mountains: Implications for subsidence mechanisms, age of breakup and crustal thinning: Geological Society of America Bulletin, v. 95, p. 155–173.
———, ———, AND DEVLIN, W. J., 1983, Thermal subsidence and eustasy in the lower Paleozoic miogeocline of western North America: Nature, v. 306, p. 775–779.
———, ———, AND GROTZINGER, J. P., 1988, Cambro-Ordovician eustasy: Evidence from geophysical modeling of subsidence in Cordilleran and Appalachian passive margins, in Paola, C., and Kleinspehn, K., eds., New Perspectives in Basin Analysis: Springer-Verlag, New York, p. 129–161.
———, NICKESON, P. A., AND KOMINZ, M. A., 1984, Breakup of a supercontinent between 625 Ma and 555 Ma: New evidence and implications for conntinental histories: Earth and Planetary Science Letters, v. 70, p. 325–345.
BOVA, J. A., AND READ, J. F., 1986, Incipiently drowned facies within a cyclic peritidal ramp sequence, Early Ordovician Chepultepec interval, Virginia Appalachians: Geological Society of America Bulletin, v. 98, p. 714–727.
BRADY, M. J., AND KOEPNICK, R. B., 1979, A Middle Cambrian platform-to-basin transition, House Range, west of central Utah: Brigham Young University Geology Studies, v. 26, p. 1–7.
———, AND ROWELL, A. J., 1976, Upper Cambrian subtidal blanket carbonate of the miogeocline, eastern Great Basin: Brigham Young University Geology Studies, v. 23, p. 153–163.
BUCK, W. R., 1986, Small-scale convection induced by passive rifting: The cause for uplift of rift shoulders: Earth and Planetary Science Letters, v. 77, p. 362–373.
CAZENAVE, A., LAGO, B., AND DOMINH, K., 1983, Thermal parameters of the oceanic lithosphere estimated from geoid height data: Journal of Geophysical Research, v. 88, p. 1105–1118.
CELERIER, B., 1986, Models for the evolution of the Carolina Trough and their limitations: Unpublished Ph.D. Dissertation, Massachusetts Institute of Technology, Cambridge, 206 p.
CHILINGARIAN, G. V., AND WOLF, K. H., 1975, Compaction of Coarse-Grained Sediments, I: Elsevier, Amsterdam, 552 p.
———, AND ———, 1976, Compaction of Coarse-Grained Sediments: II: Elsevier, Amsterdam, 8808 p.
CHOW, N., AND JAMES, N. P., 1987, Cambrian Grand Cycles: A northern Appalachian perspective: Geological Society of America Bulletin, v. 98, p. 418–429.
COCHRAN, J. R., 1981, Simple models of diffuse extension and the pre-sea floor spreading development of the continental margin of the northeastern Gulf of Aden, in Blanchert, R., and Montadert, L., eds., Geology of Continental Margins: 26th International Geological Congress, Proceedings, Oceanologica Acta, p. 259–277.
DEMICCO, R. V., 1983, Wavy and lenticular-bedded carbonate ribbon rocks of the Upper Cambrian Conococheague limestone, central Appalachians: Journal of Sedimentary Petrology, v. 53, p. 1121–1132.

DEVLIN, W. J., AND BOND, G. C., 1988, The initiation of the early Paleozoic Cordilleran miogeocline: Evidence from the uppermost Proterozoic-Lower Cambrian Hamill Group of southeastern British Columbia: Canadian Journal of Earth Sciences, v. 25, p. 1–19.

DOELLING, H. H., 1980, Geology and mineral resources of Box Elder County, Utah: Utah Geological and Mineral Survey, Bulletin 115, 251 p.

FISCHER, A. G., 1964, The Lofer cyclothems of the Alpine Triassic, in Merriam, D. F., ed., Symposium of cyclic sedimentation: State Geological Survey of Kansas, Bulletin 169, v. 1, p. 107–150.

FORTEY, R. A., 1984, Global earlier Ordovician transgressions and regressions and their biological implications, in Bruton, D. L., ed., Aspects of the Ordovician System: Palaeontological Contributions, University of Oslo, Norway, Universitetsforlaget, p. 37–50.

FUCHTBAUER, H., 1974, Sediments and Sedimentary Rocks, Part II: Halsted Press, New York, 464 p.

GROTZINGER, J. P., 1986, Cyclicity and paleoenvironmental dynamics, Rocknest platform, northwest Canada: Geological Society of America Bulletin, v. 97, p. 1208–1231.

HALLEY, R. B., HARRIS, P. M., AND HINE, A. C., 1983, Bank margin, in, Scholle, P. A., Bebout, D. G., and Moore, C. H., eds., American Association of Petroleum Geologists Memoir 33, p. 463–506.

HARLAND, W. B., COX, A. V., LLOWLELLYN, P. G., PICHON, C. A. G., SMITH, A. G., AND WALTERS, R., 1982, A Geologic Time Scale: Cambridge University Press, Cambridge, England, 131 p.

HECKEL, P. H., 1974, Carbonate buildups in the geologic record: A review, in Laporte, L. F., ed., Reefs in Time and Space: Society of Economic Paleontologists and Mineralogists Special Publication 18, p. 90–154.

HINE, A. C., AND NEUMAN, A. C., 1977, Shallow carbonate-bank-margin growth and structure, Little Bahama Bank, Bahamas: American Association of Petroleum Geologists Bulletin, v. 61, p. 376–406.

HINTZE, L. F., 1973, Geologic History of Utah: Brigham Young University Geology Studies, v. 20, 181 p.

———, 1974, Preliminary geologic map of the Wah Wah Summit Quadrangle, Millard and Beaver Counties, Utah: U.S. Geologic Survey Miscellaneous Field Studies Map MF-637, 1:48:000.

———, AND ROBISON, R. A., 1975, Middle Cambrian stratigraphy of the House, Wah Wah, and adjacent ranges in western Utah: Geological Society of America Bulletin, v. 86, p. 881–891.

JACOBI, R. D., 1981, Peripheral bulge—A causal mechanism for the Lower/Middle Ordovician unconformity along the western margin of the Northern Appalachians: Earth and Planetary Science Letters, v. 56, p. 245–251.

JOHNSON, J. G., KLAPPER, G., AND SANDBERG, C. A., 1985, Devonian eustatic fluctuations in Euramerica: Geological Society of America Bulletin, v. 96, p. 567–587.

KEEN, C. E., 1982, The continental margins of eastern Canada: A review, in Scrutton, R. A., ed., Dynamics of Passive Margins: American Geophysical Union, Geodynamics Series, v. 6, p. 45–58.

KENDALL, C. G., AND SCHLAGER, W., 1981, Carbonates and relative sea level: Marine Geology, v. 44, p. 181–212.

KEPPER, J. C., 1972, Paleoenvironmental pattern in middle to lower Upper Cambrian interval in eastern Great Basin: American Association of Petroleum Geologists Bulletin, v. 56, p. 503–527.

———, 1976, Stratigraphic relationships and depositional facies in a portion of the Middle Cambrian of the Basin and Range Province: Brigham Young University Geology Studies, v. 23, p. 75–91.

LOCHMAN-BALK, C., 1970, Upper Cambrian faunal patterns on the craton: Geological Society of America Bulletin, v. 81, p. 3197–3224.

LOHMANN, K. C., 1976, Lower Dresbachian (Upper Cambrian) platform to deep-shelf transition in eastern Nevada and western Utah: An evaluation through lithologic cycle correlation: Brigham Young University Geology Studies, v. 23, p. 111–122.

MAGARA, K., 1980, Comparison of porosity-depth relationships of shale and sandstone: Journal of Petroleum Geology, v. 3, p. 175–185.

MARKELLO, J. R., AND READ, J. F., 1982, Upper Cambrian intrashelf basin, Nolichucky Formation, southwest Virginia Appalachians: American Association of Petroleum Geologists Bulletin, v. 66, p. 860–878.

MCKENZIE, D. P., 1978, Some remarks on the development of sedimentary basins: Earth and Planetary Science Letters, v. 40, p. 25–32.

MORRIS, H. T., AND LOVERING, T. S., 1961, Stratigraphy of the East Tintic Mountains, Utah: U.S. Geological Survey Professional Paper 361, 145 p.

MOUNTJOY, E. W., AND AITKEN, J. D., 1963, Early Cambrian and late Precambrian paleocurrents, Banff and Jasper National Parks: Bulletin of Canadian Petroleum Geology, v. 11, p. 161–168.

MULLINS, H. T., NEUMANN, A. C., WILBUR, R. J., AND BOARDMAN, M. R., 1980, Nodular carbonate sediment on Bahamian slopes: Possible precursors to nodular limestones: Journal of Sedimentary Petrology, v. 50, p. 117–131.

MUSSMAN, W. J., AND READ, J. F., 1986, Sedimentology and development of a passive- to convergent-margin unconformity: Middle Ordovician Knox unconformity, Virginia Appalachians: Geological Society of American Bulletin, v. 97, p. 282–295.

NEUMANN, A. C., AND LAND, L. S., 1975, Lime mud deposition and calcareous algae in the Bight of Abaco, Bahamas: A budget: Journal of Sedimentary Petrology, v. 45, p. 763–786.

PALMER, A. R., 1971a, The Cambrian of the Great Basin and adjacent areas, western United States, in Holland, C. H., ed., Cambrian of the New World: Wiley-Interscience, London, p. 1–78.

———, 1971b, The Cambrian of the Appalachian and eastern New England regions, eastern United States, in Holland, C. H., ed., Cambrian of the New World: Wiley-Interscience, London, p. 1–78.

———, 1981a, On the correlatability of Grand Cycle tops, in Taylor, M. E., ed., Short Papers for the Second International Symposium on the Cambrian System: U.S. Geological Survey Open-File Report 81-743, p. 156–157.

———, 1981b, Subdivision of the Sauk sequence, in Taylor, M. E., ed., Short Papers for the Second International Symposium on the Cambrian System: U.S. Geological Survey Open-File Report 81-743, p. 160–163.

———, AND HALLEY, R. B., 1979, Physical stratigraphy and trilobite biostratigraphy of the Carrara Formation (Lower and Middle Cambrian) in the southern Great Basin: U.S. Geological Survey Professional Paper 1047, 131 p.

PARSONS, B., AND SCLATER, J. G., 1977, An analysis of the variation of ocean floor bathymetry and heat flow with age: Journal of Geophysical Research, v. 82, p. 803–828.

PITMAN, W. C., AND GOLOVCHENKO, X., 1988, Sea-level changes and their effect on the stratigraphy of Atlantic-type margins, in Sheridan, R. E., and Grow, J. A., eds., The Atlantic Continental Margin: U.S.: The Geology of North America, Volumes 1–2: Geological Society of America, p. 429–436.

POOLE, F. G., 1974, Flysch deposits of Antler foreland basin, western United States, in Dickinson, W. R., ed., Tectonics and Sedimentation: Society of Economic Paleontologists and Mineralogists Special Publication 22, p. 58–82.

———, SANDBERG, C. A., AND BOUCOUT, A. J., 1977, Silurian and Devonian paleogeography of the western United States, in Stewart, J. H., Stevens, C. H., and Fritsche, A. E., eds., Paleozoic Paleogeography of the Western United States: Pacific Coast Paleogeography Symposium I: Society of Economic Paleontologists and Mineralogists, p. 39–65.

PORTER, J. W., PRICE, R. A., AND MCCROSSAN, R. G., 1982, The western Canada sedimentary basin: Philosophical Transactions of the Royal Society of London, v. A305, p. 169–192.

PRATT, B. R., AND JAMES, N. P., 1982, Cryptalgal-metazoan bioherms of early Ordovician age in the St. George Group, western Newfoundland: Sedimentology, v. 33, p. 543–569.

PRICE, R. A., 1980, The Cordilleran foreland thrust and fold belt in the southern Canadian Rocky Mountains: Geological Society of London Special Publication No. 9, p. 1–22.

———, AND MOUNTJOY, E. W., 1970, Geologic structure of the Canadian Rocky Mountains between Bow and Athabasca Rivers—A progress report, in Wheeler, J. O., ed., Structure of the Canadian Cordillera: Geological Association of Canada Special Paper 6, p. 7–25.

PUGH, D. C., 1971, Subsurface Cambrian stratigraphy in southern and central Alberta: Geological Survey of Canada Paper 70-10, 33 p.

RASSETTI, F., 1965, Upper Cambrian trilobite faunas of northeastern Tennessee: Smithsonian Miscellaneous Collections, v. 148, 140 p.

READ, J. F., 1982, Carbonate platforms of passive (extensional) continental margins: Types, characteristics and evolution: Tectonophysics, v. 81, p. 195–212.

———, 1985, Carbonate platform facies models: American Association of Petroleum Geologists Bulletin, v. 69, p. 1–21.
———, GROTZINGER, J. P., BOVA, J. A., AND KOERSCHNER, W. F., 1986, Models for generation of carbonate cycles: Geology, v. 14, p. 107–110.
REES, M. N., 1986, A fault-controlled trough through a carbonate platform: The Middle Cambrian House Range embayment: Geological Society of America Bulletin, v. 97, p. 1054–1069.
ROBINSON, R. A., 1964, Upper Middle Cambrian stratigraphy of western Utah: Geological Society of American Bulletin, v. 75, p. 995–1010.
ROGERS, J., AND KENT, D. F., 1948, Stratigraphic section at Lee Valley, Hawkins County, Tennessee: State of Tennessee Division of Geology, Bulletin 55, 19 p.
ROSS, R. J., Jr., 1977, Ordovician paleogeography of the western United States, in Stewart, J. H., Stevens, C. H., Fritsche, A. E., eds., Paleozoic Paleogeography of the Western United States: Pacific Coast Paleogeography Symposium 1: Society of Economic Paleotologists and Mineralogists, p. 19–38.
ROYDEN, L., AND KEEN, C. E., 1980, Rifting process and thermal evolution of the continental margin of eastern Canada determined from subsidence curves: Earth and Planetary Science Letters, v. 51, p. 343–361.
SANDBERG, C. A., GUTSHICK, R. C., JOHNSON, J. G., POOLE, F. G., AND SANDO, W. J., 1982, Middle Devonian to Late Mississippian geologic history of the overthrust belt region, western United States, in Powers, K. E., ed., Geologic Studies of the Cordilleran Thrust Belt, v. 2: Rocky Mountain Association of Geologists, Denver, p. 691–719.
SANDBERG, C. A., AND MAPEL, W. J., 1967, The northern Rocky Mountains and plains, in Oswald, D. H., ed., International Symposium on the Devonian System, v. I: Alberta Society of Petroleum Geologists, John McAra Ltd., Calgary, p. 843–877.
SCHLAGER, W., 1981, The paradox of drowned reefs and carbonate platforms: Geological Society of America Bulletin, v. 92, p. 197–211.
SCHLAGER, W., AND GINSBURG, R. N., 1981, Bahama carbonate platforms—The deep and the past: Marine Geology, v. 44, p. 1–24.
SHAMMUGAM, G., AND LASH, G., 1982, Analogous tectonic evolution of the Ordovician foredeeps, southern and central Appalachians: Geology, v. 10, p. 562–566.
SIMPSON, E. L., AND SUNDBERG, F. A., 1987, Early Cambrian age for synrift deposits of the Chilhowee Group of southwestern Virginia: Geology, v. 15, p. 123–126.
SLEEP, N. H., 1971, Thermal effects of the formation of Atlantic continental margins by continental breakup: Geophysical Journal of the Royal Astronomical Society, v. 24, p. 325–350.
SLOSS, L. L., 1963, Sequences in the cratonic interior of North America: Geological Society of America Bulletin, v. 74, p. 93–113.
STECKLER, M. S., 1981, Thermal and mechanical evolution of Atlantic-type margins: unpublished Ph.D. Dissertation, Columbia University, New York, 261 p.
———, 1985, Uplift and extension at the Gulf of Suez: Indications of induced mantle convection: Nature, v. 317, p. 135–139.
———, AND WATTS, A. B., 1978, Subsidence of the Atlantic-type continental margins off New York: Earth and Planetary Science Letters, v. 41, p. 1–13.
———, WATTS, A. B., AND THORNE, J. A., 1988, Subsidence and basin modeling at the U.S. Atlantic passive margin, in Sheridan, R. E., and Grow, J. A., eds., The Atlantic Continental Margin: U.S.: The Geology of North America, Volumes 1–2: Geological Society of America, p. 399–416.
STEPHENSON, R. A., EMBRY, A. F., NAKIBOGLU, S. M., AND HASTAOGLU, M. A., 1987, Rift-initiated Permian to Early Cretaceous subsidence of the Sverdrup Basin, in Beaumont, C., and Tankard, A. J., eds., Sedimentary Basins and Basin-Forming Mechanisms: Canadian Society of Petroleum Geologists Memoir 12, p. 213–231.
VAIL, P. R., MITCHUM, R. M., AND THOMPSON, S., III., 1977, Seismic stratigraphy and global changes of sea level, Part 4: Global cycles of relative changes of sea level, in Payton, C. E., ed., Seismic Stratigraphy–Applications to Hydrocarbon Exploration: American Association of Petroleum Geologists Memoir 26, p. 83–97.
WALKER, R. W., SHANMUGAM, G., AND RUPPEL, S. C., 1983, A model for carbonate to terrigenous clastic sequences: Geological Society of America Bulletin, v. 94, p. 700–712.
WATTS, A. B., 1981, The U.S. Atlantic continental margin: Subsidence history, crustal structure and thermal evolution, in Geology of Passive Continental Margins: History, Structure and Sedimentologic Record: American Association of Petroleum Geologists Education Course Note Series No. 19, chapter 2, 75 p.
———, AND STECKLER, M. S., 1979, Subsidence and eustasy at the continental margin of eastern North America: American Geophysical Union Maurice Ewing Series, v. 3, p. 218–234.
———, BODINE, J. H., AND STECKLER, M. S., 1980, Observations of flexure and the state of stress: Journal of Geophysical Research, v. 85, p. 6369–6376.
WERNICKE, B., 1985, Uniform-sense normal simple shear of the continental lithosphere: Canadian Journal of Earth Science, v. 22, p. 108–125.
WILLOUGHBY, R., 1969, Lower and Middle Cambrian fossils from the Shady Formation, Austinville, Virginia: Geological Society of America, Abstracts with Programs, v. 8, p. 301–302.
WILSON, J. E., AND JORDAN, C., 1983, Middle Shelf, in Scholle, P. A., Bebout, D. G., and Moore, C. H., eds., Carbonate Depositional Environments: American Association of Petroleum Geologists Memoir 33, p. 297–344.
WILSON, J. L., 1975, Carbonate Facies in Geologic History: Springer-Verlag, New York, 471 p.
YOUNG, F. G., 1979, The lowermost Paleozoic McNaughton Formation and equivalent Cariboo Group of eastern British Columbia: Piedmont and tidal complex: Geological Survey of Canada Bulletin 288, 60 p.

JUDY CREEK: A CASE STUDY FOR A TWO-DIMENSIONAL SEDIMENT DEPOSITION SIMULATION

DAVID M. SCATURO, JOHN S. STROBEL AND CHRISTOPHER G. ST. C. KENDALL
Department of Geology, University of South Carolina, Columbia, South Carolina 29208

JACK C. WENDTE
Esso Resources Canada, Calgary, Alberta, T2P-01

GAUTAM BISWAS
Department of Computer Science, Vanderbilt University, Nashville, Tennessee 37235

JAMES BEZDEK
Boeing High Technology Center, Seattle, Washington 98124-6269

AND

ROBERT CANNON
Department of Computer Science, University of South Carolina, Columbia, South Carolina 29208

ABSTRACT: This paper describes a computer program developed at the University of South Carolina to simulate the evolution of carbonate geometries and their facies responding to: (1) varying rates of accumulation; (2) eustatic sea-level variation; and (3) tectonic movement of the crust. The simulation creates two-dimensional plots of synchronous depositional sequences within sediment bodies. Rates of carbonate accumulation are modeled as a function of water depth and lateral position across the shelf. Carbonate accumulation includes *in situ* organic production and transport by hydrodynamic processes. Influx of clastic sediments is modeled to cause an exponential decrease in the rate of carbonate accumulation. Rates of carbonate accumulation can be further diminished with a user-defined depth-controlled wave-damping function. Eustatic variation is modeled by a fourth-order linear change in sea level over time (as described by Vail and others, 1977), with a higher order sinusoidal oscillation of sea level superimposed upon it. Reef margin and interior lagoonal facies on platforms and shelves are predicted. Modeling of these facies zones includes aggradation, progradation, backstepping, shoal development, and drowning.

The Devonian Judy Creek reef complex of the western Canada Alberta basin was used both as an aid in constructing the carbonate simulation model and as an example on which to test the completed program. The Judy Creek model was constructed by the fourth co-author (J.C.W.) from the study of 100 cores and the correlation of a systematic grid of wireline log-core cross sections. Judy Creek consists of five overall shoaling-upward depositional cycles. Superimposed upon each cycle are several subcycles. Marine hardgrounds locally cap the first three cycles at the margin of the buildup. The top of the fourth cycle is a subaerial cemented surface. The top of the upper cycle is a widespread marine hardground. Subcycles consist of minor shoaling-upward sequences of lagoonal facies.

The results of two versions of the carbonate model include: (1) a site-specific version of the Judy Creek area, which shows the facies location and movement within Judy Creek; and (2) a more generalized carbonate model. Both versions use similar inputs.

The geometry and facies distributions of Judy Creek are simulated using a stairlike fourth-order eustatic sea-level curve (as described by Vail and others, 1977) and a low-amplitude higher order sea-level oscillation with a varying period in which sea-level fall is matched by tectonic subsidence. Maximum rates of accumulation average around 3.0 to 5.0 m/10 ka. The fourth-order sea-level variation consists of a rapid 3.5–5.0 m/10 ka rise followed by a gradual rise of 0.7–1.5 m/10 ka, followed by another rapid rise of the same magnitude, followed by another gradual rise. The higher order sea-level oscillation consists of 0.5 to 1.5-m amplitudes with periods ranging from 20 to 40 ka. Tectonic subsidence averages between 0.5 and 1.0 m/10 ka.

INTRODUCTION

The computer simulation of carbonate and clastic cycles provides a quantitative means to understand and interpret these sequences (Lerche and others, 1987). In order to help explain the geometries we observe in the rock record, empirical relations can be assigned to the important variables that control these processes, and a wide range of values for these variables can be tested iteratively, both independently and in concert.

In this simulation, the sedimentary fill of a basin is determined by the interaction of three controlling parameters: rates of sediment accumulation, tectonic movement of the depositional surface, and changes in eustatic sea level. The effect of each of these variables can be tested and their interaction simulated to produce a computer-generated two-dimensional plot of the time lines within sediment bodies. These plots can be compared to similar geometries seen in seismic and well cross sections.

JUDY CREEK REEF COMPLEX

The Devonian reefs in the western Canada basin provide an excellent data base for modeling ancient carbonate depositional settings. It is helped by the easy access to the numerous cores of Devonian carbonates stored in the Calgary core warehouse and by the many detailed facies studies that have been made of this area. Judy Creek is an Upper Devonian reef complex of the Swan Hills Formation in central Alberta. It has been possible to define independent variables that exert significant controls on reef growth, geometry, and facies distribution within the Judy Creek reef complex. We were able to suggest relations between these variables and use them in the computer simulation as it was developed.

The computer simulation of the Judy Creek field is based on a sequence facies model constructed by J. C. Wendte. The model incorporates 24 systematic east–west sequence facies cross sections across the field and six tie north–south stratigraphic cross sections. Previous study of the field by Murray (1966) described the distribution of lithofacies and biofacies along selected transects across the field. A later paper by Wendte and Stoakes (1982) discussed the recognition and correlation of repeated shoaling-upward sequences along a transect across the field. The present model refines the interpretations of Wendte and Stoakes (1982) and provides a comprehensive model over the entire field.

The Judy Creek field is one of several Upper Devonian (Frasnian) reef complexes that are located, along with a reef-fringed bank, on the Swan Hills platform in central Alberta (Fig. 1; Wendte and Stoakes, 1982). The platform, bank, and reef complexes interfinger with, and are overlain by, basinal limestones of the Waterways Formation. The Swan Hills platform lies along the southwest edge of the Alberta basin, covers approximately 13,000 km^2, and has a structural dip to the southwest of 8 m/km (Wendte and Stoakes, 1982).

The Judy Creek field is situated toward the eastern margin of the Swan Hills platform. The reef is as much as 67 m thick, covers 122 km^2, and lies at a burial depth of 2,600 m (Wendte and Stoakes, 1982). Judy Creek reef is bounded to the north, south, and west by contemporaneous reef complexes that are separated from each other by deep-water surge channels. The northeastern margin faced the prevailing southwesterly paleo-winds (Murray, 1966). An isopach map (Fig. 2) and cross section (Fig. 3) show that the greatest amount of progradation occurred on the eastern side of the reef complex. Topographically high areas of the underlying platform may have served as sites for the initiation of the reef complexes and carbonate bank (Wendte and Stoakes, 1982).

The primary reefal frame builders within Swan Hills are massive, encrusting, tabular stromatoporoids, with solenoporoid algae and tabulate corals locally important as frame builders (Murray, 1966). Dendroid and bulbous stromatoporoids are an important constituent of the rocks, but are

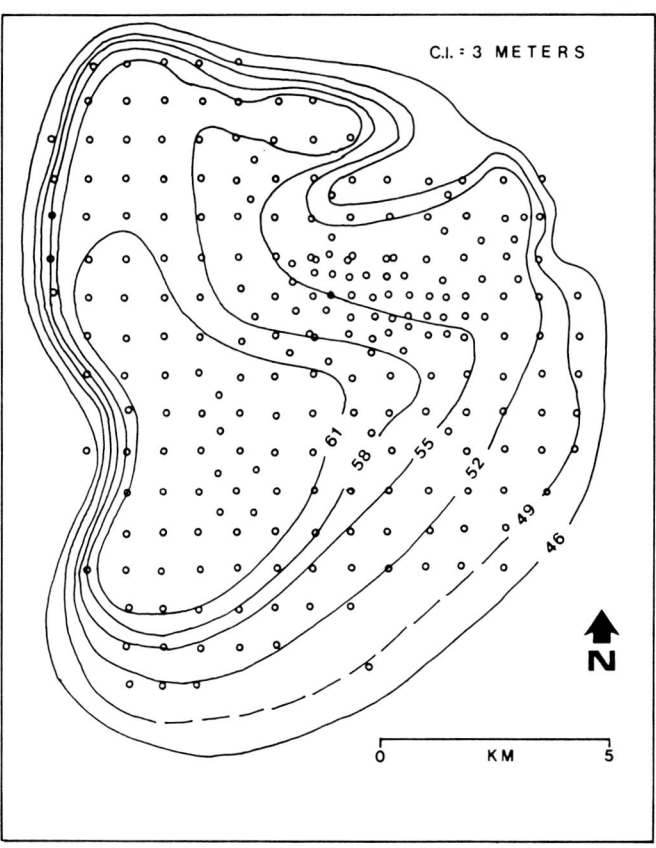

FIG. 2.—Isopach map of the Swan Hills platform underlying Judy Creek reef complex.

not considered by Murray (1966) to be framework builders. Also present, but to a lesser degree, are brachiopods, crinoids, gastropods, ostracods, and calcispheres.

Wendte and Stoakes (1982) have described and interpreted 11 environmental facies zones in the Judy Creek reef complex on the basis of textures, sedimentary structures, fossil composition, and comparisons with modern analogs present. Most of the facies they have outlined are depth

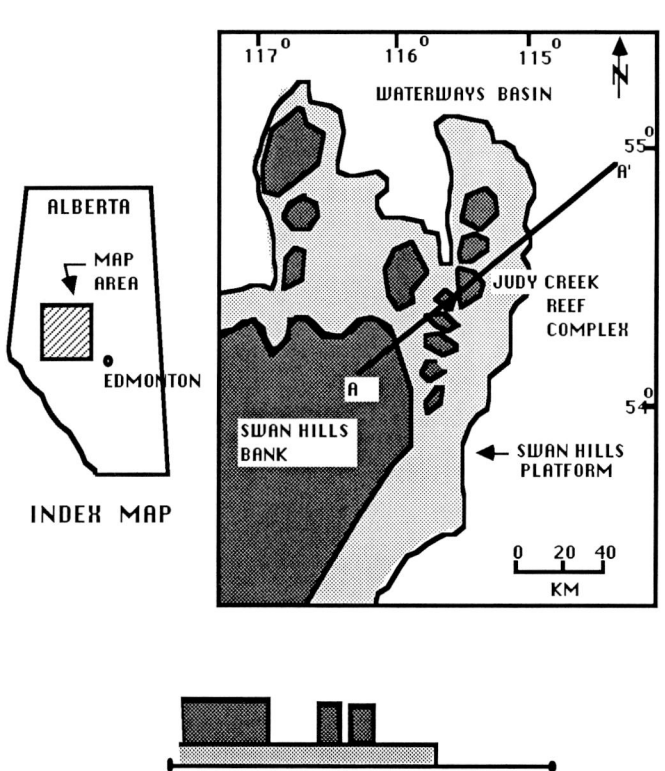

FIG. 1.—Location map of the Judy Creek reef complex (from Wendte and Stoakes, 1982, fig. 1).

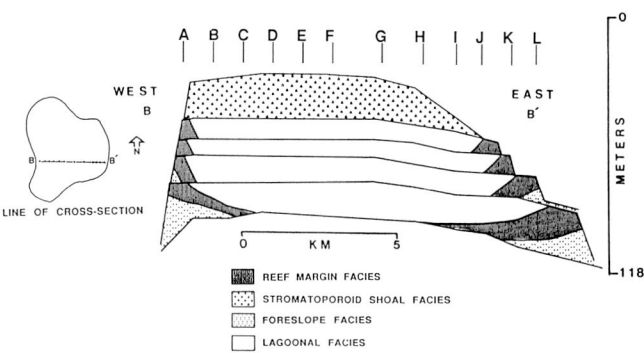

FIG. 3.—Stratigraphic cross section of the Judy Creek reef complex. Well numbers are: (A) 2-28-63-11W5; (B) 4-27-63-11W5; (C) 2-27-63-11W5; (D) 4-26-63-11W5; (E) 2-26-63-11W5; (F) 4-25-63-11W5; (G) 2-25-63-11W5; (H) 2-30-63-10W5; (I) 2-29-63-10W5; (J) 4-28-63-10W5; (K) 2-28-63-10W5; (L) 4-27-63-10W5.

dependent and coexist as a function of position across the reef complex (Fig. 4). Their findings were used to construct a sequence facies model of the Judy Creek reef complex.

Thirty cross sections of Judy Creek that had been constructed by J. C. Wendte were used in this study. Twenty four of these cross sections include distributions of the 11 facies zones summarized by Wendte and Stoakes (1982). The cross section shown in Figure 3 was used to test the completed simulation.

The Judy Creek reef complex can be divided into five time-synchronous sequences, or cycles (Fig. 3), which Wendte (pers. commun.) estimates were deposited over 600 to 700 ka. Each of these major cycles he proposes matches a 100-ka Milankovitch cycle. Each cycle can be traced across the reef complex from the leeward foreslope to the windward foreslope and consists of an overall shoaling-upward sequence and several subcycles. Cycle boundaries are defined by localized submarine hardgrounds and abrupt shifts to deeper water facies (Wendte and Stoakes, 1982).

Cycles may have formed in response to rapid rises in relative sea level (Wendte and Stoakes, 1982), based on the absence of evidence of prolonged subaerial exposure or erosion at the tops of these cycles (—all except the top of the fourth cycle, which is a subaerially cemented surface that corresponds to a lowering of sea level). Subsequent work by Wendte has shown that the top of the fourth cycle is a subaerially cemented surface that formed by a slight drop in relative sea level. Their reconstruction of the evolution of Judy Creek suggests that each cycle was initiated following a rapid sea-level rise that outpaced reef growth, resulting in a phase where carbonate accumulation lagged behind sea-level rise and a submarine hardground surface formed in downslope settings. As the rate of sea-level rise slowed, reef growth caught up to and kept pace with, or outpaced, sea level, producing a shoaling-upward sequence. Superimposed on each of these fourth-order cycles are minor, higher order subcycles that can be observed in the interior lagoon sediments. A complete lagoonal cycle consists of deeper, restricted, *Amphipora* rudstones with a mud-rich matrix that grades into shallower, less restricted, *Amphipora* rudstones with a sand-size carbonate matrix, which locally passes into a laminated tidal-flat or beach deposit. Core data indicate several of these higher order lagoonal subcycles occurring within each of the five major depositional cycles. We interpret these lagoonal cyclic sequences as resulting from low-amplitude sea-level oscillations, which can be observed in these sediments because of their slower rates of accumulation but are masked in the faster growing reef margins and reef flats.

The first cycle corresponds to the drowning of the underlying Swan Hills platform and the initiation of a north–south-trending basal stromatoporoid shoal. This formed in relatively deep water but subsequently grew upward to sea level. Once this shoal built to sea level, shallow-water high-energy conditions enabled the colonization of encrusting stromatoporoids, which formed reef margins along the edges of this shoal. The creation of these wave-resistant marginal structures fostered the formation of a sheltered interior lagoon.

During the first phase of this cycle, the rate of accumulation eventually outpaced the rate of sea-level rise, and the reef margins prograded, with the greatest progradation taking place in the east–northeast (windward direction) over approximately 5 km. As the margins prograded, the foreslopes progressively steepened, reducing the rate of advance and producing a shift to the vertical aggradation of the reef margins.

Depending on position in the reef complex, in cycles 2 through 4, the reef margin facies exhibit geometries that may vary from aggradational cycles to progradational cycles to backstepping. The backstepping phenomenon, which occurs in the carbonate provinces of different ages, has been attributed by some to a rapid rise in sea level (Playford, 1980; Wendte and Stoakes, 1982). This sea-level rise causes the flanks of a reef margin to reestablish themselves toward the interior of a reef complex, presumably due to the lower energy, more sheltered conditions that exist there. Studies of the cross sections at Judy Creek indicate the average distance of backstepping to be between 0.4 and 0.8 km. Figure 5 provides an aerial view of the breaks in slope of the reef margins for each cycle. Subsequent to each backstepping event, the reef margins built in an aggradational style, keeping pace with sea-level rise. The top of the fourth cycle was terminated by a slight drop in relative sea level.

The last cycle is composed of a number of cycles, but these cannot be easily distinguished due to a lack of well-defined reef margins or widespread hardgrounds. Presumably, the rise in sea level precluded the development of well-established, wave-resistant reef margins, resulting in a low-relief, windward-sloping stromatoporoid ramp. Therefore, this last cycle is characterized by undifferentiated, higher energy facies, particularly along the northeastern windward reef edge, where the growth of encrusting stromatoporoids was fostered more than in the lower energy conditions toward the interior of the reef complex.

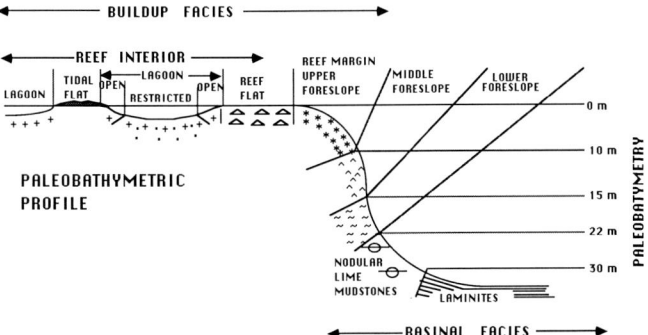

FIG. 4.—Paleobathymetric profile showing environmental facies zones at the Judy Creek reef complex. Rock types are: (1) tidal flat—pelletal mudstones/packstones, cryptalgal laminations; (2) lagoon—light to dark brown carbonaceous rudstones/floatstones containing *Amphipora*; (3) reef flat—brecciated/well-washed *Amphipora*, cylindrical and massive stromatoporoid rudstones, lesser amounts lime grainstones; (4) reef margin/upper foreslope—*in situ*, thick, tabular stromatoporoid boundstones and rubble; (5) middle foreslope—cylindrical stromatoporoid rudstones, increasing micritic matrix with depth; (6) lower foreslope—micritic boundstones/rudstones, abundant *in situ* thin wafer stromatoporoids, common crinoids and brachiopods; (7) nodular lime mudstones—bioturbated micritic limestone with nodular to banded aspect; (8) laminites—dark, unfossiliferous lime mudstones (from Wendte and Stoakes, 1982, fig. 3).

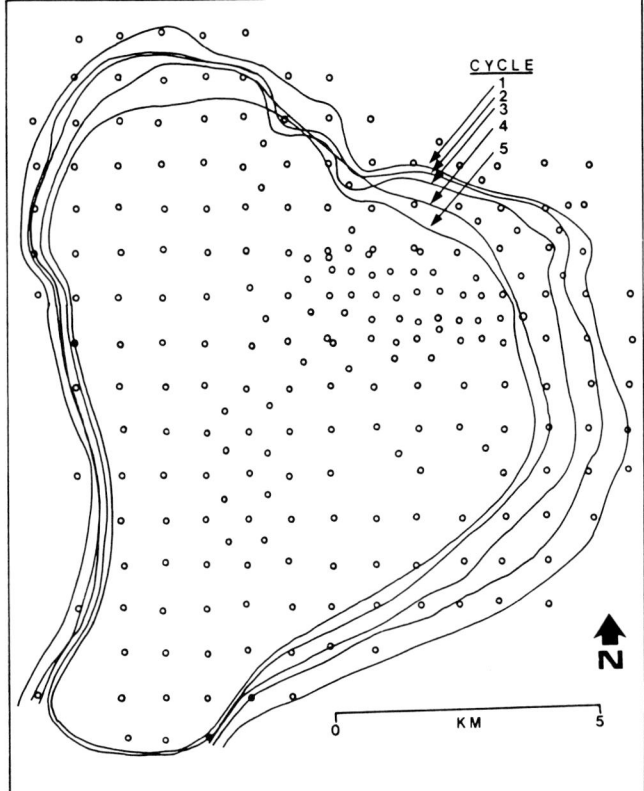

FIG. 5.—Map of Judy Creek showing the location of reef margin breaks in slope for all five stratigraphic cycles.

Reef growth is interpreted to have terminated after another rapid rise in sea level and possible nutrient stress related to clastic influx to the northeast. The reef complex ceased to grow and remained uncovered for a considerable period of time (Wendte and Stokes, 1982). Nutrient poisoning, as described by Hallock and Schlager (1986), may be a common cause of reef failure through geologic time. Prograding basinal limestones and argillaceous limestones of the Waterways Formation eventually downlapped onto and buried the drowned Judy Creek reef complex.

SIMULATION ALGORITHM

The simulation allows for the deposition of both carbonate and clastic sediments in a basinal setting. For the simulation of the Judy Creek reef complex, the carbonate subalgorithm was dominant and is described below. For a description of the full algorithm, see Strobel and others (1987) and Helland-Hansen and others (1988).

Two versions of the computer program were developed. In one version, the backstepped shelf edge is located by a statistical distribution function that simulates the pattern of dispersion of reefal organisms and calculates the probability of finding the backstepped shelf edge at a location platformward of the previous shelf edge. This distance is determined from the relation between the slope of the underlying platform, rate of sea-level rise, and rate of carbonate production.

In the other version, the backstepped shelf edge is located by assuming a relation between wave energy, rates of carbonate accumulation, and the wave shadow produced by a shelf. At the edge of a buildup, at sea level, rates of accumulation exceed the damping effect of waves, but below a certain depth, wave energy may damp carbonate production so reef growth will eventually cease. Below the effect of waves, however, carbonates can resume accumulation. Thus, it is our contention that, following a sea-level rise across a shelf, wave damping of carbonate growth occurs as a wedge-shaped zone in which wave energy suppresses reef growth, causing it to backstep.

In both models carbonate accumulation responds to variation in sea-level rise, rate of carbonate accumulation, and rate of clastic sedimentation. Each of these components is described in detail.

Rates of Accumulation

Carbonate rates of accumulation are user-defined functions in both versions of the program. Carbonate rate of accumulation in the Judy Creek model is a function of water depth and position across the platform (i.e., framework, or non-framework-building position), whereas in the general model carbonate accumulation is a function of water depth only. Allowing the user to define rates of carbonate accumulation with respect to depth gives the flexibility to construct a wide range of facies models and iteratively test them against changes in one or several variables. The Judy Creek model calculates the rate of carbonate accumulation at any water depth by linearly interpolating between user-defined coordinate pairs of depth and rate as follows:

$$A = A'' - (Z'' - Z)/((Z'' - Z') * (A'' - A')) \quad (1)$$

where A is the rate of accumulation at a specified depth Z, and (Z',A') and (Z'',A'') are the user-defined values for water depth and accumulation rate. Figure 6 provides a graphic representation of this. In the general case model, the same basic algorithm is used, except that it divides the time step into fourths, determines how much carbonate is accumulated in a fourth of a time interval, recalculates the water depth and, using the new water depth, deposits carbonates for another fourth of the time step.

The carbonate accumulation rate used in the Judy Creek model is actually the net accumulation rate of the carbonate system. It equals both the *in situ* gross productivity and the transported sediment across the back reef and fore reef of the margin. Since each time line in the simulation ranges from 5 to 100 ka, we assume that the erosion of the gross accumulation and the transportation of those sediments have already occurred, resulting in a net accumulation of sediment. The general case model handles the transportation of reef talus off the margin separately. Excess carbonate accumulation, which would cause the margins to rise above sea level, is stored as talus, transported off the buildup and deposited. The user defines what percentage of talus is transported into the lagoon and what percentage of talus is transported into the basin. The user defines the length to which the talus extends away from the reef and into the

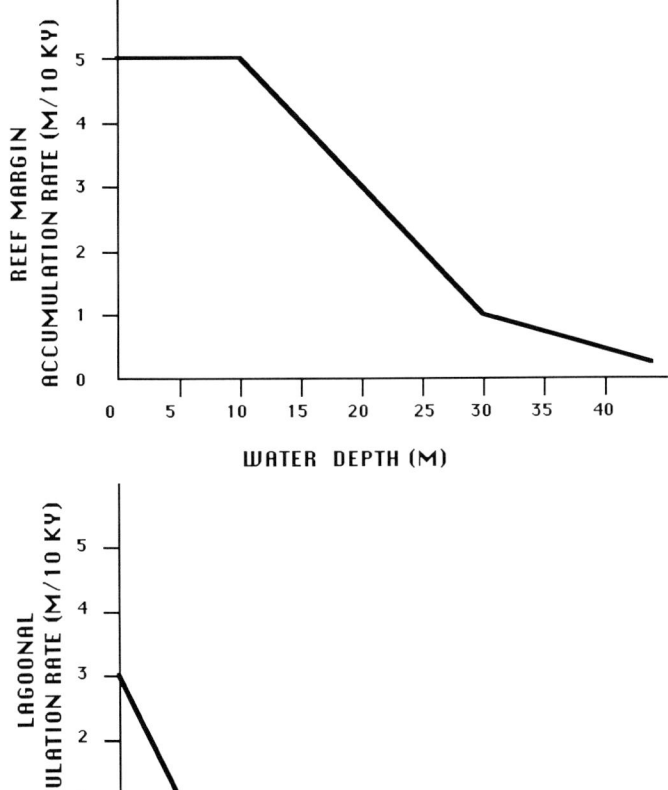

FIG. 6.—Carbonate rates of accumulation.

building) and an interior lagoon (non-framework building). These two settings have distinct sediment types, rates of accumulation, and geometric facies, responses to changes in relative sea level and clastic influx. The general model uses only one rate of accumulation, but allows this user-supplied rate to be damped by a user-defined function of distance from the reef margin toward the interior of the platform, recreating the presence of a lagoon. Figure 7 shows the results of a simulation run with two sets of rates defined, framework building and non-framework building, and a relatively rapid sea-level rise. The reef margins are able to keep pace with sea-level rise, while the interior lagoon does not, resulting in a deep-lagoon geometry.

Facies

The Judy Creek simulation concentrates on modeling the distribution of reef margin and lagoonal facies zones throughout time. The program does this by placing symbols at both the reef margin/lagoon boundary and the reef margin/foreslope boundary for each time step, thereby tracking the movement of reef margin and lagoon facies over time (Fig. 8). The reason for tracking the reef margin zone is that this is where the best primary porosity occurs. This is because high-energy reef margins are likely to have less micrite than lower energy lagoonal or foreslope positions.

Symbols are placed at both boundary locations. The reef margin facies zone consists of the reef crest, backreef, and upper forereef sediments. The interior lagoon facies zone consists of tidal-flat, shallow-lagoon skeletal sands, and deeper, mud-prone lagoonal sediments.

Siliciclastic Mixing

The amount of siliciclastic sediment available for deposition is defined in the program by the areas of two right triangles, a sand and a shale triangle. The program distributes sediment across the depositional surface, with sand and shale being deposited together. As sediment is being deposited, the area of the triangles is reduced by decreasing their heights but preserving their lengths. Sediment supply can be varied through time by defining different triangle heights for specific time steps. For a complete description

basin. These differences are important to consider when entering the accumulation rate values.

The Judy Creek model defines rates for both the reef framework-building and non-framework-building elements. This is done because both modern and ancient examples show platforms as consisting of a reef margin (framework

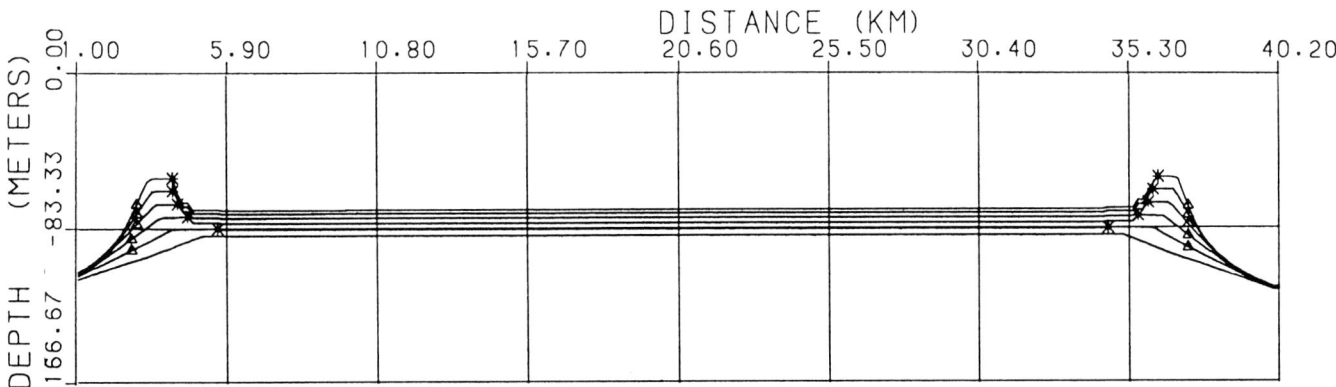

FIG. 7.—Simulation of raised-rim platform evolution where the reef margin is able to keep up with sea-level rise but the lagoon does not.

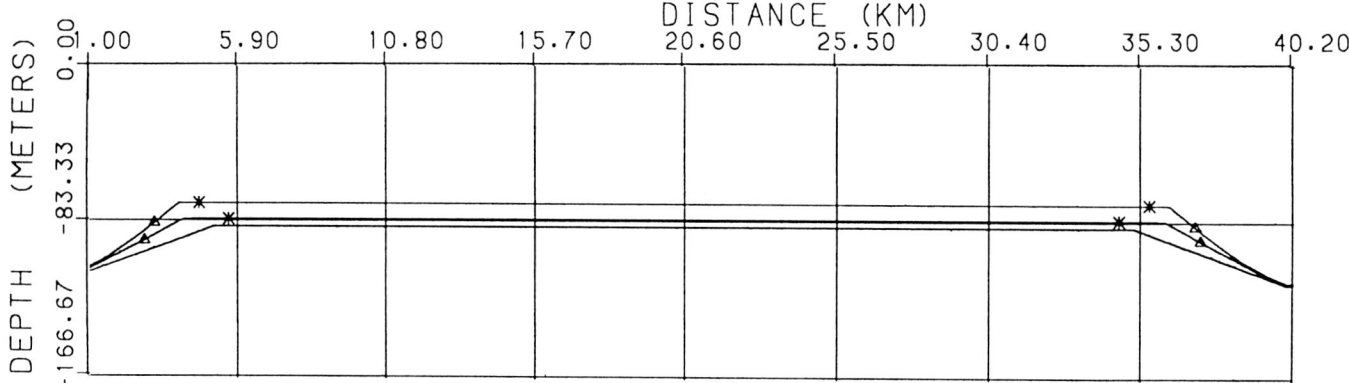

Fig. 8.—Simulation of reef margin facies zones at time step = 2.

of the siliciclastic deposition algorithm, the reader is referred to Helland-Hansen and others (1988).

Numerous authors (Wilson, 1975; Stoakes, 1980; Hallock and Schlager, 1986; Hubbard, 1986) have studied the effects of the influx of siliciclastic sediments into carbonate environments. Carbonate rates of accumulation are partially or completely suppressed by siliciclastics depending on the amount of siliciclastic sediment input into the overlying water column. This may either be a result of the reduction of water transparency or an increase of nutrients and particulate organic matter that accompanies terrestrial runoff and stresses the carbonate biological community.

Nutrient poisoning could prove to be a major contributing factor in the drowning of a reef (Kinsey and Davies, 1979). Typically, reefs are adapted to nutrient-poor environments (Birkeland, 1977). Influx of nitrates and phosphates, either from runoff or from upwelling, can result in the displacement of coelenterates and the increase in filamentous algae and bioeroders (Highsmith, 1980; Hallock and Schlager, 1986).

Turbidity has also been attributed to the diminishing of reef growth due to the attenuation of light by suspended sediment within the water column (Hubbard and Scaturo, 1985). Turbidity could play an important role in the drowning of a platform or shelf, by reducing the rate of carbonate productivity to values low enough to be significantly outpaced by a rapid rise in sea level.

The amount of suppression the influx of clastic sediments has on a reef can be expressed by first calculating the fraction of clastic sediment in the water column overlying the carbonate depositional surface. We do this by

$$S(x,t) = H(x,t)/Z(x,t) \qquad (2)$$

where S is the fraction of clastics in the water column, x, at time t, H is the amount of clastic sediment, and Z is the water depth.

If we then assume clastic suppression to be a function of S, the clastic fraction within the water column, we can choose to model this by using a simple exponential decay function. The equation

$$dC/dS = -kC \qquad (3)$$

(where k is greater than zero) shows C, the amount of clastic suppression, which ranges between 0 and 1 in value, will decrease as a function of the clastic fraction in the water column, and

$$C = 1 * \exp(-kS) \qquad (4)$$

shows C to be decreasing exponentially, where k is a scaling constant, and S is the fraction of clastics in the water column. In our simulation runs, k ranged between 1.3683 and 138.6300. This range of k gave the best fit to our geologic models. Figure 9 shows a simulation run with clastics prograding from the left, suppressing carbonate production where clastics are deposited, while the areas not influenced by clastic sediments continue to accumulate. This simulation run uses a decay rate $k = 13.8630$.

Relative Sea Level

Water depth plays a major role in controlling the rate of carbonate accumulation. This is a result of an increase in light attenuation with an increase in depth, culminating at a point where light levels are incapable of supporting photosynthetically driven calcification (Graus and Macintyre, 1976).

Eustatic variation in the simulation is calculated for each time step from user-defined input data for two orders of sea-level change over time. Coordinates of height of sea level (against a fixed datum) versus time are specified by the user. The program linearly interpolates between fourth-

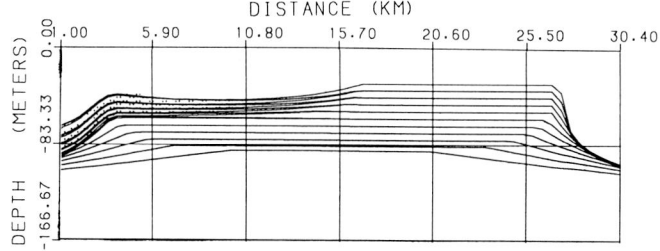

Fig. 9.—Simulation showing 50 percent suppression of carbonate growth with 5 percent clastics in the water column.

order points in the 100-ka to 1-ma time range (after terminology of Vail and others, 1977), and sinusoidally interpolates between higher order points (Fig. 10). This feature allows sea-level fluctuations of various frequencies to be modeled.

The amplitude and timing of several orders of sea-level cycles have been documented by datable unconformities and lateral positions of sedimentary sequences by Vail and others (1977), Haq and others (1987), and others. The operators use sea-level curves from the published literature as input data, or develop their own curves based on iterative trial and error, in order to match the simulation geometry to the actual cross section geometry.

Tectonic movement of the crust occurs in either a positive or a negative direction and can be varied in time as well as space. It is modeled by the user defining a tectonic rate (subsidence or uplift) against horizontal positions for specific time intervals. The program linearly interpolates the coordinates in both the time and space dimensions (Fig. 11). Tectonic subsidence varies from place to place. Fischer (1975) estimates average rates of tectonic subsidence in the

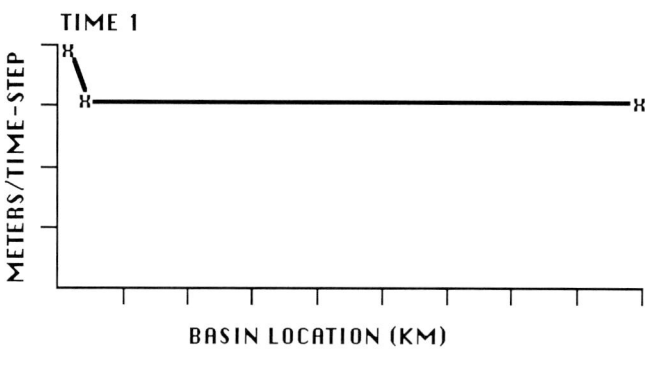
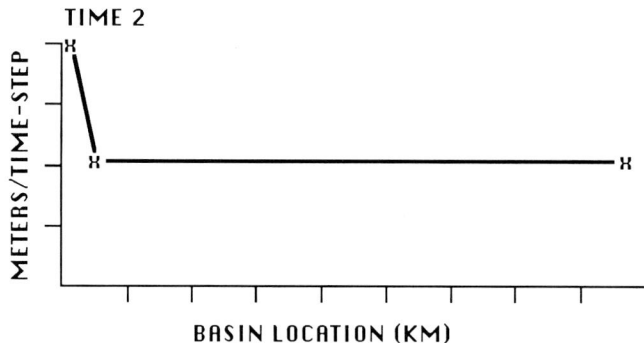
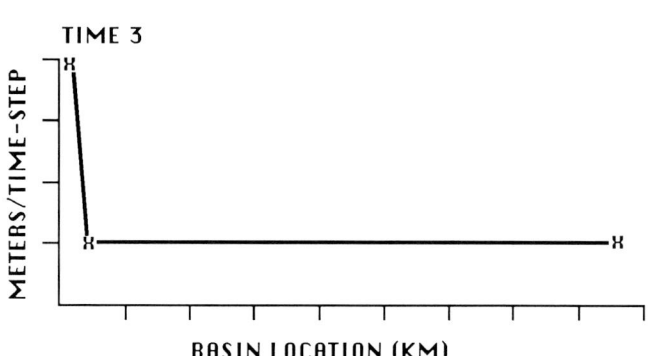

Fig. 11.—Calculation of subsidence rate at user-defined locations across the basin (denoted by "x"). Tectonic movement in the model is independent of sediment loading.

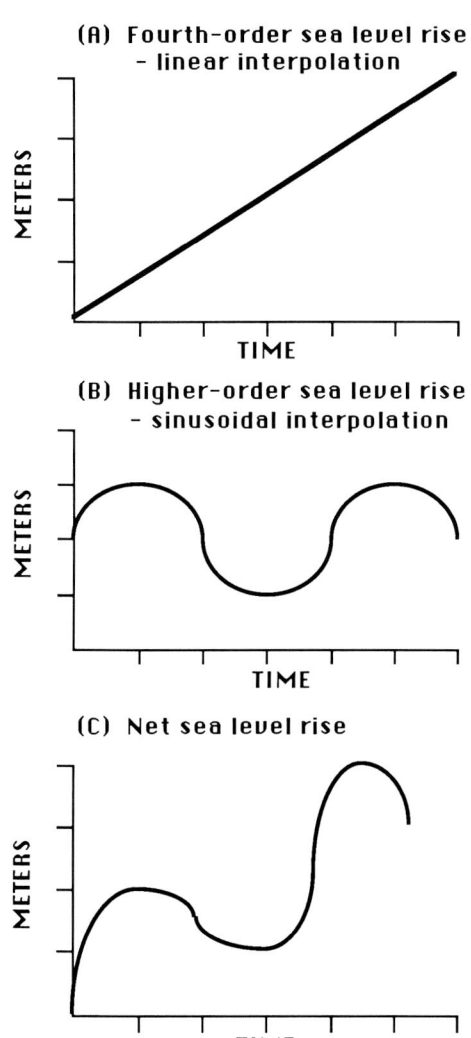

Fig. 10.—Eustatic sea-level calculation in the model.

Alpine Triassic at 0.1–1.0 m/10 ka. Guidish and others (1984) estimate the rate of basement subsidence in the North Sea during the Lower Paleocene to be between 0.7 and 1.3 m/10 ka. We assume tectonic subsidence at Judy Creek in the Upper Devonian to average between 0.5 and 1.0 m/10 ka.

The response of carbonate shelves and platforms to changes in relative sea level has been discussed by Kendall and Schlager (1981). Their study concludes that the response is a function of the difference between the growth potential of the reef and the rate of sea-level rise. These three responses include: (1) keep-up reef, where the platform or shelf remains at sea level; (2) catch-up reef, where only the faster growing reefal fauna are able to match sea-level rise, while the rest of the platform or shelf is drowned; (3) give-up reef, where the potential rate of carbonate growth is exceeded by the rate of sea-level rise, and the platform or shelf is drowned.

Adey (1978) suggests that there can be a lag period on the order of 500–1,000 years after which an incipiently drowned reef (i.e., one where sea-level rise outpaces reef growth potential but the depositional surface remains within the euphotic zone) is able to switch back on its full growth potential and reach sea level.

The various facies geometries the platform can take in response to sea-level rise are: (1) progradation of the reef margin and lagoon; (2) aggradation of the reef margin and lagoon; (3) backstepping of the reef margin and lagoon; (4) shoal development on top of platform; and (5) drowning of the platform (Fig. 12).

Progradation.—

The carbonate reef margin and lagoon will prograde in an upbuilding and outbuilding fashion when the growth potential exceeds the rate of sea-level rise. The margin will prograde out over a gently sloping substrate, and the foreslope of the margin will progressively steepen until it reaches a critical angle, at which point the foreslope is too steep for stable deposition and the platform can only build in a vertical direction. To illustrate this concept, Figure 13 shows a platform where one margin has a steep slope and one margin has a gentle slope. The platform is able to prograde out over the gentle slope but is unable to prograde on the steep slope.

Aggradation.—

The carbonate platform will aggrade vertically under two different conditions. It will aggrade if the angle of the foreslope is too steep for stable deposition (Fig. 13), or it will aggrade if the rate of sea-level rise matches the potential

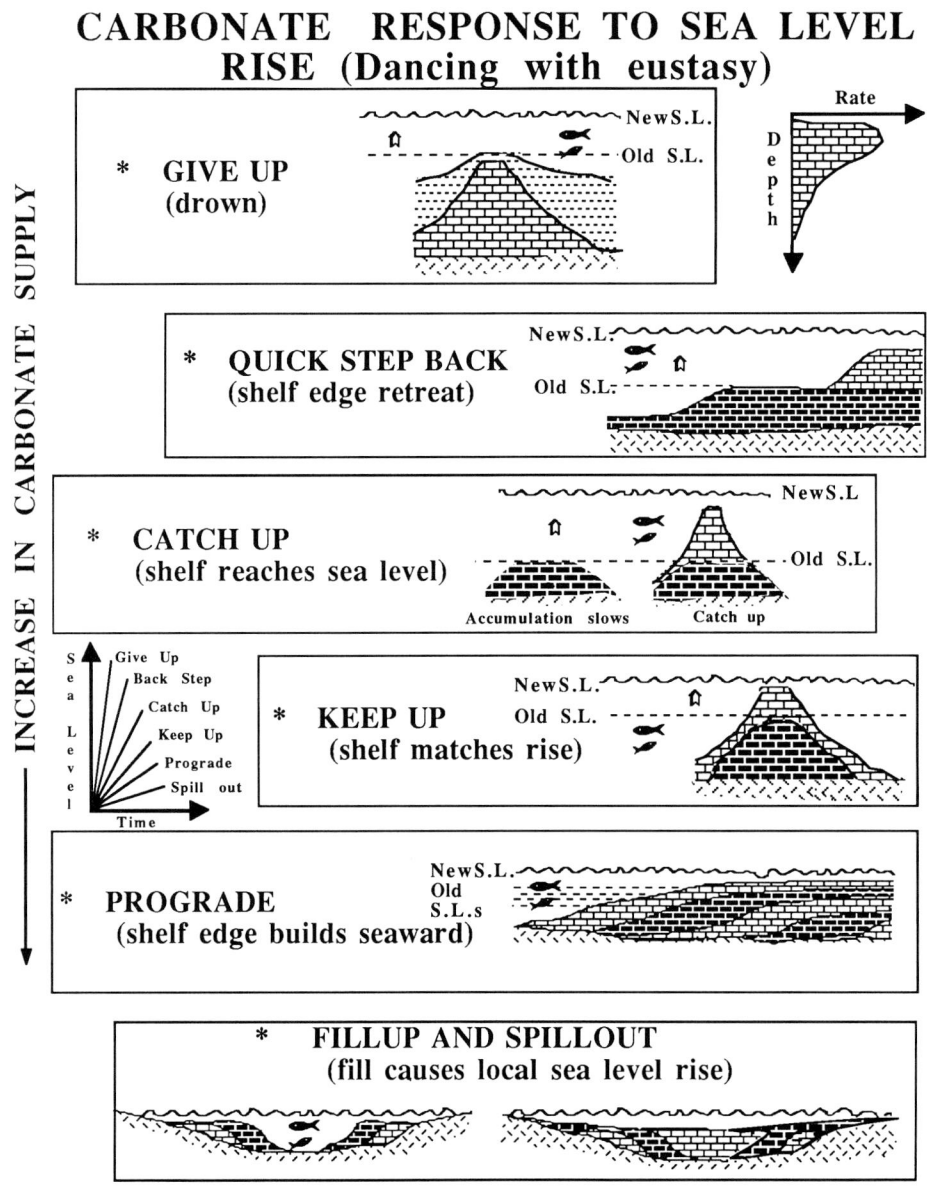

FIG. 12.—Carbonate facies geometries.

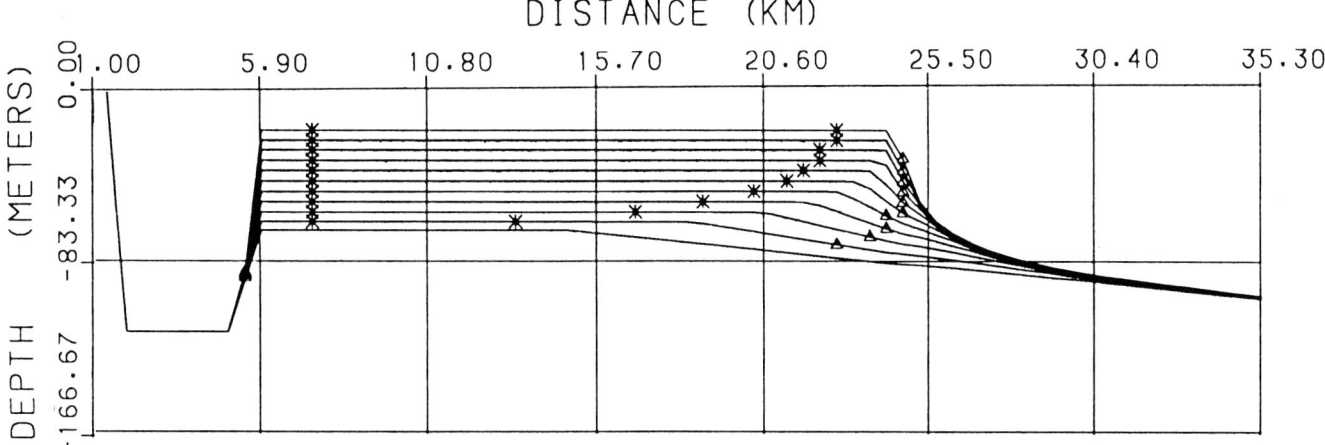

Fig. 13.—Simulation showing progradation of the reef margin as a function of steepness of the foreslope.

rate of growth (Fig. 14). If the rise in sea level equals the potential growth rate of the reef margin but is greater than that of the lagoon, the interior of the platform will develop into a deep lagoon (Fig. 7). Depending on lagoonal growth rates and the rate of sea level rise, however, the interior lagoon may eventually fill with sediment and establish a flat top, again with a shallow lagoon.

Backstepping.—

Backstepping is a term used to describe the movement and reestablishment of the reef margin toward the interior of the shelf or platform. This phenomenon has been recognized in Devonian reefs in the Canning Basin (Playford, 1980) and in the Alberta Basin of western Canada (Wendte and Stoakes, 1982) and has been attributed to rapid rise in relative sea level. During such a rise, shallow-water, faster growing reefal faunas are subjected to a sudden increase in depth that only allows the growth of deeper water, slower growing faunas. The rise may, in fact, cause a lag, during which the depositional surface of the platform may become covered by hardgrounds on its downslope margins. If the rate of sea-level rise slows and the platform surface is still within the euphotic zone, the reef margin will reestablish itself in a backstepping manner. Wendte and Stoakes (1982) observed a backstepping distance in the Judy Creek reef complex ranging from 0.4 to 0.8 km.

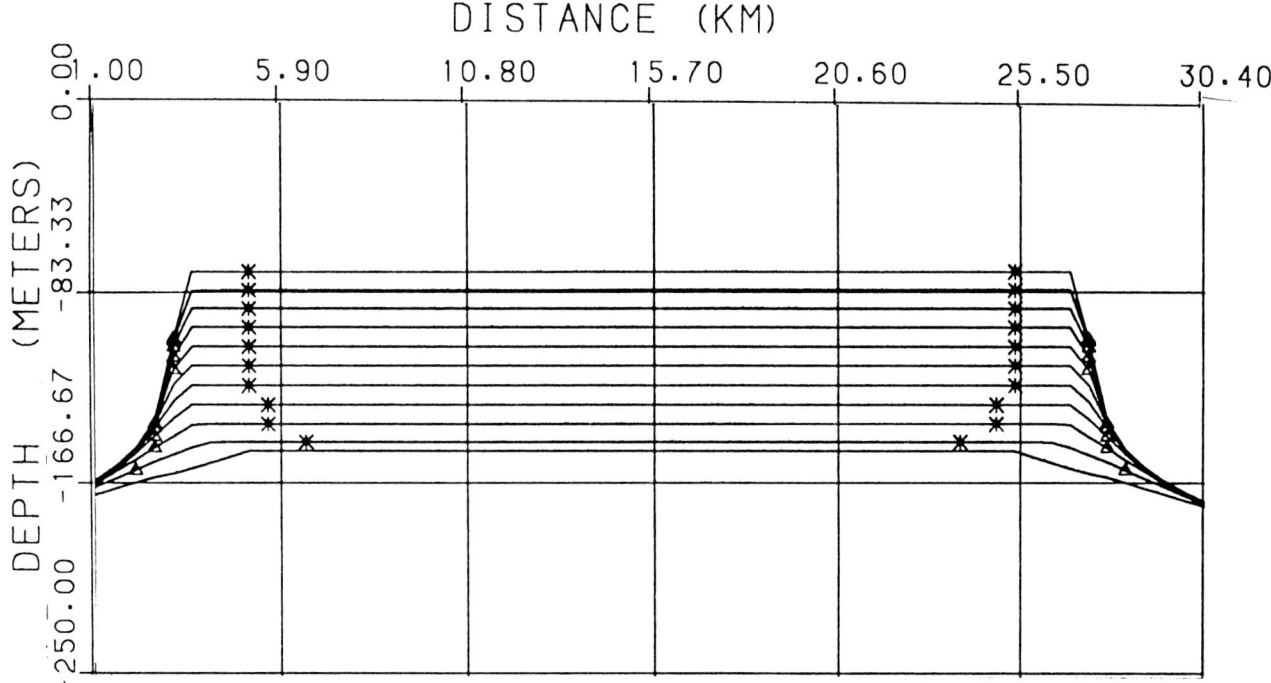

Fig. 14.—Simulation showing vertical aggradation of the reef margin, where rate of carbonate growth equals sea-level rise.

In the Judy Creek model, empirical relations are made between the distance of reef margin backstepping in Judy Creek, the rapidity of sea-level rise, and the slope of the underlying platform. Sea-level rise will move more quickly up a gradual sloping surface than up a steep sloping surface. Figure 15 shows a plot of backstepping distance against platform slope from Judy Creek.

From Figure 15, we can assume a relation between backstepping and slope, which satisfies the equation

$$dX/dY = -Y \qquad (5)$$

where X is backstepping distance and Y is slope. The solution to this equation is

$$X = \exp(-Y/p) \qquad (6)$$

where p is a scaling constant used to fit the empirical curve to the actual data. In the simulation runs, we found that a value of $p = 0.25$ produced the best geologic outputs. As p increases, the value of X increases, and the curve moves higher up the y axis.

The Judy Creek model assumes an empirical relation between backstepping and rapid sea-level rise. In this model, we assume backstepping to be an exponential function of the rapidity of sea-level rise

$$X = 1 - \exp(-w * Z) \qquad (7)$$

where Z is the ratio of rate of maximum carbonate growth to rate of sea-level rise, and w is a scaling constant used to fit the empirical curve to a "realistic" shape (Fig. 16). In the simulation runs, a value of $w = 1.0$ was used. As w increases, so does the value of X.

Therefore, we can calculate the distance of backstepping to be a function of both sea-level rise, Z, and platform slope, Y. Backstepping distance is calculated by

$$X(Z,Y) = (1 - \exp(-w * Z)) + \exp(-Y/p) \qquad (8)$$

where X is the distance the reef margin will backstep, Y is

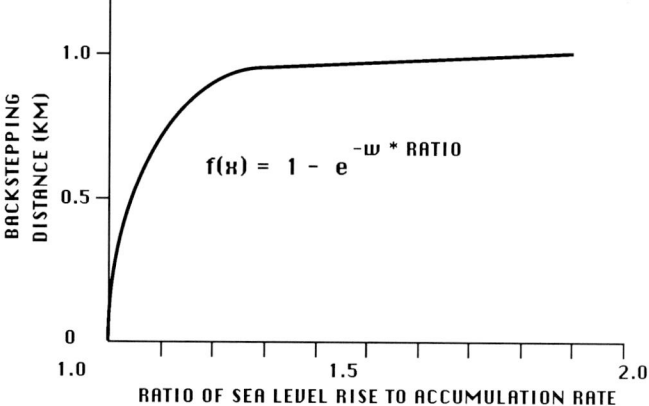

Fig. 16.—Backstepping as a function of sea-level rise.

the underlying platform slope, Z is the ratio of carbonate growth rate to rate of sea-level rise, and p and w are user-defined scaling constants.

We use this value, X, in the backstepping function B, which represents the depositional surface response to the rapid change in depth and the foreslope. The backstepping function is calculated by

$$B = 0.5 * (1 + \tanh((x - m)/b)) \qquad (9)$$

where x is each column in the depositional array starting from the previous shelf edge and ending at a distance X platformward of the shelf edge; m is the location of the midpoint of the backstepping distance X; and b is a scaling parameter of X (Fig. 17). This equation is a logistic statistical distribution function (Hastings and Peacock, 1974) and represents the probability that the backstepping distance X is less than or equal to the column distance x. The backstepping function, which ranges between 0 and 1, is then multiplied to the growth rate of the reef margin, which will switch on its full growth potential at a distance X from the drowned shelf edge ($B = 1$).

In the second case, we model backstepping to be a function of wave depth. Wave energy is considered to start damping carbonate growth potential when the sea-level change is less than the depth to wave base. From this point,

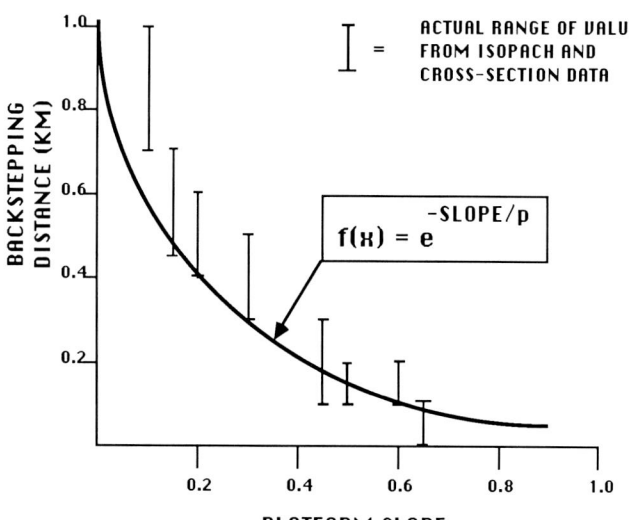

Fig. 15.—Backstepping as a function of platform slope.

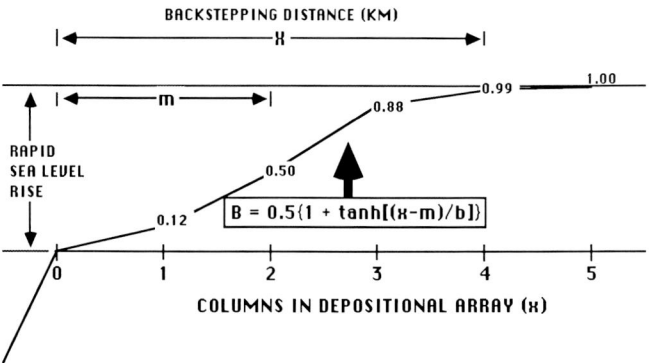

Fig. 17.—Diagram showing carbonate accumulation being damped by the backstepping function, B.

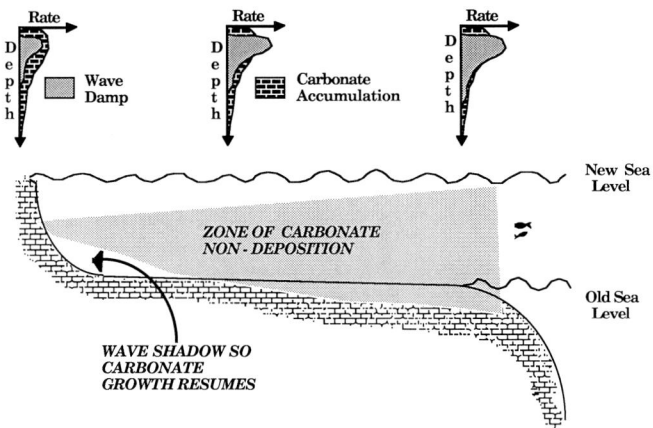

FIG. 18.—Diagram showing carbonate accumulation being damped by the effect of waves, causing the reef margin to backstep. Note how the wave damping increases, then decreases with depth. When damping is exceeded by accumulation rate, growth resumes.

at which waves start to break, to a certain distance shoreward, carbonate potential is reduced by a user-defined function that takes into account the depth of the sea and this distance. This simulates the increasing attenuation of wave energy as waves move shoreward. This is expressed by

$$A = P(z) - W(d,z) \qquad (10)$$

where A is the rate of accumulation, $P(z)$ is the potential carbonate accumulation at depth z, and $W(d,z)$ is the wave damping value at distance d and depth z. This is illustrated in Figure 18.

Shoal Development.—

Backstepping occurs when there is a rapid rise in sea level followed by a stillstand. The reef margin is able to reestablish itself as the rate of sea-level rise decreases. If, however, the rate of sea-level rise does not fall below the rate of carbonate growth, but the depositional surface is still within the upper limits of the euphotic zone, the platform will be capped with a shoal in which it is not possible to differentiate between a reef margin or a lagoon.

Drowning.—

There are two types of drowning, or give-up (Neumann, pers. commun.), responses a platform may exhibit following a rapid rise in relative sea level: (1) incipient drowning and (2) complete drowning. Incipient drowning is defined by Kendall and Schlager (1981) to occur when the rate of sea-level rise is greater than the rate of carbonate growth, and the platform is subjected to deeper water conditions but still remains within the euphotic zone. This usually results in a raised rim with deep-lagoon profile. If a platform is subjected to a sea-level rise in which the surface of the platform is below the lower limit of the euphotic zone, the platform will be completely drowned (give-up reef).

Damping

The user has the added capability of defining a damping function, which will reduce carbonate rates of accumulation across the platform for user-specified time intervals. Damping can be extended across the entire platform or, if the user so desires, damping can be restricted to the interior lagoonal positions on the platform. This enables the user to model the cumulative effects of waves, wind, and currents by damping the entire platform in the direction of the wave energy, or the user may model the cumulative effects of restricted circulation within the lagoon by damping only the platform lagoon.

RESULTS

The program was tested on actual cross sections of Judy Creek reef, and the simulated analogy is shown in Figure 19. In all of the simulation runs, rate of carbonate accu-

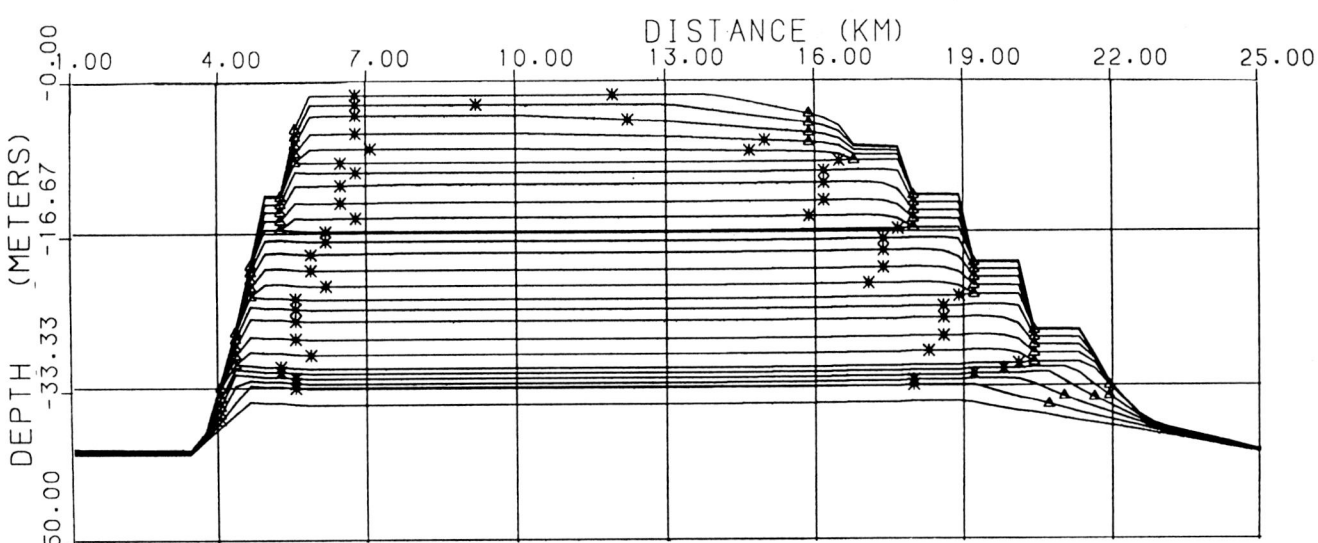

FIG. 19.—Simulation of the Judy Creek reef complex.

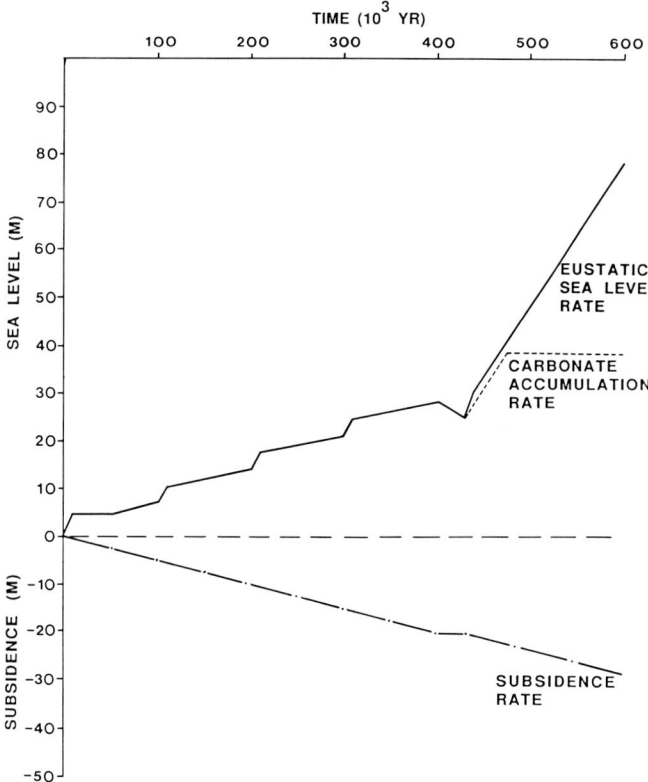

FIG. 20.—Relative sea-level curve used to simulate the Judy Creek reef complex.

mulation was assumed to remain constant through time. There is a maximum rate of accumulation over a specific depth range, which does not change. The actual thickness of the accumulation is a function of water depth. Maximum carbonate rate of accumulation for these simulations was taken to be 3.0 m/10 ka.

The fourth-order sea-level rise in the simulations is assumed to be a stairlike shape, with a rapid rise in sea level followed by a more gradual rise, followed by another rapid rise, and so on (Fig. 20). Rapid sea-level rises ranged between 3.5 and 5.0 m/10 ka. Higher order sea-level oscil-lations consisted of 0.5- to 1.5-m amplitudes with periods ranging from 20 to 40 ka. In all cases, long-term linear subsidence was matched against rates of sea-level fall. Linear subsidence rate ranges from 0.5 to 1.0 m/10 ka.

Shape of the underlying platform affected the subsequent geometries of the cycles. The steeper sloping leeward edge was not able to prograde as far as the ramplike windward edge of the initial basal platform. Slopes of the leeward edge average around 0.25°, whereas slopes of the windward edge average around 0.05° (Fig. 21).

For cycle 1, we define the sea-level rise to be 0.12 m/10 ka over a 100-ka period. This allowed the reef margin to prograde about 3 km, where slopes were less than 0.25° on the windward edge of the platform (Fig. 22). After the prograding and steepening foreslope reached a slope of 0.25°, it began to build vertically. Cycle 1 reached a maximum thickness of 12 m.

A rapid sea-level rise of 3.5 m/10 ka at the end of cycle 1 outpaced the maximum carbonate growth rate by a factor of 1.2. After sea-level rise slowed to 0.9 m/10 ka, the reef margins backstepped 0.4 km on the leeward edge and 1.4 km on the windward edge (Fig. 23). These backstepping distances are related to the relief of the margins (i.e., the windward edge has a gentler slope and therefore moved a farther distance in the platform direction). This relocation of the reef margins begins cycle 2, in which the platform aggraded for 100 ka to a thickness of 10 m. This pattern of rapid relative sea-level rise, followed by backstepping of the margins and aggradation of the platform, is repeated for cycles 3 and 4. At the end of the fourth cycle, sea level drops by about 1 m/10 ka, for a total of 3.0 m, resulting in subaerial exposure and cementation of the platform surface.

Cycle 5, the last cycle, starts after a sea-level rise of 4.0 m/10 ka, which resulted in a hardground surface and shoal-like conditions across the platform. The rate and size of sea-level rise were such that the reefs were prevented from building to sea level. It is also possible that the rate of reef growth was further damped by nutrient poisoning associated with the influx of clastics into the basin to the northeast. The platform remained under water and was subjected to wave damping from the northeast (direction of the pre-

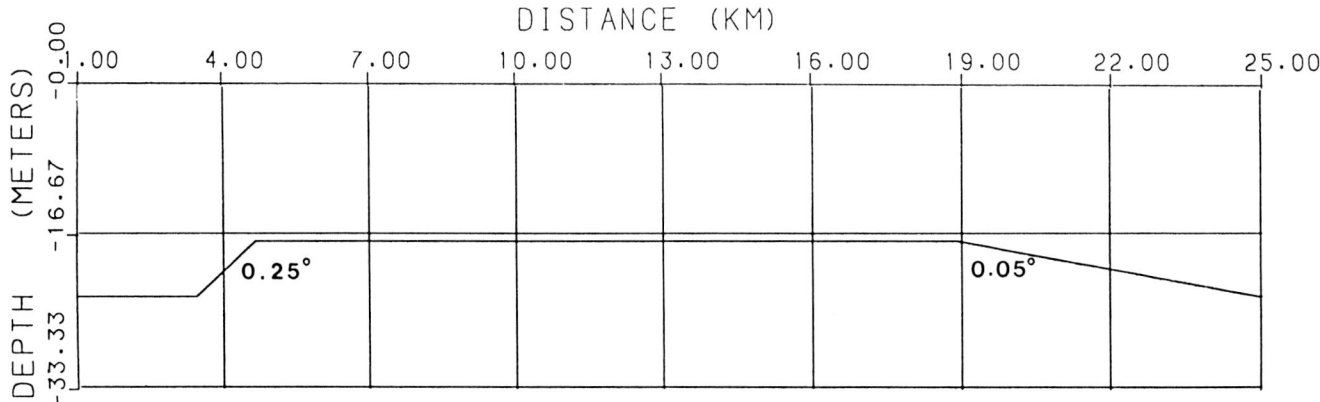

FIG. 21.—Underlying platform slope used in the simulation of the Judy Creek reef complex.

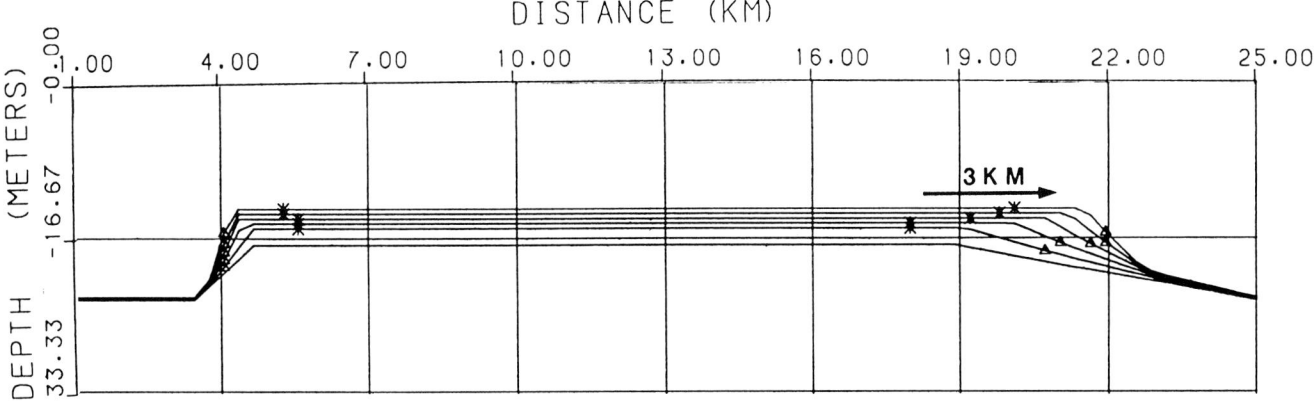

Fig. 22.—Simulation showing reef margin progradation at the end of cycle 1.

vailing winds), producing a windward-sloping ramp. Carbonate growth was damped 90 percent at the windward edge, where wave energy was strongest, and this diminished to zero half way across the platform (see Fig. 19).

DISCUSSION

We have assumed that the Judy Creek reef complex cyclicity is controlled by eustasy. We have no proof of this and, indeed, the sea-level changes may be relative in origin. Our modeling results cannot be used as evidence for eustasy. We really do not know what the short-term rates of accumulation and lag times were at Judy Creek and have therefore assumed them for our model. Obviously, other combinations of the variables and, therefore, other mechanisms, are equally possible. The model mimics the geologic record accurately only because we have fixed the variables to do so. Our results should therefore not be confused with evidence or proof that the size of the variables we use in the model is correct. In fact, we recognize that in forward modeling, there is not a unique solution set for a given geometry, but rather a family of solutions. We do feel, however, that this modeling approach helps geologists to be more objective in their understanding of sediment geometry evolution. For a complete discussion of this topic, the reader is referred to Burton and others (1987).

CONCLUSIONS

(1) Progradation of the Judy Creek reef margin occurred when the rate of carbonate accumulation was greater than the rate of relative sea-level rise, and the slope of the marginal escarpment was gentle enough to allow for stable deposition (less than 0.25°).

(2) The reef margin aggraded vertically, even though the rate of carbonate accumulation was greater than the rate of sea-level rise, provided that: (A) the foreslope was too steep for stable deposition, or (B) the rate of carbonate accumulation equaled the rate of relative sea-level rise.

(3) Backstepping of the reef margin occurred when there was a rapid relative sea-level rise that was greater than the rate of carbonate growth, but the platform remained within the euphotic zone, subjecting the shallow-water reefal fauna to deeper water conditions. The system was able to recover when the rate of sea-level rise decreased with respect to rate of carbonate growth. This caused the reefal framework builders to become reestablished platformward of the drowned shelf edge. This may have been the result of the

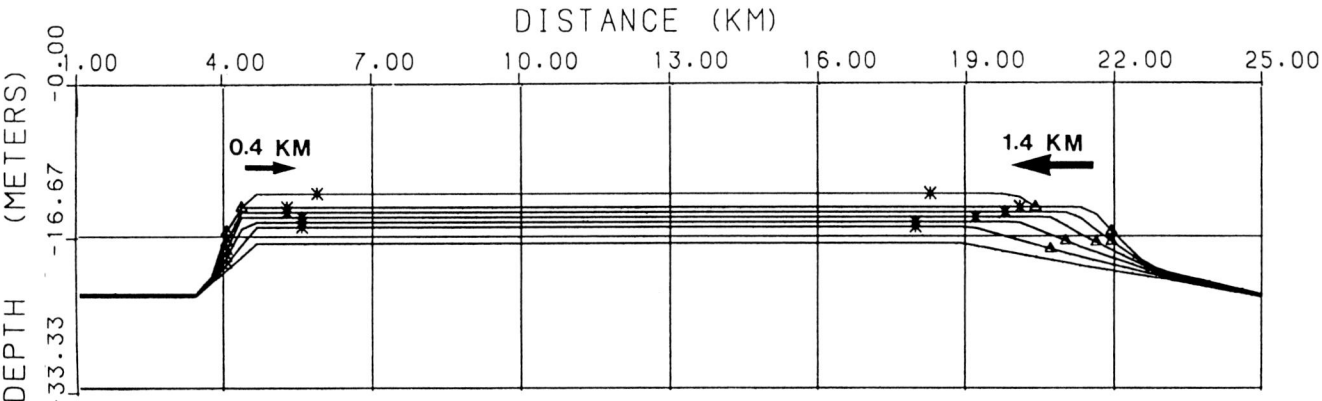

Fig. 23.—Simulation showing backstepping of the reef margins at the beginning of cycle 2.

attenuation of wave energy platformward of the drowned shelf edge or the recolonization of topographic highs behind the drowned shelf edge.

(4) When rapid sea-level rise did not decrease relative to the rate of carbonate accumulation, but the platform still remained within the euphotic zone, a shoal of reefal carbonates and skeletal sands capping the surface of the platform formed.

(5) Clastic carbonate influx into the basin of deposition away from the carbonate environment caused an exponential decrease in the rate of carbonate production.

(6) The Devonian Judy Creek reef complex of western Canada, which exhibits all of the earlier mentioned facies geometries, can be reproduced using computer simulation in response to the following variables: maximum potential carbonate growth rates for the reef margin of 3.0 m/10 ka, rapid rates of relative sea-level rise of 3.5 to 5.0 m/10 ka followed by gradual sea-level rises of 0.5 to 1.5 m/10 ka, and linear subsidence rates of 0.5 to 1.0 m/10 ka, which match the fall in sea level except at the end of cycle 4, where sea level drops by about 3.0 m. The rapid sea-level rise could either have been a result of margin downfaulting or subsidence, or glacio-eustatic rises, or a combination of both.

ACKNOWLEDGMENTS

We acknowledge the support of Esso Resources Canada in constructing the Judy Creek geologic model and express our appreciation for their approval to publish the model. We also thank the Japanese National Oil Company, Standard, and Unocal for their financial support of the development of the simulation. Interested readers should contact Christopher Kendall at the University of South Carolina for details of the code, which is written in FORTRAN 77 and runs on an IBM mainframe. Currently, a newer version is being developed in 'C' for an engineering workstation.

REFERENCES

ADEY, W. E., 1978, Coral reef morphogenesis: A multidimensional model: Science, v. 202, 831–837.
BIRKELAND, C., 1977, The importance of rate of biomass accumulation in early successional stages of benthic communities to the survival of coral recruits, in Proceedings, Third International Coral Reef Symposium, Miami, Florida, v. 1, p. 16–21.
BURTON, R., KENDALL, C. G. ST. C., AND LERCHE, I., 1987, Out of our depth: On the impossibility of fathoming eustasy from the stratigraphic record: Earth Science Reviews, v. 24, p. 237–277.
FISCHER, A. G., 1975, Tidal deposits, Dachstein Limestone of the North Alpine Triassic, in Ginsburg, R. N., ed., Tidal Deposits: Springer-Verlag, New York, p. 235–242.
GRAUS, R. R., AND MACINTYRE, I. G., 1976, Light control of growth form in colonial reef corals: Computer simulation: Science, v. 193, p. 728–734.
GUIDISH, T. M., LERCHE, I., KENDALL, C. G. ST. C., AND O'BRIEN, J. J., 1984, Relationship between eustatic sea level changes and basement subsidence: American Association of Petroleum Geologists Bulletin, v. 68, p. 164–177.
HALLOCK, P., AND SCHLAGER, W., 1986, Nutrient excess and the demise of coral reefs and carbonate platforms: PALAIOS, v. 1, p. 389–398.
HAQ, B. U., HARDENBOL, J., AND VAIL, P. R., 1987, Chronology of fluctuating sea levels since the Triassic: Science, v. 235, p. 1156–1167.
HASTINGS, N. A. J., AND PEACOCK, J. B., 1974, Statistical Distributions: Butterworth & Co., Ltd., New York, 130 p.
HELLAND-HANSEN, W., KENDALL, C. G. ST. C., LERCHE, I., AND NAKAYAMA, K., 1988, A simulation of continental basin margin sedimentation in response to crustal movements, eustatic sea level change and sediment accumulation rates: Mathematical Geology, v. 20, p. 777–802.
HIGHSMITH, R. C., 1980, Geographical patterns of bioerosion: A productivity hypothesis: Journal of Experimental Marine Biology and Ecology, v. 37, p. 105–125.
HUBBARD, D. K., 1986, Sedimentation as a control of reef development: St. Croix, U.S.V.I.: Coral Reefs, v. 5, p. 117–125.
———, AND SCATURO, D., 1985, Growth rates of seven species of scleractinian corals from Cane Bay and Salt River, St. Croix, USVI: Bulletin of Marine Science, v. 36, p. 325–338.
KENDALL, C. G. ST. C., AND SCHLAGER, W., 1981, Carbonates and relative changes in sea level: Marine Geology, v. 44, p. 181–212.
KINSEY, D. W., AND DAVIES, P. J., 1979, Effects of elevated nitrogen and phosphorous on a coral reef growth: Limnology and Oceanography, v. 24, p. 935–940.
LERCHE, I., DROMGOOLE, E., KENDALL, C. G. ST. C., WALTER, L. M., AND SCATURO, D., 1987, Geometry of carbonate bodies: A quantitative investigation of factors influencing their evolution: Carbonates and Evaporites, v. 2, p. 15–42.
MURRAY, J. W., 1966, An oil-producing reef-fringed carbonate bank in the Upper Devonian Swan Hills Member, Judy Creek, Alberta: Bulletin of Canadian Petroleum Geology, v. 14, p. 1–103.
PLAYFORD, P. E., 1980, Devonian "Great Barrier Reef" of Canning basin, western Australia: American Association of Petroleum Geologists Bulletin, v. 89, p. 1389–1403.
STOAKES, F. A., 1980, Nature and control of shale basin fill and its effect on reef growth and termination: Upper Devonian Duvernay and Ireton Formations of Alberta Canada: Bulletin of Canadian Petroleum Geology, v. 28, p. 345–410.
STROBEL, J. S., KENDALL, C. G. ST. C., BISWAS, G., AND BEZDEK, J. C., 1987, Preliminary description of program SEDFIL with carbonate module added: Proceedings of Geotech, 1987, Denver, Colorado, p. 341–349.
VAIL, P. R., MITCHUM, R. M., TODD, R. G., WILDMIER, J. M., THOMPSON, S., SANGREE, J. B., BUBB, J. N., AND HATFIELD, W. G., 1977, Seismic stratigraphy and global changes in sea level, in Payton, C. E., ed., Seismic Stratigraphy—Applications to Hydrocarbon Exploration, American Association of Petroleum Geologists Memoir 26, p. 49–212.
WENDTE, J. C., AND STOAKES, F. A., 1982, Evolution and corresponding porosity distribution of the Judy Creek reef complex, Upper Devonian, Central Alberta, in Cutler, W. G., ed., Canada's Giant Hydrocarbon Reservoirs: Canadian Society of Petroleum Geologists 1982 Core Conference, p. 63–81.
WILSON, J. L., 1975, Carbonate Facies in Geologic History: Springer-Verlag, Berlin, 471 p.

PART II
EXAMPLES OF CARBONATE RIMMED PLATFORMS AND RAMPS DEVELOPED ON PASSIVE (RIFTED) CRATONAL MARGINS

FACIES AND EVOLUTION OF PRECAMBRIAN CARBONATE DEPOSITIONAL SYSTEMS: EMERGENCE OF THE MODERN PLATFORM ARCHETYPE

JOHN P. GROTZINGER

Department of Earth, Atmospheric and Planetary Sciences, Massachusetts Institute of Technology, Cambridge, Massachusetts 02139

ABSTRACT: Precambrian carbonates generally have not been examined from the perspective of platform evolution. The major facies and stratigraphic relations of several carbonate sequences ranging from the early Archean through late Proterozoic are examined and discussed in terms of platform construction and the dominant factors that influenced patterns of sedimentation and the production of carbonate.

Early to middle Archean carbonate sedimentation was rare and generally restricted to brief interludes between episodes of tectonism and volcanism. Inferred carbonate facies are largely replaced by chert, and individual occurrences were less than a few meters thick. Most accumulated in deeper water settings, although local shallow-marine (or lacustrine) evaporitic settings also were present. Carbonates of the middle Archean Nsuze Group contain the first evidence for deposition on cratonic masses and form a thin (30 m) veneer of shallow-water facies. Late Archean carbonate sedimentation was locally substantial, forming units as thick as 500 m that may have been areally extensive, but which have since been mostly removed by uplift and erosion. Significant facies include major stromatolite buildups, grainstone belts, lagoonal to peritidal cyclic facies, and calcite pseudomorphs after possible giant aragonite fans.

Proterozoic carbonates formed platforms that are strikingly similar to Phanerozoic platforms despite the absence of metazoans. Homoclinal and distally steepened ramps developed fringing stromatolite reefs and ooid shoals, stromatolitic barrier reefs, and isolated buildups including pinnacle reefs as thick as 300 m. Rimmed shelves also formed, including accretionary margins, bypass margins with both gullied slopes and escarpments, and intra-shelf basins. Incipient and terminal drowning, rim backstepping, and subaerial exposure were also important in the evolution of Proterozoic platforms.

Globally diachronous, cratonic stabilization associated with the Archean/Proterozoic "transition" strongly influenced the long-term evolution of carbonate platforms by creating spatially extensive, stable continental masses on which carbonates formed. Widespread early Proterozoic carbonate platforms document this event. Ramps generally formed in Proterozoic extensional and foredeep basins, and often preceded development of rimmed shelves in more stable, possibly thermally subsiding basins. Rimmed shelves locally formed margins with as much as 1,000 m of relief relative to the adjacent basin and contained stromatolitic barrier reefs that were extensive for hundreds of kilometers.

The composition of Precambrian sea water was favorable for calcium carbonate production, and evidence for saturation with respect to calcite and aragonite includes well-preserved pseudomorphs of "abiotic" aragonite and calcite (high Mg?) precipitated in open-marine (non-evaporitic) shallow-water settings. Saturation probably existed to depths of at least 1 km as shown by the presence of carbonate pseudomorphs after (high Mg?) calcite in toe-of-slope sheet cracks and breccias. Evidence for possibly elevated Archean and early Proterozoic saturation values (relative to average Phanerozoic values) includes development of giant aragonite fans in the late Archean and development of extensive tidal-flat tufas in the early Proterozoic.

Proterozoic carbonate production may have been strongly regulated by the presence of stromatolite-forming benthic microbial communities. A possible mechanism was removal of CO_2 from the water column during photosynthesis. This might have induced *in situ* carbonate precipitation on and adjacent to stromatolites and provides a mechanism for carbonate mud production, which forms the major non-stromatolitic facies of many Proterozoic platforms. Thick successions of grainstone and other coarse carbonate sediments, comparable to the Jurassic of the United States Gulf Coast and Saudi Arabia, are uncommon in most platforms.

Other controls on Proterozoic platform development possibly included the effects of local stromatolite zonation according to paleoenvironment. Conical stromatolites tended to occur in basinal, slope, and deeper ramp settings and may have retarded shelf/basin differentiation and have influenced the progradation of tidal flats, promoting advance in situations where subtidal carbonate production was impeded.

Eustasy apparently influenced the stratigraphic architecture of many platforms, causing rim backstepping, incipient shelf drowning, and the development of third-order sequences or "Grand Cycles." It also was the likely cause of parasequences or "small-scale cycles" that characterize peritidal environments in late Archean through late Proterozoic sequences.

INTRODUCTION

In order to be fully appreciated, studies of carbonate platforms should have an historical perspective. Patterns deduced from their long-term development help document the evolution of carbonate-producing organisms, the succession of tectonic and subsidence regimes, the effects of eustasy, fluctuations in climate, and secular changes in the ocean and atmosphere. Description of facies, formulation of models, and analysis of variability in Phanerozoic carbonate platforms have been highly successful in fulfilling these objectives (Heckel, 1974; Ginsburg and James, 1974; Wilson 1975; Longman, 1981; James, 1984a; Read, 1982, 1985).

In contrast, studies of Precambrian carbonate platforms are uncommon and models for their facies relation and evolution are absent. Most studies of Precambrian carbonates have not been comprehensive, instead tending to focus on specific, usually biological, aspects of carbonate occurrences. An example of this concerns stromatolite morphologies and microfabrics and their potential use in global biostratigraphic correlation (Cloud, 1942; Raaben, 1969; Preiss, 1976; Donaldson, 1976a). Another involves the search for and validation of fossil microbial remnants that are often preserved in early-silicified stromatolites (Tyler and Baarghorn, 1954; Cloud, 1965; Schopf and Walter, 1983). Less commonly, geochemical studies of carbonates involving a whole-rock, facies-independent approach also have been undertaken (Veizer and Compston, 1976; Veizer and Hoefs, 1976).

Only within the last decade or so have Precambrian carbonates been studied from the perspective of platform evolution, considering the entire depositional system, and using modern approaches and comprehensive analogs (Hoffman, 1974; Cecile and Campbell, 1978; Bertrand-Sarfati and Moussine-Pouchkine, 1983; Grotzinger, 1986a,b; Beukes, 1987). The growth, spatial arrangement of facies, and evolution of many of these carbonate sequences are strikingly similar to those of younger platforms. In some instances substantial thicknesses of carbonate (over 1.5 km) accumulated over areas of at least 200,000 km^2, forming platforms with strong paleoenvironmental gradients. Barrier reefs composed exclusively of stromatolites constructed major buildups comparable in size and extent to some of the great coral-dominated barriers of the modern world. Such a degree of uniformity is perhaps an unexpected surprise given the lack of carbonate-secreting skeletal metazoa prior to Cambrian time, commonly regarded as the major producers of Phanerozoic carbonate. This, together with other similar important new discoveries, warrants a review and discussion of the occurrence and significance of Precambrian carbonate platforms.

The principle objectives of this paper are to: (1) identify the principle constituents of several Precambrian carbonate sequences ranging from the early Archean through late Proterozoic; (2) characterize and evaluate the basic facies relation observed in those sequences as they pertain to platform construction and long-term evolution; (3) assess the relative impact of major platform-regulating processes (biospheric, hydrospheric, tectonic, eustatic); and (4) develop a set of models for comparison with younger platforms and other less well understood Precambrian platforms. The models are developed using the classification scheme of Read (1982, 1985) but are modified to accommodate the specific arrangement of facies, especially unique types of buildup, that characterize the Precambrian record.

The platforms outlined here should generate an enhanced understanding of Precambrian carbonate depositional systems, reflecting the first-order features they share in common with Phanerozoic platforms, while highlighting the more specific features associated with only Precambrian platforms. This paper is not an inventory of all the known Precambrian carbonate sequences described to date. Rather, it is based on consideration of several reasonably well-documented platforms that serve to illustrate the continuity of theme as well as the diversity of style in the evolution of these earliest carbonate platforms.

ARCHEAN CARBONATES

Because of the fragmentary preservation of the Archean carbonate record, it is difficult to group the individual occurrences according to common elements for the purposes of comparison and discussion. Consequently, selected occurrences of Archean carbonates are discussed on a case history basis, emphasizing facies characteristics and interrelation and their significance in the history of platform evolution.

Early Archean

Occurrences of well-preserved early Archean carbonates are rare. A few early Archean carbonates are known from the Warrawoona Group of Western Australia and Swaziland Supergroup of South Africa (Fig. 1). These are all extremely thin, consisting of generally discontinuous limestone and dolomite units that are often extensively replaced by chert. None of these occurrences is thought to have accumulated on stable platforms, and they usually represent brief interludes in the otherwise tectonically active history of greenstone belts.

Warrawoona Group, Pilbara Block, Western Australia.—

Possible silicified carbonates from the lower and upper Warrawoona Group (approximately 3.5 Ga) have been described by Lowe (1983) and Buick and Dunlop (1987). The occurrences consist primarily of several varieties of chert, chalcedony, and quartz that are interpreted to have replaced primary carbonate mud, "diagenetic" carbonate crystals, rare stromatolites, possible sulfate evaporites, and possible paleospeleothems. Buick and Dunlop (1987) describe barite

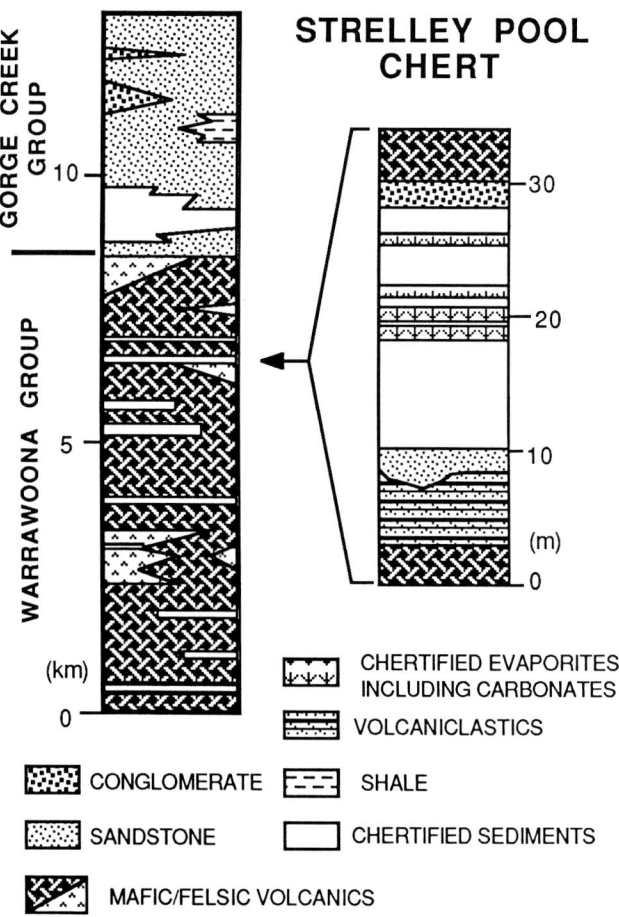

FIG. 1.—Generalized stratigraphy of eastern Pilbara block greenstone belts, Western Australia illustrating position of Strelley Pool Chert and organization of facies within the Chert. Modified after Lowe (1983).

pseudomorphs after gypsum in the lower Warrawoona occurrence.

Significance.—The Warrawoona cherts are intimately associated with thick sequences of mafic volcanics, interfingering laterally with volcaniclastic alluvial facies. The units are thought to have formed in shallow, evaporative basins adjacent to eroding volcanic sources, and it seems likely that sedimentation of these orthochemical deposits, now replaced by chert, occurred during brief quiescent periods that punctuated volcanic episodes (Lowe, 1983).

Buick and Dunlop (1987) consider the lower Warrawoona chert/carbonate/sulfate sediments to be marine in origin. Lowe (1983) also considers the upper Warrawoona orthochemical sediments as marine, but suggests that a lacustrine interpretation is also possible given the difficulty in distinguishing restricted (evaporitic) marine from lacustrine deposits. This problem is often encountered in studies of environmentally restricted Precambrian deposits (Muir and others, 1980; Rowlands and others, 1980; Grotzinger, 1986d; Southgate, 1986; Jackson and others, 1987), and also besets the Phanerozoic evaporite record (Hardie, 1984; Decima and others, 1988). It should be emphasized that carbonate and calcium sulfate deposits are characteristic of many non-marine evaporitic sequences (Tucker, 1978; Hardie, 1984), and there is potential danger in interpreting their presence as evidence for a marine origin.

Onverwacht and Fig Tree Groups, Barberton Mountainland, South Africa.—

Carbonates occur in several units of the Onverwacht and overlying Fig Tree Groups that form the lower two units of the Barberton Greenstone Belt (approximately 3.5 Ga). Carbonates in the Onverwacht Group occur as largely chert-replaced layers, consisting of as much as 20 percent finely disseminated carbonate as thick as 2 m. These are closely associated with pure cherts, shale, and very fine volcaniclastic detritus, and are interpreted as deep-water hemipelagic deposits (Lowe and Knauth, 1977). The upper Onverwacht Group and basal Fig Tree Group contain beds of dolomitic, clastic carbonate with well-preserved cross-stratification and rare desiccation cracks (Lowe and Knauth, 1977; Heinrichs and Reimer, 1977). Carbonate beds in the Fig Tree Group form units as thick as 10 m and are interpreted as shallow-water facies.

Significance.—The thin, discontinuous, cross-stratified clastic dolomite units are the best-preserved evidence for early Archean carbonate sedimentation in shallow-water environments. The paucity of early Archean carbonates may be the result of several factors. First, extensive cratonic shelves, clearly the most important habitat of Proterozoic and Phanerozoic carbonates, were apparently absent in the early Archean. Second, the lack of extended periods of tectonic quiescence in greenstone belts, coupled with high siliciclastic influx, would not have been conducive to carbonate production even during shallow-water conditions. The lack of continental shelves may be the most important factor because, as is subsequently shown, the development of large cratonic masses in the middle and late Archean permitted the production of major carbonate units that possess certain features of younger platforms.

Middle Archean

The middle Archean is significant in that it contains the first record of sedimentation on probably small but stable continents. Diachronous but widespread cratonization related to the Archean-Proterozoic "transition" initially began in South Africa, resulting in stabilization of the southern Kaapvaal Province toward the end of middle Archean time (Tankard and others, 1982). Consequently, the development of quartzite-carbonate "blanket" deposits, characteristic of the Proterozoic and Phanerozoic record, occurred there first. The Pongola Supergroup of South Africa is the best-preserved example of significant sedimentation that occurred outside of greenstone belts and consists of sedimentary and volcanic rocks that formed platform cover and intracratonic rift sequences (Matthews, 1967; Von Brunn and Mason, 1977; Tankard and others, 1982; Burke and others, 1985).

Nsuze Group, Pongola Supergroup, South Africa.—

The Nsuze Group (approximately 3.0 Ga) rests unconformably on middle and early Archean basement and consists of a heterogeneous assemblage of shales, arkosic grits, quartz arenites, and clastic-textured and stromatolitic carbonates, interspersed with two volcanic units. Carbonates are closely associated with texturally mature quartz arenites (as thick as 100 m) that have been interpreted as shallow-marine tidal-sand bodies (Von Brunn and Hobday, 1976).

Carbonates are as thick as 30 m, forming a thin, laterally discontinuous unit of clastic-textured, stromatolitic and recrystallized (massive) dolomite. Clastic-textured dolomites display well-developed herringbone crossbedding, contain large stromatolite-derived intraclasts, and are believed to have formed in tidally influenced environments similar to those of associated quartz arenites (Von Brunn and Mason, 1977). Ooid and intraclast grainstones are reported in Walter (1983). Stromatolites include low-relief structures that generally form undulatory surfaces and are characterized by fine layering, fenestral fabrics, and uncommon radial-fibrous fabrics (Von Brunn and Mason, 1977; Walter, 1983). These carbonates have high-strontium concentrations compared to those of other carbonates of this age (Veizer, pers. commun.; cited in Walter, 1983).

Significance.—The Nsuze Group carbonates are possibly the oldest sequence of sediments that can be regarded as evidence for the inception of stable carbonate platform sedimentation. Unfortunately, no information is available pertaining to either the reconstruction of various platform facies or stages of evolution. It seems likely, however, that this earliest platform may have been a relatively high-energy ramp, based on the evidence for tidal activity in both carbonate and siliciclastic sediments and the comparatively sparse distribution of stromatolites. No continuous stromatolitic intervals or buildups have been noted.

Late Archean

Significant carbonate sedimentation occurred during the late Archean, forming sequences as thick as 500 m that were deposited over at least 125 km^2. Available evidence suggests that some of these sequences probably are small remnants of formerly extensive regional systems that formed true platforms, with well-developed facies zonation.

Late Archean carbonates formed in a variety of settings, including continental platforms, greenstone belts, and possible foredeep or successor basins. Only a few of these sequences are well enough preserved for study, having survived the effects of metamorphism and deformation related to widespread "cratonization" at this time. To date, late Archean carbonate sequences containing a variety of decipherable sedimentologic facies are known from the Fortescue Group and Hamersley Group of Western Australia (Walter, 1983; Simonson, 1987), the Ventersdorp Supergroup of South Africa (Winter, 1963; Buck, 1980), the Ngezi Group of Zimbabwe (Martin and others, 1980), and the Yellowknife Supergroup and Steep Rock Group of Canada (Henderson, 1975; Wilks and Nisbet, 1985; Wilks, 1986). Of these, studies of Ngezi Group, Steep Rock Group, and Yellowknife Supergroup carbonates have produced data on the sedimentology and stratigraphy of the occurrence, in addition to geochemical and paleontologic information.

Yellowknife Supergroup, northwest Slave Province, Canada.—

Stromatolitic carbonates occur in the Yellowknife Supergroup (approximately 2.5 Ga) near Snofield Lake, northwest Slave Province, Canada (Henderson, 1975; Walter, 1983). The carbonates are developed at the transition between the lower part of the greenstone belt sequence (Yellowknife Supergroup), which consists dominantly of intermediate and felsic volcanics, and the upper part of the sequence, which consists primarily of greywacke turbidites. This stratigraphic position of the carbonates, between volcanics and sediments, is typical of other regions in the Slave Province where similar transitions occur (McGlynn and Henderson, 1970). Such transitions may represent the cessation of subaerial felsic volcanism and development of stromatolitic carbonates as atoll-type fringing barriers (cf. Darwin, 1842), followed by atoll collapse during rapid subsidence, and then deposition of basinal turbidites (Henderson, 1975). The volcanic-sediment transitions are also associated with deposition of carbonaceous pyritic mudstones and the localization of base-metal mineral deposits (Henderson, 1975).

The Snofield Lake carbonates occur as a discontinuous dolomite unit as thick as 40 m that extends along strike for about 6 km (Henderson, 1975). Dolomites are interbedded with black carbonaceous mudstones that grade upward and laterally into volcanogenic sandstones and conglomerates. The unit internally consists of thinly laminated, stromatolitic, intraclastic, and oncolitic dolomite.

Stromatolites include irregular to wavy-laminated sheets, low-relief domes, partially linked columns, and microdigitate columns (Henderson, 1975; Walter, 1983). Stromatolite form varies laterally along strike from more stratiform types in the south to higher relief forms in the north. The latter are as wide as 30 cm with 20 cm of synoptic relief.

*Significance.—*The microdigitate stromatolites reported by Walter (1983) may be some of the oldest known and are a prelude to the widespread occurrence of these types in the early Proterozoic (Grotzinger, 1987). It is noteworthy that Walter (1983) reports a palimpsest microfabric consisting of remnant acicular crystallites. On the basis of development of this microfabric type, early Proterozoic microdigitate stromatolites have been interpreted as tufas (Grotzinger and Read, 1983; Grotzinger, 1986b, 1987), formed primarily in response to precipitation of acicular carbonate, probably aragonite. By analogy then, the Snofield Lake microdigitate stromatolites may contain some of the oldest evidence for accretion of sediment by *in situ* precipitation of carbonate, rather than "trapping and binding."

The Snofield Lake carbonates probably formed during a brief interlude in a tectonically active environment. A variety of models has been proposed for the origin of greenstone belts. Regarding those in the Slave Province, earlier models propose entirely intracratonic settings, invoking continental extension producing volcanism, deep-basin development, and subsequent collapse (McGlynn and Henderson, 1970, 1972; Henderson, 1981). More recently, Kusky (1989) and Hoffman (1986) have proposed that Yellowknife supracrustal rocks are allochthonous and represent, in part, shingled oceanic crust accreted to sialic basement. In this model, the carbonates may have formed as atoll deposits that were incorporated into an accretionary prism during subduction of oceanic crust and offscraping of seamounts (Hoffman, 1986). In either model, the carbonates are considered to have formed as thin, short-lived reefs that fringed subsiding volcanos.

Steep Rock Group, southwest Superior Province, Canada.—

Carbonates in the Steep Rock Group (approximately 2.7 Ga) have been studied for almost 100 years from regional, economic, paleontologic, and sedimentologic perspectives (Smyth, 1891; Lawson, 1913; Walcott, 1913; Jolliffe, 1955; Hofmann, 1971; Walter, 1983; Wilks and Nisbet, 1985; Wilks, 1986). It is one of the most substantial Archean carbonate sequences.

The carbonates are as thick as 500 m and extend continuously along strike for 12 km, with the exception of one locality where the entire unit has been cut out beneath a major karstic unconformity developed at the top of the sequence (Fig. 2). The Steep Rock carbonate unit has been extensively faulted and was probably more extensive originally. The carbonates were deposited on siliciclastic sands and conglomerates that, in turn, lie unconformably on much older crystalline basement. Above the upper karstic carbonates is a unit of goethite, gibbsite, hematite, and kaolinite, including lenses of pisolitic ferruginous bauxite. The mineralogy and stratigraphic position of this unit are consistent with its interpretation as a carbonate soil profile, the result of intense weathering (Jollife, 1955; Wilks, 1986). This unit is overlain by a sequence of altered probable banded-iron formation, now present mostly as goethite and hematite, and is thought to have accumulated subaqueously, following transgression of the paleosol and foundering of the platform (McIntosh, 1972; Wilks, 1986). Overlying units consist of "typical" greenstone belt sequences of thick pyroclastic ashrock, mafic volcanics, and turbiditic sediments. The contact between the ashrock and overlying thick mafic volcanics is strongly sheared, sug-

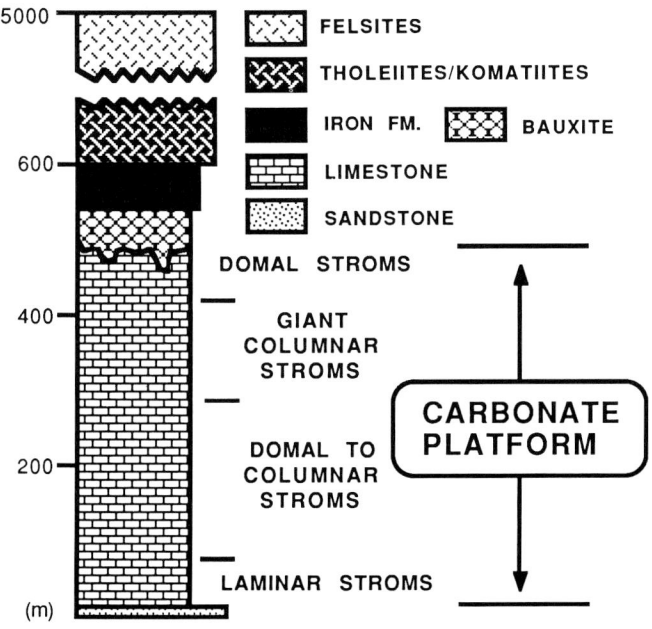

FIG. 2.—Stratigraphy of the Steep Rock Group illustrating relations between stromatolites of the carbonate unit. After Wilks and Nisbet (1985) and Wilks (1986).

gesting that the volcanics have been thrust over Steeprock platform sediments (Hoffman, 1986).

Carbonate Stratigraphy.—A crude stratigraphy (Fig. 2) based on stromatolite morphology has been defined for the Steeprock carbonates (Wilks and Nisbet, 1985; Wilks, 1986). Microbial laminites[1] and small domal stromatolites occur throughout the sequence and are best developed near the base. Large domal mounds occur abundantly in the upper 50 m.

Small domal stromatolites as wide as 15 cm with as much as 10 cm of synoptic relief occur near the base of the sequence and are associated with units containing microbial laminites and irregular to wavy-laminated stromatolites. Digitate columnar stromatolites and linked conical stromatolites may be developed, and the pseudofossil *Atikokania* (Hofmann, 1971) occurs. This structure consists of neomorphic carbonate that is pseudomorphic after a fibrous, radiating, precursor mineral. Hofmann (1971) suggested gypsum or aragonite, but later authors have favored the gypsum interpretation on the basis of the large size of the crystals that extended as much as 25 cm (Walter, 1983; Wilks, 1986).

[1] The term "microbial laminite" is used here to replace the term "cryptalgalaminite" (cf. Aitken, 1967), because, as urged by Burne and Moore (1987, p. 242), "cyanobacteria are no longer regarded as algae, and other microbial components are present alongside cyanobacteria in most microbial communities." A "microbial laminite," however, is considered here to be similar to a "cryptalgalaminite" in a rock-textural sense and with regard to facies designation.

Upper Steeprock carbonates contain large, elongate stromatolite mounds as much as 3 m wide, 5 m long, and with as much as 1 m of synoptic relief (Wilks, 1986). Stromatolites form biostromes that extend across most outcrops. Lamination in stromatolites is defined by variations in the organic content of the carbonate, as well as fenestrae that are as long as 3 cm (Wilks, 1986). Walter (1983) noted that the laminae, in part, are defined by the presence of acicular crystal fibers, similar to those of *Atikokania*. This relation further suggests that such carbonate pseudomorphs may have replaced primary aragonite rather than gypsum, because the large, elongate stromatolite mounds are more likely an open-marine rather than restricted evaporitic facies. It also suggests that precipitation of carbonate may have been an important contribution to the construction of stromatolites and, by implication, may have been a significant factor in platform construction.

Significance.—The lower Steeprock Group contains many of the elements of later Proterozoic and Phanerozoic carbonate platforms. Carbonates were deposited following deposition of a veneer of fluvial (Wilks, 1986) and possibly shallow-marine siliciclastic sediments that covered the exposed crystalline basement. It is interesting to note the apparent complete lack of grainstone in the sequence, which is dominated by stromatolitic lithologies.

The overall stratigraphy of the carbonate sequence has been interpreted by Wilks (1986) to be shallowing upward, with smaller stromatolites at the base and larger stromatolites at the top. This is the converse of many documented Proterozoic shallowing-upward sequences, however, where larger scale stromatolites are overlain by smaller varieties (Cecile and Campbell, 1978; Grotzinger, 1986a; Beukes, 1987). Furthermore, recently discovered modern subtidal stromatolites are considerably larger than known intertidal forms (Dill and others, 1986). Alternatively, the Steeprock sequence may be one of deepening upward, followed by a subsequent fall in relative sea level.

This relative fall of sea level was clearly of great amplitude as well as duration, in that parts of the 500-m-thick platform have been entirely eroded, and all areas are overlain by a bauxitic regolith. Given a 500-m minimum relative fall in sea level, based on the depth of erosion, a eustatic cause is considered to be unlikely. The largest changes in sea level in Phanerozoic time are associated with fluctuations in the volume of seafloor spreading centers as related to changes in the rate of spreading (Hays and Pitman, 1973; Pitman, 1978). Kominz (1984) has shown, however, that the likely change in sea level associated with the Cretaceous transgression, perhaps the greatest in Phanerozoic time, was only on the order of 250 m. Alternatively, changes in relative sea level related to tectonic causes, such as block faulting, may occur on the order of 500 m (Brown, 1970). On this basis, a tectonic cause is likely, although it is unclear as to whether this was rift-induced (Wilks, 1986) or possibly related to passage of a peripheral bulge during collision (cf. Hoffman, 1987).

Ngezi Group, Belingwe Greenstone Belt, Zimbabwe.—

The Ngezi Group of the Belingwe Greenstone Belt contains three distinct sequences (Fig. 3a): a basal sedimentary

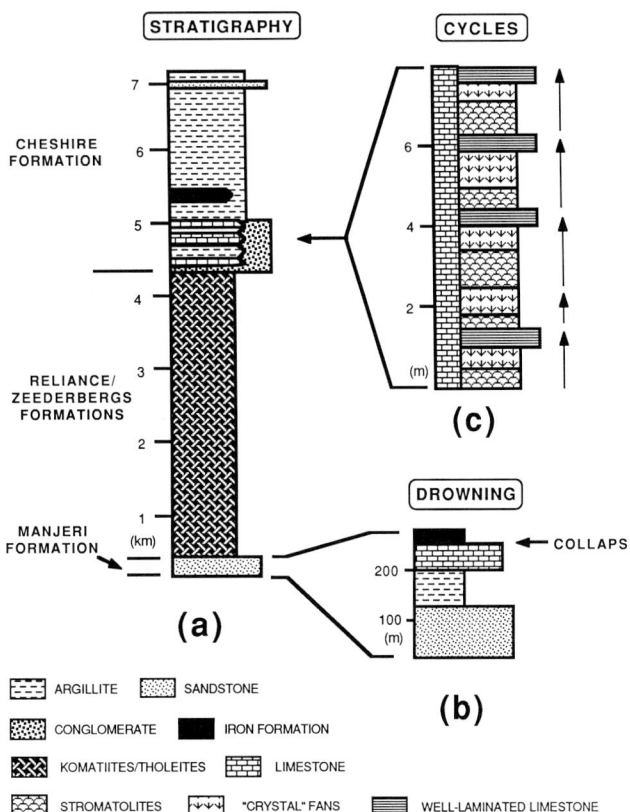

FIG. 3.—Stratigraphy of the Ngezi Group. (a) Generalized stratigraphy of the Ngezi Group in the Belingwe Greenstone Belt illustrating relative positions of the Manjeri and Cheshire Formations. (b) Stratigraphy of the Manjeri Formation showing development of thin carbonate platform followed by drowning and draping by deep-water iron formation. (c) Representative section illustrating relations between cyclic carbonate facies of the Cheshire Formation. After Martin (1978) and Martin and others (1980).

sequence (Manjeri Formation, as thick as 300 m), succeeded by almost 5 km of high-magnesia basalts and tholeiites (Reliance and Zeedersbergs Formations), and capped by 2.5 km of sediments (Cheshire Formation). Carbonates are known from the Manjeri and Cheshire Formations.

Manjeri Formation.—The Manjeri Formation is very similar to the lower Steep Rock Group of Canada, consisting of a lower transgressive siliciclastic sequence (as thick as 250 m) that unconformably overlies older crystalline basement, and an upper carbonate sequence (as thick as 50 m) that locally contains abundant stromatolites (Fig. 3b). The top of the formation is capped by banded-iron formation and then is overlain by volcanics of the Reliance Formation. As such, the Manjeri Formation can be interpreted as a platform sequence that was tectonically drowned and overlain by deep-water banded-iron formation (essentially pelagic deposits) and then volcanics. The carbonates (limestones) of the upper Manjeri Formation are not well preserved due to locally high strain (Martin, 1978) and are not discussed further here. Well-preserved areas do occur locally, and these are described in Martin (1978) and Martin and others (1980).

Significance.—It is noted that, in addition to the Steeprock carbonates, this is the other well-documented area where it has been demonstrated that carbonates formed as a continuous unit, capping an onlap sequence deposited on sialic basement. In both areas, sedimentation on relatively "stable" continental platforms was comparatively short-lived, abruptly terminated by drowning during probable major subsidence associated with greenstone belt evolution. This environment was distinct from that of intra-greenstone belt carbonates (Snofield Lake occurrence) and supra-greenstone belt carbonates, such as the Cheshire carbonates discussed below.

Cheshire Formation.—The Cheshire Formation (as thick as 2.5 km) is a heterogeneous succession of siltstone, sandstone, conglomerate, carbonate, and minor iron formation (Fig. 3a). The Cheshire is thought to overlie the Zeedersbergs volcanic sequence "conformably" (Martin, 1978), but this is only in the sense that there is no angular discordancy between the two successions. The basal and eastern parts of the sequence are conglomeratic, with abundant clasts of the underlying Zeedersbergs volcanics, indicating that uplift of the volcanics occurred during deposition of the Cheshire Formation. This suggests that a significant time gap may be present between the two sequences and that the Cheshire may have accumulated in a possible foredeep or successor basin. As such, Cheshire sedimentation postdates the main volcanic phase of the greenstone belt and represents a new basin environment in which Archean carbonates formed.

The Cheshire Formation contains some of the best-developed and best-preserved Archean carbonates in the world owing to the relatively low metamorphic grade, minimal structural disruption, and locally good outcrop (Bickle and others, 1975; Martin 1978; Martin and others, 1980; Abell and others, 1985). Initial studies by Martin (1978) and Martin and others (1980) have shown that the Cheshire carbonates compose a highly diverse, complex assemblage facies that can be organized into a generalized paleoenvironmental model. The model presented by Martin and others (1980) mainly highlights stromatolitic facies.

Cheshire carbonates are developed in western outcrops of the Belingwe Greenstone Belt and pinch out to the east, passing laterally into thick sandstones and conglomerates of the eastern outcrop belt (Martin, 1978). The carbonates occur in a mixed siliciclastic/carbonate interval approximately 500 m thick. One principle outcrop belt was studied by Martin and others (1980), who were able to recognize stromatolitic limestones, locally interbedded with rippled and mudcracked siltstone and sandstones, and coarse limestone breccia. Stromatolitic limestones (approximately 25 m thick) and interbedded clastics pass laterally into the coarse limestone breccia facies (approximately 50 m thick).

The stromatolitic facies belt consists of 22 shallowing-upward cycles (Martin and others, 1980) defined by alternation of 33 stromatolite beds and associated interstratified facies (Fig. 3c). Cycles are one to a few meters thick and extend laterally for at least 7 km. The lower unit of most cycles consists of well-laminated stromatolitic limestone and dolomitic limestone, with minor chert and rare argillaceous limestone. Stromatolites within the lower unit include crinkly laminated stratiform types, smooth-laminated domal types, and clotted varieties that form small columns. Single stromatolite types are generally specific to each cycle, and individual stromatolites may have synoptic relief of as much as 1 m. The clotted fabric consists of small, in places vertically elongate, masses of cherty and/or black earthy material that fills intercolumn spaces and occurs parallel to bedding (Martin and others, 1980).

The middle unit of many cycles consists of unlaminated limestone with well-preserved, upward-diverging, botryoidal crystal bundles and dark-weathering clots of chert. In thin section, the "crystals" are composed of a calcite mosaic that is slightly more coarse grained and "cleaner" than the adjacent groundmass. The "crystals" are regarded as calcite pseudomorphs after aragonite (Martin and others, 1980), or calcitized sulfates (Martin and others, 1980; Walter, 1983), and bear a striking resemblance to the *Atikokania*-type crystal fans of the Steeprock Group.

The upper parts of cycles consist of thin, well-laminated layers of dolomitic limestone that contains fine siliciclastic detritus. Some layers develop into small stromatolitic domes. The contact relations between the tops of cycles and bases of succeeding cycles are unspecified.

Martin and others (1980) propose a lagoonal origin for the cycles, in which relative water-level changes had a strong effect on each facies. Basal stromatolites are thought to have formed under restricted conditions in which the lagoon had limited exchange with the open ocean. The intermediate "fibrous crystal" facies is considered to have formed during a time of complete restriction of the lagoon, in which detrital influx was low and evaporitic conditions allowed precipitation of aragonite, or alternatively, gypsum. The upper part of each cycle is suggested to represent breaching of the lagoonal barrier and rapid influx of terrigenous sediment that smothered crystal growth.

The limestone breccia facies is only briefly mentioned by Martin and others (1980). Breccia clasts range in size from 2 to 20 cm, are subangular to subrounded, and consist largely of stromatolitic limestone and chert. The facies is considered to have formed adjacent to the cyclic facies as a possible back-barrier facies, produced during desiccation and brecciation of algal mats during exposure and subsequent reworking during barrier flooding events.

Significance.—The Cheshire carbonates are significant for a number of reasons. First, they display a diverse and complex assemblage of facies that is typical of younger platforms. Second, they exhibit strong, meter-scale cyclicity of facies in a shallowing-upward motif that is characteristic of both the Proterozoic (Grotzinger, 1986b,c) and Phanerozoic (James, 1984b) record. Third, they contain well-preserved pseudomorphs after possible marine carbonate precipitates, probably aragonite (supporting arguments are discussed below). Finally, the presence of thick sequences of large, angular carbonate breccia chips suggests that cryptocrystalline intraparticle early marine cementation also occurred.

PROTEROZOIC CARBONATES

Beginning with the oldest known major Proterozoic carbonate sequence, platforms were established with all of the important components of Phanerozoic platforms. Proterozoic platforms include both ramps and rimmed shelves, with many of the more specific variants, including ramps with fringing reefs, barrier reefs, and isolated buildups. Rimmed shelves show accretionary phases with flanking talus aprons, bypass phases including escarpments and gullied slopes, and development of intra-shelf basins. Sequences also are developed related to "incipient" as well as terminal drowning. Meter-scale cyclicity is evident in the oldest platforms. Platforms developed in all major tectonic environments, including stable cratons, rifts, passive margins, foreland basins, and possibly strike-slip basins.

In the following section, Proterozoic platforms are grouped (Fig. 4) and discussed in terms of the classification developed for Phanerozoic platforms by Read (1985). It is clear that this classification can be successfully applied to Proterozoic carbonates, thereby emphasizing that the fundamental controls on the gross stratigraphic architecture of platforms have been operating since at least 2.5 Ga.

Ramps

Ramps are probably the most abundant Proterozoic platform type owing to the large number of thin (few tens of meters) stromatolitic sheets intercalated within many siliciclastic shelf sequences. Several of the more substantial and better-documented Proterozoic ramps are shown in Figure 4.

As in Phanerozoic platforms, Proterozoic ramps are important precursors to the development of rimmed shelves (Grotzinger, 1986a; Beukes, 1987). They are often characterized by low gradients inherited from shallow antecedent topography of siliciclastic shelves, and show well-developed regional facies zonation from deep-ramp (below wave base) through shallow-ramp (intertidal) settings. Deeper ramp facies may include shale, concretionary shale, banded-iron formation, shale/carbonate rhythmite, laminated carbonate rhythmite, and uncommon rhythmite breccias. Shallower facies often are characterized by extensive stromatolite reefs of low local relief consisting of strongly elongate mounds that may be tens of meters in length, a few meters wide, and have up to a few meters of synoptic relief (Beukes, 1987; Ricketts and Donaldson, 1988; Pelechaty and Grotzinger, 1988). Generally, continued shallowing results in systematic decrease in the size of stromatolites in the case of ramps with fringing reefs (Logan and others, 1974; Beukes, 1987; Ricketts and Donaldson, 1988). For ramps with barrier reefs, however, the size of stromatolites may not change significantly (Pelechaty and Grotzinger, 1988). Tidal-flat facies often consist of small domal stromatolites, microbial laminites, and/or microbial tufas that form "microdigitate" stromatolites. The latter facies generally include the stromatolite types that have been designated as *Cryptozoon, Irregularia, Stratifera, Asperia,* and *Pseudogymnosolen*. Although stromatolites generally abound in

PROTEROZOIC PLATFORMS

	FRINGING REEFS	BARRIER REEFS	ISOLATED BUILDUPS
RAMPS	MONTEVILLE FM. (1) ----- REIVILO FM. (1) ----- UPPER KIMEROT GRP (7) ----- BASAL ROCKNEST FM. (8) ----- BELCHER GROUP (10)	BEECHEY FM. (2) ----- EAST RIVER FM. (13) **DISTALLY-STEEPENED** UTSINGI FM. (11) ----- LOWER NON-CYCLIC UNIT, ROCKNEST FM. (8)	LITTLE DAL (6) ----- DISMAL LAKES GRP. (14) ----- MIETTE GRP. (5)
	ACCRETIONARY	**BYPASS**	**INTRASHELF BASIN**
RIMMED SHELVES	REIVILO FM. (1) ----- KUUVIK FM. (3) ----- LOWER ROCKNEST FM. (8) ----- COWLES LAKE FM. (9) ----- TALTHEILEI FM. (11) ----- WILDBREAD FM. (11) ----- ABNER/DENAULT FMS. (12)	UPPER CAMPBELLRAND SGRP. (1) ----- MIDDLE TO UPPER ROCKNEST FM. (8) ----- CARBONATES OF GOURMA BASIN (4)	DENAULT FM. (15) ----- UPPER NON-CYCLIC UNIT, ROCKNEST FM. (8)

FIG. 4.—Classification of several Proterozoic carbonate platforms following the scheme of Read (1985). Sources of data: (1) Beukes, 1987; (2) Grotzinger and others, 1987, and Pelechaty and Grotzinger, 1988; (3) Cecile and Campbell, 1978; (4) Bertrand-Sarfati and Moussine-Pouchkine, 1983; (5) Teitz and Mountjoy, 1985, 1988; (6) Aitken, 1981, 1988; (7) Grotzinger and Gall, 1986, and Grotzinger and McCormick, 1988; (8) Grotzinger, 1986a; (9) Jackson, 1988; (10) Ricketts and Donaldson, 1981, 1988; (11) Hoffman, 1974, 1988a; (12) Donaldson, 1963, Wardle and Bailey, 1981, and Hoffman and Grotzinger, 1988; (13) Kerans and others, 1981, and Ross and Donaldson, 1988; (14) Kerans and others, 1981, and Kerans and Donaldson, 1988b; (15) Wardle and Bailey, 1981.

tidal-flat sequences, in some cases, stromatolite-barren mudflats may be developed as peritidal facies.

Ramps with fringing stromatolitic reefs.—

Beukes (1987) describes ramp facies of the Monteville Formation, Campbellrand Subgroup, South Africa (Fig. 5). Transitions between various facies occur on the scale of tens to hundreds of kilometers. A fringing, stromatolite reef passes landward into fenestral tidal-flat facies and basinward into ooid/intraclast grainstone, which interfingers with deeper ramp euxinic shales and banded sideritic iron formation. Conophyton-type stromatolites occur as an intermediate facies between grainstones and banded-iron formations in the lower part of the Monteville. The width of the platform-to-basin transition is on the order of 150 km for the lower Monteville and 75 km for the upper Monteville (Fig. 5), and water depths above the basin floor are judged to have been on the order of 60 to 80 m (Beukes, 1987). These constraints indicate that the average gradient was on the order of 50 cm/km for the earliest stages of the ramp, evolving with time and progradation to a value of 100 cm/km. These values are directly comparable to the low gradients reported by Read (1985) for Phanerozoic ramps.

Other Proterozoic fringing ramp sequences may include some units within the Reivilo Formation (Campbellrand Subgroup; Beukes, 1987), the Peg Formation (Goulburn Supergroup, Canada; Grotzinger and Gall, 1986; Grotzinger and McCormick, 1988), and the Tukarak through Laddie Formations (Belcher Group, Canada; Ricketts and Donaldson, 1981, 1988).

Ramps with barrier reefs.—

Stromatolitic barrier reefs are developed in the Beechey Formation (Goulburn Supergroup, Canada; Pelechaty and Grotzinger, 1988) and possibly the East River Formation (Hornby Bay Group, Canada; Ross and Donaldson, 1988). The Beechey Formation is an unconformity-bounded, shallowing-upward sequence of deeper ramp shales and siltstones grading upward into shallower ramp sandstones, oncolitic sandstones, and capped by stromatolite reef facies. Reefal facies grade basinward into laminated shales/siltstones and concretionary shales/siltstones (Fig. 6). Stromatolite reef facies grade cratonward into shallow-shelf hummocky cross-stratified dolarenites and dolosiltites with

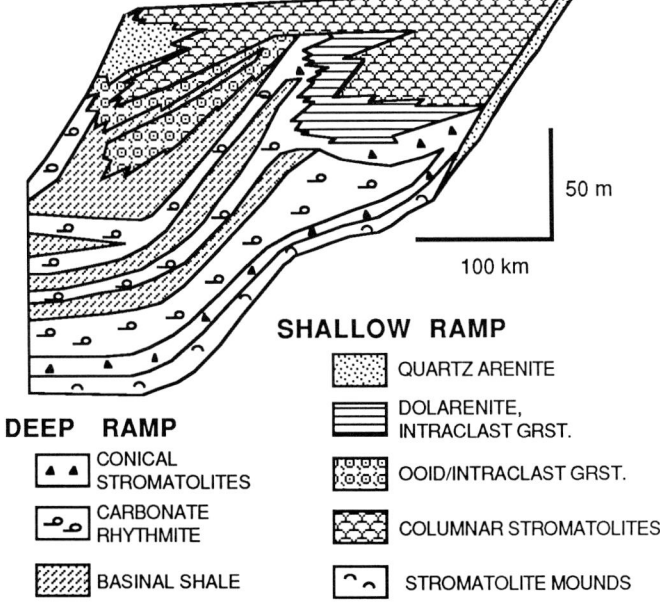

FIG. 5.—Stratigraphic cross section of Monteville Formation illustrating relations within a ramp containing fringing stromatolite reef and ooid facies. See text for description and Figure 13 for location of ramp facies with Campbellrand Subgroup. After Beukes (1987).

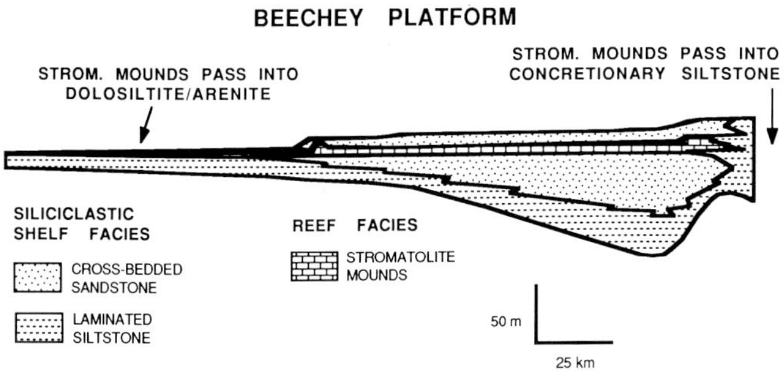

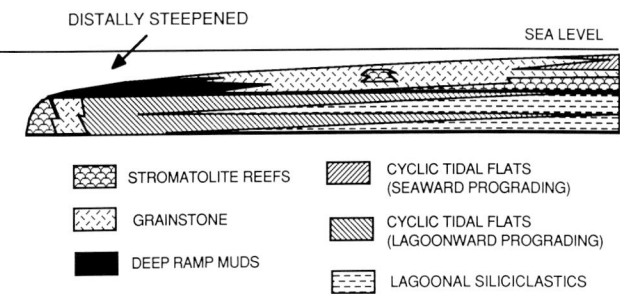

FIG. 6.—Stratigraphic cross section of Beechey Formation illustrating relations within a ramp characterized by a stromatolitic barrier reef. See text for description. After Grotzinger and others (1987) and Pelechaty and Grotzinger (1988).

FIG. 7.—Reconstructed depositional profile and stratigraphic relations within a distally steepened ramp (lower non-cyclic interval of Rocknest Formation), formed by drowning and backstepping of an earlier reefal rim. See text for description and Figure 12 for location of interval within cross section of Rocknest Formation. After Grotzinger (1986a).

minor shale and siltstone; tidal-flat facies are not observed, indicating the reefal complex was an isolated barrier.

Stromatolites of the Beechey ramp are developed in front and along the crest of a regional flexural arch that parallels the axis of a major foreland basin (Grotzinger and McCormick, 1988). The arch was probably a high during deposition of the Beechey Formation (Grotzinger and others, 1987) and may have promoted stromatolite growth in its vicinity.

In the East River Formation (Coppermine Homocline, Canada), biostromes of domal stromatolites may have formed a discontinuous barrier separating deeper ramp facies (mixed siliciclastics and carbonates?) from back-barrier ooid/intraclast grainstones and shallow subtidal/intertidal bioherms of digitate stromatolites (Ross and Donaldson, 1988). These facies grade landward into tidal-flat facies consisting of red mudstones and microbial tufas. The East River ramp developed during shallow marine onlap of a fluvio-deltaic coastal plain.

The basal Rocknest Formation is also an example of a barrier reef developed on a ramp (Grotzinger, 1986a, Fig. 16a).

Distally steepened ramps.—

Proterozoic examples of distally steepened ramps include the lower non-cyclic interval of the Rocknest Formation (Wopmay Orogen, Canada; Grotzinger, 1986a) and the Utsingi Formation of the Pethei Group (Athapuscow Basin, Canada; Hoffman, 1974, 1988a). In both cases, the ramps were formed following submergence of an earlier rimmed shelf, with deeper water, finer grained facies juxtaposed against underlying shallow-water stromatolite and grainstone facies.

In the lower non-cyclic interval of the Rocknest Formation, deep-ramp rhythmites directly overlie elongate stromatolites of an older barrier rim, and pass landward into a grainstone/bioherm facies in which isolated stromatolite patch reefs grew in a belt of intraclast/ooid grainstone (Fig. 7). These facies pass laterally into cyclic peritidal facies. The transition between the three major facies belts occurs over approximately 150 km, and ramp slopes were estimated to be on the order of 25–50 cm/km over the ramp top (Grotzinger, 1986a).

The Utsingi Formation of the Pethei Group consists dominantly of an unusual facies of probably stromatolitic, columnar limestone and dolomite (Hoffman, 1974, 1988a). The facies is characterized by closely-spaced subvertical columns, 1 to 5 cm in diameter, composed of sparry/micritic limestone having poorly defined convex-upward laminations. Adjacent intercolumn sediment consists of vertical rods of brown dolomite with well-defined concave-upward laminations. The facies is developed as monotonous intervals, devoid of other facies, tens of meters thick that are distinct only in the proportions of limestone to dolomite, and in the prominence of the columnar structure and bedding.

The columnar dolomitic limestone facies described above grades basinward, with gentle slope, into basinal limestone-argillite rhythmite, sandstone turbidites, and columnar stromatolitic marlstone (Fig. 8). The latter basinal facies consists of calcareous argillite that contains columnar to nodular, calcareous, conical stromatolites 1 to 3 cm in diameter. Shoreward equivalents of the Utsingi ramp facies are not preserved but may have been stromatolite reefs, as deduced from the paleogeographic juxtaposition of similar facies at other stratigraphic levels (Fig. 8).

Ramps with isolated buildups.—

Ramps with well-developed isolated buildups are known from the middle Proterozoic Dismal Lakes Group (Coppermine Homocline, Canada; Donaldson, 1976b; Kerans and others, 1981; Kerans and Donaldson, 1988a), from the late Proterozoic Little Dal Group (Mackenzie Mountains, Canada; Aitken, 1981, 1988), and from the late Proterozoic Miette Group (Windermere Supergroup, Canada; Teitz and Mountjoy, 1985, 1988).

*Dismal Lakes buildups.—*The Sulky Formation of the Dismal Lakes Group contains a variety of facies developed in upslope as well as downslope positions. A spectacular upslope buildup is formed by giant conophyton-type stromatolites developed as a core 25 to 30 m thick with as much as 12 m of synoptic relief on individual columns that have widths of a few meters. Giant conical stromatolites are succeeded by a sequence of giant domal stromatolites

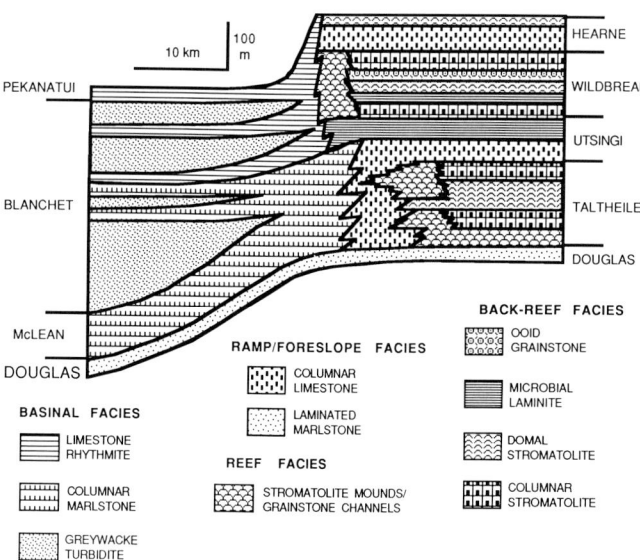

Fig. 8.—Stratigraphic cross section of Pethei Group. Note drowning of early accretionary rim (Taltheilei Formation) and development of distally steepened ramp facies (Utsingi Formation), followed by rim recovery (Wildbread Formation). Also note extensive development of stromatolitic (conical) carbonate in basinal, foreslope, and ramp settings. See text for additional descriptions. After Hoffman (1974, 1988a).

with widths of as much as 40 m and synoptic relief of 10 to 15 m. Giant domal stromatolites pass upward into smaller domal stromatolites that form laterally continuous sheets, marking the top of the buildup.

Areas adjacent to the conical columns are devoid of coarse sediment, and intercolumn fill is rare due to the extreme enveloping growth of individual stromatolites. There is little evidence for synsedimentary erosion of stromatolites, and the only fragmental debris in the buildup occurs adjacent to domes near the top. Masses of recrystallized fibrous marine cement as thick as a few tens of centimeters are developed locally throughout the buildup as early fill between columns (Kerans, 1982; Kerans and Donaldson, 1988b).

It is unlikely that much detrital sediment was utilized in the construction of the stromatolites, given the morphology of the depositional surface during the conical-column stage of the buildup (Donaldson, 1976b). This surface consisted of nested pinnacles and inverse cone-shaped depressions with very steep margins extending as far as 10 m; the absence of intercolumn detrital fill indicates that sediments were incapable of penetrating far beyond the margins of the buildup. Therefore, allochthonous carbonate within the stromatolites would have been restricted to a small component derived from settling of suspended micrite. Petrographic work, however, indicates that stromatolitic laminae are generally recrystallized to massive textures, but relic areas contain lamina-perpendicular radial fibrous crystals, suggesting that the primary layering may have formed by precipitation (Kerans, 1982). On this basis, the most probable mechanism for stromatolite growth would have been by precipi-

tation, possibly induced by the photosynthetic activity of the local benthic biocenosis. Walter (1983) has stressed the evidence in favor of conical stromatolites having been constructed by a distinct phototactic community of possibly gliding microbes.

Downslope buildups of the Dismal Lakes Group (Fig. 9) are located in a more basinal position relative to the main shallow-ramp buildup (Kerans and Donaldson, 1988b). These buildups are 2 to 40 m thick, forming bioherms and biostromes enclosed in deep-water dololutite units. The internal structure of most buildups is characterized by a curious assemblage of concave-up, dish- or bowl-shaped "algal plates" ranging in size from 0.1 to 2 m. Algal plates rest within a micrite matrix or are separated by fibrous dolomite cement crusts. Locally, the cement crusts compose over 50 percent of the bioherms.

Little Dal Pinnacle Reefs.—Isolated buildups of the Little Dal Group (Mackenzie Mountains, Canada; >.77, <1.2 Ga) are true pinnacle reefs, attaining heights of 300 m above their foundation, with horizontal dimensions of 1 to 8 km and synoptic relief of >50 m (Aitken, 1981, 1988). In scale, the buildups are comparable to the Silurian pinnacle reefs of the Illinois and Michigan basins (Lowenstam, 1950; Mesolella and others, 1974). The Little Dal reefs are contemporaneous with a surrounding basinal facies that is equivalent to an upslope platform, and were probably established on a regionally extensive oolite sheet (Oolite Member of Mudcracked Formation; Aitken, 1981). As the oolite sheet subsided, a platform and basin were formed, with local highs providing the substrate for growth of the pinnacle reefs. Basinal facies that enclose the reefs consist of nodular limestones, and thin-bedded limestone-shale/siltstone rhythmites, interpreted as deep-water, fine-grained turbidites (Aitken, 1981). Rhythmites abut bioherms of the reef core and interfinger with tongues of coarse reef talus.

The reef core is built of stromatolites (Fig. 10) and possible calcareous algae (Aitken, 1988). Columnar and domal stromatolites have been recognized in addition to new forms, similar to stromatolites, but designated informally as "Gaynia" and "Parastratifera" (Fig. 10). Aitken (1988) empha-

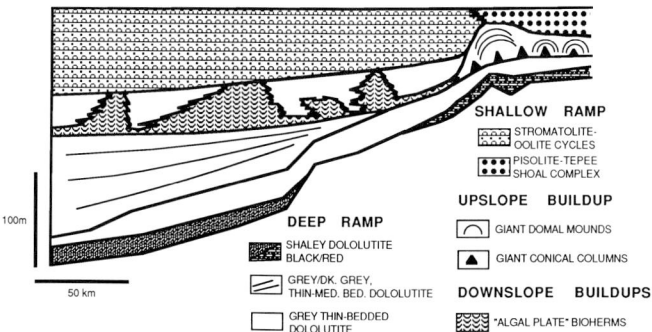

Fig. 9.—Stratigraphic cross section of Sulky Formation (Dismal Lakes Group). Note development of giant conical stromatolite bioherms followed by giant domal bioherms as an upslope buildup and "algal plate" bioherms as downslope buildups. See text for description. After Kerans and others (1981) and Kerans and Donaldson (1988b).

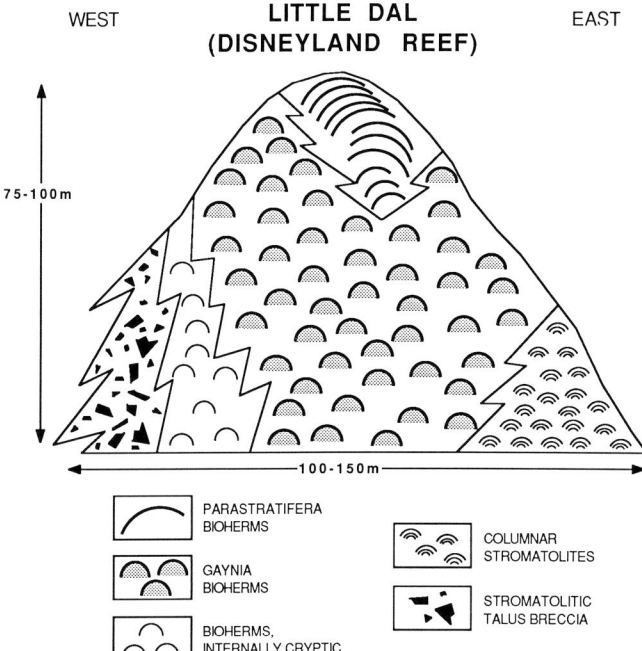

FIG. 10.—Schematic cross section of "Disneyland reef," Little Dal pinnacle reef complex. Cross section is constructed from map data of Aitken (1988, fig. 5). See text for description.

sizes that the microfabric of Gaynia and Parastratifera is "not typically stromatolitic" and notes that sediment trapping may not have been the dominant process in their formation (thereby implying that precipitation is). Gaynia is reported as a digitate form, characterized by a cellular fabric that includes thin-walled tubes of about 100 μm in diameter, suggesting a resemblance to younger calcareous algae. The framework of Gaynia had much pore space that has been occluded by infilling of internal sediment and precipitation of cement. A second variety of Gaynia apparently is characterized by a more poorly developed, non-cellular microstructure. Gaynia formed bioherms ranging to as much as 8 m in diameter.

Parastratifera displays "clearly spaced, tissue-thin laminae of dark microcrystalline calcite joined by complete or incomplete, somewhat irregular septae or walls of the same material, resembling ladder rungs" (Aitken, 1988, p.). The fabric may be formed by a meshwork of tubes, on the order of 125–160 μm in diameter. The cellular fabric is filled with sparry calcite and lacks internal sediment. Parastratifera may be more common in the upper parts of the reefs (Fig. 10) and tends to "encrust" other bioherms and steep to vertical surfaces with relief of several meters.

Columnar and domal stromatolites are interspersed throughout the framework of reefs (Aitken, 1988). Columns are branched and unbranched, with and without wall structure, and domes are commonly well laminated with meter-scale lateral linkage.

The general organization of the pinnacle reefs consists of stacks of bioherms with adjacent fill; subordinate large domes and steep-dipping stromatolitic sheets; and a mantle of boulder-size reef talus (Fig. 10). Bioherms are as much as several meters high with similar widths and have steep margins with relatively flat tops. Inter-bioherm material is thin-bedded argillaceous lime mud and silt similar to the rhythmites that enclose the reefs (Aitken, 1988). Large domes (excluding "domal" stromatolites) are present and have diameters of as much as 10 m with lower relief than bioherms and are almost exclusively formed of Parastratifera. Parastratifera domes are restricted to the upper parts of reefs (Fig. 10). In some localities, masses of Parastratifera are at least 10 m thick and encrust reef flanks, showing primary inclinations of 40° and depositional relief of at least 10 m (Aitken, 1988).

Early (probably marine), isopachous, fibrous calcite cements are ubiquitous within the reefal framework (Aitken, 1988). Cements are uncommonly interlaminated with geopedal internal sediment and may be followed by final pore-occluding blocky spar cements.

Yellowhead Platform.—The Yellowhead platform (Miette Group, Canada) is developed on and interfingers with wave-rippled and cross-laminated arkosic sandstone and laminated mudstone (Fig. 11; Teitz and Mountjoy, 1985, 1988). The platform has two basic subdivisions; a lower unit of stromatolite bioherms and related talus, and an upper, dominant unit of pisolitic and intraclastic grainstone. The stromatolite buildups are isolated and their lower parts consist of lenticular to dome-shaped mounds that average 10–15 m thick, ranging to 30 m, and are spaced 5 to 20 m apart. The upper parts of the buildups consist of stromatolitic units as thick as 40 m with laterally coalesced and vertically superimposed mounds. The buildup tops consist of thickly bedded, laterally continuous stromatolitic lamination with crinkly to wavy-laminated textures.

The pisolitic/intraclastic facies composes over 70 percent of the platform (Teitz and Mountjoy, 1985, 1988). It is 65 to 300 m thick and has a lower grainstone unit with trough to planar cross-stratification, but passes upward into interbedded grainstone and increasing wackestone. Thin stromatolitic layers are also present. The uppermost part of the platform is a paleokarstic surface beneath unconformably overlying Cambrian sandstones.

Rimmed Shelves

Proterozoic platforms showing phases of rimmed-shelf development are common and include both accretionary and

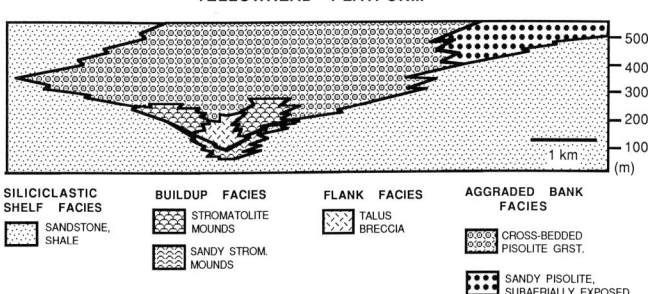

FIG. 11.—Stratigraphic cross section of carbonate facies within Miette Group (Yellowhead platform). See text for description. After Teitz and Mountjoy (1985, 1988).

bypass types (Fig. 4). Proterozoic rimmed shelves are characterized by development of linear belts of stromatolite barrier reefs that isolated inner-shelf lagoons from the deep ocean. These reefal rims show evidence of significant increase in slope into adjacent foreslope and basinal areas and in some examples constructed vertical escarpments with hundreds of meters of relief.

Accretionary margins.—

Possible examples of Proterozoic platforms with accretionary margins are shown in Figure 4. Shelf-edge stromatolite barrier reefs containing minor grainstone generally pass downslope into periplatform carbonate rhythmites, rhythmite breccias, and rare allodapic megabreccias. Basinal facies generally consist of shale, concretionary shale, and carbonate/shale rhythmites. Periplatform or foreslope facies generally intertongue with, rather than abut, shelf-edge reef facies similar to Phanerozoic accretionary margins (Read, 1985).

Facies transitions "landward" of the reefal rim are often highly variable. In some platforms reefal facies pass laterally, through minor grainstone, into shoal-complex facies consisting of peritidal small domal stromatolites, microbial laminites, microbial tufas, and disrupted zones with tepees (Rocknest Formation; Grotzinger, 1986a,b). Shoal-complex facies may then pass into areally extensive, but environmentally restricted lagoonal facies. In some platforms, reefal facies pass into laterally extensive back-reef biostromes (Taltheilei Formation; Hoffman, 1974, 1988a). In others, shelf-edge reef facies may pass abruptly into narrow lagoonal facies, and then into prograding alluvial fan facies (Kuuvik and Brown Sound Formations; Cecile and Campbell, 1978). Finally, where reefal barriers may have been highly discontinuous, thick belts of grainstone are developed as high-energy back-reef facies (Cowles Lake Formation; Jackson, 1988; Wildbread Formation; Hoffman, 1974, 1988a).

Some accretionary margins evolved from earlier ramps developed on siliciclastics, as in the case of the lower Rocknest Formation (Grotzinger, 1986a) and the upper part of the Reivilo Formation (Beukes, 1987). Other accretionary margins developed from ramps associated with phases of incipient platform drowning, such as the upper platforms of the Rocknest Formation (Grotzinger, 1986a,b), and the Wildbread Formation (Hoffman, 1974). Finally, some apparently developed as the first phase in the development of the platform, apparently bypassing a preliminary ramp phase (Kuuvik Formation; Cecile and Campbell, 1978; Jackson, 1988). In the case of the Rocknest Formation and the Campbellrand Subgroup, accretionary margins evolved into steep bypass margins.

Bypass margins.—

Bypass margins are developed in at least two major Proterozoic platforms: the Campbellrand Subgroup (Transvaal Supergroup, South Africa; Beukes, 1987) and the Rocknest Formation (Grotzinger, 1986a). In contrast to most other Proterozoic platforms discussed so far, these platforms attain significant thickness, as much as 1 km for the Rocknest and 1.5 km for the Campbellrand. Development of bypass margins occurred as the most "mature," aggradational phase in each platform, following earlier progradational phases of ramp and accretionary margin growth.

*Rocknest Formation: Escarpment Margin.—*The top of the Rocknest platform (Fig. 12) had as much as 600 m of relief above the top of the adjacent talus apron formed at the base of what probably was, at least locally, a near-vertical escarpment (Grotzinger, 1986a). This apron tapers basinward, and the relief of the top of the platform relative to the more distal slope environments was on the order of 1,000 m.

Slope facies are consistent with the evidence for a bypass margin of the escarpment type. In addition to shale, laminated rhythmites with slumps and slide scars, and rhythmite breccias, the upper portion of the Rocknest apron is characterized by spectacular megabreccias composed of shelf-edge lithologies. Individual debris sheets are 2 to 75 m thick, generally massive internally, and composed of blocks that are as long as 150 m. Most blocks are composed of stromatolitic shelf-edge reef facies, with minor grainstone breccia; small blocks of massive marine cement also occur.

In addition to the makeup of slope facies, Rocknest reefal rim successions may also reflect development of an escarpment. Neptunian dikes and sills have been shown to cut the upper part of the thickest (lower) Rocknest rim sequence, which aggraded during the most prolonged phase of escarpment development. It may be possible that gravitationally induced fracturing and failure of the rim occurred during phases of rapid rim upbuilding. Swarms of neptunian dikes may have formed at that time, leading to weakening and spalling of coherent portions of the rim downslope.

*Campbellrand Subgroup: Gullied Slope?—*The top of the Campbellrand platform (Fig. 13) may have had as much as 950 m of relief above the adjacent basin floor at the time of terminal drowning. Beukes (1987) describes basinal facies consisting of carbonaceous shale, banded chert, carbonate-type iron formation, mafic tuff, and thin-bedded carbonate rhythmites. Slope facies consist of Conophyton-

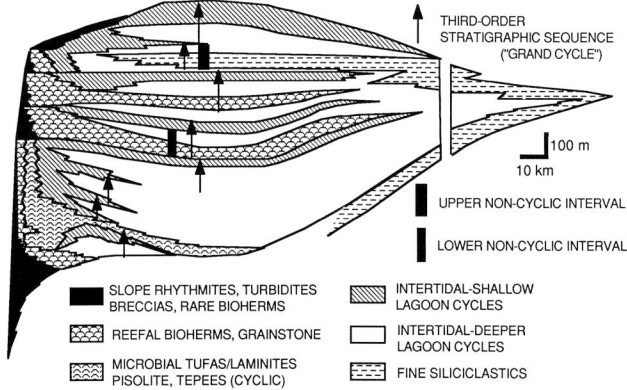

FIG. 12.—Stratigraphic cross section of Rocknest Formation, palinspastically corrected for tectonic shortening. Note location of lower noncyclic interval (distally steepened ramp; see Fig. 7) and upper non-cyclic interval (=intra-shelf basin; see Fig. 14a). After Grotzinger (1986a).

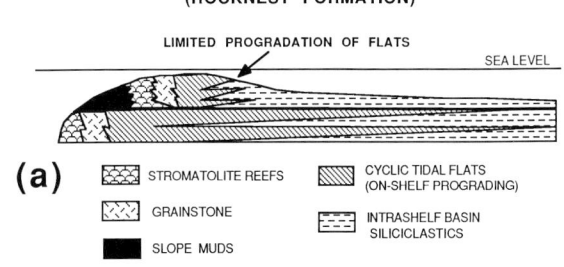

FIG. 13.—Stratigraphic cross section of the Campbellrand Subgroup. Note development of conical stromatolite facies in basinal and foreslope settings. Also note location of Monteville Formation, which contains fringing ramp facies illustrated in Figure 5. After Beukes (1987).

type stromatolites interbedded with rhythmites and chaotic rhythmite breccias. No allodapic megabreccias are described, but there is evidence for downslope transport of grainstone and possible fine fragmental stromatolitic debris. Shelf-edge facies consist of stromatolite mounds and columns that formed a reefal barrier, behind which extensive lagoonal and tidal-flat facies developed.

The reconstructed platform shows much evidence for vertical aggradation and upbuilding of the shelf edge relative to the adjacent basin (Fig. 13). Progradation of the margin was only important during the earliest phases of its evolution. On the basis of the relations described earlier, particularly the lack of allodapic breccias, it is inferred that the Campbellrand platform probably lacked escarpments but may have been characterized by gullied-slope systems that contained downslope allochthonous debris in discrete channels, rather than sheets.

Another example of a possible bypass margin with gullied slopes may be in carbonates of the Gourma Basin, west Africa (Bertrand-Sarfati and Moussine-Pouchkine, 1983). Stromatolitic reefal rim facies pass downslope into fine-grained carbonate rhythmites containing lenses of rhythmite breccia as well as allodapic breccia. Wedges of coarse megabreccia are absent, and allochthonous carbonates are restricted to discontinuous (parallel to shelf strike) lenses.

Intra-shelf basins.—

Intra-shelf basins developed within rimmed shelves are known from the Rocknest Formation (Wopmay Orogen, Canada; Grotzinger, 1986a,b), and the Denault Formation (Labrador Trough, Canada; Wardle and Bailey, 1981; Hoffman and Grotzinger, 1988). Both platforms have well-developed rims consisting of reefal stromatolites, flanked by back-reef grainstones, and cyclic peritidal facies developed a narrow shoal complex. Although the architecture of the Denault rim and shoal complex is less well known than the Rocknest rim/shoal complex, the overall assemblage of facies is very similar.

The Rocknest intra-shelf basin is represented by the upper non-cyclic interval described by Grotzinger (1986a,b). During development of this member, the reefal rim and shoal complex were very narrow (Fig. 14a), and cyclic progradation of tidal flats away from the rim, toward the craton, was not as extensive as in overlying and underlying members (Figs. 12, 14a). As much as 70 m of "non-cyclic" lagoonal facies were deposited, consisting of argillaceous dololutite, intraclast packstone, ooid packstone, and hummocky cross-stratified very fine sandstone, siltstone and dolosiltite. These sediments represent prolonged subtidal sedimentation near wave base and pass basinward into cyclic tidal-flat facies of the narrow shoal complex. Intra-shelf basin facies are probably equivalent to siliciclastic shallow-shelf and fluvio-deltaic sediments along the innermost part of the shelf (Burnside Formation; McCormick and Grotzinger, 1988).

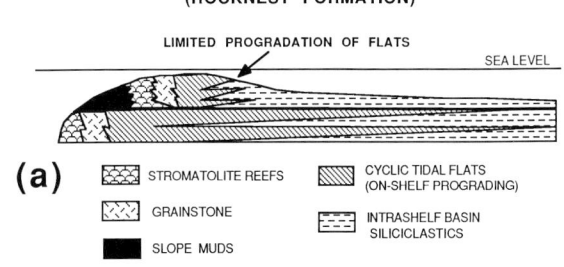

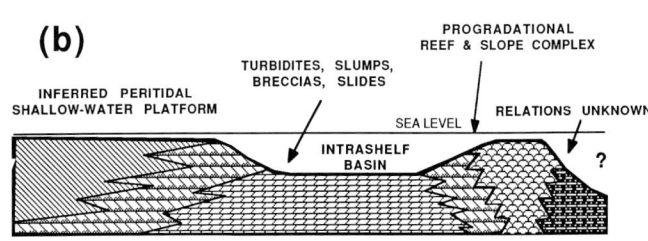

FIG. 14.—Intra-shelf basin facies. (a) Reconstructed depositional profile and stratigraphic relations for the upper non-cyclic interval of the Rocknest Formation. Following rim backstepping and recovery, fringing tidal flats did not prograde extensively as in underlying members of the Rocknest Formation (see Fig. 12); this produced an intra-shelf basin. After Grotzinger (1986a). (b) Inferred reconstruction and possible relations within the Denault Formation. The platform is poorly documented, but limited evidence suggests it may have contained a major intra-shelf basin, perhaps continuous for 600 km along strike. After Wardle and Bailey (1981) and Hoffman and Grotzinger (1988).

The Denault intra-shelf basin facies (Fig. 14b) consist of dololutite, dolosiltite, and minor shale (Wardle and Bailey, 1981). Slump folds, detachment scars, and meter-scale synsedimentary overthrust structures are common. These facies pass into outer-shelf progradational reefal rim and shoal-complex facies (Wardle and Bailey, 1981; Hoffman and Grotzinger, 1988). Intra-shelf basin facies pass cratonward into dololutites containing graded dolarenite units and lenses of coarse (allodapic?) breccia. This suggests the former presence, subsequently stripped by erosion, of a shallow-water platform on the cratonic side of the intra-shelf basin (Fig. 14b; Wardle and Bailey, 1981).

CONTROLS ON PLATFORM DEVELOPMENT

Tectonic Controls

Tectonic controls on the development of Precambrian carbonate platforms are divided into long- and short-term processes. Long-term processes involve the development of cratonic masses, or continental platforms, and what effect this had on the production of carbonate veneers. Short-term processes include continental extension, thermal subsidence, collision and flexural subsidence, the general evolution of orogens, and their influence on the evolution of carbonate platforms.

Archean/Proterozoic "transition".—

The growth of large cratonic masses of continental lithosphere, often associated with the Archean/Proterozoic "transition" (Veizer and Compston, 1976) was probably the single greatest event in the evolution of carbonate platforms. This transition was, on a global basis, highly diachronous and is reflected in the history of carbonate platforms. Some of the Earth's oldest cratons (e.g., Kaapvaal Province) are located in South Africa (Tankard and others, 1982) on which the first occurrence of cratonic carbonate sedimentation (approximately 3.0 Ga) is recorded (Von Brunn and Mason, 1977). Significantly, this same craton was also the site of the oldest major early Proterozoic carbonate platform (Campbellrand-Malmani platform, >2.3, <2.6 Ga; Beukes, 1987; Eriksson, 1977). In contrast, the oldest cratons of the Canadian shield (e.g., Slave and Superior Provinces) are significantly younger than those in South Africa (Hoffman, 1988b) and the oldest cratonic carbonates are approximately 2.7 Ga (Steeprock Group; Wilks and Nisbet, 1985). Similarly, the oldest well-developed carbonate platforms of comparable size to the Campbellrand-Malmani platform are only about 1.9 Ga (e.g., Rocknest Formation; Grotzinger, 1986a).

Stable cratonic environments were critical to growth of carbonate platforms, as shown by their rapid appearance in the early Proterozoic. Significantly, the first Proterozoic platform of regional extent (Campbellrand-Malmani), perhaps the most prominent in all the Proterozoic, was almost of Archean age. On this basis, it is suggested that the chemical and biological potential for significant carbonate production may have been present during the Archean, but the growth of major platforms was suppressed until substantial continental masses developed. These created spacious, stable (or smoothly subsiding), shallow-water platforms on which carbonates formed. Because this process was diachronous on a global scale, the initial development of carbonate platforms was also diachronous.

It should be noted that the Archean record cannot be freely interpreted without consideration of possible preservational biases. This problem has been discussed by Knoll (1984), who emphasizes that shelf-type sediments are present in "high-grade" (metamorphic) as well as "low-grade" (greenstone belt) terrains, but that only sediments of the latter provide useful sedimentologic information due to better fabric preservation. On this basis, it cannot be discounted that other (perhaps more extensive?) carbonate sequences, in addition to siliciclastic sediments, were the protoliths in some "high-grade" terrains. This is not considered to be a serious problem, however, given that the geodynamic settings in which carbonates form most extensively are generally not overwhelmed by high thermal regimes during orogeny (e.g., cratons, passive margins).

Subsidence regimes and basin development.—

Proterozoic carbonate platforms were subject to the same short-term tectonic controls that have regulated the periodicity and growth of Phanerozoic platforms. Episodes of continental extension and compression produced a diversity of sedimentary basins, each characterized by differing styles of subsidence including active crustal thinning (block faulting), passive subsidence following extension, and flexural subsidence associated with tectonic loading. It can be shown that the style of basin subsidence often influenced the development and distribution of facies, especially the position of the shelf edge in passive-margin and foreland-basin platforms.

Proterozoic strike-slip basins have been documented, but are known to contain only minor lacustrine marls (Aspler and Donaldson, 1986) with rare occurrences of calcareous stromatolites (Hoffman, 1976). These occurrences are not discussed further.

*Passive-Margin Platforms.—*The better understood Proterozoic orogenic belts indicate that the most prolific growth of carbonate platforms occurred during inferred stages of passive subsidence (probably mostly thermal in form) following episodes of continental extension.[2] This geodynamic setting produced the thickest and most extensive platforms. Carbonates may have formed on passive margins attached to oceanic lithosphere (Rocknest platform; Hoffman, 1973; Hoffman and Bowring, 1984), or on attenuated continental crust in back-arc or other marginal basins (Kimerot platform; Grotzinger and McCormick, 1988), or on minimally stretched crust in entirely intracratonic settings (Pethei platform; Hoffman and others, 1974). Most rimmed shelves of the accretionary type and all known bypass margins developed in this environment, similar to Phanerozoic analogs (cf. Read, 1982).

---

[2]An inference that passive or thermal subsidence was effective in the development of a sedimentary basin requires only that cooling of heated and extended (thinned) continental lithosphere occurred. It does not specify the actual amount of lithospheric extension, nor imply that seafloor spreading and creation of oceanic crust necessarily took place.

The location of platform margins in passive-margin settings was probably influenced by larger scale crustal structural patterns created during rift-related extension. For example, the position of the Rocknest shelf edge was controlled by the location of the underlying siliciclastic shelf edge, which was, in turn, controlled by the transition zone between unfaulted and faulted upper crust (Grotzinger, 1986a). Neither shelf edge prograded significantly beyond the location of the first few fault blocks related to initial extension, and in the case of the Rocknest platform, the shelf margin shifted from progradation to aggradation at the increase in depositional slope formed by antecedent topography of the underlying siliciclastic shelf break. This process has been documented for Phanerozoic platforms where the volume of periplatform sediment must be progressively increased in order for continued progradation of the platform top to occur. When the rate of carbonate production becomes insufficient for the slope to prograde effectively, then shelf-margin facies must build up rather than out (Schlager and Camber, 1981), resulting in aggradation and creation of bypass margins (Read, 1986).

Examples of other possible passive-margin carbonates include the Campbellrand-Malmani platform (Tankard and others, 1982; Beukes, 1987), the Fairweather-Mcleary platform (Belcher Group, Canada; Ricketts and Donaldson, 1981), and the Gourma carbonates of west Africa (Bertrand-Sarfati and Moussine-Pouchkine, 1983). Some of the thick Riphean (middle to late Proterozoic) carbonate-bearing sequences of the Siberian Platform/Baikal Fold Belt (Salop, 1983) also may have formed in thermally subsiding basins.

Foreland-Basin Platforms.—Platform development also occurred in foreland basin settings, subsiding flexurally in response to continental convergence. During growth, platforms were of more limited extent and included many ramps and some accretionary rimmed shelves. The low-platform slopes often permitted extensive offlap and vertical stacking of facies belts so that basinal facies may be overlain by shelf/shallow-ramp facies.

Examples of progradational foreland-basin ramps include the Cowles Lake platform (Wopmay Orogen, Canada; Hoffman, 1973; Jackson, 1988), the Beechy platform (Grotzinger and McCormick, 1988; Pelechaty and Grotzinger, 1988), the Kuuvik platform (Cecile and Campbell, 1978; Grotzinger and McCormick, 1988). In these platforms, basinal or deeper ramp facies are typically overlain by shallow-ramp facies. The Tukarak-Mavor platform of the Belcher Group (Ricketts and Donaldson, 1981, 1988), however, is an example of an onlap package where basinal facies overlie shallow-ramp facies; this sequence is similar to the lower part of the Middle Ordovician ramp sequence, southern Appalachians (Read, 1980; Ruppel and Walker, 1984).

The differential subsidence of foreland basins probably exerted a major control on the location of platform edges. For example, the edge of the Beechey platform (Grotzinger and McCormick, 1988) closely coincides with the zone of most rapid thickening of strata into the basin axis (Grotzinger and others, 1988), suggesting that an increase in tectonic subsidence rate in that area prohibited outbuilding of the platform beyond that point. This effect, in concert with other factors such as probably restricted seawater circulation, may have reduced the rate of carbonate accumulation so that the platform was overwhelmed by the rate of subsidence-related deepening.

Rift-Related Platforms.—The thinnest and least well-developed platforms developed in extensional (rift) settings where the intermittent influx of siliciclastic sediment, related to episodic movement of fault blocks, restricted their areal extent. Episodic fault movement also caused uplift and exposure of platforms, which would have terminated carbonate production. Examples include the late Proterozoic Yellowhead, Astoria, and "Dogtooth" platforms (Teitz and Mountjoy, 1985, 1988; Poulton, 1973, 1988), the 1.2 to 1.4-Ga Beck Spring platform (Pahrump Group, western United States; Gutstadt, 1968; Marion and Osborne, 1980), the approximately 1.2-Ga Mescal platform (Apache Group, western United States; Shride, 1967), and the various carbonate units of the approximately 1.2-Ga Borden Basin (Baffin Island, Canada; Jackson and Ianelli, 1981, 1988).

It is suggested that, as with Phanerozoic platforms, the type of basin in which Proterozoic carbonates developed had a strong control on their growth, diversification, zonation, and expansion. The most stable, probably thermally subsiding, basins were conducive to major platform development. Rates of siliciclastic influx were comparatively low, and relative changes in sea level would have been minimized. Craton interiors also would have been excellent sites for carbonate accumulation, but sequences were probably thin and most have been subsequently stripped due to postdepositional uplift of most shield areas.

In contrast, tectonically active, episodically subsiding rift or foreland basins were associated with siliciclastic influx, which inhibited diversification and expansion of most platforms. Periodic rifting or thrust loading resulted in pulses of rapid subsidence and/or uplift, testing and often exceeding the capacity of the system to produce carbonate. Subaerial exposure surfaces and karsts (Beeunas and Knauth, 1985; Teitz and Mountjoy, 1985, 1988; Grotzinger and others, 1987), in addition to drowning (Hoffman, 1987; Grotzinger and McCormick, 1988) and "smothering" by influx of siliciclastics (Cecile and Campbell, 1978; Jackson, 1988) record platform destruction under those conditions.

Eustatic Controls

The possible effects of eustasy have been related to the development of stratigraphic sequences and parasequences (Grotzinger, 1986b,c), incipient and terminal drowning of platforms (Grotzinger, 1986a; Beukes, 1987), and subaerial exposure and karst development (Kerans and Donaldson, 1988a; Pelechaty and others, 1988).

Third-Order Sequences.—Depositional sequences occur in the Rocknest platform (Grotzinger, 1986a,b) that are similar to the "Grand Cycles" of Aitken (1978). These are third-order, onlap/offlap sequences expressed by rim drowning and backstepping near the platform shelf edge and by systematic changes in the arrangement of cyclic facies in parasequences of the inner platform (Fig. 12). At least one of these sequences ("lower non-cyclic interval") is known to be correlative over a broad area, greatly exceeding the width of the platform, and affecting depositional systems tracts in other time-equivalent basins (Grotzinger, 1986a; McCormick and Grotzinger, 1988). In this example, open-shelf grainstone, packstone, and reefal stromatolite facies were deposited over restricted shelf-interior cyclic facies (Figs. 7, 12) following a likely rise in sea level. With time, the inner shelf recovered and cyclic facies prograded over open-shelf facies, establishing a new reefal rim in a back-stepped position; the previous rim was terminally drowned and draped with deep-ramp rhythmites. Third-order sequences also appear to be present in the Campbellrand Subgroup (Beukes, 1987), the Pethei Group (Hoffman, 1974, 1988a), and the Dismal Lakes Group (Kerans, 1982).

Prolonged subaerial exposure that is possibly related to eustasy occurred during development of the Dismal Lakes platform (Kerans, 1982; Kerans and Donaldson, 1988a). Exposure resulted in development of a major karst horizon that probably extended at least 400 km into the correlative Parry Bay Formation of the Bathurst Inlet area (Kerans and Donaldson, 1988a; Pelechaty and others, 1988). The karst is thought to be predominantly a eustatic feature based on the evidence for its regional extent on the order of 400 km, but locally modified by minor block faulting (Kerans, pers. commun., 1987; Kerans and Donaldson, 1988a).

Features associated with the karst are described by Kerans and Donaldson (1988a). These include collapse breccias, cross-stratified cave-floor sediments, grike systems, fibrous flowstones, cave popcorn, and regolith. Similar features have been noted along strike in the correlative Parry Bay Formation (Campbell, 1979; Pelechaty and others, 1988).

Parasequences.—Parasequences, or small-scale shallowing-upward cycles of the Rocknest platform, are considered to be eustatic in origin based on evidence including: subaerial exposure and sediment disruption at the tops of many cycles; incomplete shallowing of many cycles; extreme lateral continuity of cycles and individual beds on a regional scale; occurrence of average cycle period within the Milankovitch band; and close comparison of actual stratigraphic data with results of forward modeling using a eustatic driving mechanism (Grotzinger, 1986b,c).

It should be noted that these parasequences, or meter-scale shallowing-upward cycles, are important in many Proterozoic carbonate platforms (Grotzinger, 1986c) and even late Archean platforms (Martin and others, 1980). Therefore, the mechanism responsible for the cyclicity has been operative since the inception of the first major carbonate platforms. The eustatic model, perhaps governed by Milankovitch climate forcing and sea-level changes, is strengthened by the results of forward modeling (Read and others, 1986), and the establishment of stratigraphic equivalence between cycles with and without erosional caps (Grotzinger, 1986b).

One of the most critical contributions toward establishing the potential connection between Milankovitch forcing and platform cyclicity has been the application of time-series analysis to cyclic sequences. In Phanerozoic sequences, this method demonstrates that the distribution of cycle periods occurs in discrete clusters that closely correspond to the predicted Milankovitch periods (Olsen, 1984; Goldhammer and others, 1987; Hinnov and Goldhammer, 1988). With regard to Precambrian strata, preliminary results of time-series analysis on Rocknest cyclic carbonates indicate that cyclicity is also distributed in discrete periods, and that within the errors associated with dating, the periods may well be close to the Milankovitch band (J. Grotzinger and A. Holmes, unpubl. data).

Precambrian Sea Water and Carbonate Production

Changes in the composition of Precambrian sea water, particularly factors influencing carbonate equilibria, have been discussed by Ronov (1968), Holland (1972, 1984), Walker (1983), Kempe and Degens (1985), and Kasting (1987). In general, it is thought that no great changes have occurred to significantly alter the precipitation of chemical sediments relative to Phanerozoic and modern analogues. Holland (1972) employed a method that brackets the likely range in seawater compositions through geologic time based on constraints imposed by the assemblage of preserved and inferred evaporative minerals in the stratigraphic record. The relative abundance of minerals in marine evaporites is used to set limits on excursions in the composition of sea water. The apparent similarity of the mineralogy of marine sediments during the last 3,500 Ma suggests that the chemistry of sea water has been equally conservative (Holland, 1972). See Kempe and Degens (1985), however, for an alternative view.

Faith in the apparent similarity between Precambrian and Phanerozoic sea water is strongly dependent on the acceptance that Precambrian evaporative sequences bear evidence for precipitation of carbonate, followed by gypsum and then halite. This relation is partially supported by the Precambrian rock record, which contains preserved mineral assemblages or pseudomorphs of "vanished" assemblages (Walker and others, 1983). These relations have not been firmly established, however, particularly for the Archean and early Proterozoic record. Reports of "evaporite" pseudomorphs, most notably gypsum, are often unsubstantiated by reliable criteria. Furthermore, the context in which documented pseudomorphs occur is seldom discussed in terms of paleoenvironments, and also to what extent the mineralogy may have been influenced by influx of non-marine waters (*sensu* Hardie, 1984).

In the rare instance where it has been possible to document convincingly the former presence of primary, probably evaporative gypsum (now preserved as barite; Buick and Dunlop, 1987), it is not clear whether or not this occurrence was marine or lacustrine. Furthermore, if the occurrence was marine, it is likely that there was a cosiderable influence on the local composition of sea water through influx of "continental"-derived calcium-rich waters from erosion of highly basaltic surrounding source areas. As pointed out by Hardie (1984), even if a marine depositional setting is established independently, say on sedimentologic evidence from enclosing deposits, the hydrologic restriction required to precipitate salts ". . . cannot fail to put a strong non-marine stamp on both the geochemistry and the sedimentology of the deposit" (Hardie, 1984, p. 199). Consequently, evaporite deposits interlayered with marine sediments may not necessarily yield the correct sequence of evaporite minerals predicted by precipitation from normal sea water.

In addition to the barite- and silica-replaced gypsum in the Warawoona Group, the only other widely cited reports of evidence for former Archean gypsum are the large, fibrous calcite fans of the Cheshire Formation and Steeprock Group. These are scrutinized further later and regarded as neomorphic calcite after former aragonite rather than gypsum. If true, then this discounts the possibility of former gypsum in the two major Archean carbonate platforms, given the lack of evidence for pseudomorphic mineral casts or other suggestive features. Because the Cheshire platform was characterized in part by sedimentation in restricted environments, it is surprising that no other evidence of former gypsum is present if the fans were indeed gypsum.

Archean calcium carbonate saturation.—Evidence for saturation of Archean sea water with respect to calcite and aragonite relies almost exclusively on the presence of local carbonate beds, or in some cases the development of significant carbonate platforms when suitable tectonic conditions prevailed. More direct evidence for precipitation of carbonate also may be present, however. Large calcite fans, previously interpreted as calcite-replaced gypsum (Walter, 1983; Hofmann, 1971; Wilks, 1986; Martin and others, 1980), alternatively may be calcite pseudomorphs after giant aragonite fans. This interpretation has been historically disfavored, with one exception (Martin and others, 1980), and is considered here as the best option. These fans, which have radii as long as 1 m (Hofmann and others, 1985), appear to be characteristic of many late Archean sequences but are rare in younger sequences. If the fans were originally aragonite, then they may constitute evidence for the relative ease of precipitation of abiotic calcium carbonate, perhaps under highly supersaturated conditions.

The criteria for the determination of a gypsum precursor to the Archean calcite fans have not been specified beyond the opinion that the large crystal size of the precursor precludes an aragonitic origin (Walter, 1983). The likelihood of such excellent fabric preservation following calcitization of Archean sulfates seems improbable, however. In general, fabric preservation is poor following calcitization of Phanerozoic sulfates (Pierre and Rouchy, 1988). Such replacement involves constant flushing of the sediments by water in order to transport the reaction products. Also, a substantial reduction in the space occupied by solids takes place because gypsum has a molar volume of 74.31 cm^3/mole versus 37.94 cm^3/mole for calcite (Berner, 1971). This nearly 50 percent reduction in volume often results in brecciation and collapse of bedding and obliteration of primary fabrics (Pierre and Rouchy, 1988).

In contrast, the transformation of aragonite to calcite, involving minimal flushing by fluids and only a 10 percent increase in molar volume, results in far better fabric retention. The fabric preservation in the calcite cement fans of the widely separated and unique Cheshire and Steeprock carbonate units is excellent, consistent with calcitization of aragonite. Neomorphism of an aragonitic precursor is also supported by limited geochemical data, which argue against an extended diagenetic history as would be required in the case of calcitized sulfates. Carbonates of the Cheshire Formation were apparently stabilized in a nearly closed system and have not been subsequently modified in a significant way (Abell and others, 1985). Cheshire carbonates contain strontium concentrations on the order of several hundred ppm, with occasional values as much as 1,000 ppm (Abell and others, 1985).

Furthermore, the acicular, botryoidal habit of the original crystals, and square to feathery crystal terminations, are all suggestive of primary aragonite (Loucks and Folk, 1976; Mazzullo, 1980; Davies, 1977). Massive submarine aragonite cements with large crystal sizes are confidently known from many Phanerozoic sequences (Ginsburg and James, 1976; Yurewicz, 1977; Davies, 1977) and are frequently composed of coarse aragonite and formerly aragonitic crystal arrays, now replaced by calcite or dolomite.

The interpretation of calcitized Archean gypsum has strongly influenced models for the composition of Precambrian sea water. Past recognition of these assumed "gypsum" pseudomorphs in part forms the basis for justification of a uniformitarian model for sea water composition in which precipitation of calcium carbonate is sequentially followed by calcium sulfate and halite (Walker, 1983). If the aragonite interpretation for the fibrous fans is correct, however, true gypsum precipitation may have been more rare than previously thought and perhaps did not follow as the "normal" precipitate after calcium carbonate (cf. Grotzinger, 1987; see later discussion).

Proterozoic calcium carbonate saturation.—The precipitation of carbonate was a common phenomenon in Proterozoic marine settings. It occurred in "normal" marine environments where there is no evidence or need to invoke evaporation to concentrate dissolved components. Precipitation occurred in all environments represented by the Phanerozoic carbonate record, including the pelagic realm in certain cases (e.g., Pethei Platform; Hoffman, 1974). Supersaturation with respect to calcite and aragonite is documented by the widespread occurrence of marine cements in shallow- as well as deep-marine environments. It is important to note that fibrous, isopachous, dolomitized marine cements occur filling sheet cracks in toe-of-slope rhythmites, deposited in water depths of as much as 1,000 m (Grotzinger, 1986a). This is considered evidence for supersaturation of sea water with respect to calcite to depths of at least 1 km, suggesting that as a minimum estimate, saturation was not significantly less than that of modern sea water (cf. Li and others, 1969).

Supersaturation of Proterozoic sea water with respect to calcite and aragonite is further supported by the presence of abundant tidal-flat tufas (Grotzinger, 1986a,b; Hofmann and Jackson, 1988) in many early Proterozoic carbonate platforms (Fig. 15). The significance of these tufas has not been previously recognized, and it is suggested here that these facies represent the precipitation of mesoscopically crystalline calcium carbonate (probably aragonite; Grotzinger and Read, 1983; Hofmann and Jackson, 1988) on tidal flats under normal to mildly evaporative conditions. The apparent demise of this facies after about 1.7 Ga (Fig. 15) may suggest that early Proterozoic sea water had a higher degree of saturation than in younger times when carbonate precipitation may have been less abundant (Grotzinger, 1987).

Refinement of the Precambrian seawater model.—The Archean and early Proterozoic record bears evidence for surplus precipitation of "abiotic" carbonate as facies that are generally not represented in the middle and late Proterozoic or Phanerozoic record. In particular, this includes precipitation of probably aragonitic fibrous fans as a major facies in late Archean platforms (Cheshire and Steeprock), and precipitation of spatially and volumetrically important tidal-flat tufas in the early Proterozoic. These facies do not replace, but are developed in addition to, the usual variety of facies that characterize middle Proterozoic through Phanerozoic platforms.

As a tentative working hypothesis, it is proposed that Archean, early Proterozoic, and to a lesser extent, middle to late Proterozoic sea water favored surplus "abiotic" carbonate precipitation, as aragonite and (high Mg?) calcite, in comparison to younger times. "Normal" sea water possibly was highly oversaturated, and any perturbation such as microbially induced uptake of CO_2 might have caused immediate and prolific precipitation of calcium carbonate.

The ratio of $[HCO_3-]$ to $[Ca^{2+}]$ may also have been increased, relative to "Phanerozoic" sea water, so that $2[HCO_3-]$ was close to or even greater than $[Ca^{2+}]$. Sea water might have precipitated abundant carbonate in "excess" quantities and induced the "carbonate factory" to exploit new realms, such as tidal flats where tufas might form. In the process, most or all available calcium would be extracted simply by precipitation of carbonate and, during an evaporative situation, little or no calcium sulfate could have precipitated except near sites of continental runoff (e.g., deltas), where influx of additional calcium might be expected.

This model is compatible with theoretical studies, which suggest that atmospheric CO_2 was greatly increased in the Precambrian (Walker, 1983; Kasting, 1987). By implication, the concentration of total dissolved inorganic carbon in the oceans would have also been increased. Precipitation of carbonate would have been favored, provided that the excess negative charge represented by increased bicarbonate was balanced by an increase in the concentration of major cations. Very high concentrations of sodium in Precambrian sea water have been suggested by Kempe and Degens (1985).

The model is also supported by geologic data. Bedded or massive gypsum formed in primary evaporative environments is absent in the Archean and early Proterozoic record. These sediments first appear in the middle Proterozoic (about 1.5 Ga) MacArthur Basin (Jackson and others, 1987), with others in the approximately 1.2 Ga Borden Basin (Jackson and Ianelli, 1981) and approximately 1.2 Ga Amundsen Basin (Victoria Island, Canada; Young, 1981). Furthermore, the best documented early Proterozoic gypsum casts are not associated with the major evaporitic carbonate platforms such as the Rocknest or Campbellrand platforms. Rather, they are often developed in sequences that may have formed adjacent to deltaic regimes, where locally high concentrations of calcium might occur. For example, the widely cited gypsum casts of the Great Slave

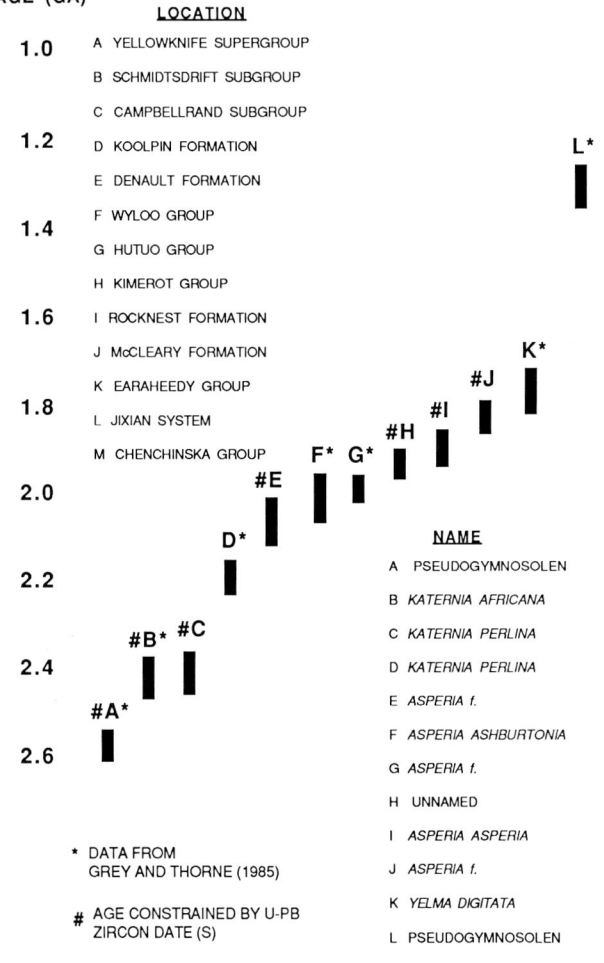

FIG. 15.—Distribution of microdigitate stromatolites in time. Note dominance in early Proterozoic, with only one example in the late Archean (occurrence A), and two in the middle Proterozoic (L,M; these are dated by the K/Ar method on glauconite and therefore could be much older). This time dependence has been noted by Grey and Thorne (1985), who suggest that biologic evolution may have been important in the development and demise of the general stromatolite form. They therefore suggest that the form has biostratigraphic significance. It is suggested here, however, that the form is largely an inorganic precipitate, as recognized by Grotzinger and Read (1983); its general absence in Archean rocks probably reflects the lack of thick tidal-flat sequences in which it generally occurs, and its rarity in middle Proterozoic to Phanerozoic rocks is inferred to record the long-term evolution of sea water toward lower degrees of calcium carbonate saturation. This model is further supported by paleontologic and textural evidence presented in Hofmann and Jackson (1988). The model also recognizes the time dependence of the form and its potential use in crude global lithostratigraphic correlation, but because of evolution of ocean composition rather than microorganisms.

Data for occurrences B, D, F, G, K, L, M are from Grey and Thorne (1985). Other data include: A (Hofmann, 1981); C (Beukes, 1987; pers. commun., 1987); E (Hoffman and Grotzinger, 1988; Donaldson, 1963; Grey and Thorne, 1985); H (Grotzinger and Gall, 1986; Grotzinger and McCormick, 1988); I (Hoffman, 1975; Semikhatov, 1978; Grotzinger, 1986a,b); and J (Hoffman and Jackson, 1988).

Supergroup (Badham and Stanworth, 1977) occur almost exclusively in siliciclastic facies, some of which are deltaic (Hoffman, 1969). None are present in the regionally extensive Pethei carbonate platform (Great Slave Supergroup), and it should be noted that halite casts (and *not* gypsum) are present as the transitional "evaporite" facies between deep basinal precipitates and an overlying evaporite solution-collapse breccia (Stark Formation; Hoffman, 1969). These relations further underscore the possibility that calcium sulfate was not a common phase in early Proterozoic evaporative sequences.

Predictions of the model.—Significantly, the demise of the early Proterozoic tidal-flat tufas coincides with the first appearance of middle Proterozoic bedded-sulfate deposits. This relation might be considered as a prediction of the model outlined above and suggests that, as total calcium carbonate saturation and the $[HCO_3-]/[CA^{2+}]$ ratio decreased with time toward average Phanereozoic values, less precipitation of carbonate occurred, leaving more calcium available to form sulfates during development of evaporative sediments.

The proposed decrease in carbonate precipitation during the middle and late Proterozoic may have been gradual, and there may be evidence for elevated precipitation from normal (unmodified by evaporation) sea water. The carbonates that are commonly associated with the globally extensive late Proterozoic glacial deposits are one possibility. Historically, this has posed a perplexing problem given the difficulties of precipitation of carbonates in apparently frigid climates (e.g., Schermerhorn, 1974). The severity of the problem would be significantly diminished, however, if the saturation state of "normal" sea water with respect to calcium carbonate was elevated relative to younger values. It has been suggested that precipitation of carbonates occurred during warming trends related to interglacial periods (Tucker, 1986; Singh, 1987); this mechanism would be greatly enhanced by increased sea water saturation. Further evidence for elevated late Proterozoic calcium carbonate saturation in sea water includes calcitized aragonite fans formed in basinal facies of the Bambui Group (Brazil; T. Peryt, pers. comun., 1988), and abundant "giant" ooids in the late Proterozoic of Greenland (A. H. Knoll, pers. commun., 1988).

Finally, the model also in part helps to explain the preponderance of Proterozoic dolomite (Chilingar, 1956; Holland, 1984). On account of their abundance and because many Proterozoic dolomites show excellent fabric retention (Tucker, 1982; Grotzinger and Read, 1983), it has been speculated that they were possibly primary (Chilingar, 1956; Tucker, 1982). Although it was first suggested by Daly (1907) that Precambrian sea water was "different" and that the ubiquitous Proterozoic dolomites may have been primary, it is now recognized that many of the better understood fabric-retentive dolomites are probably replacive after primary aragonite and/or calcite (Grotzinger and Read, 1983; Ricketts, 1983; Tucker, 1983; Zempolich and others, 1988). Unfortunately, a suitable mechanism that incorporates all aspects of the problem has not yet been identified.

An implication of the proposed increase in the $[HCO_3-]/[Ca^{2+}]$ ratio, coupled with increased total calcium carbonate saturation, is that more calcium would be extracted from surface sea water during precipitation of aragonite or calcite than occurs at present. It may have been possible to generate surface sea water with a very high Mg/Ca ratio simply by precipitating calcium carbonate under mildly evaporative or even normal conditions, depending on the initial proportions of calcium, magnesium, and bicarbonate. In many situations dolomitization might have followed immediately after enough calcium carbonate had been precipitated. It is possible that dolomites would have been widespread, forming wherever limestones were precipitated, and then replacing them shortly after sedimentation; tidal flats would have been particularly susceptible.

Biologic Controls and Carbonate Production

The single most important difference between Precambrian and Phanerozoic carbonate platforms is the lack of carbonate-secreting metazoa and calcifying algae. Surprisingly, this seems to have had little effect on the gross facies mosaics and morphologies of platforms, which closely mimic those of Phanerozoic platforms. This relation is perhaps the most striking conclusion of this paper, and demonstrates that zones of Proterozoic carbonate production and accumulation must have been nearly identical to Phanerozoic counterparts.

The Precambrian carbonate factory.—

Organic buildups.—Stromatolites, most likely constructed by sediment accreted through the trapping-and-binding and/or precipitation-inducing activities of microbial communities, occupied every major ecological niche known to be important in the construction of Phanerozoic platforms by more complex and environmentally sensitive organisms. Stromatolites probably represent prolific *in situ* carbonate production in the form of major buildups and reefs, some of which may have been comparable in scale to the largest of Phanerozoic barrier and pinnacle reefs (see later discussion and Grotzinger, 1988). Stromatolites may have also been responsible for the production of "intraclast" grainstone that occurs adjacent to many buildups and can be accounted for through their erosion and degradation.

Thus, certain emphasis can be placed on stromatolitic buildups not only as carbonate "depositories" but also as actual "factories." Other evidence for this model includes locally abundant marine cementation in the vicinity of stromatolites, as cements between component particles of stromatolites, and also in the direct precipitation of stromatolitic laminae (Kerans, 1982). The presence of substrate-parallel layers of neomorphic fibrous cement that coat and form adjacent to stromatolites are evidence for *in situ* precipitation of accretionary layers. Similarly, the preservation of radial fibrous fabrics in neomorphic carbonates (usually dolomite) that constitute "microdigitate" stromatolites (microbial tufas) of early Proterozoic tidal-flat facies is also evidence for *in situ* sediment accumulation by precipitation. Other, more indirect arguments also apply with a correspondingly lower degree of confidence. For example, the minimal incorporation of sand grains into stromatolitic laminae where stromatolites are strongly admixed with siliciclastics in mixed settings is a common field observation

(Donaldson, 1963; Serebryakov and Semikhatov, 1974). In addition, the fine lamination of Proterozoic stromatolites has been cited as evidence of possible precipitation, in contrast to Phanerozoic "trapping and binding" stromatolites which often show poor lamination (Cloud and Semikhatov, 1969; Walter, 1972). In general, however, the degree of autochthoneity of most stromatolitic carbonate remains to be shown.

Grainstones.—In general, grainstone sheets are relatively uncommon in facies tracts isolated from stromatolite buildups. Accumulations of open-marine grainstone hundreds of meters thick, comparable to the Jurassic oolite-dominated deposits of the United States Gulf Coast (Smackover Formation) and the "Arab Zone" lime sands of Saudi Arabia (Wilson, 1975), are rare in the Proterozoic record. The late Proterozoic Yellowhead platform (Fig. 11; Teitz and Mountjoy, 1985, 1988) may be the best example of a grainstone-dominated platform. In part, the peculiar scarcity of abundant Proterozoic grainstone must relate to the lack of segmented benthic metazoa and pellet-producing organisms responsible for much coarse-grained Phanerozoic carbonate sediment. It probably also attests, however, to the robust nature of most stromatolites, which develop hydrodynamically stable forms that are fundamentally resistant to normal erosional processes, thus minimizing the total yield of intraclastic carbonate grains derived through erosional processes. Suitable nucleii for ooid growth would have been relatively uncommon given the absence of pellet-producing organisms and bioclastic metazoan fragments. The uncommon and generally spatially-restricted Proterozoic oolites that have been documented probably formed wherever erosion-derived intraclast production was locally high (e.g., back-reef setting of Rocknest platform; Grotzinger, 1986a). Furthermore, where Proterozoic grainstones have been documented (e.g., Campbellrand, Rocknest, and Pethei platforms), these deposits are commonly overlain by stromatolitic units. Consequently, accumulation of thick Proterozoic open-marine grainstones also may have been impeded by the rapid growth of stromatolitic sheets, which could have quickly overwhelmed the space available for grainstone production. In comparison, Phanerozoic metazoan-dominated communities may not have been capable of such effective growth. This is supported by studies of Recent Bahamian grainstone shoals that, during migration, smother and destroy incipient metazoan buildups but preserve and promote growth of associated stromatolites (Dill and others, 1986; Griffin, 1988). This appears to be another example of how the environmentally impervious nature of stromatolites may have exerted a strong influence on the construction and differentiation of Proterozoic platforms.

Carbonate muds.—A related feature that also serves to distinguish Proterozoic from Phanerozoic platforms is the dominance of fine-grained carbonate in Proterozoic platforms. A mechanism for the lack of grainstone as discussed above does not in itself provide a mechanism for the presence of carbonate mud, which must have been generated independently of grainstones; as with siliciclastic sediments, the muds are not likely to have formed by communition and disintegration of sand-sized carbonate grains. It also seems unlikely that mud was produced in abundance independent of stromatolites in one part of the platform, and then transported to another part where it became trapped and bound. Most Proterozoic carbonate mud production was apparently aided by the presence of stromatolites, in that many Proterozoic carbonate mud units are commonly associated with stromatolitic units.

By virtue of the demonstrable involvement of ancient microbial communities in stromatolite construction (Walter, 1983; Klein and others, 1987), and because the volume of stromatolitic sediment is generally equal to or greater than the volume of non-stromatolitic carbonate (Grotzinger, 1986a), it is likely that micritic carbonate was indirectly precipitated through the activity of stromatolite-forming benthic microorganisms. This probably occurred during photosynthesis, when CO_2 was extracted from the water column, producing an increase in pH and decrease in carbonate solubility. Such a mechanism might have been effective on a platformwide scale, inducing carbonate precipitation wherever stromatolitic buildups and reefs were developed. On many platforms, stromatolites probably formed veneers spanning deeper subtidal to upper intertidal environments, and it is conceivable that the "factory" spanned the entire platform, was operative on a diurnal basis, and its performance was dependent only on the rate of turnover of bank water.

It is also possible, given the high calcium carbonate saturation levels postulated above, that "whitings" (Cloud, 1942) may have been a common or even diurnal phenomena. Although it is reasonable to expect that this may have occurred inorganically, the effect would have been greatly enhanced through the activities of benthic and planktonic photosynthetic microbiotas.

Stromatolites and reefs.—

Stromatolites are not always considered as legitimate components of "true" reefs. Early investigations of stromatolites and their role in possible reef construction (Twenhofel, 1919; Fenton and Fenton, 1931) initiated a controversy over the viability of stromatolites as "true" reef builders (compare Cumings, 1932, with Fenton and Fenton, 1933). As a result, the acceptance of stromatolites as reef builders has been slow, and studies evaluating their role in the construction of buildups and Precambrian carbonate platforms are scarce, as pointed out by James (1984c).

The question of the role of stromatolites in Precambrian reef development is partly a semantic problem and must be addressed with respect to the modern definitions of a reef. Most recent definitions (Dunham, 1970; Heckel, 1974) embrace the concepts outlined by Lowenstam (1950) during his classic study of the Niagaran reefs of the Great Lakes area. According to Lowenstam (1950), the essential criteria for establishing the presence of a "true" reef are demonstration of the *structure-building potential* of the community by: (1) growth of the structure from deeper, quiet-water settings upward into the shallow zone of continual wave agitation; and (2) demonstration that the structures were resistant to wave action. Thus, a reef, as defined by Lowenstam (1950, p. 433) in terms of ecologic principles, is ". . . the product of the actively building and sediment-binding biotic constituents, which, because of their potential wave resistance, have the ability to erect rigid, wave-resistant topographic structures." Lowenstam also stated that those principles applied to both modern and ancient reefs, independent of time, and allowed for specific evolutionary adaptation; it is clear that this definition is independent of the specific organism responsible for reef growth. The culminating stage of reef development, following growth to sea level, was lateral expansion of the reef habitat to ". . . influence its surroundings, locally effecting circulation and salinity and effecting sedimentation actively through contribution [of sediment] and passively through creating a zone of turbulence" (Lowenstam, 1950, p. 433).

Lowenstam's ecologic definition of a reef has major significance for stromatolite-constructed buildups in that it *does not require* that the reef-forming organisms be skeletal metazoa *per se*. The emphasis of Lowenstam's definition was not to specify any particular biota as being responsible for the development of a reef and, as is stated (p. 432–433), "the presence and abundance of corals, algae, and bryozoans as such do not constitute proof of reef-building activity. As we go further back in geologic history, we must examine each case critically to determine whether or not the organisms in question were potentially able to erect structures and to bind sediments and, if so, whether or not they actually did."

The essential function of a rigid, metazoan skeleton in reef building is to strengthen the substrate so it can build up into shallow water, withstand wave activity, and thereby influence other environments. A "framework" results from a combination of the intertwining growth and bioerosion of the calcified benthic organisms. If, however, the essence of the definition of a reef is biologically-induced upward growth of a rigid substrate into the zone of wave activity, then stromatolites must be considered as viable reef builders given their demonstrated ability to produce upward-shallowing, wave-resistant structures. Stromatolites and stromatolite bioherms generally streamline themselves and become elongate during upbuilding into the zone of intense wave action.

Stromatolites as "True" Reef Builders.—Individual stromatolites may be regarded as the "framebuilding" constituents in a stromatolitic buildup, provided that it is legitimate to allow the substitution of stromatolites for calcifying metazoa in the role of upbuilding. Because most stromatolites show evidence of having been produced primarily through the trapping and binding and/or precipitation-inducing activity of benthic microbial communities, they can be regarded as having had the same function as individual metazoans had during the development of Phanerozoic reefs—*they are directly responsible for the vertical accretion of the structure upward into the zone of physical destruction (wave-action)*. In many cases, reef development subsequently caused strong paleoenvironmental zonation of carbonate platforms.

The platforms illustrated in this paper, in addition to several other studies of Precambrian stromatolite buildups (Cecile and Campbell, 1978; Ricketts and Donaldson, 1981), demonstrate that stromatolites did form true reefs that are directly comparable, in a paleoecological context, to younger metazoan-constructed counterparts. These stromatolite buildups possess all the critical criteria cited by Lowenstam to be indicative of a true reef, including evidence for growth from deeper to shallower water wave-agitated settings, and continued growth and flourishing within the zone of active physical destruction. Although it can be inferred, on these grounds, that early cementation of the substrate and/or direct precipitation must have taken place in order to form these wave-resistant structures built by presumably non-calcifying microbes, there is also direct petrographic evidence (fibrous marine cements) that early cementation was indeed important in the development of these reefs (Grotzinger, 1986a).

Stromatolite reefs occupied a variety of different niches, similar to younger counterparts. These include major barrier reefs adjacent to large seaways, patch reefs that formed in restricted lagoons located behind these barriers, patch reefs located on gentle ramps facing open seaways, pinnacle reefs analogous to the Silurian reefs of the Great Lakes region, and even downslope bioherms that grew entirely within a deeper, quieter water setting. The research of the past two decades has taken a significant step toward establishing that many stromatolite buildups are, in fact, true reefs. This provides an important focal point for future work that can be used to help unravel the evolution of microbial communities in Precambrian time. Given that the role of stromatolites may be more accurately understood within a paleoecological context (i.e., reef building), then this in turn may help assign new significance to results of paleontological investigations concerned with microbiotic evolution. What, for example, is the relation between the microbial community that colonized an open-marine barrier-reef complex and that which formed the stromatolites of a more restricted lagoonal patch reef, or a highly restricted tidal flat? If nothing more, this distinction should help to characterize the bias introduced by selective sampling and preservation of lithologies that tend to be selectively silicified.

Finally, the recognition of Precambrian stromatolite buildups as true ecological reefs creates a new context for understanding the genesis and evolution of Phanerozoic reefs. The relative effects of extrinsic versus intrinsic controls on the cyclic succession of biofacies within Phanerozoic reefs continues to be a subject of much debate (Precht, 1987). Because Precambrian reefs were constructed by biologically less complex microbial communities, it may be possible to assess more accurately the relative physical versus biological controls on reef development than is possible for metazoan-dominated reefs, where the biologic complications introduced by the probably strong influence of ecologic community succession (*sensu* Walker and Alberstadt, 1975) must be confronted.

It is suggested that the physical controls on Precambrian reef development might be expected to dominate over biologic controls and that recognizable differences may exist between the architecture of Precambrian and Phanerozoic reef complexes. With further study of Precambrian reefs, it might be possible to derive a relation that can be used to help isolate the various influences that complicate Phanerozoic reef development. The strong vertical succession of stromatolite types in certain Proterozoic and even Archean reefs (Hoffman, 1988a; Nisbet and Wilks, 1988; Cecile and Campbell, 1978), similar to the succession of biofacies in Phanerozoic reefs, seems to be reason enough to justify further studies with this concept in mind.

Finally, it is suggested that the "performance capacity" (*sensu* Lowenstam, 1950) of stromatolites in forming reefs was probably exceptionally high because of the great tolerance of stromatolite-building microbes to environmental pollutants that would have been lethal to metazoans. Such effects include the influx of siliciclastics and introductions of inimical hypersaline, brackish, or nutrient-depleted bank water. The stratigraphic record shows that Precambrian stromatolite reefs may be considered as some of the most successful and widespread of all geologic time, and it is noteworthy that during times of Phanerozoic environmental crisis, stromatolites often replaced metazoans as the dominant reef builders. Examples of this include the Famennian of the Canning Basin (Playford, 1980; C. Kerans, pers. commun., 1988) and several stages of the Silurian in the Michigan Basin (Mesolella and others, 1974).

Stromatolite paleoecology.—

No firm relation has yet been established between the inferred variety of microbial benthic communities and their potential effect on development of specific facies in Precambrian carbonate platforms. Although it is now generally accepted that stromatolite morphology is affected mainly by environmental factors and that the microfabric is controlled primarily through biological interactions (Semikhatov, 1976), few studies attempt to relate specific stromatolite types to specific settings. Some exceptions include the work by Donaldson (1976b), Hoffman (1976), and Grey and Thorne (1985). Only a few generalizations can be made at this point, but it appears that certain stromatolite types recur in specific settings (Fig. 16). These types include conical columnar stromatolites and asperiaform or "microdigitate" stromatolites. The development of conical and microdigitate stromatolites probably had locally important effects on the evolution of particular carbonate platforms where these forms were present, especially with regard to carbonate production. It is noteworthy that these two stromatolite groups represent paleoenvironmental end members, and that they also contain the most clear-cut evidence for a precipitational origin. The great majority of stromatolite types occur somewhere in between these end members, including domes, linked domes, most columnar stromatolites, and their elongate equivalents; these types are the dominant Precambrian reef builders.

Conical columnar stromatolites.—Conical columnar stromatolites apparently developed in deep- to shallow-subtidal paleoenvironments. The potential subtidal restriction

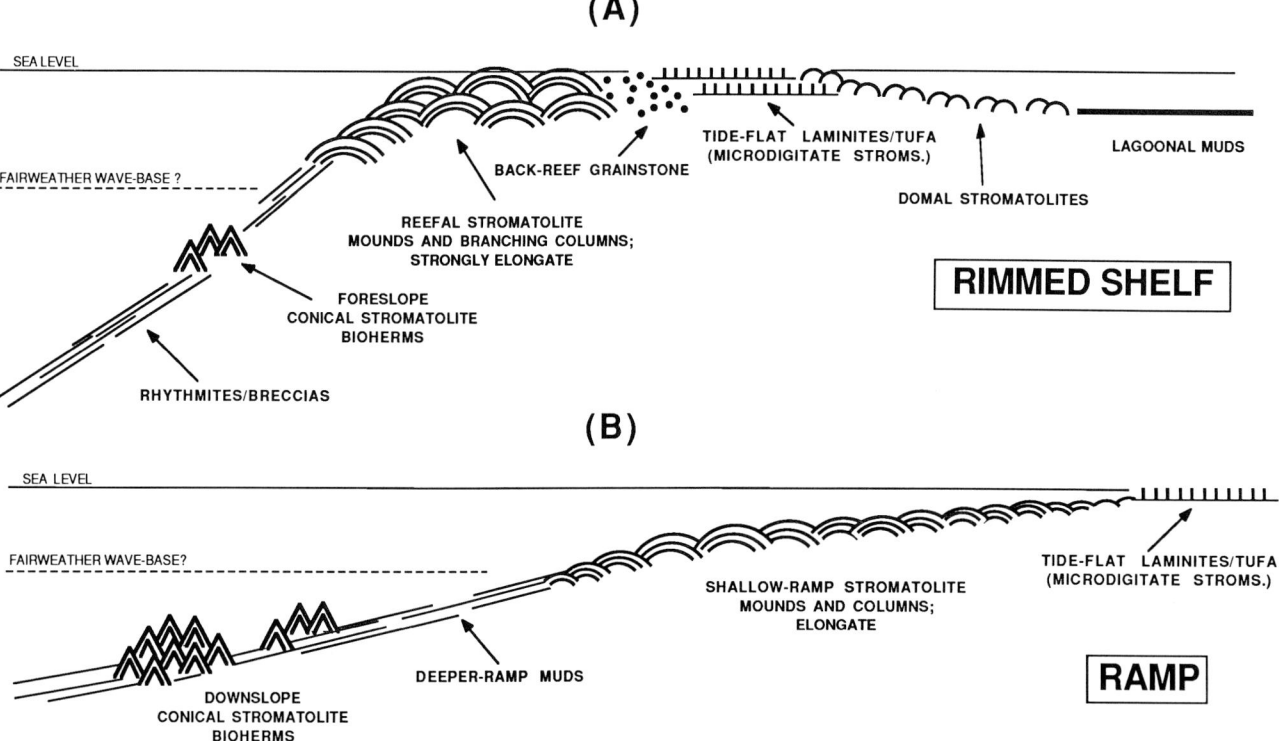

FIG. 16.—General distribution of specific stromatolite types across Proterozoic carbonate platforms. (A) Rimmed shelf as high-energy (windward) margin characterized by barrier-reef complex of strongly elongate stromatolite mounds and columns. Conical stromatolites may form below wave base as foreslope deposits; domal stromatolites of the inner-shelf lagoon are weakly to non-elongate as a result of their restricted, low-energy setting; tufas (including microdigitate stromatolites) form by precipitation on tidal flats. (B) Ramp as moderate-energy platform. Elongation of stromatolite mounds and columns is dependent on relative amount of wave surge and/or tidal strength; for low-energy settings, elongation will be minimal. Note decrease of stromatolite mound size toward both deeper and shallower settings; elongation may show a similar relationship.

of "conophyton" was first recognized by Donaldson and Taylor (1972). Subsequently, it has been recognized as the dominant and often exclusive stromatolitic component of basinal and slope environments (Fig. 16), and as a transitional facies in incipient to terminally drowned platform sequences (Donaldson, 1976b; Hoffman, 1976; Grotzinger, 1986a; Beukes, 1987). In addition, conical columnar stromatolites occur in probable deep-ramp facies of the middle to late Proterozoic Beck Spring Formation (Pahrump Group, western United States; Grotzinger, unpubl. data), and also in the late Proterozoic Bambui Group (Brazil; Cloud and Dardene, 1973), which contains well-developed basinal facies (T. Peryt, pers. commun., 1988).

In the Pethei platform (Hoffman, 1974, 1988a), conical stromatolites were important sources of carbonate precipitation in basinal and deep- to shallow-slope and ramp facies where carbonate production might not have occurred otherwise. This may have been an important control in maintaining the relatively gentle gradients of the Pethei platform (Taltheilei and Wildbread Formations), in contrast to other platforms where the lack of contemporaneous carbonate production in basin and slope environments resulted in rapid slope steepening (e.g., Rocknest platform; Grotzinger, 1986a). This relation is apparently supported by the lack of evidence for steep platform slopes in the other two platforms where conical stromatolites had a significant role in slope and basinal carbonate production (Campbellrand and Dismal Lakes platforms; Beukes, 1987; Kerans and Donaldson, 1988b).

Microdigitate stromatolites.—In contrast, asperiaform or "microdigitate" columnar stromatolites appear to be almost exclusively restricted to peritidal environments (Fig. 16). These are generally less than 1 cm wide, less than 10 cm high, and tend to form units as thick as a few meters that extend laterally for hundreds of kilometers. They are particularly well developed as capping facies in shallowing-upward tidal-flat cycles (Hoffman, 1975; Hoffman, 1975; Kerans, 1982; Grey and Thorne, 1985; Grotzinger, 1986a,b; Grotzinger and Gall, 1986). They also tend to occur as encrusting layers that drape unconformity-related erosional surfaces (Grotzinger and others, 1987).

The likely precipitated origin of microdigitate stromatolites may have had a significant effect on the evolution of tidal-flat sequences. The ability to produce carbonate directly probably enabled tidal flats to shallow and prograde more efficiently during the development of shallowing-upward peritidal cycles (Grotzinger, 1986b), rather than having to depend on shoreward transport of subtidally produced carbonate (cf. James, 1984a).

Other Types of Stromatolite.—Most other types of stromatolite probably formed in a diverse range of deeper subtidal- to intertidal-reef, open-shelf, lagoonal and tidal-flat settings (Fig. 16). In general, inferred low-energy coastlines are characterized by non-elongate domal or linked domal stromatolites with less common non-elongate columnar stromatolites. These facies are typically developed behind barrier-reef complexes of rimmed shelves in protected inner-shelf peritidal or lagoonal environments (Fig. 16a), or as deeper subtidal to intertidal facies of low-energy ramps.

In contrast, inferred high-energy coastlines typically had non-linked to partially linked columnar stromatolites with elongation aspect ratios of > 5:1. Non-elongate domes or linked domes are rare. Most shoal-water reef complexes are also characterized by strongly elongate columnar stromatolites (Fig. 16). These commonly are the basic component of much larger, strongly elongate mounds on the scale of 10 to 100 m in length. Stromatolite mounds may interfinger with lenses of grainstone to form "mound-and-channel" belts along platform margins of rimmed shelves (Fig. 16a). Buildups of strongly elongate stromatolites are characteristic of many platform positions, including barrier reefs of rimmed shelves and deeper subtidal to intertidal settings of high-energy ramps (Fig. 16b).

The variation in form of these types of stromatolite as a function of platform position is generally limited to changes in the synoptic profiles of mounds and/or individual stromatolites and in their elongation aspect ratios. There is an excellent correlation between inferred paleo-water depth and stromatolite form such that shallow-water forms often have lower synoptic relief (Grotzinger, 1986a; Hoffman, 1988a; Ricketts and Donaldson, 1988). The correlation between inferred paleo-water depth and elongation aspect is less clear, owing to other external factors such as wave interference patterns that also influence elongation (cf. Logan and others, 1974). Whereas in some cases the elongation aspect decreases directly with inferred paleo-water depth (Grotzinger, 1986a), it apparently decreases inversely with inferred paleo-water depth in others (Hoffman, 1988a). One possible explanation of this apparent contradiction may be that an initial increase in elongation during shallowing from below wave base is followed by a decrease in elongation during continued shallowing into the zone of dampened wave surge in the upper tracts of tidal flats. Consequently, platforms that were not submerged below wave base may only show sequences in which stromatolites are less elongate upward. On the other hand, platforms that never fully shallowed may only contain sequences with stromatolites that are more elongate upward.

CONCLUSIONS

(1) Carbonate platforms with most of the essential features of Phanerozoic platforms were well developed by 2.6–2.3 Ga. It is possible that some late Archean carbonates (2.5–2.7 Ga) were parts of major platforms, but the few structurally preserved remnants are insufficient to reconstruct platform morphology and paleogeography accurately.

Proterozoic carbonate platforms include ramps and rimmed shelves. These generally show strong paleogeographic zonation as reflected by the distribution of stromatolitic buildups, including reefs, relative to other deeper and shallower water facies. Because these relations recur in specific platform settings, it permits classification of Proterozoic platforms according to Phanerozoic facies models. Ramps may be homoclinal or distally steepened and have fringing reefs, barrier reefs, or isolated buildups. Rimmed shelves are developed as accretionary margins, bypass margins of both the escarpment and gullied slope subtypes, and may contain intra-shelf basins.

(2) The evolution of many Proterozoic platforms is strikingly similar to that of younger counterparts and indicates that the critical stages of development, including growth, diversification, and expansion, *were not dependent* on the presence of carbonate-secreting or other metazoan organic communities. Therefore, Phanerozoic platforms can be viewed from a new perspective as Proterozoic *templates,* on which complex organic diversification and evolution took place. This is potentially a very valuable relation because Proterozoic platforms, as simple systems, can provide an opportunity to evaluate the complex extrinsic factors on carbonate production such as eustasy or tectonism, but which have been difficult to separate from the likely intrinsic effects inherent to metazoan-dominated Phanerozoic platforms.

As an example, Proterozoic microbial communities were probably less susceptible to possible environmental factors (e.g., salinity changes, nutrient supply, and so forth) and the short-term pressures of complex community succession that are likely inherent to metazoan buildups (cf. Walker and Alberstadt, 1975). Consequently, the problem of the relative roles of community succession versus eustasy in the development of stratigraphic carbonate sequences can perhaps be more clearly evaluated within the context of the more simple Proterozoic platform, where the potential biologic effect of community succession can be downplayed.

(3) Precambrian platforms were strongly influenced by many of the same controls that have affected the development of Phanerozoic platforms. These controls include: the mode and tempo of basin subsidence and the rate of siliciclastic influx; eustatic effects on rim backstepping; incipient to terminal platform drowning; the hierarchical packaging of stratigraphy; and the potential for subaerial exposure and karsting.

(4) In addition, the evolution of Precambrian platforms was profoundly influenced by the non-reversible tectonic, biologic, and physico-chemical processes associated with the longer term evolution of the Earth. These changes are recorded as events in the history of these platforms and include: (1) emergence of "full blown" carbonate platforms in the earliest part of Proterozoic time, following the growth and stabilization of cratons during Archean time; (2) the introduction and very successful involvement of (photosynthetic?) microorganisms in the production of stromatolitic and perhaps even non-stromatolitic carbonate; and (3) the apparent abundance of carbonate precipitation as shown by time-restricted facies, including giant cement (primary aragonite?) fans in the late Archean and widespread tidal flat tufas in the early Proterozoic. It is proposed that calcium carbonate saturation and possibly the $[HCO_3^-]/[Ca^{2+}]$ ratio were much greater in the Precambrian as a mechanism to account for the apparent abundance of calcium carbonate precipitation, the paucity of calcium sulfate precipitation, and some potentially early dolomitization.

ACKNOWLEDGMENTS

I am grateful to P. F. Hoffman, K. A. Eriksson, and J. F. Read for stimulating my interest in Precambrian geology and sedimentology, and to the Geological Survey of Canada for providing the means for eight seasons of research on various aspects of Precambrian carbonate depositional systems. Many of the ideas presented here have benefited from discussions with N. J. Beukes, N. Christie-Blick, S. L. Dorobek, K. A. Eriksson, K. Griffin, L. A. Hardie, P. F. Hoffman, H. J. Hofmann, N. P. James, C. Kerans, S. M. Pelechaty, J. F. Read, B. C. Schreiber, and M. Walter. Financial support during preparation of the manuscript was provided by the Lamont-Doherty Geological Observatory of Columbia University and through National Science Foundation Grants EAR 86-14670 and EAR 88-06040. The final manuscript benefited from critical reviews by C. Kerans, A. H. Knoll, and J. F. Read.

REFERENCES

ABELL, P. I., MCCLORY, MARTIN, A., NISBET, E. G., AND KYSER, T. K., 1985, Archean stromatolites from the Ngezi Group, Belingwe greenstone belt, Zimbabwe; Preservation and stable isotopes–preliminary results: Precambrian Research, v. 27, p. 357–383.

AITKEN, J. D., 1967, Classification and environmental significance of cryptalgal limestones and dolomites, with illustrations from the Cambrian and Ordovician of southwestern Alberta: Journal of Sedimentary Petrology, v. 37, p. 1163–1178.

———, 1978, Revised models for depositional grand cycles, Cambrian of the southern Rocky Mountains, Canada: Bulletin of Canadian Petroleum Geology, v. 26, p. 515–542.

———, 1981, Stratigraphy and sedimentology of the upper Proterozoic Little Dal Group, Mackenzie Mountains, Northwest Territories, *in* Campbell, F. H. A., ed., Proterozoic Basins of Canada: Geological Survey of Canada Paper 81-10, p. 47–71.

———, 1988, Giant algal reefs, middle/upper Proterozoic Little Dal Group (>770, <1200 Ma), Mackenzie Mountains, N.W.T., Canada, *in* Geldsetzer, H., James, N. P., and Tebbutt, G., eds., "Reefs—Canada and Adjacent Areas": Canadian Society of Petroleum Geologists Memoir 13, p. 13–23.

ASPLER, L. B., AND DONALDSON, J. A., 1986, The Nonacho Basin (Early Proterozoic), Northwest Territories, Canada: Sedimentation and deformation in a strike-slip setting, *in* Biddle, K. T. and Christie-Blick, N., eds., Strike-slip Deformation, Basin Formation and Sedimentation: Society of Economic Paleontologists and Mineralogists Special Publication 37, p. 193–210.

BADHAM, J. P. N., AND STANWORTH, C. W., 1977, Evaporites from the lower Proterozoic of the East Arm, Great Slave Lake: Nature, v. 268, p. 516–518.

BEEUNAS, M. A., AND KNAUTH, L. P., 1985, Preserved stable isotopic signature of subaerial diagenesis in the 1.2-b.y. Mescal Limestone, central Arizona: Implications for the timing and development of a terrestrial plant cover: Geological Society of America Bulletin, v. 96, p. 737–745.

BERNER, R. A., 1971, Principles of Chemical Sedimentology: McGraw Hill Book Company, New York, 240 p.

BERTRAND-SARFATI, J., AND MOUSSINE-POUCHKINE, A., 1983, Platform-to-basin facies evolution: The carbonates of late Proterozoic (Vendian) Gourma (west Africa): Journal of Sedimentary Petrology, v. 53, p. 275–293.

BEUKES, N. J., 1987, Facies relations, depositional environments and diagenesis in a major early Proterozoic stromatolitic carbonate platform to basinal sequence, Campbellrand Subgroup, Transvaal Subgroup, southern Africa: Sedimentary Geology, v. 54, p. 1–46.

BICKLE, M. J., MARTIN, A., AND NISBET, E. G., 1975, Basaltic and peridotitic komatiites, stromatolites and a basal unconformity in the Belingwe Greenstone Belt, Rhodesia: Earth and Planetary Science Letters, v. 27, p. 155–162.

BROWN, G. F., 1970, Eastern margin of the Red Sea and the coastal structures in Saudi Arabia: Philosophical Transactions of the Royal Society of London, series A, v. 267, p. 75–87.

BUCK, S. G., 1980, Stromatolite and ooid deposits within the fluvial and lacustrine sediments of the Precambrian Ventersdorp Supergroup of South Africa: Precambrian Research, v. 12, p. 311–330.

BUICK, R. AND DUNLOP, J. S. R., 1987, Early Archean evaporitic sediments from the Warrawoona Group, North Pole, Western Australia: Geological Society of America, Abstracts with Programs, v. 19, p. 604.

BURKE, K., KIDD, W. S. F., AND KUSKY, T. M., 1985, The Pongola structure of southeastern Africa: the world's oldest preserved rift?: Journal of Geodynamics, v. 2, p. 35–49.

BURNE, R. V., AND MOORE, L. S., 1987, Microbialites: Organosedimentary deposits of benthic microbial communities: PALAIOS, v. 2, p. 241–254.

CAMPBELL, F. H. A., 1979, Stratigraphy and sedimentation in the Helikian Elu Basin and Hiukitak Platform, Bathurst Inlet-Melville Sound, Northwest Territories: Geological Survey of Canada Paper 79-8, 18 p.

CECILE, M. P., AND CAMPBELL, F. H. A., 1978, Regressive stromatolite reefs and associated facies, middle Goulburn Group (Lower Proterozoic) in Kilohigok Basin, N.W.T.: An example of environmental control on stromatolite forms: Bulletin of Canadian Petroleum Geology, v. 26, p. 237–267.

CHILINGAR, G. V., 1956, Ca/Mg ratio and geologic age: American Association of Petroleum Geologists Bulletin, v. 40, p. 2256–2266.

CLOUD, P. E., 1942, Notes on stromatolites: American Journal of Science, v. 240, p. 363–379.

———, 1965, Significance of the Gunflint Precambrian microflora: Science, v. 148, p. 27–35.

———, AND DARDENNE, M., 1973, Proterozoic age of the Bambui Group in Brazil: Geological Society of America Bulletin, v. 84, p. 1673–1676.

———, AND SEMIKHATOV, M. A., 1969, Proterozoic stromatolite zonation: American Journal of Science, v. 267, p. 1017–1061.

CUMINGS, E. R., 1932, Reef or bioherms: Geological Society of America Bulletin, v. 43, p. 331–352.

DALY, R. A., 1907, The limeless ocean of Precambrian time: American Journal of Science, v. 23, p. 93–115.

DARWIN, C., 1842, The structure and distribution of coral reefs, in 1962 paperback edition: University of California Press, Berkeley, 214 p.

DAVIES, G. R., 1977, Former magnesian calcite and aragonite submarine cements in upper Paleozoic reefs of the Canadian Arctic: Geology, v. 5, p. 11–15.

DECIMA, A., MCKENZIE, J. A., AND SCHREIBER, B. C., 1988, The origin of "evaporative" limestones: An example from the Messinian of Sicily (Italy): Journal of Sedimentary Petrology, v. 58, p. 256–272.

DILL, R. F., SHINN, E. A., JONES, A. T., KELLY, K., AND STEINEN, R. P., 1986, Giant subtidal stromatolites forming in normal salinity waters: Nature, v. 324, p. 55–58.

DONALDSON, J. A., 1963, Stromatolites in the Denault Formation, Marion Lake, coast of Labrador, Newfoundland: Geological Survey of Canada Bulletin 102, 33 p.

———, 1976a, Aphebian stromatolites in Canada: Implications for stromatolite zonation, in Walter, M. R., ed., Stromatolites: Elsevier, New York, p. 371–380.

———, 1976b, Paleoecology of Conophyton and associated stromatolites in the Precambrian Dismal Lakes and Rae Groups, Canada, in Walter, M. R., ed., Stromatolites: Elsevier, New York, p. 523–534.

———, AND TAYLOR, A. H., 1972, Conical-columnar stromatolites and subtidal environment (abs.): American Association of Petroleum Geologists Bulletin, v. 56, p. 614.

DUNHAM, R. J., 1970, Stratigraphic reefs versus ecologic reefs: American Association of Petroleum Geologists Bulletin, v. 54, p. 1931–1950.

ERIKSSON, K. A., 1977, Tidal flat and subtidal sedimentation in the 2250 M.Y. Malmani Dolomite, Transvaal, South Africa: Sedimentary Geology, v. 18, p. 223–244.

FENTON, C. L., AND FENTON, M. A., 1931, Algae and algal beds in the Belt series of Glacier National Park: Journal of Geology, v. 39, p. 670–686.

———, AND ———, 1933, Algal reefs or bioherms in the Belt series of Montana: Geological Society of America Bulletin, v. 44, p. 1135–1142.

GINSBURG, R. N., AND JAMES, N. P., 1974, Holocene carbonate sediments of continental shelves, in Burk, C. A., and Drake, C. L., eds., The Geology of Continental Margins: Springer-Verlag, New York, p. 137–155.

———, AND ———, 1976, Submarine botryoidal aragonite in Holocene reef limestones, Belize: Geology, v. 4, p. 431–436.

GOLDHAMMER, R. K., DUNN, P. A., AND HARDIE, L. A., 1987, High frequency glacio-eustatic sealevel oscillations with Milankovitch characteristics recorded in middle Triassic platform carbonates in northern Italy: American Journal of Science, v. 287, p. 853–892.

GREY, K., AND THORNE, A. M., 1985, Biostratigraphic significance of stromatolites in upward-shallowing sequences of the early Proterozoic Duck Creek Dolomite, Western Australia: Precambrian Research, v. 29, p. 183–206.

GRIFFIN, K. M., 1988, Sedimentology and Paleontology of Thrombolites and Stromatolites of the Upper Cambrian Nopah Formation and their modern Analogue on Lee Stocking Island, Bahamas: Unpublished M.Sc. Thesis, University of California, Santa Barbara, 147 p.

GROTZINGER, J. P., 1986a, Evolution of early Proterozoic passive-margin carbonate platform, Rocknest Formation, Wopmay Orogen, N.W.T., Canada: Journal of Sedimentary Petrology, v. 56, p. 831–847.

———, 1986b, Cyclicity and paleoenvironmental dynamics, Rocknest platform, northwest Canada: Geological Society of America Bulletin, v. 97, p. 1208–1231.

———, 1986c, Upward-shallowing platform cycles: A response to 2.2 billion years of low-amplitude, high-frequency (Milankovitch band) sea level oscillations: Paleoceanography, v. 1, p. 403–416.

———, 1986d, Shallowing-upward cycles of the Wallace Formation (Belt Supergroup), northwestern Montana and northern Idaho, in Roberts, S. M., ed., Belt Supergroup: A Guide to Proterozoic Rocks of Western Montana and Adjacent Areas: Montana Bureau of Mines and Geology Special Publication 94, p. 143–160.

———, 1987, Sediment production and evolution of Proterozoic carbonate platforms (abs.): American Association of Petroleum Geologists Bulletin, v. 71, p. 561.

———, 1988, Introduction to Precambrian Reefs in Geldsetzer, H., James, N. P., and Tebbutt, G., eds., "Reefs—Canada and Adjacent Areas": Canadian Society of Petroleum Geologists Memoir 13, p. 9–12.

———, AND GALL, Q., 1986, Preliminary investigations of early Proterozoic Western River and Burnside River Formations: Evidence for foredeep origin of Kilohigok Basin, N.W.T., Canada: in Current Research, Part A: Geological Survey of Canada Paper 86–1A, p. 95–106.

———, GAMBA, C., PELECHATY, S. M., AND MCCORMICK, D. S., 1988, Stratigraphy of a 1.9 Ga foreland basin shelf-to-slope transition: Bear Creek Group, Tinney Hills area of Kilohigok Basin, District of Mackenzie, in Current Research, Part C: Geological Survey of Canada Paper 88-1C, p. 313–320.

———, AND MCCORMICK, D. S., 1988, Flexure of the early Proterozoic lithosphere and the evolution of Kilohigok Basin (1.9 Ga), Northwest Canadian Shield, in Kleinspehn, K. and Paola, C., eds., New Perspectives in Basin Analysis: Springer-Verlag, New York, p. 405–430.

———, ———, AND PELECHATY, S. M., 1987, Progress report on the stratigraphy, sedimentology, and significance of the Kimerot and Bear Creek groups, Kilohigok Basin, District of Mackenzie, in Current Research, Part A: Geological Survey of Canada Paper 87-1A, p. 219–238.

———, AND READ, J. F., 1983, Evidence for primary aragonite precipitation, lower Proterozoic (1.9 Ga) dolomite, Wopmay orogen, northwest Canada: Geology, v. 11, p. 710–713.

GUDSTADT, A. M., 1968, Petrology and depositional environments of the Beck Spring Dolomite (Precambrian), Kingston Range, California: Journal of Sedimentary Petrology, v. 38, p. 1280–1289.

HARDIE, L. A., 1984, Evaporites: Marine or non-marine?: American Journal of Science, v. 284, p. 193–240.

HAYS, J. D., AND PIMAN, W. C., III, 1973, Lithospheric plate motion, sea-level changes, and climatic and ecological consequences: Nature, v. 246, p. 18–22.

HECKEL, P. H., 1974, Carbonate buildups in the geologic record: A review, in Laporte, L. F., ed. Reefs in Time and Space: Society of Economic Paleontologists and Mineralogists Special Publication 18, p. 90–154.

HEINRICHS, T. K., AND REIMER, T. O., 1977, A sedimentary barite deposit from the Archean Fig Tree Group of the Barberton Mountain Land (South Africa): Economic Geology, v. 72, p. 1426–1441.

HENDERSON, J. B., 1975, Archean stromatolites in the northern Slave province, Northwest Territories, Canada: Canadian Journal of Earth Science, v. 12, p. 1619–1630.

———, 1981, Archean basin evolution in the Slave province, Canada, in Kroner, A., ed., Precambrian Plate Tectonics: Elsevier, New York, p. 213–236.

HINNOV, L. A., AND GOLDHAMMER, R. K., 1988, Identification of Milankovitch signals in Middle Triassic platform carbonate cycles using a super-resolution spectral technique (abs.): American Association of Petroleum Geologists Bulletin, v. 72, p. 197.

HOFFMAN, P. F., 1969, Proterozoic paleocurrents and depositional history of the east arm fold belt, Great Slave Lake: Canadian Journal of Earth Science, v. 6, p. 441–162.

———, 1973, Evolution of an early Proterozoic continental margin: The Coronation Geosyncline and associated aulacogens of the northwestern Canadian Shield: Philosophical Transactions of the Royal Society of London, series A, 273, p. 547–581.

———, 1974, Shallow and deep-water stromatolites in lower Proterozoic platform-to-basin facies change, Great Slave Lake, Canada: American Association of Petroleum Geologists Bulletin, v. 58, p. 856–867.

———, 1975, Shoaling-upward shale-to-dolomite cycles in the Rocknest Formation, Northwest Territories in Ginsburg, R. N., and Klein, G. de V., eds., Tidal Deposits: Springer-Verlag, New York, p. 257–265.

———, 1976, Environmental diversity of middle Precambrian stromatolites, in Walter, M. R., ed., Stromatolites: Elsevier, New York, p. 599–612.

———, 1986, Crustal accretion in a 2.7–2.5 Ga "granite-greenstone" terrane, Slave Province, NWT: A prograding trench-arc system?: Geological Association of Canada, Program with Abstracts, v. 11, p. 82.

———, 1987, Early Proterozoic foredeeps, foredeep magmatism, and Superior-type iron-formations of the Canadian Shield in Kroner, A., ed., Proterozoic Lithospheric Evolution: American Geophysical Union Geodynamics Series, v. 17, p. 85–98.

———, 1988a, Pethei reef complex (1.9 Ga), Great Slave Lake, N.W.T.: in Geldsetzer, H., James, N. P., and Tebbutt, G., eds., "Reefs—Canada and Adjacent Areas": Canadian Society of Petroleum Geologists Memoir 13, p. 38–48.

———, 1988b, United plates of America, the birth of a craton: Early Proterozoic assembly and growth of proto-Laurentia, in Annual Reviews of Earth and Planetary Science, v. 16, p. 543–603.

———, AND BOWRING, S. A., 1984, Short-lived 1.9 Ga continental margin and its destruction, Wopmay orogen, northwest Canada: Geology, v. 12, p. 68–72.

———, DEWEY, S. F., AND BURKE, K., 1974, Aulacogens and their genetic relation to geosynclines, with a Proterozoic example from Great Slave Lake, Canada, in Dott, R. H., and Shaver, R. H., eds., Modern and Ancient Geosynclinal Sedimentation: SEPM Special Publication 19, p. 38–55.

———, AND GROTZINGER, J. P., 1988, Abner/Denault reef complex (2.1 Ga), Labrador Trough, NE Quebec, in Geldsetzer, H., James, N. P., and Tebbutt, G., eds., "Reefs—Canada and Adjacent Areas": Canadian Society of Petroleum Geologists Memoir 13, p. 49–54.

HOFMANN, H. J., 1971, Precambrian fossils, pseudofossils, and problematica in Canada: Geological Survey of Canada Bulletin, v. 189, 146 p.

———, 1975, Stratiform Precambrian stromatolites, Belcher Islands, Canada: Relations between silicified microfossils and microstructure: American Journal of Science, v. 275, p. 1121–1132.

———, 1981, Precambrian fossils in Canada–The 1970s in retrospect, in Campbell, F. H. A., ed., Proterozoic Basins of Canada: Geological Survey of Canada Paper 81-10, p. 419–443.

———, AND JACKSON, G. D., 1988, Proterozoic ministromatolites with radial-fibrous fabric: Sedimentology, v. 34, p. 963–971.

———, THURSTON, P. C., AND WALLACE, H., 1985, Archean stromatolites from Uchi Greenstone Belt, northwestern Ontario, in Ayres, L. D., Thurston, P. C., Card, K. D., and Weber, W., eds., Evolution of Archean Supracrustal Sequences: Geological Association of Canada Special Paper 28, p. 125–132.

HOLLAND, H. D., 1972, The geologic history of sea water—An attempt to solve the problem: Geochemica et Cosmochemica Acta, v. 36, p. 637–651.

———, 1984, The chemical evolution of the atmosphere and oceans: Princeton University Press, Princeton, 582 p.

JACKSON, G. D., AND IANELLI, T. R., 1981, Rift-related cyclic sedimentation in the Neohelikian Borden Basin, northern Baffin Island, in Campbell, F. H. A., ed., Proterozoic Basins of Canada: Geological Survey of Canada Paper 81-10, p. 269–302.

———, AND ———, 1988, Neohelikian reef complexes, Borden rift basin, northwestern Baffin Island, in Geldsetzer, H., James, N. P., and Tebbutt, G., eds., "Reefs—Canada and Adjacent Areas": Canadian Society of Petroleum Geologists Memoir 13, p. 55–63.

JACKSON, M. J., 1988, Early Proterozoic Cowles Lake foredeep reef, N.W.T., Canada, in Geldsetzer, H., James, N. P., and Tebbutt, G., eds., "Reefs—Canada and Adjacent Areas": Canadian Society of Petroleum Geologists Memoir 13, p. 64–71.

———, MUIR, M. D., AND PLUMB, K. A., 1987, Geology of the southern McArthur Basin, Northern Territory: Bureau of Mineral Resources, Geology and Geophysics Bulletin 220, Australian Government Publishing Service, Canberra, 173 p.

JAMES, N. P., 1984a, Introduction to carbonate facies models, in Walker, R. G., ed., Facies Models, second edition: Geoscience Canada Reprint Series 1, p. 209–212.

———, 1984b, Shallowing-upward sequences in carbonates, in Walker, R. G., ed., Facies Models, second edition: Geoscience Canada Reprint Series 1, p. 213–228.

———, 1984c, Reefs, in Walker, R. G., ed., Facies Models, second edition: Geoscience Canada Reprint Series 1, p. 229–244.

JOLLIFFE, A. W., 1955, Geology and iron ores of Steep Rock Lake: Economic Geology, v. 50, p. 373–398.

KASTING, J. F., 1987, Theoretical constraints on oxygen and carbon dioxide concentrations in the Precambrian atmosphere: Precambrian Research, v. 34, p. 205–229.

KEMPE, S., AND DEGENS, E. T., 1985, An early soda ocean?: Chemical Geology, v. 53, p. 95–108.

KERANS, C., 1982, Sedimentology and stratigraphy of the Dismal Lakes Group: Unpublished Ph.D. Dissertation, Carleton University, Ottawa, 404 p.

———, AND DONALDSON, J. A., 1988a, Proterozoic paleokarst profile, Dismal Lakes Group, N.W.T., Canada, in James, N. P., and Choquette, P. W., eds., Paleokarst: Springer-Verlag, New York, p. 167–182.

———, AND ———, 1988b, Deeper water conical stromatolite reef, Sulky Formation, middle Proterozoic, N.W.T., in Geldsetzer, H., James, N. P., and Tebbutt, G., eds., "Reefs—Canada and Adjacent Areas": Canadian Society of Petroleum Geologists Memoir 13, p. 81–88.

———, ROSS, G. M., DONALDSON, J. A., AND GELDSETZER, H. J., 1981, Tectonism and depositional history of the Helikian Hornby Bay and Dismal Lakes Groups, District of Mackenzie, in Campbell, F. H. A., ed., Proterozoic Basins of Canada: Geological Survey of Canada Paper 81-10, p. 157–182.

KLEIN, C., BEUKES, N. J., AND SCHOPF, J. W., 1987, Filamentous microfossils in the 2.5 to 2.1 Ga-old Transvaal Supergroup: Their morphology, significance, and paleoenvironmental setting: Precambrian Research, v. 36, p. 81–94.

KNOLL, A. H., 1984, The Archean/Proterozoic transition: A sedimentary and paleobiological perspective, in Holland, H. D., and Trendall, A. F., eds., Patterns of Change in Earth Evolution: Springer-Verlag, New York, p. 221–242.

KOMINZ, M. A., 1984, Oceanic ridge volumes and sea-level change–An error analysis, in Schlee, J. S., ed., Interregional Unconformities and Hydrocarbon Accumulation: American Association of Petroleum Geologists Memoir 36, p. 108–126.

KUSKY, T. M., 1989, (1989) Archean compressional tectonics and allochthonous greenstone belts in the Slave province: Tectonics, in press.

LAWSON, A. C., 1913, The geology of Steeprock Lake, Ontario: Geological Survey of Canada Memoir 28, p. 7–15.

LI, T. H., TAKAHASHI, T., AND BROECKER, W. S., 1969, The degree of saturation of $CaCO_3$ in the oceans: Journal of Geophysical Research, v. 74, p. 5507–5525.

LOGAN, B. W., READ, J. F., HAGAN, G. M., HOFFMAN, P. F., BROWN, R. G., WOODS, P. J., AND GEBELEIN, C. D., 1974, Evolution and diagenesis of Quaternary carbonate sequences, Shark Bay, Western Australia: American Association of Petroleum Geologists Memoir 22, 358 p.

LONGMAN, M. W., 1981, A process approach to recognizing facies of reef complexes, in Toomey, D. F., ed., European Fossil Reef Models: Society of Economic Paleontologists and Mineralogists Special Publication 30, p. 9–41.

LOUCKS, R. G., AND FOLK, R. L., 1976, Fanlike rays of former aragonite in Permian Capitan Reef pisolite: Journal of Sedimentary Petrology, v. 46, p. 483–485.

LOWE, D. R., 1983, Restricted shallow-water sedimentation of early Archean stromatolitic and evaporitic strata of the Strelley Pool Chert, Pilbara Block, Western Australia: Precambrian Research, v. 19, p. 239–283.

———, AND KNAUTH, L. P., 1977, Sedimentology of the Onverwacht Group (3.4 billion years), Transvaal, South Africa, and its bearing on the characteristics and evolution of the early Earth: Journal of Geology, v. 85, p. 699–723.

LOWENSTAM, H. A., 1950, Niagaran reefs of the Great Lakes area: Journal of Geology, v. 58, p. 430–487.

MARION, M. L., AND OSBORNE, R. H., 1980, Sedimentology of the Beck Spring Dolomite, eastern Mojave Desert, southern California, in Geological Society of America, Abstracts with Programs, v. 12, p. 117–118.

MARTIN, A., 1978, The geology of the Belingwe-Shabani schist belt: Rhodesia Geological Survey Bulletin 83, 213 p.

———, NISBET, E. G., AND BICKLE, M. J., 1980, Archean stromatolites of the Belingwe Greenstone Belt, Zimbabwe (Rhodesia): Precambrian Research, v. 13, p. 337–362.

MATTHEWS, P. E., 1967, The pre-Karroo formations of the White Umfolozi inlier, northern Natal: Geological Society of South Africa Transactions, v. 70, p. 257–272.

MAZZULLO, S. J., 1980, Calcite pseudospar replacive of marine acicular aragonite, and implications for aragonite cement diagenesis: Journal of Sedimentary Petrology, v. 50, p. 409–422.

MCCORMICK, D. S., AND GROTZINGER, J. P., 1988, Aspects of the Burnside Formation, Bear Creek Group, Kilohigok Basin, District of Mackenzie, N.W.T., in Current Research, Part C: Geological Survey of Canada Paper 88-1C, p. 321–340.

MCGLYNN, J. C., AND HENDERSON, J. B., 1970, Archean volcanism and sedimentation in the Slave structural province, in Baer, A. J., ed., Symposium on Basins and Geosynclines of the Canadian Shield: Geological Survey of Canada Paper 70-40, p. 31–44.

———, AND ———, 1972, The Slave Province, in Price, R. A. and Douglas, R. J., eds., Variations in Tectonic Styles in Canada: Geological Association of Canada Special Paper 11, p. 506–526.

MCINTOSH, J. R., 1972, The Caland Ore Company Limited deposit: A geological description: Ontario Department of Mines and Northern Affairs Geological Report 93, p. 82–105.

MESOLELLA, K. J., ROBINSON, J. D., MCCORMICK, L. M., AND ORMISTON, A. R., 1974, Cyclic deposition of Silurian carbonates and evaporites in Michigan Basin: American Association of Petroleum Geologists Bulletin, v. 58, p. 34–62.

MUIR, M., LOCK, D., AND VON DER BORCH, C., 1980, The Coorong model for penecontemporaneous dolomite formation in the middle Proterozoic McArthur Group, Northern Territory, Australia, in Zenger, D. H., Dunham, J. B., and Ethington, R. L., eds., Concepts and Models of Dolomitization: Society of Economic Paleontologists and Mineralogists Special Publication 28, p. 51–67.

NISBET, E. G., AND WILKS, M. E., 1988, The Steep Rock Lake stromatolite reef, Atikokan, northwestern Ontario, in Geldsetzer, H., James, N. P., and Tebbutt, G., eds., "Reefs—Canada and Adjacent Areas": Canadian Society of Petroleum Geologists Memoir 13, p. 89–92.

OLSEN, P. E., 1984, Periodicity of lake-level cycles in the late Triassic Lockatong Formation of the Newark basin (Newark Supergroup), New Jersey and Pennsylvania, in Berger, A. L., Imbrie, J., Hays, J., Kukla, G., and Saltzman, B., eds., Milankovitch and Climate, Part 1: D. Reidel, Hingham, p. 129–146.

PELECHATY, S. M., AND GROTZINGER, J. P., 1988, Stormatolite bioherms of a 1.9 Ga foreland basin carbonate ramp, Beechey Formation, Kilohigok Basin, Northwest Territories, in Geldsetzer, H., James, N. P., and Tebbutt, G., eds., "Reefs—Canada and Adjacent Areas": Canadian Society of Petroleum Geologists Memoir 13, p. 93–104.

———, GROTZINGER, J. P., GOODARZI, F., SNOWDON, L. R., AND STASIUK, V., 1988, Middle Proterozoic karst development and bitumen emplacement, Parry Bay Formation (dolomite), Bathurst Inlet area, District of Mackenzie: A preliminary analysis, in Current Research, Part C: Geological Survey of Canada Paper 88-1C, p. 299–312.

PIERRE, C., AND ROUCHY, J. M., 1988, The carbonate replacements after sulfate evaporites in the middle Miocene of Egypt: Journal of Sedimentary Petrology, v. 58, p. 446–456.

PITMAN, W. C., III, 1978, Relationship between eustacy and stratigraphic sequences of passive margins: Geological Society of America Bulletin, v. 89, p. 1389–1403.

PLAYFORD, P. E., 1980, Devonian, "Great Barrier Reef" of Canning Basin, Western Australia: American Association of Petroleum Geologists Bulletin, v. 64, p. 814–840.

POULTON, T. P., 1973, Upper Proterozoic 'Limestone Unit', Northern Dogtooth Mountains, British Columbia: Canadian Journal of Earth Science, v. 10, p. 292–305.

———, 1988, Stromatolite bioherms, late Proterozoic, northern Purcell mountains, B. C., in Geldsetzer, H., James, N. P., and Tebbutt, G., eds., "Reefs—Canada and Adjacent Areas": Canadian Society of Petroleum Geologists Memoir 13, p. 105–109.

PRECHT, W. F., 1987, Extrinsic or intrinsic controls on reef facies development: Which is more important? Does it Matter? (abs.): in Canadian Reef Inventory Project: Program, p. 49.

PREISS, W. V., 1976, Intercontinental correlations, in Walter, M. R., ed., Stromatolites: Elsevier, New York, p. 359–370.

RAABEN, 1969, Columnar stromatolites and late Precambrian stratigraphy: American Journal of Science, v. 267, p. 1–18.

READ, J. F., 1980, Carbonate ramp-to-basin transitions and foreland basin evolution, Middle Ordovician, Virginia Appalachians: American Association of Petroleum Geologists Bulletin, v. 64, p. 1575–1612.

———, 1982, Carbonate platforms of passive (extensional) continental margins—Types, characteristics, and evolution: Tectonophysics, v. 81, p. 195–212.

———, 1985, Carbonate platform facies models: American Association of Petroleum Geologists Bulletin, v. 69, p. 1–21.

———, GROTZINGER, J. P., BOVA, J. A., AND KOERSCHNER, W. A., 1986, Models for generation of carbonate cycles: Geology, v. 14, p. 107–110.

RICKETTS, B. D., 1983, The evolution of a middle Precambrian dolostone sequence–A spectrum of dolomitization regimes: Journal of Sedimentary Petrology, v. 53, p. 565–586.

———, AND DONALDSON, J. A., 1981, Sedimentary history of the Belcher Group of Hudson Bay, in Campbell, F. H. A., ed., Proterozoic Basins of Canada: Geological Survey of Canada Paper 81-10, p. 235–254.

———, AND ———, 1988, Stromatolite reef development on a mud-dominated platform in the middle Precambrian Belcher Group of Hudson Bay, in Geldsetzer, H., James, N. P., and Tebbutt, G., eds., "Reefs—Canada and Adjacent Areas": Canadian Society of Petroleum Geologists Memoir 13, p. 113–119.

RONOV, A. B., 1968, Probable changes in the composition of sea water during the course of geologic time: Sedimentology, v. 10, p. 25–43.

ROSS, G. M., AND DONALDSON, J. A., 1988, Depositional history and facies geometry of a high energy early Proterozoic carbonate shelf (Hornby Bay Group, N.W.T., Canada), in Geldsetzer, H., James, N. P., and Tebbutt, G., eds., "Reefs—Canada and Adjacent Areas": Canadian Society of Petroleum Geologists Memoir 13, p. 120–128.

ROWLANDS, N. J., BLIGHT, P. G., JARVIS, D. M., AND VON DER BORCH, C. C., 1980, Sabkha and playa environments in late Proterozoic grabens, Willouran Ranges, South Australia: Journal of the Geological Society of Australia, v. 27, p. 55–68.

RUPPEL, S. C., AND WALKER, K. R., 1984, Petrology and depositional history of a Middle Ordovician carbonate platform: Chickamauga Group, northeastern Tennessee: Geological Society of American Bulletin, v. 95, p. 568–583.

SALOP, L. J., 1983, Geological evolution of the Earth during the Precambrian: Springer-Verlag, New York, 459 p.

SCHERMERHORN, L. J. G., 1974, Late Precambrian mixtites: Glacial and/or nonglacial?: American Journal of Science, v. 274, p. 673–824.

SCHLAGER, W., AND CAMBER, O., 1981, Depositional, erosional and by-pass slopes on carbonate platforms, in Abstracts of Papers: International Association of Sedimentologists, Eleventh Congress on Sedimentology, Hamilton, Canada, p. 179.

SCHOPF, J. W., AND WALTER, M. R., 1983, Archean microfossils: New evidence of ancient microbes, in Schopf, J. W., ed., Earth's Earliest Biosphere: Princeton University Press, Princeton, p. 214–239.

SEMIKHATOV, M. A., 1976, Experience in stromatolite studies in the U.S.S.R., in Walter, M. R., ed., Stromatolites: Elsevier, New York, p. 337–358.

———, 1978, Nekotorye karbonatnye stromatolity afebiya Kanadskogo shchita, in Raaben, M. E., ed., Nizhnyaya granitsa Rifeya i stromatolity Afebiya: Transactions of the Geological Institute, U.S.S.R. Academy of Sciences, v. 312, p. 111–147.

SEREBRYAKOV, S. N., AND SEMIKHATOV, M. A., 1974, Riphean and Recent stromatolites: A comparison: American Journal of Science, v. 274, p. 556–574.

SHRIDE, A. F., 1967, Younger Precambrian geology in southern Arizona: U.S. Geological Survey Professional Paper 566, 89 p.

SIMONSON, B. M., 1987, 2.5 Ga carbonate turbidites in the banded iron formation-rich Hamersley Group of Western Australia, in Geological Society of America, Abstracts with Programs, v. 19, p. 846.

SINGH, U., 1987, Ooids and cements from the late Precambrian of the Flinders Ranges, South Australia: Journal of Sedimentary Petrology, v. 57, p. 117–127.

SMYTH, H. L., 1981, Structural geology of Steep Rock Lake, Ontario: American Journal of Science, v. 43, Third Series, p. 317–331.

SOUTHGATE, P. N., 1986, Depositional environment and mechanism of preservation of microfossils, upper Proterozoic Bitter Springs Formation, Australia: Geology, v. 14, p. 683–686.

TANKARD, A. J., JACKSON, M. P. A., ERIKSSON, K. A., HOBDAY, D. K., HUNTER, D. R., AND MINTER, W. E. L., 1982, Crustal evolution of southern Africa: Springer-Verlag, New York, 523 p.

TEITZ, M. W., AND MOUNTJOY, E. W., 1985, The Yellowhead and Astoria carbonate platforms in the late Proterozoic upper Miette Group, Jasper, Alberta, in Current Research, Part A: Geological Survey of Canada Paper 85-1A, p. 341–348.

———, AND ———, 1988, The late Proterozoic Yellowhead carbonate platform, west of Jasper, Alberta, in Geldsetzer, H., James, N. P., and Tebbutt, G., eds., "Reefs—Canada and Adjacent Areas": Canadian Society of Petroleum Geologists Memoir 13, p. 129–134.

TUCKER, M. E., 1978, Triassic lacustrine sediments from South Wales: Shore-zone clastics, evaporites and carbonates, in Matter, A., and Tucker, M. E., eds., Modern and Ancient Lake Sediments: International Association of Sedimentologists Special Publication 2, p. 203–222.

———, 1982, Precambrian dolomites: Petrographic and isotopic evidence that they differ from Phanerozoic dolomites: Geology, v. 10, p. 7–12.

———, 1983, Diagenesis, geochemistry, and origin of a Precambrian dolomite: The Beck Spring Dolomite of eastern California: Journal of Sedimentary Petrology, v. 53, p. 1097–1119.

———, 1986, Formerly aragonitic limestones associated with tillites in the late Proterozoic of Death Valley, California: Journal of Sedimentary Petrology, v. 56, p. 818–830.

TWENHOFEL, W. H., 1919, Pre-Cambrian and Carboniferous algal deposits: American Journal of Science, v. 48, p. 339–352.

TYLER, S. A., AND BARGHOORN, E. S., 1954, Occurrence of structurally preserved plants in Precambrian rocks of the Canadian Shield: Science, v. 119, p. 606–608.

VEIZER, J., AND COMPSTON, W., 1976, $^{87}Sr/^{86}Sr$ in Precambrian carbonates as an index of crustal evolution: Geochemica et Cosmochemica Acta, v. 40, p. 905–914.

———, AND HOEFS, J., 1976, The nature of O^{18}/O^{16} and C^{13}/C^{12} secular trends in sedimentary carbonate rocks: Geochemica et Cosmochemica Acta, v. 40, p. 1387–1395.

VON BRUNN, V., AND HOBDAY, D. K., 1976, Early Precambrian tidal sedimentation in the Pongola Supergroup of South Africa: Journal of Sedimentary Petrology, v. 46, p. 670–679.

———, AND MASON, T. R., 1977, Siliciclastic-carbonate tidal deposits from the 3000 m.y. old Pongola Supergroup, South Africa: Sedimentary Geology, v. 18, p. 245–255.

WALCOTT, C. D., 1913, Notes on fossils from limestones of Steeprock Series, Ontario, in Lawson, A.C., The geology of Steeprock Lake, Ontario: Geological Survey of Canada Memoir 28, p. 16–23.

WALKER, J. C. G., 1983, Possible limits on the composition of the Archean ocean: Nature, v. 302, p. 518–520.

———, KLEIN, C., SCHIDLOWSKI, M., SCHOPF, J. W., STEVENSON, D. J., AND WALTER, M. R., 1983, Environmental evolution of the Archean-Early Proterozoic Earth, in, Schopf, J. W., ed., Earth's Earliest Biosphere: Princeton University Press, Princeton, p. 260–290.

WALKER, K. R., AND ALBERSTADT, L. P., 1975, Ecological successions as an aspect of structure in fossil communities: Paleobiology, v. 1, p. 238–257.

WALTER, M. R., 1972, Stromatolites and the biostratigraphy of the Australian Precambrian and Cambrian: Paleontological Association of London Special Paper 11, 190 p.

———, 1983, Archean stromatolites: Evidence of the Earth's earliest benthos, in Schopf, J. W., ed., Earth's Earliest Biosphere: Princeton University Press, Princeton, p. 187-213.

WARDLE, R. J., AND BAILEY, D. G., 1981, Early Proterozoic sequences in Labrador, in Campbell, F.H.A., ED., Proterozoic Basins of Canada: Geological Survey of Canada Paper 81–10, p. 331–360.

WILKS, M. E., 1986, The geology of the Steep Rock Group, N.W. Ontario: A major Archean unconformity and Archean stromatolites: Unpublished M. Sc. Thesis University of Saskatchewan, 206 p.

———, AND NISBET, E. G., 1985, Archean stromatolites from the Steep Rock Group, northwestern Ontario, Canada: Canadian Journal of Earth Science, v. 22, p. 792–799.

WILSON, J. L., 1975, Carbonate Facies in Geologic History: Springer-Verlag, New York, 470 p.

WINTER, H. de la R., 1963, Algal structures in sediments of the Ventersdorp System: Geological Society of South Africa Transactions, v. 66, 115–128.

YOUNG, G. M., 1981, The Amundsen embayment, Northwest Territories; Relevance to the Upper Proterozoic evolution of North America, in Campbell, F.H.A., ed., Proterozoic Basins of Canada: Geological survey of Canada Paper 81–10, p. 203–218.

YUREWICZ, D. A., 1977, Origin of the massive Capitan Limestone (Permian), Guadalupe Mountains, New Mexico and West Texas, in Upper Guadalupian Facies, Permian Reef Complex, Guadalupe Mountains, New Mexico and West Texas: Society of Economic Paleontologists and Mineralogists, Permian Basin Section Publication 77-16, p. 45–92.

ZEMPOLICH, W. G., WILKINSON, B. H., AND LOHMANN, K. C., in 1988, Diagenesis of late Proterozoic Carbonates: the Beck Spring Dolomite of eastern California: Journal of Sedimentary Petrology, v. 58, p. 656–672.

TECTONIC CONTROL ON THE FORMATION OF A CARBONATE PLATFORM: THE CAMBRIAN OF SOUTHWESTERN SARDINIA

THILO BECHSTÄDT

Geologisches Institut der Universität, Albertstrasse 23 B, D-7800 Freiburg, West Germany

AND

MARIA BONI

Dipartimento Scienze della Terra, Università degli Studi di Napoli, Largo San Marcellino 10, I-80138 Napoli, Italy

ABSTRACT: The Lower to Middle Cambrian sequence of southwestern Sardinia shows different stages of platform evolution through time, from a ramp to an isolated carbonate platform: (1) terrigenous carbonate homoclinal ramp with algal-archaeocyathan mounds (*Epiphyton/Renalcis*) in the west and terrigenous, shallow-marine to tidal sequences in the east; (2) carbonate terrigenous ramp or rimmed shelf with an ooid shoal complex, prograding toward the west; the back-shoal area contains peloidal mudstones, algal-archaeocyathan biostromes (*Girvanella*), and increasingly tidal deposits (siliciclasts and carbonates) toward the east; (3) isolated platform, aggraded to sea level, with an intra-shelf basin in the southeast and slopes to the north and west; (4) isolated, flooded platform; barriers toward the open sea partly broken down; (5) isolated platform with raised margins and deep interior, often with thick breccia beds in uppermost parts; (6) segmentation and drowning of the platform with deposition of nodular limestones and intercalated limestones and shales; and (7) siliciclastic deposits covering the former platform.

Evidence of tensional tectonics (slumping, debris flows, internal breccias, neptunian dykes, intraplatform basins and ponds) is abundant. Subsidence rates, however, are relatively low; the stratigraphic horizons are largely continuous. Rifting terminated by late Early Cambrian to Mid-Cambrian time, when plate-tectonic setting changed to a drift phase.

INTRODUCTION

The Iglesiente and Sulcis areas of southwestern Sardinia (Figs. 1, 2) consist mainly of Paleozoic (especially Cambro-Ordovician) sediments and intrusives. Within the Paleozoic sequence, two tectonostratigraphic units have been found, heavily influenced by the Variscan tectonics: a lower "autochthonous" unit and an overlying "allochthonous" unit (Carmignani and others, 1982). The allochthonous unit, not discussed in this paper, consists of Paleozoic siliciclastics and is overthrust from the northeast during Variscan orogenesis. A direct paleogeographic link of these siliciclastic Cambrian to Ordovician deposits with the carbonate platform of southwestern Sardinia, as proposed by Gandin and others (1987), is highly speculative.

The autochthonous sequence (Figs. 1, 2) is dominated by a terrigenous carbonate platform (Nebida Formation), followed by a carbonate platform of several hundred meters thickness (Gonnesa Formation) that is much mineralized (Ba-Zn-Pb). Mineralizations are not discussed in this paper; they have been described (Boni, 1985; Boni and others, 1988) and are the topic of a forthcoming paper (Boni and Bechstädt, in prep.).

The terrigenous carbonate and the carbonate platform sequences are Early Cambrian. The stratigraphic position of the base of the Nebida Formation is unknown. The existence of "Eocambrian" sediments is very doubtful; most likely, the so-called "Bithia Formation" is a slightly heteropic facies of the Nebida Formation. The lower part of the Nebida Formation, the Matoppa Member, falls (at least partly) into the Atdabanian stage, due to the presence of specific trilobites and archaeocyathans. The lithologic boundary between the Matoppa and the overlying Punta Manna member coincides approximately with the Atdabanian/Botomian (ex-Lenian) stratigraphic boundary. The top of the Nebida Formation apparently is still Botomian. According to Debrenne and Gandin (1985) and Pillola (1986), the overlying Gonnesa Formation is restricted to the Lower Cambrian, to the late Botomian to Toyonian (Elankian auctorum). A Middle Cambrian age has been attributed, however, by Gross (1982) and Mostler (1985), to upper parts of the Gonnesa Formation.

The history of research within this platform sequence is treated in detail in Bechstädt and others (1988). Cocozza (1979), Carannante and others (1984), and Gandin (1986) assume an erosional area to the north-northwest of the carbonate platform. Gandin and others (1987) see a transition from an inner shelf in the northwest (Iglesiente, northern Sulcis) to an outer shelf (eastern Sulcis) and finally into a slope and basinal area in the east (represented by the overlying Variscan nappes). In contrast, Bechstädt and others (1985) present evidence of an erosional area situated in the east-southeast of present outcrops and an open-marine area in the west-northwest.

The carbonate sequence of the Gonnesa Formation is overlain by approximately 50 m of deeper water carbonate-clastic alternations or nodular limestones (Campo Pisano Formation). According to Pillola (1986), the Campo Pisano Formation is composed mainly of Middle Cambrian sediments but also contains some uppermost Lower Cambrian rocks.

The development of the Cambrian platform of southwestern Sardinia fits well into established models of platform evolution (Read, 1985). Six stages can be distinguished (compare Figs. 1, 3, and 4) outlining the development from a terrigenous carbonate ramp (Nebida Formation) to an isolated carbonate platform (Gonnesa Formation) that finally drowned (Campo Pisano Formation) and was covered by open-marine terrigenous clastics (Cabitza Formation).

The Cabitza Formation spans the uppermost Middle Cambrian to Early Ordovician. Late Cambrian trilobites occur in lower and middle parts of the formation; *Rhabdinopora* (*Dictyonema*) *flabelliformis* indicates a Tremadocian age for upper parts (Gandin and Pillola, 1985; Pillola, 1986).

After a short-lived compression with subsequent erosion (Sardic phase, probably Mid-Ordovician), alluvial fans

("Puddinga") discordantly covered the Cambro-Ordovician sequence, followed by fluviatile sediments and Late Ordovician shallow-water clastics (Oggiano and others, 1986; Laske and Bechstädt, 1987). They contain a well-known Caradocian fauna.

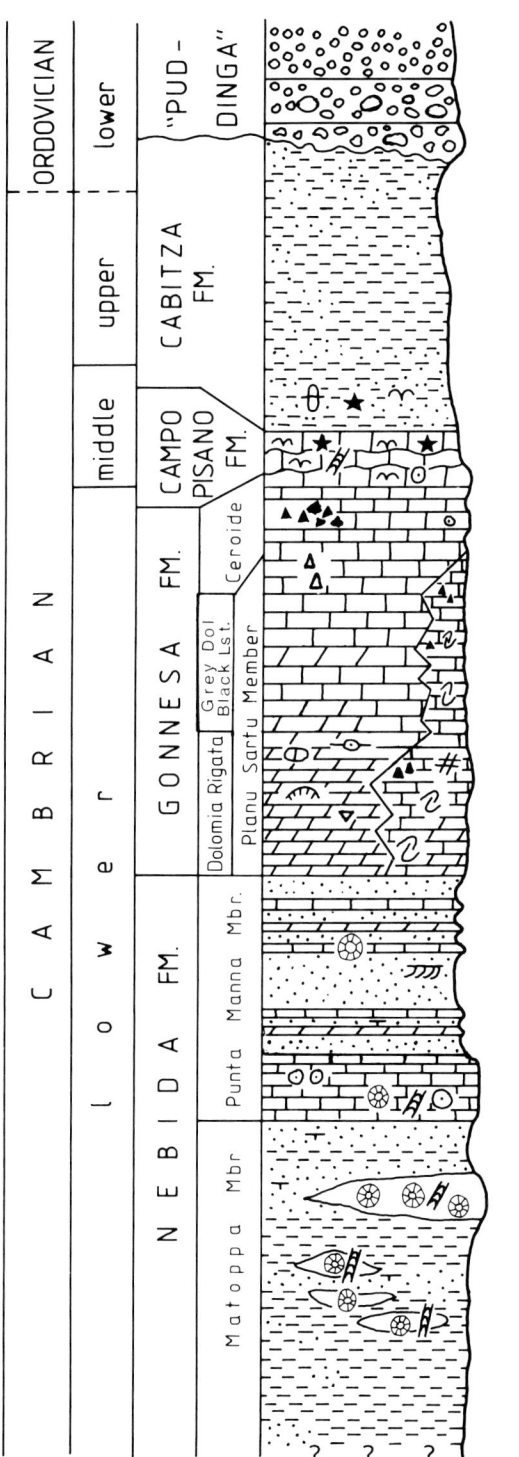

FIG. 1.—Generalized Cambro-Ordovician stratigraphic column. For legend, see Figure 4.

The whole stratigraphic interval often has been compared with Cambro-Ordovician sediments of the Spanish Meseta and the Montagne Noire of southern France. In Early Paleozoic time, Sardinia apparently was connected with these areas; movement and rotations of the island during Variscan orogenesis are debatable. In southwestern Sardinia, only scarce information is available for the late Paleozoic history. After the Variscan period with its strong low-angle thrusting, large parts of the area underwent emergence and karstification. During the Tertiary, local grabens were filled by sediments and volcanics. The present position of Sardinia is due to a separation from eastern Spain and southern France during the Tertiary.

In this paper data from two groups are presented: a group from the University of Freiburg has been active in the Sulcis and in the western Iglesiente on the Nebida and Gonnesa Formations, and a group from the University of Naples has worked mainly on the carbonate sequences of the Iglesiente. Almost 40 sections have been measured, and parts of the area have been mapped. Outcrop control is very good because of the semiarid climate and presence of many workings from old mines. Only in some areas of eastern Sulcis, is access difficult because of dense bushy vegetation. The main problem for a paleogeographic reconstruction is the heavy tectonics; several Paleozoic compressional phases, partly perpendicular to one another, are responsible for folding, faulting, and overthrusting. Therefore, only sketches (not to scale) of the paleogeography can be constructed. The main purpose of this paper is to outline the general development of the Cambrian carbonate platform, especially the tectonic control on its formation. In the discussion of the platform development, the first two stages are treated only briefly. A more detailed description of these terrigenous carbonate ramp stages at the base of the platform can be found in Bechstädt and others (1985, 1988) and in Selg (1985, 1986).

STAGE 1: HOMOCLINAL RAMP

Facies development and paleogeography.—

This stage is developed in upper parts of the Matoppa sequence. Clastics of tidal (sand flat and mud flat) origin are found in the east, whereas small algal-archaeocyathan-reef mounds occur in the west (Fig. 4). The mounds occur in several stratigraphic levels, and their dimensions increase with time. Constructed mainly by *Epiphyton-* and *Renalcis*-boundstone with some archaeocyathan floatstone, the mounds generally grew during times of strongly reduced (or interrupted) clastic supply to this western area. A low-energy environment is indicated by the abundance of carbonate mud (and early, partly radiaxial, fibrous cements) and by the archaeocyathans, which are mainly unfragmented. The contact of the siliciclastics with the carbonate mounds is very distinct, and few mixed lithologies have been found.

Because of shallowing and clastic input, the uppermost mounds were covered by terrigenous sediments. This was caused more likely by a climatic change and/or by a eu-

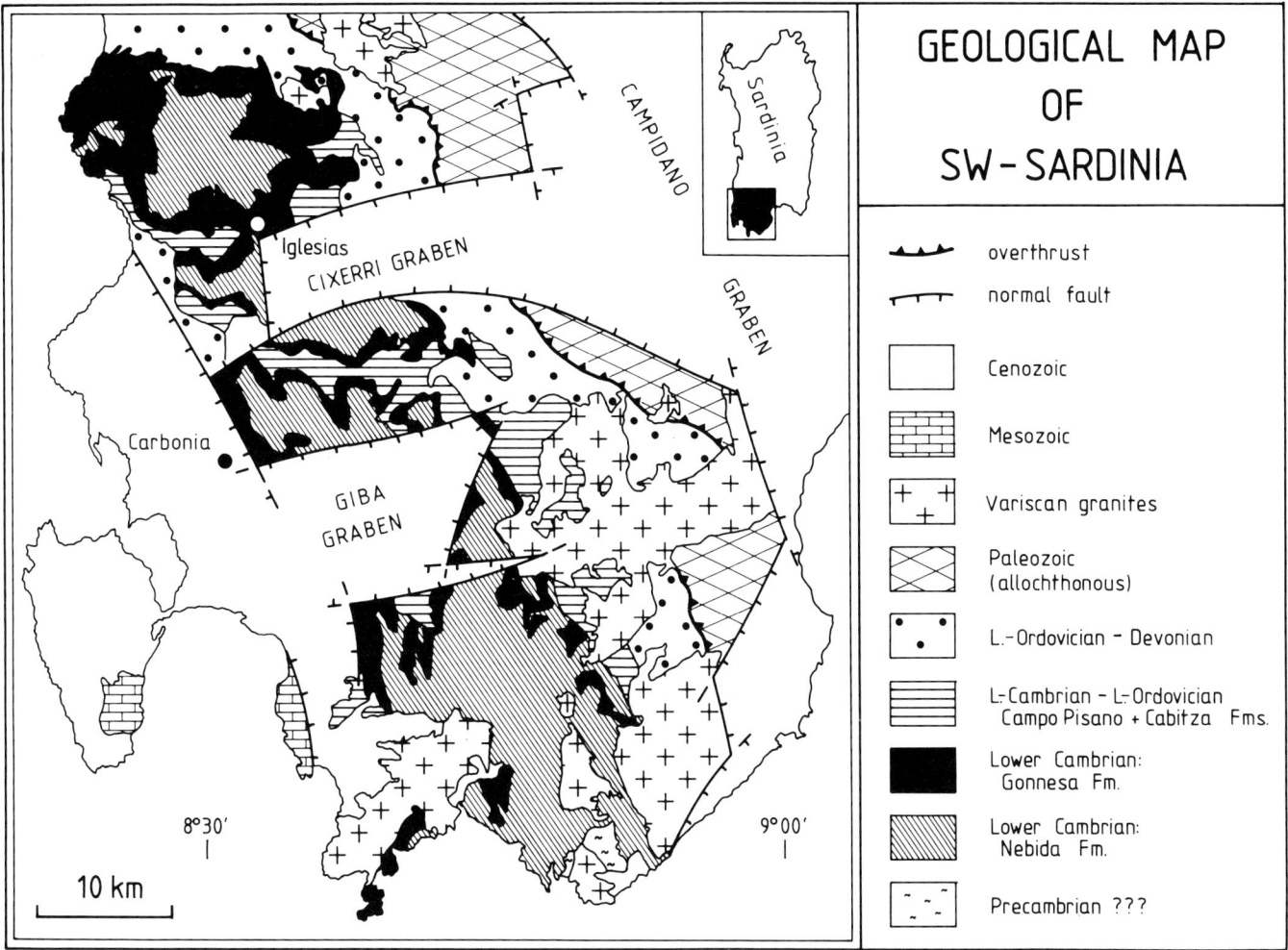

FIG. 2.—Geologic map of southwestern Sardinia.

static sea-level fall rather than by synsedimentary tectonics. This sudden influx of sands ended the homoclinal ramp stage.

STAGE 2: OOID-PELLET RAMP OR RIMMED SHELF

Facies development and paleogeography.—

Oolites, derived from ooid shoals in the western part of the area, are abundant at this stage (Figs. 4, 5). The shoals protected a back-shoal area to the east, where algal- (*Girvanella*) archaeocyathan boundstones and floatstones occur in subtidal facies. *Renalcis* and *Epiphyton* are absent. Peloidal mudstones and wackestones as well as some siliciclastics are common, together with ooid spillovers, in the shallow subtidal to intertidal parts. In back-shoal areas, three cyclic sequences occur, each approximately 20 m in thickness. They will be described by M. Selg (pers. commun., 1987). In this interval, slumping phenomena can be observed for the first time. Sediments indicate only locally restricted conditions.

Carbonates grade to the east into terrigenous clastics (Figs. 4, 5). Upper parts of this sequence are mainly cross-bedded sandstones with some shales and siltstones. Carbonate intercalations consist of local archaeocyathan framestones, but most are spillover oolites and tidal carbonates (algal laminites as well as intraclastic grapestone sediments). Number and thickness of these carbonate beds increase toward the top of this stage.

The change from high-energy deposits into tidal sequences at the top and the westward migration of the mixed clastic-carbonate tidal flats are indicative of an offlapping margin in which sedimentation rates exceeded subsidence rates. The platform increased in size beyond the present outcrop belt because of outbuilding to the west toward the more open-marine area.

The different facies belts (oolites in the west, followed by algal-archaeocyathan and peloidal limestones, passing into clastics toward the east) clearly contradict the model of Gandin and others (1987) for an eastern outer-shelf to slope area for this time.

The platform of late Nebidan time could have been either a ramp or a rimmed shelf (Read, 1985). For the Sardinian

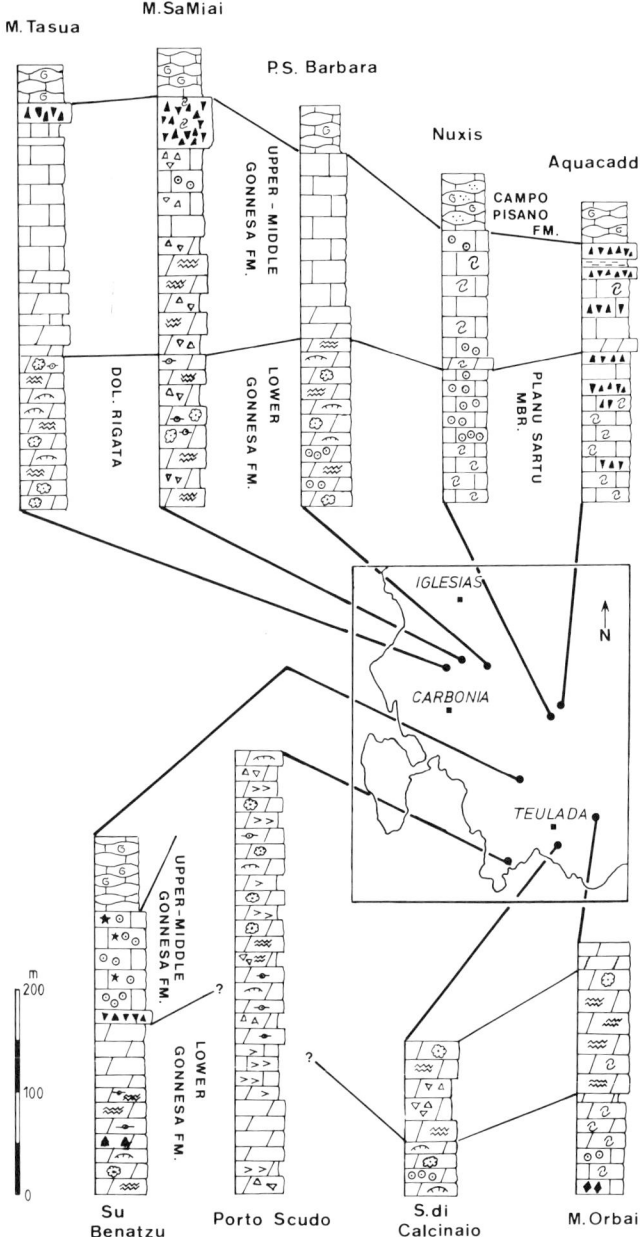

FIG. 3.—Stratigraphic sections of the Gonnesa and Campo Pisano Formations, arranged in two traverses. For legend, see Figure 4. Correlations based on lithology; base of Campo Pisano Formation is a biostratigraphic event.

platform, however, it is not possible to distinguish between these two types, because the slope and shelf break of that time are located outside of the present outcrop belt.

STAGE 3: ISOLATED PLATFORM, AGGRADED TO SEA LEVEL

With the onset of Gonnesa Formation deposition, the former clastic-carbonate platform became isolated; different basins developed because of an increase in tectonic activity.

Platform sediments.—

Tidal and supratidal carbonates with stromatolitic laminated fabrics, birdseyes, oncolites, grapestones, peloids and (vadose?) pisolites dominate the platform sequence of this stage (Figs. 3, 6, 7D). Black dolomitic tufa beds (Fig. 7C) of probably subaerial origin, showing upward- and downward-directed growth of microbial encrustations, have been observed quite frequently at the base of the sequence. Oncolitic overgrowths are common in this horizon. This facies, although much smaller in area, is almost identical to that described by Grotzinger (1986) from the Proterozoic Rocknest shelf. During subaerial periods, cavities and fissures formed within the tufa beds and later infilled with white micritic dolomite.

In the overlying sediments, often rich in stromatolitic mats, molds after anhydrite or gypsum and some idiomorphic gypsum crystals occur; gypsum platelets have been found in southern most areas (Schledding, 1985). An arid, restricted tidal environment is well documented by this sequence of the "Dolomia Rigata" (laminated dolomite).

Platform margins.—

Both high- and low-energy margins occur. This is important for interpretation of the type of basins and for elucidation of their former connections to the open sea.

The high-energy margin is oolitic and occurs in the Su Benatzu area (between Carbonia and Teulada) outlining the southwestern margin of the platform. In some places, high-energy grainstones with oolites can also be traced between the platform and the eastern slope (discussed later). In the Nuxis area (Schledding, 1985) south of Acquacadda, 40 m of cross-bedded and bioturbated oolitic grainstones were shed on slope sediments (Fig. 3). In some sections of this area, the sequence apparently is shallowing upward (flaser bedding, crossbedding), indicating a short-term local outbuilding of the platform toward the east.

FIG. 4.—Northwest-southeast paleogeographic cross sections from Nebidan to Cabitzan time, showing main stages of platform evolution: (1) terrigenous-carbonate homoclinal ramp with algal-archaeocyathan mounds (*Epiphyton/Renalcis*) in the west and terrigenous, shallow-marine to tidal sequences in the east; (2) carbonate terrigenous ramp or rimmed shelf with an ooid-shoal complex, prograding westward; the back-shoal area contains peloidal mudstones, algal-archaeocyathan biostromes (*Girvanella*), and increasingly tidal deposits (siliciclasts and carbonates) toward the east; (3) isolated platform, aggraded to sea level, with an intra-shelf basin in the southeast and intraplatform basin (Western Iglesiente Inlet) in the northwest; hypothetically, a barrier and the open sea were located farther west; (4) isolated, flooded platform; due to breccia masses bearing shallow-water debris, a barrier is still assumed farther to the west, outside present outcrops; (5) isolated platform with raised margins and deep interior; the proposed barrier toward the open sea was partly broken down because a high-energy margin can be observed in the northwest for the first time; (6) drowning and segmentation of the platform; small horsts and grabens develop with deposition of nodular limestones and intercalated limestones and shales; and (7) siliciclastic deposition covers the former platform.

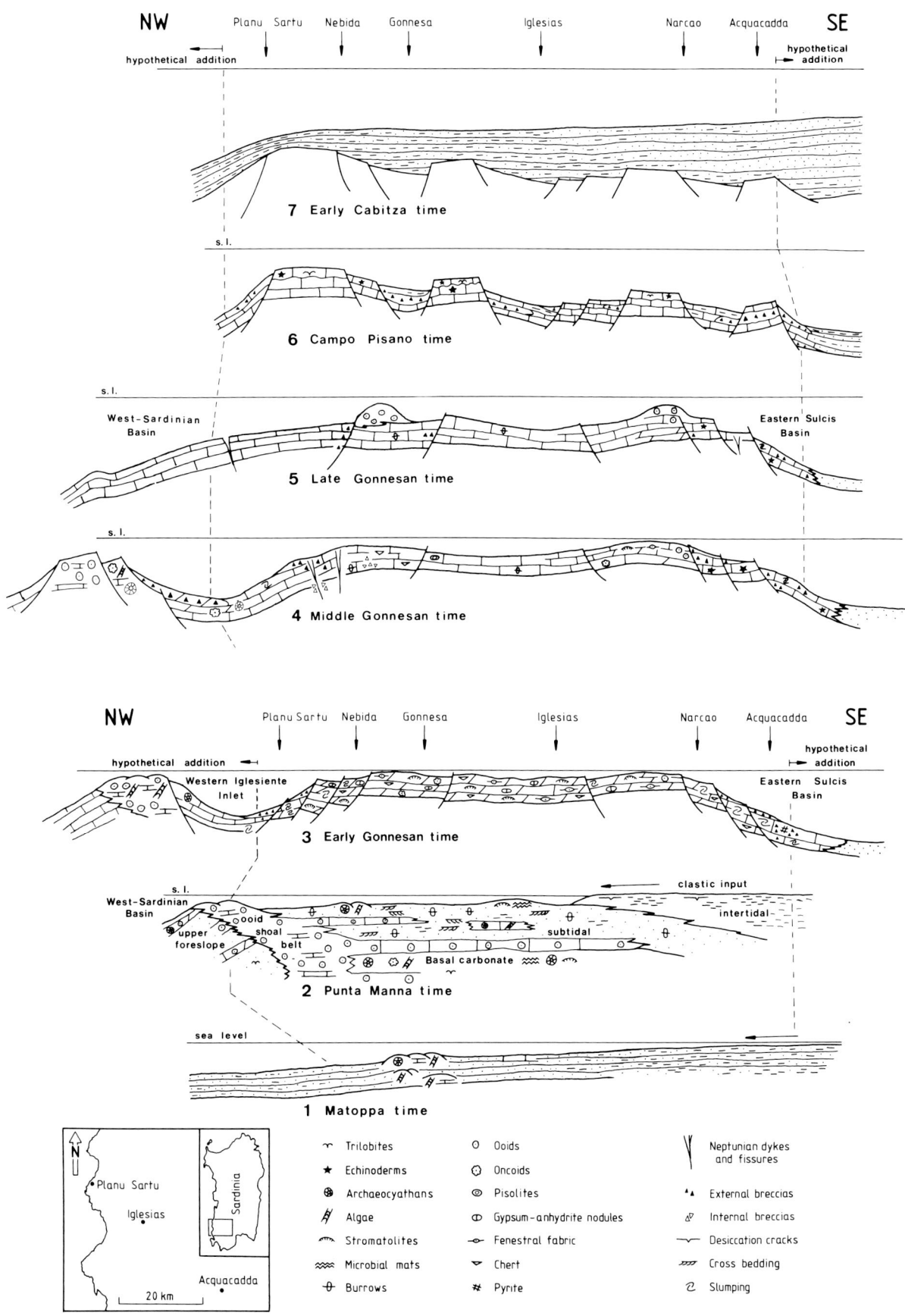

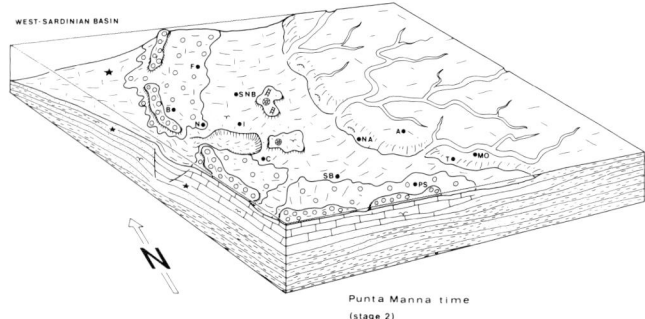

FIG. 5.—Paleogeography at late Nebidan time (stage 2, Punta Manna Member). An ooid-shoal complex is located to the west. Percentages of oolites clearly diminish when going east and are replaced by peloid, stromatolitic, and algal-archaeocyathan carbonates as well as terrigenous clastics, often cross-bedded, partly with mudcracks. Fine-grained, tidal, terrigenous clastics occupy eastern parts of the area. A = Acquacadda; B = Buggerru; C = Carbonia; F = Fluminimaggiore; I = Iglesias; MO = Monte Orbai; N = Nebida; NA = Narcao; PS = Porto Scudo; SB = Su Benatzu; SNB = San Benedetto; T = Teulada.

Low-energy margins crop out at the northwestern rim of the platform (Figs. 4, 6, 8) in the Acquaresi-Buggerru area. Farther west, slope and basinal deposits occur. These margins lack high-energy sediments but have cyclic sequences instead (Bechstädt and Boni, in prep.); the cycles are 10 or more meters thick and consist of two main types of sediment interlayered with each other: (a) Millimeter-laminated carbonates, consisting of flat and continuous laminations, form beds tens of centimeters thick. In the fresh rock the laminations are dark grey to black; weathered carbonates show very distinct white to dark grey laminations. These laminated carbonates resemble the laminites of the slope deposits described later. The laminites within the cycles frequently show small-scale slumping features and microfaulting but lack debris flows (Figs. 7A, B). (b) Tidal sediments and subaerial tufas (Fig. 7C), containing microbial mats, laminated fenestrae, rip-up clasts, stromatolites, and evaporite molds, are the second type of sediment.

Laterally, the cycles cannot be followed for more than tens to hundreds of meters; thickness of the cycles is strongly variable as well. This makes it difficult to correlate sections, which are only a few hundred meters apart.

These cycles, which can be found locally in the platform interior as well, are thought to be indicative of repeated, small-scale tilting that formed half-grabens with small ponds tens to a maximum of several hundred meters in size and as deep as 10 m. The sediments in the depressions were affected by seismic shocks. The ponds were infilled with partly slumped material and were finally covered by laminated tidal to supratidal carbonates. Layers of well-sorted

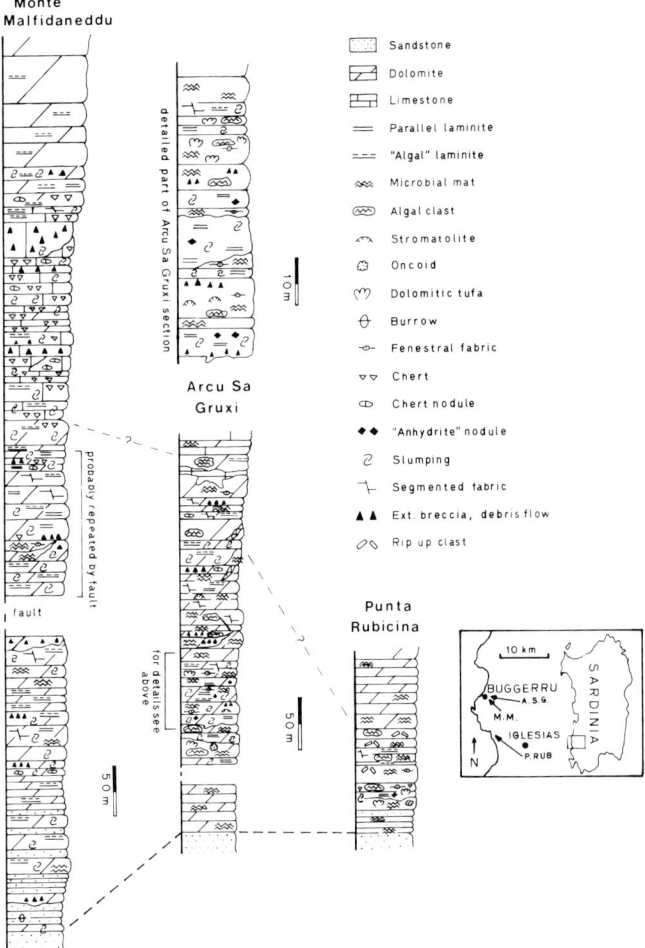

FIG. 6.—Stratigraphic sections from the northwestern area of the Iglesiente, documenting a transect from carbonate platform to slope deposits of the "Western Iglesiente Inlet." Correlations based on lithology. Lower correlation line is top of the Nebida Formation; upper line is base of the "black limestone" horizon. Inner part of the carbonate platform is documented by the Punta Rubicina section (east of Nebida), which exhibits thin Dolomia Rigata member and grey dolomite (upper half of section) deposits. Some tectonic instability is documented at the base of the section. Marginal part of the platform is represented by the Arcu Sa Gruxi section, consisting of Dolomia Rigata at the base, followed by anomalous grey dolomite and black limestone at the top of the sequence. This section is spectacular due to frequent seismites and the cyclic type of sedimentation (see enlargement and compare Figs. 7A–C and 10A.). Upper parts of a slope ("Western Iglesiente Inlet") are documented by the Monte Malfidaneddu section. The Planu Sartu member is partly repeated, followed by thicker bedded black limestones, which probably correlate with the "black limestone" of the platform sequences. This horizon is overlain by laminated limestones to dolomites. The top of the sequence (= top of Monte Malfidaneddu) consists of epigenetic "yellow dolomite." Compare also Figures 9A–C, E, and 10C, showing sediment types of this section.

FIG. 7.—(A) Slumped microlaminites from a cyclic sequence of the Dolomia Rigata at the northwestern platform margin. The slumped bed is located approximately 2 m below the sequence shown in (B) Telephone coin (2 cm) for scale. Arcu Sa Gruxi, east of Buggerru. (B) Laminated dolomite from cyclic sequence of the Dolomia Rigata at the northwestern platform margin. Very early diagenetic segmentation as well as microslumping can be observed, indicative of a seismite. Scale in inches and mm. Arcu Sa Gruxi, east of Buggerru. (C) Subaerial tufa, showing brecciations and fissures infilled by white micritic dolomite. Downward-directed growth of some tufa layers can be observed. Scale in cm. Marginal platform sequence of the Dolomia Rigata. Arcu Sa Gruxi, east of Buggerru. (D) tidal dolomites with coated grains and fenestrae. Scale in cm. Top of Dolomia Rigata at Guturru Canali Acquas, east of San Benedetto. (E) Part of seismite within finely laminated limestone of the Planu Sartu member, showing segmented zone overlain by soupy matrix containing some lithified fragments. Scale in cm. North of Acquacadda, eastern Sulcis.

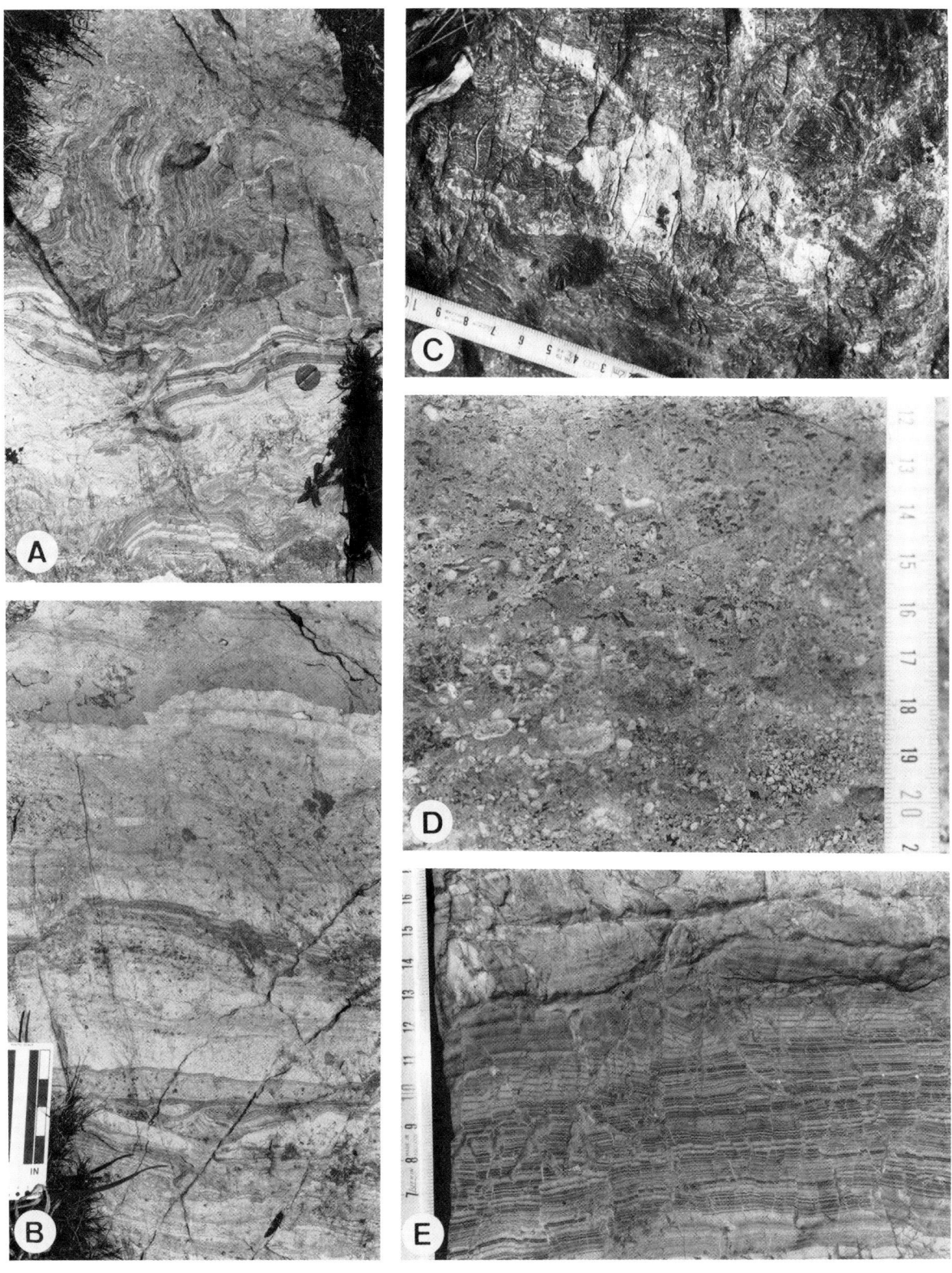

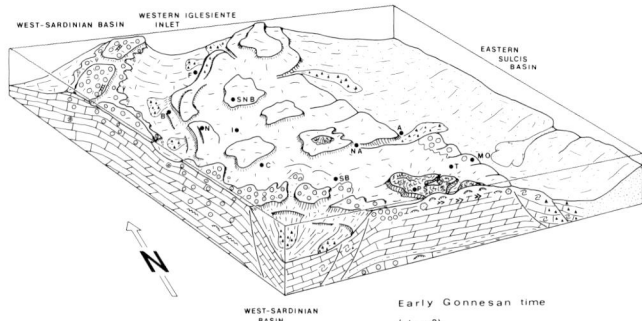

FIG. 8.—Paleogeography at Early Gonnesan time (stage 3, Dolomia Rigata/Planu Sartu Members). A peritidal, isolated carbonate platform with a high-energy southwest margin is accompanied by southeast intrashelf basin and northwest intraplatform basin (Western Iglesiente Inlet). The barrier and open sea farther northwest are hypothetical. A = Acquacadda; B = Buggerru; C = Carbonia; F = Fluminimaggiore; I = Iglesias; MO = Monte Orbai; N = Nebida; NA = Narcao; PS = Porto Scudo; SB = Su Benatzu; SNB = San Benedetto; T = Teulada.

algal debris occur within these tidal sequences. They might represent storm deposits. The tidal sediments show the impact of seismic activity as well; tidal sediments are disrupted and broken apart by many small, early diagenetic faults. In several places, neptunian dykes occur, partly infilled with calcareous or dolomitic microsparitic sediments, which are analogous to the limestones of stage 5, the "Ceroide."

Good outcrops with these facies can be found in the Punta Su Liberau–Nanni Frau area east of Buggerru. The laminites of this area are slumped, microfaulted, and show indications of seismic shocks; within microlaminites, which parallel under- and overlying algal carbonates, a soupy matrix can be observed as well as fragments of sediment in this matrix and segmented zones (Fig. 7B). All these are structures typical for seismites (Seilacher, 1969). Much of this tectonic activity might be related to mass transports on the slope, with listric shear planes that continue from the midslope to the upper slope and the bank margin.

Slope sediments.—

In southeastern and northwestern areas marginal to the platform, finely laminated limestones and dolomites ("Planu Sartu Member;" Bechstädt and others, 1985) occur together with well-bedded dark carbonates. The flat and continuous millimeter and centimeter laminations of the dark grey to black carbonates are different from the irregular laminations of the "algal" laminites from the platform. These deeper marine carbonates frequently show evidence of mass transport (slumping, debris flows, and local megabreccias). These sediments have been found in two areas to the southeast and northwest of platform sequences described earlier. They are interpreted as slope facies.

Different types of this facies have been found in the outermost northwest portion of the area (Figs. 4, 8). The slope sediments are underlain by tidal and supratidal carbonates, which are widely developed in uppermost Nebidan and lowermost Gonnesan time, in the area of the platform as well as at the base-of-slope deposits. Fine, commonly varvelike laminated limestones cover these tidal sediments and are strongly fissured locally (Fig. 9D). Seismites (Seilacher, 1969) and slumped-sediment bodies of all sizes, from millimeters to several meters, have been observed (Fig. 9A, B, C). Dips of slump folds indicate a westerly to northwesterly transport of the slump masses. During slumping, the laminites apparently were rather coherent, indicating early cementation. Debris flows, frequently containing chert breccias, become more and more prominent upward in the section. Thick-bedded dark limestones with debris flows (Fig. 9C, E) have been found on top of this lower, mostly dolomitic horizon, especially east and south of Buggerru. In the environs of Buggerru, a second horizon of dark microlaminates occurs on top of the dark limestones.

The close proximity of the slumped slope deposits and of the cyclic sequences of the platform margin (see Fig. 6) are interpreted to be similar to the clinoforms of northern Little Bahama Bank and of some Triassic carbonate flanks of the Dolomites, described recently by Kanter (1988). In these examples steeply dipping, listric shear planes of the upper slope and the bank margin turn parallel to the depositional clinoforms of the midslope area. They are interpreted as slide surfaces that developed under moderate overburden, disintegrating the upper slope and bank margins. The cyclic sequences of the Cambrian platform margin, therefore, might be the result of episodic mass transports on the slope.

The basinal sequence shallows upward only locally because of the outbuilding of the platform. In most parts of this western area, the slope sequence, which can be followed for approximately 8 km from Santa Lucia north of Buggerru to the area southeast of Canal Grande (northwest of Nebida), continues into middle Gonnesan time. The thickness of the Planu Sartu in this area is much greater than at the eastern margin of the platform (Figs. 3, 6) where it also composes parts of the overlying "grey dolomite" (450 to 600 m thick).

Slope sediments crop out in the southeast as well (compare Figs. 3, 4 and 8). Basinal deposits are missing because of an overlying Variscan nappe covering areas farther east

FIG. 9.—(A) Finely laminated, slumped dolomites covered by laminated sequence, which is in place. Planu Sartu Member from Monte Malfidaneddu, east of Buggerru. (B) Detail of left part of (A) Scale in cm. (C) Thin- to thick-bedded, cherty limestones of the Planu Sartu Member. The lower part is heavily faulted due to slumping. Monte Malfidaneddu, east of Buggerru. Hammer for scale. (D) Finely laminated dolomites from basal part of the Planu Sartu Member, showing small-scale dykes partly infilled with sediment. Scale in cm. Between Canal Grande and Cala Domestica, western Iglesiente. (E) Debris flow from dark, thick-bedded limestones of the Planu Sartu Member. This horizon corresponds to upper parts of the sequence shown in (C) Scale in cm and inches. Monte Malfidaneddu, east of Buggerru.

(Fig. 2). Intraformational mass-movements, translational slides, and slumps are common and can be followed for approximately 20 km in a north-south direction (Bechstädt and others, 1984). The slope sediments are exclusively limestones, most of them finely laminated and rich in pyrite. These commonly black limestones are less than 100 m thick, much thinner than the "normal" Dolomia Rigata (Fig. 3). Seismites are frequent in the Acquacadda area as well: all stages occur, from liquefied sediments with larger fragments of original sediment, to segmented parts of the section, to slumping and sliding (compare Fig. 7E and Bechstädt and others, 1984). Lithification during time of slumping was highly variable, and slump folds change laterally into slump breccias. In the Acquacadda area, debris flows with meter-size clasts have been observed (Bechstädt and others, 1984). Slumping and sliding were directed toward the east (Schledding, 1985), indicating an eastward-directed slope.

Isolation of the platform, paleogeography.—

During Nebidan time, no basin existed between the platform and the clastic source: clastics were delivered far to the west (compare Figs. 4 and 5). The eastern basin was established in early Gonnesan time due to a tensional phase, as shown by the slope sediments with seismites, slumping features, brecciations, and mass movements, and by ruptural events, which affected the platform and its marginal areas. This eastern Sulcis basin separated the platform from the clastic source. Almost all of the clastics delivered from an erosional area farther east were trapped in the newly formed basin. Coarse-grained clastics could not reach the western areas, and an exclusively carbonate platform developed (Figs. 4, 8).

Interpretation of the paleogeographic situation in the northwest is much more complicated. Only slope sediments and platform sediments of mostly tidal origin have been found. Platform margin high-energy deposits are lacking. Cyclic sediments with indications of strong tectonic activity occur, however. Since late Nebidan time, the platform margin was apparently located farther to the west, in front of an open-marine realm outside the onshore area of the present western Iglesiente (compare Figs. 4 and 8). The slope sediments, cropping out in the northwesternmost part of the area, did not face the Cambrian open sea. They are part of a platform internal basin or an oceanic tongue ("Western Iglesiente Inlet"). This is the reason for the absence of high-energy platform-marginal sediments in this area, in contrast to the oolite-rich late Nebidan sediments. Farther west, a barrier between this inlet and the open sea may have been developed. Reworked remnants of this proposed barrier have been found recently in younger sequences of the Planu Sartu-Cala Domestica-Canal Grande area (see stage 4). Gandin and others (1987) assume an inner-shelf position of the Iglesiente and northern Sulcis also during Gonnesan time. The high-energy oolitic sediments of the Su Benatzu area in western Sulcis (Schledding, 1985) and the slumped and brecciated slope sediments of the westernmost Iglesiente (Figs. 3, 6), several hundred meters thick, clearly contradict this paleogeographic model.

STAGE 4: FLOODED ISOLATED PLATFORM

Facies development and paleogeography.—

In eastern- and westernmost parts of the area farther from the platform, the Planu Sartu continues into middle Gonnesan time. In areas marginal to the platform, the Planu Sartu Member is frequently overlain by grey dolomites, often showing algal laminations and some reworking; these sediments also occur on top of the tidal sediments of the platform (Dolomia Rigata). The sequence will be described in more detail in Bechstädt and Boni (in prep.).

In the Iglesiente as well as the Sulcis areas (compare Figs. 1 and 4), the dolomites are followed by, or they pass laterally into, black limestones. These black limestone sequences have been described by Boni and Marinacci (1980), Boni (1985), and Schledding (1985). This type of sediment forms a distinct stratigraphic horizon. Main outcrops of the black limestones are in the eastern Sulcis ("Acquacadda Member"; Schledding, 1985) and in the Iglesiente, where they have been deposited on top of the Planu Sartu as well as on top of the Dolomia Rigata or on top of the grey dolomites. The facies mainly consists of mudstones and some peloid grainstones, which are thickly bedded and lack laminations. The mostly micritic black limestones indicate a low-energy environment. Local occurrences of lithoclast grainstones and peloid grain- to packstones (Punta Santa Barbara, north of Narcao, for instance) in the same stratigraphic level indicate higher energy conditions on local shoals, now situated within and on marginal parts of the platform (Schledding, 1985). These shallow-water sediments are sometimes dolomitized. Oolites are rare; some have been observed in the Nuxis section (Eastern Sulcis, south of Acquacadda) close to the eastern margin of the platform.

There are no direct hints for the depositional depths of the black limestones. Both the low-energy black limestones and the high-energy shallow-water facies are overlain by the same type of microsparitic limestone, the "Ceroide" (stage 5). This change to a uniform facies development is an argument for relatively shallow depths of the black limestone, not exceeding several tens of meters.

At the transition from Dolomia Rigata (stage 3) to grey dolomites and/or black limestones (stage 4), neptunian dykes have been observed in several areas (for instance, at San Benedetto east of Buggerru), sometimes filled with white microsparitic sediments similar to those of the Ceroide and/or clayey material. These, as well as frequent occurrences of breccia beds, are proof of strong tensional tectonics affecting the area at that time. The whole sequence is indicative of a relative sea-level rise that flooded the former tidal platform (Fig. 4). It is not clear whether the flooding of the platform was caused by increased subsidence or by a eustatic sea-level rise (or both). An argument for increased subsidence is the evidence of tensional tectonics. An argument for eustatic control is the uniform, widespread occurrence of the black limestones. In the preceding lower Gonnesan time, there were no shelf areas with high-carbonate production. Because of this, a distinct relative rise of sea level could not be compensated by sedimentation of carbonates.

The basinal areas in the east and west still existed (see earlier discussion). Slope sediments, breccia beds, and slumped sequences crop out in the Acquacadda area and in the westernmost Iglesiente between Cala Domestica and Planu Sartu. In this area, recently discovered breccia sequences 200 m thick occur (Fig. 10E) and are overlain by microsparitic limestones (Ceroide) of stage 5. The breccia beds contain shallow-water, algae-rich debris (unknown *in situ* so far), transported archaeocyathans, and 20 m of finely laminated, slumped-slope sequences that wedge out laterally (Fig. 10B). At Cala Domestica, the breccia masses overlie floatstones/rudstones consisting of resedimented oncolites with archaeocyathan cores, described by Debrenne and Gandin (1985). Farther south, at Canal Grande, graded beds within upper parts of the Planu Sartu Member locally contain archaeocyathan debris (Fig. 10D). No archaeocyathans have been found at the margins of the Dolomia Rigata platform; probably, the archaeocyathans were transported from a shoal area farther west, outside the present outcrops. Because of the breccia sequences, it can be assumed that this barrier between the Western Iglesiente Inlet and the open sea disintegrated in middle Gonnesan time because of tectonic activity (compare Figs. 4 and 6).

STAGE 5: PLATFORM WITH RAISED RIMS AND DEEP INTERIOR

Facies development and paleogeography.—

Peloidal mudstones to wackestones, sometimes heavily bioturbated, become abundant during stage 5 in the "Ceroide" (= waxy limestone) facies, which covers most of the area (Figs. 3, 4, 11). Deeper water, low-energy conditions are indicated by this facies; the interpretation, however, is hampered by strong recrystallization of these clean limestones. Local occurrences of high-energy sediments (oolite grainstones, packstones to grainstones containing archaeocyathans) are to be found in upper parts of some sequences in the west (G.L. Pillola, pers. commun., 1985; Schledding, 1985). In our interpretation most of the Ceroide represents a flooded platform, changing to a deeper lagoonal platform with a few local shoals in the platform interior and with shoals that form raised rims to the platform. Strong transgressive pulses could be compensated only locally by higher carbonate production, especially of the marginal facies. The local shoal facies might be due to a sea-level fall or to tectonic activity.

The high-energy deposits of Ceroide time overlie different facies: in the northwest, the platform margin is not in the same position as it was during late Nebidan and early Gonnesan time. There is no more evidence of a barrier separating an intraplatform basin (the "Western Iglesiente Inlet") from the open sea. The platform margin, instead, approximately follows the eastern margin of the former intraplatform basin. This distinct backstepping of the northwestern platform margin probably was caused by tectonic pulses of drowning of the platform. These pulses are evidenced by numerous occurrences of slumping, debris flows, and neptunian dykes (Planu Sartu area, for instance; Fig. 10F). Breccias are common and include matrix-rich debris flows and matrix-poor or matrix-lacking "internal breccias" with a good "fitting" of the components (Füchtbauer and Richter, 1983). The latter type of breccia might be caused by flexure-like deformations of the northwestern platform margin, where fissures frequently occur. The most important Pb-Zn deposits within the upper Gonnesa Formation occur in such breccia beds.

In easternmost sections, some oolitic grainstones have been found at the platform margin, as well as oolite intercalations within fine-grained slope carbonates and breccia beds. In contrast to the northwestern platform margin, which shifted in time, the positions of the eastern and the southwestern (Su Benatzu) margins were relatively stable. At Su Benatzu, the high-energy belt is located on tidal sediments of early Gonnesan time (Figs. 3, 8, 11).

In southernmost parts, in southern Sulcis, tidal sediments locally bearing gypsum platelets can be found throughout the Gonnesan sequence. A paleo-high is assumed for this area (Figs. 8, 11).

Breccia sediments are especially frequent at the top of the sequence (see stage 6); the platform was dissected heavily in late Ceroide time. Slumping and debris flows occur not only at the western and eastern rims of, but also within, the platform.

STAGE 6: SEGMENTATION AND DROWNING OF THE PLATFORM

Facies development and paleogeography.—

The Ceroide is overlain by nodular limestones and calcareous slates of the Campo Pisano Formation. Breccias, the clasts consisting of Ceroide limestone, can be found in several places in uppermost parts of the Ceroide. Some of them are internal breccias, whereas others have the characteristics of debris flows. According to Gandin (1985), tensional tectonics caused short-lived local emergence and karstification due to rotation of blocks. This might have been possible, although no evidence of real karstic facies and/or vadose diagenesis was found in the upper Ceroide sediments. The red matrix of some breccia horizons and within many fissures, taken as an indiciator of karstification, might be related to the mostly red facies of the Campo Pisano. This assumption is corroborated by slumping and debris flows at the change from Ceroide to Campo Pisano deposits. These fabrics occur mainly at the western and eastern rims of the platform but also locally within the platform. At Acquacadda, blocks of nodular limestone have been incorporated into the breccia beds (Boni and others, 1981). Breccia components, as well as the fabrics of the breccia horizons, indicate a subaqueous origin.

Nodular limestones as well as calcareous slates are typical of the fossil-rich Campo Pisano Formation (Pillola, 1986), which often has a distinct contact on top of the Gonnesa Formation. In other places (north of San Benedetto, for instance), the uppermost parts of the Ceroide contain several intercalations of red calcareous Campo Pisano slates as well as breccia beds.

The Campo Pisano Formation is a condensed sequence spanning the uppermost Lower Cambrian to uppermost Middle Cambrian. The sediments consist of bioclastic wackestones to packstones and bioclastic floatstones. Echinoderm debris, trilobite exuviae, sponge spicules (Mostler, 1985), *Chancelloria*, hyolites, and brachiopods indicate an

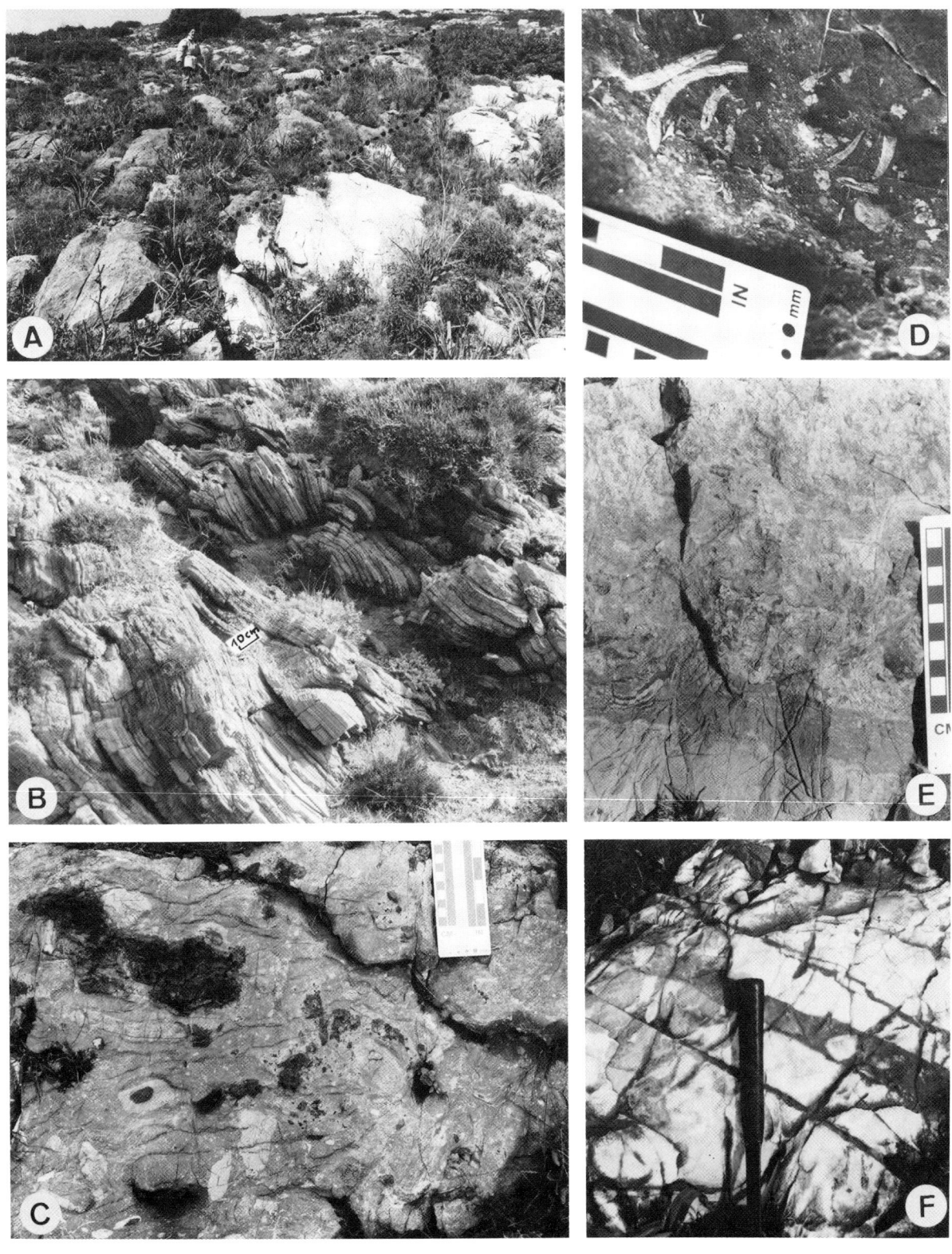

open-marine environment. This type of facies is similar to the "ammonitico rosso" facies of Tethyan sequences. The origin of the carbonate mud of the Campo Pisano sediments is problematic. No remnants of the platform able to deliver carbonate ooze are known at that time.

The calcareous sediments of the Campo Pisano Formation are followed in turn by the siliciclastics of the Cabitza Formation, which are often finely laminated. Thin carbonate layers are to be found only locally within the open-marine Cabitza sediments. This siliciclastic sedimentation lasted well into Ordovician time.

CONTROLS ON THE PLATFORM EVOLUTION

The different stages of evolution of the Cambrian platform of Sardinia fit well into the model of Read (1985). (1) A homoclinal ramp (Nebida Formation, Matoppa Member) at the base changed into (2) an ooid-pellet shoal-type of ramp and/or a rimmed shelf (Nebida Formation, Punta Manna Member) that prograded toward the west. (3) Subsequently, the platform became isolated, at first aggrading to sea level in early Gonnesan time. In the northwest, a barrier separated a small intraplatform basin or an oceanic tongue from the open sea. (4) Later, in middle Gonnesan time, this isolated platform was flooded and the barrier toward the open sea to the west disintegrated. (5) Raised margins developed in late Gonnesan time; only the margins were able to keep pace with the transgressive pulses. (6) With the onset of Campo Pisano deposition, the isolated platform was segmented and drowned. Nodular limestones and intercalated limestones shales were deposited on the platform. (7) Finally, the siliciclastics of the Cabitza Formation buried the former carbonate platform.

The Early Cambrian carbonate deposits of southwestern Sardinia can be compared quite well with Triassic/Lower Jurassic sequences of the Alps and of Italy (compare also Cocozza and Gandin, 1987), situated on the margin of the Mesozoic Tethys. The carbonate platforms of both times were extremely unstable throughout their evolution. Lead-zinc mineralizations can be found in the Triassic platforms of the Alps (Wetterstein limestone, Schlern dolomite) as well as in the Cambrian platform described. Most ore bodies in the Cambrian of Sardinia, both barite and sulfides, are contained in well-determined stratigraphic horizons. The ores are more or less directly related to the Cambrian tectonics and the instability of the platform. Facies and paleogeography strongly conditioned distribution and localization of the ore bodies. Much of the mineralization is of the Mississippi Valley-type, especially in upper parts of the Gonnesa Formation (compare Boni, 1985). Some of the deposits apparently are sedimentary exhalative; they are of the

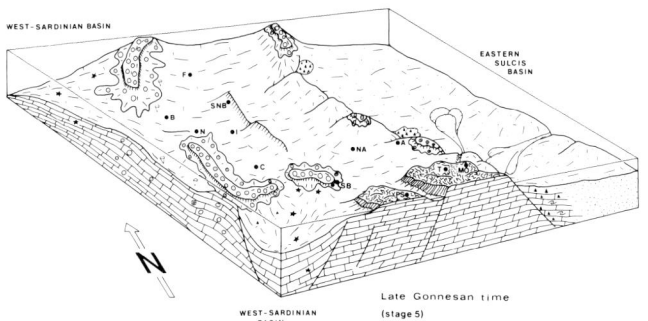

Fig. 11.—Paleogeography at Late Gonnesan time (stage 5, Ceroide). In contrast to early Gonnesan time (Fig. 4), the carbonate platform now was no longer aggraded to sea level; raised platform margins enclosed a deeper lagoonal interior. The platform margins of the southwest and east areas were located in almost the same position as in early Gonnesan time. To the northwest, however, a clear backstepping of the platform margin occurred, from a (hypothetical) position outside present outcrops into the area cropping out today. A = Acquacadda; B = Buggerru; C = Carbonia; F = Fluminimaggiore; I = Iglesias; MO = Monte Orbai; N = Nebida; NA = Narcao; PS = Porto Scudo; SB = Su Benatzu; SNB = San Benedetto; T = Teulada.

Irish/McArthur-type of ore deposits (compare Williams, 1980). These massive sulfides are related to the tectonic event initiating the isolated Gonnesa platform: the ores are restricted to uppermost parts of platform stage 2 and to stage 3. The massive sulfides generally consist of pyrite and sphalerite, with lesser amounts of galena. This type of deposit is common from the Iglesias syncline to northern and northeastern areas of the platform. The sulfide layers are often disrupted and slumped and locally contain different generations of breccias. It is quite clear that these ores formed in an environment of strong synsedimentary and/or syndiagenetic instability.

We interpret the ore genesis in southwestern Sardinia as an evolving process, resulting in a series of deposits ranging from Sedex (sedimentary exhalative) to Mississippi Valley-types.

The carbonate platform of southwestern Sardinia was heavily influenced by at least four main tensional events, each characterized by slumping, debris flows, brecciation phenomena, and often by mineralization: (1) During deposition of the uppermost Nebida/lowermost Gonnesa Formations, the main events were the formation of an intrashelf basin in the east and an intraplatform basin and ponds in the northwest. Seismites were typical for early Gonnesan time. (2) In middle Gonnesan time, the tectonic event was less prominent than event 1. It is characterized by neptunian dykes and local debris flows at the base of the "black limestone." (3) During Ceroide deposition, breccias and

Fig. 10.—(A) Light and dark micritic carbonates of the Dolomia Rigata at the northwestern platform margin, indicating mass transport. Dark beds (inside stippled area) are strongly folded and thin completely laterally due to gravitational transport. Other horizons show strong variations in thickness. Person for scale. Arcu Sa Gruxi, east of Buggerru. (B) Slumped, finely laminated dolomites from a discontinuous interlayer within thick breccia masses from upper parts of the Planu Sartu Member at Is Tres Cannabis, north of Cala Domestica, western Iglesiente. (C) Detail of partly silicified debris flow from Planu Sartu Member, Monte Malfidaneddu, east of Buggerru. (D) Transported archaeocyathans at the base of a graded bed from upper parts of Planu Sartu Member, Canal Susuia, north of Canal Grande (western Iglesiente). (E) Detail of breccia masses from upper parts of the Planu Sartu Member containing fragments of algal limestones. North of Cala Domestica (western Iglesiente). (F) Ceroide with small neptunian dyke infilled by grey calcareous matrix. East of Planu Sartu, west of Buggerru.

neptunian dykes accumulated in many places (especially of the Iglesiente) in different stratigraphic levels. (4) At the top of the Ceroide Member, event 4 started with Ceroide breccias in uppermost parts of the Gonnesa Formation and culminated in the flooding and breakdown of the platform and deposition of widespread condensed Campo Pisano Formation Sediments.

Events 1 and 4 are the most important. According to the stratigraphic data mentioned earlier event 1 fell into the Botomian; events 2-4 occurred in the early to late Toyonian.

Because of the strong tensional tectonics, estimates of overall sea-level changes are difficult. A transgression is manifested in basal parts of the sequence (Matoppa Member, Atdabanian), whereas in upper parts of the Matoppa and in the overlying Punta Manna Member, a general regressive trend can be observed (lower part of the Botomian). Its maximum was reached during early Gonnesan time, during the formation of the "Dolomia Rigata" (late Botomian). The local deeper water sequences of this time are due to tectonic subsidence of intra-shelf and intraplatform basins.

In middle Gonnesan time (probably early Toyonian), the "black limestone" manifested a transgression. The flooding of the platform at this time might have been caused by stronger subsidence or, alternatively, by a eustatic sea-level rise. Neptunian dykes and breccias at the base of this level, together with the development of local shoals, indicate tectonic events (event 2) active at that time, whereas the widespread, relatively uniform development of the black limestones and of the Ceroide indicates eustatic control of this transgression. Sediments of the subsequent middle to (?)late Toyonian show an overall trend of platform deepening. Only in late Gonnesan time (late Toyonian), in upper parts of the Ceroide, did oolite shoals develop at the platform margin. The timing of this facies development roughly coincides with the "Hawke Bay Event" of Palmer and James (1980), a circum-Iapetus regression near the Lower-Middle Cambrian boundary. In southwestern Sardinia, however, no terrigenous clastics are known from this interval. Our group could not find undebatable evidence for the local emergence and karstification mentioned by Gandin (1985) and Cocozza and Gandin (1987) from the base of the Campo Pisano Formation (uppermost Lower Cambrian, lowermost Middle Cambrian). The strong transgression in the Middle Cambrian (Campo Pisano Formation) is linked with strong tectonic instability (breccias at Acquacadda, for instance). The uniform widespread development of the Campo Pisano facies however, points to a eustatic control as well.

According to Gandin (1982), Cocozza and Gandin (1987), and Gandin and others (1987), the climate changed from arid (Dolomia Rigata) to humid tropical conditions (Ceroide facies), whereas the depth of deposition did not change significantly within Gonnesan time. We agree fully on the arid climate of late Nebidan/early Gonnesan time; indications of evaporites have been described from stage 3 especially (Schledding, 1985; Bechstädt and others, 1988). Evidence for a climatic change to humid conditions in late Gonnesan time is rather weak, however. The assumption of no changes in water depth during Gonnesan time is at odds with the sedimentary fabrics, described in this paper.

Because of the mostly subtidal conditions, estimates of the climate are difficult. Neither the facies nor the fauna (presence of archaeocyathans) give a distinct hint for a climatic change.

The tensional regime of the Cambrian ended in Ordovician time. A short-lived compression ("Sardic phase") of probable Mid-Ordovician age affected the area, leading to erosion and development of alluvial fans and fluviatile conditions, followed again by shallow-marine sequences (Oggiano and others, 1986; Laske and Bechstädt, 1987). The timing of this event largely coincides with the Knox unconformity of the Appalachians (Mussman and Read, 1986).

PLATE-TECTONIC SETTING

Subsidence rates influence the evolution of carbonate platforms and are indicative of the tectonic regime. Calculation of the subsidence rate of the Gonnesan platform is rather difficult, however. The problems are the ongoing discussion on Cambrian stratigraphic terminology and the uncertainty on the absolute ages of the Cambrian stages, as well as calculation of sea-level changes during the Cambrian. Whereas, for instance, the Proterozoic/Cambrian boundary has been set by most authors at approximately 570 Ma (Harland and others, 1982; Palmer, 1983), other authors suggest 560 ± 10 Ma (Keller and Krasnobayev, 1983); Gale (1982) suggests 530 ± 10 Ma for the base of the Cambrian. The Cambrian/Ordovician boundary is set by Palmer (1983) at 510 Ma and by Gale (1982) at 495 + 10 − 5 Ma. The Cambrian, therefore, should have lasted from between 35 and 60 m.y. There seem to be no reliable radiometric dates available that allow estimates of the dates of the Lower-Middle or Middle-Upper Cambrian boundaries. Thus, calculations on the rates of sedimentation and rates of subsidence are highly speculative. The level of error might be more than 50 percent; consequently, the following data are very rough estimates only.

The base of the Nebida Formation is under dispute; its thickness (see Fig. 1) is a rough guess only and thus, the Formation cannot be used for calculation of subsidence. The time span for deposition of the Gonnesa Formation (Botomian to Toyonian) is estimated to have lasted approximately 4 to 8 m.y. (the early Cambrian has been assumed to be 15 to 30 m.y.); the formation in the western and northwestern parts of the area is composed of approximately 500 m of sediment, locally more; thicknesses in the eastern part are mostly much lower (Fig. 3).

If we decompact 500 m of Gonnesan sediments for the end of Gonnesan time, following the procedures of Bond and Kominz (1984), which are based on the basin model of McKenzie (1978), and considering the different lithotypes present, approximately 710 m of decompacted thickness will result. The average porosity at the end of Gonnesan deposition is 30 percent. Isostatic loading correction, assuming an Airy-type crust, results in approximately 340 m of thermo-tectonic subsidence. The range of calculated thermo-tectonic subsidence is 4.25–8.5 cm/ka. For this rough calculation, differential depths at the beginning and end of the Gonnesa Formation can be neglected, because at least parts of the sequence stayed in shallow water.

These subsidence rates are comparable, although higher, than the thermo-tectonic subsidence rates reported by Bond and others (1984) for lower Cambrian rocks from the Appalachians of southwestern Virginia (roughly 3 cm/ka). They are lower than the thermo-tectonic subsidence of the northwest African continental margin in the early Jurassic: after the onset of spreading, a thermal-tectonic subsidence of 8 cm/ka occurred, as reported by Hardenbol and others (1981).

Sedimentation rates for the Campo Pisano clearly do not correspond with subsidence rates; they are much lower. The Campo Pisano, according to Pillola (1986), is mainly Middle Cambrian, some lowermost parts are early. The Middle Cambrian is roughly estimated to have lasted 10 to 15 m.y. A strong compaction rate of these slaty 50 m of carbonates has to be assumed; according to the curves of Bond and Kominz (1984), the decompacted thickness should be approximately 125 m of sediment. Sedimentation rate, per 1,000 years, should be about 1.25 to 0,8 cm, much less than the sedimentation rate during the preceding Gonnesan time (9–18 cm/ka). Assuming the calculated thermo-tectonic subsidence of the Gonnesa Formation (4.25–8.5 cm/ka) for Middle Cambrian time and adding the strong tensional impact at the beginning of the Campo Pisano (or a strong sea-level rise) of at least 100 m, about 600–1,000 m of water depth would result for the beginning of the Late Cambrian (onset of Cabitza Formation deposition). This calculated value is geologically reasonable and gives support to the calculation of the Gonnesan subsidence.

The broad-scale tectonic events of this time are the opening of the Iapetus and the possible development of a Proto-Tethys. No well-established plate-tectonic model exists for the Early Paleozoic of the western Mediterranean. The strong tensions in the Cambrian of Sardinia, followed by compression ("Sardic phase"), have been interpreted as rifting, from time to time affected by transcurrent movements (Carannante and others, 1984; Vai and Cocozza, 1985; Minzoni, 1985). Cocozza and Gandin (1987) assume early-rifting deposition for the "Bahamian-type" Ceroide carbonate platform and the overlying nodular limestones of the Campo Pisano Formation. The calcalkaline volcanism or Ordovician time (Minzoni, 1985) poses one of the problems for these models. Another problem is the large extent of the compression (Sardic phase) in Mid-Ordovician time: effects of transpression should be localized. Effects of "Sardic" compressions, however, are well known from different parts of Spain as well as from the Montagne Noire. Timewise, this compression can even be related with the Knox unconformity of the Appalachians (compare Mussman and Read, 1986).

The widespread blanket of platform carbonates in late Gonnesan time and the rates of thermo-tectonic subsidence are compatible with a late Early Cambrian drifting stage; they are certainly no argument for early, or active rifting at that time (in contrast to the model of Cocozza and Gandin, 1987). The subsidence rates during the late Early Cambrian (corrected for sediment loading) are compatible with thermal-tectonic subsidence rates of drifting stages. The values indicate, however, a higher amount of stretching (β), according to the model of McKenzie (1978). The values are in the same range as Jurassic thermo-tectonic subsidence rates at the initial drifting of marginal areas of the Atlantic (compare Hardenbol and others, 1981).

On the other hand, the formation of intra-shelf and intraplatform basins, neptunian dykes, and slumps and debris flows within the Gonnesa Formation are indicative of strong tensional effects from time to time; they point at an ongoing rifting process. Both facts can be explained, if we place the change from rifting to drifting, from extension to thermal subsidence and cooling into the Gonnesan and Campo Pisano time. In the Appalachians, the Iapetus rift-drift transition is assumed to have occurred in the Early Cambrian (Bond and Kominz 1984; Bond and others, 1984; Williams and Hiscott, 1987). In Sardinia, the change from rifting to drifting most probably occurred in the late Early Cambrian to Middle Cambrian. We assume that the evolution of a passive margin during the Cambrian, which continued into the early Ordovician, was terminated by the onset of subduction in Mid-Ordovician time by the change from a passive to a convergent margin. Subduction would be responsible for compression (Sardic phase) in Mid-Ordovician time, possibly due to arc-continent or microplate-continent collision. Until Ordovician time, the geodynamic history of Sardinia is largely compatible with the sequence of events on the Appalachian side of the Iapetus.

ACKNOWLEDGMENTS

We are indebted to former student members of our research projects in southwestern Sardinia (especially Th. Schledding and M. Selg) for their assistance; to R. Maass, Freiburg, for many discussions on plate-tectonic models for the Early Paleozoic; to K. Bitzer, Freiburg, for help with isostatic models; to Societá Italiana Miniere and Societá Bariosarda for granting access to the mineral properties; to Paul Crevello, Fred Read, and Jim Wilson for their thorough reviews of an early draft of the manuscript.

REFERENCES

BECHSTÄDT, THILO, BONI, MARIA, AND SCHLEDDING, THOMAS, 1984, Slope-sediments in the Cambrian Gonnesa Formation of the Sulcis area, SW-Sardinia: Neues Jahrbuch für Geologie und Paläontologie, Monatshefte, v. 1984/3, p. 129–138.

———, ———, AND SELG, MATTHIAS, 1985, The Lower Cambrian of SW-Sardinia: From a clastic tidal shelf to an isolated carbonate platform: Facies, v. 12, p. 113–140.

———, SCHLEDDING, THOMAS, AND SELG, MATTHIAS, 1988, Rise and fall of an isolated, unstable carbonate platform: The Cambrian of southwestern Sardinia: Geologische Rundschau, v. 77, p. 389–416.

BOND, G. C., AND KOMINZ, M. A., 1984, Construction of tectonic subsidence curves for the early Paleozoic miogeocline, southern Canadian Rocky Mountains: Implications for subsidence mechanism, age of breakup, and crustal thinning: Geological Society of America Bulletin, v. 95, p. 155–173.

———, NICKESON, P. A., AND KOMINZ, M. A., 1984, Breakup of a supercontinent between 625 and 555 Ma: New evidence and implications for continental histories: Earth and Planetary Science Letters, v. 70, p. 325–345.

BONI, MARIA, 1985, Les gisements de type Mississippi Valley du Sud-Ouest de la Sardaigne (Italie): Une synthèse: Chronique recherche miniere, v. 479, p. 7–34.

———, COCOZZA, TOMMASO, GANDIN, ANNA, AND PERNA, GIULIANO, 1981, Tettonica, sedimentazione e mineralizzazioni delle brecce al bordo sud-orientale della piattaforma carbonatica Cambrica (Sulcis, Sardegna): Memorie della Società Geologica Italiana, v. 22, p. 111–122.

———, IANNACE, ALESSANDRO, AND PIERRE, CATHERINE, 1988, Stable isotope compositions of lower Cambrian Pb-Zn-Ba deposits and their host carbonates, southwestern Sardinia, Italy: *in* Chivas, A. R. ed., Isotopes in Paleoenvironments: Chemical Geology, v. 72, p. 267–282.

———, AND MARINACCI, PAOLO, 1980, Analisi stratigrafico-strutturale della zona di Buggerru (Iglesias) con particolare riguardo alla posizione delle mineralizzazioni della Formazione di Gonnesa: Bolletino della Società Geologica Italiana, v. 99, p. 35–55.

CARANNANTE, GABRIELE, COCOZZA, TOMMASO, AND D'ARGENIO, BRUNO, 1984, Late Precambrian-Cambrian geodynamic setting and tectono-sedimentary evolution of Sardina (Italy): Bolletino della Società Geologica Italiana, v. 103, p. 121–128.

CARMIGNANI, LUIGI, COCOZZA, TOMMASO, GHEZZO, CLAUDIO, PERTUSATI, P. C., AND RICCI, C. A., 1982, Lineamenti del basamento sardo, *in* Guida alla geologia del Paleozoico sardo: Guide geologiche regionali, Società Geologica Italiana, p. 11–23.

COCOZZA, TOMMASO, 1979, The Cambrian of Sardinia: Memorie della Società Geologica Italiana, v. 20, p. 163-187.

———, AND GANDIN, ANNA, 1987, Early rifting deposition: Examples from carbonate sequences of Sardinia (Cambrian) and Tuscany (Triassic-Jurassic), Italy: An analogous tectono-sedimentary and climatic context: American Association of Petroleum Geologists Bulletin, v. 71, 5, p. 540.

DEBRENNE, FRANÇOISE, AND GANDIN, ANNA, 1985, La formation de Gonnesa (Cambrien, SW Sardaigne): Biostratigraphie, paléogéographie, paléoécologie des Archéocyathes: Bulletin de la Societé Géologique de France, tome I, p. 531–540.

FÜCHTBAUER, HANS, AND RICHTER, D. K., 1983, Relations between submarine fissures, internal breccias and mass flows during Triasssic and earlier rifting periods: Geologische Rundschau, v. 73, p. 53–66.

GALE, N. H., 1982, Numerical dating of Caledonian times (Cambrian to Silurian), *in* Odin, G. S., ed., Numerical Dating in Stratigraphy: Wiley, New York, p. 467–486.

GANDIN, ANNA, 1982, Tectono-sedimentary evolution of an epicontinental shelf: Lower-Middle Cambrian of Sardinia (Italy): Abstracts of Papers, Eleventh International Congress on Sedimentology, McMaster University, Hamilton, Canada, p. 43.

———, 1985, Anomalie al passaggio formazione di Gonnesa-Formazione di Cabitza (Cambriano inferiooree medio; Sardegna sud-occidentale), *in* Gruppi di Lavoro del Centro Nationale Ricerche, "Paleozoico" e "Evoluzione Magmatica e Metamorfica della Crosta Fanerozoica": Riunione Scientifica, Siena, Note Brevi e Riassunti, p. 28–29.

———, 1986, Depositional-paleogeographic trend of Cambrian in S-W Sardinia: A review: International Geological Correlation Program, Project 5, Correlation of Prevariscan and Variscan events in the Alpine-Mediterranean mountain belts, Cagliari, Sardinia, Abstracts of Papers, p. 24.

———, MINZONI, NELLO, AND COURJAULT-RADÉ, PIERRE, 1987, Shelf to basin transition in the Cambrian-Lower Ordovician of Sardinia (Italy): Geologische Rundschau, vol. 76, p. 827–836.

———, AND PILLOLA, G. L., 1985, Biostratigrafia e sedimentologia della Formazione di Cabitza nell' Iglesiente, *in* Gruppi Di Lavoro del Centro Nationale Ricerche, "Paleozoico" e "Evoluzione Magmatica e Metamorfica della Crosta Fanerozoica"; Riunione Scientifica, Siena, Note Brevi e Riassunti, p. 30–31.

GROSS, ULI, 1982, Die Stratigraphie der unterkambrischen Nebida Formation im Iglesiente (Südwestsardinien) unter besonderer Berücksichtigung ihrer Trilobitenfauna: Unpublished Ph. D. Dissertation, University of Heidelberg, West Germany, 123 p.

GROTZINGER, J. P., 1986, Cyclicity and paleoenvironmental dynamics, Rocknest platform, northwest Canada: Geological Society of America Bulletin, v. 97, p. 1208–1231.

HARDENBOL, J., VAIL, P. R., AND FERRER, J., 1981, Interpreting paleoenvironments, subsidence history and sea-level changes of passive margins from seismic and biostratigraphy: Oceanologica Acta, Proceedings, 26th International Geological Congress, Geology of Continental Margins Symposium, Paris, p. 33–44.

HARLAND, W. B., COX, A. V., LLEWELLYN, P. G., PICTON, A. C. G., SMITH, A. G., AND WALTERS, R., 1982, A Geological Time Scale: Cambridge University Press, Cambridge, England, 128 p.

KANTER, J. A. M., 1988, On the interpretation of clinoforms in carbonate platform flanks: Ancient and recent examples, Dolomites (Italy) and northern Little Bahama Bank (Bahamas): Terra Cognita, v. 8, p. 13.

KELLER, B. M., AND KRASNOBAYEV, A. A., 1983, Late Precambrian geochronology of the European USSR: Geological Magazine, v. 120, p. 381–389.

LASKE, RAINER, AND BECHSTÄDT, THILO, 1987, Alluviale Fächer und flachmarine Sedimentabfolgen im Hangenden der Sardischen Diskordanz (Ordovizium SW-Sardiniens): Heidelberger Geowissenschaftliche Abhandlungen, v. 8, p. 141–143.

MCKENZIE, D. P., 1978, Some remarks on the development of sedimentary basins: Earth and Planetary Science Letters, v. 40, p. 25–32.

MINZONI, NELLO, 1985, Magmatismo Paleozoico ed evoluzione geodinamica in Sardegna, *in* Gruppi di Lavoro del Centro Nationale Ricerche, "Paleozoico" e "Evoluzione Magmatica e Metamorfica della Crosta Fanerozoica"; Riunione Scientifica, Siena, Note Brevi e Riassunti, p. 87–88.

MOSTLER, HELFRIED, 1985, Neue heteractinide Spongien (Calcispongea) aus dem Unter- und Mittelkambrium Südwestsardiniens: Berichte des naturwissenschaftlich-medizinischen Vereins Innsbruck, v. 72, p. 7–32.

MUSSMAN, W. J., AND READ, J. F., 1986, Sedimentology and development of a passive- to convergent-margin unconformity: Middle Ordovician Knox unconformity, Virginia Appalachians: Geological Society of America Bulletin, v. 97, p. 282-295.

OGGIANO, GIACOMO, MARTINI, I. P., AND TONGIORGI, MARCO, 1986, Sedimentology of the Ordovician "Puddinga" Formation (SW-Sardinia): International Geological Correlation Program, Project 5, Correlation of Prevariscan and Variscan Events in the Alpine-Mediterranean Mountain Belts. Cagliari, Sardinia, Abstracts of Papers, p. 61.

PALMER, A. R., 1983, The decade of North-American geology: 1983 geologic time scale: Geology, v. 11, p. 503–504.

———, AND JAMES, N. P., 1980, The Hawke Bay Event: A circum-Iapetus regression near the Lower-Middle Cambrian boundary, *in* Wones, D. R., ed., The Caledonides in the U.S.A.: Blacksburg, Virginia Polytechnic Institute and State University Memoir 2, p. A3.

PILLOLA, G. L., 1986, Biostratigraphy of the Campo Pisano and Cabitza Formations: Preliminary report: International Geological Correlation Program, Project 5, Correlation of Prevariscan and Variscan Events in the Alpine-Mediterranean Mountain Belts, Cagliari, Sardinia, Abstracts of Papers, p. 67–68.

READ, J. R., 1985, Carbonate platform facies models: American Association of Petroleum Geologists Bulletin, v. 69, p. 1–21.

SCHLEDDING, THOMAS, 1985, Fazies, Geochemie und Paläogeographie der unter- bis mittelkambrischen Gonnesa Formation sowie der basalen Cabitza Formation des Sulcis (SW-Sardinien, Italien): Der Zerfall einer Karbonatplattform: Unpublished Ph.D. Dissertation, University of Freiburg, West Germany, 154 p.

SEILACHER, ADOLF, 1969, Fault-graded beds interpreted as seismites: Sedimentology, v. 13 (1969), p. 155–159.

SELG, MATTHIAS, 1985, Die siliziklastisch-karbonatische Wechsellagerung der unterkambrischen Nebida Formation (SW-Sardinien): Entstehung einer Karbonat-Plattform: Unpublished Ph. D. Dissertation, University of Freiburg, West Germany 105 p.

———, 1986, Algen als Faziesindikatoren: Bioherme und Biostrome im Unter-Kambrium von SW-Sardiniens: Geologische Rundschau, v. 75, p. 693–702.

VAI, G. B., AND COCOZZA, TOMMASO, 1985, Tentative schematic zonation of the Hercynian chain in Italy: International Geological Correlation Program, Project 5, Correlation of Prevariscan and Variscan Events in the Alpine-Mediterranean Mountain Belts. Cagliari, Sardinia, Abstracts of Papers, p. 15.

WILLIAMS, HAROLD, AND HISCOTT, R. N. 1987, Definition of the Iapetus rift-drift transition in western Newfoundland: Geology, v. 15, p. 1044–1047.

WILLIAMS, NEIL, 1980, Precambrian mineralization in the McArthur-Cloncurry region with special reference to stratiform lead-zinc deposits, *in* Henderson, R. A., and Stephenson, P. J., eds., The Geology and Geophysics of Northeastern Australia; Geological Society of Australia, Queensland Division, p. 89–108.

EVOLUTION OF A LOWER PALEOZOIC CONTINENTAL-MARGIN CARBONATE PLATFORM, NORTHERN CANADIAN APPALACHIANS

NOEL P. JAMES
Department of Geological Sciences, Queen's University, Kingston, Ontario, K7L 3N6
ROBERT K. STEVENS
Department of Earth Sciences, Memorial University, St. John's, Newfoundland, A1B 3X5
CHRISTOPHER R. BARNES
Geological Survey of Canada, 508 Booth Street, Ottawa, Ontario K1A 0E8
AND
IAN KNIGHT
Department of Mines, Government of Newfoundland and Labrador, St. John's, Newfoundland, A1C 3T7

ABSTRACT: The northwestern margin of the northern Appalachian Orogen and adjacent craton is a Lower Paleozoic, low-latitude carbonate platform that originally lay along the northern margin of the Iapetus Ocean. Parts of the platform interior are now exposed in Quebec, but much lies beneath the Gulf of St. Lawrence. Outer-shelf and deep-water deposits crop out in western Newfoundland. The shelf break and upper slope are nowhere exposed, but their nature has been determined from numerous clasts in sediment gravity flows redeposited on the lower slope and now stacked in allochthonous thrust complexes.

The relatively thin part of the platform (approximately 2 km) preserved in western Newfoundland lay well inboard of the shelf edge and accumulated on rigid crust that underwent relatively little thermal subsidence. Four separate phases of platform evolution can be differentiated, reflecting the interplay between tectonics, eustasy, and the evolving Lower Paleozoic biota. Phase 1: *preplatform shelf*, reflects initial siliciclastic deposition and volcanism on rifted crystalline basement, followed by a short period of carbonate sedimentation with archaeocyathan buildups and ooid-sand shoals that was terminated by offlap of thick quartz arenites. Phase 2: *narrow, high-energy platform*, is characterized by mixed siliciclastic-carbonate, non-bioclastic, peritidal sedimentation throughout and is locally manifest as three Grand Cycles. Contemporaneous deep-water sediment composes basal welded conglomerate overlain by quartzose carbonate turbidites. Phase 3: *wide, low-energy platform*, is an onlap package of muddy, bioclastic, subtidal and peritidal carbonates arranged in the form of two unconformity-bounded sequences. The adjacent deep-water slope was a narrow belt of debris flows and a wide apron of carbonate and shale, deposited in the lower part of an oxygen-minimum zone. Phase 4: *foundered platform*, documents the initial uplift, faulting, subsidence, and fragmentation of the platform in a sequence of peritidal to subtidal to deep-water carbonates, reflecting the initial stages of the Taconic Orogeny. The entire sequence is buried by synorgenic flysch and overlain by thrust sheets containing deep-water sediments of phases 1, 2, and 3, as well as oceanic lithosphere and other exotic assemblages.

The main control governing platform initiation, thickness, and demise was tectonics. Internal stratigraphy, arrangement of shallow-water facies, and style of deep-water sedimentation were governed by eustasy with a local tectonic overprint. Climate was temperate tropical. The sediments themselves, dominated by periods of bioclastic versus non-bioclastic (ooids, peloids) sedimentation, reflects temporal changes in the early Paleozoic global biosphere.

INTRODUCTION

The northwestern margin of the northern Appalachian Orogen (Humber Zone of Williams, 1979) and adjacent St. Lawrence Platform together compose an early Paleozoic, low-latitude miogeocline that originally lay along the northern margin of the Proto-Atlantic or Iapetus Ocean. This miogeocline is presently considered to be the Paleozoic passive margin of eastern North America, which was partially destroyed by closing of an early Paleozoic ocean with accompanying accretion of outboard terranes. Similar miogeoclinal facies are recognized along the western margin of most of the Appalachian-Caledonian Orogen (Williams and Stevens, 1974; Williams, 1978, 1979). Only fragments of this once extensive shelf are now exposed in eastern Canada. Most of the platform interior lies beneath the Gulf of St. Lawrence, except for small but important outcrops on the Mingan Islands, Quebec (Fig. 1). Outer platform and slope-to-basin strata are present in western Newfoundland; shallow-water sediments are autochthonous to parautochthonous, little deformed, and crop out over wide areas; deep-water deposits occur in thrust complexes of the Humber Arm and Hare Bay allochthons, which were transported westward in Middle Ordovician time, during the Taconic Orogeny. Nowhere is the ancient shelf break or upper slope exposed, but their nature has been deduced from the composition of clasts in redeposited, deep-water, sediment gravity flows. Although only specific facies are exposed, excellent outcrops and detailed biostratigraphy have allowed us to formulate a detailed reconstruction of this platform.

General attributes of the western Newfoundland platform have been documented by Rodgers, (1968) Stevens, (1970), Williams and Stevens, (1974) and James and Stevens, (1982). This study is the first to integrate results of detailed stratigraphy, sedimentology, and paleontology undertaken over the last decade and to present a coherent picture of platform evolution. Our interpretation draws heavily on our own studies as well as those of our colleagues and graduate students. Correlation and event stratigraphy have been accomplished primarily through biostratigraphy, utilizing conodonts, graptolites, and trilobites. The development of the platform reflects changes in carbonate sedimentation as a function of synsedimentary tectonics, eustatic sea-level fluctuations, and the changing character of the evolving benthos.

STRATIGRAPHY AND GEOLOGIC SETTING

The carbonate platform is preserved in two major geologic provinces. The western, generally flat-lying strata in western Newfoundland, beneath the Gulf of St. Lawrence and in Quebec, are part of the Interior Platform, locally referred to as the St. Lawrence Platform. The eastern, deformed strata are continuous with those to the west but lie

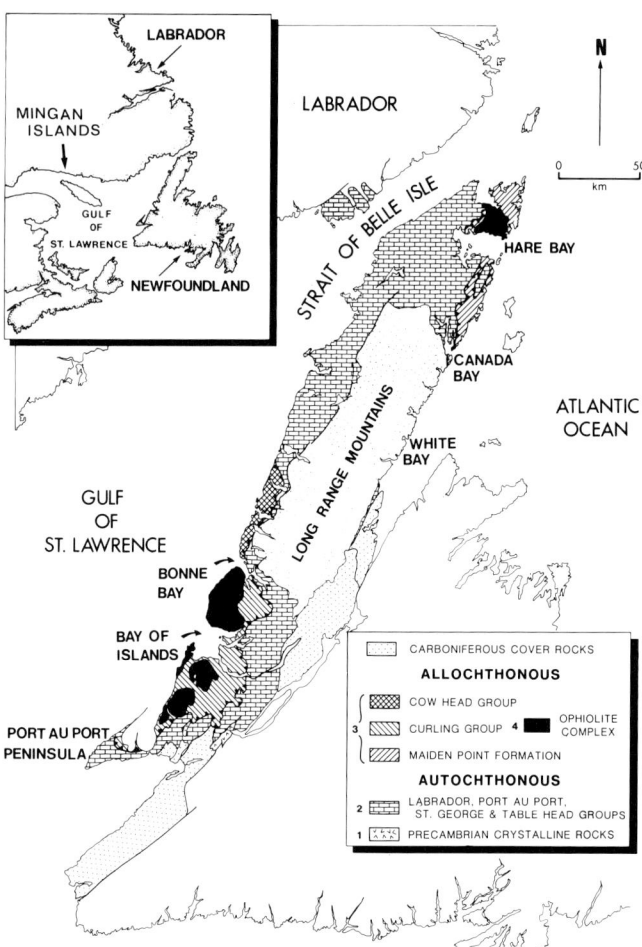

FIG. 1.—Geologic map of the Humber Zone in western Newfoundland with index map showing location of Mingan Islands, Quebec (after James and Stevens, 1986).

within the Appalachian Orogen, more specifically in the tectonostratigraphic Humber Zone (Williams, 1979). Here, the autochthonous shallow-water sediments are locally overlain by allochthonous deep-water strata and ophiolites. The eastern margin of the Humber Zone is drawn at the Baie Verte-Brompton "Line," more properly a "zone" (Williams and St. Julien, 1982), a steep structural belt marked by deformed ophiolites and mafic volcanic rocks.

Autochthonous Rocks

Crystalline Grenvillian (about 1,000 Ma) basement, which underlies the platform, is intruded by late Precambrian dikes that feed tholeiitic lava flows. These rocks are covered by a late Precambrian to early Middle Cambrian succession (Fig. 2) of siliciclastic and minor-carbonate sediment of the Labrador Group (Schuchert and Dunbar, 1934; Cumming, 1983). The bulk of the succeeding shallow-water carbonate platform sediments is encompassed by the Port-au-Port Group (Chow, 1986), St. George Group (Knight and James, 1987), and the Table Head Group (Klappa and others, 1980; Figs. 1, 2). The carbonates are overlain by easterly derived terrigenous clastic rocks of the Mainland Sandstone (Schillereff and Williams, 1979) and similar northeasterly derived sediments of the Goose Tickle Formation–deposits within an evolving foreland basin. These rocks are imbricated by thrust faults, progressively deformed and metamorphosed along the eastern margin of the Humber Zone.

Allochthonous Rocks

The Humber Arm Allochthon, located north and south of the Bay of Islands (Fig. 1), is a series of thrust complexes separated by intervening melange. The thrust complexes are imbricate slices of upper Proterozoic to middle Ordovician deep-water sedimentary rocks, called the Cow Head Group (Kindle and Whittington, 1958; James and Stevens, 1986) and the Curling Group (Stevens, 1970; Figs. 1, 2); igneous rocks and metamorphic rocks of the Skinner Cove Volcanics and Little Port Complex; and an ophiolite suite, the Bay of Islands Complex (Williams, 1975). Stratigraphic nomenclature of the transported sedimentary rocks, particularly those called the Curling Group, is under revision (J. Botsford and L. Quinn, pers. commun., 1987). The sequence in the Hare Bay Allochthon is dominated by deep-water sediments of the Maiden Point Formation, volcanics of the Cape Onion Formation, and volcanics and peridotites of the St. Anthony Complex (Williams and Smyth, 1983). Extensively deformed rocks east of White Bay, considered to be equivalent to the Humber Arm Allochthon, are called the Fleur de Lys Supergroup.

The first structures are of Early to early Middle Ordovician age, as confirmed by an unconformity beneath the Caradocian Long Point Group (Fig. 2). Later structures are related to deformation during the Devonian (Acadian; Williams and others, 1986) and Carboniferous (Alleghenian).

The following synthesis comes from study of the least deformed and best exposed of these rocks, the autochthonous succession and the Cow Head and equivalent units in the Humber Arm Allochthon. The history of platform development is most conveniently described as four separate phases.

PHASE 1—PREPLATFORM SHELF

The continental shelf was initiated on rifted and block-faulted Grenvillian basement, intruded by late Precambrian mafic dikes and sills (Strong and Williams, 1972; Williams and Stevens, 1974; Strong, 1974). The irregular topography was veneered and then buried by an onlap-offlap sequence of siliciclastic and carbonate sediments called the Labrador Group (Fig. 2).

Basal Sandstones

The oldest sediments are plutonic-boulder conglomerate, quartz arenite, arkose, siltstone, and slate (Bateau Formation). Minor volcanics are localized in the east. Synsedimentary faulting and continued extension are clearly demonstrated by deformation of these sediments prior to intrusion by numerous cross-cutting dikes that feed overlying plateau basalts and pyroclastics (Lighthouse Cove Formation). These volcanics probably filled structurally controlled topographic lows.

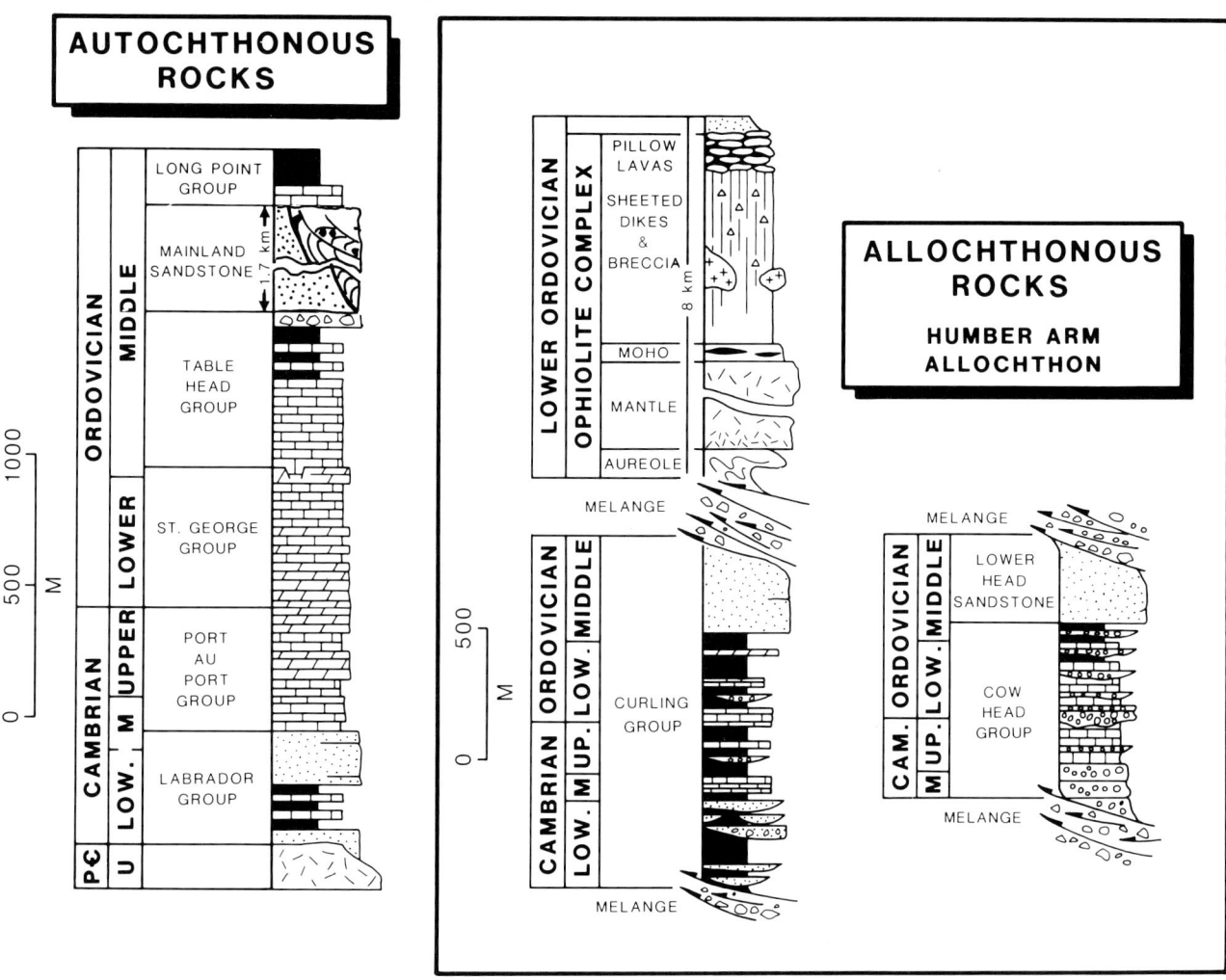

Fig. 2.—Sketch illustrating the principal rock units in shallow-water autochthonous and deep-water allochthonous successions in western Newfoundland (after James and Stevens, 1986).

An upper subaerially weathered surface is everywhere blanketed by transgressive fluvial and strandline sandstone and minor conglomerate (Bradore Formation; Fig. 3A). The sediments, however, are laterally variable. In Labrador basal sediments are red, pebbly, subarkosic, trough and planar-tabular cross-bedded sandstone (Fig. 3B) and pebble conglomerate, deposited in a complex series of braided streams that flowed eastward off the craton. Braid channels built directly to the coast, and the river sands were extensively reworked by vertical burrowers in distributary-mouth estuaries. Upper units are mineralogically more mature than earlier sediment and consist of sands that filled laterally migrating tidal inlets along a barrier coastline, the upper parts of which were removed by subsequent shoreface migration (Hiscott and others, 1984). Eastward, around the northern end of the Long Range Mountains, sandstones were deposited in either fluvial or tidal-flat settings (Knight, in prep.). Sediments south and east of the Long Range are just a veneer of green-grey sandstone and thin shale or red, white, and grey quartz arenite over basement.

Northeastward paleocurrents around the northern end of the Long Range and marked thickness variations of the Bradore probably reflect structural topographic control (Knight, in prep.). North-south thickness variations suggest that the complex was part of a faulted, northward-tilted upland that was only onlapped as sea level finally rose above the newly rifted continental margin, initiating the development of a narrow open shelf.

The top of this sequence at some localities in insular Newfoundland is capped by laminated and pisolitic hematite, indicating brief subaerial exposure and lateritic weathering.

Open-Shelf Sediments

Everywhere north of Bonne Bay (Fig. 4) the basal sandstone is overlain by limestone, shale, siltstone, and minor sandstone (Forteau Formation), which become progressively more shale-dominated southward. Farthest away to the south and east, dark shale and minor-ribbon limestone, intraclastic lime breccia, and oolite-oncolite lime grainstone prevail. Poorly studied allochthonous deep-water sed-

Fig. 3.—Sedimentary rocks deposited during phase 1 (Late Precambrian to Middle Cambrian–preplatform shelf). (A) Precambrian crystalline rocks overlain (at arrow) by a thin pebble conglomerate and trough cross-bedded sandstones (Bradore Formation) in southern Labrador. (B) Bedding plane view of trough cross-beds in the Bradore Formation indicating a southeasterly transport direction; southern Labrador. (C) Small patch reef surrounded by limestone and shale in the Forteau Formation, southern Labrador. (D) Ooid grainstone capped by a hardground and overlain by fossiliferous oncolitic grainstone in the Forteau Formation, western Newfoundland. (E) An 8-m-thick section of massive, well bedded orthoquartzite in the Hawke Bay Formation, Port-au-Port Peninsula, western Newfoundland. (F) Deep-water conglomerate of limestone clasts (recessively weathering) and quartz pebbles (white) filling a channel eroded into quartzose turbidites in the allochthonous Curling Group, western Newfoundland (scale divisions 2 cm).

iments, mainly laminated green and red shale (upper Summerside Formation, Curling Group), suggest tranquil deep-water sedimentation under conditions of varying oxygenation. Together, these facies indicate a narrow shelf with clean quartz sand at the strandline, a narrow belt of carbonate close to shore in the west, and a sea floor covered with silt and mud that sloped gently basinward to the east and south (Fig. 4) in the form of a ramp (Ahr, 1973; Read, 1982).

Trilobites and archaeocyathans from different facies indicate deposition entirely within the *Bonnia-Olenellus* Zone of the Early Cambrian (Debrenne and James, 1981). The presence of *Wanneria* in these sediments suggests that most deposition was restricted to early and middle parts of this zone (Palmer and James, 1979).

North of Bonne Bay, the sequence exhibits many attributes of a Grand Cycle (Aitken, 1966), dominated in the north by the upper carbonate half-cycle and in the south and east by the lower shaly half-cycle. The base of the Grand Cycle is everywhere marked by subtidal, transgressive carbonate (Devils Cove Member of the Forteau Formation; James and Debrenne, 1980). Shaly, nodular, red to white fossiliferous lime wackestone with solitary archeocyathans in Newfoundland grades westward and northward onshelf into archaeocyathan bioherm complexes (Fig. 3C) in Labrador (James and Kobulk, 1978).

The subtidal, shale-dominated lower half-cycle varies from burrowed calcareous siltstone with nodules and layers of *Salterella* and trilobite-rich limestone to dark shale punctuated by storm layers rich in shallow-water carbonate clasts. These mudrocks thicken eastward and southward offshelf and become progressively more deep water in aspect, with some units deformed by synsedimentary slump folds. They are metamorphosed to slate-grade equivalents at Bonne Bay and Canada Bay.

The upper, carbonate, half-cycle represents a narrow belt of high-energy shoals and tidal flats that thins rapidly eastward and southward. The extent of the belt is illustrated in Figure 4. Early deposits are skeletal and ooid calcarenites or archaeocyathan biostromes (Hughes, 1977), which colonized the shelf as far south as Hawke Bay (Knight and Boyce, 1984). The rest of the cycle is typified by cross-bedded ooid grainstone and oncolite rudstone (Fig. 3D) with dolostone lenses and oolitic sandstones that display rapid lateral variation and only minor-shale interbeds (Knight, in prep.).

Offlap Sandstones

Terrigenous clastics and minor carbonates (Hawke Bay Formation) characterize the final phase of offlap. Contained trilobites range in age from the upper part of the *Bonnia-Olenellus* Zone of the Early Cambrian to the *Bathyuriscus-Elrathina* Zone of the early Middle Cambrian (Boyce, 1977; A. R. Palmer, pers. commun., 1978; Knight, 1977; Knight and Boyce, 1987). Carbonates contain trilobites from all intervening zones (Knight and Boyce, 1987).

Basal deposits are subtidal to intertidal shallowing-upward shale to sandstone cycles on the inner shelf and subtidal fine-grained thin sandstone and shale on the outer shelf.

Thick overlying sandstones on the inner shelf (Fig. 3E) are mostly massive cross-bedded quartz arenite deposited in tidal-dominated strandline and barrier-island settings (Knight, in prep.). Interbeds of glauconitic, phosphatic sandstone, burrowed mudstone and sandstone, and oolitic and pisolitic hematite represent bioturbated intertidal-sand and mudflat deposits, which suffered lengthy periods of exposure. On the outer part of the shelf, sediments are meter-scale shallowing-upward sequences of subtidal to intertidal shale and channel-to-sheet sands, which grade upward into ooid-sand bodies containing stromatolite banks and tidal-flat dololaminites (Bridge Cove Member; Knight and Boyce, 1987). Sequences are capped by fenestral carbonates, quartz, and phosphate-rich dolostone or paleokarst.

Equivalent outer shelf-upper slope, parautochthonous strata in Canada Bay are grey, pyritiferous shale with ribbon-quartz sandstone and minor limestone (Knight, 1987). Allochthonous lower slope sediments (Irishtown Formation; Sellars Formation) are a succession of conglomerate (Fig. 3F), sandstone, siltstone, and shale, mainly turbidites which together exhibit many characteristics of a submarine fan complex (e.g., Walker, 1984). Conglomerates at the top are polymict and contain clasts of all older, shallow-shelf lithologies and basement, indicating profound erosion accompanying deposition (Fig. 5). It is difficult to explain why the shallow-water clastics lack basement clasts, whereas they are common in the fan complex.

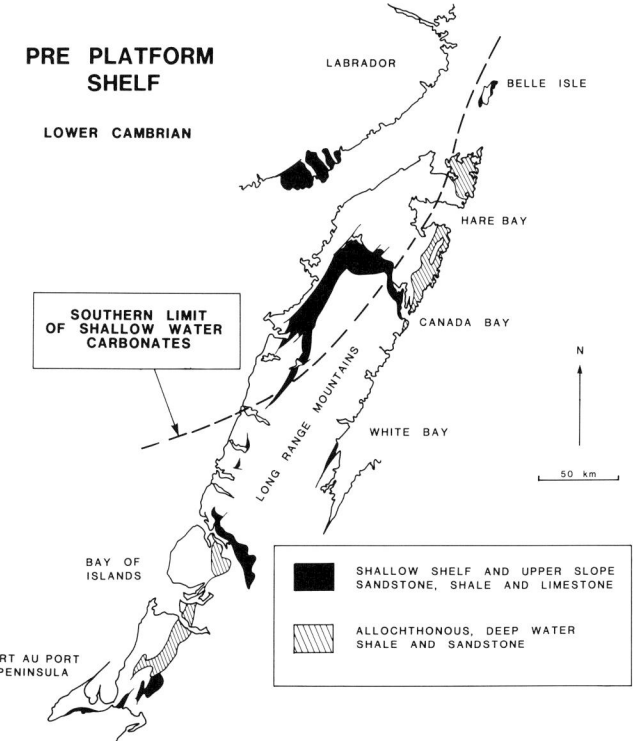

Fig. 4.—A geologic map of Lower Cambrian strata in western Newfoundland, illustrating the present distribution of autochthonous and parautochthonous facies (black) and allochthonous facies (shaded). The southern limit of shallow-water carbonates refers to the major extent of ooid sands, archaeocyathan biostromes, and tidal-flat facies in the upper carbonate half-cycle of phase 1.

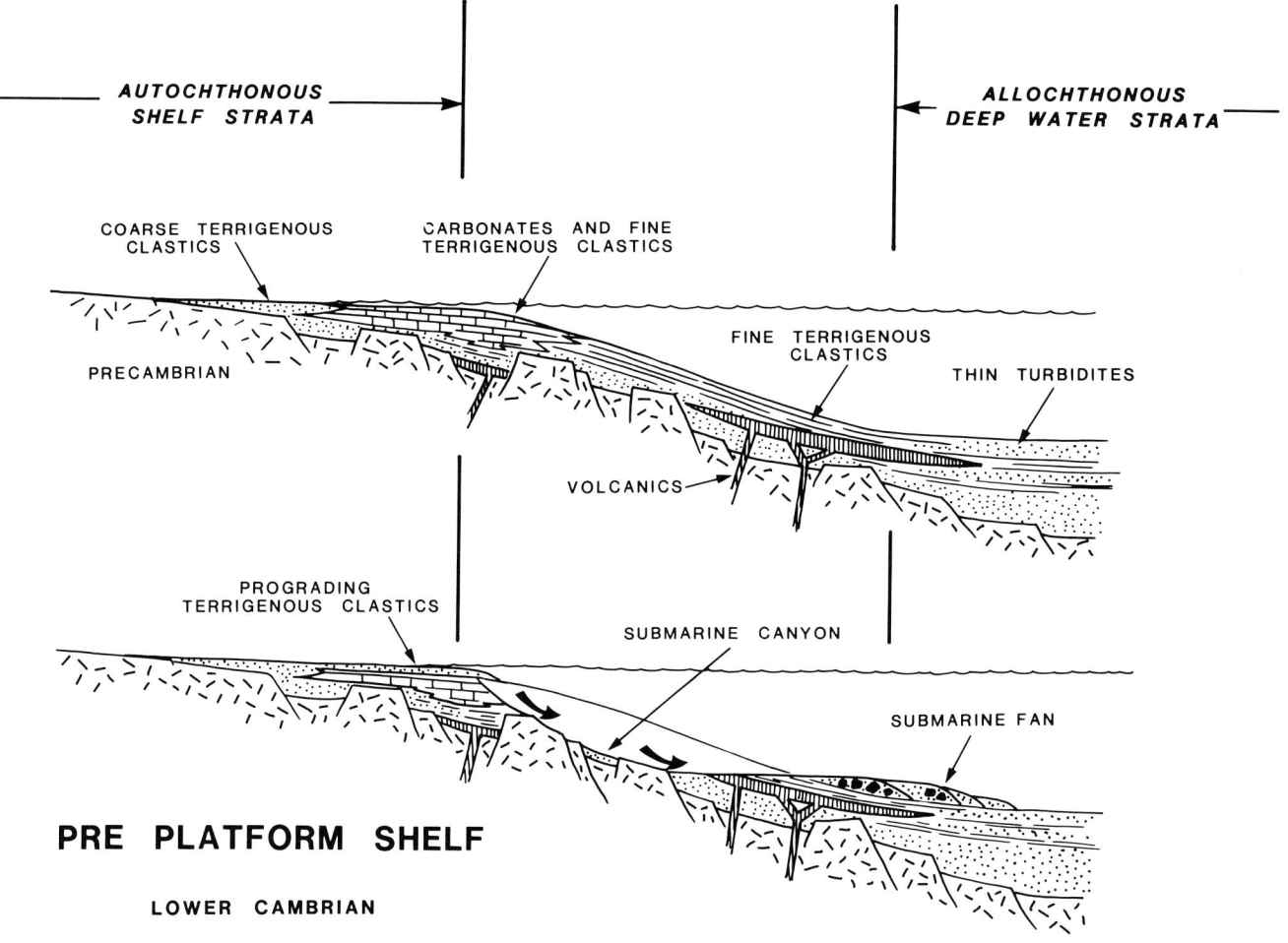

FIG. 5.—Interpretative cross sections of the restored Lower Cambrian preplatform shelf in western Newfoundland. (A) the shelf during maximum sea-level highstand (Forteau Formation–Summerside Formation; phase 1, Event 2 of Fig. 16). (B) the shelf during maximum offlap "Hawke Bay Event" (Hawke Bay Formation–Irishtown Formation). Carbonate sedimentation was continuous at the edge; submarine canyons may have been eroded locally; phase 1, Event 3 of Figure 16.

PHASE 2—NARROW, HIGH-ENERGY CARBONATE PLATFORM

Onlap of carbonate sediments (Port-au-Port Group) across the platform was rapid during late Middle Cambrian time (*Ehmaniella* Zone; just below the base of the Marjumian Stage, i.e., Late Cambrian, *sensu* Ludvigsen and Westrop, 1985). Lack of sandy strandline facies points to a shoreline somewhere west of present insular Newfoundland, but not as far west as Quebec, i.e., beneath the Gulf of St. Lawrence, suggesting that the platform was less than 200 km wide. Although dominated by carbonates, the sequence still contains a high proportion of terrigenous clastic sediments. Three northeast-southwest-trending facies belts can be recognized (Fig. 6), all of which are dominated by peritidal sedimentation.

Inner Belt

Middle Cambrian sedimentation was mainly subtidal in the form of silty mud, muddy and burrowed carbonate with local oolite shoals and stromatolite-dominated tidal flats (Knight, 1977; Knight and Boyce, 1987). Upper Cambrian strata are now mainly dolostone, but sediments were originally deposited in a complex mosaic of low-energy muddy tidal flats and intervening shallow subtidal environments (Knight, 1980; Knight and Saltman, 1980).

Mixed Belt

This belt contains sediments similar to those in the north (Fig. 7B, 7D), but intercalated with thick sequences of oolite (Fig. 7A).

Outer Belt

This narrow belt is surmised from sections at Goose Arm, Bonne Bay, and White Bay, where dolomitized massive carbonates with relic ooid fabrics and no shale are believed to represent stacked ooid-sand shoals. Stronger evidence is provided by equivalent redeposited conglomerates of the Shallow Bay Formation (Cow Head Group), which indicate that the margin was characterized by ooid- and peloid-sand shoals with numerous buildups of *Epiphyton* and *Girva-*

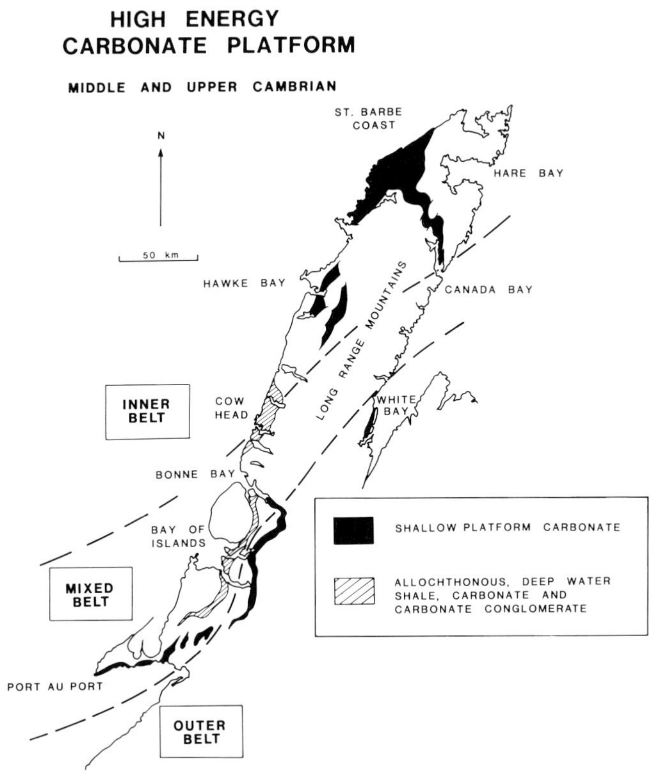

FIG. 6.—A geologic map of Middle and Upper Cambrian strata in western Newfoundland illustrating the distribution of three facies belts: (i) an inner mixed siliciclastic carbonate belt, (ii) an intermediate belt in which inner facies and ooid sands are mixed, and (iii) an outer belt of stacked ooid-sand shoals. Grand Cycles are best developed in the mixed belt.

nella between shoals and on the upper slope (James, 1981; James and Stevens, 1986).

Cyclicity

Cyclicity is a feature of Cambrian sedimentation in much of this area. A widely correlative Grand Cycle of Middle Cambrian age dominated by subtidal sediments is recognizable in the inner belt. Cycles in the overlying Upper Cambrian in the St. Barbe region are not obvious, although two sequences may be present. Four cycles are present in the parautochthon at Canada Bay, but lack of fauna precludes their dating and integration into the regional picture. Three Grand Cycles (Fig. 7A, 7C, 8), similar to those which typify Cambrian sedimentation in western North America, are recognized in the mixed belt from Bonne Bay south to Port-au-Port Peninsula (Chow and James, 1987).

Integration of platform sequences, as represented by Grand Cycles, with detailed stratigraphy of deep-water lower slope sediments indicates a linked history of platform growth (Fig. 9). Deep-water sediments equivalent to the lower two Grand Cycles are a succession of limestone conglomerates (Fig. 7F) with few calcarenite interbeds (Fig. 7E) in the Cow Head region (Downes Point Member, Shallow Bay Formation) and conglomerate with interbedded calcarenite turbidites in the Humber Arm region (Cooks Brook Formation). Such a succession indicates that carbonate sedimentation at the platform margin generally exceeded the relative rate of sea-level rise and resulted in more or less continuous margin progradation or offlap (James and Mountjoy, 1983), with blocks of early lithified sediment and calcified algal buildups shed repeatedly into deep water, forming a wide debris apron of amalgamated conglomerates and breccias. The upper Grand Cycle, which is thicker than the lower two cycles, indicates vertical rather than lateral accretion. This is reflected in lower slope facies of the Cow Head region by a sequence of overlapping beds composed of quartzose carbonate-sand turbidites (Fig. 7E; Broom Point Member, Shallow Bay Formation), representing rapid production and delivery of ooid/peloid sand to the slope and quartz sand bypassing the outer shelf. This bypassing appears to have taken place during periods of exposure, because muddy tidal-flat caps on ooid-sand shoals commonly contain quartz sand. Increasing relief between shelf and basin led to progressive narrowing of the deep-water carbonate-sand apron and onlap of anaerobic to dysaerobic basinal shales upslope (Martin Point Member, Green Point Formation). These sediments in the Humber Arm region (Cooks Brook Formation) are also quartzose calcarenite and shale (J. Botsford, pers. commun., 1986).

PHASE 3—WIDE, LOW-ENERGY, RIMMED PLATFORM

A dramatic change in the style of sedimentation, to widespread muddy carbonates (St. George Group), coincides roughly with the Cambro-Ordovician boundary. The east-west facies belts of the Cambrian are no longer discernible. Subtidal facies of burrowed, fossiliferous, muddy carbonate are now present everywhere as thin- to medium-bedded, rubbly to nodular limestone. These are intercalated with coarse-grained storm deposits of fossiliferous to intraclast rudstone. Conspicuous throughout the succession are horizons of thrombolite-metazoan (sponges and minor corals) buildups and associated calcarenites. The platform margin was rimmed by a sequence of calcified algal buildups with varying proportions of sponges, primitive corals, and, locally, ancestral stromatoporoids.

Superimposed on the general continuous rise in eustatic sea level, which reached its peak near the end of Early Ordovician time, are two unconformity-bounded sequences (Fig. 10). Each package is composed of a basal succession of shallowing-upward peritidal cycles, a middle subtidal carbonate, and an upper succession of shallowing-upward peritidal cycles (Knight and James, 1987).

Lower Sequence

The lower sequence is mostly peritidal, roughly equivalent to the Tremadoc Series, as defined in Britain, or the lower Canadian (Ibexian) Series, as defined in North America, and composes the Watts Bight Formation and most of the Boat Harbour Formation (Fig. 10). In southern and eastern localities, such as the Port-au-Port Peninsula and Goose Arm in the inner part of the Bay of Islands, the basal beds are largely peritidal, whereas farther onshelf in the St. Barbe region the sediments are more subtidal. The middle subtidal portion of the sequence is dolomitized in the north

Fig. 7.—Sedimentary rocks deposited during phase 2 (Middle and Upper Cambrian–high-energy carbonate platform). (A) Massive, interbedded brown and grey oolite sands in the Upper Cambrian Felix Member, Petit Jardin Formation, Port-au-Port Peninsula (person for scale at arrow). (B) Prism dessication cracks in laminated dolostone, Upper Cambrian Campbells Member, Petit Jardin Formatiion, Port-au-Port Peninsula. (C) Shales in the lower part of Grand Cycle A (Fig. 8) forming cliffs 20 m high, Middle Cambrian Cape Ann Member, Petit Jardin Formation, Port-au-Port Peninsula. (D) Exhumed large thrombolite mounds and buildup complexes in the Upper Cambrian Petit Jardin Formation, northwest coast. (E) Cross-bedded quartzose calcarenite turbidites in the Upper Cambrian portion of the allochthonous, deep-water, Cow Head Group, north of Bonne Bay. (F) Conglomerate composed of lime mudstone and peloid/ooid grainstone clasts in a quartzose calcarenite matrix, Upper Cambrian portion of the deep-water, allochthonous Cow Head Group, north of Bonne Bay.

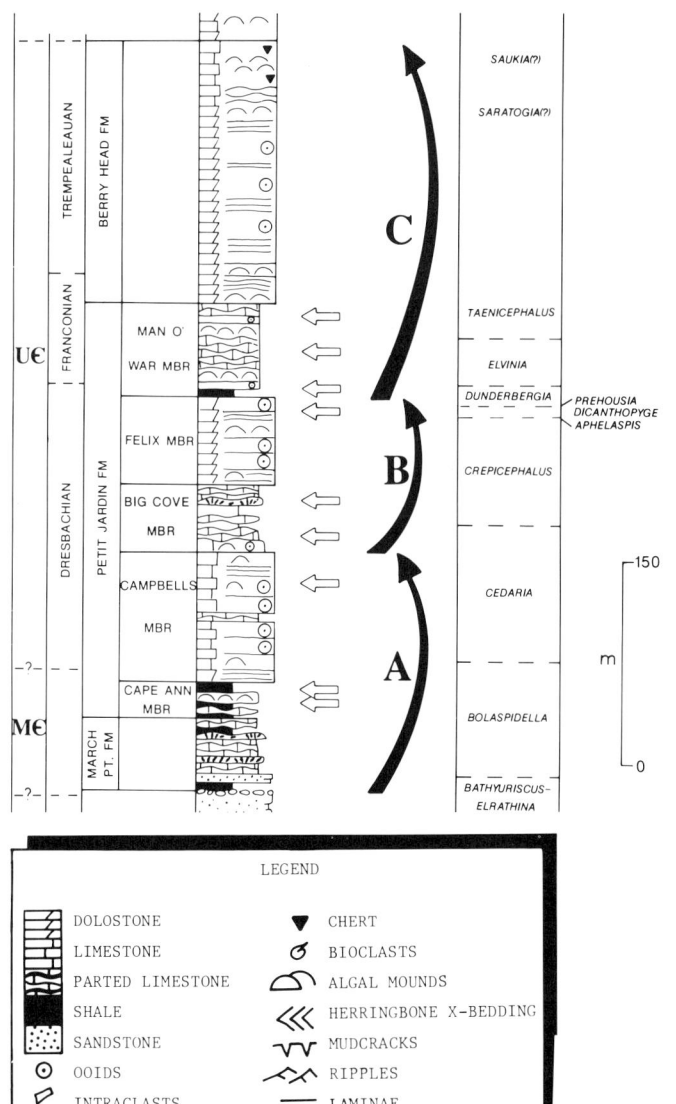

FIG. 8.—A stratigraphic column illustrating the Middle and Upper Cambrian autochthonous platform stratigraphy in the mixed carbonate-siliciclastic belt of central western Newfoundland, in which the three Grand Cycles are highlighted; arrows indicate paleontological control (after Chow and James, 1987). Phase 1, Events 4 and 5 of Figure 16.

but is limestone in the south. Small *Renalcis*-thrombolite-coral mounds and reefs surrounded by skeletal sands occur throughout but are particularly prominent near the base (e.g., Green Head Complex of Pratt and James, 1982) and the top (Knight, 1987). The generally subtidal sediments are interrupted by local peritidal horizons. Shallowing-upward cycles of burrowed lime wackestone, wavy-stratified lenticular and flaser-bedded dolomitic lime wackestone and mudstone and dololaminated or laminated lime mudstone form the upper 100 m of the sequence. Thrombolite and stromatolite mounds occur throughout the deposit. The sequences are thought to represent sedimentation in a mosaic of tidal-flat islands that accreted upon the shallow shelf (Pratt and James, 1986). Overall, the record is one of carbonate production keeping pace with relative sea-level rise. The top of the megacycle is bounded by a subaerial exposure horizon (Fig. 11B) with solution surfaces, dolomitization, silicification, intraformational dolomite and/or chert breccias, and lag deposits of intraformational quartz sand and pebbles. Paleo-caves with collapse breccias occur beneath the surface on Port-au-Port Peninsula. Sphalerite mineralization is developed locally at this horizon. Trilobites typical of the underlying sediments (Zone F) die out at this break (Boyce, 1979, 1983) and conodonts change dramatically between Faunas 2 and 3, with most of Fauna 3 interval absent (Stouge, 1982).

Coeval, deep-water, lower slope facies (Fig. 9) are encompassed within the massive limestone conglomerates of the proximal Shallow Bay Formation (Stearing Island Member) and extensive parted to ribbon limestones of the Green Point Formation (Broom Point Member) in the Cow Head region. The conglomerates, although spectacular and commonly amalgamated, are localized aerially to a relatively narrow band, suggesting a toe-of-slope debris apron fronted by a wide, gently dipping slope of carbonate silt and mud. There is abundant evidence of slope instability throughout in the form of intraformational truncation surfaces (ITS) and slumped horizons, even in the distal facies (Coniglio, 1986). The Cooks Brook Formation in the Humber Arm is characterized by similar hemipelagic limestone at this time (J. Botsford, pers. commun., 1986).

Upper Sequence

The upper sequence is a mainly subtidal facies (Fig. 11A) and corresponds roughly to the Arenig Series of Britain and the upper Canadian (Ibexian) to lower Whiterockian Series of North America. It composes the upper part of the Boat Harbour Formation (Barbace Cove Member), the Catoche Formation, and the Aguathuna Formation (Fig. 10). Onlap was extensive during this time, which represents the highest stand of sea level in the Early Ordovician (Barnes, 1984), and similar, coeval shallow-water sediments are present 300 km west of Newfoundland on the Mingan Islands, Quebec (Romaine Formation; Desrochers, 1985). Thus, the platform at this time must have had many of the attributes of an epeiric sea (*sensu* Shaw, 1964).

Peritidal sediments at the base of this sequence are distinctly coarser and more fossiliferous than most in the St. George Group. This suggests that not only was the rate of carbonate sediment production generally greater than the rate of relative sea-level rise, but also that high-energy conditions during this flooding stage extended well onto the shelf and were not dampened by an extensive barrier.

Coeval deep-water sediments (Fig. 9; Factory Cove Member, Shallow Bay Formation) are a sequence of interbedded ribbon limestone, dark laminated shale, and conglomerate, indicating platform margin progradation. Distal facies (St. Pauls Member, Green Point Formation) are mostly red bioturbated shale that migrated upslope with time. The distribution of these shales, (i.e., anaerobic to dysaerobic deposits upslope, aerobic deposits downslope) indicates that the ocean in front of the platform was distinctly stratified, with an oxygen-minimum zone impinging on the slope where

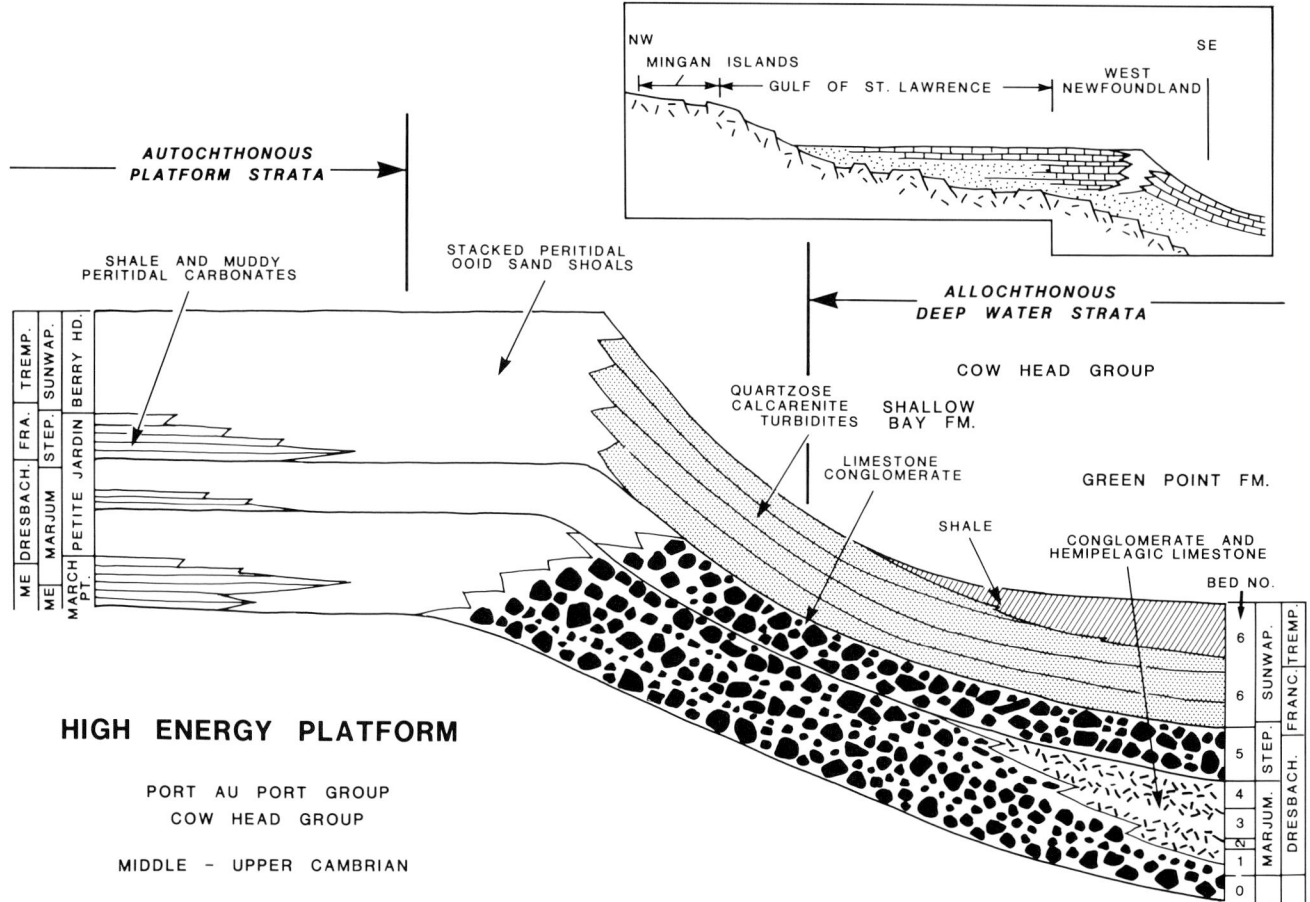

FIG. 9.—Interpretative cross section of the restored high-energy, narrow, Middle-Upper Cambrian platform and coeval deep-water sediments in western Newfoundland (Port-au-Port Group and Cow Head Group). Phase 3, Events 4 and 5 of Figure 16.

proximal deposits of the Shallow Bay Formation accumulated.

The succeeding thick sequence of shallow subtidal sediments (Catoche Formation; Fig. 11A, 11C) indicates a period when relative sea-level rise was equal to or even greater than the rate of carbonate sediment production. This subtidal facies extends to the Mingan Islands (Desrochers, 1985), indicating a vast shallow subtidal shelf of similar character throughout the area. Large bank complexes of thrombolites (Fig. 11D), *Renalcis*, sponges, primitive corals, pelmatozoans, and other skeletal invertebrates studded this shelf. Toward the margin, such mounds locally appear to form a major mound-barrier complex as seen in the parautochthonous rocks of Hare Bay (Stevens and James, 1976), Pistolet Bay, and Canada Bay (Knight, 1986, 1987).

A gradual change in the balance between sea-level rise and sediment production is reflected in the upper part of this subtidal-sediment package, as fossils become less diverse and peloidal grainstones tend to dominate (Costa Bay Member), reflecting less open-marine conditions.

Contemporaneous deep-water sediments (Fig. 12) record onlap of fine-grained facies upslope. Most proximal deepwater sediments are fine-grained ribbon limestone with a few limestone conglomerate horizons (Fig. 11E). Sediments equivalent to the middle of the subtidal (Catoche) facies are distinctive; silica replaces many ribbon-limestone layers, phosphate-pebble conglomerates are common, and graptolite zones are condensed. These characteristics together suggest starved deep-water carbonate sedimentation, likely due to the high position of sea-level and accompanied backstepping of the platform margin. Distal facies are characteristically red, bioturbated shale with minor green and black horizons and thin limestone. This sequence again indicates that proximal facies accumulated in the lower part of an oxygen-minimum zone, while those on the lower slope were deposited under aerobic seafloor conditions (James and Stevens, 1986).

The microcrystalline peritidal dolostone, dolomitic shale, and minor limestone (Aguathuna Formation) that cap this upper sequence reflect the slowing of relative sea-level rise and the onset of tectonism associated with the Taconic Orogeny. These peritidal sediments, like those beneath them, extend across the Gulf of St. Lawrence to Quebec. The predominance of dolostone, the importance of cryptalgal laminite and shale, the absence of fauna except locally in easternmost facies, and the presence of silicified evaporites suggest restricted and locally hypersaline environments. Intraformational breccias commonly containing quartz peb-

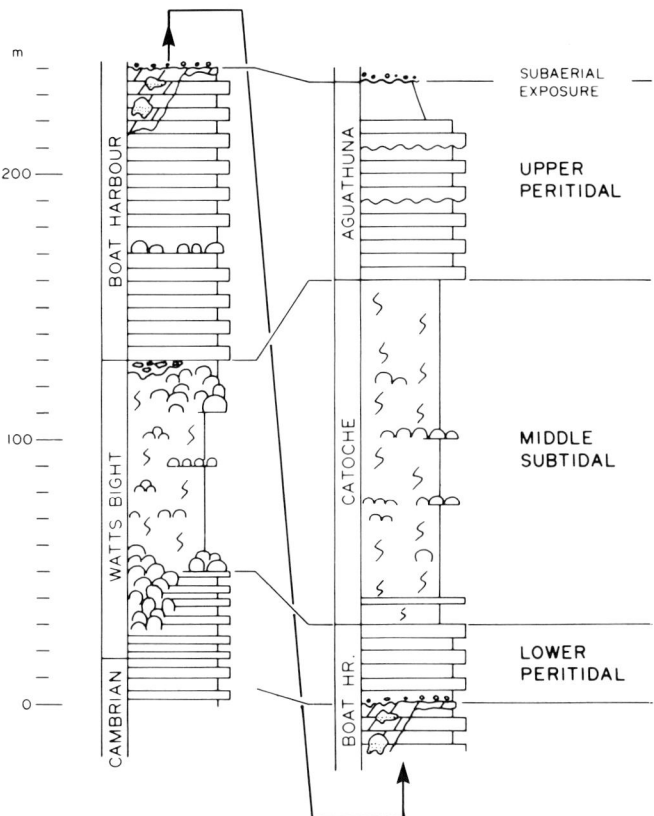

FIG. 10.—Two stratigraphic columns illustrating the two third-order sequences that compose the autochthonous Lower Ordovician St. George Group in western Newfoundland (after Knight and James, 1987). Phase 3, Events 6 and 7 of Figure 16.

bles or sand grains indicating subaerial exposure are present at several levels. The significance and duration of breaks within this peritidal succession are not presently understood. There appears to be at least one major break near the base and another near the top. This upper break, called the St. George Unconformity herein, is an erosional unconformity in many of the southern localities (Fig. 15A), whereas in some northern sections the break is recorded only by a change in the conodont fauna (K. Stait, pers. commun., 1986). Erosion has locally removed part or all of this peritidal-dolomite sequence. The dramatic, local variations in thickness reflect a combination of faulting, uplift, and erosion (Knight and James, 1987). On the Mingan Islands, the Romaine Formation was tilted before erosion so that the unconformity is a flat-karst plane which bevels several units (Desrochers and James, 1988).

Coeval deep-water facies are a particularly sensitive record of this period. Up to this time, all proximal facies of the Cow Head Group were similar, but after it, they are dramatically different (James and others, 1987). In the Cow Head region, the succession is one of hemipelagic ribbon limestones with minor calcarenites and two spectacular megaconglomerates (e.g., Fig. 11F; Beds 12 and 14; Kindle and Whittington, 1958). The abundance of limestone suggests that the margin was prograding, whereas the megaconglomerates which can be traced into the most distal deep-water facies and which sample margin facies as old as Upper Cambrian, point to wholesale margin collapse, probably the result of seismicity (James and Stevens, 1986). In the Lobster Cove area, in contrast (Fig. 13), deep-water sediments change abruptly at this time to a sequence of alternating shale and dolostone with almost no carbonate, suggesting that the margin upslope from this area was "drowned" below the zone of rapid carbonate production. This drowning is postulated to be the result of faulting (James and others, 1987).

PHASE 4—FOUNDERED PLATFORM

The demise of this carbonate platform occurred during initial stages of the Taconic Orogeny; deep-water slope deposits were rapidly buried by synorogenic flysch, the outer platform was faulted, fragmented, and subsided into deep water, and the inner platform was uplifted and underwent prolonged karstification.

The youngest deep-water carbonates of the allochthonous Cow Head Group (Middle Ordovician Whiterockian Series, *Orthidiella* Zone) are coeval with peritidal carbonates at the top of the upper sequence of phase 3. They are everywhere overlain by green lithic wackes and arenites (Lower Head Formation and equivalents; Fig. 15E), the basal beds of which are locally conglomeratic. Sole marks on the bases of graded beds in this flysch indicate a north to northeast provenance, and ophiolite detritus suggests that this source was a tectonic highland or archipelago of newly assembled thrust slices containing sediments and oceanic lithosphere, with perhaps Grenville basement slices (Stevens, 1970). The top of these sands are always in thrust contact with an overlying allochthonous slice, indicating the sediments were soon caught up in the westward-moving thrust complexes of an accretionary prism.

The autochthonous carbonates and shales (Table Head Group), which are equivalent to these foreland basin sediments, record breakup and foundering of the outer shelf (Klappa and others, 1980). Faulting, uplift, and erosion of the Lower Ordovician platform (Fig. 14) during early Middle Ordovician time (Whiterockian Series; *Orthidiella* Zone) created an irregular topography of horsts and grabens. This landscape was at first flooded only periodically, with peritidal carbonates accumulating in the lows (Springs Inlet Member, basal Table Point Formation; Ross and James, 1987), while the highs remained above water. Periods of non-deposition and prolonged subaerial exposure are reflected now by several unconformities (Knight, 1986) within this peritidal succession. Eventual complete inundation of this block-faulted terrane (Fig. 14) led to widespread subtidal carbonate deposition (Whiterockian Series, *Anomalorthis* Zone). The sediments (Fig. 15B) are monotonous dark grey, lumpy to nodular, bioturbated wackestone to packstone with numerous lenses of fossiliferous grainstone, similar to the subtidal carbonates of the Lower Ordovician below. Small reefs built by lithistid sponges (Klappa and James, 1980) dotted an otherwise relatively uniform sea floor during certain periods. Although of varying thickness, the thickest sections, over 300 m thick, span less than one

Fig. 11.—Sedimentary rocks deposited during phase 3 (Lower Ordovician–wide, low-energy, rimmed platform). (A) Erosional hoodoos of fossiliferous wackestone and packstone in the Catoche Formation, Port-au-Port Peninsula (packsack for scale at arrow). (B) Bedding plane exposure of subaerial karst unconformity separating the two sequences, strewn with quartz and chert pebbles, Boat Harbour Formation, northwest coast. (C) Dolomitized, burrowed, and fossiliferous wackestone/packstone, typical of the upper part of the Catoche Formation, northwest coast (coin at arrow is 2 cm in diameter). (D) Margin of a partially dolomitized digitate thrombolite mound (left) and partially dolomitized burrowed carbonate sediments (right), Catoche Formation, Hare Bay (scale has 2-cm divisions). (E) An 8-m-thick section of deep-water, interbedded limestone and shale and massive conglomerate (top right) in the allochthonous Cow Head Group, north of Bonne Bay. (F) Massive megaconglomerate (Bed 14) of shallow-water limestone clasts in argillaceous limestone matrix at the top of the allochthonous, deep-water, Cow Head Group, north of Bonne Bay (2-m-long measuring staff for scale, lower left).

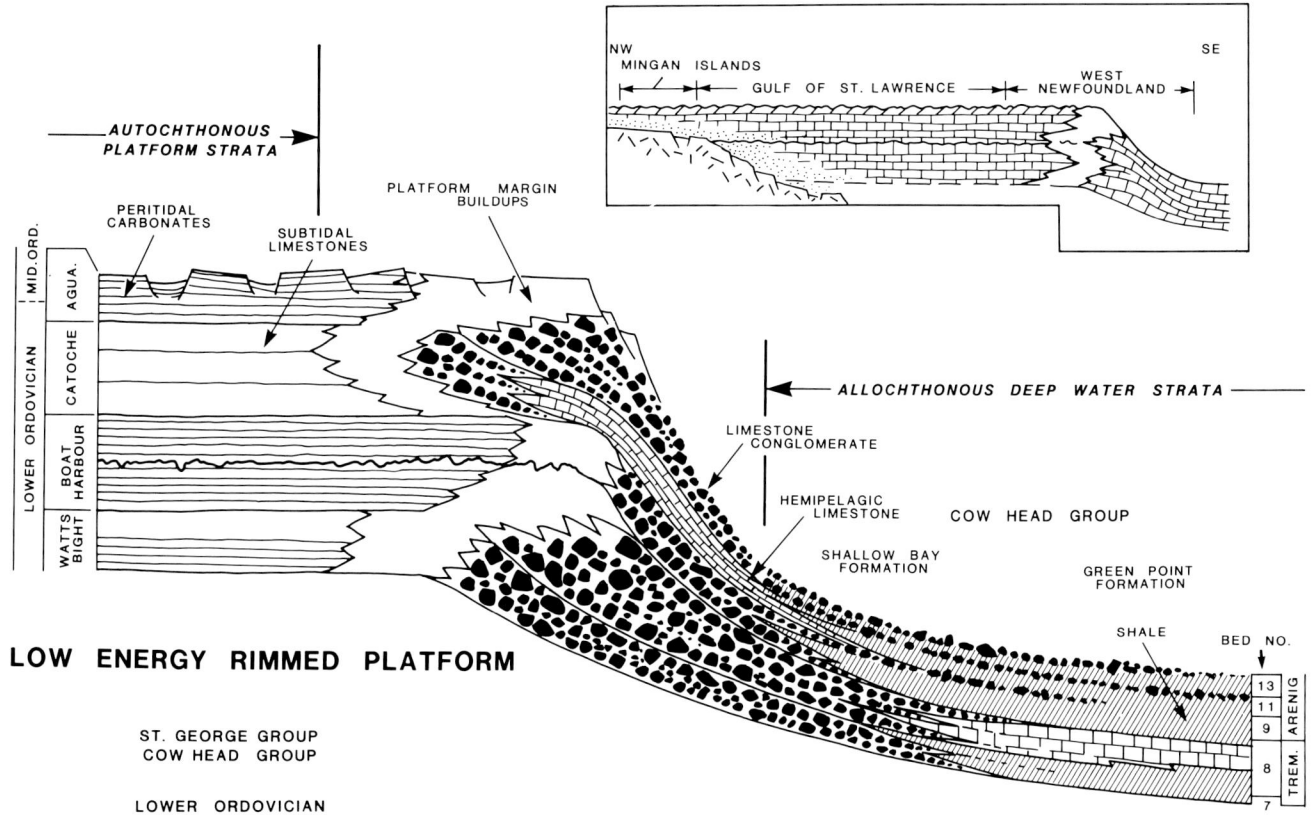

FIG. 12.—Interpretative cross section of the restored Lower Ordovician, low-energy, wide, rimmed platform and coeval deep-water sediments in western Newfoundland (St. George Group and Cow Head Group). Phase 3, Events 6 and 7 of Figure 16.

MIDDLE ORDOVICIAN

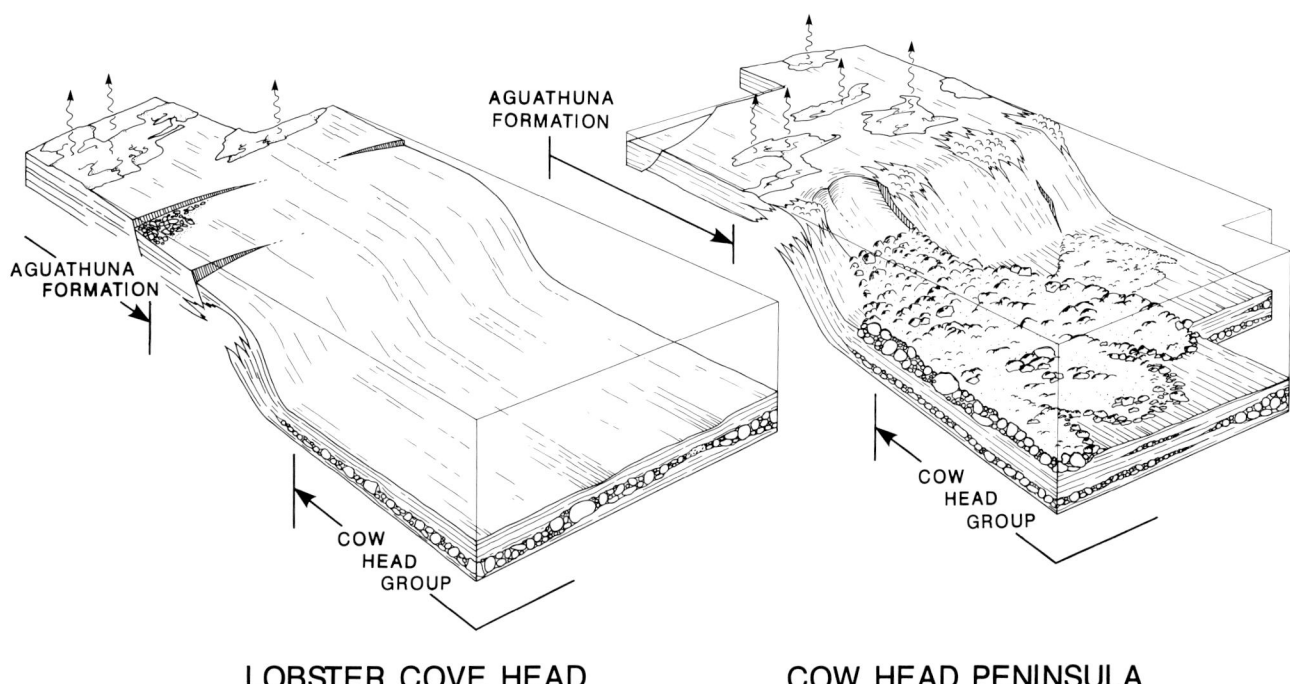

FIG. 13.—Two interpretive isometric diagrams illustrating the platform margin in the Lobster Cove Head versus Cow Head regions (1 and 40 km north of Bonne Bay, respectively) during earliest Middle Ordovician time; the difference is interpreted to reflect faulting during the initial stages of the Taconic Orogeny (from James and others, 1987).

graptolite zone (Whittington and Kindle, 1963; Williams and others, 1987), indicating that although rapid, the rate of carbonate sedimentation never caught up to relative sea-level rise. The rocks are characterized by repeated evidence of instability in the form of slumps, slides, deformed beds, and massive intraformational limestone conglomerates. The top of the sequence is everywhere abrupt and marked either by lithistid sponge biostromes, oncolite rudstones, or layers of chert (S. Stenzel, pers. commun., 1986; Knight, 1986).

Overlying sediments are variable in composition and reflect rapid subsidence within different settings. In some places the subtidal carbonates are overlain by hemipelagic slope deposits of parted to ribbon limestone (Table Cove Formation), which are extremely fossiliferous at the base, are interbedded with limestone plate conglomerates at various levels, and are characteristically deformed by synsedimentary slumps and slides (Fig. 15C). In many localities these slope deposits are capped by black, graptolitic, laminated shales (Black Cove Formation), but in others the black shales overlie the shallow subtidal carbonates (Table Point Formation) directly (Fig. 15D). In still other areas, particularly on the Port-au-Port Peninsula, the tops of the subtidal carbonates are eroded and overlain by massive conglomerates (Cape Cormorant Formation) with clasts and olistoliths of both underlying subtidal lithologies and older Cambrian and Ordovician platform carbonates. Such dramatic variations suggest that the fragmented platform foundered as separate blocks, at different rates, until the source of carbonate sediments was shut off completely (S. Stenzel, pers. commun., 1986).

Deep-water conglomerate and euxinic shale were finally buried by northeasterly derived synorogenic flysch (Mainland Sandstone, Goose Tickle Formation) as the foredeep migrated across the drowned and foundered platform (Fig. 14). Exposure and erosion of the Middle Ordovician carbonates are reflected by massive conglomerate horizons, composed almost exclusively of Table Head Group limestones (Fig. 15F), in the basal units of the flysch. The inner platform, as characterized by outcrops on the Mingan Islands, was exposed to karst erosion during most of this period. The St. George Unconformity is covered by strandline sandstone and shale, which grade upward into subtidal limestone (Mingan Formation; Chazyan Series) that are all younger than the west Newfoundland carbonates (Desrochers, 1985). Recent conodont studies by G. S. Nowlan (pers. commun., 1985) have shown the Whiterockian Series to be represented only in the uppermost strata of the Romaine Formation. This western area was probably affected by a peripheral bulge in response to filling of the westerly migrating foreland basin receiving flysch sediments and, consequently, was a region of limited, or later eroded, carbonate sedimentation during the Whiterockian.

The final phases of the Taconic Orogeny resulted in obduction of oceanic sediments and lithosphere onto the foundered platform. In western Newfoundland they are represented by the Hare Bay and Humber Arm allochthons (Figs. 1, 2) and similar ophiolite complexes in the Quebec Appalachians. Although they represent almost 10 km of sampled lithosphere, the allochthonous slices rarely exceed 2 km in thickness at any location. Various lines of evidence, including recent conodont color alteration index (CAI) studies for thermal maturity (Nowlan and Barnes, 1987a,b), indicate that the allochthons never extended much farther than their present areas (Fig. 1). The first post-orogenic phase of sedimentation is represented by the Long Point Group which, on the Port-au-Port Peninsula, overlies the Humber Arm Allochthon. The lower part of this unit is Middle Ordovician in age (early Mohawkian Series, Black Riverian Stage; possibly latest Whiterockian Series, Chazyan Stage; Bergstrom and others, 1974). It therefore tightly constrains the age of the Taconic Orogeny. Equivalent carbonates and clastics in the subsurface of Anticosti Island (Roliff, 1968), together with the carbonates of the Long Point Group to the east, represent a new phase of sedimentation in the Anti-

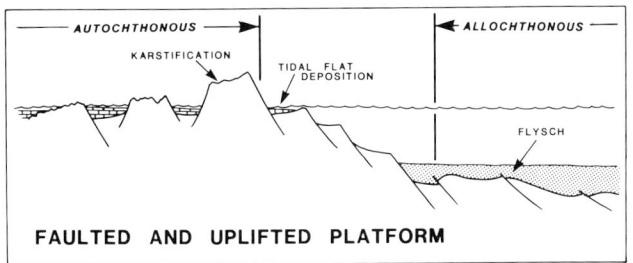

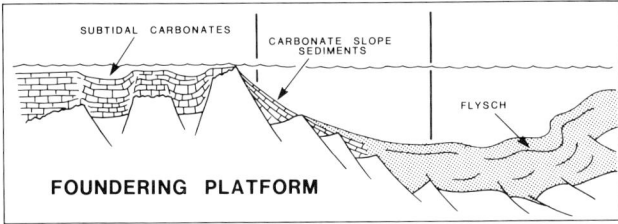

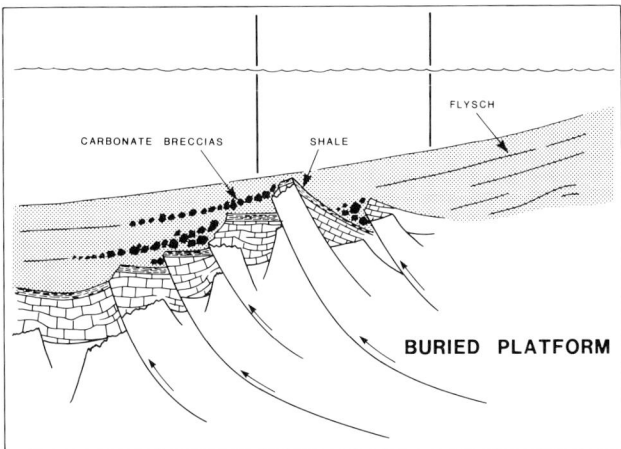

FIG. 14.—Three interpretative diagrams illustrating the evolution of the carbonate platform during the final foundering phase: (A) development of a foreland basin to the east and uplift, faulting, and karstification in the west (Aguathuna Formation, Springs Inlet Member of the Table Cove Formation, Lower Head Formation); (B) thrusting and deformation of the foreland basin flysch in the east, foundering and rapid subsidence of the faulted platform in the west (Table Point and Table Cove Formations); (C) thrusting and transport of strata in the east, subsidence into deep water below level of carbonate deposition (shale–Black Cove Formation), thrusting and collapse of scarps to form conglomerate (Cape Cormorant Formation), burial by flysch (Mainland Sandstone, Goose Tickle Formation). Phase 4, Events 8 and 9 of Figure 16.

Fig. 15.—Sedimentary rocks deposited during phase 4 (Middle Ordovician–foundered platform). (A) Erosional karst unconformity (arrows) at the top of St. George Group peritidal dolostones overlain by limestones of the Table Head Group, Port-au-Port Peninsula (section 6 m thick). (B) Sea cliffs some 22 m high sculpted into muddy fossiliferous limestone of the Table Point Formation, northwest coast. (C) Evenly bedded limestones with slumps and debris flows (arrows) in the deep-water Table Cove Formation, northwest coast. (D) A gently dipping, partially silicified limestone developed on shallow-water limestones of the Table Point Formation (T) directly overlain by deep-water shales of the Black Cove Formation (B), indicating rapid platform foundering (section is 10 m thick). (E) Interbedded massive to graded sandstones and shales of flysch, which buries the platform, Bay of Islands. (F) Massive conglomerate composed entirely of Table Head Group limestone clasts within thick flysch, which buries the platform, northwest coast.

costi Basin over the relatively low relief of the foundered and deformed Iapetus margin.

CONTROLS ON THE PASSIVE STAGE OF PLATFORM GROWTH

This early Paleozoic carbonate platform of the northern Canadian Appalachians formed mostly during a period of tectonic quiescence, with the continental margin undergoing thermal subsidence following latest Proterozoic rifting, rapid plate movement, and expanded mid-ocean ridge system. Global sea level was in a period of prolonged rise, which resulted in gradual flooding of the cratons. This second-order eustatic sea-level oscillation (Vail and others, 1977; Fortey, 1984) is reflected on most cratonic areas worldwide by deposits of the Sauk Sequence (Sloss, 1963; Sloss and Speed, 1974). The demise of the platform, brought on by continental collision, completes stages of the Wilson Cycle.

As part of the miogeocline that lay along the northern edge of the Iapetus Ocean, the western Newfoundland platform strongly resembles other parts of this continental margin now located to the south in the United States and to the north in western Europe, Spitsbergen, and Greenland. The original miogeocline does not appear to have been linear but was instead a series of promontories and reentrants (Thomas, 1977; Williams, 1978). Whereas the edge of the platform on the Newfoundland promontory ran Northeast-Southwest through the island, it probably swung sharply to a more East-West orientation south of the Port-au-Port Peninsula. These trends have yet to be proven, however, and will be considered by the Lithoprobe East deep crustal seismic transects currently being initiated.

The platform is characterized by cyclicity, as are most other platform carbonates (James, 1984), whereas coeval deep-water sediments illustrate dramatic changes in composition both laterally and vertically. In the following discussion emphasis is placed on shallow-water sequences because they are currently better understood and documented than deep-water facies. To interpret the controls on these shallow-water facies fully, however, the nature of coeval deep-water facies, which in turn reflects the nature and dynamics of the platform margin (James and Mountjoy, 1983; James and Stevens, 1986), is considered at the same time.

Since the first three phases represent development of a carbonate platform on a passive continental margin, they are discussed as a whole, followed by a treatment of the final foundering phase, during which the margin became convergent. For purposes of discussion, these phases are subdivided into major events, which are numbered on Figure 16.

It has been generally demonstrated that the rate of carbonate sediment production in early Paleozoic, shallow-water settings was roughly similar to that of modern platforms (Schlager, 1981). Thus, even though the style of sedimen-

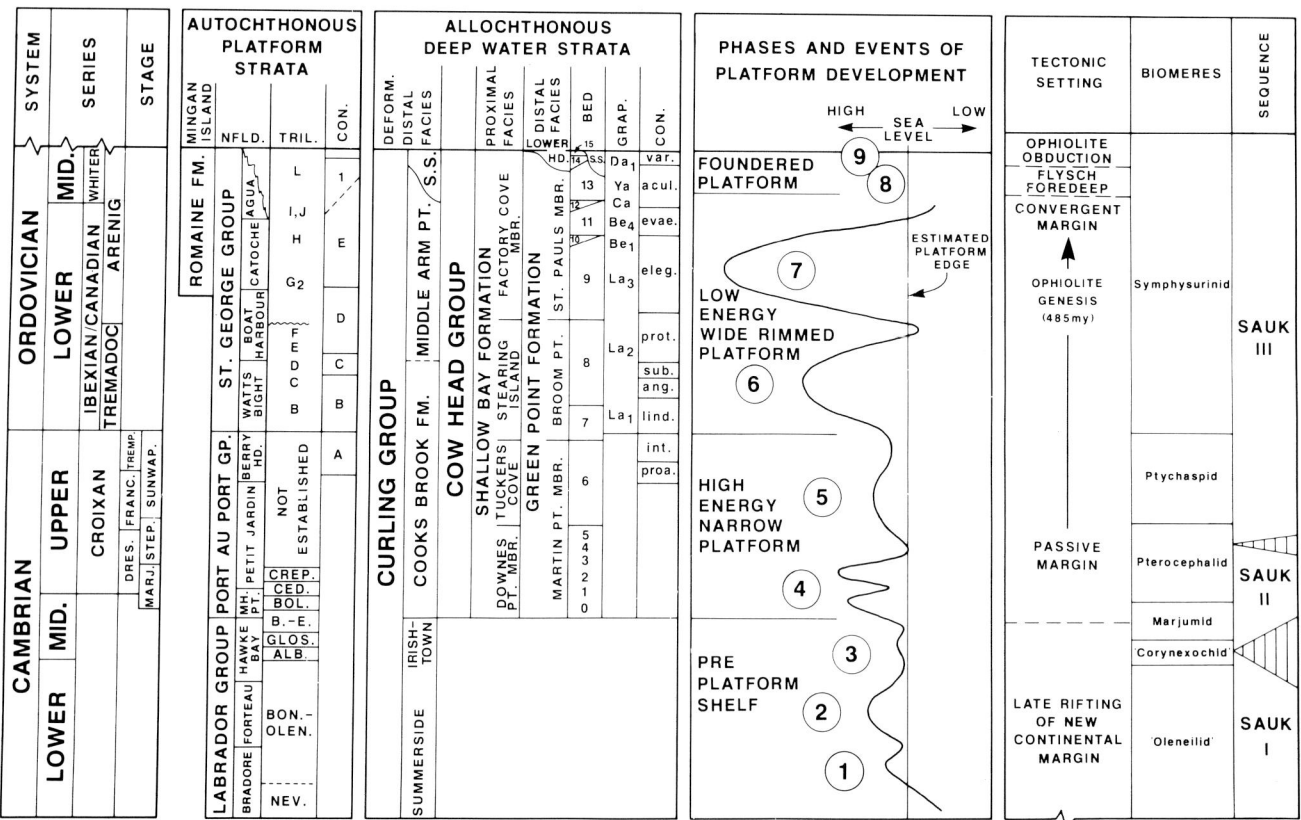

FIG. 16.—A diagram summarizing the chronostratigraphy and biostratigraphy of different autochthonous and allochthonous sequences in western Newfoundland and the Mingan Islands, Quebec, together with the four phases and nine events in the development of the platform and interpreted sea-level history.

tation may be determined by the nature of the calcareous benthos, the main controls on the development of the platform are tectonics and eustasy.

The interplay between tectonics, eustasy and climate on the western Newfoundland platform is manifest by repeated sequences of strata at several different scales. These are similar in thickness to those outlined by Vail and others (1977), although their origins may not be simply related to relative movements in sea level. To avoid confusion we use the same heirarchical ranking as Vail and others (1977). The largest scale, the second-order sequence, encompasses the whole platform succession, reflecting gradual flooding of the North American plate during prolonged eustatic sea-level rise. Superimposed third-order sequences, hundreds of meters in thickness, which have correlatives in other areas, are commonly called Grand Cycles in the Early Paleozoic. These, in turn, are made up of fourth-order and smaller meter-scale sequences, whose origin is a topic of controversy (Hardie, 1986).

The Second-Order Sequence (Sauk Sequence)

The western Newfoundland miogeocline, as part of the North American plate, is genetically related to other contemporaneous passive margins. As such, the thickness of the sedimentary sequence will reflect the interaction between tectonic subsidence and eustasy.

In general terms, post-rift tectonic subsidence of both modern and ancient passive continental margins decays exponentially with time in response to slow cooling and thermal contraction of lithosphere that was heated during rifting and continental breakup (McKenzie, 1978; Watts, 1981; Bond and others, 1984). Specifically then, Middle Cambrian time should have been a period of relatively rapid subsidence, Late Cambrian a time of slowed subsidence, and Early Ordovician a time when low subsidence rates approached those of the craton in general. The importance of these decreasing subsidence rates is reflected in the relative thickness of early Paleozoic sediments from miogeoclines in the southern Appalachians, Great Basin, and Canadian Cordillera: thick Middle Cambrian to thin Lower Ordovician sediments (Bond and others, 1984, and this volume).

The thickness of the western Newfoundland platform appears anomalous when compared to successions in other areas of North America. Both the Middle and Upper Cambrian sequences are roughly one-half the thickness of comparable miogeoclinal successions elsewhere (a total of approximately 800 m versus 1,600 m). The Middle Cambrian is complicated because of the Hawke Bay sandstone, but the Upper Cambrian shows the trend clearly, with an accumulation rate of roughly 2.5 cm/1,000 yrs. Since all areas are affected equally by the second-order eustatic sea-level rise, this anomaly must reflect the nature of subsidence.

The answer is clear when the Upper Cambrian and Lower Ordovician sequences in Newfoundland are compared. The Upper Cambrian is about 500 m thick. The Lower Ordovician also is about 500 m thick, similar to other contemporaneous North American Lower Ordovician miogeoclinal strata. Thus, it would appear, because of their similar accumulation rates, that the Middle Cambrian, Upper Cambrian, and Lower Ordovician platform successions now present in western Newfoundland accumulated on crust with a constant and not exponential rate of subsidence, more cratonic than marginal. A location inboard of the hinge zone of the inner flexural wedge (Bond and others, this volume) is most probable. Implications of such an interpretation are: (1) much of the outer platform, east of the present outcrop belt, has been removed or overridden by tectonics, and (2) the nature of most third-order sequences should be controlled by eustasy.

Third-Order Sequences (Grand Cycles)

These sequences, because they can be recognized in other parts of the world, are generally attributed to eustasy (Palmer and Halley, 1979; Aitken, 1981). Palmer (1983b) divided the second-order Sauk Sequence into three subsequences on the basis of trilobite biostratigraphy and suggested that they were separated by eustatic lowerings of sea level.

Preplatform shelf.—

The rift-drift transition, reflecting continental breakup and the onset of seafloor spreading, is interpreted to have occurred in western Newfoundland about 550 to 570 Ma (Williams and Hiscott, 1987), or stratigraphically within the Bradore Formation. The overlying, widespread limestone, shale, and minor sandstone correspond to much of the *Bonnia-Olenellus* Zone, or the upper part of Sauk I (Palmer, 1983b). Fritz (1975) extended the Grand Cycle concept of Aitken (1966, 1978) into the Lower Cambrian and differentiated three onlap-offlap cycles in western North America. The Grand Cycle represented by the Forteau Formation in western Newfoundland (Event 2, Fig. 16) is approximately, but not precisely, the same age as Cycle B of Fritz, suggesting that eustasy may be but one of the primary controls.

This interpretation of eustatic control is further strengthened by the thick sequence of offlap sandstones that caps the Grand Cycle (Event 3, Fig. 16). This period, which extends from the end of Early Cambrian time through much of Middle Cambrian time, is typified widely by reduced thicknesses of sediments that contain several trilobite zones (the Corynexochid Biomere of Palmer, 1977), or by a major clastic unit representing a regional regression. The presence of a similar event around the margins of the Iapetus Ocean led Palmer and James (1979) to postulate that it represented a major eustatic fall in sea-level, or regional uplift—the Hawke Bay Event. Recent work by Knight and Boyce (1987) indicates that sea-level did not fall below the edge of the shelf, suggesting that uplift may have occurred in the west causing influx of much coarse terrigenous detritus eastward onto the shelf. The intrusion of anorthosite near Sept Iles, Quebec, at this time (Higgins and Doig, 1981) could have created movement along the St. Lawrence rift system. The eastward progradation of terrigenous clastics smothered carbonate deposition, spilled over the shelf edge, may have cut submarine canyons, fed deep-water quartzose turbidites, and probably accentuated the overall effect of lower relative sea-level. These canyons may later have served as conduits to feed some of the Cow Head Group conglomerates.

High-energy narrow platform.—

This late Middle Cambrian to latest Late Cambrian succession is highly variable, but sediments in the transitional facies belt between inner-shelf terrigenous clastic/carbonate sediments and an outer belt of stacked ooid shoals exhibit striking cyclicity, with most of the characteristics of classic Cambrian Grand Cycles (Chow and James, 1987).

The base of Newfoundland Grand Cycle A (Event 4, Fig. 16) is within the upper part of the *Bathyuriscus-Elrathina* Zone (Palmer, 1977) or the *Ptychagnostus gibbus* Interval-Zone (Robison, 1984). Globally, the base of the *P. gibbus* Zone commonly coincides with an unconformity or an abrupt facies change possibly related to a eustatic rise in sea level (Robison, 1984). The upper boundary of Cycle A within the *Cedaria* Zone appears to be equivalent to the break between the Maryville Limestone and Nolichucky Shale in the Tennessee Appalachians (Palmer, 1983a), which has no equivalents on the western part of the craton (Palmer, 1983a). This distribution suggests that the change may be localized to the Appalachians and, like many Grand Cycles in the Great Basin, is only local in extent, and therefore not controlled by global eustatic changes in sea level.

In contrast, the top of Grand Cycle B is a break that is recognized almost everywhere in North America (Palmer, 1983a) and has been designated as the top of Sauk II subsequence (Palmer, 1983b). Recent detailed work in the Canadian Rockies (Westrop, 1986) confirms the continent-wide nature of this event. This break is interpreted, therefore, as reflecting a eustatic fall in sea level.

The overlying Cycle C (Event 5, Fig. 16) is dominated by ooid sand-shoal deposition in the outer part of the belt and tidal-flat dolostones in the inner belt. It is as thick as both of the preceding cycles combined. There are commonly two Grand Cycles in the Upper Cambrian of the southern Canadian Rockies, but to the north and east there is only one cycle that spans the entire interval (Aitken and Greggs, 1967) as it does in Newfoundland. Shoals and banks were particularly widespread in the Great Basin and southern Appalachians during this time (Palmer, 1971a,b).

Thus, from the record in the central part of western Newfoundland at least, sedimentation on this relatively narrow, high-energy shelf took place in two stages, separated by a eustatic lowering in sea level. The early stage (Event 4) is represented by two Grand Cycles, which probably reflect sedimentation during two local, long-term fluctuations superimposed on a eustatic rise and fall of sea level. The later stage (Event 5) composes only one Grand Cycle but one in which sedimentation, or vertical accretion, was much more rapid.

This two-stage development is clearly illustrated by coeval deep-water facies. The early stage (Event 4) resulted in extensive deposition of carbonate conglomerates and megaconglomerates, whose vertical persistence indicates platform margin progradation. Thus, the amplitude of the rise and fall in sea-level was probably low. The later, quite different stage (Event 5) is represented by quartzose calcarenite turbidites punctuated by limestone conglomerates and upslope onlap of shale. These attributes suggest reduced sediment delivery to the slope, most easily explained by vertical accretion of the margin and on-bank transport of most sediment, reflecting a high-amplitude fluctuation in sea level.

The end of Event 5 (the top of Grand Cycle C) at or just below the Cambro-Ordovician boundary is a regressive event, characterized by peritidal deposition, which is also recognized farther south in the Virginia Appalachians (Demicco and Hardie, 1981) and on most continents. The reasons for this regression are currently the topic of much debate, with some suggesting it may be eustatic (Miller, 1984; Fortey, 1984), possibly related to glaciation in South America. Others argue that it is of local significance (Ludvigsen and others, 1988). In many sections of the world there is evidence of shoaling, shallowing, or emergence during all or part of the *Cordylodus proavus* conodont Zone (= lower conodont Fauna A on Fig. 16). The initiation of a subsequent transgressive event appears to correlate with the overlying *C. lindstromi* and *C. angulatus* Zones (= conodont Fauna B); the *C. intermedius* Zone or *C. lindstromi* Zone will probably be selected to define formally the base of the Ordovician System.

Low-energy, wide, rimmed platform.—

The Early Ordovician is marked by a rapid and extensive flooding of the shelf, representing the climax phase of the Sauk Sequence. The submergence of the North American craton is extensive, exceeded only in the late Middle and Late Ordovician during the climax of the Tippecanoe Sequence of Sloss (1963). The Early Ordovician transgression is well documented for the North American craton (Ross, 1976; Barnes, 1984), is evident on other cratons, and is interpreted as a global eustatic event (Jaanusson, 1979; McKerrow, 1979; Leggett and others, 1981; Fortey, 1984).

In western Newfoundland the Low-Energy, Wide, Rimmed Platform phase includes two transgressive and two regressive events or two third-order sequences (Events 6 and 7, Fig. 16). These all-carbonate sequences lack the fine-grained terrigenous clastics of true Grand Cycles, yet are the same scale.

The tripartite nature of the lower sequence (Event 6) is identical to the Chepultepec interval to the south in the Virginia Appalachians (Bova and Read, 1987). The thickness and composition of the three divisions (basal peritidal, including unnamed beds above the Upper Cambrian Berry Head Formation = 120 m; middle subtidal unit = 80 m, upper peritidal unit = 110 m) in both areas are remarkably similar, indicating a strong eustatic control on sedimentation. The precise response of carbonate sedimentation to this fluctuation in sea level is argued in James (1984), Knight and James (1987), and Bova and Read (1987). The second transgressive phase (Event 7, Fig. 16) with a wide range of lithofacies and more diverse faunas as well as identical outcrops at Mingan Islands, Quebec, all suggest a more extensive platform than that for Event 6 (cf. Barnes, 1984, figs. 6,7). A more rapid deepening and possibly development of a more continuous mound-barrier back from the platform margin during deposition of the lower Catoche Formation probably reduced the amount of carbonate reaching distal-slope facies, as reflected in fine sediments of the coeval Cow Head Group (James and Stevens, 1986).

A conclusion of this analysis of third-order sequences is a confirmation that eustasy is a primary control. Nevertheless, the effect of local tectonics is still important, as demonstrated by the fact that these sequences, while present in many localities, are not ubiquitous. Ultimately, the only way to separate tectonics from eustatic control may be to compare the sequence in the northern Appalachians with similar rocks on a different plate with a different history (e.g., northwestern Argentina; Ramos and others, 1986).

Fourth-Order (Meter-Scale) Sequences

The nature and causes of meter-scale sequences are currently a topic of controversy (James, 1984; Hardie, 1986) and new interpretations (Read and others, 1985). The following discussion is directed toward sequences developed on the platform during phases 2 and 3, in terms of their composition and arrangement within the larger third-order sequences.

High-energy narrow platform.—

The details of peritidal cyclicity within the mixed siliciclastic-carbonate belt have been described by Chow (1986) and are summarized in Chow and James (1987). Widespread, correlatable, meter-scale, parted limestone-shale cycles, which characterize the shaly parts of Newfoundland Grand Cycles, compose a basal ripple-laminated limestone (subtidal) that grades transitionally upward into parted limestone (predominantly intertidal) which, in turn, is capped by carbonate laminate and/or shale (supratidal). Parted limestones, composed of thinly interbedded limestone, dolostone, and shale exhibit upward changes from flaser bedding to wavy bedding to lenticular bedding. Flat-pebble conglomerates, skeletal wackestone and packstone, grey oolite, and horizons of algal mounds randomly punctuate the cycles. Although they vary between Grand Cycles (Chow and James, 1987), the sequences are on the whole identical in most respects to those in coeval platform sediments farther south in the United States Appalachians (Hardie, 1986). The major difference is that those farther south contain a much thicker component of subtidal facies (ooid sands, algal mounds, and flat-pebble conglomerates), probably reflecting the higher rates of tectonic subsidence and therefore accommodation interval at this time.

In contrast, the unpredictable meter-scale sequences of oolite sand (Fig. 7A), characteristic of the upper carbonate parts of the Grand Cycles, are non-cyclic and composed of three units: intertidal grey oolite, subtidal brown oolite, and thin, discontinuous supratidal laminite caps.

The parted limestone-shale cycles and other argillaceous lithologies that compose the basal parts of Grand Cycles and Cambrian platform sediments elsewhere have been interpreted to represent deposition in intra-shelf basins, in mostly subtidal settings (Markello and Read, 1981; Aitken, 1978). There is no information as to a possible intra-shelf basin west of Newfoundland beneath the Gulf of St. Lawrence. Thus, the part of the platform in western Newfoundland is either the transition zone between a western intra-shelf basin and an eastern platform margin ooid-shoal belt, or the whole platform was very shallow to peritidal throughout its history. Regardless, the meter-scale cycles are peritidal throughout, and so quite different from most of the Grand Cycles as recognized in western North America (Aitken, 1966, 1978) and modeled by Read and others (1985). Chow and James (1987) suggest that this may be the result of a much narrower shelf in this area or, more probably, a low-subsidence rate. Thus, whereas intra-shelf basins developed in those areas with a relatively high rate of thermal subsidence, in regions with a low-subsidence rate, such as Newfoundland, sediments were deposited throughout in shallow water. The peritidal cycles are, nevertheless, dominantly subtidal at the base, increasingly intertidal in the middle, and mostly supratidal at the top of the shaly half-cycles. These third-order meter-scale cycles were related to either autocyclic or allocyclic deposition superimposed on a second-order rise and fall in sea level (James, 1984) in which there were relatively small variations in amplitude of the third-order cycle (Hardie, 1986).

Low-energy wide-rimmed platform.—

Sequences on this part of the platform have been described in detail by Pratt and James (1986). Peritidal deposition characterizes the basal and upper parts of the two third-order cycles. They observed rapid lateral facies changes in some outcrops and when correlating closely spaced sections. The stratigraphic array of lithotopes appeared too irregular to be simplified into widespread, laterally continuous, shallowing-upward cycles. Instead, Pratt and James (1986) suggested that these peritidal sediments were deposited on and around small tidal-flat islands and banks that were scattered across the wide shelf, particularly during periods when sedimentation matched relative sea-level rise. Removal of some beds during burial pressure dissolution may make this phenomenon more apparent than real. Recent work indicates that, locally, some tidal-flat caps may be traced laterally over areas of 255 km^2.

Along the craton shoreline, as exposed 300 km to the west on the Mingan Islands, Quebec, Desrochers (1985) found similar coeval facies in the upper second-order sequence and was able to trace individual meter-scale cycles over 20 km with only minor changes in their thickness and internal lithofacies organization. Thus, it appears that different parts of the platform may have responded in different ways to similar sea-level-sedimentation couplets. Close to the cratonic margin (e.g., Mingan Islands) or when onlapping regional exposure surfaces (e.g., basal peritidal parts of the sequences), cycles exhibit good lateral correlation, indicating extensive progradation. In mid-platform areas remote from the craton margin or regional subaerial exposure surfaces during offlap (e.g., upper peritidal parts of the sequences), cycles appear to have rapid lateral variation, probably reflecting the tidal-flat island model.

NATURE OF THE FOUNDERED PLATFORM

The platform demise (Events 8 and 9, Fig. 16) took place in three stages: (1) extensional faulting and partial exposure; (2) rapid subsidence; and (3) thrusting, erosion, and burial with conversion to an easterly migrating foreland basin.

Faulting and exposure are expressed by the upper part of the St. George Group (Aguathuna Formation) and basal peritidal part of the Table Point Formation (Springs Inlet Member). This period is marked by peritidal deposition and erosion along most of the Appalachian miogeocline, indicating that the rate of carbonate sedimentation exceeded the rate of sea-level rise. Barnes (1984) noted that this period, in both Arctic Canada and the Cordillera, was a time of peritidal deposition, suggesting that the succession may represent a global eustatic sea-level fall. Analysis of the position of the unconformity or unconformities (Mussman and Read, 1986) worldwide indicates, however, that the exposure event in the Appalachians cannot be explained by eustatic sea-level fall alone but must also require some tectonic influences. This conclusion is also valid for the St. George Unconformity (Knight and James, 1987).

Nature of the peritidal successions and rapid changes in amount of erosion of the Aguathuna Formation all point to a gently folded and block-faulted terrane of horsts and grabens (Klappa and others, 1980; Knight, 1985, 1986; Stenzel, pers. commun., 1986). The inner part of the platform on the Mingan Islands, Quebec, however, appears to be marked by gentle warping (Desrochers, 1985). In the southern Appalachians the shelf was also deformed into a series of open folds (Thomas and others, 1980). The block faulting indicates extensional tectonics, supporting the hypothesis that the shelf was warped by a westward migration of a peripheral bulge that formed in response to eastward underthrusting, foreland flysch basin subsidence, and eventual ophiolite obduction (Jacobi, 1981; Quinlan and Beaumont, 1984).

The subsidence stage, represented by the Table Head Group, indicates that relative sea-level rise, dominated by platform subsidence, kept pace with, but then exceeded, the rate of carbonate sedimentation.

This period in the southern Appalachians is also complicated with the rocks being affected by a later Middle Ordovician phase of the Taconic Orogeny. In Virginia, the Middle Ordovician carbonate ramp sequence grades southeastward into the foreland basin facies (Read, 1979). The sequence, like that in the north, commences with an onlap package of peritidal facies, which passes upward into shallow-ramp facies (Lenoir and Lincolnshire limestones), muddy carbonate reef mounds, and deep-ramp facies. In the east these are overlain by slope to basin carbonates and shales that are buried by conglomerates, greywackes, and shales (Knobs/Martinsburg) comparable to the succession in Newfoundland. In the west the unconformity is of longer duration, and this facies is younger and thinner and overlain by an offlap package of shallow subtidal to peritidal carbonates capped by mixed deltaic/shallow-marine terrigenous clastics and peritidal carbonates. A similar succession is present in the Mingan Islands, Quebec, where a long hiatus is followed by an offlap package of peritidal siliciclastics, followed by subtidal carbonates of the Chazyan Mingan Formation (Desrochers, 1985).

Tectonics clearly play a key role during platform demise, yet similarities between widely separated parts of the Appalachian miogeocline suggest eustasy may also have an effect. In the Great Basin, where the miogeocline is still a passive margin, this period (Whiterockian) is one during which carbonate sedimentation kept up with or outpaced relative sea-level rise (Ross and others, this volume). Thus, the effect of eustasy was probably overshadowed by orogenesis in the Appalachians during this last phase of carbonate sedimentation.

Understanding of the burial phase in western Newfoundland is complicated because of synsedimentary thrusting accompanying sedimentation (Fig. 14). Burial by northeasterly derived flysch is accompanied, in the early stages, by uplift and erosion of the platform, producing numerous levels of granule to boulder conglomerates and breccias and olistoliths. In the northeast these conglomerates compose only Table Head Group clasts, whereas in the southwest much of the underlying succession is eroded. Similar events are recorded in the southern Appalachians (Mussman and Read, 1986), where lithoclasts from all of the Lower Ordovician units of the Knox and Beekmantown are found in Middle Ordovician conglomerates associated with flysch derived from the southeast.

THE EVOLVING BIOTA

Whereas carbonate sedimentation is demonstrably cyclic throughout the history of platform development, the nature of the sediments changes dramatically. The long-term trend is one in which terrigenous clastic sediments give way to carbonates, reflecting gradual flooding of the craton concommitant with progressive narrowing and isolation of the siliciclastic source area. A much more remarkable trend is the changing nature of carbonate sediments.

Early Cambrian carbonates are extremely fossiliferous and mostly bioclastic. This changes gradually toward the end of the early Cambrian, so that by comparison, middle and upper Cambrian sediments are almost abiotic, composed of peloids, ooids, and intraclasts with microbial buildups and relatively few bioclasts. These sediments are more like those of the Proterozoic than the Phanerozoic. The style of deposition changes again across the Cambro-Ordovician boundary with Lower Ordovician sediments becoming not only low energy but progressively more fossiliferous with time, so that most deposits are bioclastic, with insignificant ooid or peloid carbonates.

These changes do not reflect some peculiarity of the western Newfoundland platform, but typify other early Paleozoic miogeoclines around the North American plate, confirming that climate or water circulation were not governing factors. They clearly reflect some as yet poorly understood changes in the marine biosphere. A parallel trend is not apparent in younger intracratonic carbonate platforms but does appear in miogeoclinal carbonate platforms along the margin of the Tethys (Wilson, 1975), particularly during the late Jurassic through Cretaceous, another period of high sea-level stand (Vail and others, 1977). The importance of this difference in the biota appears to have been mainly a change from bioclastic sedimentation and microbial/metazoan buildups to a platform characterized by widespread carbonate sands and microbial buildups. It is as though, in this latter mode, the platform reverted to a simpler, more primitive style, akin to that of the Precambrian.

SUMMARY

The striking similarity between the history of the Lower Paleozoic continental margin in the northern Canadian Appalachians and in the southern and central Appalachians suggests a uniform model may be constructed for this type of carbonate platform. The history of the northern sector can be resolved into four phases and nine events summarized in Figure 16. Lack of evaporites, except for minor local peritidal remnants in latest stage 3, suggests a humid tropical setting. Whereas variations in climate may not have been important, the apparent uniform conditions resulted in normal open-marine environments, promoting rapid fixation of carbonate.

Phase 1—Preplatform Shelf

(Upper Proterozoic?—Middle Cambrian): a narrow shelf developed on rifted plutonic basement and was covered by terrigenous clastic and minor-carbonate sediments in the form of a pronounced onlap-offlap sequence; primary control was tectonic.

Event 1.—

(Upper Proterozoic-Lower Cambrian): block faulting of Precambrian basement following continental rifting; sand deposition began in depressions and eventually covered underlying topography; shelf sedimentation occurred in braided fluvial to strandline settings; primary control was eustatic sea-level rise against a background of minor tectonism and moderate thermal flux related to rifting.

Event 2.—

(Upper Lower Cambrian): onlap-offlap in the form of a Grand Cycle; variable coarse and fine terrigenous clastic and fossiliferous carbonate deposition produced pronounced facies belts; a subtidal ramp with no effective organic rim; most conspicuous carbonates were ooid-sand shoals, archaeocyathan buildups, and muddy tidalites; deep-water deposits were terrigenous muds and silts; primary control was eustasy.

Event 3.—

(Uppermost Lower Cambrian-middle Middle Cambrian): pronounced offlap; quartzose sands prograded to shelf edge; local carbonate deposition continued throughout at shelf edge; probable cutting of submarine canyons; deep-water slope muds, silts, and conglomeratic submarine fans; primary controls were tectonic and eustasy.

Phase 2–High-Energy Narrow Platform

(Middle-Upper Cambrian): evolution to a carbonate platform roughly 200 km wide and covered by terrigenous clastic and carbonate sediments with substantially fewer biotic constituents than in previous phase; distinct facies belts; shelf margin of ooid/peloid-sand shoals; primary control was eustasy.

Event 4.—

(Middle to middle Upper Cambrian): marked cyclicity throughout this time; two Grand Cycles in transition zone between outer ooid-sand shoals and inner terrigenous/carbonate belt; pronounced margin progradation; *Epiphyton-Girvanella* buildups on upper slope; extensive conglomerate-debris flows in deep water; primary control was eustasy, but generation of two Grand Cycles, rather than one, was probably due to local tectonism.

Event 5.—

(Middle Upper to Upper Cambrian); similar to preceding event, but only one grand Cycle developed; proximal quartzose carbonate turbidites and distal terrigenous muds characterized lower slope deposition; margin accretion was mostly vertical and deep-water shales prograded upslope; primary control was eustasy.

Phase 3–Low-Energy, Wide, Rimmed Platform

(Lower to basal Middle Ordovician): evolution into a huge platform more than 500 km across, overall onlap; entirely bioclastic carbonate depositions with a diverse fauna and buildups at shelf edge; primary control was eustasy, representing climax of Sauk Sequence.

Event 6.—

(Basal to middle Lower Ordovician): onlap-offlap cycle with karst unconformity at top; quiet, muddy, subtidal to peritidal deposition, local and marginal microbial-metazoan buildups; margin progradation; deep-water carbonate toe-of-slope apron with minor debris sheets, which accumulated under anaerobic conditions within, and aerobic conditions below an oxygen-minimum zone, primary control was eustasy.

Event 7.—

(Middle Lower to basal Middle Ordovician): onlap-offlap cycle with karst unconformity at top, same style as Event 6 except that onlap was more rapid and extensive, marginal buildups were larger with more diverse metazoan fauna, and platform was block faulted during offlap; margin back-stepping and then progradation; deep-water sediments formed a narrow toe-of-slope debris apron fronted by a wide shale slope deposited in the lower part of a fluctuating oxygen-minimum zone; deep-water sediments were covered by synorogenic flysch; primary controls were initially eustasy but later tectonics.

Phase 4–Foundered Platform

(Basal to middle Ordovician): collapse and burial of platform; carbonate deposition followed by flysch sedimentation; primary control was tectonic.

Event 8.—

(Basal to middle Middle Ordovician: block faulting and uplift with passage of peripheral bulge across platform; karstification of highs, peritidal carbonate depositions in lows; subsequent general subsidence and rapid subtidal deposition throughout on an irregular series of platforms and depressions subject to continuing seismicity; eventual subsidence below zone of carbonate deposition; primary control was tectonic against a background of eustatic sea-level rise.

Event 9.—

(Medial Middle Ordovician): burial by flysch; contemporaneous erosion of carbonates from highs created by thrust faulting; primary control was tectonic.

ACKNOWLEDGMENTS

This synthesis summarizes the work of many colleagues and studies, and we particularly thank W. D. Boyce, M. Coniglio, N. Chow, F. Debrenne, A. Desrochers, R. A. Fortey, D. Haywick, S. Hughes, T. Lane, R. Levesque, R. Ludvigsen, C. F. Klappa, D. R. Kobluk, A. R. Palmer, B. R. Pratt, R. J. Ross, Jr., B. A. Stait, S. R. Westrop, H. Williams, and S. H. Williams for all their help. Research students J. Botsford, S. L. Pohler, S. Stenzel, and K. Stait kindly allowed us to include results of their present research. G. Bond kindly allowed us to see several unpublished manuscripts, and the paper was improved by the comments of J. F. Read. The research of Barnes, James, and Stevens was supported by the Natural Sciences and Engineering Council of Canada and Department of Energy Mines and Resources. Knight has been supported through Canada-Newfoundland Mineral Development Subsidiary Agreements, 1976–1989, and publication is with the permission of the Newfoundland Department of Mines.

REFERENCES

AITKEN, J. D., 1966, Middle Cambrian to Middle Ordovician cyclic sedimentation, southern Rocky Mountains of Alberta: Bulletin of Canadian Petroleum Geology, v. 14, p. 405–441.

———, 1978, Revised models for depositional grand cycles, Cambrian of the southern Rocky Mountains, Canada: Bulletin of Canadian Petroleum Geology, v. 26, p. 515–542.

———, 1981, Generalizations about Grand Cycles, *in* Taylor, M. E., ed., Second International Symposium on the Cambrian System: U.S. Geological Survey Open-File Report 81-743, p. 8–14.

———, AND GREGGS, R. G., 1967, Upper Cambrian formations, southern Rocky Mountains of Alberta, an interim report: Geological Survey of Canada Paper 66-49, 91 p.

AHR, W. M., 1973, The carbonate ramp–An alternative to the shelf model: Gulf Coast Association of Geological Societies Transactions, v. 23, p. 221–225.

BARNES, C. R., 1984, Early Ordovician eustatic events in Canada, *in* Bruton, D. L., ed., Aspects of the Ordovician System: Paleontological Contributions from the University of Oslo No. 295, Universitetsforlaget, p. 51–63.

BERGSTROM, S. M., RIVA, J., AND KAY, M., 1974, Significance of conodonts, graptolites and shelly faunas from the Ordovician of western and north-central Newfoundland: Canadian Journal of Earth Sciences, v. 11, p. 1625–1660.

BOND, G. C., NICKERSON, P. A., AND KOMINZ, M. A., 1984, Breakup of a supercontinent between 625 Ma and 555 Ma: New evidence and implications for continental histories: Earth and Planetary Science Letters, v. 70, p. 325–345.

BOVA, J. A., AND READ, J. F., 1987, Incipiently drowned facies within a cyclic peritidal ramp sequence, Early Ordovician Chepultepec interval, Virginia Appalachians: Geological Society of America Bulletin, v. 98, p. 714–727.

BOYCE, W. D., 1977, New Cambrian trilobites from western Newfoundland: Unpublished B.S. Honors Thesis, Memorial University of Newfoundland, St. John's, 66 p.

———, 1979, Further developments in western Newfoundland Cambro-Ordovician biostratigraphy, *in* Current Research, Mineral Development Division, Department of Mines and Energy, Report 79-1, p. 7–10.

———, 1983, Early Ordovician trilobite faunas of the Boat Harbour and Catoche Formations (St. George Group) in the Boat Harbour-Cape Norman area, Great Northern Peninsula, western Newfoundland: Unpublished M.S. Thesis, Memorial University of Newfoundland, St. John's, 272 p.

CHOW, N., 1986, Sedimentology and diagenesis of Middle and Upper Cambrian platform carbonates and siliciclastics, Port-au-Port Peninsula, western Newfoundland: Unpublished Ph.D. Dissertation, Memorial University of Newfoundland, St. John's, 458 p.

———, AND JAMES, N. P., 1987, Cambrian Grand Cycles: A northern Appalachian perspective: Geological Society of America Bulletin, v. 98, p. 418–429.

CONIGLIO, M., 1986, Synsedimentary submarine slope failure and tectonic deformation in deep-water carbonates, Cow Head Group, western Newfoundland: Canadian Journal of Earth Sciences, v. 23, p. 476–490.

CUMMING, L. M., 1983, Part 2. Lower Paleozoic autochthonous strata of the Strait of Belle Isle area, *in* Geology of the Strait of Belle Isle Area, Northwestern Insular Newfoundland, Southern Labrador and Adjacent Quebec: Geological Survey of Canada Memoir 400, p. 75–105.

DEBRENNE, F., AND JAMES, N. P., 1981, Reef associated archaeocyathans of the Forteau Formation, southern Labrador: Palaeontology, v. 24, p. 343–378.

DEMICCO, R. V., AND HARDIE, L. A., 1981, Patterns of platform and off-platform carbonate sedimentation in the Upper Cambrian of the central Appalachians and their implications for sea level history: Second International Symposium on the Cambrian System: U.S. Geological Survey Open-File Report 81-743, p. 67–70.

DESROCHERS, A., 1985, The Lower and Middle Ordovician platform carbonates of the Mingan Islands, Quebec: Stratigraphy, paleokarst and limestone diagenesis: Unpublished Ph.D. Dissertation, Memorial University of Newfoundland, St. John's 342 p.

———, AND JAMES, N. P., 1988, Early Paleozoic surface and subsurface paleokarst: Middle Ordovician carbonates, Mingan Islands, Quebec, *in* James, N. P., and Choquette, P. W., eds., Paleokarst: Springer-Verlag, New York, p. 183–210.

FRITZ, W. H., 1975, Broad correlations of some Lower and Middle Cambrian strata in the North American Cordillera: Geological Survey of Canada Paper 75-1, Part A, p. 533–540.

FORTEY, R. A., 1984, Global early Ordovician transgressions and regressions and their biological implications, *in* Bruton, D. L., ed., Aspects of the Ordovician System: Paleontological Contributions from the University of Oslo No. 295, Universitetsforlaget, p. 37–50.

HARDIE, L. A., 1986, Stratigraphic models for carbonate tidal flat deposition, *in* Warme, J. E., and Shanley, K. W., eds., Carbonate Depositional Environments, Part 3, Tidal Flats: Colorado School of Mines Quarterly, v. 81, p. 59–74.

HIGGINS, M. D., AND DOIG, R., 1981, The Sept Iles anorthosite complex field relationship, geochronology and petrology: Canadian Journal of Earth Sciences, v. 18, p. 561–573.

HISCOTT, R. N., JAMES, N. P., AND PEMBERTON, S. G., 1984, Sedimentology and ichnology of the Lower Cambrian Bradore Formation: Fluvial to shallow-marine transgressive sequence, coastal Labrador: Bulletin of Canadian Petroleum Geology, v. 32, p. 11–26.

HUGHES, S., 1977, Facies anatomy of a Lower Cambrian archaeocyathid biostrome complex, southern Labrador: Unpublished M.S. Thesis, Memorial University of Newfoundland, St. John's, 276 p.

JAANUSON, V., 1979, Ordovician, *in* Robison, R. A., and Teichert, C., eds., Treatise on Invertebrate Paleontology, Part A, Introduction: Geological Society of America, p. 136–166.

JACOBI, R. D., 1981, Peripheral bulge–A causal mechanism for the Lower-Middle Ordovician unconformity along the western margin of the northern Appalachians: Earth and Planetary Science Letters, v. 56, p. 245–251.

JAMES, N. P., 1981, Megablocks of calcified algae in the Cow Head Breccia, western Newfoundland; vestiges of a Lower Paleozoic continental margin: Geological Society of America Bulletin; v. 92, p. 799–811.

———, 1984, Shallowing-upward sequences in carbonates, *in* Walker,

R. G., ed. Facies Models, second edition: Geological Association of Canada, p. 213–228.

———, BOTSFORD, J., AND WILLIAMS, S. H., 1987, Allochthonous slope sequence at Lobster Cove Head: Evidence for a complex Middle Ordovician platform margin in western Newfoundland: Canadian Journal of Earth Sciences, v. 24, p. 1199–1211.

———, AND DEBRENNE, F., 1980, Regular archaeocyaths from the Forteau Formation, west Newfoundland: Canadian Journal of Earth Sciences, v. 17, p. 1609–1615.

———, AND KOBLUK, D. R., 1978, Lower Cambrian patch reefs and associated sediments, southern Labrador, Canada: Sedimentology, v. 25, p. 1–35.

———, AND MOUNTJOY, E. W., 1983, The shelf slope break in fossil carbonate platforms, in Stanley, D. J., and Moore, G. T., eds., The Shelf Slope Boundary, A Critical Interface on Continental Margins: Society of Economic Paleontologists and Mineralogists Special Publication 36, p. 189–207.

———, AND STEVENS, R. K., 1982, Anatomy and evolution of a Lower Paleozoic continental margin, western Newfoundland; Field Excursion No. 2B, International Association of Sedimentologists Congress, Hamilton, Ontario, 75 p.

———, AND ———, 1986, Stratigraphy and correlation of the Cambro-Ordovician Cow Head Group, western Newfoundland: Geological Survey of Canada Bulletin 366, 143 p.

KINDLE, C. H., AND WHITTINGTON, H. B., 1958, Stratigraphy of the Cow Head region, western Newfoundland: Geological Society of America Bulletin, v. 69, p. 315–342.

KLAPPA, C. F., AND JAMES, N. P., 1980, Lithistid sponge bioherms of Middle Ordovician age, Table Head Group, west Newfoundland: Bulletin of Canadian Petroleum Geology, v. 28, p. 425–451.

———, OPALINSKI, P., AND JAMES, N. P., 1980, Stratigraphy of the Table Head Group: Canadian Journal of Earth Sciences, v. 17, p. 1007–1019.

KNIGHT, I., 1977, The Cambro-Orodovician platformal rocks of the Northern Peninsula, Newfoundland: Mineral Development Division, Newfoundland Department of Mines and Energy, Report 77-6, 27 p.

———, 1980, Cambro-Ordovician carbonate stratigraphy of western Newfoundland; sedimentation, diagenesis and zinc-lead mineralization: Mineral Development Division, Newfoundland Department of Mines and Energy, Open-File Newfoundland 1154, 43 p.

———, 1985, Geological mapping of Cambrian and Ordovician sedimentary rocks of the Bellburns (12I/5/6), Portland Creek (12I/4) and Indian Lookout (12I/3) map areas, Great Northern Peninsula, Newfoundland, in Current Research, Mineral Development Division, Newfoundland Department of Mines and Energy, Report 85-1, p. 79–88.

———, 1986, Ordovician sedimentary strata of the Pistolet Bay and Hare Bay area, Great Northern Peninsula, Newfoundland, in Current Research Mineral Development Division, Newfoundland Department of Mines and Energy, Report 86-1, p. 147–160.

———, 1987, Geology of the Roddickton (12I/16) map area, in Current Research, Mineral Development Division, Newfoundland Department of Mines and Energy, Report 87-1, p. 343–357.

———, AND BOYCE, W. D., 1984, Geological mapping of the Port Saunders (12I/11), St. John Island (12I/14) and parts of the Torrent River (12I/10) and Bellburns (12I/6) map sheets, western Newfoundland, in Current Research, Mineral Development Division, Newfoundland Department of Mines and Energy, Report 84-1, p. 114–123.

———, AND ———, 1987, Lower to Middle Cambrian terrigenous-carbonate rocks of Chimney Arm, Canada Bay: Lithostratigraphy, preliminary biostratigraphy and regional significance, in Current Research, Mineral Development Division, Newfoundland Department of Mines and Energy, Report 87-1, p. 359–365.

———, AND JAMES, N. P., 1987, The stratigraphy of the Lower Ordovician St. George Group, western Newfoundland: The interaction between eustasy and tectonics: Canadian Journal of Earth Sciences, v. 24, p. 1927–1951.

———, AND SALTMAN, P., 1980, Platformal rocks and geology of the Roddickton map area, Great Northern Peninsula, in Current Research, Mineral Development Division, Newfoundland Department of Mines and Energy, Report 80-1, p. 10–28.

LEGGETT, J. K., MCKERROW, W. S., COCKS, R. L. M., AND RICKARDS, R. B., 1981, Periodicity in the early Paleozoic marine realm: Journal of the Geological Society of London, v. 138, p. 167–176.

LUDVIGSEN, R., AND WESTROP, S. R., 1985, Three new Upper Cambrian stages for North America: Geology, v. 13, p. 139–143.

———, PRATT, B. R., AND WESTROP, S. R., 1988, The myth of a eustatic sea level drop near the base of the Ibexian Series: New York State Museum Bulletin 462, p. 65–70.

MCKENZIE, D. P., 1978, Some remarks on the development of sedimentary basins: Earth and Planetary Science Letters, v. 40, p. 25–32.

MARKELLO, J. R., AND READ, J. F., 1981, Carbonate ramp to deeper shale shelf transitions of an Upper Cambrian intrashelf basin, Nolichucky Formation, southwest Virginia Appalachians: Sedimentology, v. 28, p. 573–597.

MCKERROW, W. S., 1979, Ordovician and Silurian changes in sea level: Quarterly Journal of the Geological Society of London, v. 136, p. 137–145.

MILLER, J. F., 1984, Cambrian and earliest Ordovician conodont evolution, biofacies and provincialism, in Clark, D. L., ed., Conodont Biofacies and Provincialism: Geological Society of America Special Paper 196, p. 43–68.

MUSSMAN, W. J., AND READ, J. F., 1986, Sedimentology and development of a passive to convergent margin unconformity: Middle Ordovician Knox Unconformity, Virginia Appalachians: Geological Society of America Bulletin, v. 97, p. 282–295.

NOWLAN, G. S., AND BARNES, C. R., 1987a, Application of conodont colour alteration indices to regional and economic geology, in Austin, R. L., ed., Conodonts: Investigative Techniques and Applications; British Micropaleontological Society, Ellis Horwood Ltd., Chischester, p. 188–202.

———, AND BARNES, C. R., 1987b, Thermal maturation of Paleozoic strata in eastern Canada from conodont colour alteration index (CAI) data with implications for burial history, tectonic evolution, hotspot tracks and mineral and hydrocarbon potential: Geological Survey of Canada Bulletin 369, 47 p.

PALMER, A. R., 1971a, The Cambrian of the Great Basin and adjacent areas, western United States, in Holland, C. H., ed., Cambrian of the New World: Wiley-Interscience, London, p. 1–78.

———, 1971b, The Cambrian of the Appalachians and eastern New England regions, eastern United States, in Holland, C. H., ed., Cambrian of the New World: Wiley-Interscience, London, p. 78–143.

———, 1977, Biostratigraphy of the Cambrian System–A progress report: Annual Review of Earth and Planetary Sciences, v. 5, p. 13–33.

———, 1983a, On the correlatability of Grand Cycle tops, in Taylor, M. E., ed., Second International Symposium on the Cambrian System: U.S. Geological Survey Open-File Report 81–473, p. 156–159.

———, 1983b, Subdivision of the Sauk sequence, in Taylor, M. E., ed., Second International Symposium on the Cambrian System: U.S. Geological Survey Open-File Report 81–473, p. 160–162.

———, AND HALLEY, R. B., 1979, Physical stratigraphy and trilobite biostratigraphy of the Carrara Formation (Lower and Middle Cambrian) in the southern Great Basin: U.S. Geological Survey Professional Paper 1047, 131 p.

———, AND JAMES, N. P., 1979, The Hawke Bay event; a circum-Iapetus event of Lower Cambrian age, in Wones, D. R., ed., The Caledonides in the U.S.A.: Blacksburg, Virginia Polytechnic Institute and State University Memoir 2, p. 15–18.

PRATT, B. R., AND JAMES, N. P., 1982, Cryptalgal-metazoan bioherms of Early Ordovician age in the St. George Group, western Newfoundland: Sedimentology, v. 29, p. 543–569.

———, AND JAMES, N. P., 1986, The tidal flat island model for peritidal shallowing-upward sequences; St. George Group, western Newfoundland: Sedimentology, v. 33, p. 313–345.

QUINLAN, G. M., AND BEAUMONT, C., 1984, Appalachian thrusting, lithospheric flexure and the Paleozoic stratigraphy of the eastern interior of North America: Canadian Journal of Earth Sciences, v. 21, p. 973–996.

RAMOS, V. A., JORDAN, T. E., ALLMENDINGER, R. W., MPODOZIS, C., KAY, S. M., CORTEZ, J. M., AND PALMA, M., 1986, Paleozoic terranes of the central Andean-Chilean Andes: Tectonics, v. 5, p. 855–886.

READ, J. F., 1979, Depocenters, carbonate facies and foreland basin evolution, Middle Ordovician, Virginia, in Wones, D. R., ed., The Caledonides in the USA: Blacksburg, Virginia Polytechnic Institute and State University Memoir 2, p. 19–26.

———, 1982, Carbonate platforms of passive (extensional) continental margins: Types, characteristics and evolutions: Tectonophysics, v. 81, p. 195–212.

——, GROTZINGER, J. P., BOVA, J. A., AND KOERSCHNER, W. F., 1985, Models for generation of carbonate cycles: Geology, v. 14, p. 107–110.

ROBISON, R. A., 1984, Cambrian Agnostida from North America and Greenland: Part 1. Ptychagnostidae: University of Kansas, Paleontological Contributions, Paper 109, 59 p.

RODGERS, J., 1968, The eastern edge of the North American continent during the Cambrian and Early Ordovician, in Zen, E. A. White, W. S. Hadley, J. B., and Thompson, J. B., Jr., eds., Studies of Appalachian Geology, Northern and Maritime: Interscience Publications, New York, p. 141–149.

ROLIFF, W. A., 1968, Oil and gas exploration–Anticosti Island, Quebec: Geological Association of Canada, Proceedings, v. 19, p. 31–36.

ROSS, R. J., Jr., 1976, Ordovician sedimentation in the western USA, in Bassett, M. G., ed., The Ordovician System, Proceedings of the Palaeontological Association Symposium, Birmingham, September, 1974: University of Wales Press and National Museum of Wales, Cardiff, p. 73–105.

——, AND JAMES, N. P., 1987, Brachiopod biostratigraphy of the Middle Ordovician Cow Head and Table Head groups, western Newfoundland: Canadian Journal of Earth Sciences, v. 24, p. 70–95.

SCHLAGER, W., 1981, The paradox of drowned reefs and carbonate platforms: Geological Society of America Bulletin, v. 92, p. 197–211.

SCHUCHERT, C., AND DUNBAR, C. O., 1934, Stratigraphy of western Newfoundland: Geological Society of America Memoir 1, 123 p.

SCHILLEREFF, H. S., AND WILLIAMS, H., 1979, Geology of the Stephenville map area, Newfoundland, in Current Research, Part A, Geological Survey of Canada, Paper 79-1A, p. 327–332.

SHAW, A. B., 1964, Time in Stratigraphy: McGraw-Hill, New York, 365 p.

SLOSS, L. L., 1963, Sequences in the cratonic interior of North America: Geological Society of America Bulletin, v. 74, p. 93–114.

——, AND SPEED, R. C., 1974, Relationships of cratonic and continental margin tectonic episodes, in Dickinson, W. R., ed., Tectonics and Sedimentation; Society of Economic Paleontologists and Mineralogists Special Publication 22, p. 98–119.

STEVENS, R. K., 1970, Cambro-Ordovician flysch sedimentation and tectonics in west Newfoundland and their possible bearing on a Proto-Atlantic Ocean, in Lajoie, J., ed., Flysch Sedimentology in North America: Geological Association of Canada Special Paper 7, p. 165–177.

——, AND JAMES, N. P., 1976, Large sponge-like mounds from the Lower Ordovician of western Newfoundland: Geological Society of America, Abstracts with Programs, v. 8, p. 1122.

STRONG, D. F., 1974, Plateau lavas and diabase dykes of northwestern Newfoundland: Geological Magazine, v. 3, p. 501–514.

——, AND WILLIAMS, H., 1972, Early Paleozoic flood basalts of northwestern Newfoundland, their petrology and tectonic significance: Geological Association of Canada, v. 24, p. 43–54.

STOUGE, S., 1982, Preliminary conodont biostratigraphy and correlation of Lower to Middle Ordovician carbonates of the St. George Group, Great Northern Peninsula, Newfoundland: Mineral Development Division, Newfoundland Department of Mines and Energy, Report 80-3, 59 p.

THOMAS, W. A., 1977, Evolution of Appalachian-Ouachita salients and recesses from reentrants and promontories in the continental margin: American Journal of Science, v. 277, p. 1233–1278.

——, TULL, J. F., BEARCE, D. N., RUSSELL, G., AND ODUM, A. L., 1980, Geological synthesis of the southernmost Appalachians, in Wones, D. R., ed., The Caledonides in the USA: Blacksburg, Virginia Polytechnic Institute and State University Memoir 2, p. 91–98.

VAIL, P. R., MITCHUM, R. M., AND THOMPSON, S., III, 1977, Seismic stratigraphy and global changes of sea level, Part 3: Relative changes of sea level from coastal onlap, in Payton, C., ed., Seismic Stratigrapy–Applications to Hydrocarbon Exploration: American Association of Petroleum Geologists Memoir 26, p. 63–81.

WALKER, R. G., 1984, Turbidites and associated coarse clastic deposits, in Walker, R. G., ed., Facies Models, second edition: Geological Association of Canada, Reprint Series 1, p. 171–188.

WATTS, A. B., 1981, The U.S. Atlantic continental margin: Subsidence history, crustal structure and thermal evolution, in Geology of Passive Continental Margins; History, Structure and Sedimentologic Record: American Association of Petroleum Geologists Education Course Note Series No. 19, 75 p.

WESTROP, S. R., 1986, Trilobites of the Upper Cambrian Sunwaptan Stage, southern Canadian Rocky Mountains, Alberta: Palaeontographica Canadiana No. 3, 179 p.

WHITTINGTON, H. B., AND KINDLE, C. H. 1963, Middle Ordovician Table Head Formation, western Newfoundland: Geological Society of America Bulletin, v. 74, p. 745–758.

WILLIAMS, H., 1975, Structural succession, nomenclature and interpretation of transported rocks in western Newfoundland: Canadian Journal of Earth Sciences, v. 12, p. 1874–1894.

——, (comp.), 1978, Tectonic lithofacies map of the Appalachian Orogen, Map No. 1A: International Geological Correlation Program, Project 27, The Appalachian-Caledonides Orogen, Canadian Contribution No. 5.

——, 1979, Appalachian Orogen in Canada: Canadian Journal of Earth Sciences, v. 16, p. 792–807.

WILLIAMS, H., AND HISCOTT, R. N., 1987, Definition of the Iapetus rift-drift transition in western Newfoundland: Geology, v. 15, p. 1044–1047.

——, JAMES, N. P., AND STEVENS, R. K., 1986, Humber Arm Allochthon and nearby groups between Bonne Bay and Portland Creek, western Newfoundland, in Current Research, Part A, Geological Survey of Canada Paper 85-1A, p. 399–406.

——, AND SMYTH, W. R., 1983, Geology of the Hare Bay Allochthon, in Geology of the Strait of Belle Isle area, Northwestern Insular Newfoundland, Southern Labrador and Adjacent Quebec: Geological Survey of Canada Memoir 400, p. 109–133.

——, AND STEVENS, R. K., 1974, The ancient continental margin of eastern North America, in Burke, C. A., and Drake, C. L., eds., The Geology of Continental Margins: Springer-Verlag, New York, p. 781–796.

——, AND ST. JULIEN, P., 1982, The Baie Verte-Brompton Line: Early Paleozoic continent-ocean interface in the Canadian Appalachians, in St. Julien, P., and Beland, J., eds., Major Structural Zones and Faults of the Northern Apppalachians: Geological Association of Canada Special Paper 24, p. 177–207.

WILLIAMS, S. H., BOYCE, W. D., AND JAMES, N. P., 1987, Graptolites from the Middle Ordovician St. George and Table Head Groups, western Newfoundland, and implications for the correlation of trilobite, brachiopod and conodont zones: Canadian Journal of Earth Sciences; v. 24, p. 456–470.

CONTROLS ON EVOLUTION OF CAMBRIAN-ORDOVICIAN PASSIVE MARGIN, U.S. APPALACHIANS

J. F. READ
Department of Geological Sciences, Virginia Polytechnic Institute and State University, Blacksburg, Virginia 24061

ABSTRACT: The Cambro-Ordovician shelf was initiated during Late Precambrian rifting with breakup at 600 ± 25 Ma. The passive-margin carbonates developed over Early Cambrian rift to marine-shelf clastics. Subsidence rates were low (1 to 10 cm/ka), and driven by exponentially decreasing thermo-tectonic subsidence coupled with flexural loading. Variation in subsidence rates formed depocenters in Tennessee, Pennsylvania, and Vermont, separated by arches in New Jersey, Virginia, and Alabama. Calculated platform slopes per cycle for the carbonate sequences were 2 to 4 cm/km.

Initially, an early Cambrian bank-fringed ramp developed. By Middle Cambrian, this ramp had formed into a high-relief reef-rimmed margin fringed by thick periplatform talus deposits. This rimmed shelf persisted into the Late Cambrian. Middle Cambrian rifting formed the Rome trough inboard of the passive margin, while redbeds were deposited over much of the shelf. With sea-level rise, a huge intra-shelf basin formed within the Tennessee depocenter. Both the Rome trough and the intra-shelf basin had ceased to exist by the Latest Cambrian. The rimmed shelf developed into a ramp during initial subduction in the Early Ordovician. During Middle Ordovician collision, a widespread unconformity developed, the ramp foundered, and its leading edge was deformed.

At the formation level, the stratigraphy reflects third-order (1 to 10 m.y.) sea-level fluctuations. At a smaller scale, platform facies are mainly 1 to 5 m thick, peritidal carbonate cycles that reflect 20- to 100-ka (Milankovitchian) sea-level fluctuations with less than 10-m amplitudes. These low amplitudes allowed the shelf to track sea-level fall as it continued to subside during regressions; thus, soils or regoliths are lacking. Dolomitization during this phase was intense, however. Regional thick (30–300 m) subtidal limestones punctuate the shelf sequence and likely were due to third-order sea-level rises. Long-term sea-level falls periodically caused widespread deposition of eolian and coastal clastics over the shelf at tops of carbonate cycles. Intra-shelf basin Grand Cycles reflect gradual shallowing during third-order sea-level rise/fall coupled with superimposed Milankovitch-scale fluctuations.

INTRODUCTION

The Cambrian-Ordovician sequence in the Appalachians has long been the classic Early Paleozoic continental-shelf prism or miogeocline (Rodgers, 1969; Bird and Dewey, 1970). It contains over 3.5 km of shelf sediment (Fig. 1), dominated by shallow-water carbonates. The sequence is presently exposed in the fold-thrust belt of the Appalachians in the United States and is penetrated by numerous wells on the autochthon.

In this paper, the various factors that appear to have influenced the evolution of the shelf are examined. These include tectonic subsidence, which controlled the development of depocenters and arches, and hinge-line migration. The effects of the development of a major rift inboard of the margin some 40 m.y. after breakup are documented. Collisional tectonics appear to have influenced the margin prior to its destruction.

The evolution of the shelf from a ramp to rimmed-shelf and intra-shelf basins, into an aggraded rimmed shelf, and then into a collisional ramp is documented in terms of facies, marginal relief, and tectonic and sea-level history.

Sea-level changes at various scales affected the development of the shelf sequence. Third-order cycles are documented using qualitative and quantitative techniques. The effects of these third-order (1 to 10 m.y.) sea-level fluctuations on formational stratigraphy is examined. Superimposed on these were short-term (20,000 to over 100,000) sea-level fluctuations, which appear to have controlled deposition of small-scale cycles. Amplitude of sea-level oscillation has been an important factor in generation of the platform morphology and facies; the relatively low amplitudes involved appear to be the controlling factor making this ancient shelf different from the modern. Finally, some results of computer modeling of the sequences are outlined.

STRUCTURAL FRAMEWORK

The passive margin developed on rifted continental crust bordering eastern North America. This rifting of a Late Proterozoic supercontinent occurred between 650 and 570 Ma, while breakup occurred at about 600 ± 25 Ma (Bond and others, 1984). Long-term subsidence of the margin appears to have been from a few centimeters to 10 cm/ka. Tectonic subsidence of outer-shelf sections was dominated by thermal contraction of substantially thinned continental lithosphere with exponentially decreasing subsidence from the Early Cambrian to the Early Ordovician (Bond and others, 1984, 1988; Fig. 2). This contributed as much as 2 km of subsidence for outer-shelf sections, whereas the bulk of remaining subsidence reflects sediment loading. Thickness and facies distribution of the passive-margin sequence were controlled by depocenters in Vermont, Pennsylvania, and Tennessee (Figs. 1, 2). Subsidence rates in the Vermont depocenter were less than farther south, but are similar to rates farther north in Newfoundland (James and Stevens, 1986). The pattern of depocenters reflects regional differential subsidence, which probably was basement controlled. Depocenters were separated by arches located in New Jersey, Virginia, Alabama, and in the platform interior, the Waverly arch (Fig. 1; Woodward, 1961; Thomas, 1977). The Rome trough (Fig. 1) developed during the Middle to early Late Cambrian as an extensive graben system about 300 km inland from the shelf edge (Webb, 1980).

The passive-margin sequence appears to have been influenced by at least three hinge lines that are separated in time and show a progressive westward migration (Fig. 1). Wehr and Glover (1985) suggest that a Late Precambrian hinge zone, which they defined as a zone of major crustal attenuation, lay near the Blue Ridge, which would have lain just seaward of the Cambro-Ordovician carbonate shelf edge. This Late Precambrian hinge zone separates thick deep-

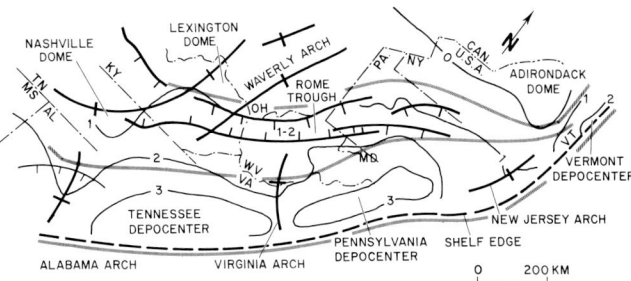

FIG. 1.—Regional tectonic map on palinspastic base, United States Appalachian orogen, showing major structural features. Hinge lines are shown in grey stippled lines; Late Precambrian hinge zone is lower line and is located along carbonate shelf edge; Early Cambrian hinge is middle line, and Middle Cambrian hinge is upper line north of Rome trough.

water clastics (Late Precambrian) from thin to absent shelf facies. An Early Cambrian hinge line, which marks the landward limit of major flexuring, is located less than 100 km landward of the restored shelf edge in Vermont and 200 km landward of the shelf edge elsewhere. The hinge line during the Middle Cambrian was located less than 100 km back from the restored shelf edge in Vermont and 400 km landward of the restored shelf edge in Pennsylvania-Virginia. Later hinge lines lie even farther landward.

Deformation of the passive margin was initiated in the early Middle Ordovician when a regional unconformity (Knox or Beekmantown unconformity) developed on the sequence. This collisional orogeny continued into the Late Ordovician. The margin was involved in further deformation during the Devonian, but major thrusting occurred during the Late Paleozoic when the region underwent as much as 50 percent shortening.

SEQUENCE STRATIGRAPHY OF THE PASSIVE MARGIN

The passive-margin succession in the Appalachians is divided into five sequences that encompass the major stratigraphic units traditionally used as major map units (Fig. 3). Few of these are bounded by unconformities that are recognizable on the basis of the paleontologic data. The stratigraphy is described more fully in Read (in prep.).

BASAL CLASTICS–SEQUENCE 1

Thick sequences of Early Cambrian to Latest Precambrian(?) clastics (Chilhowee or Cheshire interval; Fig. 3) unconformably overlie basement beneath the Cambrian shelf (Figs. 4 to 7). Seaward of the Late Precambrian hinge zone, the Cambrian clastics are deep-water facies that rest on thick Late Proterozoic deeper water metasediments and rift-volcanics (Fig. 8).

In general, the Early Cambrian section beneath the outer shelf consists of rift-related fluvial clastics and volcanics at the base, overlain by a deeper water shelf sequence of muds that pass upward into storm- and tide-dominated shelf sands (Figs. 5 to 7; Schwab, 1971; Mack, 1980; Simpson, 1987; Myrow, 1983). A thin dolomitic quartz sandstone separates the clastic shelf from the overlying carbonate shelf. These shelf clastics pass into deep basinal siliciclastics seaward of the Late Precambrian hinge.

Regionally, the Cambrian clastics thicken into embryonic depocenters along the shelf edge in Tennessee, Pennsylvania, and Vermont (Fig. 4), which were characterized by highest subsidence rates (as much as 10 cm/ka). They thin landward and pinch out near the Rome trough, which had not yet developed as a graben system (Webb, 1980). Early Cambrian sands appear to be absent from the New Jersey high.

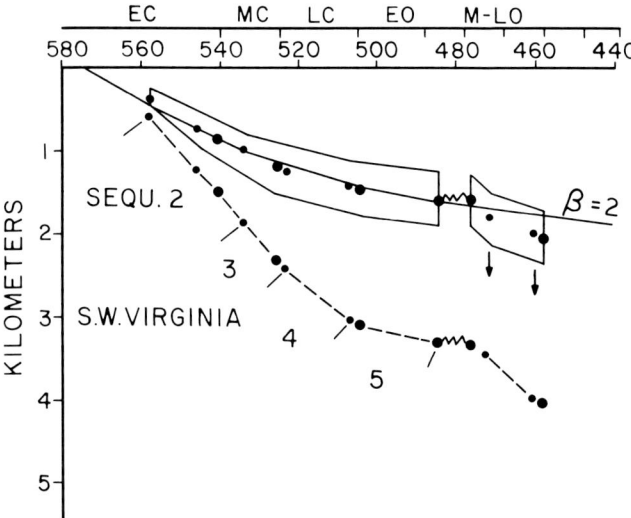

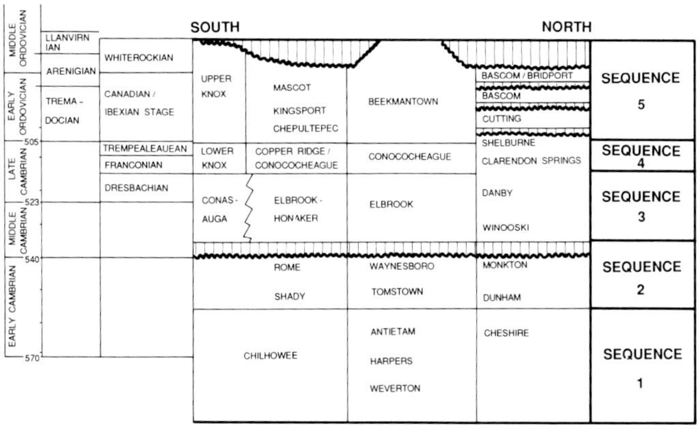

FIG. 2.—Observed sediment thickness (dashed) and tectonic subsidence (solid) plotted against time, for section from southwest Virginia (Bond, pers. commun., 1988). Times of deposition of sequences 2 to 5 shown below observed sediment-thickness curve. Large dots are stratigraphic horizons to which radiometric ages are assigned. Small dots are formation boundaries. Upper and lower lines for tectonic subsidence calculated from minimum and maximum values for delithified thicknesses. Note close correspondence of tectonic-subsidence curve to post-rift thermal-subsidence model curve (stretching factor or beta = 2) for Cambrian to Early Ordovician, suggesting that the margin underwent relatively uniform, exponentially decreasing subsidence. Departure of the tectonic-subsidence curve from the model curve in the Middle Ordovician marks time of initial collision (arrows indicate where an amount of tectonic subsidence must be added to curve owing to deep-water loading).

FIG. 3.—Chart showing major stratigraphic units that compose sequences 1 to 5, Cambro-Ordovician, U.S. Appalachians. The three columns are for the southern, central, and northern U.S. Applachians.

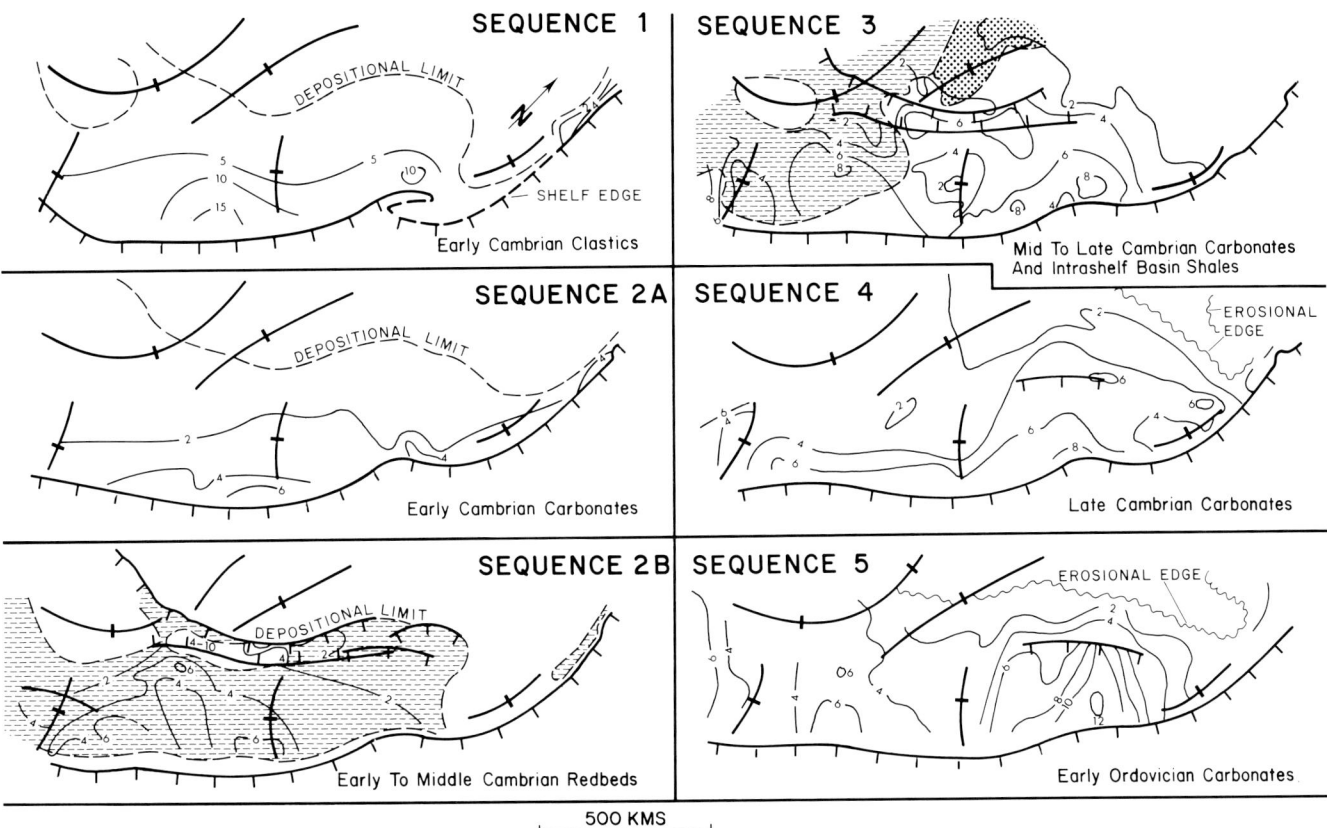

FIG. 4.—Isopach maps of sequences 1 to 5 on palinspastic base. Contours in hundreds of meters. Only shelf sequence contoured; offshelf sequences left blank. Sequence 1 clastics not patterned. For sequences 2 to 5, patterns for shelf sequence are carbonates (blank) and fine clastics (dashed); Kerbal "delta" of sequence 3 shown by dot pattern.

EARLY TO MIDDLE CAMBRIAN CARBONATES AND REDBEDS–SEQUENCE 2

The clastic shelf is overlain by a regional carbonate and redbed succession (Shady-Rome sequence) 1 to 1.5 km thick (Figs. 4 to 8). The carbonate unit (sequence 2A) locally thickens into depocenters at the edge of the shelf (Fig. 4). The overlying clastics (sequence 2B) thicken into depocenters that extend across much of the southern shelf and are locally associated with normal faulting, as in Alabama (Fig. 4). The redbeds thicken to over 1 km in the Rome trough in West Virginia-Kentucky (Webb, 1980), which began to develop at this time. Subsidence rates in the Rome trough locally matched those of the outer-shelf depocenters. Rome clastics are absent from the New Jersey arch, which appears to have been strongly positive.

Lower Carbonate Sequence 2A

In Virginia and Tennessee (Fig. 7; Byrd, 1973; Pfeil and Read, 1980; Read and Pfeil, 1983; Barnaby, pers. commun., 1987), the oldest carbonates are 300 m thick and are deep-ramp, nodular-bedded wackestone/mudstones with argillaceous seams. These deep-ramp facies are overlain by cyclic peritidal carbonates as thick as 300 m. In Maryland, these cyclic carbonates rest directly on basal clastics, are 150 m thick, and contain as many as eight shallowing-upward, limestone to dolomite sequences (Reinhardt and Wall, 1975).

In Virginia on the outer ramp, deep-ramp wackestone/mudstones pass upward into stromatactoid mudmound complexes (Fig. 7) that locally are capped by small archeocyathid reefs and are heavily marine cemented (Barnaby and Read, 1987). The mud mounds are interlayered with nodular-bedded carbonates and form several large-scale cycles. The outer-ramp mudmound sequences pass downslope into slope conglomerates and thinly bedded shaly lime muds. During regional drowning of the Early Cambrian shelf, black shales were deposited over the outer platform in both Virginia and in the St. Albans embayment, Vermont (Dorsey and Stanley, 1983; Stone and Dennis, 1964; Figs. 5, 7).

Upper Redbed Sequence 2B

Along the backstepped margin, mudmound deposition was succeeded by upbuilding of a rimmed-shelf margin constructed by skeletal algal reefs and backreef lime sands (Read and Pfeil, 1983; Barnaby and Read, 1987; Fig. 7). In Vermont (Fig. 5), oolitic dolomites dominate the margin (Rha-

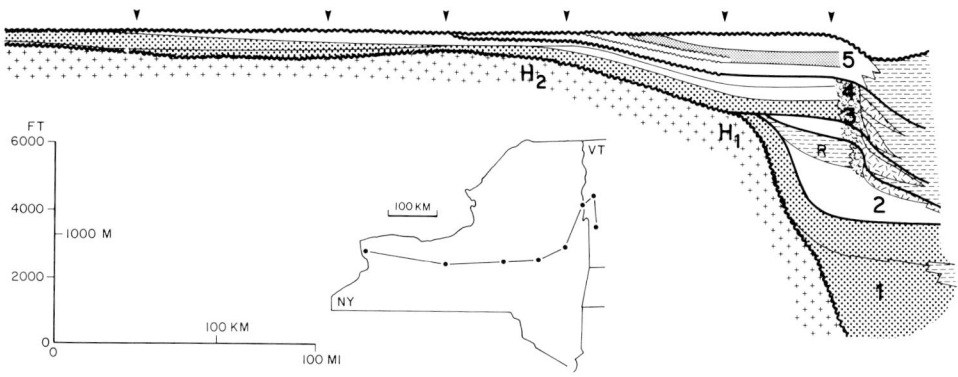

Fig. 5.—Stratigraphic cross section, New York to Vermont. Legend in Figures 5 to 8 is cyclic dolomites (blank), limestones (fine stipple), sandstones (dots), and shale (dashes); R denotes redbeds. Offshelf breccias shown by random dashes. Shelf edge facies shown by reef pattern. H1 and H2 are Early Cambrian and Middle Cambrian hinge lines, respectively.

manian, 1981). The carbonate rim at this stage formed a narrow belt fringing a clastic platform with widespread redbed deposition (Figs. 4 to 8).

The redbeds are cyclic (Harris, 1964) and consist of subtidal quartz sands passing upward into tidal-flat muds and minor carbonates (Samman, 1975; Speyer, 1983). Thick sequences of massively marine cemented periplatform talus and foreslope lime sands accumulated adjacent to the margin above earlier deep-ramp/slope, finer grained facies (Figs. 5 to 7). The talus passes seaward into turbidites and shale. A major hiatus appears to be present at or near the top of the redbed sequence (Hawke Bay Event, Palmer and James, 1979) and may be traceable biostratigraphically onto the slope in Pennsylvania, where it increases in magnitude basinward (Fig. 6; Gohn, 1976) possibly as a result of nondeposition on a starved slope.

Platform Morphology

The ramp sequence 2A developed on the gently sloping surface of the drowned clastic shelf (sequence 1). Deposition of banks on the outer ramp caused increased platform-to-basin relief and shallowing of the outer platform. This allowed establishment of algal-reef communities on the incipient shelf edge leading to formation of a high-relief rimmed shelf. At the end of deposition of sequence 2B, the platform-to-basin relief on the margin was likely to have been as many as a few hundred meters in Vermont, 600 to 800 m in Pennsylvania, and possibly over 1,000 m in Virginia to Alabama (Figs. 5 to 7).

Sea-level cycles.—

Sequence 2 formed during a 20-m.y. sea-level cycle (Fig. 9), which was responsible for initial drowning of the clastic shelf, followed by deposition of deep-ramp carbonates, then cyclic carbonates (sequence 2A), culminating in deposition of cyclic clastics (sequence 2B). At least five and possibly eight smaller scale third-order cycles are superimposed on the 20-m.y. cycle (Fig. 9) and appear to have given rise to thick carbonate buildup cycles on the ramp margin and thick (20 m) shallowing-upward sequences on the inner platform. The large influx of red clastics probably was due to the regional regression, possibly aided by local uplift along Rome trough border faults. These clastics were transported by sheet flooding onto the peritidal platform. Rimmed-shelf development appears to have accompanied regional regression, but high-subsidence rates on the outer platform must have exceeded sea-level fall rates, allowing vertical upbuilding of the rimmed-shelf margin. Abundant meter-scale cycles in the Rome Formation suggest that numerous Milankovitchian sea-level oscillations were superimposed on the larger scale events. Regional regression apparently culminated in the widespread development of the unconformity.

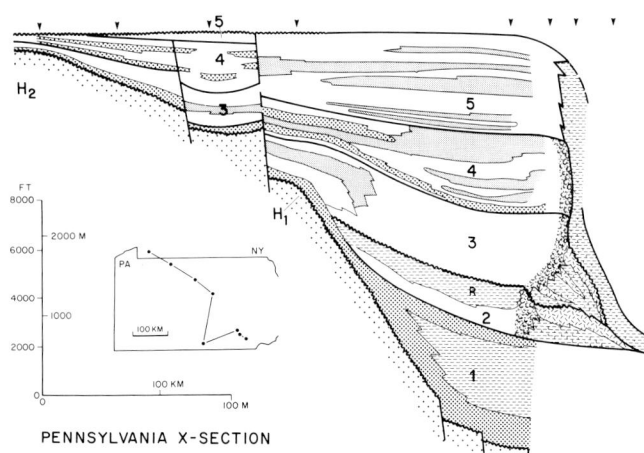

Fig. 6.—Stratigraphic cross section, Pennsylvania. See Figure 5 for explanation.

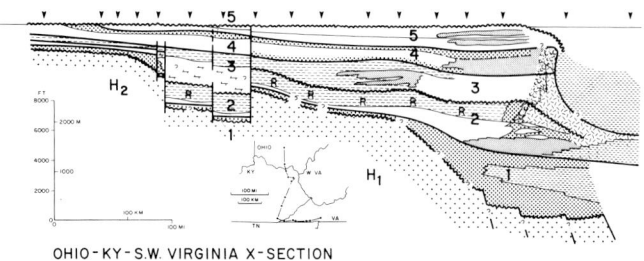

Fig. 7.—Stratigraphic cross section, Ohio-Kentucky-Virginia. Modified from Koerschner (1983). See Figure 5 for explanation.

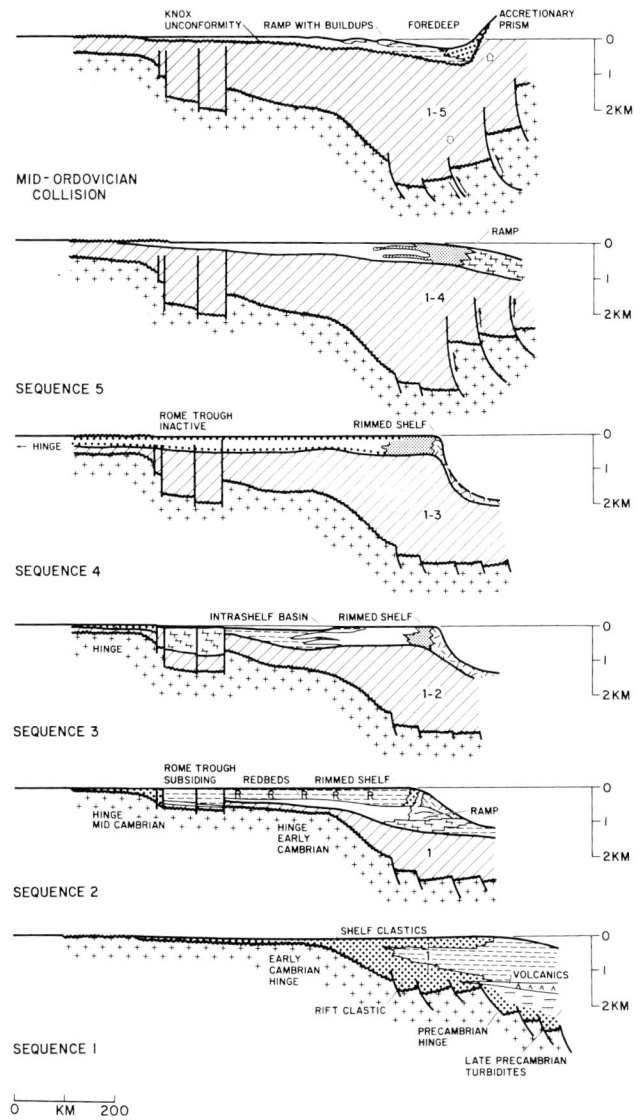

FIG. 8.—Diagrammatic evolution of passive-margin sequences 1 to 5 and subsequent collisional sequence.

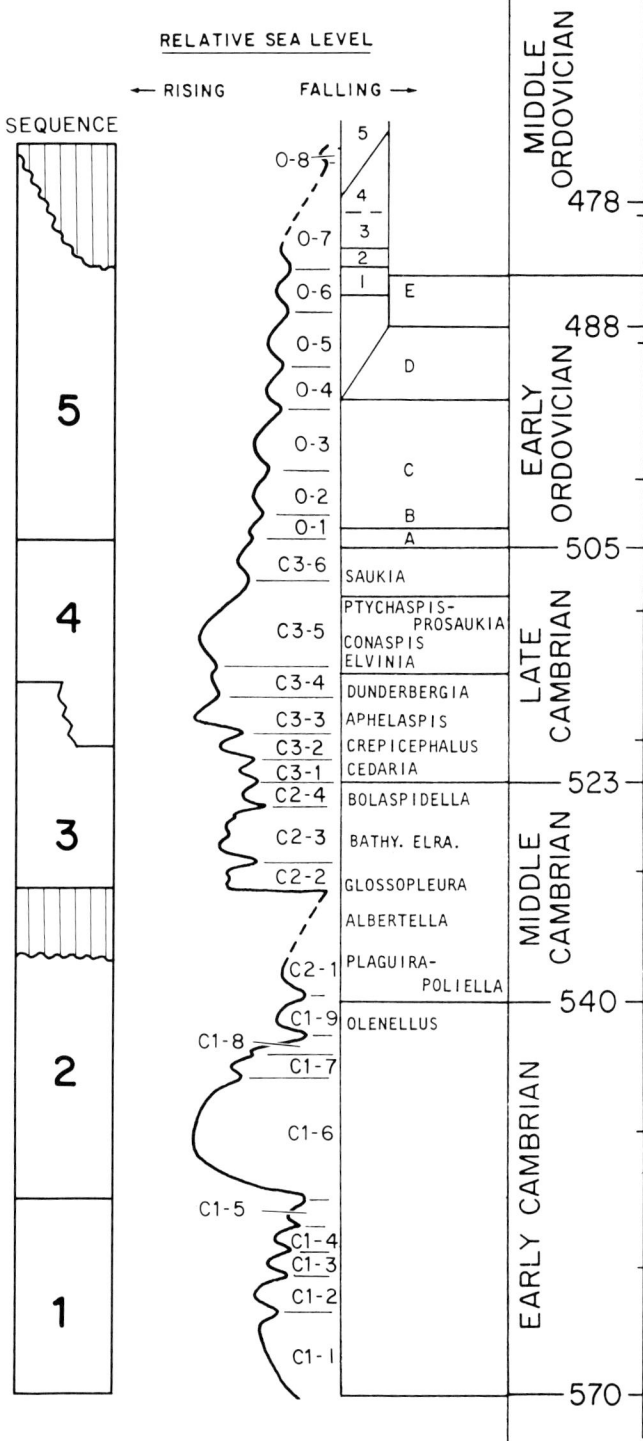

FIG. 9.—Relative sea-level curve for U.S. Appalachians, Cambro-Ordovician sequence. Note that numbered sea-level cycles are informal only and subject to modification.

MID-TO LATE CAMBRIAN INTRA-SHELF BASIN FILL AND PERITIDAL PLATFORM–SEQUENCE 3

This sequence thickens to 800 m in the Tennessee and Pennsylvania depocenters and is as thick as 1,000 m in the Rome trough (Fig. 2), which was actively subsiding (Webb, 1980). The depocenters at this stage were broad and extended well back onto the shelf (Fig. 4).

Facies and Platform Morphology

Cyclic peritidal carbonates appear to be localized over much of the outer shelf seaward of depocenters. These peritidal carbonates are similar to those in younger sequences 4 and 5 (Fig. 10; Koerschner, 1983; Markello and Read, 1982). Along the outer shelf in Pennsylvania, these formed a broad barrier and passed toward the craton into shallow-water, sandy limestone and limy shale, passing upward into deeper subtidal, dark, burrowed lime mudstones (Fig. 6; Wilson, 1952). This subtidal sequence indicates that here the rim bordered a shallow intra-shelf basin located over the Pennsylvania depocenter. This shallow-basin sequence passed into sands and sandy dolomites toward the cratonic shoreline.

The peritidal barrier narrowed along strike to the southwest, where it bordered the seaward edge of a huge intrashelf basin (Conasauga basin) located within the Tennessee depocenter (Figs. 1, 4, 7; Markello and Read, 1982; Hasson and Haase, 1988). This basin fill consisted of deposition of 0.5- to 4.5-m-thick cyclic units of deeper, quiet-water shales overlain by storm-deposited quartz and pellet siltstones, capped by glauconitic skeletal and oolitic sands and flat-pebble conglomerates (Fig. 10). Hardgrounds are abundant. The carbonate units are open-marine facies, commonly packstones, and the ooids probably were high-Mg calcite with primary radial fabrics and skeletal cores. A major delta system in Ohio (Kerbal "delta"; Janssens, 1973; Fig. 4) may have supplied some of the fine clastics. Slopes from the peritidal platform into the intra-shelf basin were oolitic ramps that passed downslope into storm-deposited, thinly bedded and burrowed muds with local renalcid bioherms.

The seaward margin of the platform at this time was a high-relief rimmed shelf bordered by calcareous algal reefs and lime sands (Figs. 5 to 8). Platform-to-basin relief was likely to have been 200 to 300 m in Vermont, 1,500 m in Pennsylvania, and 1,000 m in Virginia-Tennessee, and probably decreased to the south into Alabama.

Sea-Level Cycles

The sea-level curves for sequence 3 consist of 6 to 9 sea-level cycles ranging from 1 to 3 Ma duration (Figs. 9, 11). They are manifested in 20-m to over 100-m-thick sequences (Grand Cycles) consisting of a subtidal shale unit overlain by a regressive limestone (Aitken, 1978; Palmer, 1981; Bond and others, this volume). The sequence shows gradual onlap, reflecting long-term sea-level rise from the mid- to late Cambrian (Bond and others, 1984, and this volume).

LATE CAMBRIAN CYCLIC CARBONATES–SEQUENCE 4

Sequence 4 thickens to 800 m in the Pennsylvania depocenter, which extends some 300 km in from the shelf edge (Fig. 4). The unit is as thick as 600 m in the Tennessee depocenter, which is localized as a narrow sedimentary basin along the outer shelf. The sequence generally shows little thickening over the Rome trough (Fig. 7), which was stable by this time. There was some basement faulting in the northern extension of the trough, however, which shows thickened sections in Pennsylvania (Figs. 4, 6; Wagner, 1976).

Facies

The sequence is dominated by cyclic peritidal carbonates, which tend to be heavily dolomitized to the west. The cycles are 1 to 5 m thick with lower parts of lime conglomerates and lime sands and muds, overlain by cryptalgal heads or ribbon carbonates; these fine upward into dolomitized cryptalgal laminites (Fig. 10; Demicco, 1985; Zadnick, 1960; Major, 1976; Koerschner, 1983). In general, many outer-shelf cycles are dominated by digitate algal bioherms, thick grainstones, or ribbon carbonates. Mid- and inner-shelf cycles contain abundant thrombolites and relatively restricted subtidal facies.

Sequence 4 also contains some dominantly subtidal, so-called "non-cyclic" limestone sequences from 50 to over 200 m thick (Figs. 6, 7). They contain grainstone, bio-

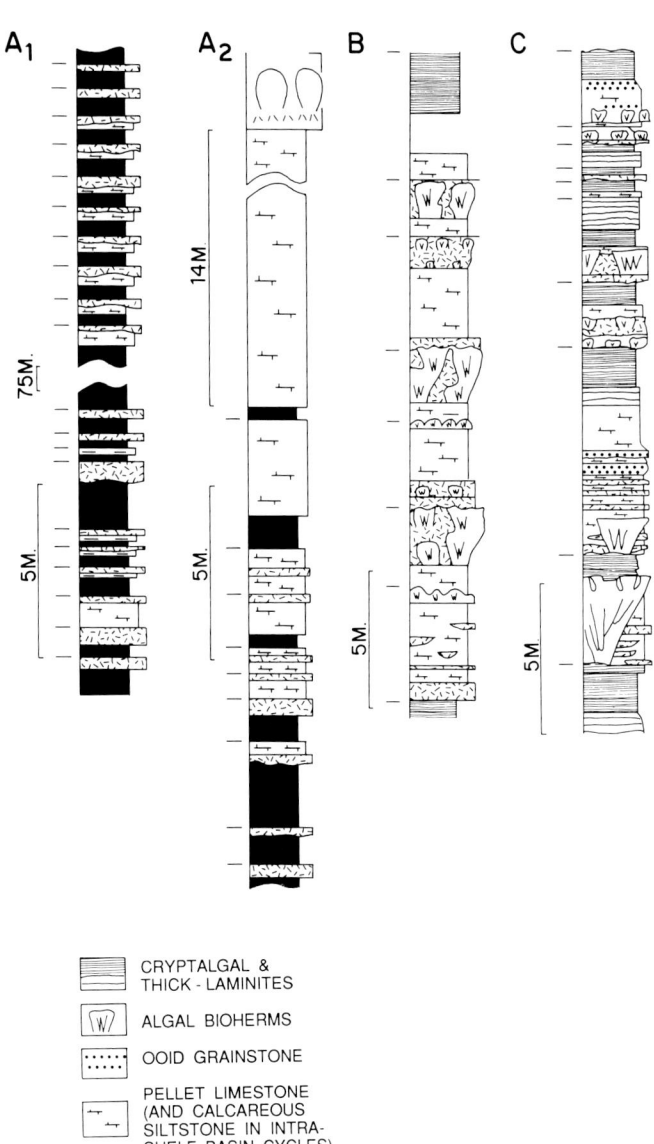

FIG. 10.—Selected parts of stratigraphic sections illustrating major facies sequences. Column A1 illustrates cycles typical of intra-shelf basin successions of sequence 3 (Conasauga Group), Duffield, Virginia (data collected by J. R. Markello). Lower part of column is from Lower Shale; upper part of column is from Upper Shale, Nolichucky Formation. Column A2 is transition from Upper Shale into Maynardville ribbon carbonates upward into peritidal cycles of the Copper Ridge Formation (sequence 4). Column B illustrates subtidal limestone sequences and is from the Chepultepec Formation (lower sequence 5), Fincastle, Virginia (from Bova and Read, 1987). Column C illustrates peritidal cyclic carbonates from the Conococheague Formation (sequence 4), Wytheville (from Koerschner and Read, in prep.).

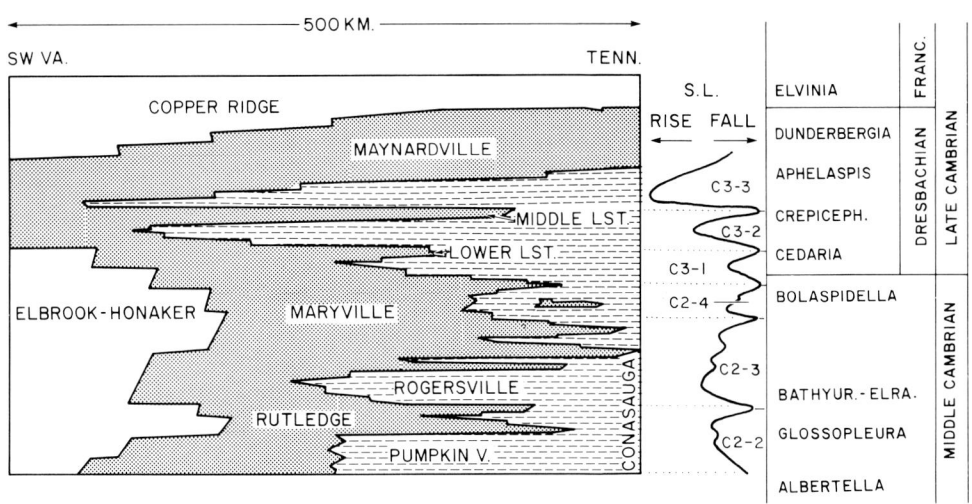

FIG. 11.—Chronostratigraphic diagram of sequence 3 from Virginia to Tennessee, showing time relations of lithologic units and relative sea-level curve. Diagram is parallel to the passive margin and runs from the shelf into the intra-shelf basin. Intra-shelf basin shales (dashes) form mainly during the rise, whereas the limestones (stippled) form mainly during the fall; cyclic peritidal dolomites are blank.

herms, and ribbon carbonates, but lack algal laminites. Facies are arranged in shallowing-upward sequences that formed in a shelf lagoon that rarely shallowed into algal flats (Demicco, 1985).

Two prominent quartz-sand zones occur in the lower and upper parts of sequence 4 (Figs. 6, 7, 12; Wilson, 1952; Wagner, 1961). Sandy zones contain from a few to over 25 sand beds that overlie mud-cracked laminites of cycle tops, but show marine reworking. Toward the shoreline in western New York, sequence 4 contains basal conglomerates overlain by sandy and silty dolomites, oolitic and stromatolitic dolomites, and local solution collapse breccias (Zenger, 1979; Selleck, 1975).

Platform Morphology

The platform at this time continued to be reef rimmed (Demicco, 1985) and probably had attained maximum platform-to-basin relief (Figs. 5 to 8). This relief was lowest in Vermont (probably less than a few hundred meters; Fig. 5) and greatest (over 1,200 m of relief) in Pennsylvania (Fig. 6; Meisler and Belcher, 1968; Cady, 1945). In Virginia, relief may have been about 800 m (Fig. 7). Shelf margin(?) and foreslope facies (up to 360 m thick) accumulated off the margin in Pennsylvania and are medium to coarse dolomites with scattered algal(?) blocks and slope clasts. To the northwest, this unit is overlain by shelf carbonates, indicating progradation of the shelf (Gohn, 1976; Meisler and Belcher, 1968). These foreslope facies are erosionally overlain by as much as 50 m of megabreccias (which are Franconian in part), which are overlain by lime sand and thin-bedded, shaly limestone. Lower slope and basin margin facies include scattered periplatform breccias, along with resedimented quartz sands (locally channelized), that were shed out onto thin-bedded limestones of the basin margin in Maryland (Reinhardt, 1974) and onto black limy shales in New York and Vermont (Kidd, pers. commun., 1986; Dorsey and Stanley, 1983).

Sea-Level Cycles

Sequence 4 is dominated by as many as five 3- to 9-m.y. sea-level cycles (Figs. 9, 12). Poor dating of the sequences makes accurate determination of the actual facies relations between sands of the inner-shelf and cyclic and non-cyclic carbonates of the outer shelf difficult. It seems likely, however, that the thick subtidal limestone sequences are related to third-order sea-level rise (Fig. 12). The peritidal sequences may be related to third-order sea-level fall (Fig. 12) with superimposed 20- to 100-ka sea-level fluctuations of less than 10 m. The sands are lowstand deposits related regionally to this third-order sea-level fall, coupled with lowstands associated with the short-term sea-level cycles (Fig. 12). Based on regional distribution of laminite caps, Demicco (1985) suggested that the peritidal flats prograded from the reefal rim toward the platform interior.

EARLY ORDOVICIAN CYCLIC CARBONATES–SEQUENCE 5

Depocenters at this time were very well developed and trended normal to the margin, extending far in from the shelf edge (Fig. 4). In the south, the sequence is truncated by the major (up to 10 m.y.) Knox-Beekmantown unconformity, which erodes deeper into the section toward the craton (Figs. 5, 6, 7). The unconformity disappears into the Pennsylvania depocenter (Figs. 1, 4), however, where de-

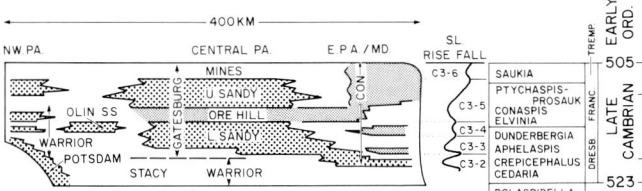

FIG. 12.—Chronostratigraphic diagram of sequence 4, Pennsylvania to Maryland, showing time relations of lithologic units and relative sea-level curves. Cross section is roughly normal to the passive margin. Subtidal limestone units (stippled) form during the rise, whereas cyclic dolomites (blank) and quartz sands (dots) typify falls.

position was continuous into the early Middle Ordovician, and as much as 1,200 m of cyclic carbonates accumulated. The Rome trough was inactive, although there appears to have been down-to-basin growth faulting in the northern extension of the trough at this time (Figs. 4, 6; Wagner, 1976).

Facies

The sequence is dominated by cyclic, peritidal dolomites with laminite caps. Regionally, the most restricted cycles consist of dolomitized muds or pellet sands, overlain by dolomitized laminites. In depocenters, the cyclic carbonates may be skeletal or oolitic and tend to have subtidal units that mainly are undolomitized (Hobson, 1963; Lees, 1967). These cyclic carbonates pinch out into areas of slow subsidence (such as New York) into unconformities, solution collapse breccias, or quartz sands (Rubin and Friedman, 1977; Braun and Friedman, 1969; Fisher, 1977; Fig. 13).

The cyclic dolomites are interbedded with six to eight regional limestones (Figs. 5, 6, 7, 13). The limestone units are composed of cycles of subtidal limestone capped by dolomitic laminites, or are composed of subtidal sequences that lack laminites (Fisher, 1954; Sando, 1957; Donaldson, 1959; Hobson, 1963; Braun and Friedman, 1969; Harris 1969; Spelman, 1966; Lees, 1967; Conway, 1977; Mazzullo, 1978; Ross and others, 1982; Bova and Read, 1987). In general, the limestones are thinly bedded wackestone/mudstone with abundant storm-deposited carbonate sands and conglomerates, thinly bedded to cross-bedded lime sands, and digitate algal and thrombolitic bioherms (Goldhammer and others, 1984; Bova and Read, 1987). Limestone units pass into cyclic dolomites toward the craton and the landward margins of depocenters. Locally, within depocenters, the limestones contain diverse biotas; facies appear to become more restricted toward landward margins of depocenters.

Platform Morphology

The platform margin during sequence 5 deposition was a carbonate ramp (Fig. 8) which, in Maryland, consists of thin-bedded grainstone passing seaward into renalcid bioherms and then into *Epiphyton* bioherms and lime sands (Goldhammer and others, 1984). The ramp morphology in Maryland and Pennsylvania is indicated by the absence of shelf edge-derived reef detritus in the slope facies, which were deposited as a deeper water limestone blanket (up to 600 m thick) over the earlier talus wedge (Gohn, 1976; Reinhardt, 1974; Demicco, 1985; Fig. 6). This deep-water blanket contains carbonaceous shales, lime muds, and re-sedimented lime sands that commonly are quartzose.

Ramp morphologies also are indicated in the southern Appalachians, where cyclic peritidal carbonates pass seaward into scattered thrombolitic bioherms and ribbon carbonates, and then into thick sequences of ribbon carbonates with storm sequences (Oder, 1934; Aadland, 1984; Bova and Read, 1987). Conversion of the high-relief rimmed shelf of the Late Cambrian into the Early Ordovician ramp could not have been done due to basin filling by sedimentation, because offshelf sections are too thin. Instead, it appears to have been due to tectonic shallowing of the basin, possibly caused by incipient collision, and reverse movement on earlier, listric normal faults (Fig. 8).

Sea-Level Cycles

The major limestone to cyclic-dolomite sequences appear to represent the effects of as many as eight third-order, 2- to 6-Ma sea-level cycles (Figs. 9, 13). Smaller scale 1- to 5-m cycles within the cyclic carbonates resulted from Milankovitch sea-level fluctuations of a few meters (Bova and Read, 1987). It is possible that some of the limestone sequences also might have been due to increased amplitude of these short-term sea-level fluctuations, perhaps in excess of 10 to 15 m (Bova and Read, 1987). The laminites pro-

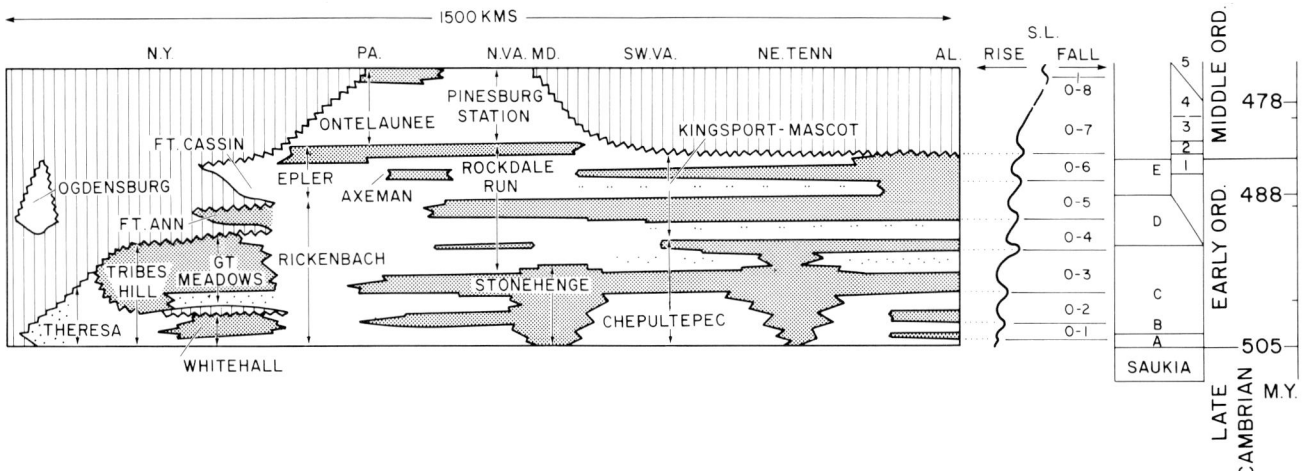

FIG. 13.—Chronostratigraphic diagram of sequence 5, extending parallel to the passive margin from New York to Alabama, showing time relations of lithologic units and realtive sea-level curve. Subtidal carbonates (stippled) form mainly during the rise, whereas cyclic dolomites (blank) and lesser sands (dots) form during the fall. Note unconformities developed along cratonic margin during sea-level falls.

graded from cratonic highs and arches out into the depocenters and toward the ramp margin, in contrast to the Late Cambrian flats, which prograded in from the rimmed-shelf margin.

TECTONIC CONTROLS

Global tectonics appear to have controlled the second-order sea-level cycle from the Late Precambrian to the Middle Ordovician. The gradual rise may have been due to breakup of a Late Proterozoic supercontinent and formation of a major spreading system that caused maximum decrease in ocean basin volume some 70 m.y. after breakup (Bond and others, 1984). The fall possibly was related to subsidence of the ridges with cooling and may have been aided by destruction of the ridges during subduction.

The early carbonate platform of sequence 2 was controlled by the location of a broad hinge zone of rifted, thinned basement (Wehr and Glover, 1985), and the fluvial- to marine-shelf cover of clastics of sequence 1 (Simpson, 1987; Fig. 8). Thomas (1977) suggested that the present shape of the fold-thrust belt was inherited from the configuration of the rifted margin, which contained a series of promontories and re-entrants. Thus, the carbonate shelf edge appears to be roughly located above the Late Precambrian hinge zone. The hinge line appears to have backstepped 100 to 200 km in from the shelf edge by Early Cambrian time (Figs. 1, 8). It backstepped 200 km from the shelf edge in Vermont to 400 to 500 km in Pennsylvania to Virginia in the Middle Cambrian to Early Ordovician (Figs. 1, 8). This Middle Cambrian backstep was synchronous with the development of the Rome trough, which may have formed a zone of weakness in the continental lithosphere. The Late Precambrian to Ordovician landward shift in the hinge line, however, also may reflect increased flexural rigidity and thickening of the lithosphere with cooling and loading following breakup (Bond pers. commun., 1988).

Long-term subsidence of the passive margin was controlled by thermal cooling of thinned continental lithosphere and tended to decrease exponentially into the Early Ordovician (Bond and others, 1984, 1988; Fig. 2). Major pulses of subsidence do not appear to have occurred, judging by geohistory plots, except in the Rome trough and the Alabama basement fault zone. Synchroneity of relative sea-level changes along the length of the passive margin (Figs. 11, 12, 13) in fact point to eustatic causes rather than jerky subsidence.

Regional variation in subsidence rate over the shelf was expressed as depocenters in Tennessee, Pennsylvania, and Vermont and intervening arches (Figs. 1, 2). Initially, depocenters were confined to the shelf edge but tended to extend onto the shelf with time. The carbonates were initiated over regional highs and seaward of the cratonic shoreline, or along the rimmed-shelf edge, and migrated toward more rapidly subsiding depocenters. Thicker sections were deposited in depocenters, where more rapid subsidence also favored deeper, more open-marine conditions on the shelf during transgressions.

Rifting in the Middle Cambrian formed the Rome trough some 300 km inboard of the passive margin (Figs. 1, 4), and subsidence continued into the early Late Cambrian. Uplifts associated with rifting may have shed clastics to form the redbeds of sequence 2. Major down-to-basin faulting of the shelf during this period also is evidenced by development of a half-graben in Alabama, and perhaps by the St. Albans re-entrant in Vermont (Fig. 4), which appears to have controlled the backstepping of the shelf edge and distribution of the Parker Slate and breccias above the drowned carbonate shelf of sequence 2. Wagner (1976) suggests that down-to-basin faulting continued into the Late Cambrian and early Ordovician in Pennsylvania along the northern extension of the Rome trough (Olin Basin; Figs. 1, 6).

The Waverly arch, which is oblique to the trend of the shelf (Figs. 1, 4), appears to have been activated in the Late Cambrian and possibly slightly earlier. It may have influenced the direction of progradation of the Kerbal "delta" (sequence 3) in Ohio and caused thinning of sequence 4 and possibly sequence 5.

Tectonic deformation appears to have controlled the transition from the Late Cambrian rimmed shelf, which had a maximum relief of 2.5 km, into a ramp. In the absence of thick basinal sequences adjacent to the margin in Pennsylvania-Maryland (Gohn, 1976; Reinhardt, 1974), shallowing of the basin probably resulted from uplift of the basin floor due to reverse movement along earlier listric normal faults during incipient collision (Fig. 8). This was followed in the early Middle Ordovician by regional uplift of the shelf in the southern Appalachians to form the Knox unconformity (Fig. 8), which has been ascribed to development of a peripheral bulge as the margin was subducted (Jacobi, 1981), possibly in association with a eustatic sea-level fall (Sloss, 1963; Mussman and Read, 1986). Subduction in the north appears to have been accompanied by broad folding and normal faulting of the shelf (Fisher, 1954; Cady, 1945). In the Pennsylvania depocenter, this unconformity is not evident, because subsidence continued into the Middle Ordovician. Regionally, subsidence rates started to increase into the early Middle Ordovician, reflecting subduction and tectonic loading of the margin (Rowley and Kidd, 1981; Shanmugam and Lash, 1982; Bond and others, this volume). This downwarping of the shelf also may have been synchronous with a eustatic sea-level rise. Deepening promoted local development of large carbonate buildups on the foundering carbonate platform, followed by drowning of the ramp beneath black shale (Fig. 8; Benedict and Walker, 1978; Read, 1980).

WAVE-ENERGY, FETCH, AND SEAFLOOR SPREADING

The opening and closing of the oceanic basin during the Cambrian to Middle Ordovician probably was accompanied by a change in wave-energy conditions as fetch changed. Given that the passive margin formed at about 570 Ma, then it seems likely that during the early breakup, wave-energy would have been relatively low due to the narrow basin width which limited fetch. This is supported by the muddy, low-energy Early Cambrian ramp with its mud mounds and scarcity of well-developed carbonate sands.

Considering that about 100 m.y. were available to generate and destroy an oceanic basin, then with active spreading rates an ocean basin with large fetch could have developed by the Middle to Late Cambrian. This basin could have been underlain by either oceanic crust, thinned continental or transitional crust. If a large basin had developed by the Middle to Late Cambrian, then the rimmed platform effectively blocked the effects of oceanic waves on the platform interior, where muddy cycles are relatively common landward of the rim. Evidence of periodic high-energy storm events, however, are common in the shelf sequences of this age in the form of abundant flat-pebble conglomerates with well-rounded platy clasts of early lithified carbonate (Whisonant, 1987). These flat-pebble conglomerates were formed in outer-shelf settings both above and below fair-weather wave base characterized by multidirectional currents and bordering tidal flats, where they locally formed gravelly beach ridges or storm layers veneering the tidal flats. The inner-shelf conglomerates associated with tidal flats reflect onshore winds and waves, which also probably promoted transport of fine sediment into tidal flats, enabling them to prograde rapidly. These conglomerates decrease in abundance into the Early Ordovician, when energy appears to have been lower, possibly reflecting incipient collision and onset of closure of the basin. Low-energy conditions in the southern Appalachians are indicated by presence of a mud-dominated ramp with shoal complexes of scattered bioherms and thin grainstones. Relatively high-energy conditions associated with outer-ramp grainstone and bioherm complexes appear to have continued farther north in Maryland, where the stratigraphic evidence of the orogenic clastic wedges suggests closure at a later date than areas along strike.

PLATFORM SLOPES, SHORELINE MIGRATION, TIDAL RANGE

Slopes

Slopes on top of the platform during deposition of each cycle can be estimated roughly by the rate of thickening of a stratigraphic sequence (m/km) divided by the maximum number of cycles within the interval. Calculated slopes are about 2 cm/km for the platform top during the Middle Cambrian to early Ordovician. By comparison, the Bahama platform has slopes of 4 cm/km, whereas most modern shelves have slopes that are much higher. The effects of these low slopes was to allow even small sea-level fluctuations to cause widespread transgressions of several hundred kilometers. During 20- to 100-ka regressions, the low slopes caused rapid seaward migration of the shoreline as sea level fell. Thus, only relatively low-amplitude sea-level oscillations would have favored regional tidal-flat deposition, although because we cannot assess the maximum potential progradation rates of the flats, we cannot be sure of the actual amplitudes involved. Relatively large-amplitude fluctuations (above 10 to 20 m), coupled with third-order sea-level rise, would have tended to cause the strandline to move off the platform faster than tidal flats could prograde during 20- to 100-km sea-level falls; thus, tidal flats would not be deposited over the whole shelf. At these times, dominantly subtidal cycles capped by erosion surfaces on the outer shelf would be developed (Bova and Read, 1987).

Slopes on the platform increased toward the margin. Slopes on the outer part of the early Cambrian ramp were about 1° (10 m/km) but increased greatly during rimmed-shelf development, when slopes may have reached tens of degrees and thick deposits of periplatform talus developed. Slopes decreased into the Early Ordovician as the rimmed shelf evolved back into a ramp.

Tidal Range

The thickness of tidal-flat caps of cycles may provide clues as to tidal range. Average thickness of tidal-flat caps in the Conococheague Formation in Virginia exceed 1 m. In many cycles, the caps compose more than half the cycle thickness. Computer modeling suggests that tidal ranges of 0.5 and 1.0 m are too low to generate these thick caps, unless unreasonably high-sedimentation rates and very short lag times are used (Koerschner and Read, in prep.). Decreasing the amplitude forms thick caps, but it also greatly increases the number of cycles formed as well as decreasing their thickness. Using 2-m tidal ranges generates cycle caps averaging 1.4 m, close to the average value for the Conococheague caps. A relatively large tidal range also is suggested by the mechanical sedimentary structures of the ribbon carbonates, which have many features in common with tidal-flat facies in the high-tidal-range flats of the North Sea (Demicco, 1983). If the volume of water on the shelf at low and high tides is taken into account, however, unrealistically high tidal currents would be required to generate 2-m tides on a diurnal or semidiurnal cycle, given the shelf water depths of less than 3 m and its great width of up to 400 km. Consequently, the 2-m tidal range on the Cambrian flats probably was due to wind tides. Wind tides would have allowed a more gradual transfer of water across the shelf. Because much tidal-flat deposition takes place during storms, the sedimentary record would seem likely to preserve the effects of wind tides over astronomic tides.

ALLOCYCLIC OR AUTOCYCLIC CONTROLS

The peritidal cycles at outcrop scale show no evidence that the tidal-flat caps are merely the product of random, mosaic shallowing under stable sea level. Tidal-flat caps are continuous as far as the outcrops extend. Walker (pers. commun., 1987) has documented that individual tidal-flat caps of cycles containing well-defined marker beds in the Knox Group could be traced in the subsurface in nine cores that extend over 2.4 km, with no evidence of pinchout. Markello (1979) showed that individual cycles in the Middle Limestone (Conasauga Group) were traceable over an area of 80 by 140 km. Thus, the evidence suggests that the cycles are shelf-wide events.

There is little evidence that jerky subsidence could be causing these regional cycles given the short periods of the cycles. Nor is it likely that the cycles formed under stable sea level by cessation of sedimentation due to shrinking of the subtidal factory because of tidal-flat progradation (autocyclic model of Ginsburg, 1971). The low-subsidence rates would not allow the platform to be flooded to the required depths, except by having lag times of tens of thousands of years, which seems unreasonable. Furthermore, shallowing-upward cycles are well developed in the intra-shelf basin, yet these cycles were formed completely in subtidal settings.

Thus, the most likely mechanism for the cycles appears to be sea-level fluctuations in the range of 20 to over 100 ka, possibly coupled with longer term third-order fluctuations. Shrinking of the subtidal carbonate factory during tidal-flat progradation, however, probably decreased progradation rates as flats migrated toward depocenters. Thus, tidal-flat deposition tended to be inhibited in depocenters and locally along the platform margin, especially during times of third-order sea-level rise.

SEA-LEVEL CONTROLS

Third-Order Sea-Level Cycles

The major stratigraphic sequences were controlled by third-order sea-level cycles with durations of 1 to 3 m.y. The third-order cycles may be defined qualitatively on the basis of relative water depth of lithofacies composing large-scale shallowing-upward or deepening-upward sequences (Figs. 11, 12, 13). They also may be defined using subsidence modeling techniques applied to Grand Cycles. In this, systematic differences between the tectonic subsidence curves and best-fit exponential cooling curves are interpreted as changes in eustatic sea level (Bond and others, this volume). They also may be defined using Fischer plots of peritidal sequences (Fischer, 1964; Goldhammer, 1987; Read and Goldhammer, in prep.; Figs. 14, 15). Fischer plots graph cumulative cycle thickness (corrected for linear subsidence) against time. On the plots, the cycles are assigned an average period, and linear subsidence is assumed (valid for the relatively short times involved). Thus, eustatic events (or major changes in subsidence rate) plot as departures relative to the horizontal axis (Figs. 14, 15).

These third-order sea-level changes typically resulted in sequences tens to a few hundred meters thick. During third-order sea-level rise in depocenters, accommodation space was increased, and slightly deeper water shale or shaly limestones, carbonate banks, and algal bioherms were deposited. During third-order sea-level fall, accommodation

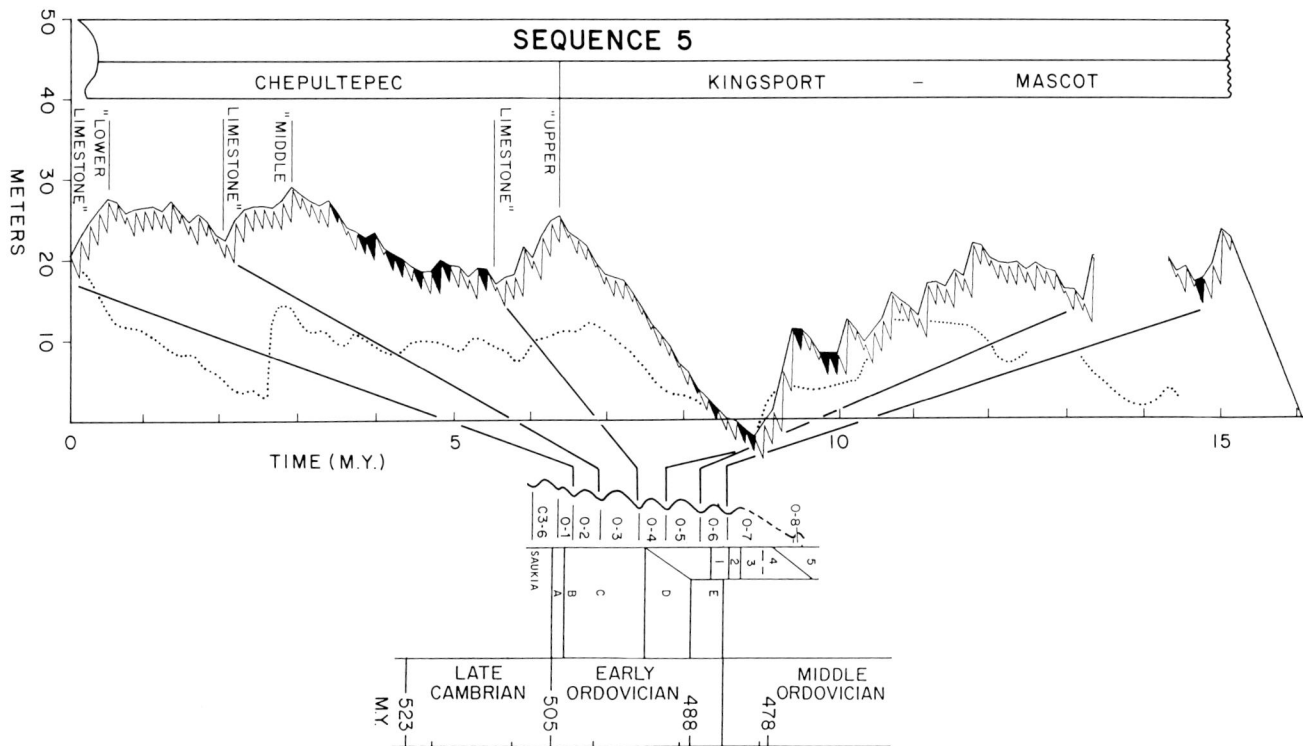

FIG. 14.—Fischer plot of sequence 5, cyclic upper Knox Group (Early Ordovician), Avens Bridge, Virginia, based on data in Bova (1982) and Read (unpubl. data). Plot when turned on side is in normal reading position, with time along horizontal axis and thickness along vertical axis. Inclined lines sloping to lower right mark linear subsidence path of each cycle, arranged with oldest (base of section) on left. Thickness of each cycle is plotted vertically after platform is allowed to subside for one cycle period. Heavy top curve marks path of sea level (or subsidence greater than average linear subsidence); upward slope to right suggests rise in sea level, whereas downward slope to right indicates fall in sea level. Sandstone-bearing cycles marked in black. Curve suggests net eustatic fluctuations of 30 m maximum over periods of as much as 4 m.y. Lower part of diagram shows how sea-level curves from Fischer plots relate to qualitative curves defined by relative water depths of lithofacies (Fig. 13). Thick subtidal cycles that lack tidal-flat caps (lower, middle, and upper limestones of the Chepultepec Formation, and much of the Kingsport Formation) were developed during the rises, whereas thin dolomite cycles and quartz sands occur during the sea-level falls. Dotted line is trace of Fischer plot from Goodwins Ferry, Virginia, which has a lower subsidence rate than Avens Bridge; note this results in apparent lower amplitudes. Cycles O2 to O6 have been well documented in Pennsylvania by Goldhammer, Nguyen, and Hardie (in Hardie and Shinn, 1986).

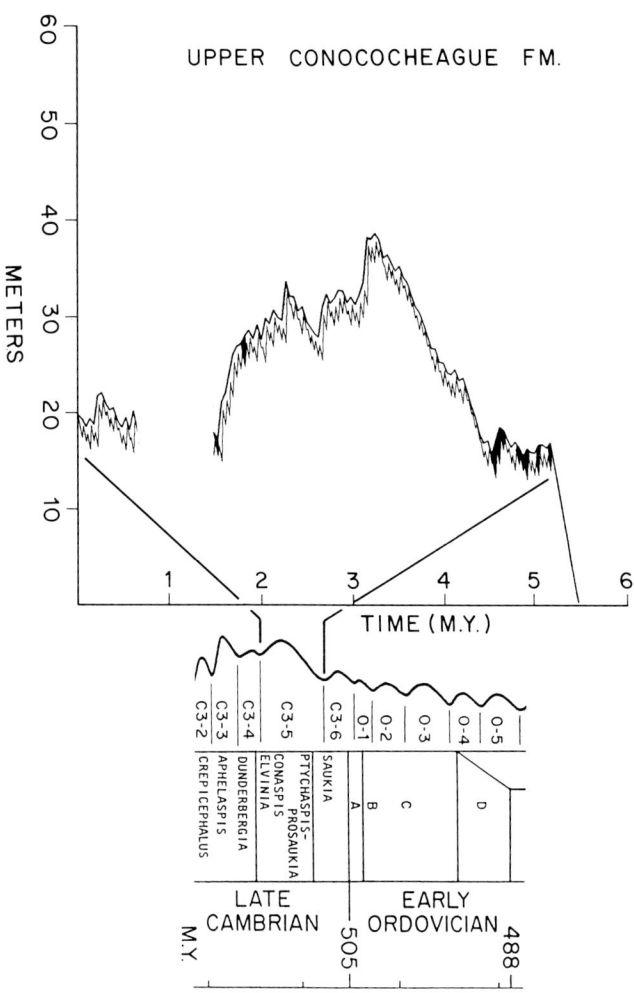

Fig. 15.—Fischer plot of part of sequence 4, Late Cambrian, upper Conococheague Formation, Wytheville, Virginia (from Koerschner, 1983), compared with qualitative sea-level curve derived from relative water depths of lithofacies (Fig. 12). For details of plot, see Figure 14. Fischer plot suggests eustatic departure of about 25 m over 4 m.y.

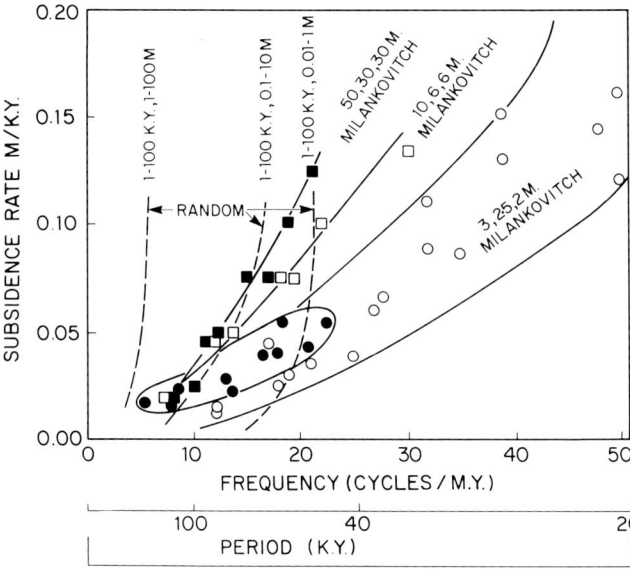

Fig. 16.—Plot of cycle frequency versus subsidence rate from various measured sections (solid circles), along with data derived from synthetic cyclic sequences using Milankovitch sea-level curves (Fig. 17) of various amplitudes; for the Milankovitch synthetic data sets, the 100-, 40-, and 20-ka amplitudes are labelled 50, 30, 30 m; 10, 6, and 6 m; and 3, 2.5, and 2 m. Three dotted lines mark trend of data from sea-level curves generated by random numbers between 1 and 100 ka and amplitudes of as much as 100 m, as much as 10 m, and as much as 1 m, respectively (data points not shown). Note Cambro-Ordovician cycles lie on trend with low-amplitude Milankovitch cycles and are discordant to trends formed by high-amplitude Milankovitch fluctuations and with random fluctuations.

space decreased, and peritidal dolomites and quartz sands were deposited, and erosional unconformities developed where the sea-level-fall rate exceeded platform subsidence. Constraints can be put on the amplitudes of third-order sea-level cycles by the fact that for conformable sequences (which are dominant), these sea-level fall rates would have had to be less than subsidence rates. Thus, for 0.05 m/ka-subsidence rates (typical for much of the shelf), and assuming a symmetrical oscillation, maximum amplitudes would have been less than 50 m for a typical 2-m.y. cycle. In fact, most of the third-order amplitudes appear to have been from 10 to 50 m, based on Fischer plots (Figs. 14, 15). These values are similar to those calculated by Bond and others (this volume) using subsidence modeling techniques.

Random versus Milankovitch Sea-Level Fluctuations

Average frequencies of cycles from the various stratigraphic sections increase into areas with high-subsidence rates (Fig. 16), an effect noted elsewhere by Kendall and Schlager (1981) and Goodwin and Anderson (1985). This effect is reproduced by the modeling, in that more sea-level fluctuations are preserved in the sedimentary record in areas of higher subsidence rate because the platform is subsiding sufficiently rapidly to allow most of the sea-level fluctuations cover the platform to form cycles. The field data have maximum subsidence rates of below 0.06 m/ka and maximum frequencies of 22 cycles/m.y., equivalent to 44-ka periods. With decreasing subsidence rates, frequencies decrease to less than 10 cycles/m.y. (periods greater than 100 ka).

The synthetic cycles produced by the sea-level curve shown in Figure 17, and various subsidence rates and third-order sea-level rise/fall rates, can be used to form a synthetic data set extending over the range of naturally occurring subsidence rates on the platform, and into the range above 0.06 m/ka. The synthetic data, when plotted, show a trend that lies along the same trend as the naturally occurring data set, and extend down through the 40-ka to the 20-ka range, with increasing subsidence values. This suggests that the natural cycles and the synthetic cycles formed under similar sea-level fluctuations of similar periods and perhaps amplitudes.

To test whether the natural data set could be the result of purely random sea-level fluctuations, cycles were generated for various subsidence rates, using a data set formed by random periods (1 to 100 ka) and random amplitudes of 1 to 100 m, another set for random amplitudes of 0.1 to

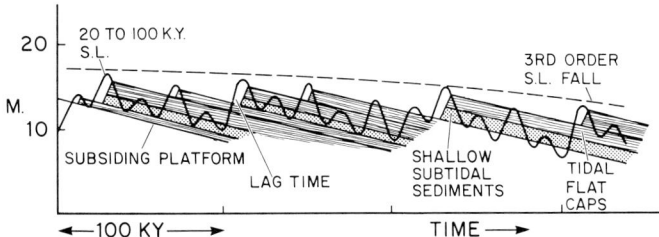

Fig. 17.—Computer model showing formation of peritidal cycles such as those shown in Figure 10C. Vertical scale is meters; horizontal scale is time. Sinusoidal curve is track of Milankovitch sea-level fluctuations with time. These are superimposed on gradually falling (third-order) sea level (dotted line). Inclined lines to lower right are paths of subsiding facies boundaries, whereas lines extending upward to upper right are paths of aggrading sediment surface. Note the 2-ka lag time between time when platform is initially flooded and time when it begins aggrading. Subsidence rate is 0.05 m/ka, fall rate of third-order sea level averages 0.02 m/ka; 100-, 40-, and 20-ka amplitudes of sea-level curve are 3 m, 2.5 m, and 2 m, and water depths and sedimentation rates are tidal flat (2-m tidal range; 0.4 m/ka) and shallow subtidal (2 to 3 m below high water; 0.5 m/ka).

10 m, and a third set for 0.01- to 1-m amplitudes. The data set for 1- to 10-m amplitudes and 1- to 100-ka periods most likely would duplicate the data set, if they were the result of random sea-level (or subsidence) fluctuations, judging by histograms of cycle thicknesses (Koerschner and Read, in prep.) and considerations of maximum amplitudes allowable to generate tidal-flat capped cycles of regional extent (Read and others, 1986). In fact, the random data sets form trends, which are roughly parallel to each other, but highly discordant with the actual data set, and the synthetic Milankovitch data set. Note that the natural data set only overlaps the lower range of the random data set. The natural data set could not be generated by these random processes, because it would require sea levels on one part of the platform to be fluctuating with amplitudes of as much as 10 m, while on the more rapidly subsiding part of the platform, sea levels would have to fluctuate only as much as 1 m, which is impossible. The random data sets differ from the Milankovitch data sets in that they do not have a built in 100-, 40-, and 20-ka signal. In the Milakovitch data set, the higher frequency signals become increasingly more important with increasing subsidence rate. Consequently, with highest subsidence rates (above 0.1 m/ka), 20-ka cycles tend to be preserved. In contrast, for the randomly generated sea-level fluctuations, the highest average periods preserved only are about 50 to 60 ka at the high-subsidence rates of 0.1 m/ka. This cutoff at about 50-ka average period in the random data set reflects the absence of an underlying higher frequency signal that is able to be preserved at high-subsidence rates.

Spectral analysis supports non-random Milankovitch fluctuations for these rocks (Coruh and Read, in prep.). The record of sea-level fluctuations is best preserved in intra-shelf basin sequences of sequence 3, where ratios of the periods of the spectral peaks are roughly 1:2:5:40, which on the basis of long-term accumulation rates, may correspond to cycle periods of 20, 40, 100, and 800 ka, although large-error bars on subsidence rates make accurate estimates of cycle periods imprecise. A poorer record is preserved in the peritidal sequences, because many of the sea-level fluctuations are not preserved in the sedimentary record (Fig. 17).

Development of peritidal sequences.—

The small-scale, 1- to 5-m cycles that typify much of the shallow shelf appear to have resulted from low-amplitude (probably a few meters) sea-level oscillations with periods of from 20 to over 100 ka (Fig. 17; Bova and Read, 1987; Koerschner and Read, in prep.). Early attempts at modeling these used a simple sinusoidal or zig-zag sea-level curve of fixed amplitude (Read and others, 1986; Bova and Read, 1987). This fails to create the variation in cycle thickness and facies that is present in the stratigraphic record. The cycles are more realistically modeled using Milankovitch-like sea-level fluctuations (Hays and others, 1976; Goldhammer and others, 1986; Koerschner and Read, in prep.), with a superimposed third-order (1 to 10 Ma) sea-level fluctuation (Fig. 17), the magnitude of which can be obtained from the Fischer plots. Note that we cannot accurately constrain the amplitudes of the 20-ka fluctuations, in that sea-level curves shown in Figures 17 and 19 would form similar peritidal sequences (perhaps differing slightly in diagenesis related to emergence) and would tend to have the same values for frequency versus subsidence rate (Fig. 16).

A lag time following platform inundation during the sea-level rise associated with the 20-ka to 100-ka fluctuations allowed deepening to occur with little sediment upbuilding. This lag time is basically a non-depositional period following initial flooding of the platform, until carbonate production reaches its full potential. Once the carbonate production reached maximum values, however, shallowing upward would have occurred in a relatively short time (a few thousand years or so) given likely sedimentation rates for these facies. With sea-level fall associated with the 20-ka to 100-ka cycles, tidal flats migrated rapidly over distances of as much as 400 km from cratonic highs or the rimmed-shelf edge into the subsiding depocenters. Much of the cycle period (probably many tens of thousands of years for 100-ka cycles) would have been non-depositional, as the supratidal platform slowly subsided, as sea level fell. Sedimentation would have restarted only following the next transgression of the platform.

During third-order sea-level rise, if not too rapid (say below 0.05 m/ka), thick, relatively open-marine, peritidal cycles formed. The third-order rise amplified the low-amplitude 20-ka to 100-ka rises and tended to suppress the sea-level falls associated with these short-term oscillations. It also favored formation of some conformable cycle boundaries, because the Milankovitch lowstands on the outer shelf rarely fell much below the surface of the subsiding platform. Because rise rates are low near the position of the lowstands, flooding of the platform occurs relatively slowly during the initial part of the Milankovitch rise. This allowed deposition of a deepening-upward tidal-flat sequence above the tidal-flat cap of the previous cycle.

During third-order sea-level falls, thin, relatively restricted cycles formed over the shelf due to decreased accommodation space. This was because the long-term fall decreased the space created by subsidence for cycle deposition and tended to suppress the short-term sea-level rises (20-ka to 100-ka amplitudes; compare Figs. 17 and 18). The third-order falls tended to amplify the Milankovitch falls, causing widespread regression and tidal-flat progradation across the shelf. The modeling shows that fewer cycles are deposited during third-order falls compared to the rise. Consequently, average cycle periods are long during the third-order falls. This results in long periods of emergence of the tidal flats, favoring brecciation, and development of disconformities. Furthermore, rates of sea-level rise at the times of flooding of the platform are high as a result of the Milankovitch lowstands falling some distance below the platform top. Consequently, by the time the sea finally starts to flood the platform, Milankovitch rise rates are approaching maximum values, and together with lag time, prevent a deepening-upward sequence from being deposited. Consequently, disconformable tops of cycle caps are sharply overlain by a thin, transgressive, subtidal lag or by a subtidal shallowing-upward sequence. In landward areas toward the craton, where subsidence rates were less than the third-order fall rates, sea level would have dropped below the level of the platform for long periods (1 ma or more), resulting in unconformity formation and quartz sands being supplied to the shelf (Fig. 14).

Cycles decrease in number onto the inner shelf. The overall decrease in cycle number into the shelf interior reflects the fact that areas of higher subsidence rate tend to record more of the Milankovitch fluctuations than do areas of lower subsidence, discussed previously. In some cases, however, there also is a slight decrease in the number of cycles toward the outer shelf. This may have resulted from condensed sequences on the outer shelf during sea-level rise, which prevented shallowing to sea level during each Milankovitch fluctuation. It may also be due to tidal flats being unable to prograde all the way onto the outer shelf, especially during times of third-order sea-level rise.

Development of thick subtidal-limestone sequences.—

Bova and Read (1987) suggested that dominantly subtidal-limestone sequences that lack tidal-flat caps may have been due to Milankovitch sea-level fluctuations with amplitudes over 10 or 15 m. This was based on estimates of maximum progradation rates of 5 km/ka for Holocene tidal flats and the low slopes on the platform. Given the low (1 to 2 cm/km) slopes of the platform, amplitudes above 10 to 20 m would have caused rapid regression during sea-level falls, leaving tidal flats stranded on the inner platform (Bova and Read, 1987), while outer-shelf cycles were capped by karstic surfaces, rock platforms, or hardgrounds. It seems likely, however, that tidal-flat progradation rates were underestimated and may have exceeded 20 km/ka (Hardie and Shinn, 1986). Bova and Read (1987) also used 140-ka average cycle periods in the modeling. During third-order sea-level rise, however, almost all of the Milankovitch fluctuations leave a record in the cyclic stratigraphy (Fig. 18), resulting in average periods during the third-order rise of 20 to 30 ka. These result in little time for emergence of the platform but cause rapid regression. Thus, unless periods and maximum progradation rates can be better constrained, then estimation of amplitudes of Milankovitch sea-level fluctuations is approximate at best.

The fact that thick subtidal cyclic sequences formed during third-order rise superimposed on Milankovitch fluctuations (Bova and Read, 1987) is borne out by relative sea-level curves and Fischer plots of cyclic sequences (Figs. 12, 13, 14). The third-order rises amplify the Milankovitch rises, causing deposition of open-marine facies. The modeling suggests, however, that if the 20- to 40-ka fluctuations are symmetric sine waves, then the amplitudes of the Milankovitch fluctuations have to be higher than 7.5 m in order to generate the facies observed. If the 20- to 100-ka fluctuations are asymmetric (Fig. 19), however, then open-marine facies will be produced, even with low-amplitude Milankovitch fluctuations of less than 10 m.

It is possible that during third-order sea-level rises, the outer shelf became sediment starved. This would result in condensed sequences that would be equivalent to several peritidal carbonate cycles on the inner platform (B. Ross, pers. commun., 1987). During long-term fall, the outer shelf would initially become the site of thick, shallowing-upward subtidal units that filled the space above the condensed interval. Although this is possible, the modeling does not suggest that the outer shelf became sediment starved for long periods. It is not possible to rule this out on the basis of outcrop data, however. The presence of downlap surfaces on the outer shelf, if observed in seismic cross sections, might suggest sediment starvation for long periods.

The modeling suggests that tidal-flat sedimentation rates have to be suppressed to produce these thick subtidal sequences. The third-order rises decrease the magnitude of the Milankovitch sea-level falls and the duration of emergence of the platform (Fig. 18). Progradation of tidal flats

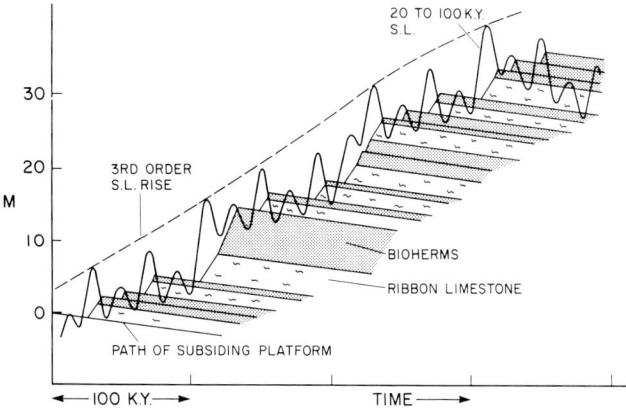

FIG. 18.—Computer model showing formation of dominantly subtidal sequences such as those shown in Figure 10B. Milankovitch curve (100-, 40-, and 20-ka fluctuations of 6 m, 4 m, and 4 m, respectively) is superimposed on third-order sea-level rise (average 0.125 m/ka). Subsidence rate is 0.025 m/ka, and water depths and sedimentation rates are tidal flats (1-m tidal range, 0.05 m/ka), shallow subtidal bioherms (1 to 3.5 m below high water, 0.5 m/ka), and storm-influenced deeper subtidal (over 3.5 m below high water, 0.3 m/ka). Combination of long-term rise and Milankovitch rise results in deeper subtidal facies in lower parts of cycles. Long-term rise also inhibits progradation of tidal flats across platform, hence suppressed tidal-flat sedimentation rates. Cycles may be capped by tidal- or minor subaerial-erosion surfaces. Note that cycles are much thicker than those formed during long-term fall in previous figure.

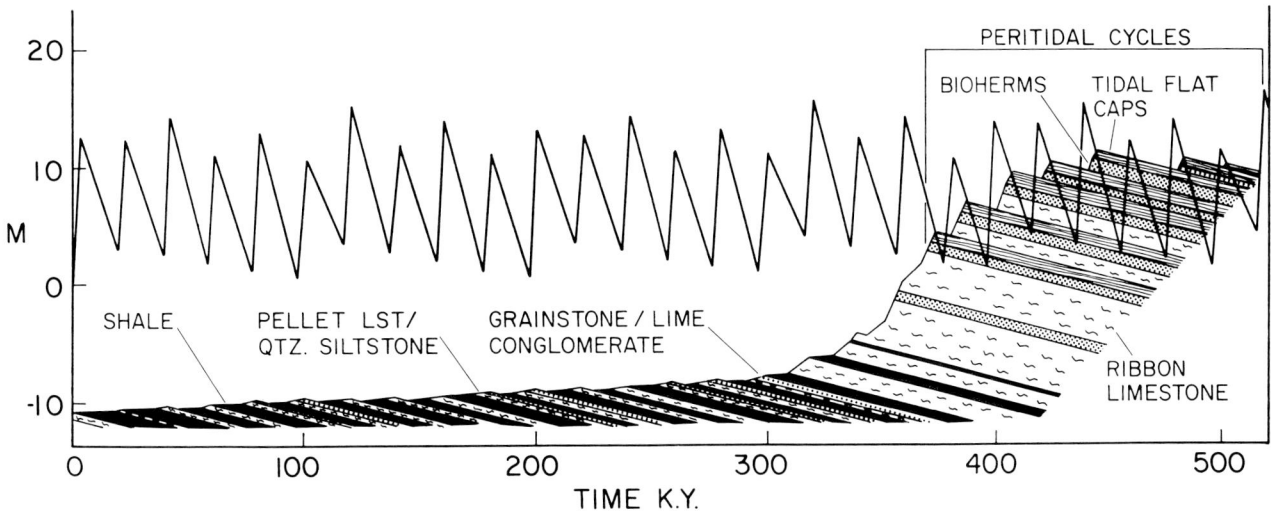

FIG. 19.—Formation of shale-limestone-conglomerate sequences of intra-shelf basin, and overall shallowing-upward sequence from intra-shelf basin into peritidal platform (Grand-Cycle; Fig. 10A). Starting water depth is 13 m below the low sea-level stand and results from rapid sea-level rise exceeding sedimentation rate. Amplitudes for 100-, 40-, and 20-ka fluctuations are 3 m, 2.5 m, and 10 m. Sediment types, water depths, and sediment rates for the basin are shale (below 15 m, 0.06 m/ka), siltstone/pellet limestone (11 to 15 m, 0.08 m/ka) and lime sand/conglomerate (4 to 11 m, 0.10 m/ka). On the platform parameters used were tidal flats (2 m tidal range, 0.4 m/ka), shallow subtidal bioherms and grainstones (2 to 4 m below high water, 0.5 m/ka), ribbon carbonates (4 to 15 m water depth, 0.60 m/ka decreasing to 0.20 m/ka below 8 m), and shale (greater than 15 m water depth, 0.06 m/ka). Sequence formed consists of shale-silt-lime sand/conglomerate cycles of the intra-shelf basin, overlain by thick ribbon carbonates of the deep-ramp, and then peritidal cycles. Compare section developed with Figure 10A.

across the shelf is inhibited because the tidal flats tend to be pushed farther landward by successive Milankovitch transgressions that reach successively higher elevations due to the superimposed third-order sea-level rise (Fig. 18), and because the magnitude and duration of the effective sea-level fall is decreased.

Development of intra-shelf basin sequences.—

Given likely rise rates for the third- to fifth-order sea-level rises, we need to examine the likely mechanisms involved in deepening over the shelf. The third-order sea-level rises, which probably spanned at least a few hundred thousand years and were mainly less than 50 m in amplitude, likely had rise rates of no more than 0.2 m/ka, assuming 20- to 30-ka cycles developed on the rise. For these third-order rise rates, deepening would only occur if average sedimentation rates were kept low (below 0.2 m/ka) and slightly less than subsidence rate plus long-term rise rate for many tens of thousands of years. This could be achieved by having carbonate production on the shelf decrease because of high turbidity, variable salinities due to influxing fine siliciclastics under humid conditions, increased water depths of 10 to 20 m, and perhaps cooler waters.

If Milankovitch rises were superimposed on the long-term rise, then these would amplify the third-order sea-level rise, whereas the short-term sea-level falls would be offset by the third-order rise. This would result in a series of short-term (20 to 100 ka) rises interspersed with relative stillstands. If this occurred in subtidal settings with highly suppressed sedimentation rates, then water depths would tend to increase over the platform during third-order rises.

Gradual shallowing would take place once sedimentation rate started to exceed relative sea-level rise rate slightly (subsidence plus eustatic sea-level rise), or where third-order sea levels started to fall (Fig. 19). The modeling suggests that average, long-term, sedimentation rates in the basin must have only slightly exceeded subsidence rates (0.02 to 0.10 m/ka). If sedimentation rates exceed these values by only a few centimeters/ka, shallowing to peritidal depths occurs rapidly, and thick intra-shelf basin sequences are not developed. Where initial deepening was to subwave base depths, then shallowing upward would form a sequence with a basal quiet-water shale. As long-term shallowing progressed, the bottom would come into the zone periodically affected by storm reworking during Milankovitch lowstands, and quiet-water shale deposition would occur only during Milankovitch highstands (Fig. 19), possibly under the influence of humid conditions which favored increased supply of fine siliciclastics to intra-shelf basins. During the Milankovitch sea-level falls (Fig. 19), storm-wave base would be lowered onto the sea floor, causing storm reworking of the bottom and successive deposition of silts capped by coarse carbonate sands and conglomerates deposited during the lowstand. Sea-level fall also appears to have been accompanied by more arid conditions and decreased fine siliciclastic supply to the basin, which became the site of clear-water carbonate deposition with storm reworking.

Modeling indicates that the dominance of coarsening-upward shale-silt-lime sand/conglomerate cycles in the basin cannot be produced by symmetrical sinusoidal 20- to 100-ka sea-level oscillations. These form symmetrical coarsening-upward fining cycles. The typical coarsening-upward cycles are most easily modeled by asymmetric 20- to 100-ka sea-level fluctuations (Fig. 19). The modeling also suggests that the abundant 20- to 40-ka cycles in the basin are most easily produced where 20-ka amplitudes exceed 10 m, but the 40- and 100-ka amplitudes are only a few meters. The relatively large 20-ka fluctuations enable the vertical migration of the three arbitrary depth zones of shale (greater than 15 m), silt (11 to 15 m), and sand/conglomerate (less than 11 m) to occur with sufficient frequency to form the many three-component, shale-based cycles. These depth zones are compatible with short-period waves that would typify a shallow protected intra-shelf basin. The modeling suggests that the time equivalent peritidal facies that would have formed under these elevated amplitudes would be little different from peritidal facies that developed under lower amplitudes, assuming tidal flats were capable of migrating out across the shelf. Consequently, at this stage it is not possible for us to determine the exact form of the sea-level fluctuations that formed these sequences.

Third-order sea-level fall favored shallowing of the basin, and intra-shelf basin shale-storm sequences were succeeded by deposition of widespread, thick (10 to 20 m), deeper ramp, ribbon limestone with thin relatively fine-grained storm sequences; this sequence is overlain by shallow-water bioherms or lime sands, and finally by peritidal cyclic carbonates (Fig. 19). The modeling suggests that shallowing of the basin was accompanied by a decrease in wave base, otherwise the ramp would be dominated by lime sands; the shallowing also was coupled with an increase in sedimentation rate.

Effects on periplatform sedimentation.—

It is not clear whether sea levels influenced development of periplatform talus deposits. The well-developed talus at the foot of the rimmed shelf of sequence 2 appears to have formed during a time of long-term sea-level fall (Hawke Bay Event of Palmer and James, 1979). This might have favored oversteepening of the margin as organic growth was concentrated at the shelf edge. This also, however, was a time of considerable seismic activity associated with Rome trough rifting. Consequently, earthquakes may have caused frequent failure of the margin and breccia deposition.

Other regional breccias that might relate to sea-level fall include those in the Franconian (Late Cambrian), when regressive sands developed on the shelf. The limited biostratigraphic control on the timing of the breccias, however, does not allow them to be tied in accurately to the sea-level curves. Other breccias in the sequence, which do not appear to be as regional, may relate to local oversteepening of the margin at various times and may be little influenced by sea-level history. Some have been interpreted as submarine fan deposits associated with canyons incised into the margin (Keith and Friedman, 1977; Gohn, 1976).

CLIMATIC CONTROLS

It is possible that the change from the clastic shelf (sequence 1) to the carbonate shelf of sequence 2 reflects warming of the global climate from the Late Precambrian glaciation and migration of North America toward the equator during the Late Precambrian to Early Cambrian (Bond and others, 1984).

Climate during much of the Cambro-Ordovician appears to have been semiarid (Mazzullo and Friedman, 1975; Reinhardt and Hardie, 1976; Pfeil and Read, 1980). Consequently, this limited the supply of clastics to the shelf from the low-relief hinterland. Clastics brought in by ephemeral streams and by sheet flooding tended to be trapped at the cratonic shoreline or deposited as widespread thin shales and silts within a cratonic moat landward of the carbonate platform. Clastics also were transported by eolian processes onto the shelf during emergence.

Semiarid climate likely favored development of algal flats, because high salinity of inner-platform waters inhibited browsing and burrowing. High salinities also tended to restrict subtidal skeletal assemblages, favoring widespread deposition of largely non-skeletal sediments (muds, peloids, intraclasts, ooids) over the platform.

Climatic changes appear to have influenced intra-shelf basin deposition. The overall evaporative setting might have promoted oxygenation of the shelf bottom waters by sinking of more dense surface waters (Purser, 1973), which would have reinforced surface-to-bottom mixing effects of waves. There is a possibility, however, that the overall semiarid climate was interrupted by more humid periods when increased runoff provided siliciclastics to the shelf and intra-shelf basins. This influx would have decreased salinity of surface waters, promoting stratification, and decreased oxygen levels in bottom waters, as well as making the water column more turbid, thus decreasing carbonate production by photosynthetic organisms (Bottjer and others, 1986). This might account for the relatively poorly fossiliferous intra-shelf basin shales of the Conasauga Group, interpreted to have formed during highstand conditions. Dalrymple and others, (1985) suggested that paleo-trade winds were important in transporting these fine siliciclastics south and west of the source areas. The quiet-water shales contrast with the interbedded, highly fossiliferous, storm-deposited limestones and silt/fine sandstones believed to have formed during lowstands in relatively clear, well-oxygenated waters (Read and others, 1986).

The cause of the third-order sea-level changes is not clear. Because of their 1- to 10-ma relatively short periods, they have been ascribed to climatically induced glacio-eustatic changes (Vail and others, 1977; Olsen, 1986). Shorter term climate changes typical of the Milankovitch range (20 ka to 100 ka) probably were the cause of the small-scale cyclicity of the shelf sequences (Goodwin and Anderson, 1985; Read and others, 1986; Grotzinger, 1986; Hardie and Shinn, 1986). If these were related to glacio-eustatic effects, then they they must have involved low-ice volumes associated with small-scale continental or alpine glaciations, because of the low amplitudes involved. This would be compatible with the prevailing view of the Cambro-Ordovician as a time of little (or no?) continental glaciation. Fluctuations in sea level also may have been aided by expansion and contraction of the oceanic water column during heating and cooling.

Possible short-term, regular climatic changes with periods of 1 to 3 ka may have been responsible for the even banding of clay-rich and clay-poor layers in deep-ramp facies (e.g., the Maynardville Formation, Late Cambrian) and in many thin-bedded deep-water facies. Regular interbedding of shale and limestone is common in many deep-shelf and ramp settings and has been ascribed purely to effects of storms (Kreisa, 1981). These commonly show periodicities in the range of 1 to 3 ka, however, with alternation of clear-water, storm-influenced sedimentation and quiet, deeper water deposition in a turbid basin. A 2.5-ka climatic signal is evident in Permian evaporites (Anderson, 1986). It might be this climatic signal that is causing the limestone-shale couplets by alternation of arid and humid conditions, which influenced the amount of fine clastics being transported out onto the shelf and slope.

ACKNOWLEDGMENTS

Much of the literature search and thickness compilation was done by R. Brent Bray, an undergraduate student at Virginia Polytechnic Institute, whose help was invaluable. Computer programming was done by S. Sriram and V. Phuah. Unpublished information was provided by W. A. Thomas, J. Tull, K. O. Hasson, C. S. Haase, R. V. Dolfi, C. Mehrtens, L. Sternbach, A. R. Palmer, F. L. Schwab, G. Lash, L. Glover, III, R. Demicco, P. Geiser, R. Hatcher, E. Simpson, W. S. F. Kidd, A. Drake, and L. A. Hardie. The manuscript was read by R. Barnaby, D. Osleger, M. Elrick, B. Ross, and R. Ross, who provided valuable criticism. E. A. Anderson and P. Goodwin suggested that the intra-shelf basin sequences may reflect eustatic sea-level changes. L. A. Hardie and R. K. Goldhammer suggested the use of Fischer plots. I also owe a great debt to my students R. Pfeil, J. Markello, J. Bova, W. Koerschner, W. Mussman, J. P. Grotzinger, I. P. Montanez, R. Barnaby, D. Osleger, and M. Elrick, who were the source of many of the ideas in this paper. Typing was done by Belinda Pauley, and drafting was done by Tony Agliata. Financial support was provided by National Science Foundation Grants from 1980 to 1987.

REFERENCES

AADLAND, R. K., 1984, Petrology of the Upper Knox carbonate sequence, Shelby and Talledega counties, Alabama: Southeast and North-Central Section Meeting, Geological Society of America, Lexington, Kentucky, p. 121.

AITKEN, J. d., 1978, Revised models for depositional grand cycles, Cambrian of the southern Rocky Mountains: Bulletin of Canadian Petroleum Geology, v. 26, p. 515–542.

ANDERSON, R. Y., 1986, The varve microcosm: Propogator of cyclic bedding: Paleoceanography, v. 1, p. 373–382.

BARNABY, R. J., AND READ, J. F., 1987, Evolution of Early to Middle Cambrian carbonate platform, southwest Virginia Appalachians: American Association of Petroleum Geologists, Book of Abstracts, p. 24.

BENEDICT, G. L., III, AND WALKER, K. R., 1978, Paleobathymetric analysis in Paleozoic sequences and its geodynamic significance: American Journal of Science, v. 278, p. 578–607.

BIRD, J. M. AND DEWEY, J. F., 1970, Lithosphere plate-continental margin tectonics and the evolution of the Appalachian orogen: Geological Society of America Bulletin, v. 81, p. 1031–1060.

BOND, G. C., KOMINZ, M. A. AND GROTZINGER, J. P., 1988, Cambro-Ordovician eustasy: evidence from geophysical modeling of subsidence in Cordilleran and Appalachian passive margin, in Kleinspehn, K., and Paola, C., eds., New Perspectives in Basin Analysis: Springer-Verlag, New York, p. 129–161.

———, NICKESON, P. A. AND KOMINZ, M. A., 1984, Breakup of a supercontinent between 625 Ma and 555 Ma: New evidence and implications for continental histories: Earth and Planetary Science Letters, v. 70, p. 325–345.

BOTTJER, D. J., ARTHUR, M. A., DEAN, W. E., HATTIN, D. E., AND SAVRDA, C. E., 1986, Rhythmic bedding produced in Cretaceous pelagic carbonate environments: Sensitive recorders of climatic cycles: Paleoceanography, v. 1, p. 467–481.

BOVA, J. P., AND READ, J. F., 1987, Incipiently drowned facies within a cyclic peritidal ramp sequence, Early Ordovician Chepultepec interval, Virginia Appalachians: Geological Society of America Bulletin, v. 98, p. 714–727.

BRAUN, M., AND FRIEDMAN, G. M., 1969, Carbonate lithofacies and environments of the Tribes Hill Formation (Lower Ordovician) of the Mohawk Valley, New York: Journal of Sedimentary Petrology, v. 39, p. 113–135.

BYRD, W. J., 1973, Petrology of the Cambrian Shady dolomite in North Carolina, northeast Tennessee, and southwest Virginia: Unpublished Ph.D. Dissertation: University of North Carolina, Chapel Hill, 152 p.

CADY, W. M., 1945, Stratigraphy and structure of west-central Vermont: Geological Society of American Bulletin, v. 56, p. 515–587.

CONWAY, S. W., 1977, Depositional environments and diagenesis of a sequence in a borehole: The Gailor Formation (Lower Ordovician) of the Mohawk Valley-Saratoga region, New York: Unpublished M.S. Thesis, Rennselaer Polytechnic Institute Troy, New York, 125 p.

DALRYMPLE, R. W., NARBONNE, G. M., AND SMITH, L., 1985, Eolian action and the distribution of Cambrian shales in North America: Geology, v. 13, p. 607–610.

DEMICCO, R. V., 1983, Wavy and lenticular bedded carbonate ribbon rocks of the Upper Cambrian Conococheague Limestone, Central Appalachians: Journal of Sedimentary Petrology, v. 53, p. 1121–1132.

———, 1985, Patterns of platform and off-platform carbonates of the Upper Cambrian of western Maryland: Sedimentology, v. 32, p. 1–22.

DONALDSON, A. L., 1959: Stratigraphy of Lower Ordovician Stonehenge and Larke Formations in central Pennsylvania: Unpublished Ph.D. Dissertation, Pennsylvania State University, State College, Pennsylvania, 393 p.

DORSEY, R. J., AND STANLEY, R. S., 1983, Bedrock geology of the Milton Quadrangle, northwest Vermont: Vermont Geological Survey Special Publication 3, p. 1–14.

FISCHER, A. G., 1964, The Lofer cyclothems of the Alpine Triassic: Geologic Survey of Kansas Bulletin v. 169, p. 107–149.

FISHER, D. W., 1954, Lower Ordovician (Canadian) stratigraphy of the Mohawk Valley, New York: Geological Society of America Bulletin, v. 65, p. 71–96.

———, 1977, Correlation of the Hadrynian, Cambrian and Ordovician rocks in New York State: The University of the State of New York, New York State Museum Map and Chart Series No. 25, Albany, 75 p.

GINSBURG, R. N., 1971, Landward movement of carbonate mud: A new model for regressive cycles in carbonates (abs.): American Association of Petroleum Geologists Bulletin, v. 55, p. 340.

GOHN, G., 1976, Sedimentology, stratigraphy, and paleogeography of Lower Paleozoic carbonate rocks, Conestoga Valley, southeastern Pennsylvania: Unpublished Ph.D. Dissertation, University of Delaware, Newark, 315 p.

GOLDHAMMER, R. K., 1987, Platform carbonate cycles, Middle Triassic of northern Italy: The interplay of local tectonics and global eustasy: Unpublished Ph.D. Dissertation, The Johns Hopkins University, Baltimore, 468 p.

———, DUNN, P. A., AND HARDIE, L. A., 1986, High frequency glacio-eustatic sea level oscillations with Milankovitch characteristics recorded in Middle Triassic platform carbonates in N. Italy: American Journal of Science, v. 278, p. 853–892.

———, NGUYEN, C., AND HARDIE, L., 1984, Compactional features in Lower Ordovician carbonates of the Central Appalachians and their significance: Appalachian Basin Industrial Associates, Lexington, Kentucky, p. 16–17.

GOODWIN, P. W., AND ANDERSON, E. J., 1985, Punctuated aggradational cycles: A general hypothesis of episodic stratigraphic accumulation: Journal of Geology, v. 93, p. 515–533.

GROTZINGER, J. P., 1986, Cyclicity and paleo-environmental dynamics, Rocknest Platform, northwest Canada: Geological Society of America Bulletin, v. 97, p. 1208–1231.

HARDIE, L. A., and SHINN, E. A., 1986, Carbonate Depositional Environments, Modern and Ancient, Part 3: Tidal Flats, *in* Warme, J. E., and Shanley, K. W., eds., Colorado School of Mines Quarterly, v. 81, p. 1–74.

HARRIS, L. D., 1964, Facies relations of exposed Rome Formation and Conasauga Group of northeastern Tennessee with equivalent rocks in the subsurface of Kentucky and Virginia: U.S. Geological Survey Professional Paper 501-B, p. B25–B29.

──────, 1969, Kingsport Formation and Mascot Dolomite, (Lower Ordovician) of east Tennessee, *in* Papers on the Stratigraphy and Mine Geology of the Kingsport and Mascot Formations (Lower Ordovician) of East Tennessee: Tennessee Department of Conservation, Report of Investigations No. 23, p. 1–44.

HASSON, K. O., AND HAASE, C. S., 1988, Lithofacies and paleogeography of the Conasauga Group (Middle and Late Cambrian) in the Valley and Ridge Province of east Tennessee: Geological Society of America Bulletin, v. 100, p. 234–246.

HAYS, J. D., IMBRIE, J., AND SHACKLETON, N. J., 1976, Variations in the earth's orbit: Pacemaker of the ice ages: v. 194, p. 1121–1132.

HOBSON, J. P., 1963, Stratigraphy of the Beekmantown Group in southeastern Pennsylvania: Pennsylvania Geological Survey Bulletin G37, 331 p.

JACOBI, R. D., 1981, Peripheral bulge–A causal mechanism for the Lower/Middle Ordovician unconformity along the western margin of the Northern Appalachians: Earth and Planetary Science Letters, v. 56, p. 245–251.

JAMES, N. P., AND STEVENS, R. K., 1986, Stratigraphy and correlation of the Cambro-Ordovician Cow Head Group, western Newfoundland: Geological Survey of Canada Bulletin 366, 143 p.

JANSSENS, A., 1973, Stratigraphy of the Cambrian and Lower Ordovician rocks in Ohio: Ohio Geological Survey Bulletin 64, 197 p.

KEITH, B. D., AND FRIEDMAN, G. M., 1977, A slope-fan-basin-plain model, Taconic sequence, New York and Vermont: Journal of Sedimentary Petrology, v. 47, p. 1220–1241.

KENDALL, G. ST. C., AND SCHLAGER, W., 1981, Carbonates and relative changes in sea level: Marine Geology, v. 44, p. 181–212.

KOERSCHNER, W. F., III, 1983, Cyclic peritidal facies of a Cambrian aggraded shelf: The Elbrook and Conococheague Formations, Virginia Appalachians: Unpublished M.S. Thesis, Virginia Polytechnic Institute and State University, Blacksburg, 181 p.

KREISA, R. D., 1981, Storm-generated sedimentary structures in subtidal marine facies with examples from the Middle and Upper Ordovician of southwestern Virginia: Journal of Sedimentary Petrology, v. 51, p. 823–848.

LEES, J. A., 1967, Stratigraphy of the Lower Ordovician Axeman Limestone of Central Pennsylvania: Pennsylvania Geological Survey Bulletin G52, 78 p.

MACK, G. H., 1980, Stratigraphy and depositional environments of the Chilhowee Group (Cambrian) in Georgia and Alabama: American Journal of Science, v. 280, p. 497–517.

MAJOR, R. P., 1976, Petrology and stratigraphy of the Allentown Dolomite (U. Cambrian), northwestern New Jersey: Unpublished M.S. Thesis, University of Connecticut Storrs 148 p.

MARKELLO, J. R., 1979, Carbonate ramp to deeper shale shelf transitions of an Upper Cambrian (Dresbachian) shelf embayment, Nolichucky Formation, southwest Virginia: Unpublished M.S. Thesis, Virginia Polytechnic Institute and State University, Blacksburg, 162 p.

──────, AND READ, J. F., 1982, Upper Cambrian intrashelf basin, Nolichucky Formation, southwest Virginia Appalachians: American Association of Petroleum Geologists Bulletin, v. 66, p. 860–878.

MAZZULLO, S. J., 1978, Early Ordovician tidal flat sedimentation, western margin of proto-Atlantic Ocean: Journal of Sedimentary Petrology, v. 48, p. 49–62.

──────, AND FRIEDMAN, G. M., 1975, Conceptual model of tidally influenced deposition on margins of epeiric seas: Lower Ordovician (Canadian) of eastern New York and southwestern Vermont: American Association of Petroleum Geologists Bulletin, v. 59, p. 2123–2141.

MEISLER, H., AND BELCHER, A. E., 1968, Carbonate rocks of Cambrian and Ordovician age in the Lancaster Quadrangle, Pennsylvania *in* Contributions to Stratigraphy, U.S. Geological Survey Professional Paper 1254-G, p. G1–G14.

MUSSMAN, W. J., AND READ, J. F., 1986, Sedimentology and development of a passive-to-convergent margin unconformity: Middle Ordovician Knox unconformity, Virginia Appalachians: Geological Society of America Bulletin, v. 97, p. 282–295.

MYROW, P. M., 1983, A paleo-environmental analysis of the Cheshire Formation in west central Vermont: Vermont Geological Society Bulletin, v. 9, p. 12–13.

ODER, C. R. L., 1934, Preliminary subdivision of the Knox Dolomite in east Tennessee: Journal of Geology, v. 42, p. 469–497.

OLSEN, P. E., 1986, A 40-million-year lake record of Early Mesozoic orbital climatic forcing: Science, v. 234, p. 842–848.

PALMER, A. R., 1981, On the correlatibility of Grand Cycle tops, *in* Taylor, M. E., ed., Second International Symposium on the Cambrian System: U.S. Geological Survey Open-File Report 81-743, p. 156–157.

──────, AND JAMES, N. P., 1979, The Hawke Bay event: A circum-Iapetus regression near the Lower to Middle Cambrian boundary, *in* Wones, D. R., ed., Proceedings, Caledonides in the U.S.A.: Blacksburg, Virginia Polytechnic Institute and State University Memoir 2, p. 15–18.

PFEIL, R. W., AND READ, J. F., 1980, Cambrian carbonate platform margin facies, Shady Dolomite, southwestern Virginia, U.S.A.: Journal of Sedimentary Petrology, v. 50, p. 91–116.

PURSER, B. W. (ed.), 1973, The Persian Gulf, Holocene Carbonate Sedimentation and Diagenesis in a Shallow Epicontinental Sea: Springer Verlag, New York, 471 p.

RAHMANIAN, V. D., 1981, Mixed siliciclastic-carbonate tidal sedimentation in the Lower Cambrian Monkton Formation of west central Vermont: Northeastern Section, Geological Society of America, Abstracts with Programs, p. 170.

READ, J. F., 1980, Carbonate ramp to basin transitions and foreland basin evolution, Middle Ordovician sequence, Virginia: American Association of Petroleum Geologists Bulletin, v. 64, p. 1575–1612.

──────, GROTZINGER, J. P., BOVA, J. A., AND KOERSCHNER, W. F., 1986, Models for generation of carbonate cycles: Geology, v. 14, p. 107–110.

──────, AND PFEIL, R. W., 1983, Fabrics of allochthonous reefal blocks, Shady Dolomite (Lower to Middle Cambrian), Virginia Appalachians: Journal of Sedimentary Petrology, v. 53, p. 761–778.

REINHARDT, J., 1974, Stratigraphy, sedimentology and Cambro-Ordovician paleogeography of the Frederick Valley, Maryland: Maryland Geological Survey Report of Investigations, No. 23, 74 p.

──────, AND HARDIE, L. A., 1976, Selected examples of carbonate sedimentation, Lower Paleozoic Maryland: Maryland Geological Survey Guidebook No. 5, 53 p.

──────, AND WALL, E., 1975, Tomstown Dolomite (Lower Cambrian), cental Appalachian Mountains, and the habitat of *Salterella conulata*: Geological Society of America Bulletin, v. 86, p. 1377–1380.

RODGERS, J., 1969, The eastern edge of the North American continent during the Cambrian and Early Ordovician, *in* Zen, E-an, White, W., Hadley, J., and Thompson, J. Jr., eds., Studies of Appalachian Geology: Northern and Maritime: Interscience Publishers, New York, p. 141–149.

ROSS, R. J., AND 34 OTHERS, 1982, The Ordovician System in the United States: International Union of Geological Sciences, Publication No. 12, 73 p.

ROWLEY, D. B., and KIDD, W. S. F., 1981, Stratigraphic relationships and detrital composition of the Medial Ordovician flysch of western New England: Implications for the tectonic evolution of the Taconic orogeny: Journal of Geology, v. 89, p. 199–218.

RUBIN, D. M., AND FRIEDMAN, G. M., 1977, Intermittently emergent shelf carbonates: An example from the Cambro-Ordovician of eastern New York State: Sedimentary Geology, v. 19, p. 81–106.

SAMMAN, N. F., 1975, Sedimentation and stratigraphy of the Rome Formation in east Tennessee: Unpublished Ph.D. Dissertation, University of Tennessee, Knoxville, 337 p.

SANDO, W. J., 1957, Beekmantown Group (Lower Ordovician) of Maryland: Geological Society of America Memoir 68, 161 p.

SCHWAB, F. L., 1971, The Chilhowee Group and the Late Precambrian-Early Paleozoic sedimentary framework in the central and southern Appalachians, *in* Lessing, P., Hayhurst, R. I., Barlow, J. A., and Woodfork, L. D., eds., Appalachian Structures, Origin, Evolution, and Possible Potential for New Exploration Frontiers: West Virginia University and West Virginia Geological and Economic Survey, p. 59–86.

SELLECK, B., 1975, Paleoenvironments and petrography of the Potsdam

Sandstone, Theresa Formation and Ogdensburg Dolomite (Upper Cambrian-Lower Ordovician: Unpublished Ph.D. Dissertation, The University of Rochester, New York, 162 p.

SHANMUGAM, G., AND LASH, G., 1982, Analogous tectonic evolution of the Ordovician foredeeps, southern and central Appalachians: Geology, v. 10, p. 562–566.

SIMPSON, E. L., 1987, Sedimentology and tectonic implications of the Late Proterozoic to Early Cambrian Chilhowee Group in southern and central Virginia: Unpublished Ph.D. Dissertation, Virginia Polytechnic Institute and State University, Blacksburg, 298 pp.

SPELMAN, A. R., 1966, Stratigraphy of Lower Ordovician Nittany Dolomite in central Pennsylvania: Pennsylvania Geological Survey, Fourth Series Bulletin G 47, 187 p.

SPEYER, S. E., 1983, Subtidal and intertidal clastic sedimentation in a Lower sequence Monkton Quartzite, northwestern Vermont: Northeastern Geology, v. 5, p. 29–39.

STONE, S. W., AND DENNIS, J. G., 1964, The geology of the Milton quadrangle, Vermont: Bulletin of the Vermont Geological Survey 26, 78 p.

SLOSS, L. 1963, Sequences in the cratonic interior of North America: Geological Society of America Bulletin, v. 75, p. 93–114.

THOMAS, W. A., 1977, Evolution of Appalachian-Ouachita salients and recesses from reentrants and promontories in the continental margin: American Journal of Science, v. 277, p. 1233–1278.

VAIL, P. R., MITCHUM, R. M., JR., AND THOMPSON, S., III, 1977, Seismic stratigraphy and global changes of sea level, Part 4: Global cycles of relative changes of sea level, in Payton, C. E., ed., Seismic Stratigraphy–Applications to Hydrocarbon Exploration: American Association of Petroleum Geologists Memoir 26, p. 83–97.

WAGNER, W. R., 1961, Subsurface Cambro-Ordovician stratigraphy of northwestern Pennsylvania and bordering states: Pennsylvania Geological Survey, Fourth Series Progress Report 156, 22 p.

———, 1976, Growth faults in Cambrian and Lower Ordovician rocks of western Pennsylvania: American Association of Petroleum Geologists Bulletin, v. 60, p. 414–427.

WEBB, E. J., 1980, Cambrian sedimentation and structural evolution of the Rome Trough in Kentucky: Unpublished Ph.D. Dissertation, University of Cincinnati, Cincinnati, 98 p.

WEHR, F., AND GLOVER, L., III, 1985, Stratigraphy and tectonics of the Virginia-North Carolina Blue Ridge: Evolution of a Late Proterozoic-Early Paleozoic hinge zone: Geological Society of America Bulletin, v. 96, p. 285–295.

WHISONANT, R. C., 1987, Paleocurrent and petrographic analysis of imbricate clasts in shallow-marine carbonates, Upper Cambrian, southwest Virginia: Journal of Sedimentary Petrology, v. 57, p. 983–994.

WILSON, J. L., 1952, Upper Cambrian stratigraphy in the central Appalachians: Geological Society of America Bulletin, v. 63, p. 275–322.

WOODWARD, H. P., 1961, Preliminary subsurface study of southeastern Appalchian interior plateau: American Association of Petroleum Geologists Bulletin, v. 45, p. 1634–1655.

ZADNICK, V. E., 1960, Petrography of the Upper Cambrian dolomites of Warren County, New Jersey: Unpublished Ph.D. Dissertation, University of Illinois, Urbana, 155 p.

ZENGER, D. H., 1979, Stratigraphy and petrology of the Little Falls dolostone (Upper Cambrian), east-central New York: The University of the State of New York, New York State Museum Map and Chart Series No. 34, Albany, 138 p.

ARCHITECTURE AND EVOLUTION OF A WHITEROCKIAN (EARLY MIDDLE ORDOVICIAN) CARBONATE PLATFORM, BASIN RANGES OF WESTERN U.S.A.

REUBEN J. ROSS, JR.
Colorado School of Mines, Golden, Colorado 80401

NOEL P. JAMES
Queens University, Kingston, Ontario, K7L 3N6 Canada

LEHI F. HINTZE
Brigham Young University, Provo, Utah 84602

AND

FORREST G. POOLE
U.S. Geological Survey, Denver, Colorado 80225

ABSTRACT: The Ordovician Whiterock Series of western Utah, Nevada, and southern California constitutes a vast carbonate platform that grew during a single offlap-onlap cycle, 12 m.y. in duration, in late Arenig through Llandeilo time, when most of the North American continent was exposed to subaerial weathering. Evolution of the platform took place in four phases during which carbonate sedimentation initially kept pace with, then generally exceeded, the rate of relative subsidence.

The margin of the platform was initially occupied by a carbonate-shoal complex rich in *Nuia* and calathid algal reefs, but at its climax it became composed mostly of *Girvanella*-rich oncolites and receptaculitid algae. Seaward of the margin, slope deposits consist of fine-grained carbonate and terrigenous clastic sediments, largely as parted to ribbon limestones with hardgrounds and slumped horizons, punctuated by limestone conglomerates. Rapid accretion of massive prograding oncolite-sand shoals during the second phase impeded water circulation to the open platform shoreward, or east, of them. The result was lagoonal accumulation of euxinic black shales and intercalated storm deposits as tongues of coquina, oncolite, and fossiliferous rudstone in an intra-shelf basin. Slowing of relative sea-level rise during phase 3 is indicated, especially in the south, by extensive peritidal deposits, which prograded westward precisely over the most substantial parts of the previous oncolite-rimming facies and assumed their function as the shelf margin. Shoreward and northward, the shallow shelf was covered by burrowed carbonate mud.

Quartz sands prograded from the northeast toward the southwest and west, eventually overwhelming most of the carbonate platform. Dolomitization and karstification in a seemingly anomalous southwest-northeast, narrowly elongate area from the Talc City Hills to the Sheep Range appear to have taken place prior to deposition of quartz sands.

The vast scale of this carbonate platform is such that all its elements may not be seen at a single location, in contrast, for example, to the Permian Basin sediments of the Guadalupe Mountains of New Mexico. Rather, the three-dimensional geometry of the Whiterockian platform is evident only through careful measurement of stratigraphic sections and interpretation of depositional facies patterns, which are then tied together through detailed biostratigraphic correlation. The unique nature of the sediments that compose this platform is a function of its equatorial setting on a leeward continental margin and the lack of any large skeletal metazoans during this interval of Ordovician time.

INTRODUCTION

The early part of middle Ordovician time, in what is now North America, was a period of continental exposure and subaerial weathering, reflected in the geologic record by a major unconformity (Ross and others, 1982; Mussman and Read, 1986) or widespread orthoquartzite sheets with few fossils. Margins of the early Paleozoic continent, however, remained submerged through much of this time. Rocks of this age, referred to as the Whiterockian Series (Ross and others, 1982), contain a distinctive fossil assemblage that defines the Toquima-Table Head Faunal Realm (Ross and Ingham, 1970) that is recognized as a marginal carbonate belt on several continents. Carbonates of this faunal realm form a yoke around modern North America and are preserved in the Appalachians, the Arctic, the Canadian Rockies, and the Great Basin. In the Basin Ranges much of the original carbonate shelf is still preserved intact.

The purpose of this paper is to present an integrated synthesis of the sedimentology and stratigraphy of Whiterockian rocks in the Basin Ranges, based upon detailed paleontologic and stratigraphic study for over 30 years. The time interval represented is 12 m.y. The Whiterockian strata are conveniently divided into four time slices on the basis of fossils. The picture that emerges is one of a vast carbonate platform that accreted in response to gradual platform margin subsidence in a humid equatorial setting adjacent to a low-lying landmass when few reef-building organisms had evolved.

PREVIOUS STUDIES

Middle Ordovician stratigraphy has been vital to tectonic interpretations of the Basin Ranges for 45 years. The juxtaposition of correlative but contrasting Ordovician facies in the Roberts Creek Mountains (Fig. 1, loc. 12) led Merriam and Anderson (1942) to recognize the Roberts Mountains Thrust. The history and general facies relation of Ordovician carbonates in the Basin Ranges, south of the Tooele Arch of Webb (1958) and Hintze (1973a), have been recorded by Hintze (1973b) and by Ross (1976, 1977). Lowell's (1960) regional east-west interpretation of stratigraphy added the interrelation of the basinal western graptolite shales with the platform carbonates to Webb's (1958) report on Utah and eastern Nevada. Studies by Hintze (1951, 1952, 1973a, 1973b), by Ross (1951, 1964, 1967, 1970), by Ross and Shaw (1972), and by Harris and others (1979) provided paleontologic control.

In southeastern California, Stevens (1986, pp. 11–15) noted shallow-water deposits in much of the Whiterockian

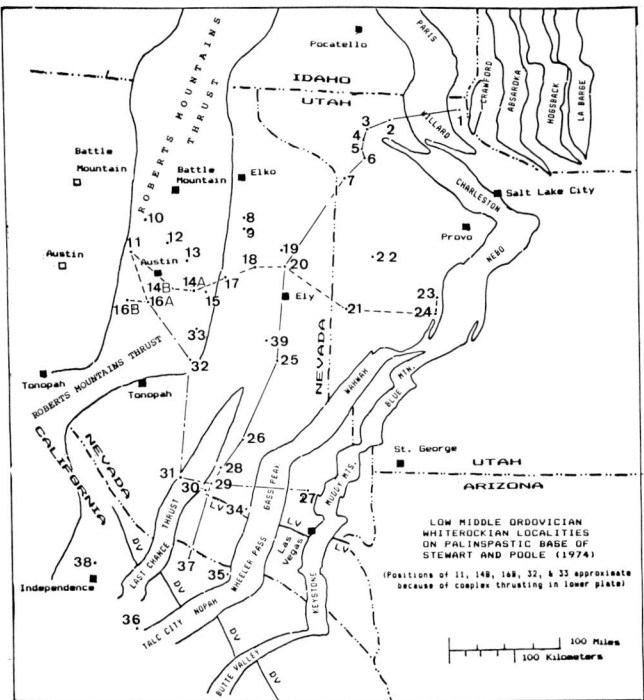

FIG. 1.—Index map showing relative positions of localities in Figures 8–11d Plotted on palinspastic base of Stewart and Poole (1974). Positions of localities 11, 14B, 16B, 32, and 33 are approximate because of complex faulting in lower plate. Key to localities:

UTAH

1. Crawford Mountains; 2. Logan area; 3. Wellsville Mountains; 4. Promontory Point; 5. Lakeside Mountains; 6. Newfoundland Mountains; 7. Silver Island Mountains; 21. Confusion Range, Fossil Mountain in Ibex area; 23. Northern Pavant Range; 24. Fillmore-Richfield area.

NEVADA

8. Pearl Peak, Ruby Range; 9. Sherman Mountain, Ruby Range; 10. Cortez face, Mountain Tenabo; 11. Callaghan Ranch; 12. Western Peak, Roberts Creek Mountains; 13. Lone Mountain; 14A. Martin Ridge, Monitor Range; 14B. Whiterock Canyon, Monitor Range; 15A. North of Ninemile Canyon; 15B. Hill 8308, South of Water Canyon; 16A. Hill 8474, lower Ikes Canyon; 16B. Northwest and Southwest of Hill 8937, upper Ikes Canyon; 17. vicinity of Eureka; 18. Pogonip Ridge; 19. Southern Cherry Creek Range; 20. Steptoe Ranch, northern Egan Range; 25. southern Egan Range, Sunnyside; 26. Pahranagat Range, west side; 27. Arrow Canyon Range, northwest; 28. Aysees Peak; 29. Ranger Mountains, Northwest side; 30. Spotted Range, Southwest end; 31. Meiklejohn Peak, Bare Mountain; 32. Rawhide Mountain, south end; 33. Clear Creek Canyon, Monitor Range; 34. Black Gate Canyon, Sheep Range; 39. Horse Range, west of Currant.

CALIFORNIA

35. Northern Nopah Range; 36. Talc City district; 37. Pyramid Peak; 38. Mazourka Canyon.

Badger Flat Formation, from the Talc City Hills in the south (Fig. 1, loc. 36) to central Mazourka Canyon in the northwest (Fig. 1, loc. 38). He concluded that the shelf slope break lay outboard (northwest) of Mazourka Canyon in Whiterockian time.

METHODS

All maps are plotted on the palinspastic base of Stewart and Poole (1974). Cross sections are prepared using four paleontologically established time zones. Two of the sections (Figs. 9, 10) trend east-west and one (Fig. 8) trends north-south. The sedimentary record is divided into four phases. In a sizeable data base of measured sections, we recognize 13 lithofacies. The geographic distribution of lithofacies for each of the depositional phases is plotted on four paleogeographic maps (Figs. 11a–d) based approximately on the mid-level of each phase.

STRATIGRAPHIC FRAMEWORK

Two generalized diagrams (Fig. 2a, b) outline the stratigraphic and lithologic framework of the Whiterockian shelf. These lower middle Ordovician rocks, covering a vast area, have a varied stratigraphic nomenclature, which reflects the natural lithofacies associations. Figure 2a is a simplified northeast-southwest cross section illustrating the disposition of the formational stratigraphic units. Figure 2b is essentially the same diagram showing various lithofacies and phases of their development.

Time slices erected on paleontologic evidence show that deposition of lithofacies was markedly diachronous. Boundaries of the phases of depositional progradation and transgression into which we have divided the sedimentary record do not coincide precisely with the boundaries of fossil zones; phase 2, for example, terminates within the upper *Anomalorthis* Zone (Zone N of Hintze, 1952).

LITHOFACIES

The overall composition of sediments on this shelf are like other lower Paleozoic epicontinental platform deposits,

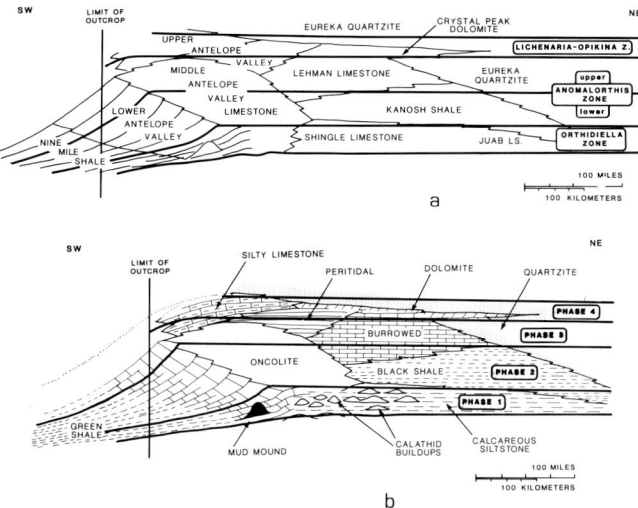

FIG. 2.—(a) Summary diagram of formal stratigraphic units relative to fossil zones, demonstrating marked diachroneity of most units. Compare with (b). Compiled from Figures 8–10. (b) Idealized summary diagram of lithofacies and phases of development of the Whiterockian carbonate platform in the Basin Ranges. Compiled from Figures 8–10. The diagram represents a theoretical northeast-southwest-trending area between Figures 8 and 9.

ranging from sandstone, siltstone, and shale to limestone and dolostone, with most being mixtures of terrigenous clastic and carbonate end members. There is a clear inner-shelf to outer-shelf, terrigenous clastic to carbonate polarity in sediment disposition. It recalls the underlying Cambrian inner-detrital to outer-carbonate belts of Palmer (1960), which mimic the distribution patterns of sediments on some modern tropical continental shelves, such as Belize and the Great Barrier Reef (Ginsburg and James, 1974). The following discussion of facies reflects this transition from terrigenous clastic to carbonate sediments.

Platform Facies

(1) Sandstone facies (Eureka, Swan Peak, and Watson Ranch Quartzites).—

*Description.—*These sediments are thin- to thick-bedded, cross-laminated quartz arenites. The well-rounded and well-sorted quartz grains are medium-sand size. Sorting improves and grain size decreases to the southwest. Collophane nodules occur in beds 10–30 m above the base in southeastern localities. Details of this lithofacies are found in Webb (1956) and Ketner (1968).

*Environment.—*These are possibly aeolian sands blown off the craton and reworked in a marine environment, in a manner similar to that of the Persian Gulf today (Shinn, 1973). Webb (1956, 1958) recognized that the combined quartzite units prograded westward, intertonguing with marine carbonates (loc. 20, 21). In addition, the basal sands at the Ranger Mountains (loc. 29) are interbedded with sandy limestone beds containing trilobites, ostracodes, and conodonts (Ross, 1964, p. C18; Harris and others, 1979, p. 18–21, fig. 9) of marine origin. Gilluly and Masursky (1965, p. 18, 20) reported corals from sandy dolomite 102–112 ft (31–34 m) below the top of the Eureka Quartzite at Cortez (loc. 10), indicating at least temporary deposition in marine waters.

(2) Siltstone facies (Copenhagen and Barrel Spring Formations).—

*Description.—*Sandy to slightly calcareous siltstones, the facies occurs in beds 5 cm to 1 m thick that weather a yellowish gray. They are burrowed and locally highly fossiliferous, containing abundant brachiopods and *Nuia* but with less common bryozoans.

*Environment.—*The facies formed in a subtidal, nearshore setting dominated by fine-grained terrigenous clastic material derived from reworking of weathered cratonic regolith (Dalrymple and others, 1985).

(3) Calcareous Siltstone facies (Shingle [Kellogg, 1963] and Juab Limestones).—

*Description.—*The siltstone is weathered brown to orange and is thin to medium bedded, locally nodular, thoroughly burrowed, and clay rich. Skeletal grainstone, silty shale, and flat-pebble conglomerate punctuate the sequence. Numerous fossils include silicified trilobites, rare brachiopods, sponges(?), and *Nuia* (usually near or stratigraphically above calathid mounds). Graptolites are locally present in shales.

*Environment.—*The burrowed, terrigenous, fossiliferous, fine-grained sediments suggest a nearshore setting like that of facies 2 (siltstones), whereas the graptolitic shales imply quiet water. Imbricated clasts and graded conglomerates, however, point to infrequent deposition by storms (Aigner, 1985). Skeletal grainstones probably reflect winnowing by similar high-energy events.

(4) Black shale facies (Kanosh Shale; Fig. 4d).—

*Description.—*The Kanosh Shale is a black, fissile, graptolite-bearing shale, which grades locally to dark green shale. Sediments occur in beds as thick as 10 m, separated by thinner limestone. Lime mudstone increases in abundance upward (Hintze, 1973b). Lenses of oncolite rudstone are common locally. Stratigraphically higher parts also contain lenses and discontinuous limestone beds of packstones, rudstone, and grainstone, composed of numerous brachiopods, large ostracodes, trilobites, and *Receptaculites,* which varies from conical to discoidal, together with local concentrations of oncoids.

*Environment.—*The organic nature of the sediments, lack of burrowing, and preservation of graptolites indicate deposition in a quiet-water, dysaerobic to anaerobic environment (Byers, 1977). Upward shallowing and more oxygenation are suggested by the increased calcareous content of the higher interbeds. Sheets of oncolites were probably washed off adjacent, contemporaneous oncolite shoals. The well-sorted bioclasts also indicate winnowing and lateral transport, possibly by storms as in facies 2 and 3. Lateral transitions into oncolite-shoal and sandstone facies, deposited under aerobic conditions, suggest deposition in a stratified water body, characterized by an anoxic deep layer and an aerobic upper layer.

(5) Silty limestone facies (Upper Antelope Valley Limestone).—

*Description.—*These sediments are similar to the calcareous siltstones (facies 3) but are separated regionally and stratigraphically from them and lack the high-clay content. They have been confirmed only in the Toquima Range, particularly at Ikes Canyon (loc. 16A, Figs. 9, 10). Sediments are highly fossiliferous, thin bedded, and bioturbated; portions weather yellowish orange or medium gray. Fossils are predominantly brachiopods.

*Environment.—*A subtidal, low-energy, normal-marine environment is postulated for these beds, which may represent nearshore or upper slope to platform margin deposits in the vicinity of an island of the Tooele Arch (Fig. 11d).

(6) Burrowed limestone facies (Lehman Limestone, Fig. 3 a,c,d).—

*Description.—*Rubbly, micritic to very fine grained, this richly fossiliferous, light gray limestone is in beds 10–25 cm thick. It is essentially a fossiliferous wackestone that has been intensely bioturbated. In the southern Egan Range (loc. 25), burrowing is so intense that bedding is virtually obliterated (Fig. 3c). In other areas there are burrows within burrows. Fossils include brachiopods, trilobites, pelmatozoans, both large and small ostracodes, and conodonts. The primitive colonial coral *Lichenaria* (Fig. 3b) was common

FIG. 3.—(a) Burrowed, micritic limestone of the Lehman Formation, representing phase 3 at Fossil Mountain, Ibex area (loc. 21). Compare fabric with correlative carbonate rocks in (c) and possibly correlative rocks in Figure 5a. (b) Primitive colonial coral *Lichenaria* in Lehman Formation, west of Steptoe Ranch (loc. 20). Phase 3-4 transition. (c) Almost completely burrowed Lehman Formation, which forms cliff in (d). This is highly fossiliferous rock composed of amalgamated micrite and bioclastic wackestone. Phase 3-4 transition, at Sunnyside (loc. 25). (d) View southward along resistant cliff of burrowed Lehman Formation, which is separated by thin, thinly bedded, fossiliferous, burrowed dolomitized limestone from overlying Eureka Quartzite. Sunnyside (loc. 25).

in the Ibex, Pogonip Ridge, and northern Ranger Mountains areas during the later stages of deposition.

Dolomitization.—With a few exceptions, these sediments are dolomitized when directly beneath or interbedded with the lowest sandstone of the Eureka Quartzite. The dolostone is finely crystalline, retains the characteristic burrowing, is thin to thick bedded, and weathers to light brown to light gray. Dolomitization may be only indirectly related to the original environment of deposition.

Environment.—The paucity of terrigenous clastic material and fossiliferous nature of this sediment suggest subtidal deposition distal from the strandline, on a wide, relatively shallow shelf (Wilson, 1975; Flugel, 1982; Wilson and Jordan, 1983). The intense burrowing (*Thalassinoides?*) is like that in Upper Ordovician dolomites in parts of the Basin Ranges (Sheehan and Schiefelbein, 1984).

Shelf Edge or Platform Margin Facies

(7) Oncolite facies (Middle Antelope Valley Limestone; Fig. 4a,b).—

Description.—These rudstones are composed of oncoids, some of which exceed 2 cm in diameter, whose cortex is composed of intertwined *Girvanella* filaments. Sediments are sorted by size, and the matrix between oncoids is bioclastic packstone to grainstone. Deposits vary from lenses (loc. 21, 18) to beds meters thick at the base (loc. 26), to massive beds tens of meters in thickness in the middle and upper parts (loc. 27). The massive beds form impressive cliffs but exhibit little obvious crossbedding. Upper units also contain (1) ooid grainstone locally in minor amounts, peritidal teepees, and dessication cracks (loc. 11, 14A), (2) the filter-feeding gastropods *Palliseria* (Fig. 4a) and *Maclurites,* and (3) *Receptaculites* in the form of large thickened discs.

Environment.—These coarse carbonate sands and gravels, with evidence of periodic exposure near the top, are interpreted as shoals. Their position between shelf and slope facies indicates a shelf edge position.

(8) Calathid buildup facies (Lower Antelope Valley and Shingle [Kellogg, 1963] Limestones; Fig. 4c).—

Description.—Buildups occur as mounds as much as 6 m in diameter and 2 m high. Complexes were formed by stacked and overlapping mounds. The structures are composed of calathid algae, receptaculitid algae, and possible sponges, making up 35% of the whole, whereas 65% is lime mud. Calcarenites surrounding the mounds are rich in *Nuia.*

Environment.—The structures are (1) a discontinuous outer-shelf belt of scattered to stacked buildups, or (2) isolated buildups on the shelf.

(9) Peritidal carbonate facies (Fig. 5a, b).—

Description.—These deposits are typically medium-bedded limestone and minor dolostone, which display dark to light, gray to pink to yellowish grey or pale orange banding in outcrop and contain numerous sedimentary structures indicating peritidal accumulation. Sedimentation is in the form of repeated partial to complete, meter-scale shallowing-upward sequences.

The most common subtidal carbonates are burrowed to mottled gray wackestones to packstones, locally very fossiliferous and variably dolomitic (facies 6–burrowed limestone). At some localities (Loc. 32), these sediments contain abundant oncolites, similar to the oncolite facies (facies 7) beneath them. Intertidal limestones vary from gray to pink to orange wackestone-mudstones, are locally bioturbated, and include the ichnofossils *Diplocraterion* and *Chondrites.* They locally contain stromatolites, fenestrae, fine parallel laminations, and dessication cracks with small teepees. Supratidal carbonates are generally rusty-weathered, laminated, microcrystalline dolomites.

Environment.—Although details of the possibly cyclic repetitions remain to be worked out, subtidal, intertidal, and supratidal lithofacies are easily differentiated. These meter-scale sequences are typical of low-energy, muddy peritidal deposition in a relatively humid climate (Shinn, 1983; James, 1984; Hardie and Shinn, 1986).

Dolomitization.—The sediments are locally dolomitized when directly beneath or interbedded with the Eureka Quartzite, similar to facies 6 (burrowed limestone).

Slope to Basin Sediments

(10) Mottled limestone facies (Lower Antelope Valley Limestone; Fig. 6d).—

Description.—This grey limestone is in beds 5–20 cm thick but rendered indistinct by pervasive burrowing, which has produced coarse mottling weathered orange. Burrows are 2–6 cm in diameter and resemble those attributed to *Thalassinoides* (Sheehan and Schiefelbein, 1984). Beds are composed of wackestone and sporadic packstone rich in pelmatozoan plates; complete articulated brachiopods are locally abundant. The brachiopods occur both in limestone and in silty orange dolomite that fills the burrows. In the burrowed, silty, upper limestone beds at Ikes Canyon (Mill Canyon thrust sequence; loc. 16A, Figs. 9, 10) there is a local concentration of sponges together with abundant trilobites and brachiopods; these are the famous sponge beds described by Bassler (1941).

Lenses of very fine, poorly rounded quartz sand, weathered orange, are present locally. From its typical development in the Ranger Mountains (loc. 29), this facies grades laterally into black shale with brachiopod-grain flows (Fig. 6c) at Little Rawhide Mountain (loc. 32).

Environment.—These sediments are similar in style to those of facies 6, the burrowed onshelf limestones, and are likewise interpreted as subtidal carbonates deposited on a well-oxygenated sea floor. Their position between shallow oncolite shoals (facies 7) and deep slope deposits (facies 11), however, suggests that they were deposited outboard of the shelf edge, on the upper part of the slope.

(11) Parted and ribbon-limestone facies (Lower Antelope Valley Limestone; Fig. 6a).—

Description.—These medium gray to dark gray, alternating limestones and shales occur in beds 2–10 cm thick.

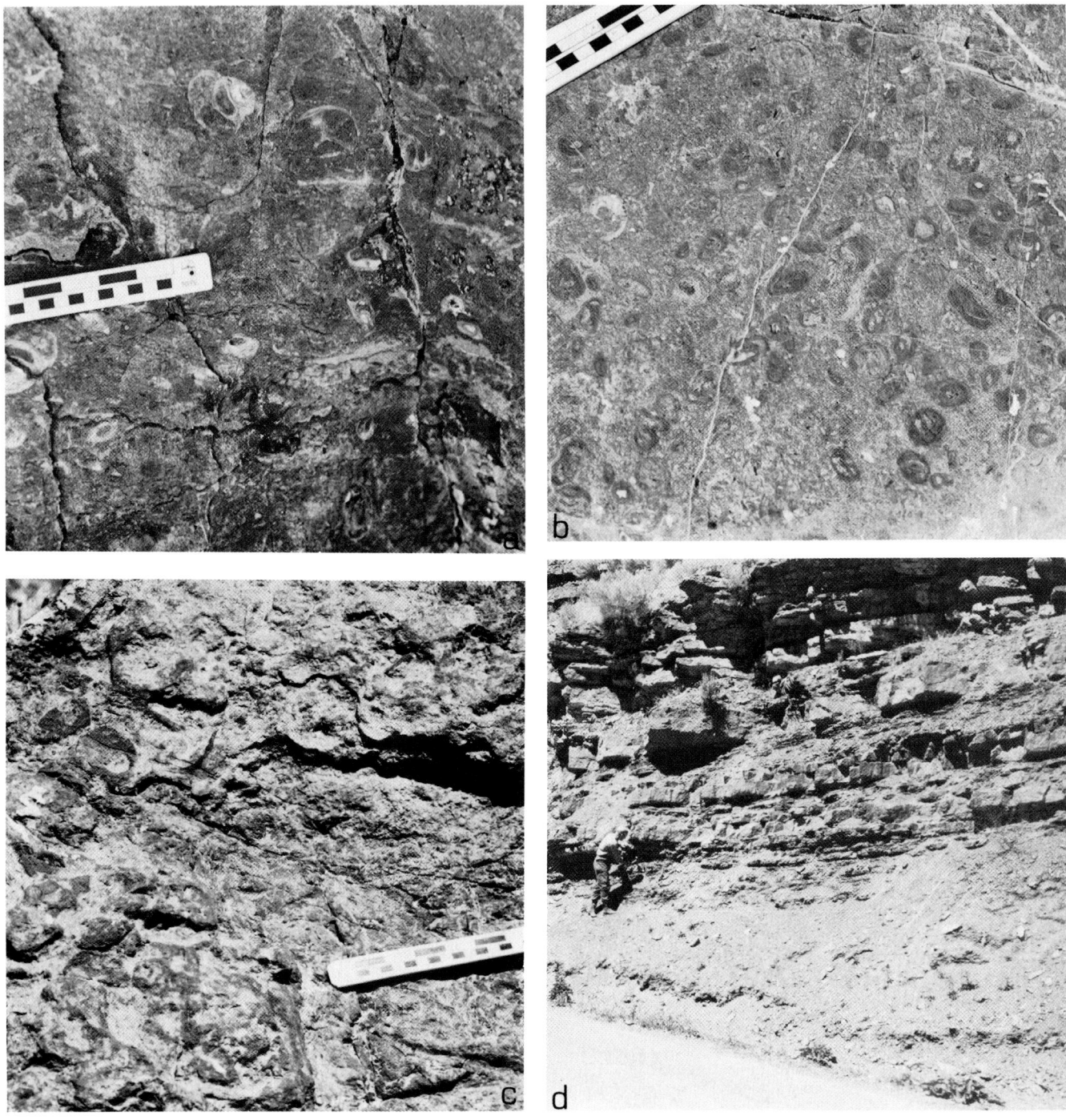

FIG. 4.—(a) Gastropod *Palliseria*, a filter feeder, in oncoid-rich middle member of Antelope Valley Limestone, northern Ranger Mountains (loc. 29). Represents platform margin shoal-water environment of phase 2. Scale in inches and centimeters. (b) Large oncoids in massive oncolitic packstone and bioclastic grainstone of middle member of Antelope Valley Limestone. Represents widespread shoal-water platform margin of phase 2. Northern Ranger Mountains (loc. 29). Scale in inches and centimeters. (c) Part of a calathid-sponge mound near base of *Orthidiella* zone, phase 1, at Sunnyside (loc. 25, Fig. 1). Similar coeval mounds are at Steptoe (loc. 20) and Pogonip Ridge (loc. 18). Scale in inches and centimeters. (d) Shale of the Swan Peak Formation overlain by quartzite, Logan Canyon (loc. 1), in 1947. Shale is lateral-facies equivalent of Kanosh Shale and quartzite corresponds to Watson Ranch and younger Eureka Quartzite of central western Utah and Nevada.

Fig. 5.—(a) Thin laminae of peritidal limestone capping burrowed (*Diplocraterion*) dolomitic micrite. These two lithologies form the core of seven repeated sequences (b) in which the base is dark-weathered burrowed wackestone and the top is white-weathered fenestral micrite. Phase 3. Arrow Canyon Range, west side, toward north end (loc. 27). (b) Peritidal beds in repeated sequences overlain gradationally by Eureka Quartzite. Only the upper four of seven sequences are visible; section repeated by faulting. Looking northwestward in Arrow Canyon Range (loc. 27). (c) Cross-bedded and channeled coarse grainstone. Outcrop is below increasingly quartzose beds at base of Eureka Quartzite at Meiklejohn Peak (loc. 31). Indicative of temporary reversal of deepening water in phase 4. (d) Karst deposit of quartzose sand in middle oncolitic member of Antelope Valley Limestone in northern Nopah Range (loc. 35). This lithology also occurs in the Sheep Range (loc. 34) and the Talc City Hills (loc. 36). Sand was derived from overlying Eureka Quartzite and suggests an Ordovician high, which predicted future position of Talc City–Gass Peak thrust. See Figures 1 and 11c.

Fig. 6.—(a) Ribbon limestone, phase 1-2 transition at this locality. Slope environment of deposition. Basal Antelope Valley Limestone, Ikes Canyon (June Canyon thrust sequence) (loc. 16B). Jacob's staff for scale, graduated in inches and feet. (b) Mud mound at Meiklejohn Peak (loc. 31). Massive mound is overlapped by flanking mottled grainstones and wackestone beds of the *Orthidiella* Zone, which are overlain by thick beds of oncolitic limestone of phase 2. Represents upper-slope deposition (see Fig. 9). View from the north. (c) Interbedded shale and nodular grainstone of phase 1, at Little Rawhide Mountain (loc. 32). Grainflow tongues are rich in *Orthidiella*. Scale in inches and centimeters. (d) Mottled limestone, consisting of thoroughly burrowed grainstone and wackestone, represents subtidal, upper slope of phase 1 and contains abundant silicified fossils of *Orthidiella* Zone at this locality. Northern Ranger Mountains (loc. 29). Scale in inches and centimeters.

They vary from ribbon to parted limestone, as defined by James and Stevens (1986, p. 27, figs. 17, 18). Texture varies from lime mudstone to local lenses of grainstone. Bioclasts are composed of abraded fragments of *Nuia* and brachiopods. Synsedimentary slumps and folds, intraformational truncation surfaces, and phosphatic rinds on hardgrounds are present throughout the facies; extensive hardgrounds covered with orthoconic cephalopods are present at two localities (14B, 32). Other fossils include trilobites, graptolites, and conodonts.

Environment.—These sediments are carbonate slope deposits (Cook and Mullins, 1983; McIlreath and James, 1984). Lack of burrowing and bioturbation suggests deposition under anaerobic conditions.

(12) Mudmound facies (Lower Antelope Valley Limestone; Figs. 6b, 7a).—

Description.—These buildups are mostly micrite, with lenses of micritic wackestone including *Sphaerocodium*, *Nuia*, abundant small ostracods, trilobites, brachiopods, and pelmatozoan plates (see Ross, 1972; Ross and others, 1975, p. 41, table 2). The mounds are rooted above encrinite beds at the base of the mottled limestone (upper slope–facies 10) and at some localities extend upward through the entire thickness of the facies.

Environment.—These mounds are similar to other muddy buildups that develop on the slopes of carbonate platforms (Wilson, 1975; James, 1983, 1984).

(13) Green shale facies (Ninemile Shale).—

Description.—This greenish, highly fossiliferous, fissile shale is interbedded with silty limestone. Limestone occurs as nodules and discontinuous beds 2–15 cm thick. Composition ranges from 25–50% carbonate to 75–50% clay shale. There are few obvious sedimentary structures and no evidence of sedimentary displacement. Locally (loc. 15a) abundant grainstone is composed mainly of branched, unabraded *Nuia*, interpreted by Ross and others (1988) as indicative of shoal water of moderate energy. Other fossils include trilobites (some enrolled), graptolites, and brachiopods.

Environment.—The lack of definitive environmental indicators prohibits assignment of this facies to any setting other than quiet water. Its position outboard of the slope sediments (facies 12) suggests, however, that a high slope to basin setting is most probable. Increase in percentage of fine siliciclastic sediment may reflect a decrease in production of biogenic carbonate in an offshelf setting.

GEOLOGIC HISTORY

In the following discussion, we base our sedimentologic interpretations on summaries by Wilson (1975), Kendall and Schlager (1981), and James and Mountjoy (1983). The term relative sea level refers to the sum of eustatic sea-level changes, sedimentation, and crustal movements. We interpret relative sea-level rise to be the cumulative effect of tectonic subsidence and eustatic rise (cf. Kendall and Schlager, 1981).

Antecendent Platform of Late Ibexian Age

Lithologies of the antecedent platform are characterized by widespread uniformity and are mostly terrigenous clastic sediments. The facies pattern is simple. A western green shale facies and an eastern calcareous facies of discontinuous calathid-sponge buildups.

Runzel marks in microlaminated siltstone at Ibex (loc. 19) indicate a strandline in a clay- and silt-rich environment, whereas essentially correlative fenestral micritic limestone in the Arrow Canyon Range (loc. 27) indicates a strand in a predominantly carbonate environment. Lack of similar indicators to the west suggests widespread subtidal conditions and absence of abrupt transition from shallow to deep water.

We interpret these depositional settings to have been on a ramp (Ahr, 1973; Read, 1985) that sloped gently seaward. The calcareous siltstone facies formed the proximal part of the ramp and was periodically exposed along a fluctuating strandline. Scattered calathid buildups flourished offshore. Distal parts of the ramp were the site of green shale deposition.

A sea-level rise or sudden shelf subsidence prompted low-energy deposition. An increase in water depth may have resulted in a maximum eastward incursion of the green shale facies onto the shelf.

Whiterockian Platform

Phase 1.—

Phase 1 marks the beginning of differentiation into discrete deep-water, platform margin, and interior platform facies, a pattern which typifies the platform throughout most of its subsequent middle Ordovician history. Phase 1 is characterized by mostly vertical accretion.

Geographic Distribution of Lithofacies (Fig. 11a).—The slope facies define a laterally continuous facies belt and are composed of, from east to west, from shallow to deep, upper-slope mottled limestone, middle-slope parted to ribbon limestone, and lower-slope green shale. Mud mounds nucleated on the slope in the lower part of the burrowed facies and accreted upward.

During its initial stages, the oncolite facies appears to have been a series of isolated elongate shoals in a belt 30–40 km wide and possibly as much as 100 km in length. Throughout most of the area, the oncolite shoals separate platform from slope facies, but where shoals are absent, platform sediments grade directly into slope sediments.

Buildups of calathid algae are localized along a north-south trend with a disposition similar to that of the oncolite shoals. The platform itself, in places more than 200 km in width, was floored by calcareous siltstone containing an open-marine fauna. In the eastern, landward region, black shale prevailed.

Stratigraphy (Figs. 8, 9, 10).—Three cross sections illustrate the changing patterns of sedimentation during this and later phases. Initial vertical accretion (Figs. 8–10) typical of phase 1 is illustrated best by the oncolite facies, calathid buildups, and the calcareous siltstone facies. The mottled and parted to ribbon limestones of the slope facies,

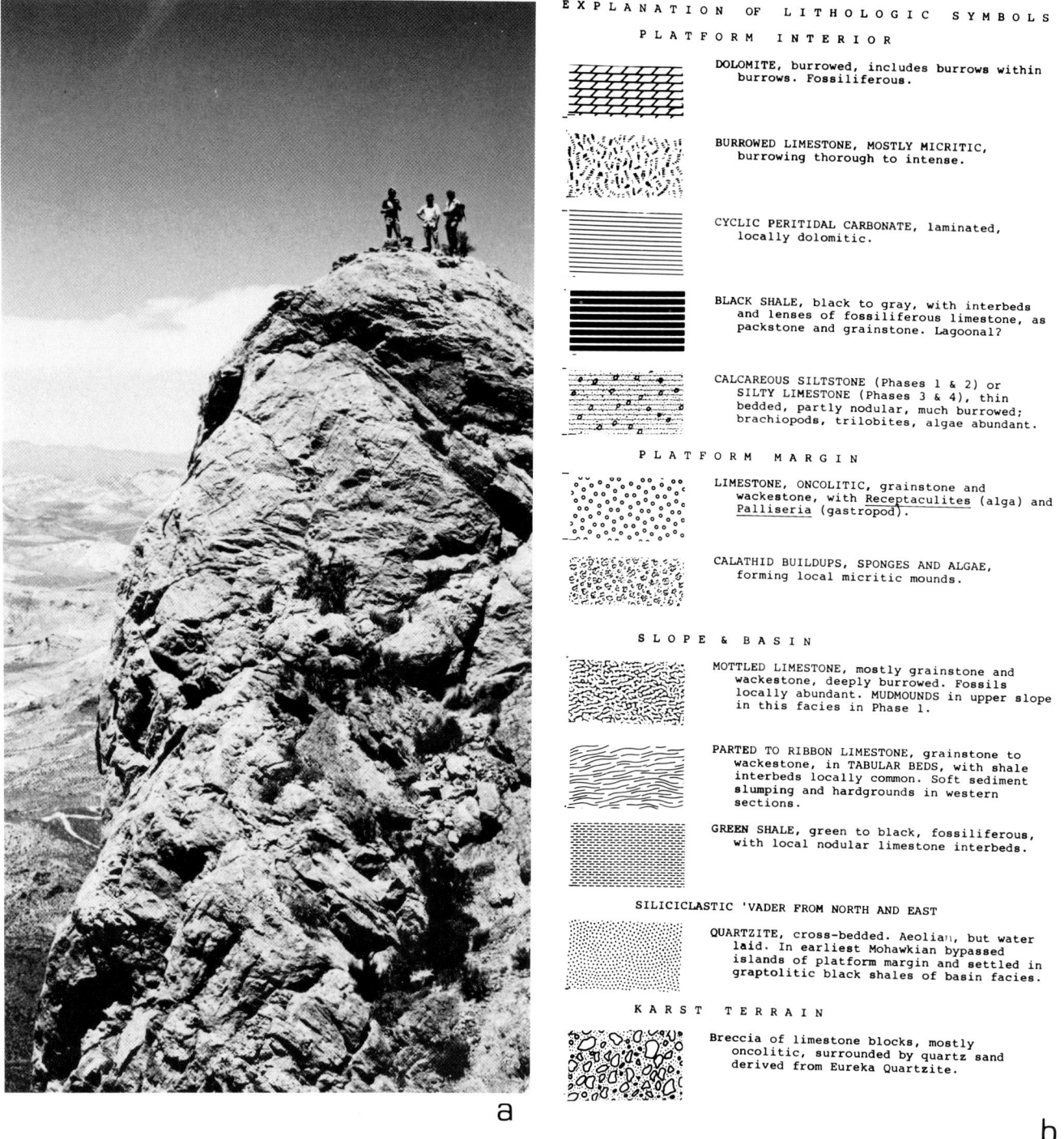

Fig. 7.—(a) Top surface and west face of micritic mud mound at Meiklejohn Peak (loc. 31). Thin beds to right of geologists contain either fenestrae or similar cryptic organic traces. Represents upper-slope environment of phase 1. (b) Explanation of the 11 principal lithologies depicted in Figures 8–11d.

however, illustrate rapid westward progradation, although accretion of mud mounds rooted thereon was also vertical. In addition, the black shale on the shelf also began progradation (Figs. 8, 10) toward the west and southwest at this time.

Interpretation.—Initiation of oncolite shoals, supplemented by calathid mounds, formed an elevation at the shelf edge, which affected water circulation on the platform to the east. Rapid accretion of the oncolite facies during this phase outpaced deep-water sedimentation and created increased relief, which in turn resulted in an outboard slope facies. This increase in relief also resulted in restriction of the platform facies.

Phase 2.—

Geographical Distribution of Lithofacies (Fig. 11b).— Phase 2 of platform development is characterized by expansion of the oncolite shoals into an areally extensive oncolite-bank complex which, during its maximum growth, was about 75 km wide in the north and 200 km wide in the south (Fig. 11b). Deep-water sediments are similar to those of phase 1, but platform interior sediments are characterized by black shales or burrowed limestone. Quartzose sands encroached on both of these facies, especially in the north. Sediments in oncolite shoals at the close of this phase provide clear evidence of periodic exposure.

*Stratigraphy.—*Phase 2 is distinguished by rapid, westward, basinward progradation of all major facies belts, as illustrated in Figures 8–10. The inner-platform facies, black shale, and burrowed limestone, however, have a complex relation. During early stages of deposition, corresponding to the time of most rapid margin progradation, the platform was covered by black euxinic muds. We interpret this setting to be similar to that of the intra-shelf basins postulated by Markello and Read (1981) in the Cambrian of the Appalachians and recognized in the Jurassic on the Arabian Platform (Murris, 1980; Alsharhan and Kendall, 1986). These shales become progressively more carbonate rich upward, reflecting more open circulation and probably shallowing. Later stages of deposition record the increased expansion of burrowed lime mud eastward until, at the end of this phase, the platform is covered with carbonates. This change to platform carbonate deposition coincides with a slowing of westward platform margin progradation.

*Interpretation.—*This phase records the interplay between rising sea level and carbonate sedimentation. At first, carbonate sedimentation at the margin just matched or exceeded the rate of sea-level rise, but onshelf sedimentation lagged behind, resulting in development of a deep lagoon with impeded water circulation. Subsequently, as illustrated by pronounced westward progradation of the margin, the relative rise in sea level slowed, or carbonate sedimentation in the lagoon began to catch up with it. The shelf became progressively shallower. Finally, the margin shoaled to sea level, probably as sea-level rise slowed even more, and the shelf floor accreted into very shallow water, as reflected by the uniform widespread burrowed limestones (facies 6).

Phase 3 (Fig. 11c).—

Phase 3 is characterized by carbonate deposition in all environments except that of the encroaching Eureka Quartzite (Fig. 11c). The distinctive platform margin oncolite facies are restricted to the northern and southern parts of the area, while the platform is covered by subtidal burrowed limestone in the north and a wide peritidal carbonate complex in the south. Where oncolite shoals are absent, upper-slope mottled limestone passes directly into shallow-water burrowed limestone or peritidal carbonate. The whole inner part of the shelf is mantled by quartz sand.

*Stratigraphy.—*This period records an abrupt change both in facies patterns and style of accretion. The platform margin facies either narrow dramatically or disappear upward and are abruptly overlain by peritidal or burrowed subtidal sediments, or locally (loc. 16A) by silty limestone limestone of facies 5. Quartz sands continued to prograde westward over the platform interior.

*Interpretation.—*These facies patterns imply that carbonate sedimentation was, in general, keeping pace with or exceeding relative sea-level rise. It should be emphasized that the peritidal facies are not linked to the shoreline but are a vast complex on the outer part of the shelf.

Phase 4 (Fig. 11d).—

By this time much of the platform was covered by quartzose sand, which had prograded from the north and east (Fig. 11d). Carbonate facies were restricted to a narrow peritidal complex and subtidal burrowed carbonates, which are characteristically dolomitized. The most seaward facies was a series of silty limestones, developed on top of the now buried shelf margin. Neither platform margin nor slope facies are present in the area; they presumably lay to the west and are now buried beneath thrust sheets.

*Stratigraphy.—*During the early part of phase 4, there was a rapid eastward onshelf shift of carbonate facies onlapping quartzose sands. This event was short lived, however.

Channeled grainstone (Fig. 5c) in planar beds of the siltstone facies (facies 2) at Meiklejohn Peak (loc. 31) reflects temporary slowing of subsidence, also suggested by transitional calcareous siltstone above Antelope Valley Limestone and below sandstone of the Eureka Quartzite. Similarly, the lowest beds of the Copenhagen Formation at Martin Ridge (loc. 14A) and Clear Creek Canyon (loc. 33) belong in this phase, as do the upper shale and mudstone beds of the Barrel Spring Formation (Ross, 1964, p. C37, C40) in the Northern Inyo Range (loc. 38). In the northern part of the study area, rapid westward progradation of a tongue of the sand approximates the end of this phase. In the north (loc. 14A, 15B), this progradation is dated accurately within the conodont zone of *Pygodus anserinus,* of latest Whiterockian age, i.e., equivalent to the upper Chazy Group of New York. Thereafter, deposition of the silty limestone of the Copenhagen Formation records early Mohawkian subsidence, which is terminated by progradation of the main body of the Eureka Quarzite in the conodont zone of *Phragmodus undatus* (A. G. Harris, pers. commun., 1987).

In the south at Meiklejohn Peak (loc. 31), however, the quartz sand arrived in the lower *Amorphagnathus tvarensis* conodont zone slightly later than in the north, a time equivalent to the Ashbyan Stage, lowest in the Mohawkian Series.

*Interpretation.—*This phase records a short-lived rapid rise in relative sea level, reflected by an onshelf shift in carbonate facies followed by massive progradation of quartzose sands.

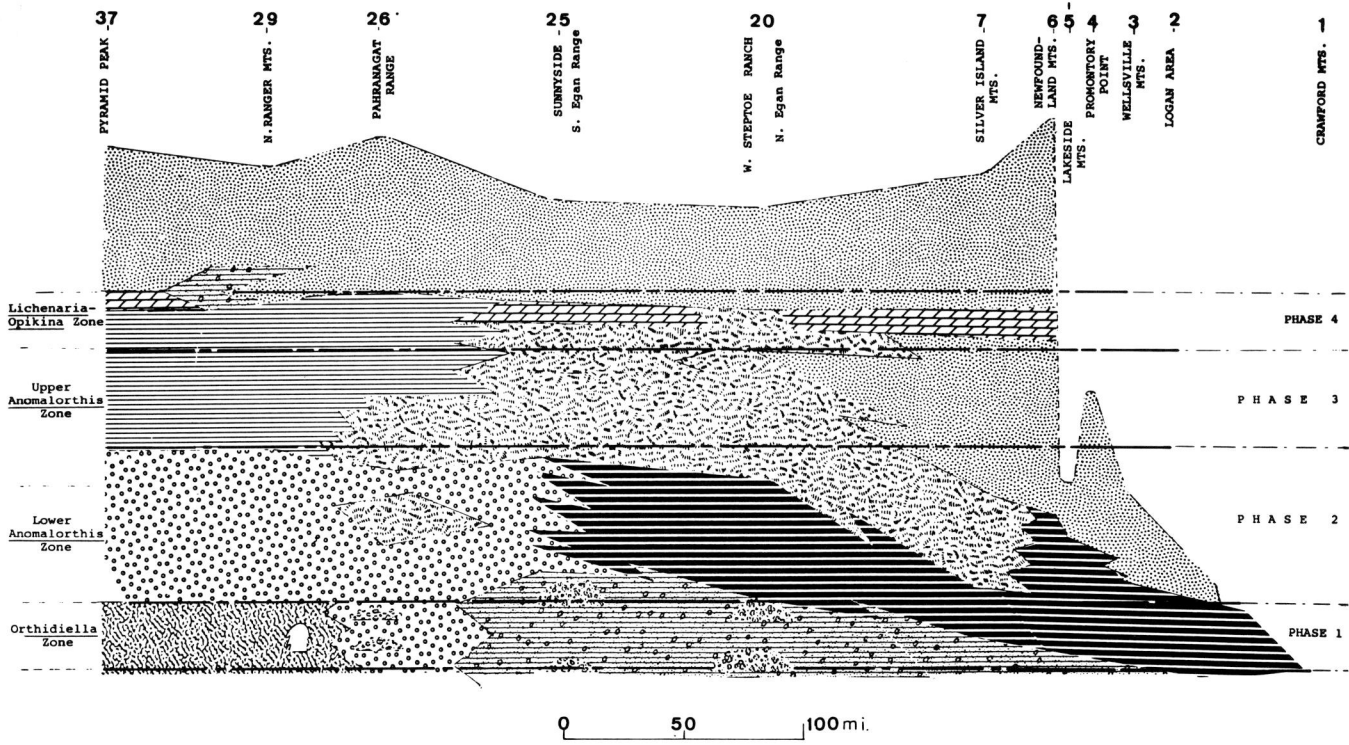

Fig. 8.—North-south cross section from Logan, Utah, (loc. 1) on the north to Pyramid Peak (loc. 37) on the south. Anomalous thicknesses between localities 5 and 6 are related to tectonic complication along the Tooele Arch.

Fig. 8, 9, 10.—Cross sections of lithic facies based on four fossil (time) zones: *Orthidiella*, lower *Anomalorthis*, upper *Anomalorthis*, and *Lichenaria-Opikina* Zones, which are indicated on left of each diagram. Phases 1-4 are indicated on right of each diagram. Thicknesses of most measured sections are in remarkably close agreement with time zones, except that type section of Antelope Valley Limestone (loc. 14A, 15) is anomalously thick. In all sections progradation of carbonate facies dominated through phases 1 and 2. Top of phase 2, *within* upper *Anomalorthis* Zone, marks change to transgression in north (Fig. 9).

CONTROLS ON PLATFORM DEVELOPMENT

Primary Controls

The thick sequence of sediments that composes the Whiterockian platform necessitates a continuing rise in relative sea level during deposition to accommodate the ever-increasing sediment wedge composed largely of shallow-water carbonates. Postulated variations in the rate of relative sea-level rise, as reflected in the sediments, have been discussed above. Changes in the rates of sea-level rise are the result of some combination of eustasy, continental margin subsidence, and local tectonics.

Eustasy.—

The record of eustasy during this period is poorly understood, because only continental-margin facies were deposited and many of these are tectonized in orogenic belts. Data from Canada (Barnes, 1984) suggest a major fall in sea level during the *Orthidiella* Zone followed by a sea-level rise in the following *Anomalorthis* Zone. As noted above, however, the upward growth of sponge and algal mounds and accretion of oncolite shoals indicate an increase in water depth over the western carbonate ramp and developing platform in phase 1 (*Orthidiella* Zone). In Australia in parts of New South Wales and west central Victoria (Webby and others, 1981), an unconformity beneath Mohawkian or younger strata suggests that the continent experienced falling relative sea level in Whiterockian time. In southwest Europe (Hamman and others, 1982), in the precordillera of Argentina (Acenolaza and Baldis, 1987), and in Kazakhstan (Nikitin and others, 1986), Whiterockian deposition took place in deep or deepening water in probable shelf or shelf edge facies.

Regional subsidence.—

The early to middle Paleozoic margin exposed in the Basin Ranges is a long-lived feature, with facies similar to the middle Ordovician, under discussion here, present from Cambrian through Devonian time (Stewart and Poole, 1974). Thus, subsidence would have been relatively uniform, similar to that of other trailing continental margins (Pitman, 1978; Keen and Hyndman, 1979) and not a major factor in changing rates of sea-level rise.

Local tectonics.—

The effect of local tectonics is reflected by three anomalies in the sedimentary sequence.

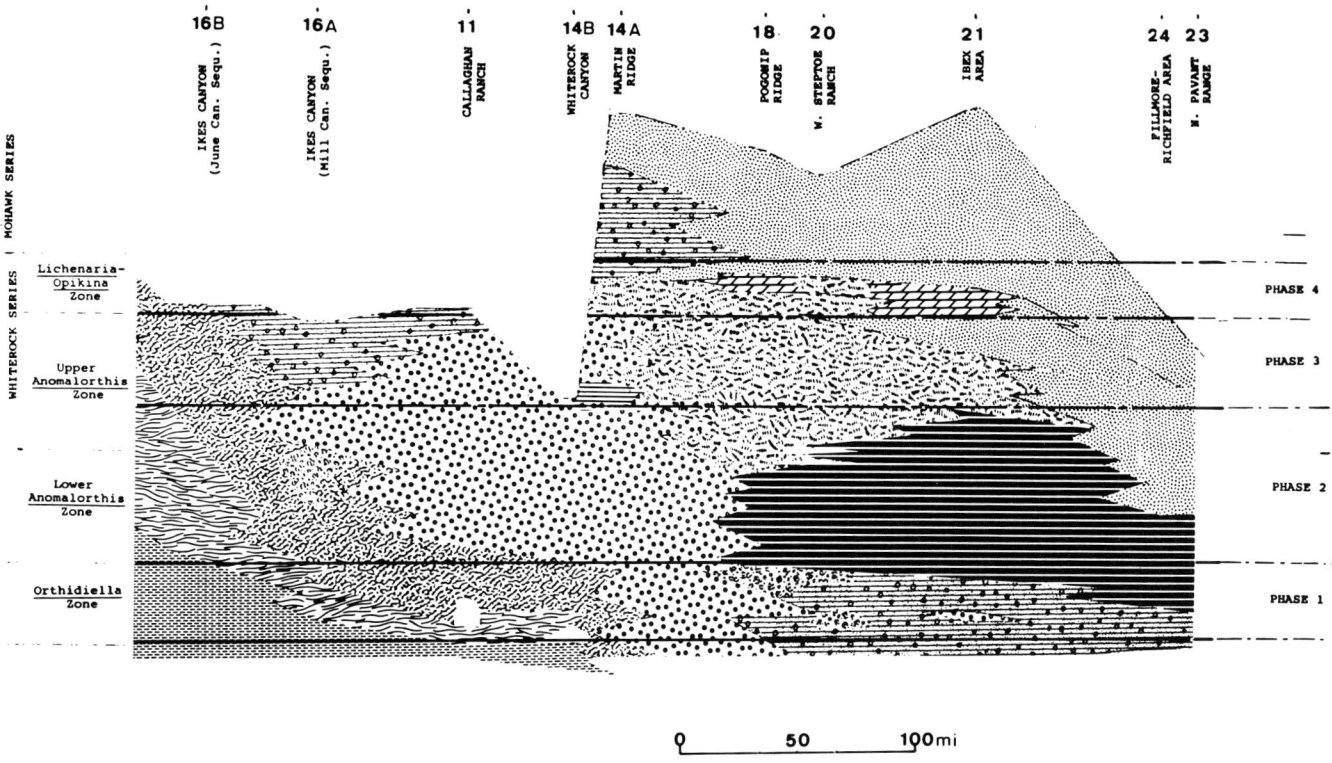

FIG. 9.—East-west cross section from Pavant Range, Utah (loc. 23), on the east, to Ikes Canyon, June Canyon thrust sequence (loc. 16B), in Toquima Range on the west. Note marked shift from progradational to transgressive deposition at beginning of phase 3.

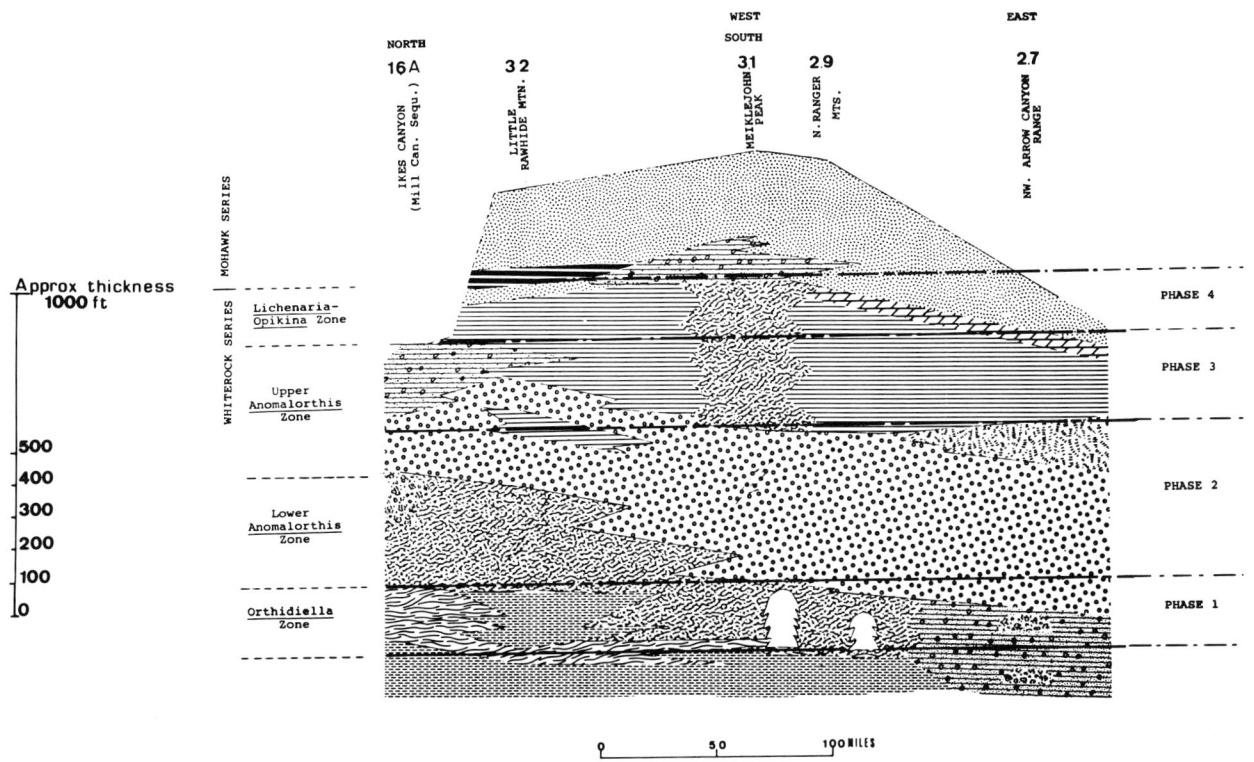

FIG. 10.—East-west-north "dog-leg" cross section from Arrow Canyon Range (loc. 27) on the east to Meiklejohn Peak (loc. 31) on the west to Ikes Canyon, Mill Canyon thrust sequence, (loc. 16A) on the north. Note lower-slope shale in phase 1 (Fig. 2b) at Little Rawhide Mountain (loc. 32), overlain by platform margin oncolite and peritidal carbonates.

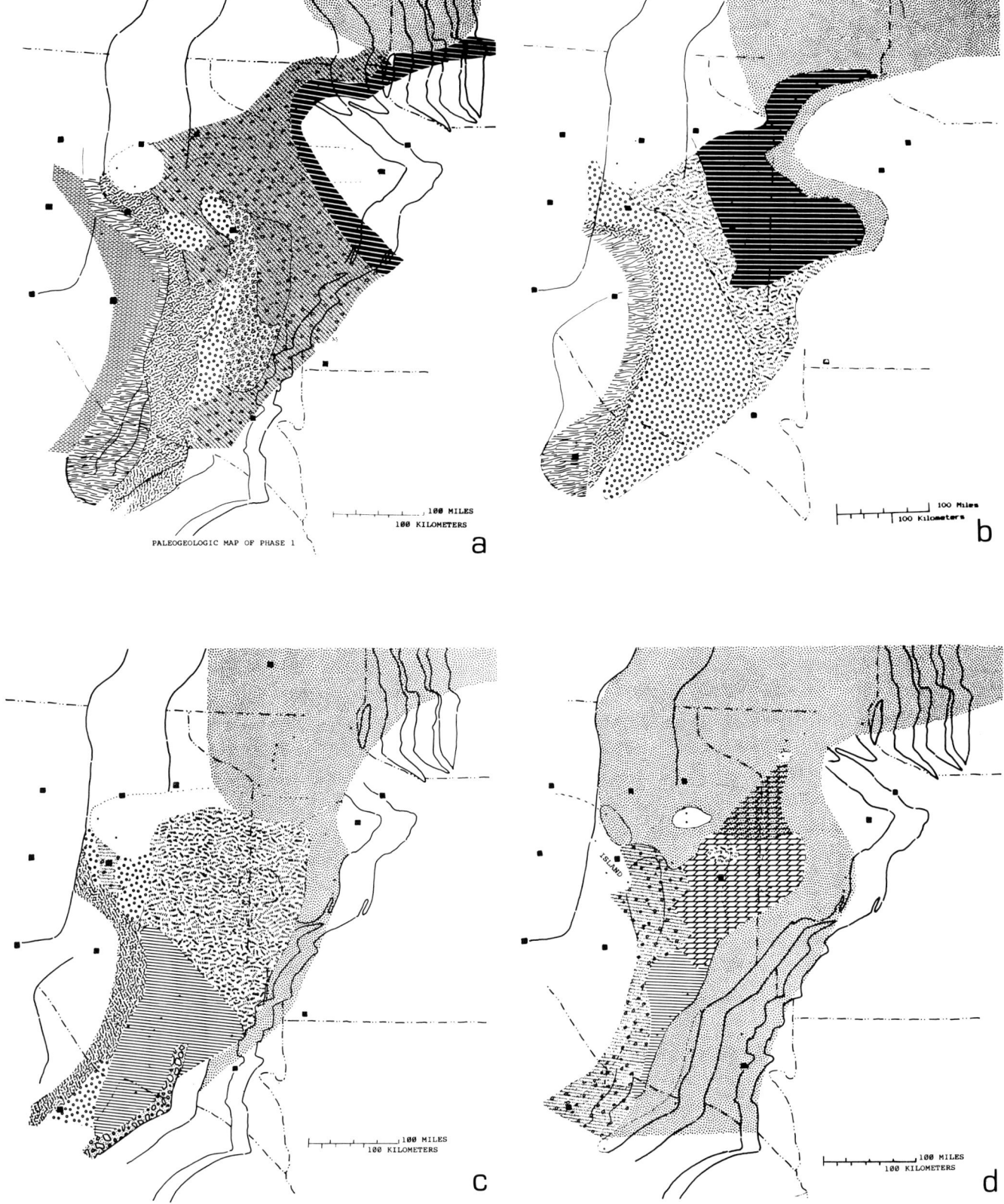

FIG. 11.—Paleogeologic maps of the four depositional phases during Whiterockian time. Phases 1-4 are the same as those portrayed in Figures 8–10. Progradational building of the carbonate platform in phases 1-2 (a, b) was reversed in phase 3 (c), except in the south where peritidal beds and local karst are developed. In Phase 4 (d), continued submergence and transgression of the platform carbonates is in conflict with increased progradational deposition of quartzose sand along the eastern strand.

(1) The Toole Arch (Webb, 1958) was a positive element in northwestern Utah and adjacent Nevada throughout Paleozoic time. The lack of Whiterockian sediments at Cortez (loc. 10) and in the Roberts Creek Mountains (loc. 12), together with the absence of the upper 244 m of Whiterockian strata at Lone Mountain (loc. 13), indicate the arch extended across the shelf in northern Nevada.

Several lines of evidence suggest that the area of the platform margin was tectonically active during Whiterockian time, especially during phase 4. McKee (1976) demonstrated the absence of different parts of the Whiterockian sequence in stacked thrust sheets in the Toquima Range (loc. 16A, 16B, and others). The Eureka Quartzite is absent or anomalously thin in the Lakeside Mountains (loc. 5), the Ruby Range (loc. 8, 9), the Toquima Range (loc. 14B, 16A, 16B) and at Callaghan Ranch (loc. 11). In addition, Harris and others (1979) have shown that part of the shelf edge underwent erosion during Mohawkian time (post-phase 4). Tectonic activity could not have been extensive during the deposition of phases 2 and 3, however, because slope sediments do not contain megaconglomerate debris flows (cf. James and Stevens, 1986).

(2) There is some evidence that the underlying Cambrian reentrant called the "House Embayment" (Rees, 1986), postulated to be fault bounded, may have controlled the location of the Whiterockian intra-shelf basin in which the Kanosh Shale accumulated. In addition, there is an abrupt lateral facies change in phase 3 from burrowed subtidal limestones in the southern Egan Range (loc. 25; Fig. 3c, d) and Horse Range (loc. 39) on the north to peritidal carbonates in the Arrow Canyon Range (loc. 27; Fig. 5a, b), Pahranagat Range (loc. 26), northern Ranger Mountains (loc. 29), and Little Rawhide Mountain (loc. 32) on the south and west. This pattern indicates that the rate of subsidence was greater in the northern part of the platform, over the "House Embayment," suggesting some underlying control. Comparison of paleogeographic maps for phases 2 and 3, however, shows that the peritidal facies lies above the widest and most substantial part of the oncolite shoal-platform rim, suggesting a sedimentologic rather than tectonic control. Furthermore, presently available crude isopachs of Ordovician sediments in the area (Hintze, 1973a; Ross, 1964, 1977) suggest that faulting was not important during Ordovician time.

(3) The third important feature is a sub-Eureka karst network (Fig. 5d) in a narrow belt from the Talc City Hills (loc. 36; Eberz, 1985) through the northern Nopah Range (loc. 35) and the Sheep Range (loc. 34; Ross, 1964, p.C15). Breccia beds composed of the oncolite facies and higher beds of the Antelope Valley Limestone are surrounded by cross-bedded quartz sand, traceable upward into the base of the overlying Eureka Quartzite. This area lies on the west side of the Talc City–Gass Peak thrust (Figs. 1, 11c). We suggest that this "Local Karst" (Choquette and James, 1988) formed on a locally uplifted area subjected to solution by meteoric water prior to or at the same time as the arrival of the prograding quartz sand. We are aware of the unlikely possibility, however, that this extensive stratigraphically positioned breccia may have been the result of much younger Basin Range extensional tectonics.

On balance, it seems that tectonics were of minor importance on a regional scale but locally were very significant.

Summary.—When the shelf sequence is considered as a whole, it has many of the attributes of Cambrian Grand Cycles (Aitken, 1966), with a lower shaly half-cycle (phases 1 and 2) overlain by an upper carbonate half-cycle (phase 3); phase 4 is the beginning of a different cycle. Many of the arguments for the first-order controls of these Grand Cycles (Aitken, 1978; Palmer and Halley, 1979; Chow and James, 1987) are equally applicable to the Whiterock platform and have been utilized in our interpretation. Such similarities argue strongly for the universality of these first-order controls.

Secondary Controls

Within the overall sedimentary sequence, climate and nature of the carbonate-producing biota are the most important factors governing the style of carbonate sedimentation.

Paleoclimate.—

This carbonate platform lay close to the middle Ordovician equator (Ross, 1976) and thus should have been outside the region of annual violent cyclonic storms (Schopf, 1980), which should have occurred to the north and south. Nevertheless, the presence of storm deposits (Aigner, 1985) throughout the sequence indicates that the platform was periodically swept by deep low-pressure systems.

Ross (1976) has speculated that during the early Ordovician, the area between the Great Basin and the Canadian Shield could have been subjected to severe chemical and biochemical weathering, converting feldspars into clays and releasing quartz as sand. Subsequent continental movement could have put the clays and silts and later the reworked sands within the belt of the equatorial easterly (blowing toward the west) trade winds (Van Dorn, 1974). These winds would also have blown large amounts of silt and clay offshore (Dalrymple and others, 1985) downwind, onto the Whiterockian platform. Thus, the silty and shaly nature of much of the shelf sediment is probably a function of this specific combination of paleoclimate and paleolatitude. In some shallow-water environments, production of biogenic carbonate may have completely masked the fine siliciclastics.

The almost incalculable volume of well-rounded, frosted, multicycled quartz sand found in the Eureka and contemporaneous Kinnikinnic, Mount Wilson, and St. Peter sandstones must have had a similar or common cratonic source. Sands blown seaward off the coast in what is now Idaho-Utah-Nevada must have been reworked by marine currents (Van Dorn, 1974). It is possible that, at times, quartz sand was supplied so much faster than the rise in relative sea-level that carbonate production was smothered.

The lack of evaporites in the peritidal deposits of phases 3 and 4 indicate that the climate was humid to possibly semiarid (James, 1984; Hardie and Shinn, 1986), a situation which appears to have prevailed into the Upper Ordovician and Silurian (Dunham and Olson, 1980). This was a function of paleolatitude, since arid conditions, conducive to evaporite precipitation, are most common 30° to 40° north and south of the equator (Schopf, 1980). Problematic molds of selenite, however, are found in facies 6 at locality 20 in phase 4.

Most sediments on the platform are muddy and, except for the infrequent storm deposits, seem to have been deposited under relatively tranquil conditions. The presence of peritidal deposits argues against a deep-water shelf, at least in phases 3 and 4. Thus, it seems that the coast during deposition was generally low, similar to the type II settings envisaged by Wilson (1975, p. 361, fig. XII-3).

The widespread normal-marine sediments, with a rich and diverse benthic fauna, suggest good water circulation and adequate nutrient supply. The organic-rich black shales, however, clearly signal deposition under conditions of depleted oxygen, in an intra-shelf basin. Circulation in this basin was probably restricted by the platform margin shoal facies. Euxinic conditions were likely dramatically enhanced by the presumed continuous offshore winds, which probably set up an estuarine (lagoonal) circulation pattern (Berger, 1970), leading to water stratification.

Paleobiology.—

Between middle Cambrian and the end of Whiterockian or beginning of Mohawkian time, there were no skeletal organisms capable of constructing large reefs. Even with the appearance of primitive colonial corals and early bryozoans in the *Lichenaria-Öpikina* Zone (phase 4), no organisms were present to form a platform rim composed of reefs, although skeletal metazoan reefs first appear elsewhere in the correlative Chazyan (Webby, 1984).

Buildups in the form of reef mounds (James, 1983) did form, however, in protected settings during the major period of subsidence and progradation (phases 1 and 2), on the shelf in the lee of oncolite shoals and in deep water on the upper slope. Shelf buildups, essentially reef mounds composed of lime mud rich in calathids, receptaculitids, and sponges and surrounded by *Nuia* sands, are similar to other onshelf reef mounds of Early Ordovician (Church, 1974; Toomey and Nitecki, 1979; Pratt and James, 1982) and pre-Chazyan Middle Ordovician (Klappa and James, 1980) age. The huge mud mounds, which grew on the slope (Ross and others, 1975), are not known elsewhere in the Ordovician, but their setting and internal structure are like those of slightly younger mud mounds rich in ramose bryozoans (Virginia: Read, 1982; Sweden: Jaanusson, 1982).

In the absence of a marginal reef, rimmed carbonate platforms generally have a shallow, carbonate sand-shoal facies (Ginsburg and James, 1974; James and Mountjoy, 1983; Read 1985) usually composed of bioclastic sands or ooids. This facies commonly contains abundant sedimentary structures indicative of high-energy deposition and periodic exposure. The massive oncolite limestones, interpreted as a shalf edge facies, do not exhibit such sedimentary structures, except near the very top, in marking the change from phase 2 to phase 3 at localities 11, 14A, and 32. In addition, oncolites are generally regarded as a moderate but not high-energy accumulation (Wilson, 1975; Flugel, 1982; Peryt, 1983). The absence of sedimentary structures indicative of high energy is in accordance with the generally muddy, low-energy features of the other carbonates. The cortex of the oncolites is formed by *Girvanella,* a common occurrence in most early Paleozoic reefs, in many reef mounds, and in blocks from the platform margin resedimented in deep water (James, 1981; Pfiel and Read, 1980; Coniglio and James; 1985).

On the basis of the above evidence, we interpret the oncolite sediments as a series of shallow-water but not strandline shoals. Instead, we suggest that they accumulated under permanently submerged conditions in waters that may have been as deep as 10 m. Only when relative sea-level rise slowed considerably at the inception of phase 3 did these shoals become awash and at that time were capped by oolite shoals and intertidal islands. To our knowledge, this is a unique style of margin facies.

SUMMARY AND CONCLUSIONS

(1) An epicontinental carbonate platform, some 400 m thick and at its climax over 250 km wide, now located in the southwestern United States, formed a substantial part of the middle Ordovician (Whiterockian) margin.

(2) The 12-m.y. span of deposition can be divided on the basis of detailed biostratigraphy into four time slices and phases, with the sedimentary facies in each reflecting variations in the rate of relative sea-level rise.

During phase 1 (*Orthidiella* Zone) and phase 2 (lower to mid-upper *Anomalorthis* Zone), the shelf evolved from a flooded ramp to a rimmed shelf as a belt of calathid-sponge mounds and oncolite bars coalesced into an oncolite-shoal complex that was 75 km wide in the north, 200 km wide in the south, and over 600 km in length from north to south. Vertical accretion was followed by westward progradation of margin facies as relative sea-level rise slowed. Ocean-facing carbonate slope facies contain large carbonate mud mounds, whereas shelf facies are mostly burrowed siltstones or black shales deposited in an intra-shelf basin.

The rate of sea-level rise slowed dramatically in phase 3 (upper *Anomalorthis* Zone) and is reflected by extensive peritidal sediments on top of the oncolite-shoal complex and widespread carbonate deposition on the platform.

Phase 4 (*Lichenaria–Öpikina* Zone) records first a short period of onlap, followed rapidly by profound offlap and burial of the platform by prograding quartzose sands.

(3) Local tectonics appear to have been most important during phase 3, localizing peritidal facies in the south and subtidal facies over the underlying House Embayment in the north.

(4) The unique style of carbonate sedimentation on this platform is a function of its equatorial location during the Ordovician, in a humid climate, covered by seas that were rarely disturbed by large storms, in a belt of offshore winds. Of equal importance is the nature of the biota, which at this time lacked any large skeletal metazoans. Therefore, the marginal facies was a massive oncolite shoal, which accumulated in a shallow subtidal, but not intertidal, environment.

ACKNOWLEDGMENTS

This study is the result of a research project by Ross and James, funded by National Science Foundation Grant EAR-8117322 to Ross and Natural Sciences and Engineering Council of Canada Grant A-9159 to James. We acknowledge with gratitude assistance in the field from D. L. Schmidt in the Arrow Canyon and Sheep Ranges, from G. L. Dixon in the Ranger Mountains, and from Noel Eberz in the Talc City and Mazourka Canyon areas. Eberz called our attention to the karst breccias in the Talc City district. J. E. Valusek and Ronald McDowell not only helped measure several stratigraphic sections, but also provided stratigraphic information in connection with their dissertations. We have benefited from comments of and discussions with numerous visitors who have joined in the field. We gratefully received critical comments on the manuscript from J. B. Dunham, J. L. Wilson, and J. F. Read, although we have not taken all their advice.

REFERENCES

ACENOLAZA, F. G., AND BALDIS, BRUNO, 1987, The Ordovician System in South America, Correlation Chart and Explanatory Notes *in* Ross, R. J., Jr., ed., International Union of Geological Sciences Publication No. 22, 64 p, 4 figs., 3 pls.

AHR, W. M., 1973, The carbonate ramp–An alternative to the shelf model: Gulf Coast Association of Geological Societies Transactions, v. 23, p. 221–225.

AIGNER, T., 1985, Storm depositional systems: Lecture notes in Earth Sciences: Springer-Verlag, Heidelberg, 174 p.

AITKEN, J. D., 1966, Middle Cambrian to middle Ordovician cyclic sedimentation, southern Rocky Mountains of Alberta: Bulletin of Canadian Petroleum Geology, v. 14, p. 405–441.

———, 1978, Revised models for depositional grand cycles, Cambrian of the southern Rocky Mountains, Canada: Bulletin of Canadian Petroleum Geology, v. 26, p. 515–542.

ALSHARHAN, A. S., AND C. G. ST. C. KENDALL, 1986, Precambrian to Jurassic rocks of Arabian Gulf and adjacent areas, their facies, depositional setting, and hydrocarbon habitat: American Association of Petroleum Geologists, v. 70, p. 977–1002.

BARNES, C. R., 1984, Early Ordovician eustatic events in Canada, *in* Bruton, D. L., ed., Aspects of the Ordovician System: Paleontological Contributions, University of Oslo, Universitetsforlaget, No. 295, p. 51–63.

BASSLER, R. S., 1941, The Nevada early Ordovician (Pogonip) sponge fauna: U. S. National Museum Proceedings, v. 91, no. 3126, p. 91–102.

BERGER, W. H., 1970, Biogenous deep-sea sediments; Fractionation by deep sea circulation: Geological Society of America Bulletin, v. 81, p. 1385–1402.

BYERS, C. W., 1977, Biofacies patterns in euxinic basins, a general model, *in* Cook, H. E., and Enos, Paul, eds., Deep-Water Carbonate Environments: Society of Economic Paleontologists and Mineralogists Special Publication 25, p. 5–18.

CHOQUETTE, P. W., AND JAMES, N. P., 1988, Introduction, *in* James, N. P., and Choquette, P. W., eds., Paleokarst: Springer-Verlag, New York, p. 1–21.

CHOW, N., AND JAMES, N. P., 1987, Cambrian grand cycles, a northern Appalachian perspective: Geological Society of America Bulletin, v. 98, p. 418–429.

CHURCH, S. B., 1974, Lower Ordovician patch reefs in western Utah: Brigham Young University Geological Studies, v. 21, p. 41–62.

CONIGLIO, MARIO, AND JAMES, N. P., 1985, Calcified algae as sediment contributors to early Paleozoic limestones; Evidence from deep-water sediments of the Cow Head Group, western Newfoundland: Journal of Sedimentary Petrology, v. 55, p. 746–755.

COOK, H. E., AND MULLINS, H. T., 1983, Basin margin environment, *in* Scholle, P. A., Bebout, D. G., and Moore, C. H., eds., Carbonate Depositional Environments: American Association of Petroleum Geologists Memoir 33, p. 539–618.

DALRYMPLE, R. W., NARBONNE, G. M., AND SMITH, L., 1985, Eolian action and the distribution of Cambrian shales in North America: Geology, v. 13, p. 607–610.

DUNHAM, J. B., AND OLSON, E. R., 1980, Shallow subsurface dolomitization of subtidally deposited carbonate sediments in the Hanson Creek Formation (Ordovician-Silurian) of Central Nevada, *in* Zenger, D. H., Dunham, J. B., and Ethington, R. L., eds., Concepts and Models of Dolomitization: Society of Economic Paleontologists and Mineralogists Special Publication 28, p. 139–162.

EBERZ, NOEL, 1985, Geology of the Badger Flat Limestone (Middle Ordovician), Southeastern California: Unpublished M.S. Thesis: San Jose State University, San Jose, California, 91 p.

FLUGEL, ERIC, 1982, Microfacies Analysis of Limestones: Springer-Verlag, New York, 633 p.

GILLULY, JAMES, AND MAZURSKY, HAROLD, 1965, Geology of the Cortez Quadrangle, Nevada: U. S. Geological Survey Bulletin 1175, 117 p., geologic map.

GINSBURG, R. N., AND JAMES, N. P., 1974, Holocene carbonate sediments of continental shelves, *in* Burk, C. A., and Drake, C. L., eds., The Geology of Continental Margins: Springer-Verlag, New York; p. 137–155.

HAMMAN, WOLFGANG, ROBARDET, MICHEL, AND ROMANO, MICHAEL, 1982, The Ordovician System in southwestern Europe (France, Spain, and Portugal), Correlation Chart and Explanatory Notes: International Union of Geological Sciences Publication No. 11, 47 p.

HARDIE, L. A., AND SHINN, E. A., 1986, Carbonate depositional environments, Part 3. Tidal flats, *in* Warme, J. E., and Shanley, K. W., eds., Colorado School of Mines Quarterly, v. 81, 74 p.

HARRIS, A. G., BERGSTROM, S. M., ETHINGTON, R. L., AND ROSS, R. J., Jr., 1979, Aspects of middle and upper Ordovician conodont biostratigraphy of carbonate facies in Nevada and southeast California and comparison with Appalachian successions: Brigham Young University Geology Studies, v. 26, pt.3, p. 7–33, 5 pls.

HINTZE, L. F., 1951, Lower Ordovician Detailed Stratigraphic Sections for Western Utah: Utah Geological and Mineralogical Survey, Bulletin 39, 98 p.

———, 1952 (1953), Lower Ordovician Trilobites from Western Utah and Eastern Nevada: Utah Geological and Mineralogical Survey, Bulletin 48, 249 p., 28 pls.

———, 1973a, Geologic History of Utah: Brigham Young University Geology Studies, v. 20, pt. 3, 181 p., 83 text figs., 46 stratigraphic charts.

———, 1973b, Lower and middle Ordovician stratigraphic sections in the Ibex area, Millard County, Utah: Brigham Young University Geology Studies, v. 20, pt. 4, p. 3–36.

JAANUSSON, VALDAR, 1982, The Siljan district, *in* Bruton, D. L., and Williams, S. H., eds., Field Excursion Guide, Fourth International Ordovician Symposium: Paleontological Contributions, University of Oslo, Universitetsforlaget No. 279, p. 15–42.

JAMES, N. P., 1981, Megablocks of calcified algae in the Cow Head Breccia, western Newfoundland, vestiges of a lower Paleozoic continental margin: Geological Society of America Bulletin, v. 92, p. 799–811.

———, 1983, Reef environment, *in* Scholle, P. A., Bebout, D. G., and Moore, C. H., eds., Carbonate Depositional Environments: American Association of Petroleum Geologists Memoir 33, p. 345–440.

———, 1984, Reefs, *in* Walker, R. G., ed., Facies Models: Geological Association of Canada Reprint Series No. 1, p. 229–244.

———, AND MOUNTJOY, E. W., 1983, Shelf-slope break in fossil carbonate platforms, an overview, *in* Stanley, D. J., and Moore, G. T., ed., The Shelf Break, Critical Interface on Continental Margins: Society of Economic Paleontologists and Mineralogists Special Publication 33, p. 189–206.

———, AND STEVENS, R. K., 1986, Stratigraphy and Correlation of the Cambro-Ordovician Cow Head Group, Western Newfoundland: Geological Survey of Canada, Bulletin 366, 143 p., maps and charts.

KEEN, C. E., AND HYNDMAN, R. D., 1979, Geophysical review of the continental margins of eastern and western Canada: Canadian Journal of Earth Sciences, v. 16, p. 712–747.

KELLOGG, H. E., 1963, Paleozoic stratigraphy of the southern Egan Range, Nevada: Geological Society of America Bulletin, v. 74, p. 685–708, 4 pls. Geologic map in v. 75, p. 949–968.

KENDALL, C. G. ST. C., AND SCHLAGER, W., 1981, Carbonates and relative changes in sea level: Marine Geology, v. 44, p. 181–212.

KETNER, K. B., 1968, Origin of Ordovician quartzite in the Cordilleran miogeosyncline: U. S. Geological Survey Professional Paper 600-B, p. B169–B177.

KLAPPA, C. F., AND JAMES, N. P., 1980, Small lithistid sponge bioherms, early middle Ordovician Table Head Group, western Newfoundland: Bulletin of Canadian Petroleum Geology, v. 28, p. 435–451.

LOWELL, J. D., 1960, Ordovician miogeosynclinal margin in central Nevada: Twenty-first International Geological Congress, Norden, Part VII, Ordovician and Silurian Stratigraphy and Correlations, p. 7–17.

MARKELLO, J. R., AND READ, J. F., 1981, Carbonate ramp-to-deeper shale-shelf transitions of an upper Cambrian intrashelf basin, Nolichucky Formation, southwest Virginia Appalachians: Sedimentology, v. 28, p. 573–597.

MCILREATH, I. A., AND JAMES, N. P., 1984, Carbonate Slopes, in Walker, R. G., ed., Facies Models: Geological Association of Canada Reprint Series No. 1, p. 245–258.

MCKEE, E. H., 1976, Geology of the Northern Part of the Toquima Range, Lander, Eureka, and Nye Counties, Nevada: U. S. Geological Survey Professional Paper 931, 49 p., geologic map.

MERRIAM, C. W., AND ANDERSON, C. A., 1942, Reconnaissance survey of the Roberts Mountains, Nevada: Geological Society of America Bulletin, v. 53, p. 1675–1728.

MURRIS, R. J., 1980, Middle east, stratigraphic evolution and oil habitat: American Association of Petroleum Geologists Bulletin, v. 64, p. 597–618.

MUSSMAN, W. J., AND READ, J. F., 1986, Sedimentology and development of a passive to convergent margin unconformity, middle Ordovician Knox unconformity, Virginia Appalachians: Geological Society of America Bulletin, v. 97, p. 282–295.

NIKITIN, I. F., APPOLONOV, M. K., TZAJ, D. T., KOROLJOV, V. G., KIM, A. I., ERINA, M. V., LARIN, N. M., AND GOLIKOV, A. N., 1986, The Ordovician System in Kazakhstan and middle Asia, Correlation Charts and Explanatory Notes: International Union of Geological Sciences Publication No. 21, 34 p.

PALMER, A. R., 1960, Some aspects of the early upper Cambrian stratigraphy of White Pine County and vicinity: Intermountain Association of Petroleum Geologists, Eleventh Annual Field Conference, p. 53–58.

———, AND HALLEY, R. B., 1979, Physical Stratigraphy and Trilobite Biostratigraphy of the Carrara Formation (Lower and Middle Cambrian) in the Southern Great Basin: U. S. Geological Survey Professional Paper 1047, 130 p.

PERYT, T., 1983, ed., Coated Grains: Springer-Verlag, New York, 655 p.

PFIEL, R. W., AND READ, J. F., 1980, Cambrian carbonate platform margin facies, Shady Dolomite, southwestern Virginia, U.S.A.: Journal of Sedimentary Petrology, v. 50, p. 91–116.

PITMAN, W. C., III, 1978, Relationship between eustasy and stratigraphic sequences of passive margins: Geological Society of American Bulletin, v. 89, p. 1389–1403.

PRATT, B. R., AND JAMES, N. P., 1982, Cryptalgal-metazoan bioherms of early Ordovician age in the St. George Group, western Newfoundland: Sedimentology, v. 29, p. 543–569.

READ, J. F., 1982, Geometry, facies and development of Middle Ordovician carbonate buildups, Virginia Appalachians: American Association of Petroleum Geologists Bulletin, v. 66, p. 189–209.

———, 1985, Carbonate platform facies models: American Association of Petroleum Geologists Bulletin, v. 69, p. 1–21.

REES, M. N., 1986, A fault-controlled trough through a carbonate platform, the Middle Cambrian House Range embayment: Geological Society of America Bulletin, v. 97, p. 1054–1069.

ROSS, R. J., Jr., 1951, Stratigraphy of the Garden City Formation in Northeastern Utah, and its Trilobite Faunas: Peabody Museum Natural History, Bulletin 6, 155 p., 36 pls.

———, 1964, Middle and Lower Ordovician Formations in Southernmost Nevada and Adjacent California: U.S. Geological Survey Bulletin 1180-C, 101 p.

———, 1967, Some Middle Ordovician Brachiopods and Trilobites from the Basin Ranges, Western United States: U. S. Geological Survey Professional Paper 523-D, 43 p., 11 pls.

———, 1970, Ordovician Brachiopods, Trilobites, and Stratigraphy in Eastern and Central Nevada: U. S. Geological Survey Professional Paper 639, 103 p., 22 pls.

———, 1972, Fossils from the Ordovician Bioherm at Meiklejohn Peak, Nevada: U. S. Geological Survey Professional Paper 685, 43 p., 18 pls.

———, 1976, Ordovician sedimentation in the western United States, in Basset, M. G., ed., The Ordovician System: Proceedings, Palaeontological Association Symposium, University of Wales Press and National Museum, Cardiff, p. 73–105.

———, 1977, Ordovician paleogeography of the western United States, in Stewart, J. H., Stevens, C. H., and Fritsche, A. E. eds., Pacific Coast Paleogeography Symposium I, Paleozoic Paleogeography of the Western United States: Pacific Section, Society of Economic Paleontologists and Mineralogists, p. 19–38.

———, ADLER, F. J., AMSDEN, T. W., BERGSTROM, DOUGLAS, BERGSTROM, S. M., CARTER, CLAIRE, CHURKIN, MICHAEL, CRESSMAN, E. A., DERBY, J. R., DUTRO, J. T., Jr., ETHINGTON, R. L., FINNEY, S. C., FISHER, D. W., FISHER, J. H., HARRIS, A. G., HINTZE, L. F., KETNER, K. B., KOLATA, D. L., LANDING, E. D., NEUMAN, R. B., SWEET, W. C., POJETA, JOHN, Jr., POTTER, A. W., RADER, E. K., REPETSKI, J. E., SHAVER, R. H., THOMPSON, T. L., AND WEBERS, G. F., 1982, The Ordovician System in the United States of America, Correlation Chart and Explanatory Notes: International Union of Geological Sciences Publication No. 12, 73 p.

———, AND J. K. INGHAM, 1970, Distribution of the Toquima-Table Head (middle Ordovician Whiterock) faunal realm in the northern hemisphere: Geological Society of America Bulletin, v. 81, p. 393–408.

———, JAANUSSON, V., AND FRIEDMAN, I., 1975, Lithology and Origin of Middle Ordovician Calcareous Mudmound at Meiklejohn Peak, Southern Nevada: U. S. Geological Survey Professional Paper 871, 48 p.

———, AND SHAW, F. C., 1972, Distribution of the Middle Ordovician Copenhagen Formation and its Trilobites in Nevada: U. S. Geological Survey Professional Paper 749, 33 p., 8 pls.

———, VALUSEK, J. E., AND JAMES, N. P., 1988, Nuia and its environmental significance, in Wolberg, D. L., compiler, Contributions to Paleozoic Paleontology and Stratigraphy in honor of Rousseau H. Flower: New Mexico Bureau of Mines and Mineral Resources Memoir 44, p. 115–121.

SCHOPF, T. J. M., 1980, Paleoceanography: Harvard University Press, Cambridge, Massachusetts, 341 p.

SHEEHAN, P. M., AND SCHIEFELBEIN, D. R. J., 1984, The trace fossil Thalassinoides from the upper Ordovician of the eastern Great Basin, deep burrowing in the early Paleozoic: Journal of Paleontology, v. 58, p. 440–447.

SHINN, E. A., 1973, Sedimentary accretion along the leeward, SE coast of Qatar Peninsula, Persian Gulf, in Purser, B. H., ed., The Persian Gulf: Springer-Verlag, New York, p. 199–210.

———, 1983, Tidal flat environment, in Scholle, P. A., BEBOUT, D. G., AND MOORE, C. H., eds., Carbonate Depositional Environments: American Association of Petroleum Geologists Memoir 33, p. 171–210.

STEVENS, C. H., 1986, Evolution of the Ordovician through mid-Pennsylvanian carbonate shelf in east-central California: Geological Society of America Bulletin, v. 97, p. 11–25.

STEWART, J. H., AND POOLE, F. G., 1974, Lower Paleozoic and uppermost Precambrian, Cordilleran miogeocline, Great Basin, western United States, in Dickinson, W. R., ed., Tectonics and Sedimentation: Society of Economic Paleontologists and Mineralogists Special Publication 22, p. 28–57.

TOOMEY, D. F., AND NITECKI, M. H., 1979, Organic Buildups in the Lower Ordovician (Canadian) of Texas and Oklahoma: Fieldiana, Series 2, 181 p.

VAN DORN, W. G., 1974, Oceanography and Seamanship: Dodd, Mead, and Company, New York, 481 p.

WEBB, G. W., 1956, Middle Ordovician detailed stratigraphic sections for western Utah and eastern Nevada: Utah Geological and Mineralogical Survey Bulletin 57, 77 p.

———, 1958, Middle Ordovician stratigraphy in eastern Nevada and western Utah: American Association of Petroleum Geologists Bulletin, v. 42, p. 2335–2377.

WEBBY, B. D., 1984, Ordovician reefs and climate: A review, *in* Bruton, D. L., ed., Aspects of the Ordovician System: Paleontological Contributions, University of Oslo, Universitetsforlaget, No. 295, p. 89–100.

———, VANDENBERG, A. H. M., COOPER, R. A., BANKS, M. R., BURRETT, C. F., HENDERSON, R. A., CLARKSON, P. D., HUGHES, C. P., LAURIE, J., STAIT, B., THOMSON, M. R. A., AND WEBERS, G. F., 1981, The Ordovician System in Australia, New Zealand, and Antarctica, Correlation Chart and Explanatory Notes, *in* Webby, B. D., compiler and ed., International Union of Geological Sciences Publication No. 6, 64 p.

WILSON, J. L., 1975, Carbonate Facies in Geologic History: Springer-Verlag, New York, 471 p.

———, AND JORDAN, CLIFFORD, 1983, Middle shelf environment, *in* Scholle, P. A., Bebout, D. G., and Moore, C. H., eds., Carbonate Depositional Environments: American Association of Petroleum Geologists Memoir 33, p. 297–344.

REEFAL PLATFORM DEVELOPMENT, DEVONIAN OF THE CANNING BASIN, WESTERN AUSTRALIA

PHILLIP E. PLAYFORD
Geological Survey of Western Australia, Perth, 6000, Australia

NEIL F. HURLEY
Marathon Oil Company, Littleton, Colorado 80160

CHARLES KERANS
Bureau of Economic Geology, University of Texas, Austin, Texas 78713

AND

MICHAEL F. MIDDLETON
Geological Survey of Western Australia, Perth, 6000, Australia

ABSTRACT: The Devonian "Great Barrier Reef" of the Canning Basin developed beside a mountainous landmass of Precambrian rocks (the Kimberley Block) and around islands of Precambrian and Ordovician rocks. Basement topography, commonly fault controlled, was important in localizing the reefal platforms.

Growth of the platforms was nearly continuous from late Givetian to late Famennian times, with only rare intervals of brief emergence. The earliest (Givetian) platforms were low-relief banks; later Frasnian and Famennian platforms were usually reef rimmed, with high relief. Upright reef margins predominated in the Frasnian, with intervals of backstepping in the late Frasnian associated with widespread drowning and the development of pinnacle reefs. A brief regression, with minor subaerial erosion of platforms (including mild karstification), occurred at the Frasnian-Famennian boundary, and the succeeding Famennian platforms advanced basinward over their equivalent marginal-slope and basin facies. Platform extinction in the late Famennian resulted from abrupt drowning.

The hypothetical curve of relative sea level for the Canning Basin Devonian shows only partial resemblance to the Euramerican eustatic curve. In both areas, rapid rises in relative sea level in the late Frasnian caused widespread drowning of reefal platforms, followed by a brief regression at the Frasnian-Famennian boundary, which coincided with the world-wide mass extinction of metazoan reef builders. The Famennian regressions identified in Euramerica are not recognized in the Canning Basin, where the limestone platforms advanced continuously under conditions of steady rise or stillstand in relative sea level.

Earthquakes associated with faulting during growth of the complexes resulted in extensive fracturing of rigid submarine-cemented limestones along the platform margins and upper marginal slopes, leading to the development of neptunian dikes and the collapse of platform margins to form massive debris flows. Contemporary faulting also influenced platform–basin morphology and was responsible for the mountainous topography of the adjoining landmass, which shed masses of boulder conglomerate and other terrigenous sediments interfingering with the reef complexes.

Seismic stratigraphic modeling, in comparison with observed seismic records and well data, suggests that the evolutionary model for the Devonian reef complexes deduced from outcrop studies can also be recognized in the subsurface.

Three small oil fields are associated with or overlie a well-defined subsurface Famennian platform margin. Recent research involving outcrop, core, seismic, and geochemical studies suggest, however, that the best prospects for future oil discoveries are likely to be in Givetian-Frasnian platforms, which have yet to be adequately tested.

Mississippi Valley-type zinc-lead orebodies have been found in Givetian-Frasnian platforms in the outcrop area, and one of these is now being developed. The area is believed to have a good potential for further zinc-lead mining developments.

INTRODUCTION

The well-exposed Devonian "Great Barrier Reef" of the Canning Basin (Figs. 1–3) is renowned as one of the world's best-preserved Paleozoic reef sequences, and is also of considerable economic interest because of its oil and mineral potential. Consequently, it has been the subject of a great deal of research and exploration activity over the past 25 years. This research has largely been directed or coordinated by the Geological Survey of Western Australia, in association with other institutions in Australia and overseas.

The research program has been designed to assist petroleum and mineral exploration, by seeking to elucidate the stratigraphy, evolutionary development, paleontology, structure, diagenetic history, geochemistry, and paleomagnetics of the reefal carbonates and their associated facies. Earlier references to the geology are quoted by Playford and Lowry (1966), Petersen (1975), and Playford (1980). The principal contributions since then have been by Playford (1981, 1984); Cockbain (1984); Alexander and others (1985); Kerans (1985); Kerans and others (1986); Rigby (1986); Hurley (1986); Hurley and Van der Voo (1987); and Wallace (1987). The regional geology of the basin is described by Towner and Gibson (1983), and much detailed information is contained in the Canning Basin Symposium volume edited by Purcell (1984), including papers on evolution of outcropping Devonian reef complexes by Benn (1984), Cooper and others (1984), and Hall (1984).

Exploration for subsurface reef complexes by Home Oil succeeded during 1982 in discovering a small oil field in a Famennian reef complex and overlying dolomite of the Fairfield Group (Playford, 1982; Moors and others, 1984). Since then, two additional small fields have been found in Carboniferous-Permian sandstones above the Famennian platform margin (Figs. 2, 21).

In addition, several zinc-lead Mississippi Valley-type orebodies have recently been found by BHP in the reef complexes southeast of Fitzroy Crossing (Hall, 1984; Murphy and others, 1986). The first of these (Cadjebut) is currently under development, and the company has expressed optimism that this area will eventually prove to be a major zinc-lead province. Mineralization has also been found elsewhere in the outcropping reef complexes, notably in the Napier Range area (Buchhorn, 1986).

MORPHOLOGY OF THE REEF COMPLEXES

The barrier-reef belt developed during the Middle Devonian (Givetian) and Late Devonian (Frasnian and Famennian) as a series of reefal limestone platforms fringing a mountainous landmass of Precambrian rocks (the Kim-

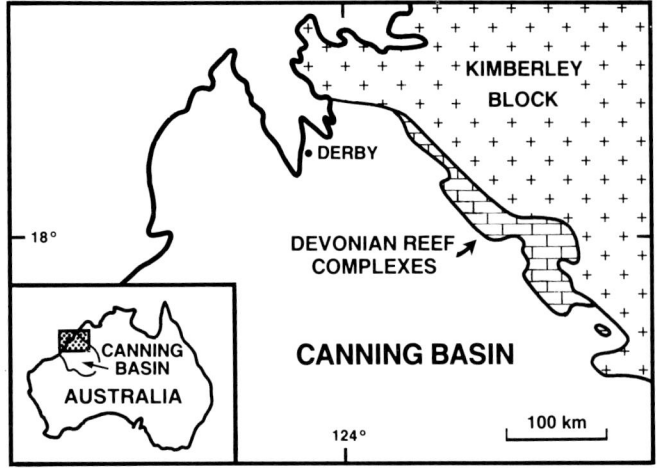

FIG. 1.—Location map.

berley Block) and around islands of Precambrian and Ordovician rocks. Basement topography, commonly fault controlled, was important in localizing platforms.

Most of the complexes grew as reef-rimmed platforms flanked by steep marginal slopes that descended to depths of as much as several hundred meters in the adjoining basins. The morphology and facies nomenclature are illustrated in Figure 4, which also shows the extent of strong early submarine cementation around the platform margins and upper marginal slopes. This cementation was very important in developing and maintaining the high-relief margins and very steep reefal slopes (Playford, 1980; Kerans and others, 1986). Some of the older Pillara platforms were not fringed by reefs, however, but were uncemented low-relief banks flanked by gentle marginal slopes and shallow basins.

Platform margins.—

The wide variety of platform margins recognized in the complexes is illustrated in Figures 5–7. The types shown are end members, which intergrade, both laterally and vertically. Upright margins are of three types: upright scarp, upright rollover, and collapsed margins (Fig. 6A). An *upright-scarp margin* formed where a well-developed reef grew vertically to form a reef scarp, with adjoining marginal-slope deposits progressively mounting up against it; an *upright-rollover margin* developed where the sea floor simply "rolled over" from a flat-topped platform to the marginal slope, without any intervening scarp (Fig. 6B); and a *collapsed margin* formed when there was submarine landsliding of a platform margin, resulting in a sheer scarp (Fig. 7).

An *advancing margin* (Fig. 6C) formed where a reefal platform expanded outward, growing over its own marginal-slope deposits, whereas a *retreating margin* (Fig. 6D) formed where a platform was shrinking steadily in extent as it grew upward. A *backstepping margin* (Fig. 14) is one where a platform retreated abruptly into the platform interior, for distances ranging from a few meters to several kilometers.

A *fault-controlled margin* (Fig. 6E) is one that formed along an active fault, downthrown to the basin. An *interfingering-bank margin* occurred where a bank, lacking any reef rim, interfingered directly with gently dipping marginal-slope deposits. Finally, *pinnacle reefs* (having breadth to height less than about 2:1; Playford, 1980) formed over the reef margins or interiors of drowned platforms, or as the final phase of drowned atolls (Figs. 6F, 8).

The different types of platform margin are believed to have formed in response to a number of factors, including basement topography, water depths in the adjoining basins, influx of terrigenous detritus, degree of submarine cementation, contemporary faulting and associated earthquakes, the types of reef-building organisms involved, and especially changes in relative sea level resulting from combined eustatism and tectonism.

The hypothetical association between the types of margin and rates of eustatic sea-level change may be summarized as follows: advancing margins–slow rise or stillstand; interfingering-bank margins–slow rise; upright-rollover margins–moderately rapid rises; retreating and upright-scarp margins–rapid rise; and backstepping margins and pinnacle reefs–very abrupt rise.

Water depths in the adjoining basins also influenced platform growth–shallow water favored interfingering-bank and advancing margins, whereas deep water favored upright margins.

Lateral gradations occur between those types of platform margin that are associated with similar rates of relative sea-level change or similar water depths.

Faulting.—

Faulting before and during growth of the complexes influenced platform development in a number of ways. It commonly controlled the basement topography, especially that of islands on which platforms were established (Hurley, 1986), whereas contemporary faulting influenced platform–basin morphologies, and associated earthquake shaking resulted in fracturing of rigid early cemented limestones and the collapse of platform margins.

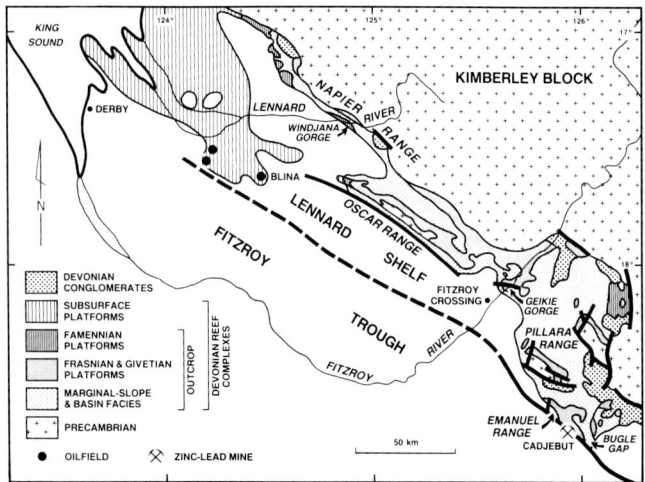

FIG. 2.—Generalized geologic map (modified after Playford, 1984).

FIG. 3.—Aerial views of the reef complexes. (A) View looking northwest over Windjana Gorge and the Napier Range, showing the exhumed fringing-reef complex. (B) View looking north over the exhumed Laidlaw Range platform atoll (right foreground) and Lloyd Hill platform atoll–pinnacle reef (left middle distance).

Growing platform margins rarely coincided with active faults; the only prominent example is the southwestern margin of the Emanuel Range platform, on the upthrown side of the Cadjebut Fault (Fig. 6E; Playford and Lowry, 1966). In addition, two small faults joining a branch of the Cadjebut Fault can be shown to have moved during the Frasnian, and they appear to have influenced growth of a reef spine at the south end of the Lawford Range platform atoll (Playford, 1981).

Logan and Semeniuk (1976, p. 2, 10, 128), on the other hand, claimed that the "boundaries between all formations and rock units are major sinuous strike faults," and that the "major rock units are bounded by parallel strike faults of major dimensions." They maintained that the rocks seen in outcrop are not reef complexes, but are the products of dynamic metamorphism, the platform margins being zones of dislocation "characterized by intense cataclastic metamorphism and metasomatism."

There is no factual basis for these conclusions, however (see discussions by Playford, 1980; Kerans, 1985; Hurley, 1986; and Wallace, 1987). Although the reef complexes are no longer "pristine" in that they have been subject to some faulting, tilting, and mild folding, the carbonates are unmetamorphosed, depositional fabrics are very well preserved, and the platform margins are, with few exceptions, unfaulted.

Submarine fracturing and debris flows.—

Contemporary fracturing of the strongly cemented limestones around the platform margins and upper marginal slopes caused periodic submarine landsliding of large sections of reef, giving rise to massive debris flows and large allochthonous reef blocks in the marginal-slope deposits, and turbidites in the basin deposits (Figs. 9–11). Where the fractured limestone remained in place, fissures were variously filled at any early stage with cement, detrital sediment, and organic growths to form extensive networks of neptunian dikes (Playford, 1984; Kerans and others, 1986).

The factors responsible for this fracturing and landsliding are summarized in Figure 10. Early cementation and contemporary earthquakes are believed to have been most important, combined with one or more of: unsupported reef

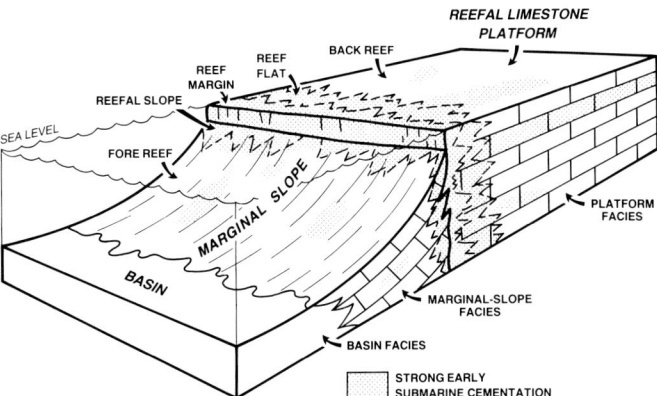

FIG. 4.—Block diagram illustrating morphology of the reef complexes, facies subdivisions, and distribution of early submarine cementation (modified after Playford, 1984).

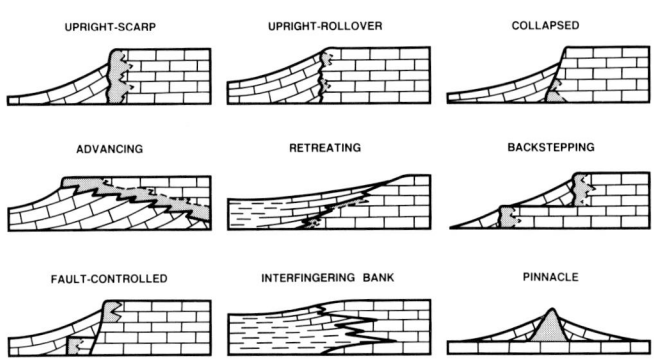

FIG. 5.—Diagrammatic sections illustrating the principal types of platform margin.

Fig. 6.—Types of platform margin. (A) Upright-scarp platform margin at Nadji Cave, Bugle Gap, showing well-bedded forereef subfacies (Sadler Limestone) on the right, abutting massive reef margin subfacies (Pillara Limestone) on the left. (B) Upright-rollover platform margin in the Pillara Range, 1 km northwest of Menyous Gap, showing marginal-slope facies (Sadler Limestone) on the left, and reef-flat subfacies (Pillara Limestone) on the right. (C) Advancing margin in the Napier Range, 5.5 km northwest of Windjana Gorge, showing massive reef (Windjana Limestone) overlying and interfingering with well-bedded marginal-slope facies (Napier Formation). (D) Retreating platform margin in the Pillara Range, 1 km southeast of Menyous Gap, showing well-bedded marginal-slope facies (Sadler Limestone) mounting over thick-bedded to massive reef-flat subfacies (Pillara Limestone). (E) Aerial view of the Cadjebut Fault, which was active during the Devonian, forming a fault-controlled margin of the Emanuel Range reefal platform. (F) Aerial view of Wade Knoll pinnacle reef (Pillara Limestone), encircled by marginal-slope facies (Sadler Limestone) and basin facies (Gogo Formation).

FIG. 7.—Panoramic view of the eastern wall of Windjana Gorge, 1 km from the eastern entrance (opposite the "Classic Face, Fig. 14), showing a collapsed platform margin, formed by submarine landsliding along a steep fracture subparallel to the reef front. Dipping marginal-slope deposits (Napier Formation) on the right abut a steep scarp of flat-lying reef margin and reef-flat subfacies (Pillara Limestone) on the left.

scarps, compaction of basin sediments, and periodic slippage along marginal-slope bedding. In some areas (for example, parts of the Napier and Oscar Ranges), discrete marginal-slope beds with abundant debris flows and allochthonous beds can be traced for many kilometers. It is thought that they reflect periods of strong seismicity causing extensive platform margin collapse and associated neptunian fissuring. None of the fracturing or debris flow development can be linked with platform emergence, which contrasts with the interpretation of Reeckman and Sarg (1986) for major debris flows in the Permian reef complex of West Texas, but is in agreement with the views of Schlager and Camber (1986) on self-erosion of limestone escarpments.

Terrigenous conglomerates.—

Contemporary faulting is believed to have led to mountainous topography in the adjoining landmass of the Kimberley Block, which was the source of the enormous masses of terrigenous-boulder conglomerate interfingering with some Frasnian and Famennian reef complexes (Playford and Lowry, 1966; Playford, 1984). These fanglomerate complexes reflect environments ranging from proximal alluvial fans through fan deltas to submarine fans. In some places the conglomerates extended right through the reefal platforms, reaching into the deep water of the adjoining basins. Sandstones and finer terrigenous deposits also interfinger with all facies of those reef complexes that directly adjoin the Kimberley Block.

The amount of terrigenous influx was important in limiting the extent of some Pillara platforms. As a result, a series of isolated reef-rimmed platforms, rather than a continuous fringing reefal platform, formed beside the shoreline in the Napier Range area during the Frasnian, when large amounts of terrigenous detritus were being shed from the adjoining Kimberley Block. These isolated platforms were succeeded in the Famennian, however, by a continuous fringing reefal platform, broken at only one locality by a major conglomerate body, the Behn Conglomerate (Playford and Lowry, 1966).

EVOLUTION OF THE REEF COMPLEXES THROUGH TIME

The evolution of the reef complexes through time is illustrated diagrammatically as a model cross section in Figure 12, and the relative change in sea-level through the same period is shown in Figure 18. Two cycles of platform development are recognized: the Givetian-Frasnian Pillara cycle, characterized by essentially vertical platform growth followed by widespread drowning and backstepping; and the Famennian Nullara cycle, characterized by strongly ad-

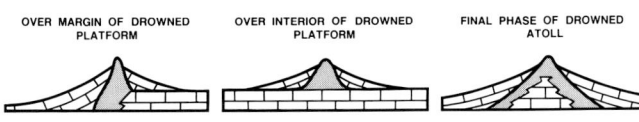

FIG. 8.—Diagrammatic sections illustrating the genesis of pinnacle reefs.

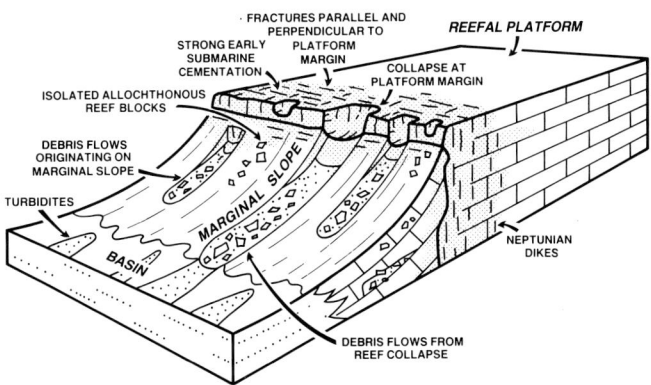

FIG. 9.—Block diagram illustrating the development of debris flows, allochthonous reef blocks, turbidites, and neptunian dikes (modified after Playford, 1984).

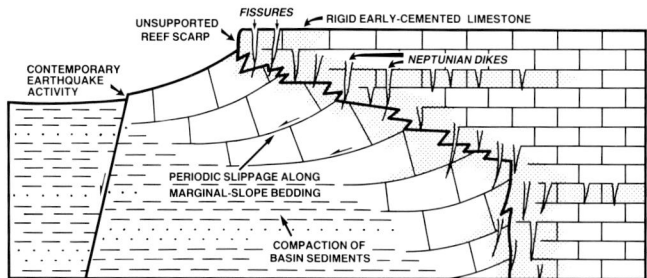

FIG. 10.—Diagrammatic cross section illustrating factors responsible for neptunian fracturing and submarine landsliding in early cemented limestones (modified after Playford, 1984).

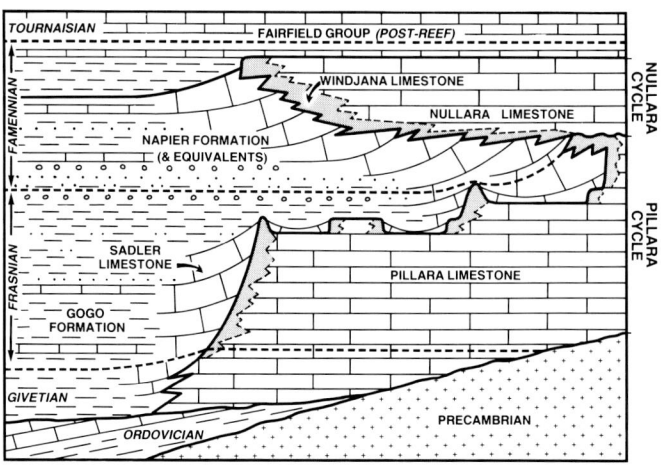

FIG. 12.—Diagrammatic section illustrating development of the complexes through time (modified after Playford, 1984).

vancing platforms. The total maximum thickness of the platform deposits of the two cycles is about 2,000 m, whereas seismic data suggest that contemporary basinal facies, composed of terrigenous clastics and turbidites, are as thick as 2,500 m.

The growth history of the reef complexes now exposed on the Lennard Shelf was one of almost continuous relative rise in sea level, reflecting major subsidence during deposition. Eustatic sea-level changes were also thought to have been important, but their overall effects were subordinate to those of basin subsidence. Emergence of the platforms occurred only briefly and rarely; the sole really significant regression that has been identified was at the Frasnian-Famennian boundary, when the platforms emerged briefly for not more than a few tens of meters above sea level, while deposition continued without break in the adjoining basins. Although it seems probable that there were a number of other minor episodes of platform emergence, few have been recognized from outcrop studies (Playford, 1980, 1982; Hurley, 1986).

Pillara cycle.—

The oldest platforms of the Pillara cycle began growth during the mid- to late Givetian. They developed on a basement of igneous-metamorphic Precambrian rocks or (in the Emanuel Range area) Ordovician dolomite and shale. These early Pillara platforms were low-relief stromatoporoid, coral, and cyanobacterial banks and biostromes, with little or no reef around their margins. They were flanked by, and interfingered with, gently dipping marginal-slope limestone and shale, which extended into shallow basins (water depths a few tens of meters) containing dark shales. Some banks interfingered directly with these basinal shales.

The rate of relative rise in sea level increased significantly during the early to mid-Frasnian. The platforms generally grew vertically upward, with well-developed reef margins, adjoining progressively deeper basins, in which water depths were as much as several hundred meters (Playford, 1980). This was followed in the late Frasnian by two major and several lesser episodes of abrupt relative sea-level rise, associated with widespread drowning of platforms, backstepping of platform margins, and the development of isolated platform atolls and pinnacle reefs (Fig.

FIG. 11.—(A) Debris flow deposit in marginal-slope facies (Napier Formation) at the southern entrance to Dingo Gap, Napier Range. (B) Allochthonous block of reef limestone in forereef deposits (Napier Formation), 0.5 km south of the east entrance to McSherry's Gap, Napier Range.

12). Figure 13 shows two examples of drowning and the associated development of pinnacle reefs in the late Frasnian of the Bugle Gap area, at Teichert Hills and northern Laidlaw Range. In each case, pinnacle reefs developed along the line of the old (pre-drowning) platform margin, while others formed over the interior of the old platform (cf. Kendall and Schlager, 1981). This late Frasnian drowning period is potentially important to oil explorers because the drowned platforms were, at least in some cases, sealed beneath basinal shales.

A small backstepping episode in the late Frasnian is evidenced at the "Classic Face" in Windjana Gorge (Fig. 14), where the margin retreated by only some 10 m, compared with hundreds of meters in examples at other localities. The interesting feature of the well-exposed Windjana Gorge example, however, is that the backstepped margin overlies an erosion surface in the underlying platform, with a total relief of about 8 m (Playford, 1980, 1981). This erosion surface is only evident near the platform margin—it disappears passing back into the platform interior. The most likely explanation is that it marks a very brief period when the platform emerged a few meters above sea level and its edge was eroded. Sea level then rose abruptly and the margin stepped back a short distance to be reestablished on a high point near the edge of the older eroded platform (Fig. 14). It is not known whether other backstepping events are similarly preceded by platform emergence; exposures elsewhere are not as good as in this, the "Classic Face."

The intraplatform erosion surface at the "Classic Face" is continuous with the platform margin unconformity that is so well exposed there, so it is likely that part of this near-vertical margin was subject to subaerial exposure. Similar platform margin unconformities are usual features of other upright margins, however, where they have apparently formed by normal submarine processes of reef growth, contemporary erosion, and progressive onlap by marginal-slope deposits, without any evidence of subaerial exposure.

In latest Frasnian times, when most of the platform areas had been drowned, there was a marked deceleration in the rate of relative rise in sea level, and the remaining platforms advanced nearly horizontally (Figs. 12, 18). This short interval was followed, at the Frasnian-Famennian boundary, by a very brief sea-level fall of a few tens of meters, when the upper surfaces of the platforms were exposed to subaerial erosion. The total amount of section eroded away at that time probably amounted to only a few meters, while deposition, although condensed, continued without break on the adjoining marginal slopes and basin floors (Druce, 1976; Playford and others, 1984). The erosion surface on the platforms is generally planar in the northwestern Oscar Range (Hurley, 1986), but at the only other locality where it is exposed, in the Napier Range, it shows evidence of mild karstification, with the development of small sinkholes as deep as about 1 m.

The Famennian platforms of the Nullara cycle were established on the eroded Pillara platforms, maintaining the same style of reefal sedimentation, with comparable platform, marginal-slope, and basin facies. The main change was in the platform-building reefal biota, as discussed below.

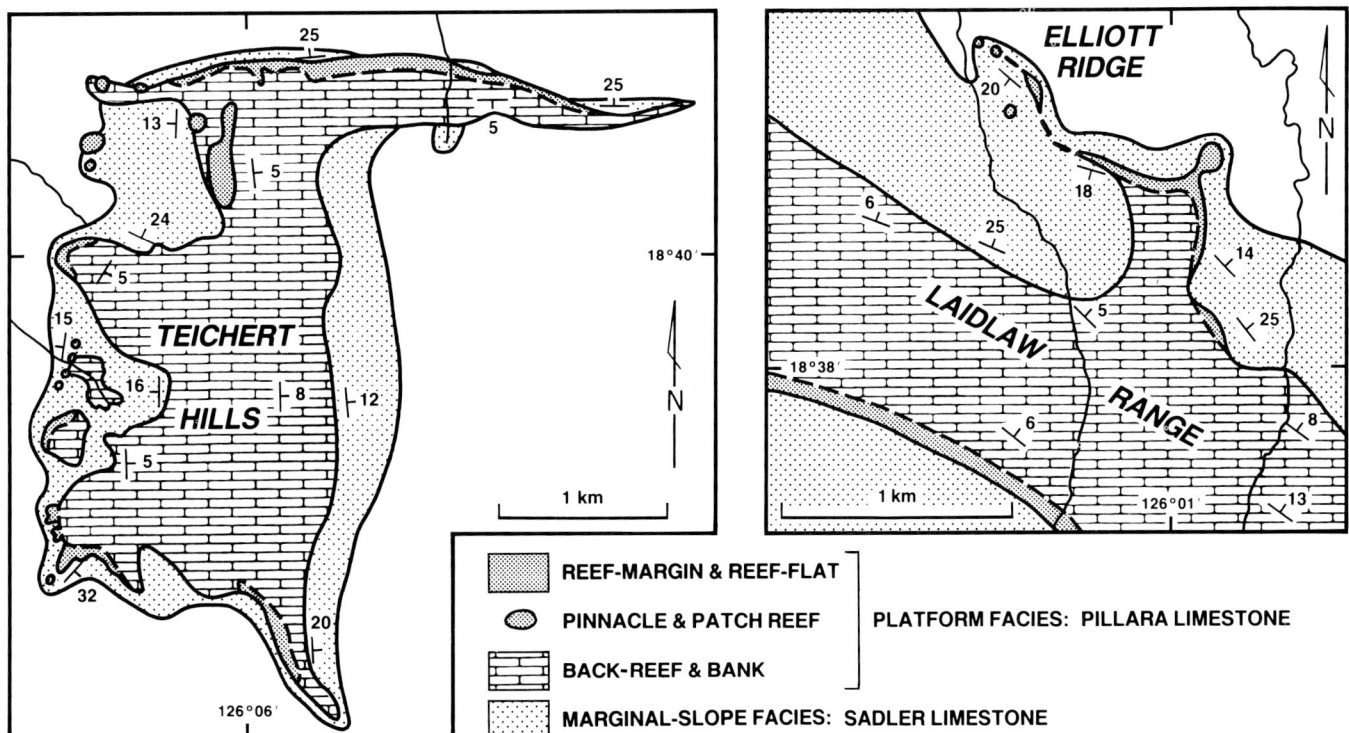

FIG. 13.—Geologic maps of the Teichert Hills and northern Laidlaw Range (Bugle Gap area) illustrating late Frasnian platform drowning and the development of pinnacle reefs over the old platform margins.

Fig. 14.—The "Classic Face" at Windjana Gorge. (A) Panoramic view (the wall is about 80 m high) and (B) closer view showing: (1) the intraplatform erosion surface, (2) the platform margin unconformity, and (3) associated minor backstepping of the platform margin.

Frasnian-Famennian mass extinction.—

The regression at the close of the Frasnian marked the end of the Pillara cycle of reef development. It coincided with the world-wide mass extinction of metazoan reef-builders and other shallow-water animal life (McLaren, 1970; House, 1975; Playford and others, 1984). The reef-building stromatoporoid-cyanobacteria-coral association of the Pillara cycle (Cockbain, 1984; Wray, 1967; Hill and Jell, 1970) was replaced almost entirely by cyanobacteria in the Nullara cycle.

Through much of the Lennard Shelf, the mass extinction is marked on the marginal slopes and drowned Frasnian platforms by spectacular developments of deep-water stromatolites (Playford and others 1976; Hurley, 1986). Playford and others (1984) found a significant iridium anomaly in such stromatolites at or near the Frasnian-Famennian boundary, and hypothesized that this could have resulted from a major meteoroid impact on earth, which in turn caused the mass extinction. Alternatively, they suggested that the iridium could have been concentrated from sea-water, together with other metals, by the iron cyanobacterium *Frutexites*, which is abundant in the stromatolites. This second hypothesis has since been confirmed through subsequent work by Playford, Hurley, R. S. Nicoll, and C. J. Orth, details of which remain to be published. The close coincidence between the widespread bloom of deep-water stromatolites and the mass-extinction event may have resulted from the abrupt decline in metazoan species that fed on cyanobacteria.

Nullara cycle.—

The Famennian platforms of the Nullara cycle were established on eroded Frasnian platforms of the Pillara cycle. During most of the Famennian, there was a slow relative rise in sea level, and the cyanobacterial platforms advanced basinward for as much as 3 km, interfingering with the marginal-slope deposits over which they grew (Figs. 6C, 12, 18). Water depths in the adjoining basins became progressively shallower, but probably still exceeded 100 m in most areas.

Extensive neptunian fissuring and platform margin collapse occurred during growth of the Nullara platforms, due partly to the compaction of basin sediments and slippage along marginal-slope bedding below the platforms (Fig. 10).

The rate of relative sea-level rise increased near the end of the Nullara cycle, so that the platform margins grew vertically upward, forming steep reef scarps. Termination of the Nullara cycle and initiation of the mixed siliciclastic-carbonate depositional sequence of the Fairfield Group probably resulted from a sudden rise in sea level. Evidence of this is seen in the Blina field, where high-energy shoal carbonates of the Nullara platform are succeeded by peloid-crinoid wackestones, which in turn grade rapidly into shales of the Fairfield Group (Fig. 22). This sequence is interpreted as indicating progressive deepening, with the rate of sea-level rise being sufficiently rapid to cause platform drowning (cf. Schlager, 1981). The steep reef scarps formed at the end of the Nullara cycle were then buried below post-reef shales and limestones of the Fairfield Group (Fig. 12).

Evolution of the reef complexes through time is illustrated in Figure 15 as a series of map and cross-sectional views of platform development in the Oscar Range area (after Hurley, 1986). Basement topography, in the form of an uplifted fault block of Precambrian metasedimentary rocks, provided the site for growth of a Frasnian reef complex; a rapid relative rise in sea level during the late Frasnian led to the development of pinnacle reefs and back-stepped platform margins; and after a brief emergence at the close of the Frasnian, the succeeding Famennian platform advanced strongly, eventually burying the Precambrian core.

Backreef cyclicity.—

Regular cyclicity in backreef deposits of the Pillara cycle has been recorded by Read (1973a, b) and Playford (1981). This cyclicity is best known in the Givetian and early Fras-

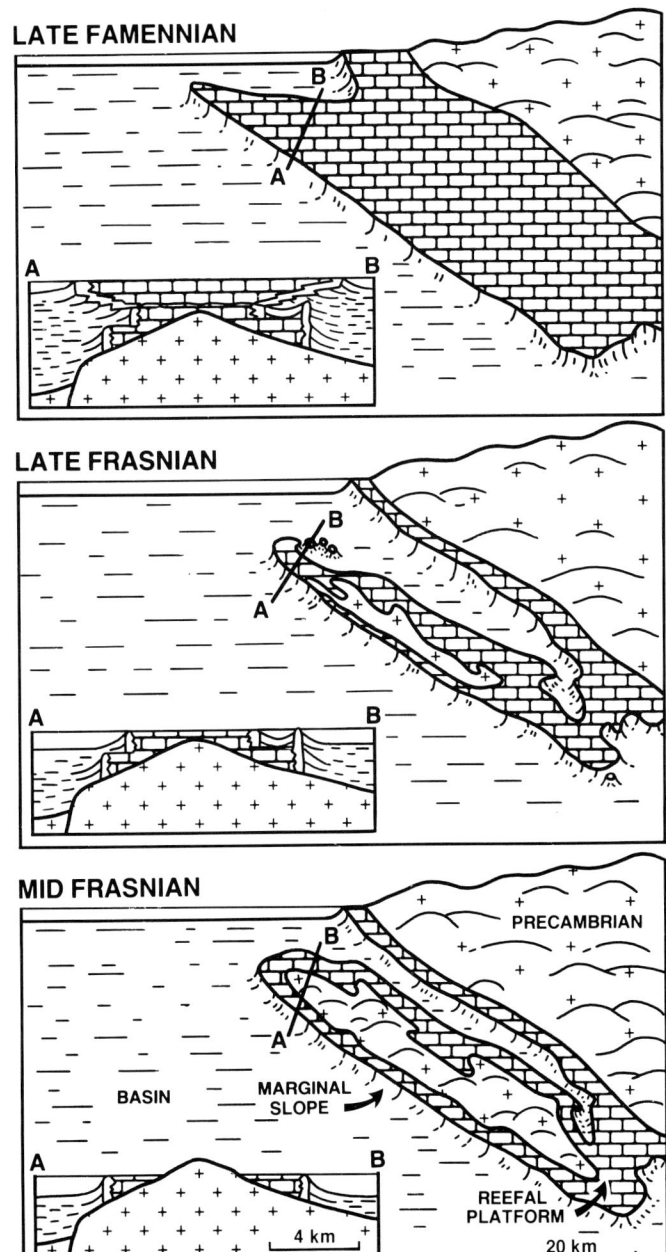

FIG. 15.—Series of diagrammatic views and cross sections illustrating evolution of the Oscar Range reef complex from mid-Frasnian to late Famennian times (modified after Hurley, 1986).

nian of the Pillara Range and the late Frasnian of Windjana Gorge, but it is not always recognizable elsewhere for example, in most of the Pillara Limestone in the Oscar Range (Hurley, 1986). Shoaling-upward cycles also occur in the Famennian Nullara Limestone in some areas, including the type section in the northwestern Oscar Range (Hurley 1986).

Cyclicity in the Pillara Limestone backreef deposits at Windjana Gorge is illustrated in Figures 16 and 17. At this locality, 31 carbonate terrigenous cycles have been measured over 153 m of section, ranging in thickness from 1.3 to 12.6 m and averaging about 5 m. They consist of shal-

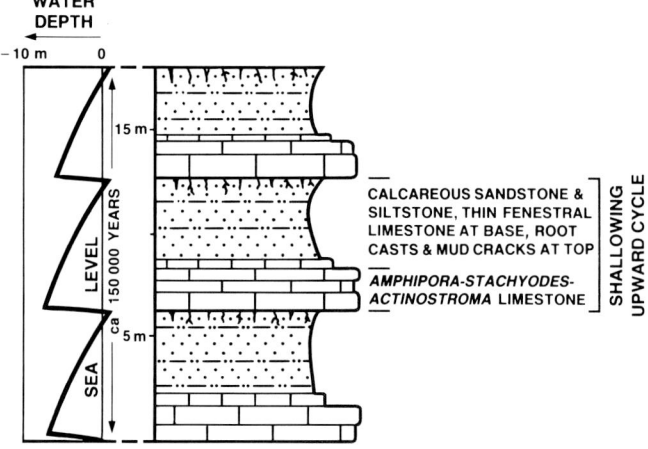

Fig. 16.—Diagrammatic columnar section illustrating the typical cyclicity developed in backreef deposits of the Pillara Limestone at Windjana Gorge. The average duration of individual cycles is approximately 50 ka.

lowing-upward cycles, with characteristics and deduced changes in water depth as shown in Figure 16. Each cycle began with an abrupt relative rise in sea level, followed by gradual shallowing through sedimentation, until the sediment surface built to sea level or slightly above.

The average duration of each cycle is estimated to be about 50 ka based on time averaging of the Frasnian section, assuming that the latest Frasnian section exposed at Windjana Gorge was deposited over an interval of about 1.5 Ma (or a little more than one-fifth of the 7 Ma estimated for the Frasnian by Harland and others, 1982).

Read (1973a, b) described carbonate and other shallowing-upward cycles in the Pillara Limestone in some detail. In the Pillara Range he measured about 70 cycles over 440 m of section in the late Givetian and early Frasnian. We estimate that this section was deposited over about 4 Ma, placing the average length of each cycle at 57 ka–reasonably close to the estimate of 50 ka for the Windjana Gorge cycles.

Eustatic versus tectonic controls.—

Figure 18 illustrates the cumulative changes in relative sea level in the Devonian of the Canning Basin that resulted from a combination of basin subsidence and global eustatism. This can be compared with the Euramerican eustatic curve shown in Figure 19. Figure 20 shows a burial geohistory plot of the Devonian and younger section encountered in the Blina 1 well (Figs. 21, 22).

The tectonic and eustatic components of relative sea-level change in the Canning Basin have not been definitively separated, although eustatism is deemed more likely to have controlled the short-term cycles shown in Figures 16 and 17. Their regularity suggests global control, possibly reflecting Milankovitch-cycle changes in ocean temperatures.

We do know that major subsidence continued in the Canning Basin from Givetian to Famennian times, in order to accommodate the thick (maximum about 2,000 m) reefal sequence, and changes in the rate of basin subsidence are presumably reflected in some features of the relative sea-level curve. Characteristics of that curve (Fig. 18), which combines changes due to both tectonism and eustatism, are the lack of significant regressive intervals, other than the brief regression at the Frasnian-Famennian boundary, and the lack of detailed correlation with most of the Euramerican Devonian eustatic curve (Fig. 19).

Fig. 17.—Low-level aerial view of cycles (each averaging about 5 m thick) in backreef Pillara Limestone, 0.5 km north of the eastern entrance to Windjana Gorge. The cycles are as illustrated in Figure 16, each beginning with a prominently outcropping unit of stromatoporoid limestone, followed by a recessive, largely terrigenous, unit.

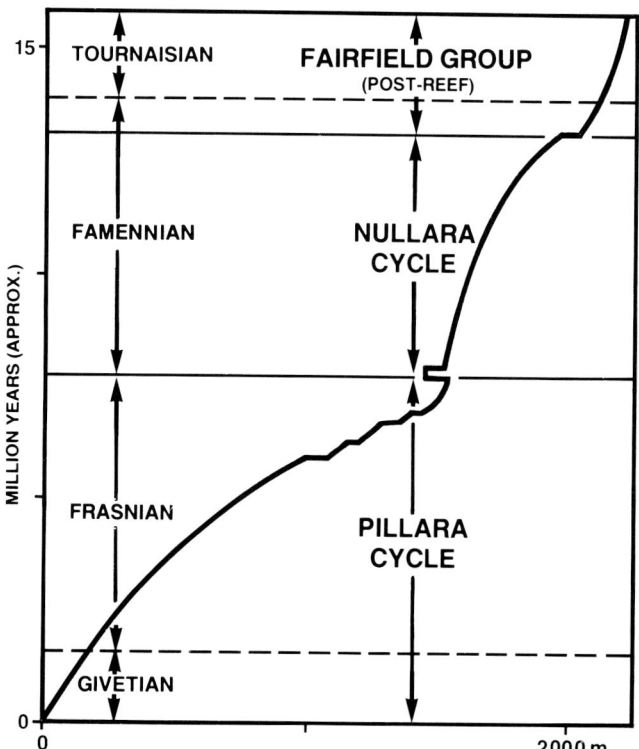

Fig. 18.—Canning Basin relative sea-level curve illustrating the combined result of tectonism (basin subsidence) and eustatism from the mid-Givetian to early Tournaisian.

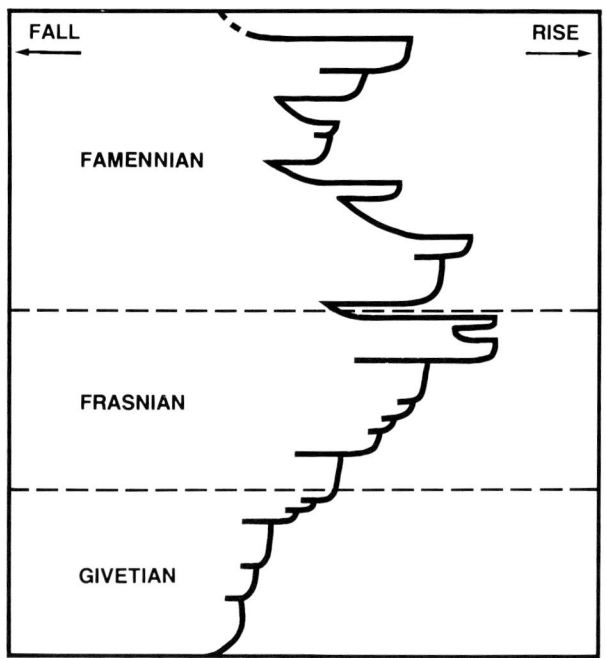

FIG. 19.—Qualitative eustatic sea-level curve for the Givetian to Famennian of Euramerica (after Johnson and others, 1985).

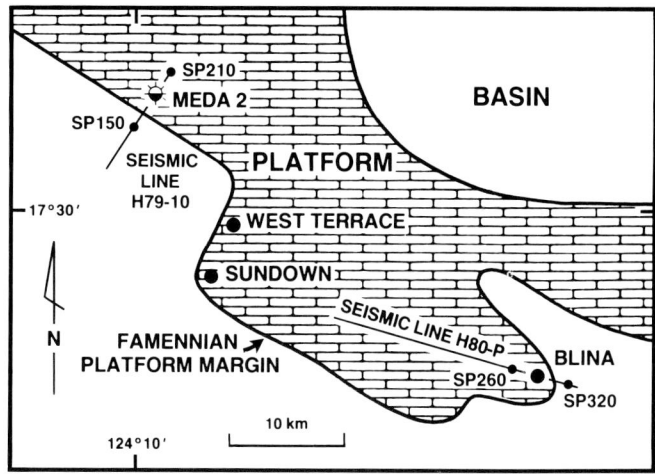

FIG. 21.—Map of the subsurface Famennian Blina-Meda limestone platform, showing the locations of oil fields and the seismic lines illustrated in Figures 23–26.

iferous, when there was significant uplift and erosion throughout the Lennard Shelf area, resulting in the removal of part of the post-reef Fairfield Group. This was followed

Some aspects of the Canning Basin curve do correlate generally with the Euramerican curve, however, suggesting eustatic control. In particular, widespread drowning of reef complexes occurred in the late Frasnian of both areas, associated with pulses of rapid rise in relative sea level. Moreover, there was a brief regression at the Frasnian-Famennian boundary in both areas.

On the other hand, the repeated sea-level falls evidenced in the Famennian of Euramerica (Johnson and others, 1985) are not recognized in the Canning Basin, where they may have been counterbalanced by continued basin subsidence. It is important to note, in this connection, that Western Australia is the only part of the world where major reef growth extended from the Frasnian into the Famennian.

The geohistory plot (Fig. 20) of the Blina 1 section shows increasing burial of the reef complexes to the mid-Carbon-

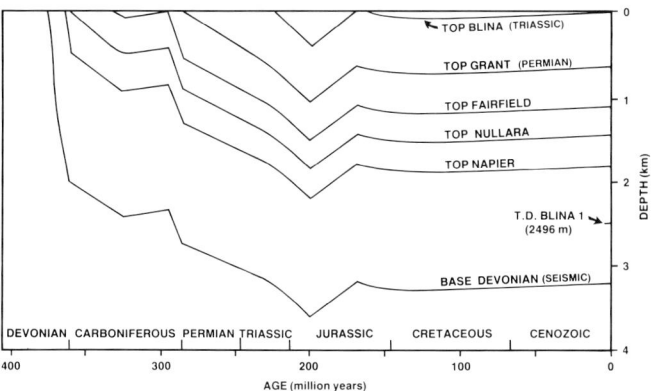

FIG. 20.—Burial geohistory plot, Blina 1 well.

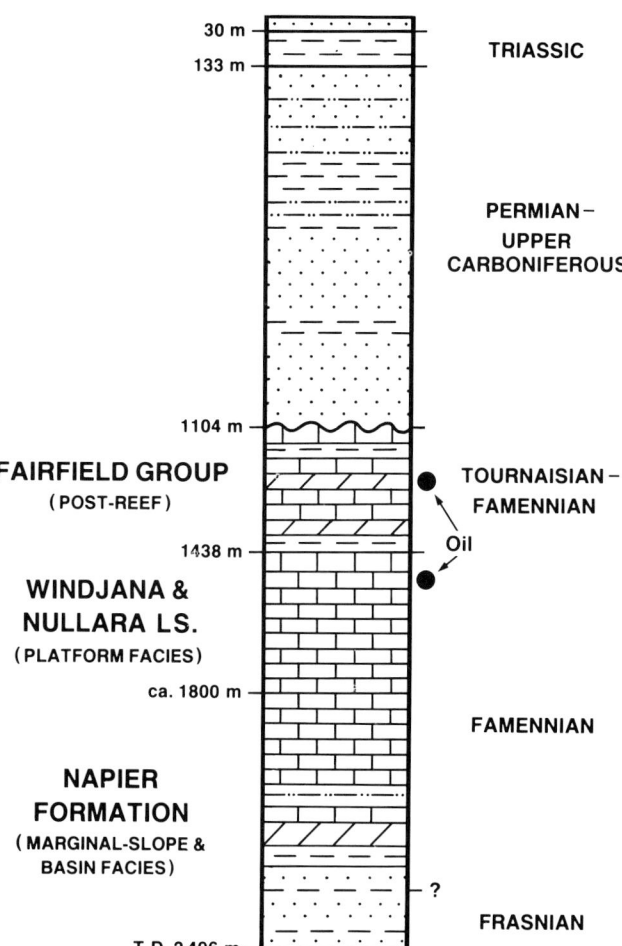

FIG. 22.—Section penetrated in Blina 1 well (for location see Fig. 21).

by renewed subsidence in the Late Carboniferous, continuing throughout the Permian and Triassic. There was a pulse of compressive tectonism in the Early Jurassic, with right lateral wrench movement along the bounding faults of the Fitzroy Trough, and associated folding of Triassic and older sediments within the trough. Minor renewed subsidence occurred during the Late Jurassic and Early Cretaceous, and since then gradual erosion has occurred. In the outcrop area this has stripped off most of the Carboniferous and younger section above the reefal platforms, but in the subsurface of the Blina-Meda area, erosion has not reached below the Triassic sequence (Fig. 20).

SEISMIC STRATIGRAPHY

The first wells to be drilled with reef objectives (Meda 1 and 2) were based on perceived reefal seismic anomalies, and both encountered significant oil and gas shows in the Fairfield Group and underlying reef complex. The first commercial oil field, Blina, was found by drilling a well-defined Famennian platform margin delineated by seismic surveys (Playford, 1982; Moors and others, 1984; Figs. 21–26). The oil field is in backreef and reefal limestone of the Nullara cycle and dolomite of the overlying Fairfield Group (Fig. 22). The Sundown and West Terrace oil fields are in Carboniferous-Permian sandstones (Grant Group) above the Famennian platform margin.

Seismic surveys (Figs. 23–26) have accurately delineated the Famennian platform margin, but they have not yet clearly defined any Givetian-Frasnian platforms of the Pillara cycle. Meda 1 and 2 are the only wells in this area that appear to have intersected a Pillara platform, and an interpretation of a seismic line passing through Meda 2 is shown as Figure 26. The Famennian platform at Blina (Fig. 25) had presumably advanced basinward beyond the edge of the older Pillara platform. Consequently, only basinal deposits were penetrated in the Frasnian section of this well.

Sediment wedges of two types, occurring several kilometers basinward of the platform margin, are displayed on the seismic sections. Large wedges, showing toplap reflections (Fig. 24, shotpoint 150, 1.5-1.6 sec), are interpreted by Middleton (1987a) as submarine fans deposited during the highstand of sea level following drowning of the Famennian platforms. This interpretation is consistent with the model of Brown and Fisher (1977). Smaller wedges, generally occurring within one kilometer of the platform margin (Fig. 23, shotpoint 304, 1.3 sec; Fig. 24, shotpoint 160, 1.5-1.6 sec), may represent either Frasnian-Famennian-boundary "lowstand wedges" (Haq and others, 1987), or debris flow deposits.

A drilling program being carried out in the Blina-Meda area during 1987–1988 should assist in determining whether prospective Pillara reefal platforms occur in the area. Such platforms are likely to have better prospects for oil than

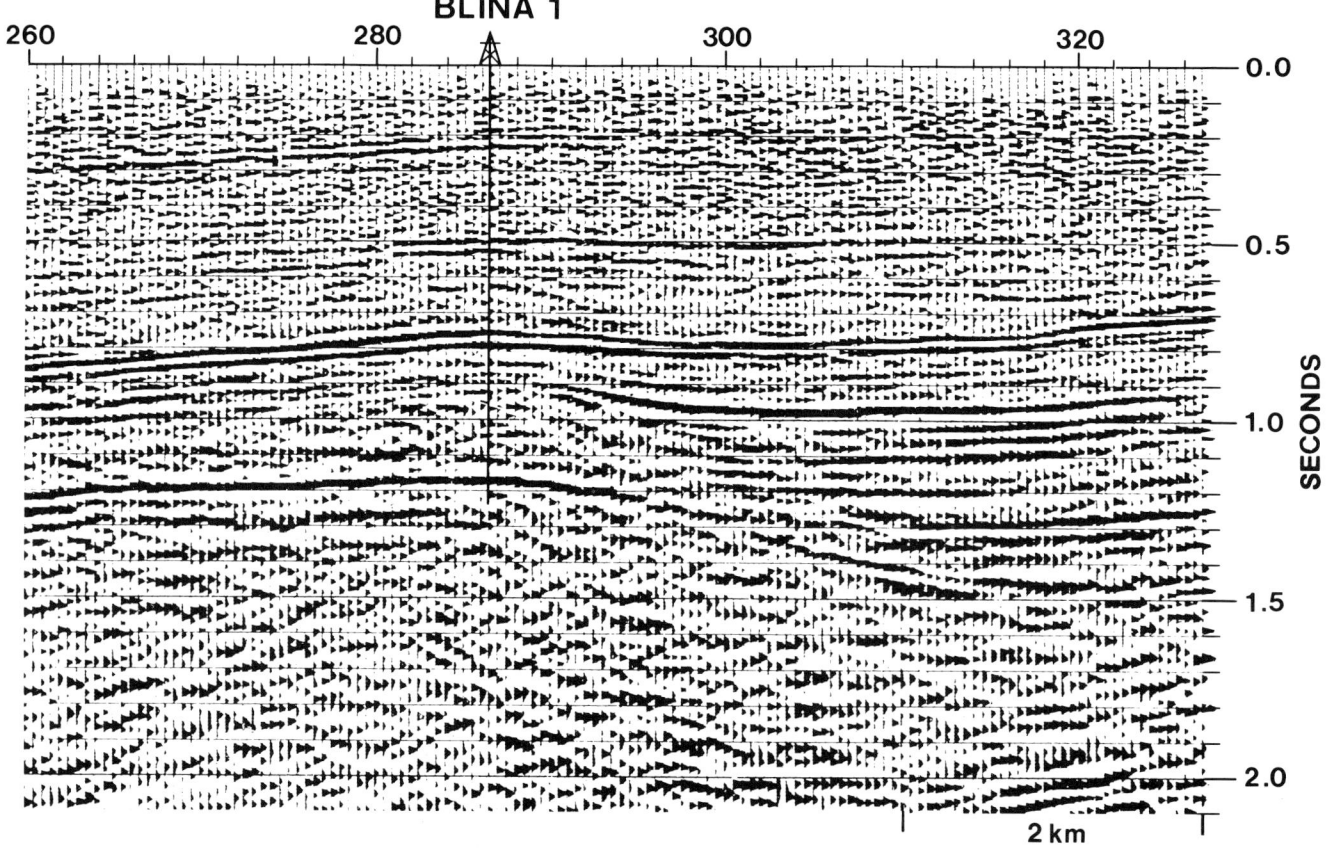

Fig. 23.—Record section, part of seismic line H80-P through the Blina 1 well.

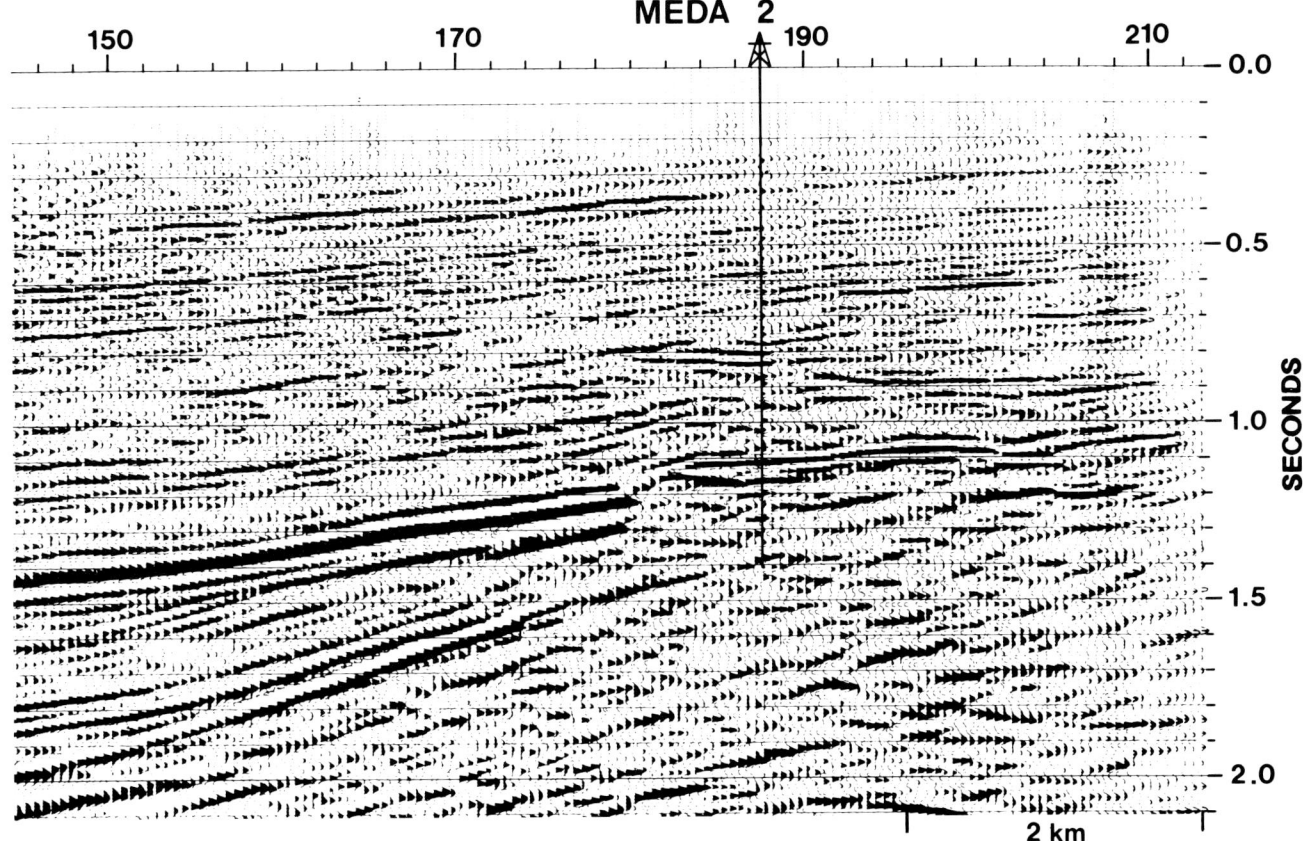

FIG. 24.—Record section, part of seismic line H79-10 through the Meda 2 well.

those of the Nullara cycle, because they are more dolomitized and consequently contain better reservoirs (Kerans, 1985). They also potentially include drowned late Frasnian pinnacle reefs and platform atolls, sealed by basinal shales with known oil-source potential (Playford, 1982; Alexander and others, 1985).

The Devonian seismic stratigraphy of the area has recently been studied by Middleton (1984, 1987a, b). He has sought to determine how the platforms of the Pillara cycle may be expressed seismically below the advancing platforms of the Nullara cycle. His seismic modeling of the outcropping reef complexes can be matched reasonably well with actual seismic records of lines shot over the subsurface Blina-Meda platform. A synthetic seismogram derived from a model cross section of the outcropping reef complexes, using observed velocities and densities of the various facies and rock types intersected in wells, is shown as Figure 27. It compares reasonably well with a seismic interpretation of the Famennian and Frasnian reefal section intersected in the Meda wells (Fig. 26).

SUMMARY AND CONCLUSIONS

This paper has outlined some of the factors controlling the evolution and morphology of the Canning Basin reefal platforms during the Middle and Late Devonian, emphasizing the importance of the rates of rise (or fall) in relative sea level due to combined eustatism and tectonism. Other significant factors are believed to include: basement topography, water depths in adjoining basins, degree of submarine cementation, and contemporary faulting and associated earthquakes.

The rate of change in relative sea level is thought to have been especially important in controlling platform morphology. Advancing platforms were associated with slow rises or stillstands; banks with slow rises (and shallow basins); upright-rollover margins with moderate rises; retreating and upright-scarp margins with rapid rises; and backstepping, pinnacle reefs, and drowning with abrupt rises. Shoaling-upward cycles in backreef limestones resulted from periodic abrupt rises (probably eustatic) of a few meters, followed by longer stillstands.

Two main cycles of platform development are recognized: the Givetian-Frasnian Pillara cycle and the Famennian Nullara cycle. The upright and backstepped platforms of the Pillara cycle formed mainly in response to rapid rises in relative sea level, punctuated by episodes of abrupt rise, whereas the advancing Nullara platforms resulted from a prolonged period of slower rise.

The earliest (Givetian and earliest Frasnian) platforms of the Pillara cycle were low-relief banks; later platforms were normally reef rimmed, with high relief. Upright-reef margins predominated in the Frasnian, with intervals of backstepping in the late Frasnian associated with widespread

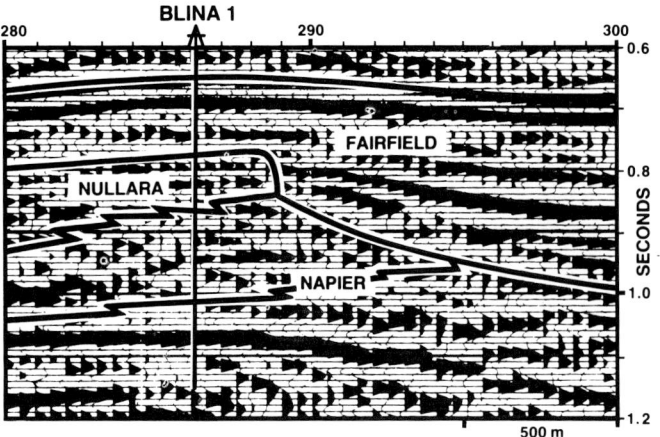

FIG. 25.—Enlargement of part of the record section shown in Figure 23, showing an interpretation of the Famennian reef complex.

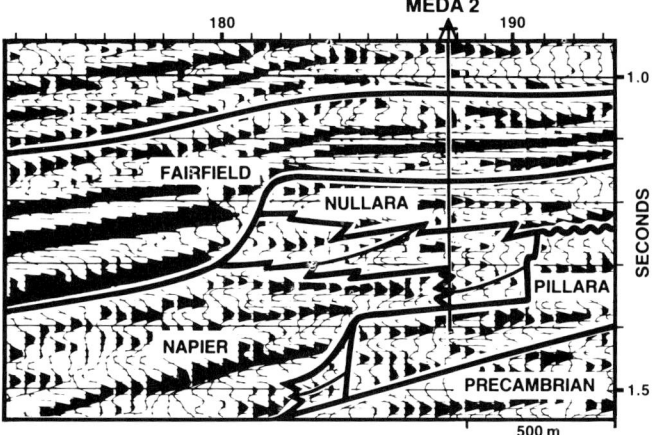

FIG. 26.—Enlargement of part of the record section shown in Figure 24, showing an interpretation of the Famennian and Frasnian reef complexes.

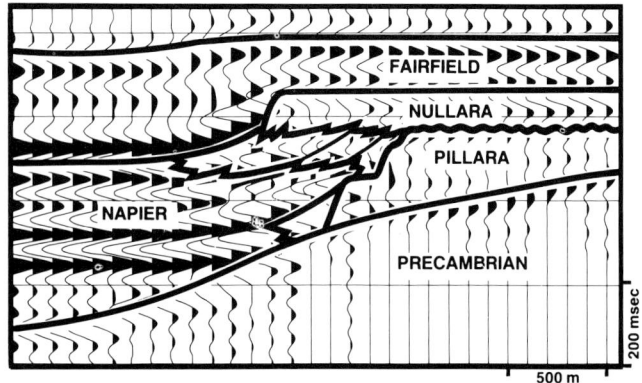

FIG. 27.—Synthetic seismogram based on the superimposed model cross section through a Devonian reef complex.

drowning and the development of pinnacle reefs. A brief emergence occurred at the close of the Frasnian, and the succeeding reef-fringed Famennian platforms of the Nullara cycle advanced basinward over their equivalent marginal-slope and basin facies. Platform extinction in the late Famennian resulted from abrupt drowning.

A world-wide mass extinction of shallow-water metazoan organisms occurred at the close of the Frasnian, and this was marked in the Canning Basin by the disappearance of nearly all reef-building metazoans. Whereas the Pillara-cycle platforms were constructed by stromatoporoid-cyanobacteria-coral associations, the succeeding Famennian platforms of the Nullara cycle were built almost entirely by cyanobacteria.

In deeper water deposits the Frasnian-Famennian mass extinction event is marked by widespread stromatolites, which may have bloomed then because of the elimination of metazoans that fed on cyanobacteria. The reason for the mass extinction has not been resolved, but previous evidence for a possible meteoroid-impact origin has now been discounted; the iridium anomaly that occurs in stromatolites at the Frasnian-Famennian boundary apparently resulted from cyanobacterial concentration of this metal and other elements.

Faulting before initiation of the reef complexes was important in controlling the topography that localized certain reefal platforms, especially in the case of islands. Continued faulting during growth of the complexes controlled or influenced a few platform margins, while associated earthquake activity caused extensive fracturing of the pervasively submarine-cemented platform margins and upper marginal slopes, leading to submarine landsliding, massive debris flows, and the development of extensive neptunian-dike networks.

The relative sea-level curve for the Devonian of the Canning Basin shows increasing rates of sea-level rise from the Givetian through the Frasnian, culminating in several abrupt rises during the late Frasnian, when wide areas of the platforms were drowned. Following a brief emergence at the Frasnian-Famennian boundary, there was a slow relative rise in sea level during most of the Nullara cycle, with an increasing rate toward the end of the cycle. Platform growth was finally terminated by abrupt drowning in the late Famennian.

There is only partial resemblance between the Euramerican sea-level curve and that of the Canning Basin, and it seems probable that continuing strong subsidence and rapid growth of reefal platforms in the Canning Basin counteracted some of the eustatic effects observed in Euramerica. Some eustatic correlation between the two areas is evidenced, however, by abrupt relative rises in sea level during the late Frasnian, followed by a brief regression at the Frasnian-Famennian boundary. There is no clear evidence in the Canning Basin of the Famennian regressions that have been identified in Euramerica.

Seismic stratigraphic studies suggest that the evolutionary model for the Devonian reef complexes deduced from outcrop studies may also apply to the subsurface Blina-Meda reef complex. A synthetic seismogram developed from simplified model cross sections of the outcropping reef complexes is comparable with actual record sections and known well intersections.

Three small oil fields have been found within and above the Famennian reef margin, which is clearly defined seismically. There has been little success to date in seismic definition of the Pillara platforms, but it is suggested that mid- to late Frasnian platform atolls and pinnacle reefs of the Pillara cycle are likely to have the best oil prospects. The recent discovery of Mississippi Valley zinc-lead orebodies in Pillara platforms in the outcrop area is also of considerable economic significance.

ACKNOWLEDGMENTS

We are grateful for help received from the many geoscientists who have participated in research sponsored by the Geological Survey on the Canning Basin Devonian carbonates. Their work has contributed greatly to the synthesis presented in this paper. We would especially like to acknowledge recent assistance and advice received from A. E. Cockbain, W. J. Meyers, V. A. Pedone, M. W. Wallace, and B. Ward.

REFERENCES

ALEXANDER, R., CUMBERS, K. M., HARTUNG, B., AND KAGI, R. I., 1985, Petroleum geochemistry of the Canning Basin: Western Australian Mining and Petroleum Research Institute, Report 22, 119 p.

BENN, C. J., 1984, Facies changes and development of a carbonate platform, east Pillara Range, in Purcell, P. G., ed., The Canning Basin, Western Australia: Geological Society of Australia and Petroleum Exploration Society of Australia, Canning Basin Symposium, Perth, p. 243-248.

BROWN, J. F., AND FISHER, W. L., 1977, Seismic-stratigraphic interpretation of depositional systems: Examples from Brazilian rift and pull-apart basins, in Payton, C. D., ed., Seismic Stratigraphy—Applications to Hydrocarbon Exploration: American Association of Petroleum Geologists Memoir 26, p. 213-248.

BUCHHORN, I. J., 1986, Geology and mineralization of the Wagon Pass prospect, Napier Range, Lennard Shelf, Western Australia, in Berkman, D. A., ed., Geology and Exploration: Council of Mining and Metallurgical Institutions, Thirteenth Congress Publications, v. 2, p. 163-172.

COCKBAIN, A. E., 1984, Stromatoporoids from the Devonian reef complexes, Canning Basin, Western Australia: Geological Survey of Western Australia Bulletin 129, 108 p.

COOPER, R. W., HALL, W. D. M., AND STYLES, G. R., 1984, The Devonian stratigraphy of the central Pillara Range in Purcell, P. G., ed., The Canning Basin, Western Australia: Geological Society of Australia and Petroleum Exploration Society of Australia, Canning Basin Symposium, Perth, p. 249-234.

DRUCE, E. C., 1976, Conodont biostratigraphy of the Upper Devonian reef complexes of the Canning Basin, Western Australia: Australia, Bureau of Mineral Resources Bulletin 158, 303 p.

HALL, W. D. M., 1984, The stratigraphy and structural development of the Givetian-Frasnian reef complex, Limestone Billy Hills, western Pillara Range, W. A., in Purcell, P. G., ed., The Canning Basin, Western Australia: Geological Society of Australia and Petroleum Exploration Society of Australia, Canning Basin Symposium, Perth, p. 235-244.

HAQ, B. U., HARDENBOL, J., AND VAIL, P. R., 1987, Chronology of fluctuating sea levels since the Triassic: Science, v. 235, p. 1156-1167.

HARLAND, W. B., COX, A. V., LLEWELLYN, P. G., PICKTON, C. A. G., SMITH, A. G., AND WALTERS, R., 1982, A Geologic Time Scale: Cambridge University Press, Cambridge, England, 131 p.

HILL, D., AND JELL, J. S., 1970, Devonian corals from the Canning Basin, Western Australia: Geological Society of Western Australia Bulletin 123, 158 p.

HOUSE, M. R., 1975, Faunas and time in the marine Devonian: Yorkshire Geological Society, Proceedings, v. 40, p. 459-490.

HURLEY, N. F., 1986, Geology of the Oscar Range Devonian reef complex, Canning Basin, Western Australia: Unpublished Ph.D. Dissertation, University of Michigan, Ann Arbor, 269 p.

———, AND VAN DER VOO, R., 1987, Paleomagnetism of Upper Devonian reefal limestones, Canning Basin, Western Australia: Geological Society of America Bulletin, v. 98, p. 138-146.

JOHNSON, J. G., KLAPPER, G., AND SANDBERG, C. A., 1985, Devonian eustatic fluctuations in Euramerica: Geological Society of America Bulletin, v. 96, p. 567-587.

KENDALL, C. ST. G., AND SCHLAGER, W., 1981, Carbonates and relative changes in sea-level: Marine Geology, v. 44, p. 181-232.

KERANS, CHARLES, 1985, Petrology of Devonian and Carboniferous carbonates of the Canning and Bonaparte Basins, Western Australia: Western Australian Mining and Petroleum Research Institute, Report 12, 223 p.

———, HURLEY, N. F., AND PLAYFORD, P. E., 1986, Marine diagenesis in Devonian reef complexes of the Canning Basin, Western Australia, in Schroeder, J. H., and Purser, B. H., Reef Diagenesis: Springer-Verlag, Berlin, p. 357-380.

LOGAN, B. W., AND SEMENIUK, V., 1976, Dynamic metamorphism; Processes and products in Devonian carbonate rocks, Canning Basin, Western Australia: Geological Society of Australia, Special Publication No. 6, 138 p.

MCLAREN, D. J., 1970, Time, life, and boundaries: Journal of Paleontology, v. 44, p. 801-815.

MIDDLETON M. F., 1984, Seismic geohistory analysis—A case history from the Canning Basin, Western Australia: Geophysics, v. 49, p. 333-343.

———, 1987a, Seismic stratigraphy of the northern Canning Basin: Exploration Geophysics, v. 18, p. 141-144.

———, 1987b, Seismic stratigraphy of the Devonian reef complexes of the northern Canning Basin, Western Australia: American Association of Petroleum Geologists Bulletin, v. 71, p. 1488-1498.

MOORS, H. T., GARDNER, W. E., AND DAVIS, J., 1984, Geology of the Blina oilfield, in Purcell, P. G., ed., The Canning Basin, Western Australia: Geological Society of Australia and Petroleum Exploration Society of Australia, Canning Basin Symposium, Perth, p. 277-283.

MURPHY, G. C., BAILEY, A., AND PARRINGTON, P. J., 1986, The Blendevale carbonate-hosted zinc-lead deposit, Kimberley region, Western Australia, in Berkman, D. A., ed., Geology and Exploration: Council of Mining and Metallurgical Institutions, Thirteenth Congress Publications, v. 2, p. 153-161.

PETERSEN, M. S., 1975, Upper Devonian (Famennian) ammonoids from the Canning Basin, Western Australia: Paleontological Society Memoir 8, 55 p.

PLAYFORD, P. E., 1980, Devonian "Great Barrier Reef" of the Canning Basin, Western Australia: American Association of Petroleum Geologists Bulletin, v. 64, p. 814-840.

———, 1981, Devonian reef complexes of the Canning Basin, Western Australia: Geological Society of Australia, Fifth Australian Geological Convention Guidebook, 64 p.

———, 1982, Devonian reef prospects in the Canning Basin: Implications of the Blina oil discovery: Australian Petroleum Exploration Association Journal, v. 24, p. 258-271.

———, 1984, Platform-margin and marginal-slope relationships in Devonian reef complexes of the Canning Basin, in Purcell, P. G., ed., The Canning Basin, Western Australia: Geological Society of Australia and Petroleum Exploration Society of Australia, Canning Basin Symposium, Perth, 1984, p. 189-234.

———, COCKBAIN, A. E., DRUCE, E. C., AND WRAY, J. L., 1976, Devonian stromatolites from the Canning Basin, Western Australia, in Walter, M. R., ed., Stromatolites: Developments in Sedimentology, Elsevier, Amsterdam, Oxford, New York, v. 20, p. 543-563.

———, AND LOWRY, D. C., 1966, Devonian reef complexes of the Canning Basin, Western Australia: Geological Survey of Western Australia, Bulletin 118, 150 p.

———, MCLAREN, D. J., ORTH, C. J., GILMORE, J. S., AND GOODFELLOW, W. D., 1984, Iridium anomaly in the Upper Devonian of the Canning Basin: Science, v. 246, p. 437-439.

PURCELL, P. G., 1984, The Canning Basin, Western Australia: Geological Society of Australia and Petroleum Exploration Society of Australia, Canning Basin Symposium, Perth, 582 p.

READ, J. F., 1973a, Carbonate cycles, Pillara Formation (Devonian), Canning Basin, Western Australia: Canadian Petroleum Geology Bulletin, v. 21, p. 38–51.

———, 1973b, Paleo-environments and paleography, Pillara Formation (Devonian), Western Australia: Canadian Petroleum Geology Bulletin, v. 21, p. 344–394.

REECKMAN, S. A., AND SARG, J. F., 1986, Foreslope deposition on a steep margin reef complex; A record of sea level fluctuation–Capitan reef complex, Guadalupe Mountains, Texas (abs.): Twelfth International Sedimentological Congress, Canberra, Australia, p. 255.

RIGBY, J. K., 1986, Late Devonian sponges of Western Australia: Geological Survey of Western Australia, Report 18, 59 p.

SCHLAGER, W., 1981, The paradox of drowned reefs and carbonate platforms: Geological Society of America Bulletin, v. 92, p. 197–231.

———, AND CAMBER, O., 1986, Submarine slope angles, drowning unconformities, and self-erosion of limestone escarpments: Geology, v. 14, p. 762–765.

TOWNER, R. R., AND GIBSON, D. L., 1983, Geology of the onshore Canning Basin, Western Australia: Australia, Bureau of Mineral Resources, Bulletin 235, 51 p.

WALLACE, M. W., 1987, Sedimentology and diagenesis of Upper Devonian carbonates, Canning Basin, Western Australia: Unpublished Ph.D. Dissertation, University of Tasmania, Hobart, Tasmania, 184 p.

WRAY, J. L., 1967, Upper Devonian calcareous algae from the Canning Basin, Western Australia: Colorado School of Mines Professional Contributions, No. 3, 76 p.

SEDIMENTARY AND TECTONIC CONTROLS ON THE DEVELOPMENT OF AN EARLY MISSISSIPPIAN CARBONATE RAMP, SACRAMENTO MOUNTAINS AREA, NEW MEXICO

WAYNE M. AHR
Department of Geology, Texas A&M University, College Station, Texas 77843

ABSTRACT: This study focuses on the controls that governed the evolution from ramp to shelf in the northern and central Sacramento Mountains during the time between first sedimentation in the Kinderhookian and the start of Waulsortian-style reef growth in the Osagian. Depositional environments and paleobathymetry are inferred from facies sections and from isopach maps of the Caballero Formation and the Andrecito Member of the Lake Valley Formation.

The isopach maps show that both intervals thin regionally from north to south. Local variations in thickness are present as loose skeletal buildups (pods), a thin deposit over a structural uplift in the northern part of the study area, and as thin and thick sediments over highs and lows on the Devonian-Mississippian unconformity in the central and southern parts of the area. There are no linear slope breaks on the ramp to indicate where the subsequent formation of the Lake Valley Shelf would take place.

Caballero and Andrecito facies contain more calcarenites and terrigenous silt-fine sand in the north; they contain more shale and lime mud in the south. The calcarenites indicate denser populations of crinoids and fenestrate bryozoans along with greater winnowing in the north, probably because the water was shallower. The greater amount of terrigenous silt and fine sand in the north is interpreted to indicate proximity to the source.

All of the Waulsortian reefs from Alamo Canyon southward are associated with concentrations of crinoid and fenestrate bryozoan skeletal remains in the underlying Andrecito, whereas the inter-reef equivalents contain comparatively fewer bioclasts. The reefs in Indian Wells and Marble Canyons rest on a crinoidal calcarenite blanket in the Andrecito and form a loosely defined semicircle around a structural uplift in central Indian Wells Canyon. From Andrecito through Tierra Blanca time, the detrital sedimentary blanket prograded and covered the ramp from Alamo Canyon northward. During Tierra Blanca time alone, as much as 60 m of crinoidal calcarenite accumulated in the growing sedimentary prism that transformed the northern Sacramento Mountains ramp into the Tierra Blanca Shelf.

The evolution from ramp to shelf during Kinderhookian to Osagian time in the Sacramento Mountains is similar to the depositional history of the Lower Carboniferous in much of Western Europe and Montana, where Waulsortian reefs grew in downslope positions on ramps and were covered by prograding detrital sedimentary packages, which created shelf slope breaks. This pattern is uncommonly similar in Europe and North America during Tournaisian-middle Visean times. It is as though the Lower Carboniferous is the premier time for ramps.

INTRODUCTION

The Sacramento Mountains of New Mexico are well known for their exposures of Early Mississippian, Waulsortian-type reefs. The environmental setting of the reefs is described in much of the geologic literature as a "shelf" and the reefs are described as being either on the shelf margin or down slope from it. The comparative scarcity of work on the less spectacular, pre-reef Mississippian strata has left important questions about the physiography and facies patterns of the pre-reef platform (*sensu* Ahr, 1985) unanswered. As a result, the ways in which sedimentation and tectonics have influenced platform development have remained relatively undocumented. As a corollary, the manner in which paleobathymetry and pre-reef sedimentation patterns may have influenced reef nucleation sites and growth patterns has been poorly understood. The difficulties in reconstructing the platform configuration are exacerbated by: (1) platform truncation on the west by a Tertiary fault that has about 2,130 m of vertical displacement; (2) many reef exposures limited to a north-south outcrop trend which follows the mountain scarp, a post-Mississippian feature; (3) exposures east of the scarp front limited to steep cliffs in a few canyons; and (4) thin, easily weathered pre-reef beds, which are poorly exposed.

Master's thesis work by Reed (1982) was followed by several similar projects on Caballero-Andrecito sedimentology (Blount, 1985; George, 1985; Morey, 1985). New interpretations of the Sacramento Mountains structural geology are presented in another thesis (Johnson, 1985). Those studies, and work by the author, form the basis for this paper. The main points are that the pre-reef, Mississippian depositional model for the Sacramento Mountains was a ramp on which Waulsortian buildups grew as patch reefs on or around local bathymetric highs, which consisted of calcarenite pods and a structural uplift. Subsequently, the ramp was partly covered by thick accumulations of bioclastic calcarenites that created a subregional shelf slope break.

The study area extends from Indian Wells to Escondido Canyons in the northern and central Sacramento Mountains (Figs. 1, 2). Sixty stratigraphic sections were measured in the Caballero Formation and the Andrecito Member of the Lake Valley Formation. Andrecito and Caballero facies descriptions are based on descriptions of measured sections, 1,600 slabbed samples, and 200 point counts per slide of about 500 thin sections. The data are on file at Texas A and M University.

STRATIGRAPHY

The stratigraphy of the Sacramento Mountains has been studied extensively. Some of the better known works include those by Darton (1917), Laudon and Bowsher (1941), Meyers (1974), Pray (1961) and DeKeyser (1983). The stratigraphic nomenclature used in this paper follows that of Pray (1961). The paper focuses on the pre-reef Mississippian strata and describes the depositional model that existed between Kinderhookian time and the start of reef growth during the Osagian, mainly, but not exclusively, in the lower Alamogordo Member of the Lake Valley Formation.

The Caballero Formation

The Caballero Formation was deposited during a transgression over a Late Devonian, low-relief erosional surface (Armstrong, 1962). It consists of silty calcareous

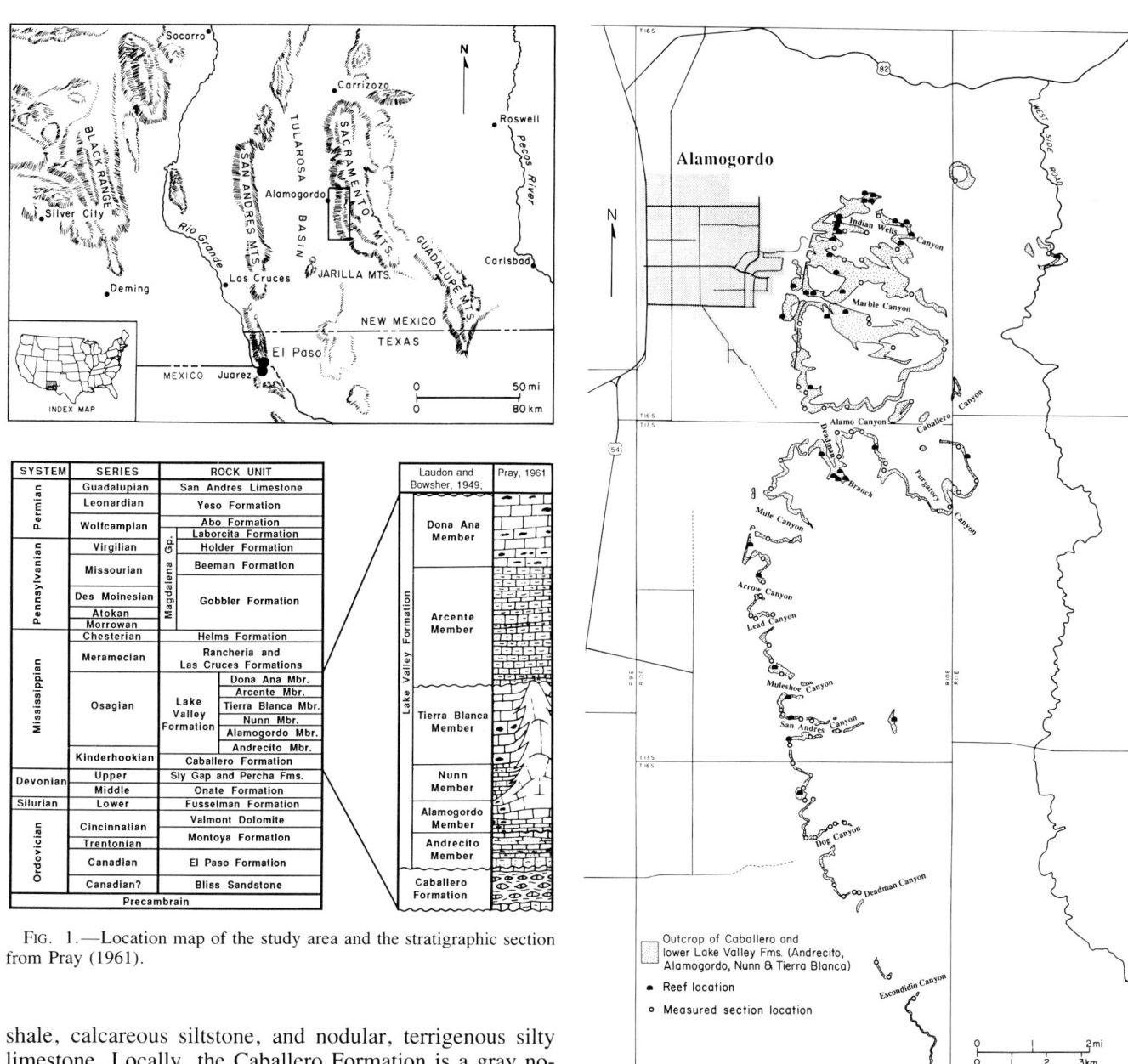

FIG. 1.—Location map of the study area and the stratigraphic section from Pray (1961).

FIG. 2.—Map of the Mississippian outcrop in the study area (adapted from Pray, 1961) showing the locations of measured sections and Waulsortian reefs. The reef locations were made during this study; no attempt is made to indicate the size or shape of the reefs on this figure.

shale, calcareous siltstone, and nodular, terrigenous silty limestone. Locally, the Caballero Formation is a gray nodular, argillaceous limestone interbedded with gray calcareous shale and contains crinoidal calcarenites and calcirudites in the upper parts of the unit (Pray, 1961). Caballero rocks are limited mainly to the northern Sacramento Mountains. A basin of Caballero deposition has been mapped from the Sacramento Mountains area westward to the vicinity of Silver City (Kottlowski, 1975; Fig. 3). Caballero strata thin toward the south in the Sacramento Mountains and pinch out completely in the Grapevine Canyon area 8 km south of Escondido Canyon (Fig. 2) (Pray, 1961). The depositional environment of the Caballero was subtidal marine (Meyers, 1975).

The Andrecito Member of the Lake Valley Formation

The contact between the Andrecito Member of the Lake Valley Formation and the underlying Caballero Formation has been described as an unconformity that occurs at the Kinderhookian-Osagian boundary (Laudon and Bowsher, 1949; Pray, 1961; Lane, 1974). It has also been described as an unconformity that straddles the conodont-defined division between the Kinderhook and the Osage (DeKeyser, 1983). In contrast, it has been described as being generally conformable with only a local discordance (Meyers, 1975). Since the Caballero is present everywhere beneath the Andrecito, any unconformity must only be local (Pray, 1961).

Andrecito rocks consist of gray thin-bedded, arenaceous

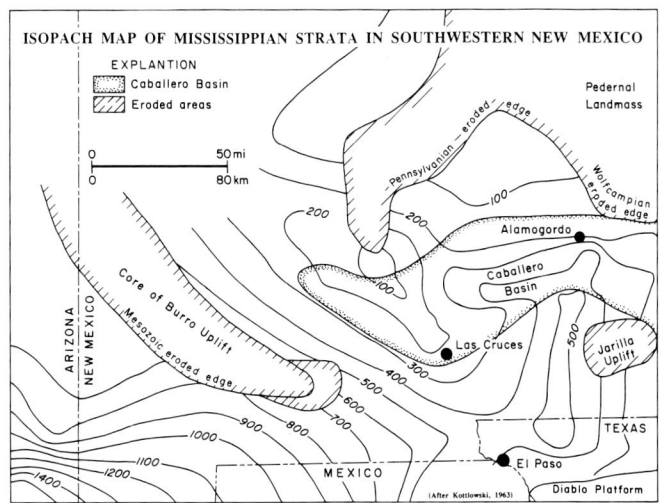

FIG. 3.—The location and extent of the Caballero Basin and isopach map of the Mississippian strata in southwestern New Mexico. Contours are in feet. After Kottlowski (1963).

limestones; dark gray thin-bedded, cherty limestones; dark gray quartz silty and argillaceous limestones; and limey, silty shale (Kottlowski, 1963). The member is more widespread than the underlying Caballero Formation and grades laterally into the Alamogordo Member or pinches out. South of the study area, the Andrecito thins and is difficult to distinguish from the Caballero. The depositional setting for the Andrecito was subtidal oxygenated marine (Meyers, 1975).

THICKNESS TRENDS AND FACIES PATTERNS

The paleobathymetry and lithofacies of the pre-reef ramp in the Sacramento Mountains are revealed by analyses of the thicknesses and petrography of the Caballero Formation and the Andrecito Member of the Lake Valley Formation.

Caballero Thickness Trends

The Caballero formation is present only in the area mapped as the Caballero Basin by Kottlowski (1963; Fig. 3). In the study area, the Caballero formation ranges in thickness from 20 m in the north-central study area to 4.5 m in the south. The formation isopach shows a general eastward thickening in the central study area, along with local variations in thickness (Fig. 4).

Local variations in Caballero thickness include an example in the central branch of Marble Canyon (Fig. 2), where a 12-m-deep erosional channel in the underlying Devonian has been described (Pray, 1961; DeKeyser, 1978). The submarine channel is filled with shale, possibly of Devonian Percha age (Pray, 1961). The middle Caballero Formation directly over this channel is 2 m thicker than at adjacent measured sections, suggesting that the channel fill sagged due to pre-middle Caballero compaction. Only the middle Caballero exhibits thickening; both the lower and upper parts of the formation do not (Blount, 1985).

In the central Indian Wells Canyon area, the top of the

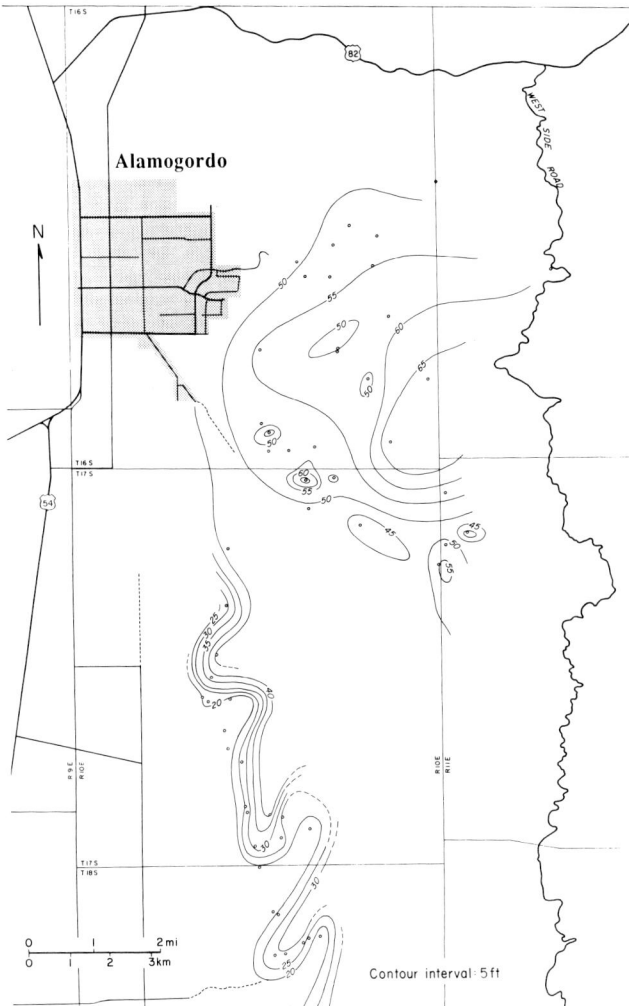

FIG. 4.—Isopach map of the Caballero Formation in the study area.

Caballero Formation is incised by a channel. This feature has been described as an angular unconformity (Meyers, 1975); however, the outline of an erosional channel has been mapped (Fig. 5). The channel is 36.5 m wide and filled to a thickness of 2 m with nodular packstones and grainstones of typical Caballero lithology (Blount, 1985). No evidence of subaerial exposure is present, but the basal 3 m of the overlying Andrecito Member are absent over the channel. About 2.4 km east-northeast of the channel, there is no angular discordance at the Caballero-Andrecito contact but 4.5 m of the basal Andrecito are missing and the first Andrecito lithology is a skeletal calcarenite-lag deposit (Blount, 1985). At surrounding sections in the northern Sacramento Mountains, *Zoophycos* burrows are present in several succeeding thin beds of the upper Caballero and thence into the Andrecito, suggesting that the contact is gradational and conformable. As there is no faunal break at the contact and the Caballero Formation was deposited in a subtidal open-marine environment (Meyers, 1975; Lane, 1982), the periods of submarine erosion must have been brief in duration and local in extent.

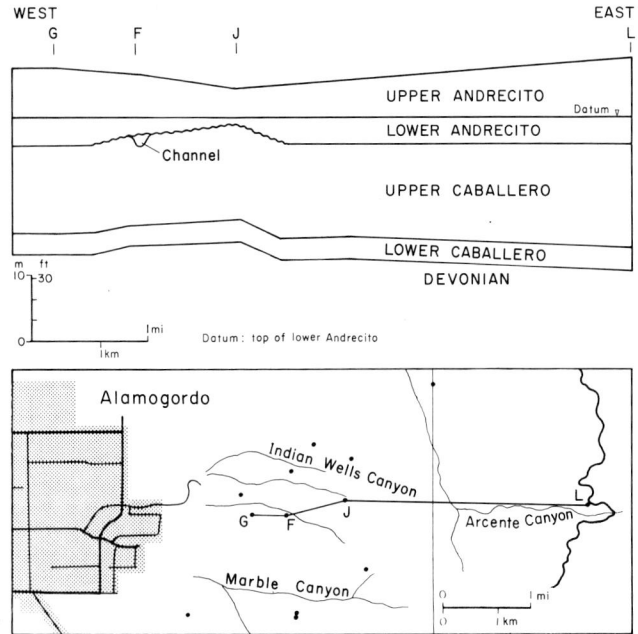

FIG. 5.—East-west stratigraphic cross section through Indian Wells and Arcente Canyons showing the thickness changes and the erosional channel in the uppermost Caballero Formation. The location of each measured section is shown on the inset map. From Blount (1985).

Caballero Facies

Simplified descriptions of four Caballero lithofacies have been derived from the detailed work mentioned earlier. The facies are designated, in vertical succession, C_b, C_1, C_2, and C_3 (Fig. 6).

The C_b, lithofacies is limited in lateral extent from Indian Wells to Mule Canyon. It is a blanket deposit with thicknesses ranging from 0 to 3.6 m, its mean carbonate allochem content (skeletal grains) is 30 percent, its mean quartz silt content is 0.30 percent, and the dominant lithology consists of olive-gray, crinoid-bryozoan wackestones to mud stones with calcareous-shale interbeds. Sedimentary structures include rare, fining-upward packstone lenses, common, discontinuous, non-parallel, wavy laminae, and vertical and horizontal burrows. Skeletal grains, except for small, whole brachiopods, are fragmented. These beds of thin to medium thickness commonly contain flattened, 5- to 8-cm nodules surrounded by shale laminae.

The C_1 facies extends over the entire study area and consists of calcareous shale along with nodular limestone. Rock colors range from olive gray to yellowish brown to dark gray. The facies ranges in thickness from about 1 to 15 m. Skeletal allochems constitute as much as 46 percent of the limestone fraction and most of the grains are broken. The mean quartz silt content is only 0.05 percent. Sedimentary structures include non-parallel, discontinuous wavy beds, burrows, and one submarine channel.

The discontinuous C_2 lithofacies is present only in the area between Mule and San Andres Canyons, where it may be as thick as 6 m in some localities. The rocks consist of olive-gray to light gray lime mudstones to packstones with coarse sand-size crinoid and bryozoan fragments. The coarse-grained limestone beds are separated by thin calcareous-shale interbeds. Sedimentary structures include wavy, discontinuous, non-parallel laminae, especially on the upper and lower surfaces of the calcarenites. Nodular bedding is present south of the Lead Canyon area and is very common in the Mule-Arrow Canyon area. The C_2 facies thickens

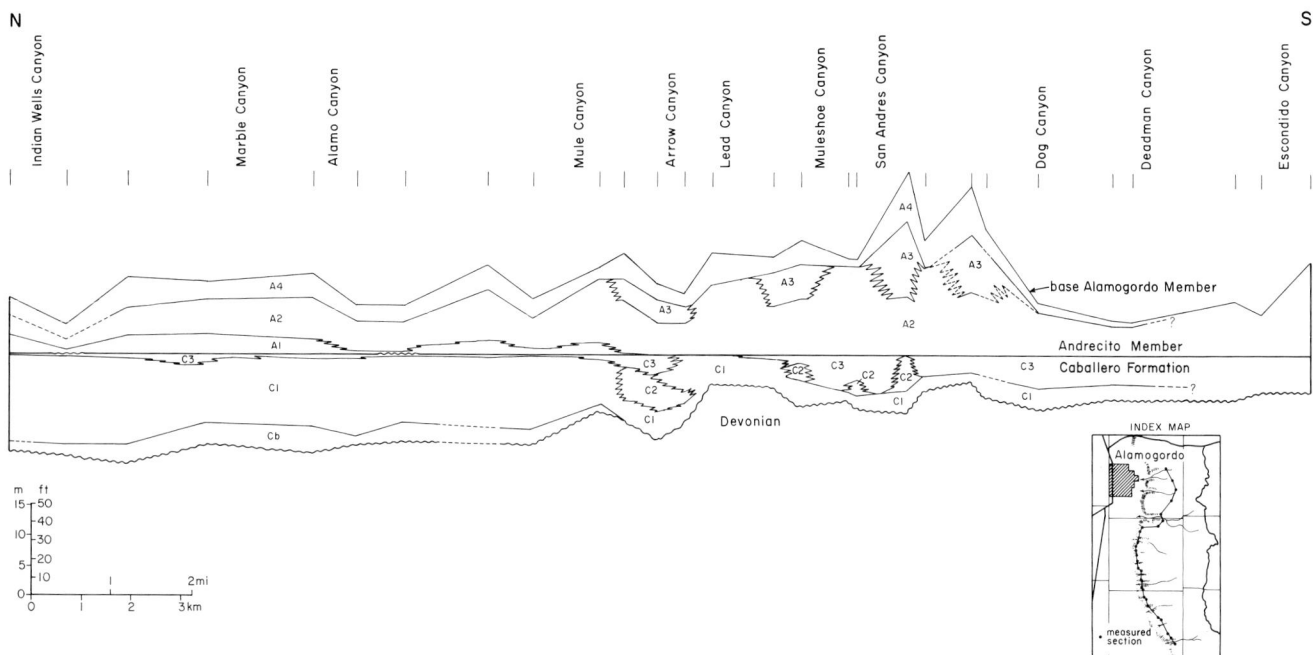

FIG. 6.—Stratigraphic cross section of the Caballero Formation and the Andrecito Member of the Lake Valley Formation in the study area. The section extends from Indian Wells to Escondido Canyons and shows the general southward thinning of the stratigraphic units. The section is hung on the top Caballero datum.

beneath Waulsortian bioherms in the Muleshoe-San Andres Canyon area (George, 1985).

The C_3 Caballero lithology ranges from argillaceous lime mudstones to packstones, light to brownish gray, to dark olive gray in color. North of Alamo Canyon, the facies consists mainly of crinoid and fenestrate bryozoan hash. To the south, brachiopods and ostracods are common as well. In the Alamo Canyon area, quartz silt ranges in abundance from 0–7 percent (Morey, 1985). Unit thicknesses range from 0 to 6 m, with the thickest intervals being south of Muleshoe Canyon. The C_3 and C_1 lithologies are difficult to separate south of Deadman Canyon.

Limestones and shales alternate in beds 7.6 to 17.8 cm thick with 2.5-cm beds common. Nodular bedding in facies C_3 is less common than in the other Caballero facies. Horizontal burrows and rare *Zoophycos* are present. The mean skeletal allochem content in the Alamo Canyon area and northward is 83 percent with a mean quartz silt content of only 0.09 percent. Fining-upward packstone lenses are common in the north. In the southern study area, the allochem content is highly variable and lower. Quartz silt is present in trace amounts.

In general, Caballero limestones contain relatively more allochems in the north than in the south, except for some skeletal calcarenite pods in the southern C_2 facies. Calcareous shale is abundant; quartz silt is relatively rare; allochems consist of broken crinoids and fenestrate bryozoans, along with whole or broken ostracods and brachiopods. The brachiopod *Cyrtospirifer lator* is common in the lower Caballero. *Rhipidomella tenuicostata* is common in the upper Caballero. Wormlike burrows are common throughout the formation; *Zoophycos* is present only in the upper C_3 facies (Blount, 1985).

Loose skeletal buildups in the C_2 facies at Muleshoe and San Andres Canyons (Fig. 6) are associated with thinner Caballero than in the other parts of the study area. The combination of skeletal pods and thin Caballero is interpreted to be linked with erosional highs on the underlying Devonian-Mississippian unconformity (George and Ahr, 1986). The calcarenite pods are located beneath sites where even larger loose skeletal buildups were deposited during Andrecito time.

Andrecito Thickness Trends

The Andrecito Member of the Lake Valley Formation is more widespread than the underlying Caballero Formation and either pinches out or grades laterally into limestones of the Alamogordo Member of the Lake Valley Formation (Kottlowski, 1963).

The Andrecito-Caballero contact is diachronous and becomes younger to the south (Meyers, 1975). Except for the channel described earlier, the contact is conformable. The Andrecito ranges in thickness from a maximum of 29 m to 0 m in the vicinity of Grapevine Canyon 8 km south of Escondido Canyon (Pray, 1961). The Andrecito isopach map in this study area shows local thick and thin strata in the northern half of the area and a more monotonous trend along the Sacramento Mountain front in the southern half of the area (Fig. 7). The absence of exposures east of the moun-

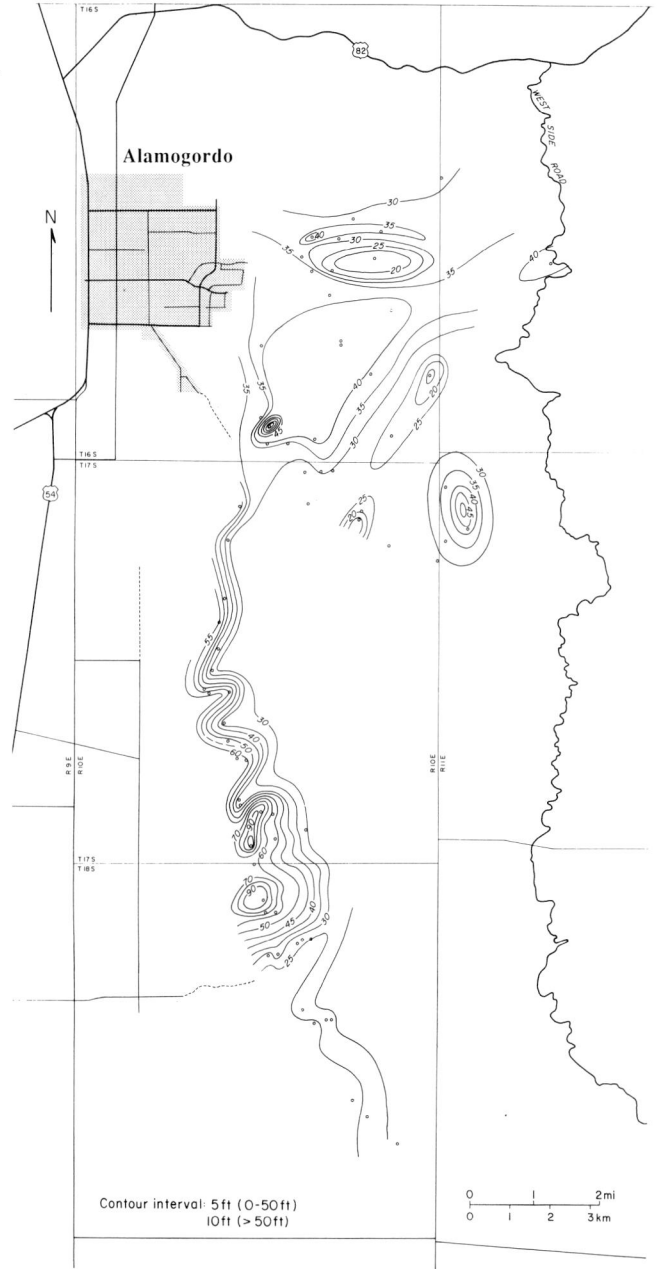

FIG. 7.—Isopach map of the Andrecito Member in the study area.

tain front in the south imposes a limitation on what can be interpreted about the thickness trends.

Locally in Indian Wells Canyon, the Andrecito is thin where it rests on the channelized Caballero surface (Fig. 5). Across the rest of the study area, local thickening typically results from skeletal buildups having accumulated as crinoid-bryozoan calcarenite pods. In general, these pods are immediately beneath large, Waulsortian biohermal buildups (George and Ahr, 1986).

In short, Andrecito paleobathymetry resulted from a combination of structural and sedimentary processes. Structural uplift at Indian Wells Canyon during Caballero time created a high over which the Andrecito thinned, because the lower part of the member was either eroded or not deposited. Accumulation of skeletal calcarenite pods created the local thick deposits. The Andrecito calcarenite pods were deposited directly above similar pods in the underlying Caballero in the Muleshoe-San Andres Canyon area. The loose, skeletal buildups were probably the result of dense crinoid-bryozoan populations having been winnowed and piled into calcarenite mounds. Otherwise, the calcarenites should occur as infill of lows in the Caballero surface and would be diluted with co-deposited fines.

The subsequent nucleation and growth of Waulsortian reefs in the study area were closely linked to the Andrecito thickness and facies trends. The reefs, which took root mainly in the Alamogordo Member, appear to have formed on the Andrecito calcarenite pods and around the uplift in Indian Wells Canyon.

Andrecito Facies

Four lithofacies have been identified in the Andrecito Member during this study. In vertical sequence, they are the A_1 through the A_4 facies (Fig. 6).

The Andrecito A_1 facies is laterally discontinuous, extending only from Indian Wells southward to Mule Canyon. The A_1 rocks consist of bryozoan packstones interbedded with silty bryozoan wackestones and calcareous quartz silt and sandstone. The facies is a sedimentary blanket of irregular thickness, which ranges from 0.6 to about 4.8 m. Skeletal allochems include crinoids, bryozoans, and brachiopods, with bryozoans being the most common grain type. Sedimentary structures include abundant vertical and horizontal burrows, among which *Zoophycos* is common. Fining-upward packstone lenses are present. Non-parallel, wavy, discontinuous laminae are abundant in this thin- to medium-bedded facies. Skeletal grains are fragmented but rarely abraded. Chert, small amounts of dolomite, and abundant calcareous, very fine terrigenous sand are present. The siliciclastics occur mainly in the upper one-third of the facies, where the mean quartz content is over 14 percent. The average skeletal allochem content is 66 percent. The lower 2/3 of the facies have a mean quartz content of only 0.29 percent, and skeletal allochems compose an average of 89 percent of the rock. The A_1 facies is absent in central Indian Wells Canyon, where a channel is present at the Andrecito-Caballero contact.

Rock types of the A_2 facies include olive-gray packstones and wackestones that form resistant beds and are interbedded with less resistant silty wackestones. Sedimentary structures include mainly discontinuous wavy laminae, *Zoophycos*, and wormlike horizontal and vertical burrows. The average skeletal allochem content is 47 percent, and quartz silt composes about 3 percent of the rock. The skeletal allochems are fragmented, but whole gastropods and horn corals are common. Whole brachiopods with geopetal sediment have been reported from the Mule Canyon area southward (George, 1985).

The A_3 facies composes discrete pods (Fig. 6). The rocks consist of fenestrate bryozoan-crinoid hash wackestones to packstones. Quantities of skeletal grains vary widely but are highest beneath the Waulsortian reefs in the overlying Alamogordo (George and Ahr, 1986). The olive-gray to dark olive-gray limestones are interbedded with thin calcareous-shale laminae. Sedimentary structures include wavy discontinuous laminae, wavy top and bottom surfaces of packstone lenses, horizontal and vertical burrows, and rare *Zoophycos*. Whole fossils are rare. The calcarenite ranges in grain size from fine to coarse sand. Terrigenous sand and silt make up less than 5 percent of the rock volume.

The basal beds of the A_4 lithofacies weather in relief and are commonly rust colored. These beds are reliable markers for the entire study area. Fresh surfaces are medium gray to olive gray. Very fine to very coarse sand-size allochems compose the lime wackestones to grainstones. Bedding surfaces are wavy and discontinuous. Burrows are uncommon. Inclined, non-parallel, curved, discontinuous beds are present and may occur along with rounded lime mud intraclasts. The lower A_4 beds contain an average quartz content of 6 percent north of Alamo Canyon. Skeletal allochems compose 30–100 percent of the rock. All skeletal grains are broken, abraded, and commonly rounded. Grainstones and intraclastic calcarenites commonly occur as discontinuous lenses. The A_4 facies marks the change from the Andrecito to the Alamogordo Member of the Lake Valley Formation. Skeletal grains and quartz are more common to the north; calcareous shale and lime mudstones are more common to the south. The A_4 and A_2 facies are indistinguishable south of Escondido Canyon (Fig. 6).

In summary, the Andrecito Member has a rich assemblage of trace fossils in addition to the crinoids, bryozoans, brachiopods, ostracods, horn corals, and trilobites present as skeletal grains. The trace fossils include *Zoophycos, Teichichnus,* and abundant wormlike burrows.

Zoophycos, known as *Taonurus* in the older literature, is especially abundant in the Andrecito rocks, but it is not indicative of a specific water depth. It and the abundant and diverse biota indicate aereated open-marine conditions, which were favorable for invertebrate life.

The terrigenous clastic content of the Andrecito ranges around only 0.5 to 2.0 percent in the south (George, 1985). Silty and sandy zones are present in the lower Andrecito of the Indian Wells-Marble Canyon areas, where the mean quartz content may reach 14.0 percent (Blount, 1985). The greater terrigenous clastic content toward the north is interpreted by Ahr and others (1986) as an indication of proximity to a source for Andrecito clastics in the north.

Andrecito rocks contain relatively more skeletal allochems and quartz silt and sand in the outcrops from Mule Canyon northward. To the south, the facies contain more calcareous shale, and the limestones typically have a higher lime mud content, except where crinoid-bryozoan loose skeletal buildups formed pods, especially in the Muleshoe-San Andres Canyon area. The pods commonly occur over older calcarenite pods in the Caballero Formation, and some of the Caballero calcarenite pods occur at breaks in slope around erosional highs on the underlying Devonian, as at the Muleshoe Canyon area (George, 1985).

DEPOSITIONAL HISTORY

Depositional Environments

The sedimentologic and paleontologic data from this study indicate that the Caballero and Andrecito were deposited in open-marine waters. The comparatively high taxonomic diversity indicates that the environment was equitable. Skeletal grains and trace fossils are present everywhere; no barren facies were found. Evaporites, flat-pebble breccias, birdseye fenestral fabrics, stromatolites, and other indicators of shoreline and subaerial exposure, desiccation or evaporation are absent. There are local accumulations of breccia clasts in Andrecito strata in the southern area, but the 5-cm- to 0.6-m- diameter clasts are encased in marine rocks. These brecciated intervals are interpreted to be submarine-debris flows associated with a local topographic high on the Andrecito sea bed near San Andres Canyon (George, 1985).

Paleobathymetry.—

The Caballero Formation rests on an erosional unconformity that cuts out rocks of Devonian and at least Early Kinderhookian age (Lane, 1974). The Devonian-Carboniferous contact in the Muleshoe and San Andres Canyon area is an irregular surface with local highs and lows (George, 1985). This irregular surface has been described previously (Pray, 1961; Armstrong and others, 1980), but it was not linked to the formation of crinoid-bryozoan calcarenite pods that began to develop as early as Caballero time and then into Andrecito deposition (George and Ahr, 1986). These pods are located immediately beneath Waulsortian reefs in several locations from Alamo Canyon southward (George, 1985; T. Byrd, pers. commun., 1987). Where the calcarenite pods are absent, there are no reefs. The Andrecito isopach map (Fig. 7) and the outcrop map with reef locations (Fig. 2) show that the reef locations in the southern and central parts of the study area correspond to areas of increased thickness in the Andrecito. A striking correspondence between thick Andrecito calcarenite pods and reef occurrences has been reported by George (1985). It has been stated previously that the Andrecito appeared to be grainier beneath some reefs than away from them (Meyers, 1975), but confirmation through isopach and point count studies has not been done.

Northward from Alamo Canyon, Caballero and Andrecito rocks contain more allochems than their equivalents to the south, and the calcarenite pods which are readily mapped in the south are absent. Instead, the entire northern substrate appears to have been covered by crinoids and bryozoans, whose remains were winnowed by a more vigorous hydrologic regime than in the south.

The Andrecito Member contains significantly more siliciclastics in outcrops from Alamo Canyon northward than it does in the south. The source of the clastics must have been from a northerly direction.

Reef locations in Indian Wells and Marble Canyons are not linked to skeletal calcarenite pods as they were in the south. On the contrary, a pronounced thin Andrecito section is present there (Fig. 7), and the reefs appear to form a crude semicircle around it. Measured sections around this area show that the lower Andrecito facies (A_1) is absent and that the Caballero surface is dissected by a submarine channel (Fig. 5), which is interpreted to have formed in Caballero time. The eroded Caballero and missing Andrecito indicate a paleobathymetric high. Because the lower Caballero Formation is of normal thickness, the paleo-high could not have existed until late Caballero time, and because the middle and upper parts of the Andrecito Member are present, the paleo-high must have been submerged to a depth beneath the threshold of vigorous current scour by that time. The local extent and the bracketed times of erosion or non-deposition indicate that the paleo-high was a structural feature no older than late Caballero or younger than middle Andrecito time. The paleo-high is interpreted to be a tectonic uplift (Ahr and others, 1986). Paleozoic tectonism is well documented in the Sacramento Mountains (Wilson, 1967, 1972). Paleozoic deformation, which was most vigorous during the Pennsylvanian, was the result of forced folding of the Paleozoic section by differential vertical movements on tectonic basement blocks (Johnson, 1985). The Marble-Indian Wells Canyon area is at a complex block boundary, whereas the remainder of the study area lies over a large, relatively stable block (Johnson, 1985). It is tempting to link the paleo-high with block movements, which occurred during Lake Valley time.

It is well known that the Caballero and Andrecito strata thin regionally to the south. This study has shown that the units do not exhibit pronounced thickness changes along depositional dip, which one could interpret as a slope break of Osage or Kinderhook time. On the contrary, the rather gradual changes in thickness with distance to the south indicate that the Caballero-Andrecito platform was a homoclinal surface with low paleoslope. Paleo-highs and -lows existed, but they were not aligned regionally and they did not form on an inherited tectonic or erosional shelf break. The paleo-highs were of two main kinds: (1) Caballero tectonic features; and (2) sedimentary buildups of crinoid and bryozoan skeletal debris. In some instances the buildups accumulated on local highs that were probably inherited from the Devonian-Missippian unconformity.

In summary, the Kinderhook-Osage platform in the Sacramento Mountains was a homoclinal ramp that developed after marine transgression and subsequent deposition on the Devonian-Carboniferous unconformity. Subsequent evolution of the ramp was affected by both tectonic and depositional controls prior to the appearance of the Waulsortian reefs.

The Tierra Blanca Shelf.—

The Mississippian section thickened significantly after Andrecito and before Arcente deposition (DeKeyser, 1978). The thickened sections occur north of the Alamo Canyon area and consist mainly of thick Tierra Blanca encrinites (Fig. 8). The post-Andrecito, pre-Arcente section north of Alamo Canyon has been interpreted to have been the site for skeletal carbonate accumulation that, along with the

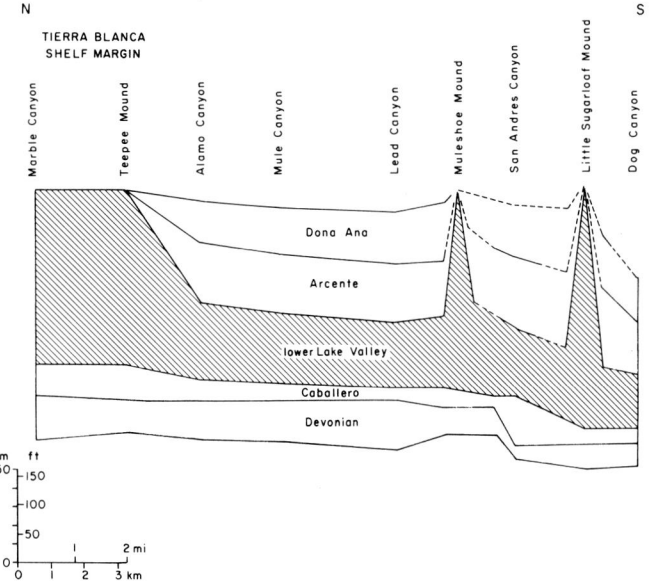

Fig. 8.—A stratigraphic section through the Tierra Blanca Shelf. Note the dramatic thickening of the lower Lake Valley Formation (Alamogordo-Nunn-Tierra Blanca) from Alamo Canyon northward to Marble Canyon (shaded). The abrupt thinning of the Alamogordo-Tierra Blanca "package" in the Alamo Canyon area marks the shelf slope break. The local thick sections at Muleshoe and Little Sugarloaf Canyons are major reef localities. Modified from DeKeyser (1978).

growth of tabular Walsortian reef clusters, created a sedimentary prism that had a definite slope break by the end of Tierra Blanca time. This is the Tierra Blanca Shelf of Meyers (1975), and it represents evolution of a ramp into a shelf by depositional processes during relative tectonic stability. The slope break was formed by passive deposition; there is no evidence that structural movements had any shaping effect on the Tierra Blanca Shelf, and there was no inherited slope break from Caballero-Andrecito times.

The age of these events is particularly interesting. The ramp existed during Kinderhook and Osage (Tournaisian: lower *typicus,*) times and evolved into a shelf during Osage (Tournaisian: upper *typicus* to *anchoralis-latus*) time (DeKeyser, 1983). Comparison of the depositional history of the Sacramento Mountain area with that of England, Wales, Ireland, Belgium, and Montana shows that various kinds of ramps were very common in the Tournaisian. By later Tournaisian or Visean time, aggrading and prograding sedimentary wedges had transformed virtually all of them into shelves.

A Comparison With Other Early Carboniferous Platforms

Western Europe.—

In Devonian-Carboniferous times, the Old Red Sandstone continent underwent a major marine transgression that moved progressively northward. Widespread carbonate sedimentation followed, extending from the Rhine to Northumberland to Kansas (Anderton and others, 1979). Apparently, virtually all carbonate sedimentation during the Tournaisian and early Visean took place on various forms of ramps. In central England, the "ramps" consist of regional tilt blocks (Miller and Grayson, 1982; Fig. 9). Southern Britain and Belgium sloped off the Precambrian Wales-Brabant massif (Lees, 1982), and gently sloping ramps existed in Wales and Ireland (Wright, 1985; MacCarthy and Gardiner, 1987). The tilt blocks do not strictly fit the definition of ramp because there is no updip end to them. The tilt-block truncated ramps and the other more conventional ramps mentioned earlier, however, characteristically have Waulsortian reefs in downslope positions, and the ramps commonly evolved into shelves after the reef phase ended. The Waulsortian reefs in central England formed on the downdip parts of the tilt blocks, and the "ramps" underwent depositional aggradation and progradation until a shelf slope break was formed. In Asbian times (late Meramec), the shelf margins became the sites for shallow-water, shelf-edge reef buildups which had shallow-water, shelf facies equivalent updip and slope and basin facies downdip (Wolfenden, 1958). Subsequently, during the Brigantian, small patch reefs developed in shallow water across the tops of some blocks, notably the Derbyshire block (Gutteridge, 1983). The blocks had become isolated platforms (*sensu* Wilson, 1975).

Montana.—

The depositional model for Kinderhookian (Tournaisian) carbonate sedimentation in Montana is described as a ramp (Smith, 1982). Smith describes vertical block faulting and downtilting of "stable" platform margins into a central, subsiding, unstable shelf in Montana. The block faulting created the Central Montana High, the Crazy Mountains Trough, and the Big Snowy Trough. The downtilting created ramps on both downtilted margins and, by latest Kin-

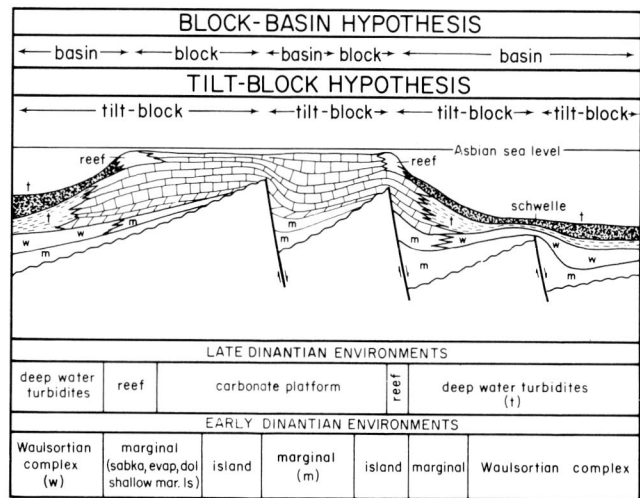

Fig. 9.—The tilt-block "ramp" model of Miller and Grayson (1982). Note the Waulsortian reefs in downslope positions on the truncated Osagian ramp. Subsequent sedimentary progradation created a shelf slope break by Meramec (Asbian) time. The shelf edge was then rimmed by a new generation of shallow-water reefs quite different from the Waulsortian ones. Reefs of this age are not present in the Sacramento Mountains but are relatively common in the subsurface of the Hardeman and Ft. Worth basins of north Texas.

derhookian times, the Big Snowy bioherms had developed on the distal parts of the ramp that extended out of Wyoming. Subsequently, during early Osagian times, the deepwater ramp facies and the reefs were buried beneath a prograding sedimentary wedge. The distal edge of the sedimentary wedge became a shelf slope break.

Similarities or Coincidences?

Early Mississippian sedimentation in the Sacramento Mountains area took place on a gently sloping ramp. Waulsortian reefs grew on the ramp and were preferentially sited on paleobathymetric highs of tectonic and depositional origin. Subsequent sedimentary infill of the updip part of the Sacramento ramp by crinoid and bryozoan calcarenites produced a shelf margin or shelf slope break by the end of Tierra Blanca time. The Early Mississippian ramp had evolved into a shelf by the more rapid deposition of detrital carbonates on the updip end of the ramp than on the distal end.

Ramps of comparable age were common in Western Europe and Montana, among other areas. In these cases, as in New Mexico, Waulsortian reefs are present in downslope positions on the ramps and, commonly, the upslope portions of the ramps, sometimes including the reefs, were blanketed with detrital carbonate sediments that prograded to form shelf slope breaks.

Are these patterns similarities or coincidences? The Early Carboniferous was world-wide one of the most profound times of carbonate sedimentation. It was marked in Europe and North America by an initial, major marine transgression over mainly terrestrial or paralic environments. It is tempting to draw the conclusion that this major transgression initially flooded more ramps than it did shelves. Perhaps the Early Carboniferous is truly the time of ramps. Ramps grew to maturity by sedimentary infill with detrital carbonates to become shelves. If this is truly a pattern and not a coincidence, then the rationale for exploration and exploitation of Early Mississippian reefs as hydrocarbon reservoirs must be quite different than one for Late Mississippian reef reservoirs, where shelves seem to be more common. The reefs on ramps would occur as individual buildups in downslope positions, typically surrounded by fine-grained, basinal deposits. In contrast, reefs of Chester and Meramec age may occur at shelf margins or as patch reefs scattered about across the shelves, as is the case of the Asbian and Brigantian reefs of England (Ahr, 1984). The shelf margin reefs are not isolated mounds. They follow regional slope breaks and are flanked updip by calcarenites and downdip by slope deposits (Wolfenden, 1958). Trans-shelf patch reefs may be small, isolated features surrounded by contemporary calcarenites in all directions (Gutteridge, 1983).

ACKNOWLEDGMENTS

Much of this study was made possible by grants from the New Mexico Bureau of Mines and Mineral Resources, thanks to Dr. Frank Kottlowski. Other funding from the American Association of Petroleum Geologists, the Gulf Coast Association of Geological Societies, and the Department of Geology at Texas A and M University is gratefully acknowledged. Discussions with L. C. Pray, W. J. Meyers, Art Bowsher, and Tom DeKeyser helped greatly, as did the 1982 field trip conducted after the Waulsortian Facies Conference in El Paso. The drafting was done by Tom Byrd, whose patient help with the facies sections was invaluable. Finally, the manuscript, which has an interesting history, benefited greatly from the critical reviews by James Lee Wilson, J. Fred Read, and Bill Meyers. Their help is gratefully acknowledged.

REFERENCES

AHR, WAYNE M., 1984, Biohermal facies in Mississippian carbonates from Texas, New Mexico, and Utah (abs.): First Meeting, European Dinantian Environments, Open University Publication, p. 131–132.
———, 1985, Limestone depositional sequences on shelves and ramps: Modern and ancient: Geology Today, v. 1, p. 84–89.
———, BLOUNT, W. M., GEORGE, P. D., AND MOREY, E. D., 1986, Paleotopography and substrate lithology as controls on initiation of Waulsortian reef growth: Examples from the Sacramento Mountains of New Mexico: American Association of Petroleum Geologists Bulletin, v. 70, p. 558–559.
ANDERTON, R., BRIDGES, P., LEEDER, M., AND SELLWOOD, B., 1979, A dynamic stratigraphy of the British Isles: George Allen and Unwin, London, 301 p.
ARMSTRONG, A. K., 1962, Stratigraphy and paleontology of the Mississippian System in southwestern New Mexico and adjacent southeastern Arizona: New Mexico Bureau of Mines and Mineral Resources, Memoir 8, 99 p.
———, MAMET, B., AND REPETSKI, J., 1980, The Mississippian System of New Mexico and southern Arizona, *in* Fouch, T. D., and Magathan, E., eds., Paleozoic Paleogeography of the West-Central United States: Rocky Mountain Paleogeography Symposium 1, Rocky Mountain Section, The Society of Economic Paleontologists and Minerologists, p. 82–99.
BLOUNT, W. M., 1985, Paleoenvironmental analysis of lower Mississippian Caballero Formation and Andrecito Member of Lake Valley Formation in northern Sacramento Mountains, Otero County, New Mexico: Unpublished M.S. Thesis, Texas A and M University, College Station, 191 p.
DARTON, N. H., 1917, A comparison of Paleozoic sections in southern New Mexico: U.S. Geological Survey Professional Paper 108-C, p. 31–55.
DEKEYSER, T., 1978, The Early Mississippian of the Sacramento Mountains, New Mexico–An ecofacies model for carbonate shelf margin deposition: Unpublished Ph.D. Dissertation, Oregon State University, Corvallis, 304 p.
———, 1983, Depositional sequences and stratigraphic revision of the Lake Valley shelf, Early Mississippian, Sacramento Mountains, New Mexico: West Texas Geological Society Bulletin, v. 23, p. 4–12.
GEORGE, P. G., 1985, A paleoenvironmental study of the lower Mississippian Caballero Formation and Andrecito Member of the Lake Valley Formation in the south-central Sacramento Mountains, Otero County, New Mexico: Unpublished M.S. Thesis, Texas A and M University, College Station, 240 p.
———, AND AHR, W. M., 1986, The effects of paleotopography and substrate lithology on the origin of Waulsortian reefs: South-central Sacramento Mountains, New Mexico: Gulf Coast Association of Geological Societies, Transactions, v. 36, p. 129–139.
GUTTERIDGE, P., 1983, Sedimentological study of the Eyam Limestone Formation in the east-central part of the Derbyshire Dome: Unpublished Ph.D. Dissertation, University of Manchester, Manchester, England, v. 1, 320 p., v. 2, 138 p.
JOHNSON, M. R., 1985, Pennsylvanian-Permian deformation at 1,000–5,000 feet overburden, Sacramento Mountains, New Mexico: Unpublished M.S. Thesis, Texas A and M University, College Station 116 p.

KOTTLOWSKI, F. E., 1963, Paleozoic and Mesozoic strata of southwestern and south-central New Mexico: New Mexico Bureau of Mines and Mineral Resources Bulletin, No. 19, 100 p.

———, 1975, Stratigraphy of the San Andres Mountains in south-central New Mexico, in Seager, W. R., Clemons, R. E., and Callender, J. F., eds., Guidebook of the Las Cruces Country: New Mexico Geological Society, p. 95–104.

LANE, H. R., 1974, The Mississippian of southeastern New Mexico—A wedge-on-wedge relation: American Association of Petroleum Geologists Bulletin, v. 28, p. 269–282.

———, 1982, The distribution of Waulsortian facies in North America as exemplified in the Sacramento Mountains of New Mexico, in Bolton, K., Lane, H. R., and LeMone, D. V., eds., Symposium on the Paleoenvironmental Setting and Distribution of the Waulsortian Facies: El Paso Geological Society and the University of Texas, El Paso, p. 96–114.

LAUDON, L. R., AND BOWSHER, A. R., 1941, Mississippian formations of the Sacramento Mountains, New Mexico: American Association of Petroleum Geologists Bulletin, v. 25, p. 2107–2160.

———, AND ———, 1949, Mississippian formations of southeastern New Mexico: Geological Society of America Bulletin, v. 60, p. 1–88.

LEES, ALAN, 1982, The paleoenvironmental setting and distribution of the Waulsortian facies of Belgium and southern Britain, in Bolton, K., Lane, H. R., and Lemone, D. V., eds., Symposium on the Paleoenvironment and Distribution of Waulsortian Facies: El Paso Geological Society and the University of Texas, El Paso, p. 1–16.

MACCARTHY, I. A. J., AND GARDINER, P. R. R., 1987, Dinantian cyclicity: A case history from the Munster Basin of southern Ireland, in European Dinantian Environments: John Wiley & Sons, New York, p. 199–237.

MEYERS, W. J., 1974, Carbonate cement stratigraphy of the Lake Valley Formation (Mississipian), Sacramento Mountains, New Mexico: Journal of Sedimentary Petrology, v. 44, p. 837-861.

———, 1975, Stratigraphy and diagenesis of Lake Valley Formation, Sacramento Mountains, in Pray, L. C., ed., A Guidebook to the Mississippian Shelf-Edge and Basin Facies Carbonates, Sacramento Mountains and Southern New Mexico: Dallas Geological Society, p. 45–66.

MILLER, JOHN, AND GRAYSON, R., 1982, The regional context of Waulsortian facies in northern England, in Bolton, K., Lane, H. R., and LeMone, D. V., eds., Symposium on the Paleoenvironmental Setting and Distribution of Waulsortian Facies: El Paso Geological Society and the University of Texas, El Paso, p. 17–33.

MOREY, E. D., 1985, A paleoenvironmental analysis of the Mississippian Caballero and lower Lake Valley formations, Sacramento Mountains, Otero County, New Mexico: Unpublished M.S. Thesis, Texas A and M University, College Station, 155 p.

PRAY, L. C., 1961, Geology of the Sacramento Mountains escarpment, Otero County, New Mexico: New Mexico Bureau of Mines and Mineral Resources Bulletin, No. 35, 114 p.

REED, R. E., 1982, Paleoenvironmental analysis of biohermal facies, Mississippian Lake Valley Formation, New Mexico: Unpublished. M. S. Thesis, Texas A and M University, College Station, 182 p.

SMITH, D. L., 1982, Waulsortian bioherms in the Paine Member of the Lodgepole Limestone (Kinderhookian) of Montana, U.S.A., in Bolton, K., Lane, H. R., and Lemone, D. V., eds., Symposium on the Paleoenvironment and Distribution of Waulsortian Facies. El Paso Geological Society and the University of Texas, El Paso, p. 51–64.

WILSON, J. L., 1967, Cyclic and reciprocal sedimentation in Virgilian strata of southern New Mexico: Geological Society of America Bulletin, v. 78, p. 805–818.

———, 1972, Influence of local structure in sedimentary cycles of Beeman and Holder Formations, Sacramento Mountains, Otero County, New Mexico, in Elam, J. C., and Chuber, S., eds., Cyclic Sedimentation in the Permian Basin, second edition: West Texas Geological Society, p. 41–54.

———, 1975, Carbonate Facies in Geologic History: Springer-Verlag, New york, 471 p.

WOLFENDEN, E. B., 1958, Paleoecology of the Carboniferous reef complex and shelf limestones in northwest Derbyshire: Geological Society of America Bulletin, v. 69, p. 871–898.

WRIGHT, V. P., 1985, Facies sequences on a carbonate ramp: The Carboniferous Limestone of South Wales: Sedimentology, v. 33, p 221–241.

SILICICLASTIC INFLUENCE ON MESOZOIC PLATFORM DEVELOPMENT: BALTIMORE CANYON TROUGH, WESTERN ATLANTIC

FRANZ O. MEYER

Bellaire Research Center, Shell Oil Company, P.O. Box 481, Houston, Texas 77001

ABSTRACT: Late Jurassic (Kimmeridgian) to Early Cretaceous (Valanginian) carbonate platforms exhibit "keep-up" and "give-up" types of growth in the Western Atlantic. Twenty seismic lines and four well ties delineate the shelf-edge geometries, depositional systems, and chronostratigraphy. Geometric relations together with paleontologic and lithologic facies data permit interpretation of the primary controlling factors on platform evolution.

Atlantic platform "keep-up" growth and development during successive tectono-eustatic fluctuations occur in two stages and vary directly with the supply of siliciclastics. Platform growth stages receiving siliciclastics prograde; those which do not aggrade. During stage I (Kimmeridgian), 8.1 km (5 mi) of margin progradation occurred near the depocenter of mixed carbonate-siliciclastic deposition, while 96 km (60 mi) northwest of the depocenter only 1.6 km (1 mi) of progradation took place across a 1,829-m-deep (6,000 ft) basin. During stage II (Portlandian-Middle Berriasian), margin progradation limited to carbonate deposition decreased to 0.5 km (0.3 mi) and the relief between the platform top and basin floor increased to about 2,134 m (7,000 ft). The switch from progradation to aggradation in the Baltimore Canyon Trough occurred after platform exposure and after the supply of siliciclastics was cut off to the basin seaward of the platform. This variation in platform evolution is found also in equivalent-age platforms in the Sable Island Delta area of the Scotian Basin, but not in coeval platform sequences of the southern Scotian Basin and Lahave Platform.

Platform "give-up"-type growth occurs as the last stage in the evolution of the Great Mesozoic Carbonate Bank in the Baltimore Canyon Trough. The cessation of growth is caused by platform drowning, not burial by siliciclastics. During stage III (Late Berriasian-Early Valanginian), more than a hundred meters of mud-dominated carbonates, rich in hexactinellid sponges and planktonic organisms, were deposited at the top of the carbonate bank. These deep-water carbonates succeeded shallow-water shelf carbonates, high-relief, shelf margin pinnacle reefs, and preceded a Valanginian siliciclastic sequence. Drowning resulted because of (1) high-amplitude tectono-eustatic fluctuations, (2) nutrient-suppressed carbonate deposition, and (3) volumetric constraints presented by a high-relief 2,440 m (8,000 ft) platform.

Siliciclastic filling of basins seaward of carbonate platforms has several important implications. First, siliciclastic bypassing of active carbonate margins may be contemporaneous with carbonate deposition. Second, the volumetric contribution of siliciclastics to basin fill can be sufficiently large to allow progradation of platforms that are otherwise limited to aggradational growth by their sedimentary carbonate budget.

INTRODUCTION

The Kimmeridgian-Valanginian platforms, underlying the continental shelf off the New Jersey coast, offer an exceptional place to study the influence of an influx of siliciclastics on the development of a carbonate platform because: (1) the area has spectacular seismic coverage and critical borehole data with which the seismic stratigraphy can be constrained; (2) the area has three successive developmental stages that feature a switch from progradational to aggradational deposition; (3) the area has an influx of siliciclastics from the nearby Appalachian hinterland; and (4) data are available on the history of eustasy and platform subsidence. This combination of circumstances provides some instructive examples of not only the influence of relative sea-level fluctuations but also the influence of siliciclastic sedimentation on the stratigraphy, geometry, and evolution of a carbonate platform deposit.

This paper (1) describes the local Kimmeridgian-Valanginian stratigraphy and (2) depositional facies; (3) details the large-scale platform margin geometries; and (4) integrates basin subsidence rates from the COST B-3 well and the sea-level changes of Haq and others (1987) to examine the contribution that supply of siliciclastics and other factors have on Mesozoic carbonate deposition in the Baltimore Canyon Trough. The main data base comes from an examination of 20 seismic lines, nine cores, cuttings, and logs from the three Shell Offshore Incorporated (SOI) and others, wells OCS-A 0336-1, OCS-A 0337-1, and OCS-A 0317-1, and logs from the Tenneco well OCS-A 0131-1. The four wells are abbreviated 0336, 0337, 0317, and 0131. A review of a similar, calibrated seismic data base from the Scotian and Lahave Platforms provides a comparison of coeval Atlantic platforms whose development occurred in areas with and without a large supply of terrigenous sediments.

REGIONAL GEOLOGIC SETTING

Carbonate strata drilled off the New Jersey coast document only a small and final segment in the history of Mesozoic carbonate deposition on the eastern North American margin. This regionally extensive carbonate deposit, here termed the Great Mesozoic Carbonate Bank (Fig. 1), stretches from the Bahamas to northern Canada (Emery and Uchupi, 1972; Mattick and others, 1974; Sheridan, 1974; Schlee and others, 1979; Schlee and Grow, 1980; Gamboa and others, 1985). The bank is thought to consist of a series of separate platform deposits whose trend of buried paleoshelf edges (Fig. 1) roughly subparallel the 1,981-m (6,500 ft) depth contour of the modern continental slope (Jansa, 1981). These platform sediments form a broad belt that extends through several basins (Scotian, Georges Bank, and Blake Plateau) and over the intervening high structural platforms (Carolina and Lahave). Seismic data from different locations along the trend of carbonate platform sequences demonstrate there is considerable inter- and intraplatform variation in growth and morphology (Schlee and others, 1979; Jansa, 1981; Schlee and Jansa, 1981; Poag, 1985). From the Mesozoic carbonate trend, Jansa (1981) distinguished a variety of platform shapes, ranging from prograding or stationary to retreating.

In the Baltimore Canyon Trough, a basin that is largely filled by terrigenous sediments off the New Jersey coast

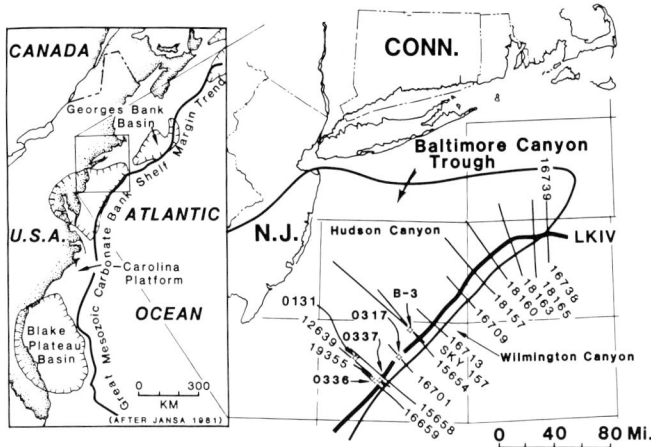

FIG. 1.—Index map showing trend of the Great Mesozoic Carbonate Banks, structural basins, and LKIV paleoshelf edge. The figure also shows the location of boreholes cited in text and a portion of the seismic lines studied.

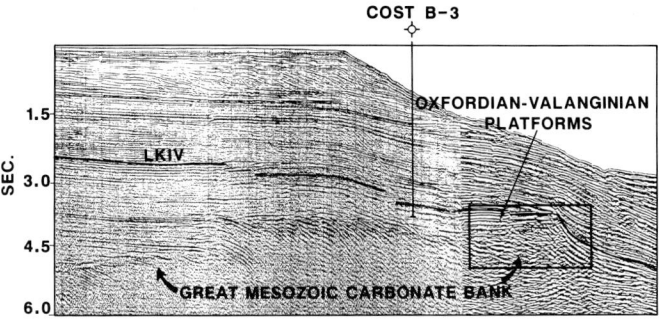

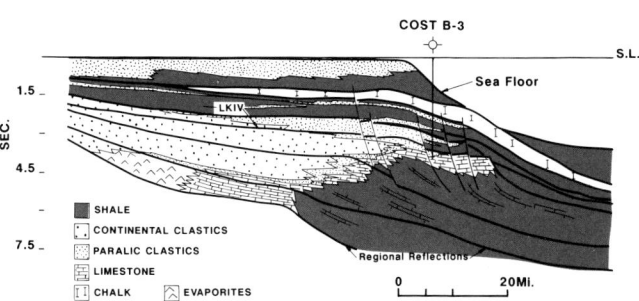

FIG. 2.—Regional Baltimore Canyon Trough cross section. Portion of JOIDES line SKY 157 (for location, see Fig. 1) with the COST B-3 well located on the line, and line drawing interpretation showing the prograded stratigraphy of the carbonate platform deposits in the Baltimore Canyon Trough. Carbonate deposits prograde across basinal shales and limestones and are buried by terrigenous clastics. The regional Lower Cretaceous reflection, LKIV, occurs at the top of the prograded carbonate complex.

(Poag, 1985), the carbonate complex is characterized by alternately prograding and aggrading growth intervals. SKY line 157 and limited well data (Fig. 2) demonstrate the lateral morphologic and stratigraphic attributes across the carbonate bank in this area (also see Schlee and others, 1979; Jansa, 1981; Schlee and Jansa, 1981). The bank, whose slope is represented by an aggraded and prograded clinoform sequence in seismic lines, is interpreted to form a narrow ribbon of limestones between the leading edge of a large, prograded, continental-shelf siliciclastic deposit, and a thick basinal deposit of interbedded carbonates and siliciclastics (Fig. 2). At least four different prograding and aggrading growth intervals and their interaction with siliciclastics can be documented over a distance of 63.4 km (40 mi) since the time of its inception. Carbonate deposition probably began some time in the early Jurassic and continued until early Cretaceous time before the deposition of carbonates ceased (Jansa and Wade, 1975; Jansa, 1981, E. R. Ringer and H. L. Patten, pers. commun., 1987). Thus, the bank is an extensive and long-lived system of interacting carbonate and siliciclastic deposition. Direct stratigraphic data on the depositional history across this prograded mixed carbonate-siliciclastic deposit are available only on the Kimmeridgian-Valanginian portion or about a 12.9-km-wide (8 mi) segment in the Baltimore Canyon Trough (Figs. 2, 3). This study limits itself to a detailed description and interpretation of controls on the evolution of this tested portion of the Great Mesozoic Carbonate Bank.

SUBDIVISION OF THE PLATFORM SEQUENCE

Stratigraphy

A combination of biostratigraphic and seismic data subdivide the tested portion of the carbonate buildup in the Baltimore Canyon Trough. Dinocyst horizons and calpionellid ecozones, supplemented by benthic foraminiferal assemblages, establish the chronostratigraphic zonation (Ringer and Patten, 1986, and pers. commun., 1987).

Chronostratigraphy and geologic ages shown are generally those recognized by Haq and others (1987). A revision of the chronostratigraphy adopted in this paper substitutes the Berriasian for the Ryazanian of Haq and others (1987). This change follows that of E. R. Ringer and H. L. Patten (pers. commun., 1987), who define a biostratigraphic zonation for the Berriasian that is the same as the one used by Haq and others (1987) for the Ryazanian interval.

Stage boundaries and biostratigraphic horizons reported for various wells in this paper are all given as depths below kelly bushing (K.B.).

The results of the integrated zonation of E. R. Ringer and H. L. Patten (pers. commun., 1987) allow recognition of Kimmeridgian-Valanginian and questionable (?)Oxfordian carbonates. Possible (?)Oxfordian strata are reported by E. R. Ringer and H. L. Patten (pers. commun., 1987) from the 0336 well. In this well, the age assignment of the basal 94 m (310 ft) are in question. Dolomitic, coral-stromatoporoid lime boundstones and mixed skeletal grainstones predominate in this basal interval. These well-cemented, shelf margin deposits contain the foraminifera *Anchispirocyclina lusitanica* and possible *Alveosepta jaccardi*, and the range top of *Ctenidodinium chondrum*, datum A, is recorded at 4,865 m (15,960 ft) from the top of this interval (E. R. Ringer and H. L. Patten, pers. commun., 1987). Datum A is now considered to be basal Kim-

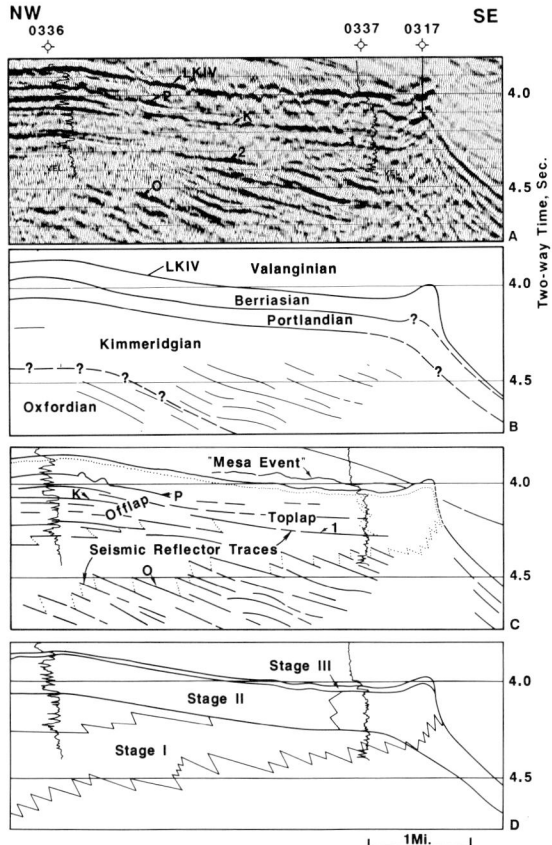

FIG. 3.—Major subdivisions of Kimmeridgian-Valanginian platform. (A) Portion of time-migrated seismic section 19355 (for location, see Fig. 1) and line drawing interpretations showing stratigraphy, seismic reflection geometries, and developmental stages. Velocity logs from 0336 and 0337 (located on line) tie well and seismic data. Well 0317 is projected 35 km (22 mi) south. Top platform carbonate (LKIV) and series boundaries (P, K, O) discussed in text are shown. (B) Chronostratigraphy recognized from drilling and seismic profiling. Sequence boundary LKIV outlines the fossilized profile from top of platform to time-equivalent slope. (C) Traces of internal reflections and geometries discussed in text; dotted lines outline inferred depositional systems. (D) Successive growth stages exhibited by Kimmeridgian-Berriasian platforms. Stage I represents a prograded "keep-up" mode, stage II an aggraded "keep-up" mode, and stage III a "give-up" mode.

meridgian and correlative with the *Rasenia cymodoce* ammonite zone on the cycle chart of Haq and others, 1987, (H. L. Patten, pers. commun., 1987). The position of datum A, coupled with an average uncompacted sedimentation rate of 12 cm/ka (379 ft/m.y.) for the Kimmeridgian interval in the 0336 well, suggest the Oxfordian strata of E. R. Ringer and H. L. Patten (pers. commun., 1987) should also be considered Kimmeridgian in this well.

Indisputable Kimmeridgian strata are recognized in both the 0336 and 0337 wells. About 884 m (2,900 ft) of shallow-water, skeletal shelf limestones and thin interbedded sandstones in the 0336 well, and 700 m (2,300 ft) of grain-supported, skeletal, shelf margin carbonates in the 0337 well characterize the two Kimmeridgian intervals. Above dinocyst horizon A, E. R. Ringer and H. L. Patten (pers. commun., 1987) report the foraminifera *Anchispirocyclina lusitanica* and possible *Alveosepta jaccardi*. They also note the range top of the dinocyst *Senoniasphaera jurassica* (datum C) at 3,971 m (13,030 ft). The top of the Kimmeridgian is placed approximately at 3,953 m (12,970 ft) in the 0336 well and at 3,735 m (12,255 ft) in the 0337 well, based on datum C (E. R. Ringer and H. L. Patten, pers. commun., 1987).

Portlandian rocks succeed the Kimmeridgian in both the 0336 and 0337 wells. In the basal 76 m (250 ft) of the 213-m (700 ft) Portlandian section at 0336, shales and argillaceous limestones predominate. In the Upper Portlandian section, lime grainstones and wackestones, rich in coated grains and a variety of skeletal material, are interbedded with thin beds of siliciclastics. The Portlandian interval in the 0337 well measures about 198 m (650 ft). Except for the skeletal lime wackestones at the base, this section consists mostly of pure, mixed-skeletal, lime grainstones.

Portlandian strata at 0336 and 0337 contain a variety of benthonic formaminifera, including *Trocholina elongata*, *T*. cf. *solecensis*, *Epistomina uhligi*, *E*. *stellicostata*, and *E*. *dneiprica* (E. R. Ringer and H. L. Patten, pers. commun., 1987). Also reported from the top of this interval is the range top of the dinocyst, *Pareodinia sp.* (datum E). This biohorizon is used to discriminate the Portlandian and Berriasian Stages. Datum E is recognized and places the Portlandian-Berriasian boundary at 3,758 m (12,330 ft) in the 0336 well. An upper boundary for the Portlandian Stage in the 0337 well is not precisely defined since *Pareodinia sp.* was not recognized there, but may be approximated from integrated biostratigraphic and seismic stratigraphic data. Well to seismic ties of datum E at 0336 and its seismic correlation suggest the position of the Portlandian-Berriasian boundary ties into the 0337 well at approximately 3,551 m (11,650 ft). The pick for this top occurs about 91 m (300 ft) below the regional Early Berriasian dinocyst *Muderongia sp.* datum recognized by E. R. Ringer and H. L. Patten (pers. commun., 1987) in this well.

Berriasian rocks are recognized in each of the three Shell wells. Except for the section at 0337, the carbonates of the Berriasian interval include the last accumulation of Mesozoic limestones before burial by siliciclastics.

Berriasian strata at 0336 have two distinctive carbonate textures. Interbedded foraminiferal grainstones and oncolitic wackestones, together with a few thin shale beds at the base, are predominant in the basal 171 m (560 ft). These shallow-water carbonates are succeeded by 58 m (190 ft) of argillaceous lime wackestones and packstones that contain a distinctive deep-marine fauna (E. R. Ringer and H. L. Patten, pers. commun., 1987). The argillaceous content, as suggested by gamma ray response, increases progressively up section in the top 15 m (50 ft) of these limestones. Hexactinellid sponges, coccoliths, *Tubiphytes,* and calpionellids are common throughout these carbonate textures.

In the 0337 and 0317 wells, the Berriasian strata exhibit the same shallow- to deep-water succession, but textural differences are noted in the shallow-water limestone intervals. At 0337, 85 m (280 ft) of mixed skeletal grainstones are interbedded with rare thin lime mudstones and foraminiferal wackestones beneath the deep-water limestones of the Berriasian sequence. At 0317, the Berriasian interval begins with a massive 131-m (430 ft) limestone section of interbedded oncolitic, skeletal, lime grainstones and coral-algal boundstones before the transition to sponge limestones is reached.

Shallow-water limestones and shales contain a variety of microfossils that suggest most of these strata are Lower Berriasian. E. R. Ringer and H. L. Patten (pers. commun., 1987) note the foraminiferal range tops for *Trocholina* cf. *solecensis, Epistomina stellicostata,* and *E. dneiprica* at 0336 and 0337. They also report *Muderongia sp.* (dinocyst datum F) near the top of the shallow-water limestones interval at 3,612 m (11,850 ft) in 0336 and at 3,493 m (11,460 ft) in 0337. At 0317, their studies show the carbonates near total depth are early Berriasian based on *Calpionella alpina* (small variety), which is found in the muddy internal sediments of boundstone cavities. In the overlying deep-water carbonates, the diversity of calpionellids increases. An analysis of these microfossils shows a late Berriasian assemblage of *C. alpina, Tintinella carpathica* (large variety), and rare *Calpionellopsis simplex* is present at 0336 and 0317 near the top of the carbonate section (E. R. Ringer and H. L. Patten, pers. commun., 1987).

Valanginian rocks are represented by interbedded shales and sandstones at 0336, carbonates at 0337, and are missing at 0317. Like the underlying late Berriasian, the Valanginian carbonates are also deep-water deposits and contain the same carbonate textures. These carbonate strata have a thickness of 86 m (282 ft) and yield specimens of *Tintinella carpathica* (large variety) and the dinocyst *Systemotophora areolata* near the top of the section. An analysis of the available microfossils indicates the carbonate strata extend upward into the lower Valanginian and are succeeded unconformably by Hauterivian strata (E. R. Ringer and H. L. Patten, pers. commun., 1987).

Correlative Valanginian strata at 0336 are a coarsening-upward sequence of which the lower 76 m (250 ft) are pure shales. The succeeding interval is predominantly shale with interbedded fine quartz sandstones. Collectively, these strata contain a diverse Valanginian microfossil assemblage. Shales of the basal section contain the dinocysts *Biobifera johnewingii, Diacanthum hollisteri, Amphorula metaelliptica,* and foraminifera of the *Lenticulina busnardoi* ecozone (E. R. Ringer and H. L. Patten, pers. commun., 1987). In the overlying siliciclastics, E. R. Ringer and H. L. Patten (pers. commun., 1987) report the dinocyst datum G and foraminifera ascribed to the *Epistomina anterior-E. carocolla* ecozone.

Figure 1 shows part of the grid of seismic lines that extends the stratigraphy of the three Shell wells beyond the borehole control. Primary reflections in these lines outline and detail the stratigraphic geometric relations that form the framework used in interpreting controls on the evolution of Kimmeridgian-Valanginian platforms in the Baltimore Canyon Trough.

Seismic line 19355 (Fig. 3) shows reflection arrangements that helped delineate the chronostratigraphy and depositional geometries. Seismic event LKIV, the JO of E. R. Ringer and H. L. Patten (pers. commun., 1987), and the reflection informally called the "mesa event" outline the top of the carbonates. Reflection LKIV is a strong, continuous, regional seismic event that can be correlated to the clinoform configuration down depositional dip. It occurs above the *Tintinella carpathica-Calpionella alpina* ecozone at 0336 and 0317, and approximates the Berriasian-Valanginian boundary. This reflection profiles a platform with a spectacular raised rim and a steeply dipping slope (Figs. 3, 4). A strike line (Fig. 5) shows this rim is highly dissected. Detailed seismic mapping of the shelf margin delineates a trend of pinnacles that now tower more than 152 m (500 ft) above the adjacent shelf floor and locally 3,048 m (10,000 ft) above a sediment-starved basin floor.

The "mesa event" is an intra-Valanginian reflection with a limited stratigraphic expression above the regionally traceable top of the platform (see Figs. 3, 4, 5). Seismic mapping demonstrates this event is restricted to a narrow zone behind the LKIV margin and forms a discontinuous belt of mesa-like features. The seismic-well tie at 0337 suggests the reflection originates at a carbonate/shale boundary.

Although unrecognized based on biostratigraphic analysis (E. R. Ringer and H. L. Patten, pers. commun., 1987), the Berriasian-Valanginian boundary is unconformable. An unconformable relation is suggested by a group of weak reflections that downlap on LKIV landward (northwest) of the "mesa event" (Fig. 4). As indicated (Fig. 4), weak reflections also downlap on the "mesa event," indicting a disconformable relation exists at this boundary. The reflection pattern is less clear, however, whether a disconformity exists at LKIV beneath the "mesa event." Termination of the "mesa event" against LKIV at the raised rim (Fig. 3) suggests an unconformable relation between Valanginian "mesa" deposits and the Berriasian platform, but this reflection pattern also may be because of spurious events derived from the pinnacles.

Figure 3A shows a portion of seismic section 19355. The 0336 and 0337 wells are located on the sesimic line. Velocity logs for these wells are used to tie reflections to specific depths in each well. Five reflections below LKIV are identified on the seismic line. Three represent chronostratigraphic boundaries; the other two help subdivide the carbonate deposit.

Reflection P (Fig. 3A, B) corresponds to the boundary between the Portlandian and Berriasian. Dinocyst horizon E is reported from this horizon at 0336. Reflection P cannot be traced throughout the study area even with additional well control. As shown (Fig. 3A), the continuity and strength of reflection P diminishes toward the shelf margin and is lost entirely in lines crossing the northern part of the study area. Correlation of reflection P across the shelf margin and into the basin is very tenuous (Fig. 3B).

Reflection K occurs at the top of the Kimmeridgian section. Dinocyst datum C is found at this level in the 0336 and 0337 wells. Reflection K is fairly persistent across the shelf, but velocity and signal problems associated with the shelf break interrupt the reflection at the margin and make it difficult to trace into the basin. Convergence of reflection 1 immediately below event K complicates, but does not prevent, tracing reflection K throughout the study area.

Reflection O is interpreted to correspond to the Oxfordian-Kimmeridgian boundary. Reflection O occurs between the 0336 and 0337 well locations on seismic line 19355 and is projected to pass at a level just beneath the depth reached by the 0336 well. At total depth, this well contains Late Oxfordian-Early Kimmeridgian microfossils. In Figure 3, the O reflection is a discontinuous inclined depositional surface and part of a classic progradational pattern. The seismic event diminishes in both directions, and no attempt was made to trace reflection O throughout the study area.

Depositional Systems

Four depositional systems are discriminated from the 0336, 0337, and 0317 borehole data and traced throughout the Kimmeridgian-Valanginian platforms with seismic lines. They are slope, shelf margin, shallow-shelf and deep-shelf systems.

Slope system.—

The Kimmeridgian-Berriasian slope system is known almost exclusively from seismic expression since only the 0337 well penetrated this portion of the Great Mesozoic Carbonate Bank in the Baltimore Canyon Trough. Slope strata form the basal 107-m (350 ft) section at 0337. The strata consist of a muddy carbonate sequence that is interbedded with widely separated, thin shale beds. Included in the carbonate sections are intervals of thrombolitic lime boundstones with mud-filled and marine-cemented stromatactis cavities (Fig. 6) and intervals of redeposited mudstone and wackestone intraclasts.

Seismic dip line 19355 and the 0337 velocity log (Fig. 3A) show that slope strata at the base of well 0337 represent the uppermost part of an extensive slope system. In line 19355 and companion dip lines, seismic expression of slope deposits is characterized by a wedge-shaped package of prograded, steeply dipping, concave-up reflections that converge and slope seaward with pronounced downdip downlap (see Figs. 4, 7). Individual reflections tend to be wavy or mounded, vary in continuity and amplitude, and decrease in dip gradually in their lower portions. Diffractions disturb reflection continuity beneath the position of the Berriasian shelf margin, and diminished amplitudes may lead to reflection discontinuities elsewhere (Figs. 3, 4). Amplitudes diminish abruptly updip (Fig. 3). Gradual decreases in amplitude occur downdip in places of reflection convergence and downlap (Fig. 4). Decreased stratal thickness probably causes the progressive downdip loss in amplitude.

The Kimmeridgian-Berriasian wedge of inclined reflections varies in breadth in the northern part of the Baltimore Canyon Trough. An analysis of dip lines shows the wedge is about 6.5 km (4 mi) wide in line 12639 (Fig. 4) and 15658 near the three Shell wells, increases to 8.1 km (5 mi) in line 16709 (Fig. 7B) about 129 km (80 mi) to the northeast, and decreases to less than 1.6 km (1 mi) in line 18165 (Fig. 7A) near the Long Island Arch.

The upper-slope deposits at 0337 show that the pattern of locally wavy or mounded strong reflections can be expected to indicate reworked and mixed siliciclastic-carbonate deposition. Wavy or mounded reflections created by changes in stratal thickness may be expressions of slope mud mounds, or redeposited sediments resulting from debris flows. The strong velocity contrast is probably the expression of interbedded terrigenous clastics and marine carbonates. Complete Mesozoic slope deposit well penetrations in the Scotian Basin contain stratigraphic sequences (Eliuk, 1978) similar to those inferred for the Kimmeridgian-Berriasian system in the Baltimore Canyon Trough. The slope sequence may also be similar to Kimmeridgian slope deposits in Portugal. In Portugal, the Kimmeridgian, mud-dominated basal Tojeira member of the Abadia Formation exhibits carbonate textures similar to those found at 0337, whereas the Mem Martins Formation features a mixed siliciclastic-carbonate slope sequence (R. C. L. Wilson, pers. commun., 1987) that is similar to the one inferred for the Baltimore Canyon Trough from seismic data.

Shelf margin system.—

A shelf margin depositional system is recognized for a massive sequence of coarse-grained carbonates in each of the three Shell wells (Fig. 6). The shelf margin deposits are mostly clean carbonate grainstones and framestones with a shallow-water assemblage of allochems that include algae, corals, mollusks, echinoderms, foraminifera, oncoids, and ooids at 0337 and 0317. At 0336 fine-grained quartzose sand and silt are interbedded and admixed with the shallow-water carbonate grainstone and framestone textures. Shelf margin deposits dominate the basal sections in each Shell well except the 0337 well. At 0337 grain-supported carbonates of backreef and reef deposits succeed a sequence of slope sediments.

Velocity logs tie the sequence of shelf margin deposits to a seismically quiet zone with poor reflection characteristics (Fig. 3). This zone of discontinuous, weak and disorganized reflections passes laterally and below into the steeply inclined reflection pattern of the slope system and into wavy, subparallel or diverging reflections in the opposite direction (Fig. 3). In the seismically quiet shelf margin zone near 0336 and 0337, the zone has climbing progradational and aggradational patterns that collectively span more than 6.5 km (4 mi) above the top of the Oxfordian event (Fig. 3D). Seismic reflection data show that the seismically quiet shelf margin zone changes progressively away from the Shell wells in the direction of Georges Bank, until 241 km (150 mi) north of 0336 and 0337 in seismic line 18165 (Fig. 7A). The Kimmeridgian-Valanginian seismic quiet zone is aggradational and less than 1.6 km (1 mi).

Shallow-shelf system.—

A shallow-water shelf system is interpreted for a sequence of interbedded, shallow-water, terrigenous clastics and carbonates occurring in the 0336 well (Fig. 8). The shallow-shelf strata contain a complete spectrum of depositional textures, but they are characterized by a predominance of muddy siliciclastics and carbonates that are rich in algae-coated grains, mollusks, foraminifera, algae and echinoderms.

Continuous subparallel and divergent low-frequency, high-amplitude reflections coincide with the interval of shallow-

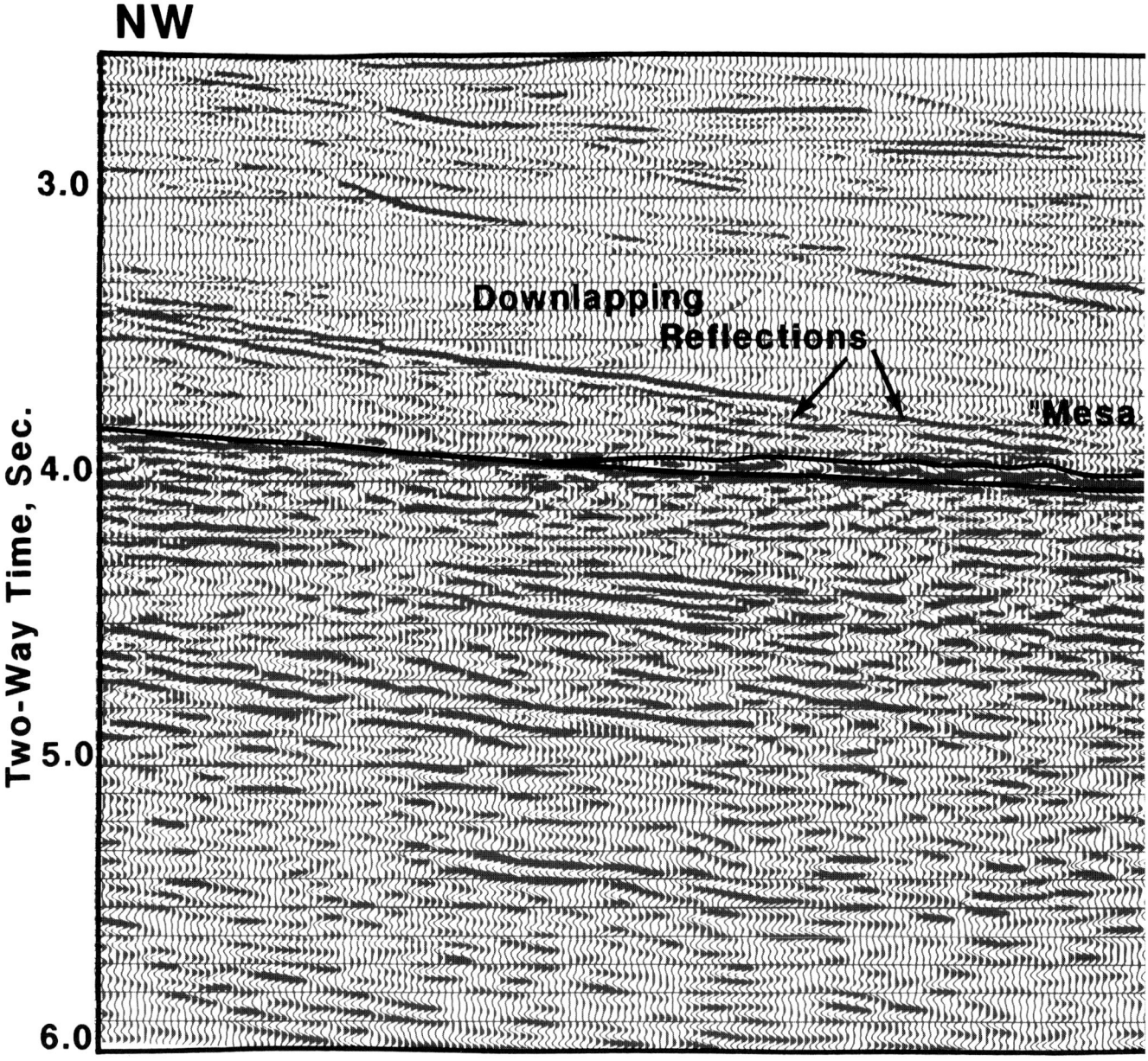

Fig. 4.—Line 12639. Seismic time section showing stratigraphic relations between LKIV, "mesa event," and succeeding LK events (for location, see Fig. 1).

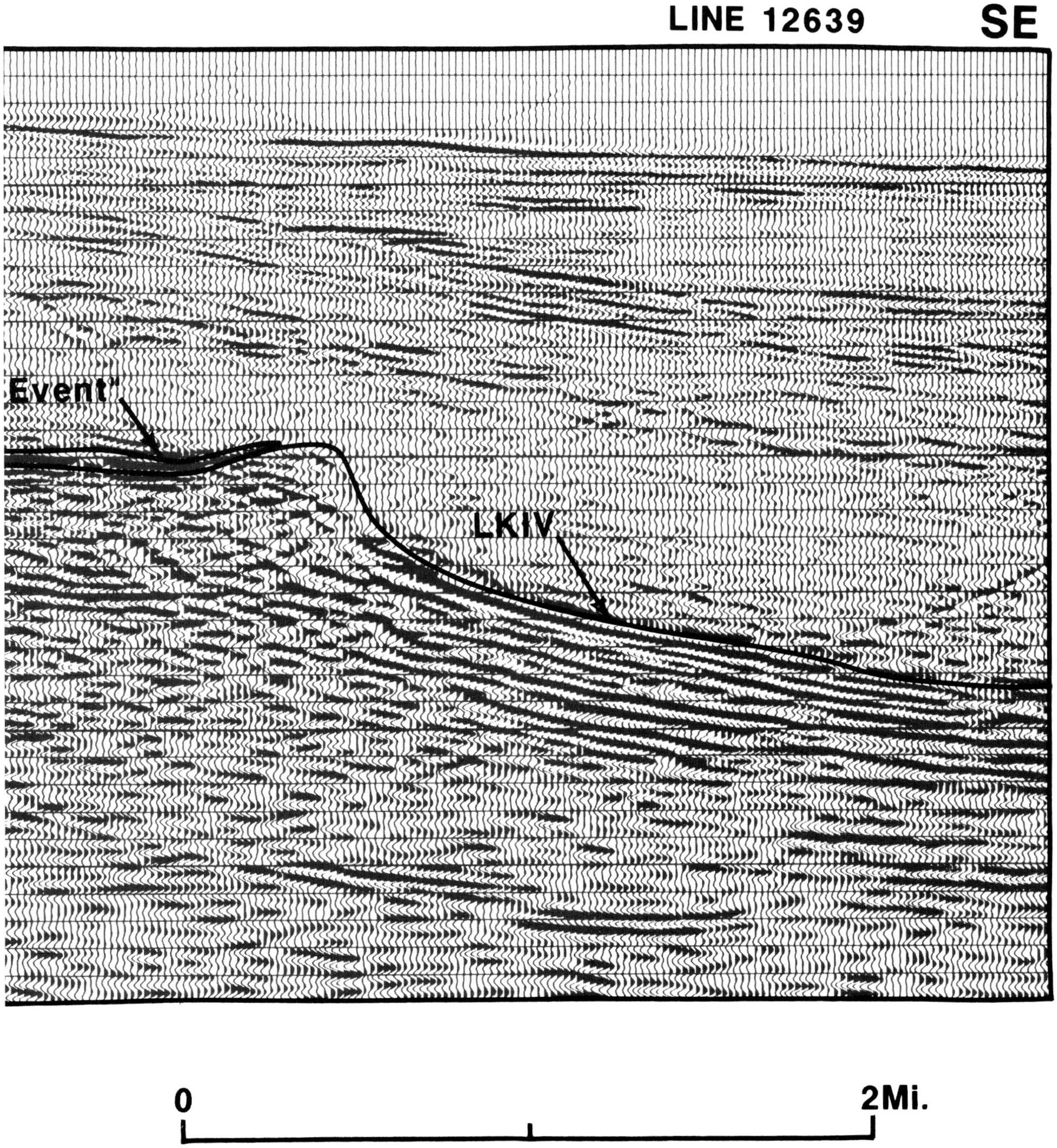

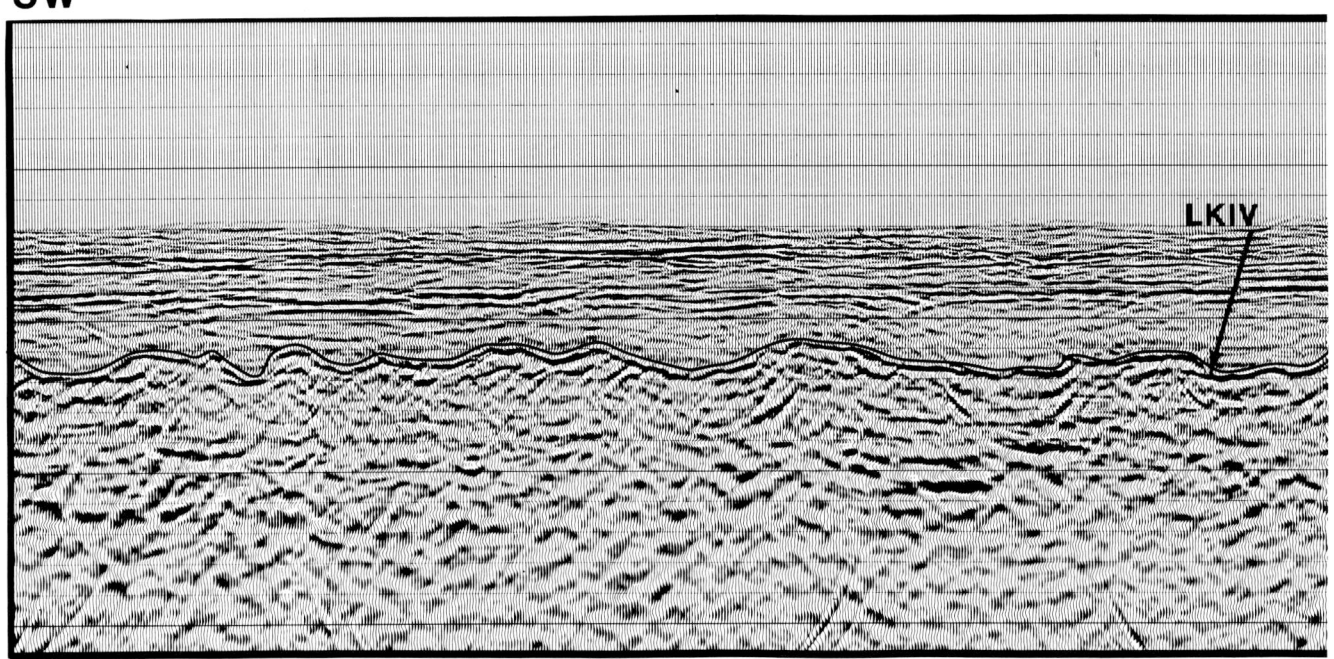

FIG. 5.—Seismic line 18904. Seismic time section showing high relief and discontinuity of LKIV carbonate margin and "mesa" deposits along depositional strike. Line 18904 parallels LKIV margin and extends between location of dip lines 16659 and 15654 in Figure 1.

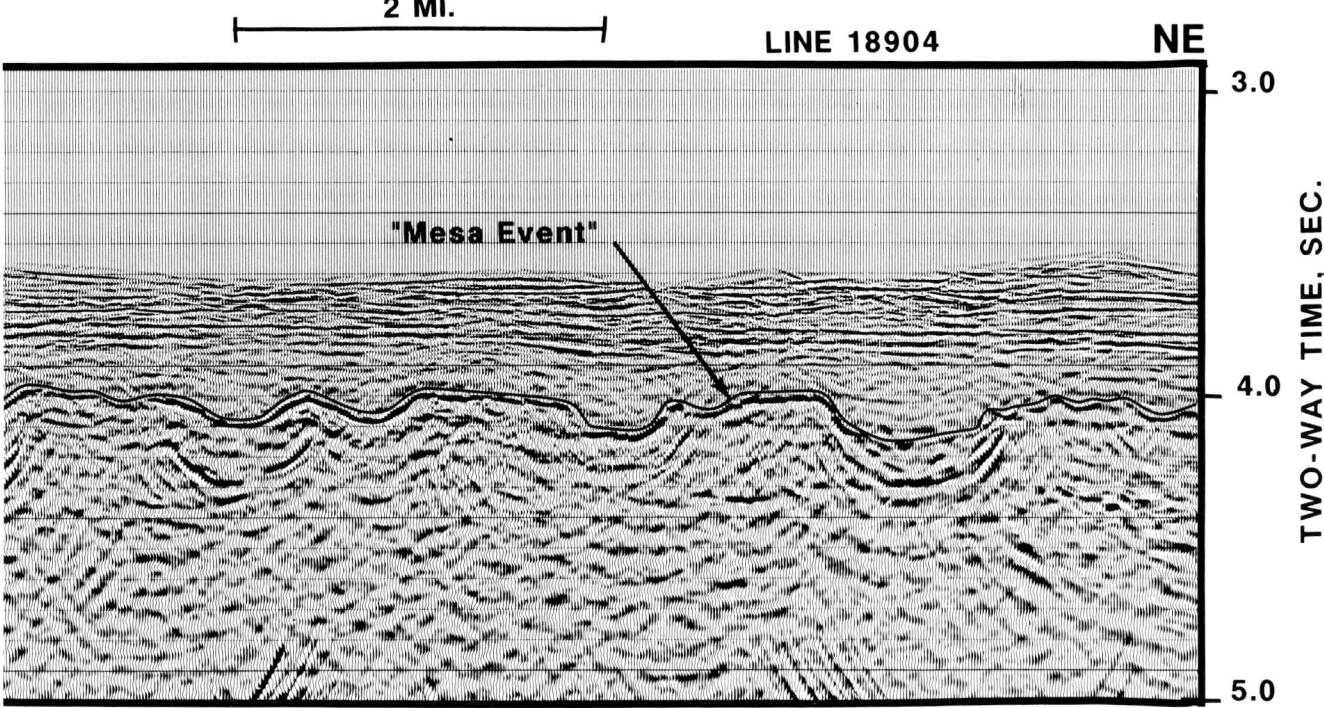

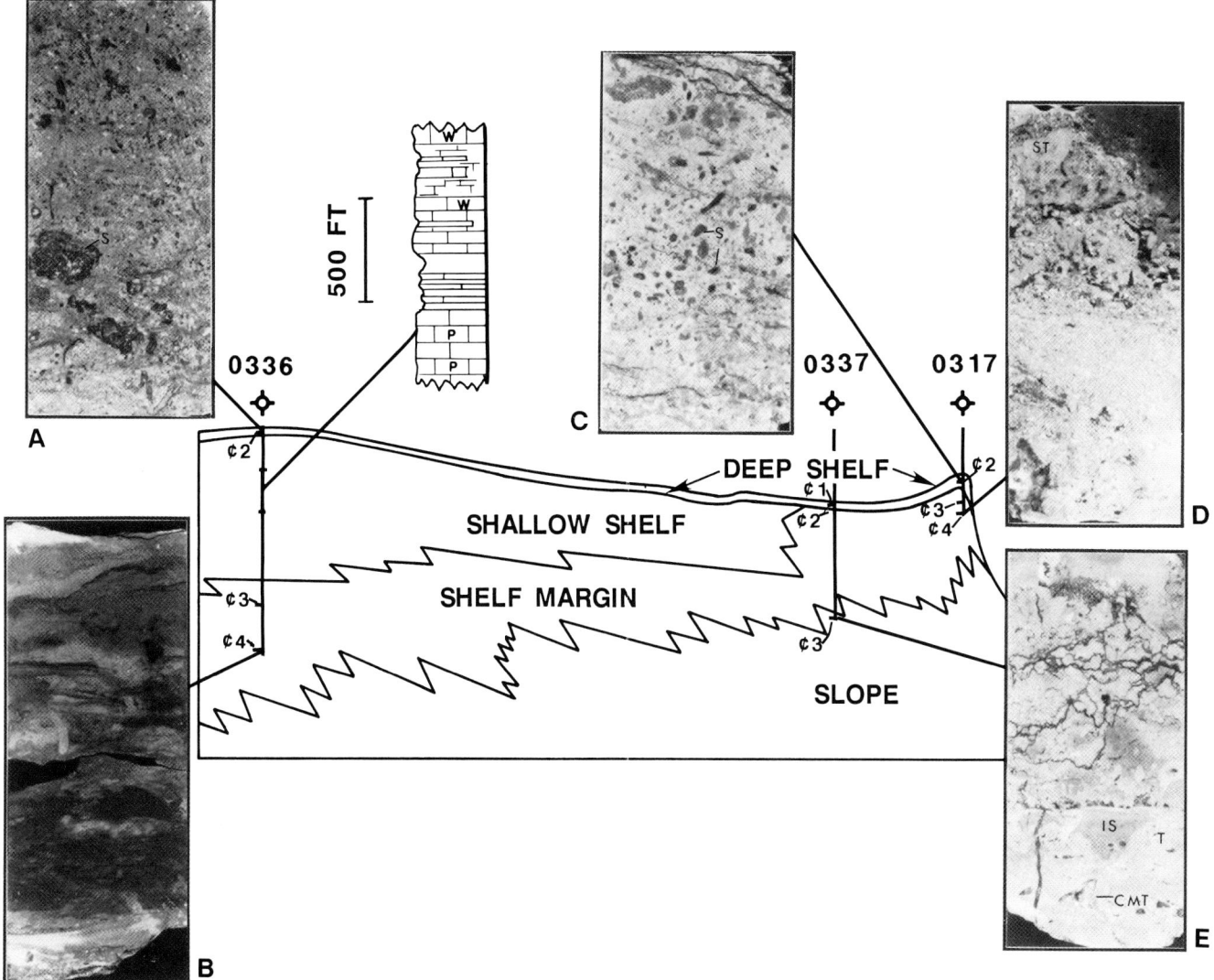

Fig. 6.—Depositional systems. Kimmeridgian-Berriasian platform depositional systems, constrained by seismic time section 19355 and borehole data from 0336, 0337, and 0317. Photographs show a portion of slabbed core (numbered ¢1, ¢2, and so forth, for each well) material from three depositional systems: (A) sponge (S) lime wackestone; (B) fine, burrowed quartz sandstone; (C) fragmented sponge-rich lime wackestone; (D) stromatoporoid (ST) boundstone and coarse skeletal lime grainstone; and (E) stylolitized thrombolite (T) boundstone with constructional cavities partially filled with internal sediment (IS) and cement (CMT). Lithologic log (follows Swanson, 1981) shows cyclic carbonate and clastic bedding of shallow-shelf facies.

shelf strata (Fig. 3). In strike section, this pattern of reflections collectively composes an elongate unit that parallels the paleomargin. Reflection patterns compose a wedge-shaped unit that thins toward and grades laterally into the seismically quiet zone of the shelf margin system in the direction of depositional dip and passes landward into a package of high-frequency discontinuous reflections (continental deposits).

Deep-shelf system.—

Overlying the shallow-shelf and shelf margin deposits occurs a deep-shelf system that forms the top of the Mesozoic carbonate sequence. It varies from a maximum of 122 m (400 ft) at 0337 to about 61 m (200 ft) thick at 0336 and 0317. This deposit is a variably porous, coccolith-bearing, mud-dominated argillaceous limestone. Cores taken in this interval contain a varied pelagic fauna that includes calpionellids, dinoflagellates, and rare foraminifera and hexactinellid siliceous sponges.

The deep-shelf deposits generally are below seismic resolution. Seismic response to this shelf system may be locally identified as an amplitude anomaly in reflection LKIV in most places. A distinctive reflection in response to a thick sequence of deep-shelf deposits at 0337 is the discontinuous, wavy, strong "mesa event." Seismic mapping of the "mesa event" delineates a series of narrow, elongated stratigraphic units that are located just behind and parallel to the Berriasian shelf break.

Depositional Stages

Reflection patterns and geometric relations of depositional systems permit a three-fold subdivision of the Kimmeridgian-Valanginian carbonate sequence in seismic dip lines (Fig. 3D). These subdivisions distinguish distinct episodes of platform growth.

The lowermost unit (stage I) is limited to the Kimmeridgian platform and is a period of prograded platform growth. Stage I is defined by shallow-shelf, shelf margin, and slope system offlaping and toplapping geometric relations at the top of the platform (Fig. 3A). Reflection 1, a weak, discontinuous intra-Kimmeridgian shelf event, whose amplitude increases near the slope break, is recognized as the top for the prograded sequence and stage I (Fig. 3A, C). Seismic correlation allows event 1 to be carried as far north as line 16709. This limit coincides with the area of maximum Kimmeridgian platform progradation in the northern Baltimore Canyon Trough. The base of stage I is established at reflection O, which approximates the Oxfordian-Kimmeridgian boundary at 0336. Reflection O does not coincide with a change in platform growth, but it does represent the oldest horizon for which there is lithostratigraphic control at the shelf margin. Reflections assigned to the base and top of stage I enclose an obliquely prograded carbonate platform segment that features a progression from offlapping to toplapping reflection trace geometries at the top of the platform, and a package of inclinded reflections whose declivity decreases during slope accretion (Fig. 3A, C). A thick basinal system is correlative with the slope system of stage I.

The middle unit (stage II) includes uppermost Kimmeridgian, Portlandian, and Early Berriasian platforms and defines a period of aggradational platform growth. Stage II is identified by a stacked sequence of shallow-shelf and shelf margin reflection patterns (Fig. 3C). Reflection 1 differentiates between stages I and II as discussed previously. Beyond the extent of reflection 1, an approximation of this horizon is afforded by reflection K. The top of stage II is defined by the abrupt intra-Berriasian transition from shallow- to deep-water carbonates in boreholes and identified as approximately equivalent to the basal inflection point of event LKIV in seismic lines. A transition analogous to that on the shelf is recognized at the shelf margin, but it is unclear whether the boundary between the two rock types is the same age at both locations. The top of stage II at the shelf margin is placed about 30 m (100 ft below the deep- and shallow-water carbonate boundary. This position is constrained by biostratigraphic data and the 0317 well to seismic tie, and by conceptual considerations. The upper boundary of stage II shows the carbonate platform developed an "empty bucket" (Kendall and Schlager, 1981) morphology that featured a prominent, raised rim of pinnacles at the margin. Significant relief at the margin by mid-Berriasian time supports the notion that shallow-water deposition persisted there, while deep-water sedimentation was initiated on the shelf.

Stage II boundries enclose an interval of shelf strata that passes laterally into a narrow zone of steeply inclined slope reflections. In contrast to the underlying prograded depositional stage, the aggraded slope system consists of only a few concave-up reflections, which merge into a single reflection a short distance down the paleoslope.

The uppermost unit (stage III) is limited to the Late Berriasian platform. It is represented by a 61 to 91-m (200–300 ft) stratigraphic interval that records, in part, the cessation of growth of the bank in Baltimore Canyon Trough. Stage III is recognized exclusively from borehole data, because its stratigraphic thickness is below seismic resolution. On the shelf, the strata are deep-shelf deposits; at the shelf margin, the strata include both shallow- and deep-water deposits. The entire stratigraphic interval assigned to stage III is included in the LKIV event and appears to represent the end of widespread carbonate deposition on the platform in the Baltimore Canyon Trough.

Because Stage III is identified from borehole data and since the stratigraphic thickness is below seismic resolution, this uppermost interval cannot be traced throughout the study area. Identification of a similar interval (Artimon member of Eliuk, 1978) at the top of the coeval carbonate platform in the Scotian Basin suggests the distribution of stage III is regional.

SILICICLASTIC CONTROL ON PLATFORM DEVELOPMENT

Siliciclastic-Carbonate Progradation, Stage I

Platform subsidence and long-term sea-level fluctuations shaped platform growth during stage I. Subsidence was rapid as evident from 945 to 1,067 m (3,100–3,500 ft) of Kimmeridgian strata. Curves displaying rates of sediment accumulation for the 0336 and 0337 wells record an average accumulation rate of 134 m/m.y. (439 ft/m.y.) throughout stage I (Fig. 9). This rate is similar to the total basin subsidence curve (including compaction and loading of sediment) calculated from the nearby COST B-3 well (Fig. 9). The average subsidence rate of 118 m/m.y. (386 ft/m.y.) is not significantly higher than the accumulation rate at 0336 and 0337. This point is important, because progradation during this interval demonstrates sedimentation rates exceeded the aggradation potential resulting from platform subsidence and a eustatic rise in sea level.

Accumulation rates did not fluctuate during the growth history of stage I (Fig. 9). This is particularly significant, since the initial offlapping and later toplapping facies relations that characterize the progradational phase of platform growth do not conform to the vertical growth trend predicted by the platform subsidence curve. Eustasy must exercise a principal control on the boundary relations at the top of the platform.

Figure 8 shows the relations of the sea-level curves and time scale of Haq and others (1987) to the Oxfordian-Berriasian portion of the platform sequence represented in the 0336 well. The sea-level curves and time scale shown have a uniformly adjusted fit to five biostratigraphic zones (Fig. 8). Four released (E. R. Ringer and H. L. Patten, pers. commun., 1987) and one proprietary biostratigraphic zones define the interval boundaries and tie the 0336 well to the biostratigraphic framework of the cycle chart of Haq and others (1987). Correlation of microfossil zones common to

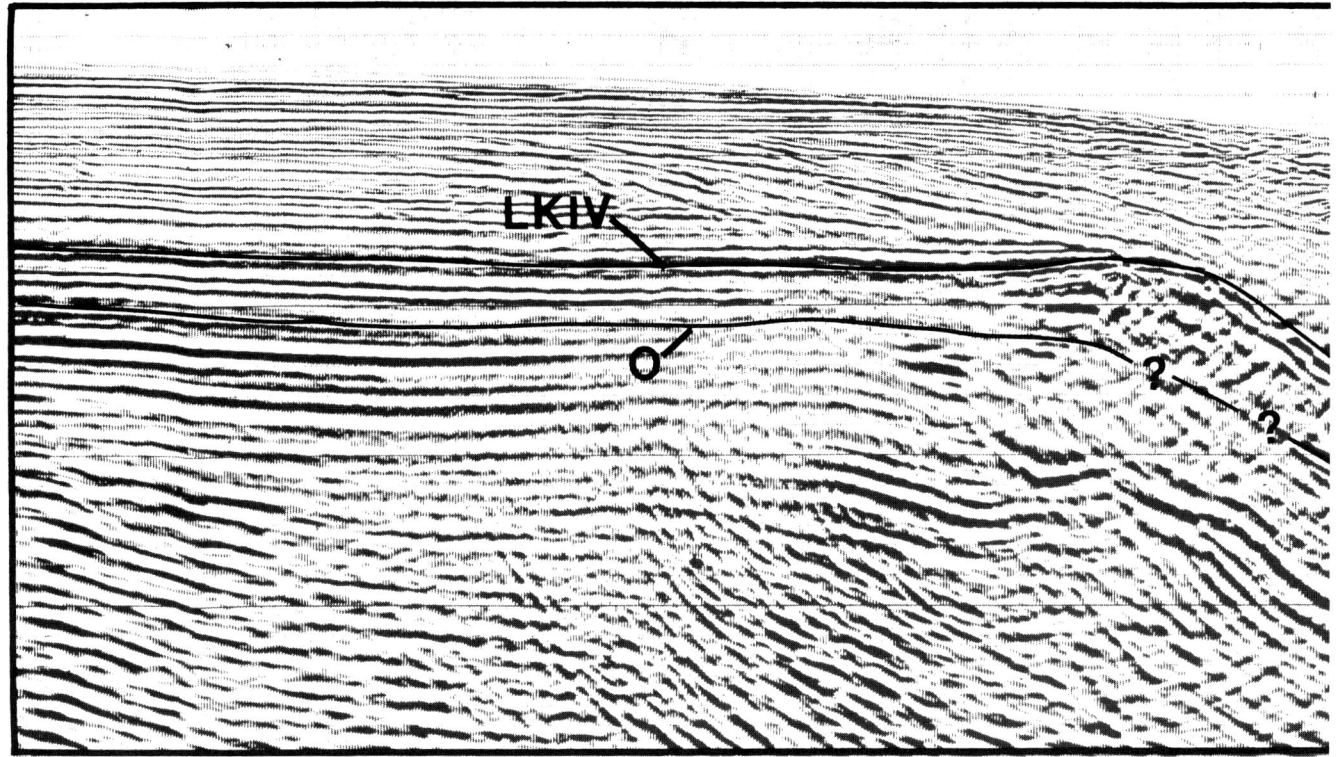

FIG. 7.—Seismic lines 16709 and 18165. Regional seismic depth sections across LKIV carbonate margin (A) approaching Long Island Arch and (B) in Baltimore Canyon Trough (for location, see Fig. 1). Profile 18165 shows predominantly aggradational growth, whereas profile 16709 shows predominantly progradational growth for the carbonate platform since the Oxfordian (O). Note that most of platform progradation occurred during the Kimmeridgian (K).

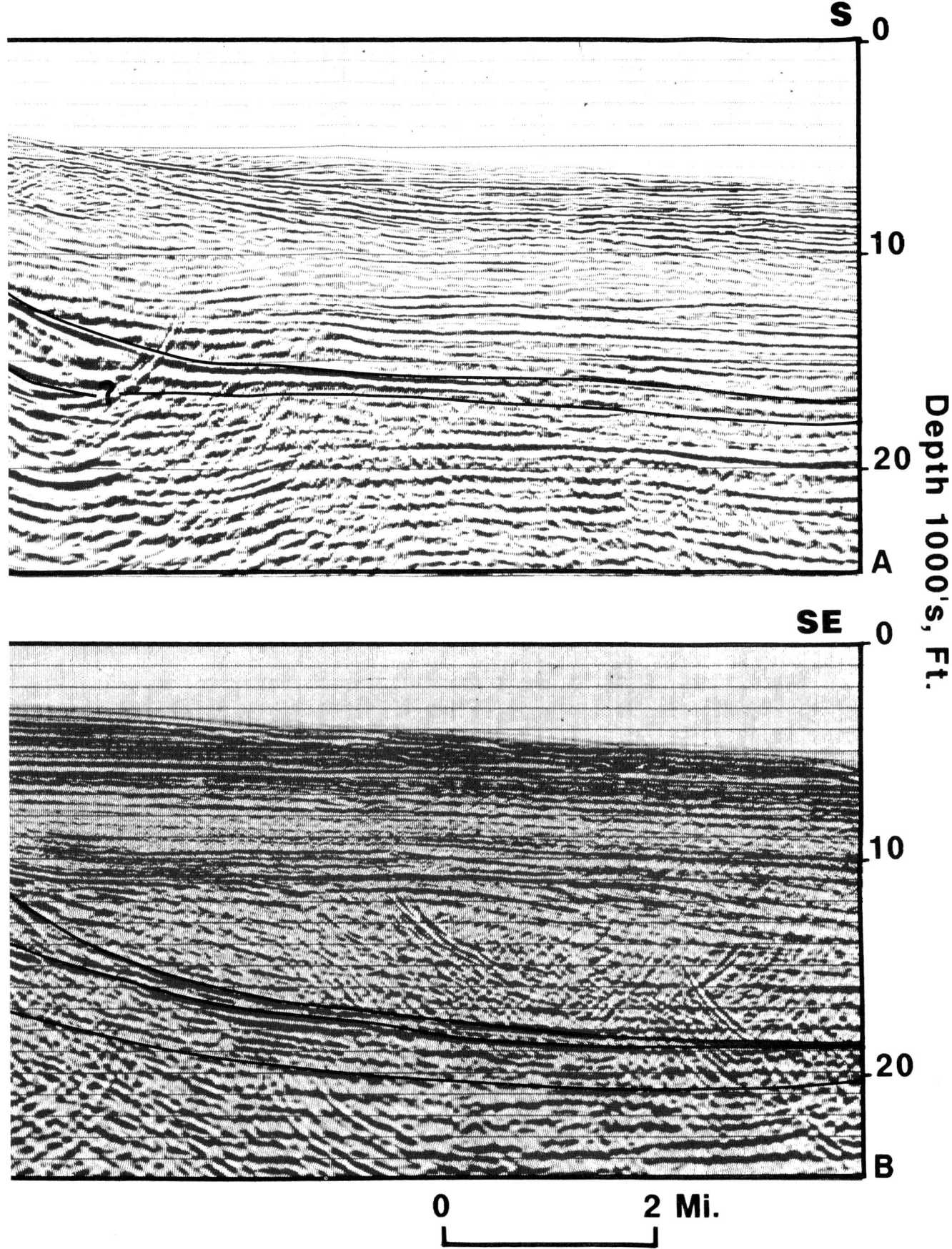

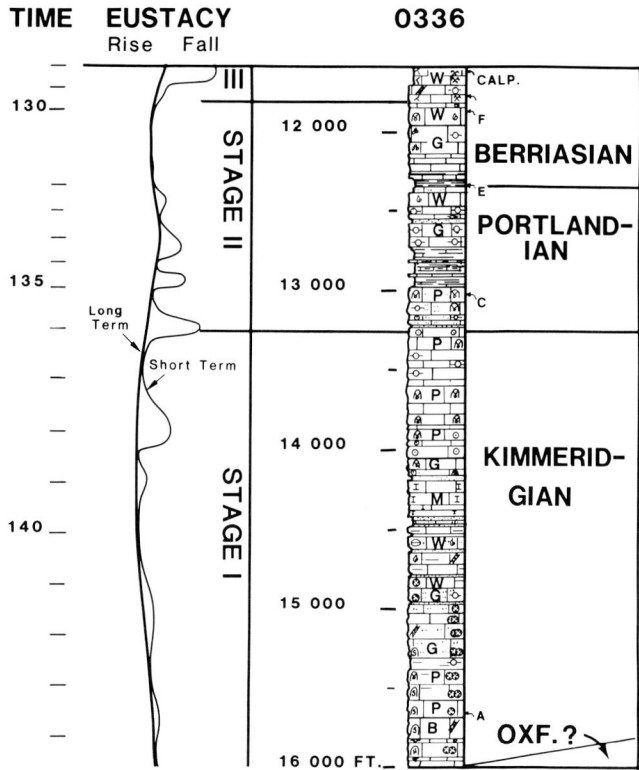

FIG. 8.—Eustasy and sediment type. Short- and long-term sea-level curve (from Haq and others, 1987) is fitted to depositional sequence of the Kimmeridgian-Valanginian platform by using the biostratigraphy and rock data of the 0336 well. Note poor correlation of Kimmeridgian sandstones and short-term lowstands. A good fit exists between Portlandian short-term lowstands and siliciclastics.

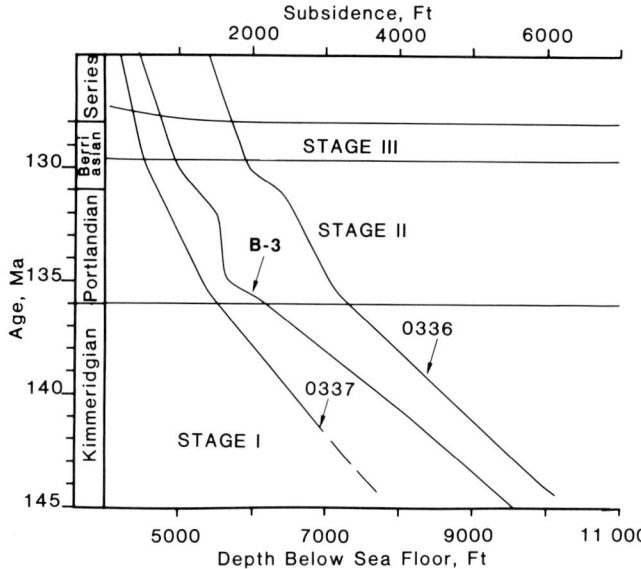

FIG. 9.—Sediment accumulation and platform subsidence. Plots for 0336 and 0337 are accumulation curves and are plotted by referencing the subsea positions of the carbonate platform top during the late Oxfordian to middle Berriasian. The COST B-3 curve charts total subsidence in Baltimore Canyon Trough during the same interval.

the 0336 well and the biostratigraphic framework of Haq and others (1987) is limited to *Senoniasphaera jurassica*, a dinocyst horizon found near 3,962 m (13,000 ft) and *Calpionellopsis* zone D (*Calpionella alpina, Tintinella carpathica* (large variety), and *Calpionellopsis simplex*) identified in core material from 3,536 m (11,600 ft) (E. R. Ringer and H. L. Patten, pers. commun., 1987). All other microfossil zones represent additional dinoflagellate horizons whose ties to the cycle chart of Haq and others (1987) are based on their ties to the European ammonite zones. Additional fine-scale constraints on the correlation of the sea-level curves or ages to specific strata are not accurately known within a stratigraphic interval. So, a uniform subdivision of the strata establishes the relation of the time scale and sea-level curves of Haq and others (1987) within each stratigraphic interval. If this fit is reasonable between the 0336 stratigraphic sequence and the cycle chart (Haq, and others, 1987), then the plotted sea-level fluctuations afford several important observations about stage I.

The effect of long-term sea-level fluctuations was to modify geometric relations between depositional systems at the margin of the prograding shelf during the Kimmeridgian. A good conformity exists between long-term sea-level fluctuations and platform growth during this growth interval. The Early Kimmeridgian long-term rise (Fig. 8) is clearly marked by offlapping progradation, and the Late Kimmeridgian long-term fall is reflected by a switch to a toplapping progradation geometry.

Further analysis of the history of falling sea level and thermal subsidence during the Kimmeridgian suggests subaerial platform exposure could not have resulted from the long-term sea-level fall. During the period of Late Kimmeridgian toplapping progradation, sea level fell at a rate of about 9 m (30 ft) m.y. (Fig. 10). This rate is significantly less than the thermal cooling subsidence rate 23 m or 75 ft/Ma calculated for the platform (M. W. Shuster, pers. commun., 1987).

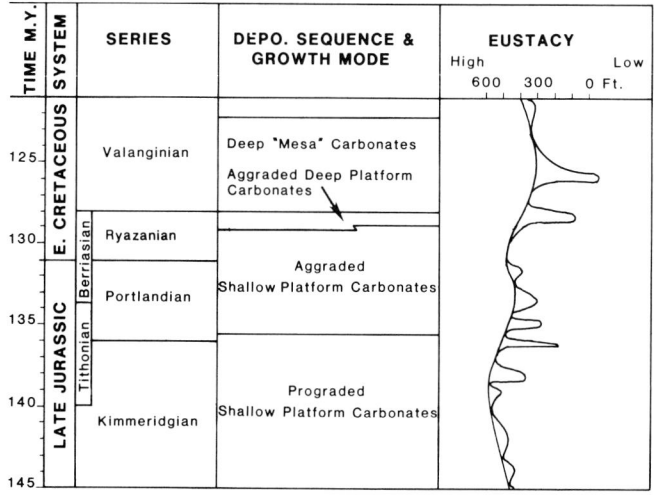

FIG. 10.—Chronology of platform growth and eustasy. Chart showing successive platform growth stages, chronostratigraphy, and sea-level fluctuations (chronology and sea-level curve are adapted from Haq and others, 1987).

Depositional and diagenetic fabrics interpreted from cuttings and seismic expression of the late Kimmeridgian strata suggest that platform growth was interrupted by periods of subaerial exposure. These interruptions in platform growth must result from short-term sea-level fluctuations.

Reflections 1 and 2 (Fig. 3A) mark two horizons interpreted as exposure surfaces on the basis of their depositional and diagenetic features. Both reflections tie to log depths at 0336, which have an interbedded sequence of carbonates and siliciclastics. Although the siliciclastics may be traced across much of the carbonate platform by their seismic expression, a coeval section of the uppermost, 0336, siliciclastic horizon is entirely pure carbonates at 0337. In this respect, the accumulation of terrigenous sediments on the Kimmeridgian shelf, but not on the shelf margin, is similar to patterns of reciprocal sedimentation described by Silver and Todd (1969) and Meissner (1972). Further evidence suggestive of subaerial exposure comes from diagenetic fabrics of carbonates associated with siliciclastics. Well cuttings from the Late Kimmeridgian interval at 0336 and 337 exhibit meteoric leaching and cementation similar to those reported in core 2 at 0337.

Although the two siliciclastic cycles (reflections 1 and 2) have features consistent with sea-level lowstands, the occurrence of Late Kimmeridgian terrigenous interbeds (Fig. 8) is not uniquely related to short-term lowstands predicted by the cycle chart (Haq and others, 1987). Multiple siliciclastic interruptions of carbonate deposition occur besides admixing of siliciclastics and carbonates during deposition of the entire Kimmeridgian interval (Fig. 8). Cyclic interruptions of carbonate deposition on the slope are also suggested by the seismic expression of this depositional system. The frequency of observed and inferred terrigenous cycles is far greater than the number of short-term sea-level fluctuations indicated by the eustasy curve of Haq and others (1987). Poor correspondence between short-term sea-level fluctuations and siliciclastic cycles suggests that other mechanisms besides short-term sea-level fluctuations may be responsible for regulating depositional variations. Meteoric diagenetic fabrics found above the youngest Kimmeridgian siliciclastic-carbonate couplet clearly indicate that similar diagenetic fabrics found beneath siliciclastic beds do not necessarily furnish evidence of platform exposure during the period of terrigenous deposition. Although the evidence is less than compelling, some cyclical interruptions of carbonate deposition probably are caused by short-term sea-level lowstands. Contemporaneous siliciclastic-carbonate deposition is an alternative process.

A high rate of basinal sedimentation and a decrease in platform height were additional factors contributing to platform progradation during growth of stage I. High-sedimentation rates allowed the platform to prograde rapidly into a basin that was 1,829 to 2,286 m (6,000–7,500 ft) deep during the Kimmeridgian (Fig. 11). Sedimentation rates varied considerably laterally. For example, maximum progradation rates calculated from seismic line 16709 are 894 m/m.y. (2,934 ft/m.y.) during this interval. This value contrasts significantly with the 201-m/m.y. (660 ft/m.y.) rate of advance during this interval in line 18165, which is located about 96 km (60 mi) northeast of line 16709.

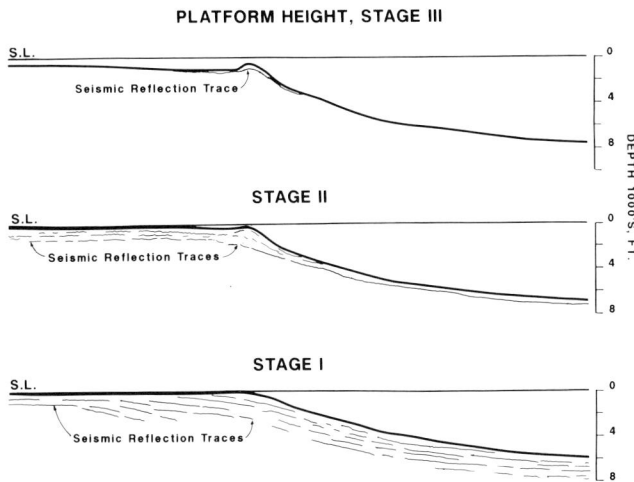

FIG. 11.—Platform height. Reflection traces from seismic depth section 16709 show platform height at the end of each growth stage. Height calculations for each stage were made at an arbitrary distance of 9 km (5.5 mi) seaward of the slope break. Tangents drawn to the slope range from 10° to 15° at these locations. Incomplete seismic coverage prevented taking measurements at a point where slope declivity fell below 1.4° (Schlager and Camber, 1986), so platform heights shown and cited are conservative figures.

The progression of progradation into the basin is recorded by a sequence of clinoforms whose relative declivity varies with the amount of progradation. Seismic lines showing several miles of progradation typically have clinoforms whose upper slope decreases over the period of progradation (Fig. 12). These declivity changes register a progressive decrease in platform height resulting from differential shelf and basin sedimentation rates. At 0336, 488 m (1,600 ft) of shelf sediment accumulated during the Kimmeridgian but during this interval 762 to 914 m (2,500–3,000 ft) of basinal sediments accumulated as shown by seismic stratigraphy in depth sections. Basin sediment accumulation exceeded that of the shelf, thereby decreasing platform height. As platform height decreased, so did the volume of sediment required for progradation.

Because Kimmeridgian basinal sediments have yet to be

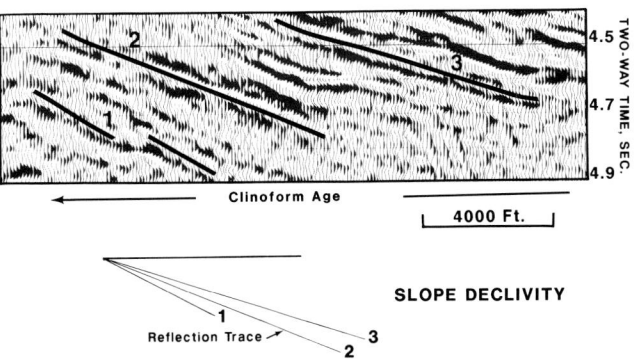

FIG. 12.—Stage I slope declivity. Kimmeridgian clinoform reflections from line 19355 show the relative angle decreases in the upper third of the platform slope (also, see Fig. 4). No absolute slope angles may be inferred because this is a time section.

drilled in the Baltimore Canyon Trough, their lithologic composition and type of vertical sequence can only be inferred. Several lines of reasoning, however, suggest the basin fill is composed of mixed siliciclastics and carbonates. Fine quartz sand and shale beds occur in shelf margin and slope facies, and there is a similarity of reflection characteristics between the basin system and parts of the slope system that are known to have a mixed lithologic composition. Both systems have high-amplitude reflections that indicate great lithologic contrast.

A narrow submerged-shelf setting appears to have existed during the period of Kimmeridgian platform progradation, because continental redbed facies are found now only a few kilometers behind the coeval carbonate shelf margin. An analysis of seismic depositional systems suggests the Kimmeridgian shelf was 3.2 to 16 km (2–10 mi) wide. The narrow shelf of stage I favors terrigenous-sediment bypassing of the shelf margin. Suspended clays that settle out within 8 km (5 mi) of the coast (Scott, 1975) would be transported to the basin margin. Coastal sands, transported and redistributed by storm-generated currents, are likely to reach shelf margin, slope and basin environments (Aigner, 1985). Deposition of siliciclastics beyond the shelf margin may occur through channeled transport across the shelf and shelf margin (Fig. 13). This episode of platform development ended in the Late Kimmeridgian, when the carbonate platform top was exposed during a short-term fall in sea level.

Carbonate Aggradation, Stage II

Conceptual arguments maintain aggraded platforms develop when their growth potential is matched by a rapid relative rise in sea level (Kendall and Schlager, 1981). In the Baltimore Canyon Trough, aggradational growth occurred in two phases. A low-sedimentation rate, a low rate of platform subsidence, and fluctuating sea-level conditions interact to shape the initial phase (Portlandian), whereas rapidly rising sea level, accelerated subsidence, and a high-sedimentation rate interact to form the final phase (early Berriasian).

Initial aggradational phase.—

The effect of slow sedimentation was to promote platform aggradation, because only enough sediment was deposited to track relative sea-level rise. Actual sedimentation rates are not known, but a comparison of the rates of platform accumulation and progradation, and basinal accumulation from stage I and stage II, suggests a drastic reduction in sedimentation occurred during the initial phase of aggradational growth, stage II. At 0336 and 0337, the calculated sediment accumulation rates average about 47 m (155 ft)/m.y. during the Portlandian (Fig. 9). Similar values are calculated for the Portlandian interval beneath other well projections in the Baltimore Canyon Trough (Poag, 1985). Compared to the Portlandian, the Kimmeridgian rates are about three times as high in areas where rapid progradation occurred.

A similar reduction in sedimentation rate is dramatically shown by differences between Portlandian and Kimmeridg-

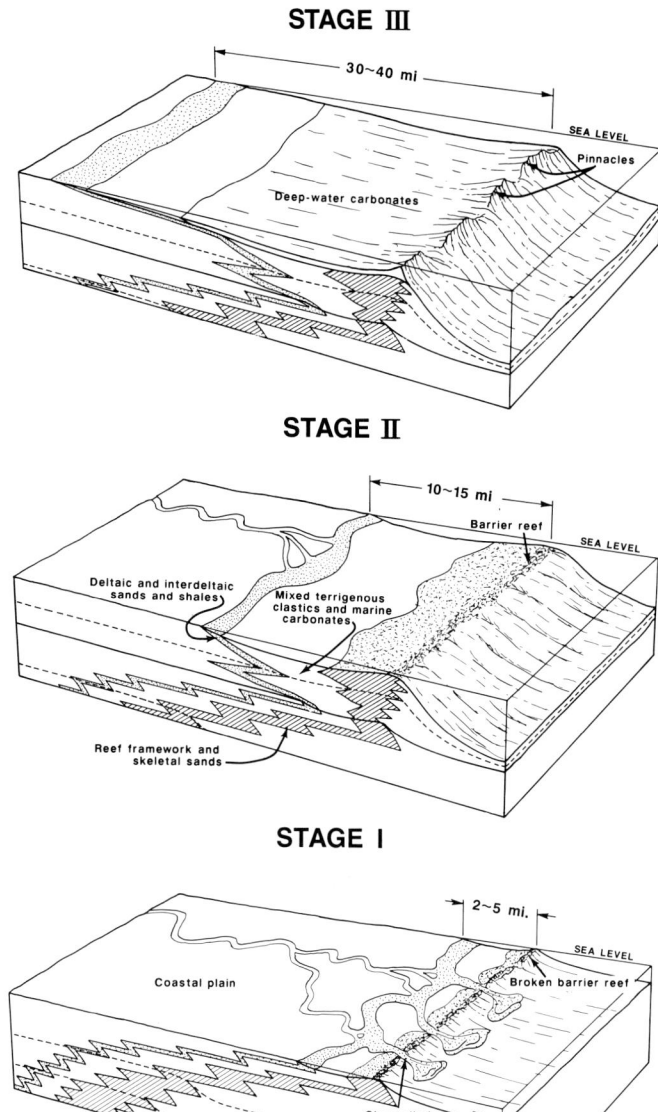

FIG. 13.—Growth stage block diagrams. Diagrammatic representation of the three stages in the evolution of the Kimmeridgian-Valanginian platform, prepared using depositional-facies types and their geometries as depicted by seismic profiling.

ian progradation rates. Maximum progradation occurred beneath the projection of seismic line 16709 (Fig. 7B) during stage II. At this location, the rate of progradation is calculated to be about 93 m (307 ft)/m.y. This rate is probably conservative. It assumes progradation occurred throughout stage II, because the seismic resolution is insufficient to distinguish between Portlandian and Berriasian episodes of progradation. Limiting progradation to the Portlandian would increase the rate, but the Kimmeridgian rate would still be nine times greater. This dramatic difference in progradation is a clear indication of the reduced relative rates of sedimentation during Portlandian time.

A thin stage II basinal depositional sequence affords additional evidence that little excess sediment was received or generated on the platform. The actual thickness of the basinal sequence is unclear, but it is significantly thinner than the platform sequence since it is limited to a single basinal reflection (Fig. 7B). This difference between platform and basinal strata is important because it documents differential sedimentation and an overall increase in platform height (Fig. 11). It also suggests a sediment-starved basin existed during stage II.

Subsidence contributed to platform aggradation by maintaining the aggradation potential of the platform. During the initial phase of stage II, the actual contribution resulting from platform subsidence is small. Following uninterrupted, moderately rapid subsidence during stage I, a Portlandian deceleration at 135 Ma reduced platform subsidence to approximately 27 m (90 ft)/m.y. at the COST B-3 well (Fig. 9). Sediment accumulation rates at 0336 and 0337 show a corresponding decline at this interval. The similarity of these two plots (Fig. 9) suggests subsidence is largely dependent on sediment loading. This is significant because changes in platform aggradation potential must result from eustatic fluctuations.

Eustasy probably is the chief factor controlling sediment distribution and production during the Portlandian. Numerous short-term sea-level fluctuations (Haq and others, 1987) were most important in determining sediment distributions. The excellent fit between the short-term sea-level history and stratigraphy of stage II supports this point (Fig. 8). At 0336, most but not all shale or sandstone intervals match the Portlandian framework of short-term sea-level lowstands, whereas limestone intervals match periods of platform flooding (Fig. 8). This conformity is strong support for the inference that short-term sea-level cycles were the chief causative agents in controlling the delivery of clastics to the platform during this interval.

Short-term sea-level fluctuations maintained carbonate production at rates sufficient to "keep-up" but not exceed the available accommodation potential of the platform. During the Portlandian, sea-level lowstands averaged about one every million years (Haq and others, 1987). These suppressed carbonate production by interrupting and reducing the size of carbonate-producing areas by introduction of terrigenous clastics (compare Walker and others, 1983). Similar disruptions in carbonate production occurred during growth of Berriasian platforms of the Baltimore Canyon. Platform exposure probably occurred since the magnitude (25 m, 82 ft to 100 m, 328 ft) of each lowstand (Haq and others, 1987) exceeded the rate of thermal subsidence. The spread of siliciclastics on the platform, as exemplified at 0336, not only documents an interruption in carbonate production but also a reduction in the size of the carbonate-producing area of the shelf. This is important because an influx of siliciclastics also can inhibit carbonate production before siliciclastics actually arrive (Hallock and Schlager, 1986). Clastics are typically delivered to carbonate shelves by nutrient-rich waters that reduce water transparency, promote the growth of fleshy algae over carbonate-producing organisms, stimulate bioerosion, and suppress carbonate production by crystal poisoning (Hallock and Schlager, 1986). This may explain why carbonate production was not significantly greater at the margin (where there are no clastics) than on the shelf during this time.

Final aggradation phase.—

The effect of a combined short- and long-term rise in sea level during latest Portlandian (131 Ma) time (Haq and others, 1987) was to greatly increase the aggradation potential and area of carbonate production on the platform. During latest Portlandian-early Berriasian time, deepening on the platform had a faunal, textural, and geometric expression. Biologically, it is expressed by the shoreward migration of benthic foraminiferal ecologic zones (E. R. Ringer and H. L. Patten, pers. commun., 1987). Texturally, a change from shallow-water grainstones to wackestones at 0336 and a switch from sand to shale at 0131 reflect increased water depths on the platform. Development of a raised rim (Schlager, 1981) at the margin documents not only deepening but also rapid platform flooding. Concurrent with flooding is an expansion of the carbonate-producing area. This is suggested by the change from shale to carbonate deposition in the basal Berriasian section at 0336 (Fig. 8).

Accelerated platform subsidence during the Portlandian-Berriasian interval was an additional factor contributing to increased space on the platform. At 0336, the relative rise is expressed by a 213-m (700 ft) section of carbonates. Deposition of these carbonates corresponds to a pronounced 2-Ma-long acceleration in platform subsidence that began at 132 Ma (Fig. 9). Tectonic subsidence curves fail to show a similar spurt in subsidence at this time in the Baltimore Canyon Trough or other Atlantic basins (Poag, 1985). A uniform deceleration is calculated for this interval instead. The variation in total subsidence must reflect the effect of a major change in sediment loading.

High rates of sedimentation enabled the platform to keep pace with an accelerated, Berriasian, relative rise in sea level. During the early Berriasian interval (130–131 Ma), a dramatic increase in accumulation rate took place beneath both COST B-3 well projections (Poag, 1985) and the 0336 projection. At 0336, the 145 m (478 ft)/m.y. accumulation rate is the highest rate calculated for the drilled platforms, but development of a raised rim at the margin during this time suggests sedimentation is only sufficient to match the rate of platform flooding. This is peculiar, because the Kimmeridgian accumulation rates are of a similar magnitude to those calculated for maximum Berriasian accumulation. Also, accumulation rates significantly higher than 145 m (478 ft)/m.y. are reported for older prograded Atlantic platforms (Poag, 1985). Several explanations could account for the lack of significant early Berriasian platform progradation. One is that nutrient poisoning (Hallock and Schlager, 1986) continued to suppress carbonate sedimentation rates. The widespread and uniform thickness of Portlandian-Berriasian strata (Poag, 1985) argues against this explanation. Another possibility is that early Berriasian sedimentation rates reflect the actual growth potential of the carbonate platform without benefit of significant influx of siliciclastics, as happened during Kimmeridgian progradation.

Cessation of Platform Growth, Stage III

Major fluctuations in sea level and nutrient poisoning (Hallock and Schlager, 1986) interacted to terminate platform growth during stage III. Rapid sea-level fluctuations with high amplitudes altered deposition across the platform and modified the shelf margin geometry. A good conformity exists between Late Berriasian short-term sea-level fluctuations and platform stratigraphy, but the lowstand (128.5 Ma) and highstand (128 Ma) of Haq and others (1987) is about 1 m.y. later than the stratigraphic expression of these events in the carbonate platform (Fig. 8). Lithologic and diagenetic data, whose age is constrained by paleontology, suggest platform flooding began approximately 129 Ma, and that the preceding lowstand was of short duration since no biotic zones are missing in the Berriasian section (E. R. Ringer and H. L. Patten, pers. commun., 1987).

The Late Berriasian lowstand terminated rapid shallow-water sediment accumulation of stage II. A 100 m (325 ft) sea-level fall (Haq and others, 1987) exposed the platform as evidenced by diagenetic fabrics. Meyer (1986) reported extensive meteoric diagenesis of grainstones in 0337 core 2 (Fig. 6) from strata immediately below stage III rocks. Absence of leaching and meteoric cementation in the overlying deep-water carbonate sequence implies these fabrics originated before platform flooding.

A rapid and large rise in sea level after exposure contributed to platform drowning by reducing the growth potential. Rapid establishment of deep-water conditions decreased shallow-water depositional sites. Stratigraphic sequences across much of the carbonate shelf are marked by an abrupt upward change from grainy Early Berriasian carbonates with a shallow-water fauna to mud-dominated Late Berriasian carbonates with a deep-water assemblage. This change is readily recognized in cuttings from the 0336 and 0337 wells and also on logs in all three Shell wells by an abrupt change in resistivity and radioactivity.

The rapidity and magnitude of flooding are dramatically documented at the shelf margin. Depositional bathymetry on the shelf behind the rim must have reached depths of 183 m (600 ft) or more since that much relief is mapped from seismic lines for pinnacles that rim the shelf margin. Similar deep-neritic paleobathymetries based on planktonic microfossils are postulated for these muddy carbonates by E. R. Ringer and H. L. Patten (pers. commun., 1987). Depths as great as 250 m (820 ft) are reasonable during maximum flooding because the rim is capped with Late Berriasian deep-water carbonates. An analysis of bathymetric and age data for this stratigraphic interval suggests the relative rate of sea-level rise averaged about 0.2 m (0.66 ft)/ka. Although the rate of rise calculated for the Late Berriasian is high, Schlager (1981) points out that a rise of at least 1 m/ka is required to drown a healthy platform.

A low-sediment accumulation rate during the Late Berriasian suggests the platform productivity was low. Accumulation curves (Fig. 9) show that sedimentation rates were lowest in the late Berriasian. At 0337, where deep-water carbonate deposition extended into the Valanginian, accumulation rates were about 19 m/m.y. (63 ft/m.y.). Similarly, reduced accumulation rates characterize this interval elsewhere in the Baltimore Canyon Trough and in other Atlantic basins (Poag, 1985). Because platform growth was aggradational, a doubling of the accumulation rate may constitute a reasonable estimate of the carbonate production rate. Compared with modern rates of carbonate production (Schlager, 1981), a rate of 37 m/m.y. (120 ft/m.y.) is low.

The Late Berriasian sea-level rise may have contributed to a reduction in carbonate productivity in two ways: (1) submergence below the euphotic zone, and (2) release of nutrients. First, after the emergent shelf was transgressed, a lag period probably occurred (Schlager, 1981) before carbonate production began. Given the high amplitude of the Late Berriasian rise (Haq and others, 1987) the widespread distribution of deep-water carbonates, and several hundred meters of pinnacle relief, a lag period of 5 ka (Read and others, 1986) may have helped to submerge the platform below the realm of high-carbonate productivity. Prolific carbonate production in the Recent is limited to the upper 50 to 100 m (164–328 ft) of the euphotic zone (Schlager, 1981). Below this zone, a reduction in the contribution of carbonate-producing algae is the most important constraint of hermatypic productivity. Calcareous algae, a common to abundant component of shallow-water stage II carbonates, are notably absent in deep-water facies of stage III.

Second, carbonate production may have been suppressed by release of nutrients (Hallock and Schlager, 1986) during transgression of coastal areas. Late Berriasian carbonates are most widespread post-Kimmeridgian platform carbonate strata and typically interfinger with coastal clastic facies on a scale of kilometers (Fig. 13). The thin Late Berriasian limestone is part of a thick carbonate sequence at 0336 (Fig. 10), but in the correlative, dominantly clastic section of the 0131 well located 15 km (9.5 mi) up depositional dip, only 12 m (40 ft) of Late Berriasian limestone occur in the sequence. Eutrophic conditions, reducing the width of the euphotic zone by 70% or more (Hallock and Schlager, 1986), may have existed, given the extensiveness of transgressed coastal areas.

Cessation of platform growth during stage III is interpreted to be a result of drowning rather than burial by terrigenous sedimentation, as suggested by Jansa (1981) and Gamboa and others (1985). An "empty bucket" platform morphology, low accumulation rates, deep-water faunas in muddy carbonates across the shelf, and extensive flooding of coastal clastics are consistent with this interpretation.

COMPARISON OF BALTIMORE CANYON, LAHAVE, AND SCOTIAN PLATFORM EVOLUTION

The Kimmeridgian-Valanginian platforms of the Baltimore Canyon Trough and coeval platforms on the Lahave Platform and the Scotian Basin are shown in Figure 14. Compared with the Baltimore Canyon Trough, the overall thickness of Kimmeridgian-Valanginian platforms is thinner in sections 2 and 3 (Fig. 14), suggesting sediment-loaded platform subsidence rates, because they are much less than in the Baltimore Canyon Trough. About 1,400 m (4,600 ft) of carbonates compose the Baltimore Canyon Kimmeridgian, Portlandian, and Berriasian platforms. The same se-

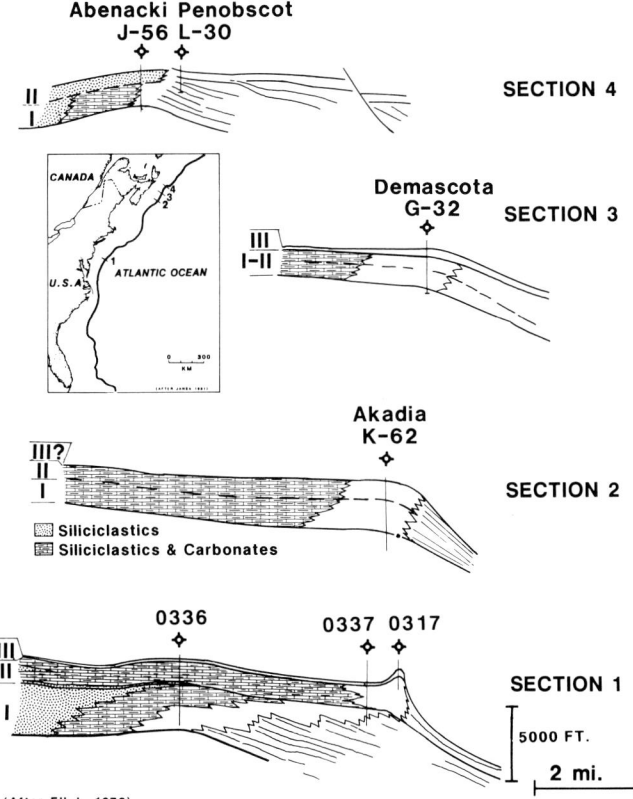

FIG. 14.—Atlantic platform evolution. Seismic reflection tracings, growth stages, and facies geometries from interpretation of United States and Canadian time sections. Vertical scale is approximate for intervals shown. Each cross section illustrates the stratigraphy of the platform based on an integration of borehole data and seismic reflection characteristics. Section 1 is based on line 19355; sections 2–4 are from Eliuk (1978, and pers. commun., 1987). Only Kimmeridgian strata are represented by stage 1 in section 1. Stage 1 in each Canadian line includes Oxfordian and Callovian rocks. Growth mode and thickness differences are discussed in text.

quence is about 790 m (2,600 ft) thick beneath the Akadia K-62 projection in Canada (Eliuk, 1978). Sediment-loaded subsidence rates calculated for the Canadian areas (Watts and Steckler, 1979) are about half the 79-m (260 ft) average Kimmeridgian-Valanginian rate calculated for the Baltimore Canyon Trough. Aggradational platform growth dominated the early phase of the Canadian development, even though the tectono-eustatic rise in sea level was reduced. Siliciclastic contribution was minimal except in section 4, located near the Sable Island Delta and possibly influenced by salt tectonics (Eliuk, 1978). Platform progradation and an influx of siliciclastics occur in the late stages of platform growth in section 4.

The difference in coeval platform development cannot be explained by variations in tectono-eustatic sea-level rise. In fact, the absence of progradation of the Scotian coast is peculiar, because the low-subsidence rate and broad platforms would seem to favor platform progradation. The variation in coeval platform development seen in these sections supports the hypothesis here that siliciclastic influx plays a major role in platform progradation.

CONCLUSIONS

Development of the Kimmeridgian-Valanginian platforms took place in three successive stages at the leading edge of a prograding wedge of continental-shelf siliciclastics. Its growth mode during sequential tectono-eustatic rises in sea level varied directly with supply of siliciclastics to the slope and basin. The switch to successive stages was punctuated by sea-level lowstands. Oblique progradation of a very narrow but high platform took place during the Kimmeridgian, while terrigenous bypassing of the carbonate shelf margin deposited quartzose sediments on the slope and in the surrounding basin. Aggradation and cessation of platform growth occurred when the supply of siliciclastics was cut off from the basin during rapid relative sea-level rise. Growth ended at the end of Berriasian time because of drowning caused by a high-amplitude tectono-eustatic fluctuation and a reduced growth potential, perhaps because carbonate production took place below the euphotic zone and was suppressed by nutrient poisoning.

Comparison of Kimmeridgian-Berriasian platforms from the Baltimore Canyon Trough, Lahave Platform, and Scotian Basin illustrates that platform progradation was greatest in those areas where there was influx of terrigenous sediments. A narrow shelf and siliciclastic bypassing of the carbonate shelf margin strongly influenced the evolution of the Kimmeridgian platform. These observations on the evolution of Kimmeridgian-Valanginian platforms and their coeval counterparts have led to the hypothesis that the evolution of these platforms is, in part, controlled by the availability of siliciclastics. Progradation of the Baltimore Canyon and Scotian platforms occurred because siliciclastics partially filled the basins and enabled the platforms to prograde at times when their carbonate potential alone was only sufficient for aggradational growth.

ACKNOWLEDGMENTS

This paper is published with the permission of Shell Oil Company. I thank Shell colleagues J. F. Karlo for planting the seeds of this study during reviews of the Atlantic data, and E. R. Ringer, D. T. Lawrence, and M. W. Shuster for helpful discussions in the development of my ideas. I also acknowledge J. H. Barwis, M. W. Shuster, R. L. Nicholas, and E. B. Picou, Jr., for their comments on an earlier version of this manuscript, C. F. Conrad for making available seismic and well data, and D. Maybry, A. W. Gallowy, and S. L. Widacki for drafting assistance with figures. I especially thank J. F. Sarg, R. C. L. Wilson, P. D. Crevello, and F. J. Read for their thoughtful review of the manuscript.

REFERENCES

AIGNER, THOMAS, 1985, Storm depositional systems, dynamic stratigraphy in modern and ancient shallow-marine sequences, in Friedman, G. M., Neugebauerm, H. J., and A. Seilacher, eds., Lecture Notes in Earth Sciences: Springer-Verlag, Berlin Heidelberg, 174 p.

ELIUK, L. S., 1978, The Abenaki Formation, Nova Scotia shelf, Canada–A depositional and diagenetic model for a Mesozoic carbonate platform: Canadian Petroleum Geology Bulletin, v. 26, p. 424–514.

EMERY, E. O., AND UCHUPI, E., 1972, Western North Atlantic ocean:

Topography, rocks, structure, water, life, and sediments: American Association of Petroleum Geologists Memoir 17, 532 p.

GAMBOA, L. A., TRUCHAN, M., AND STOFFA, P. L., 1985, Middle and Upper Jurassic depositional environments at outer shelf and slope of Baltimore Canyon Trough: American Association of Petroleum Geologists Bulletin, v. 69, p. 610–621.

HALLOCK, PAMELA, AND SCHLAGER, WOLFGANG, 1986, Nutrient excess and the demise of coral reefs and carbonate platforms: PALAIOS, v. 1, p. 389–398.

HAQ, B. U., HARDENBOL, J., AND VAIL, P. R., 1987, Chronology of the fluctuating sea level since the Triassic: Science, v. 235, p. 1156–1166.

JANSA, L. F., 1981, Mesozoic carbonate platforms and banks of the eastern North American margin: Marine Geology, v. 44, p. 97–117.

———, AND WADE, J. A., 1975, Geology of the continental margin off Nova Scotia and Newfoundland, in van der Linden, W. J. M., and Wade, J. A., eds., Offshore Geology of Eastern Canada, volume 2, Regional Geology: Geological Survey Paper 74-30, p. 51–106.

KENDALL, C. G. St. C., AND SCHLAGER, W., 1981, Carbonates and relative changes in sea level: Marine Geology, v. 44, p. 181–212.

MATTICK, R. E., FOTTE, R. Q., WEAVER, N. L., AND GRIM, M. S., 1974, Structural framework of United States Atlantic outer continental shelf north of Cape Hatteras: American Association of Petroleum Geologists Bulletin, v. 58, p. 1179–1190.

MEISSNER, F. F., 1972, Cyclic sedimentation in Middle Permian strata of the Permian Basin, West Texas and New Mexico, in Elam, J. C., and Chuber, S., eds., Cyclic Sedimentation in the Permian Basin, Second ed., p. 203–232.

MEYER, F. O., 1986, Facies specificity of megaporosity in Mesozoic shelf-edge carbonates from the Baltimore Canyon Basin: American Association of Petroleum Geologists Bulletin, v. 70, p. 621.

POAG, C. H., 1985, Depositional history and stratigraphic reference section for central Baltimore Canyon Trough, in Poag, C. H., ed., Geologic Evolution of the United States Atlantic Margin: Van Nostrand Reinhold Company, New York, p. 217–264.

READ, J. F., GROZINGER, J. P., BOVA, J. A., AND KOERSCHNER, W. F., 1986, Models for generation of carbonate cycles: Geology, v. 14, p. 107–110.

RINGER, E. R., AND PATTEN, H. L., 1986, Biostratigraphy and depositional environments of Late Jurassic and Early Cretaceous carbonate sediments in Baltimore Canyon Basin: American Association of Petroleum Geologists Bulletin, v. 70, p. 639–640.

SCHLAGER, WOLFGANG, 1981, The paradox of drowned reefs and carbonate platforms: Geological Society of America Bulletin, v. 92, p. 197–211.

———, AND CAMBER, OREN, 1986, Submarine slope angles, drowning unconformities and self-erosion of limestone escarpments: Geology, v. 9, p. 762–765.

SCHLEE, J. S., AND GROW, J. A., 1980, Buried carbonate shelf edge beneath the Atlantic continental slope: Oil and Gas Journal, v. 78, p. 148–159.

———, AND JANSA, L. F., 1981, The paleoenvironment and development of the eastern North American continental margin: Oceanologica Acta, No. SP, p. 71–80.

———, DILLON, W. P., AND GROW, J. A., 1979, Structure of the continental slope off the eastern United States, in Doyle, L. J., and Pilkey, O. H., eds., Geology of Continental Slopes: Society of Economic Paleontologists and Mineralogists Special Publication 27, p. 95–117.

SCOTT, M. R., 1975, Distribution of clay minerals on Belize Shelf, in Wantland, K. F., and Pusey, P. C., III, eds., Belize Shelf–Carbonate Sediments, Clastic Sediments, and Ecology: American Association of Petroleum Geologists, Studies in Geology No. 2, p. 97–130.

SHERIDAN, R. E., 1974, Atlantic margin of North America, in Burk, C. A., and Drake, C. L., eds., Geology of Continental Margins: Springer-Verlag, New York, p. 391–407.

SILVER, B. A., AND TODD, R. G., 1969, Permian cyclic strata, northern Midland and Delaware basins, West Texas and southeastern Mexico: American Association of Petroleum Geologists Bulletin, v. 53, p. 2223–2251.

SWANSON, R. G., 1981, Sample Examination Manual: American Association of Petroleum Geologists, Methods in Exploration Series, 31 p.

WALKER, K. R., SHANMUGAN, G., AND RUPPEL, S. C., 1983, A model for carbonate to terrigenous clastic sequences: Geological Society of America Bulletin, v. 94, p. 700–712.

WATTS, A. B., AND STECKLER, M. S., 1979, Subsidence and eustacy at the continental margin of eastern North America: American Geophysical Union, Maurice Ewing Symposium Series 3, p. 218–234.

THE EVOLUTION OF THE CARBONATE PLATFORMS OF NORTHEAST AUSTRALIA

PETER J. DAVIES, PHILIP A. SYMONDS, DAVID A. FEARY, AND CHRISTOPHER J. PIGRAM
Division of Marine Geosciences and Petroleum Geology, Bureau of Mineral Resources, Canberra, 2601 Australia

ABSTRACT: The carbonate platforms of northeast Australia, the Great Barrier Reef province and the Eastern, Queensland, and Marion Plateaus, contain a record of the complex interactions between the factors that controlled carbonate deposition over the past 60 m.y. Analysis of the extensive geological and geophysical data shows that both long-term (plate motion and subsidence) and short-term (rifting, eustasy, climate, oceanography, and collision) factors influenced platform evolution.

—The size, shape, and location of the high-standing structural features on which the carbonate platforms developed was determined by continental rifting.

—Northward plate movement controlled the distribution of climate-related facies within the Great Barrier Reef sequence, resulting in a tropical carbonate wedge that thins and becomes younger to the south and overlies temperate and subtropical facies.

—Large-scale facies distribution patterns reflect the complexities of the subsidence regimes that affected the northeast Australian platforms. The simple subsidence situation, where high-subsidence rates favoring backstepping are succeeded by lower subsidence rates favoring progradation, was complicated by episodes of accelerated subsidence.

—Sea-level variation directly controlled platform facies: rising and high sea-level periods favored increased carbonate deposition, whereas falling and low sea-levels restricted carbonate deposition, caused increased terrigenous input along the shelf, and in many cases resulted in exposure of the platforms and the formation of unconformities.

—In addition to the overall climatic consequence of northward plate motion, facies sequences show the effects of the development throughout the Cenozoic of more pronounced latitudinal climatic zonation and progressive high-latitude cooling.

—Chemical and physical oceanographic factors affected platform evolution in various ways, e.g., the inhibition of reef development by high oceanic-phosphate levels during the Early and Middle Miocene, and deposition of facies reflecting the progressive development of the east Australian current from the Miocene.

—The development of a foreland basin on the northern edge of the northeast Australian region initially caused a dramatic expansion of carbonate facies, but ultimately terminated carbonate deposition as a result of uplift and inundation by clastic detritus.

General conclusions applicable to other carbonate platforms may be deduced from analysis of the factors that controlled deposition on the northeast Australian platforms. The evolution of any particular carbonate platform will be fundamentally dependent on whether the subsidence history is simple or complex; whether plate motion is toward or away from the tropics; and whether movement from one climatic regime to another is slow or rapid. Short-term eustatic, climatic, and oceanographic factors are responsible for complexities in the facies sequences produced. The most complex and varied carbonate platform sequences will be those deposited under the influence of compound subsidence, together with plate motion through a range of climatic zones over a substantial time period. The northeast Australian carbonate platforms illustrate such a complex history and demonstrate that facies diachroneity is a fundamental characteristic of complex carbonate platform development.

INTRODUCTION

The sediments that form carbonate platforms contain a record of vertical and horizontal tectonic effects, sea-level change, and paleoclimatic and paleoceanographic variation, as a result of their dominantly biological shallow-water origin. Carbonate platforms commonly occur in a passive-margin setting, and they contain almost 50% of global oil and gas reserves. Their study is therefore of considerable economic and scientific importance.

Current knowledge of carbonate platform evolution is based almost exclusively on studies in the Caribbean region, particularly of the Blake Plateau and Bahama Platform (Hollister and others, 1972; Schlager and Ginsburg, 1981; Austin and others, 1986). In addition, seismic studies of the eastern margin of the United States have revealed Jurassic and Cretaceous carbonate platforms that extend from the latitudes of New England to Florida (Sheridan and others, 1981; Jansa, 1981) and into the Gulf of Mexico (Kauffman, 1984).

Studies of these carbonate platforms along the eastern margin of the United States and in the Caribbean have provided a model for the evolution of passive-margin platforms. Analysis of the Cenozoic evolution of the northeast Australia continental margin provides a different perspective, and one that appears to be more complete. The Great Barrier Reef and the Eastern, Queensland, and Marion Plateaus (Fig. 1) are major carbonate platforms. Studies by the Australian Bureau of Mineral Resources since the 1970s (Davies, 1983; Symonds and others, 1983; Davies and others, 1989) have provided new insights into the development of these platforms and their evolutionary relations. Furthermore, these studies have defined the major operative processes and the relations between process and product. The Cenozoic history of the northeast Australian margin illustrates a 60 m.y. history of platform development, including both initiation and demise, and demonstrates the facies diachroneity resulting from the complex interdependence of the factors controlling platform evolution. Our objectives in this paper are to describe the major features of the platforms; to describe the major processes controlling platform development; and to discuss the global implications of this study in terms of the long- and short-term factors that control carbonate platform development, emphasizing both the major features and principal stages of development and the dynamic interactions between the controlling factors.

REGIONAL TECTONIC SETTING

The passive continental margin off northeastern Australia extends over a distance of about 2,000 km between Fraser Island in the south and the Gulf of Papua in the north and covers an area of some 930,000 km^2 (Figs. 1, 2). The margin is composed of a number of marginal plateaus and rift

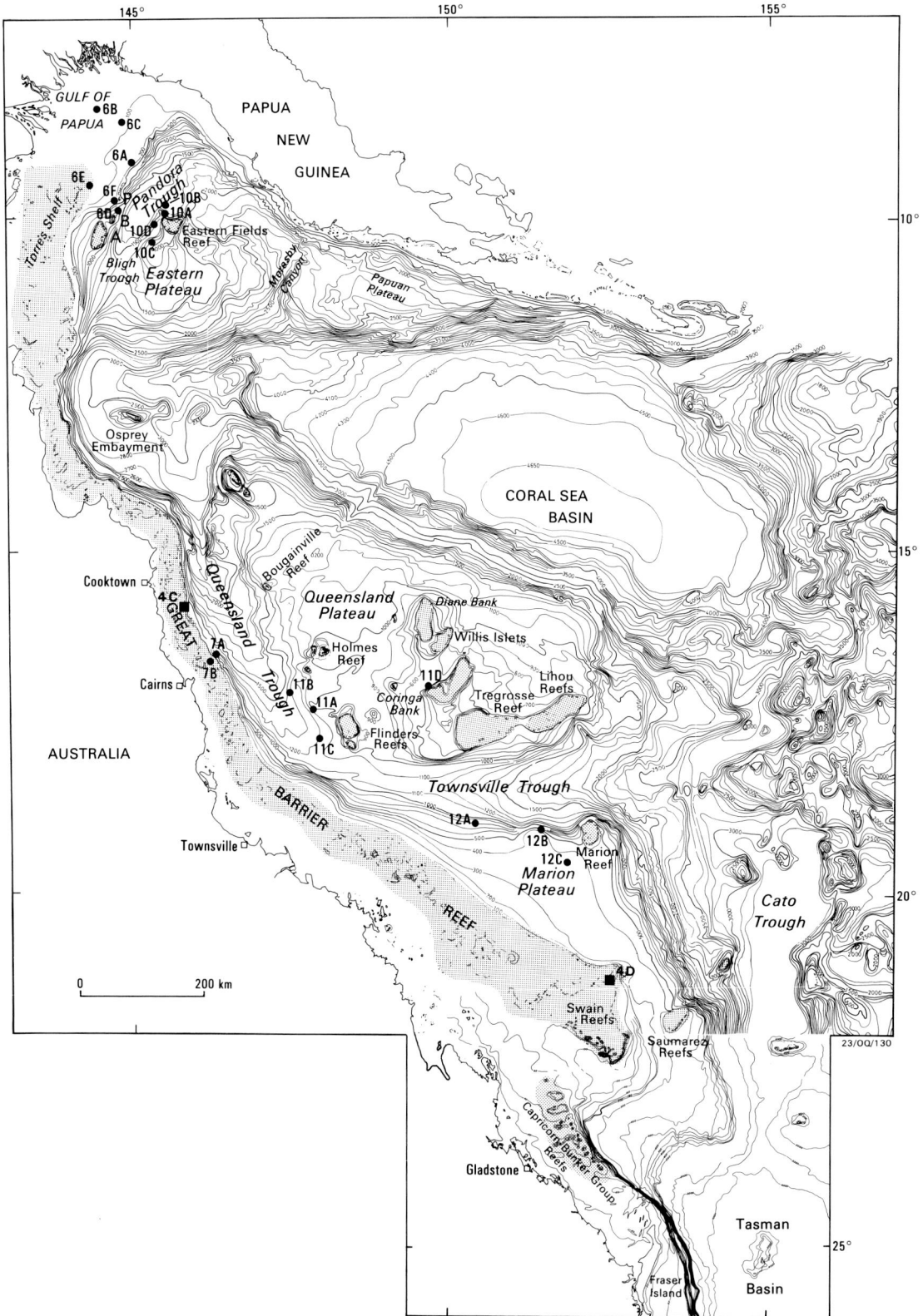

Fig. 1.—Locality map showing the principal bathymetric features of the northeast Australian continental margin (modified after Taylor, 1977, and Marshall, 1977). Areas of modern reef growth are screened; **ABP** designates the Ashmore-Boot-Portlock reef system. The locations of figured seismic sections are also shown (e.g., **6A** shows the location of Fig. 6A).

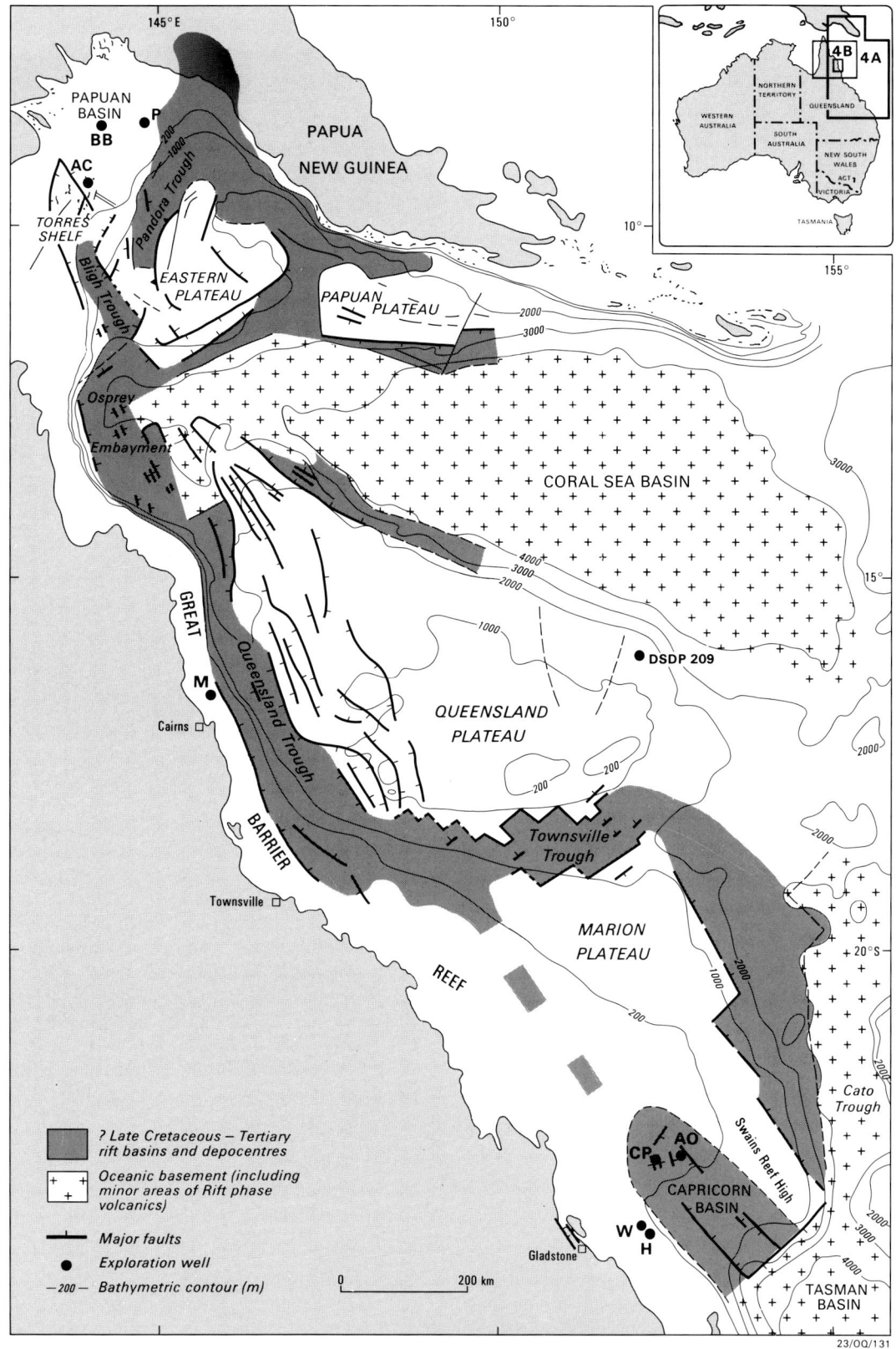

FIG. 2.—Map showing the major structural features of northeast Australia (based on Davies and others, 1988). The location of exploration wells (**BB**–Borabi 1; **P**–Pasca A1 and C1; **AC**–Anchor Cay 1; **M**–Michaelmas Cay; **AQ**–Aquarius 1; **CP**–Capricorn 1A; **W**–Wreck Reef; **H**–Heron Island) and DSDP Site 209 are also shown. The areas shown in Figures 4A and B are indicated on the inset.

troughs: the Eastern, Queensland, and Marion Plateaus; the Pandora and Bligh Troughs; the Osprey Embayment; and the Queensland and Townsville Troughs. In addition, a zone of narrow rift basins, which extend southeast from the Queensland Trough toward the Capricorn Basin, separate the Marion Plateau from the continental shelf. The entire margin is generally considered to be underlain by modified continental crust formed as a result of fragmentation of a northeastern extension of the Tasman Fold Belt (Gardner, 1970; Ewing and others, 1970; Falvey, 1972; Falvey and Taylor, 1974; Taylor, 1975; Mutter, 1977; Taylor and Falvey, 1977; Mutter and Karner, 1980; Symonds, 1983; Symonds and others, 1984).

The rift phase of margin development, which may have commenced in the Early Cretaceous but was certainly in progress by the Late Cretaceous, preceded continental breakup and the formation of small ocean basins to the east and south (Coral Sea Basin, Cato Trough, and Tasman Basin). Tasman Basin seafloor spreading commenced in the Late Cretaceous (80 Ma; Hayes and Ringis, 1973; Weissel and Hayes, 1977; Shaw, 1978) and then extended northward to form the Cato Trough and Coral Sea Basin by the Paleocene (65 Ma; Weissel and Watts, 1979). Seafloor spreading had ceased along the length of this system by the earliest Eocene (56 Ma). The nature and development of the northern end of the rift system in the Bligh and Pandora Troughs region (Fig. 2) appear to have been even more complex. These troughs were initially formed during Late Cretaceous rifting and were reactivated in the Late Oligocene by foreland basin development (Davies and others, 1989).

Although the exact structural style and development history of the rift system of the northeast Australia region is not fully understood, there is little doubt that it has controlled the gross architecture of the margin and the form of the high-standing structural elements on which the carbonate platforms have evolved.

THE CARBONATE PLATFORMS OF NORTHEAST AUSTRALIA

The northeast Australian carbonate platforms are comparable in size to any platforms known in the geologic record. The Great Barrier Reef is as large as the Cretaceous and Jurassic reef systems of the eastern United States; the Marion Plateau is almost as large as the Blake Plateau; and the Queensland Plateau is as large as the combined areas of the Great and Little Bahama Banks. The carbonate platforms of northeast Australia range from the mixed siliciclastic-carbonate Great Barrier Reef to the purely carbonate-dominated oceanic regime of the Queensland Plateau. The most comprehensively studied platform in northeast Australia is the Great Barrier Reef, where integrated geological and geophysical investigations have provided the basis for models of reef growth and platform evolution (Davies, 1983; Symonds and others, 1983). Our understanding of the structure and composition of the marginal plateaus is based on regional seismic studies and limited geologic sampling. The models derived from the detailed studies on the Great Barrier Reef have been used as the basis for interpreting carbonate platform development on the marginal plateaus.

Great Barrier Reef.—

The Great Barrier Reef is approximately 2,000 km long and is composed of about 2,500 reefs. The shelf occupied by the Great Barrier Reef is generally narrowest in the north (minimum width of 23 km occurs at 14°S) and widens to the south (maximum width of 290 km at 21°S). Reefs occupy the whole shelf in the northern region, but only the mid- to outer shelf in the central and southern parts. The Great Barrier Reef is a very diverse physiographic province (Maxwell, 1968; Hopley, 1982) and may be divided into five distinct areas using the terminology most commonly applied to carbonate platforms (Ginsburg and James, 1974; Wilson, 1975):

(1) The northernmost areas (9°–16°S) is a narrow, rimmed, high-energy platform (Fig. 3A). This zone is characterized by a shallow, narrow shelf (generally 50–75 km wide) and steep continental slope (10°–60°). Reefs have grown across the full width of the shelf, occurring as large mid-shelf platform reefs (Fig. 4A,B) and as an almost continuous line of ribbon reefs, which form the outer barrier (Fig. 4A–C).

(2) Between 16°S and 18°S, the shelf is a narrow, partially rimmed, high-energy platform (Fig. 3B). The shelf is 50–75 km wide with a steep continental slope. Reefs generally occur on the mid- to outer shelf and are separated from the coast by a channel or inner lagoon approximately 35 m deep. The outer-shelf ribbon reefs are less continuous to the south and are replaced by a line of shoals on the shelf edge.

(3) The central Great Barrier Reef (18°–20°S) is a wide, unrimmed, high-energy platform (Fig. 3C). The shelf is 90–125 km wide with a gentle continental slope (<2°). Reefs are sparse and largely restricted to the outer shelf. A drowned barrier-reef complex, which occurs on a 75-m terrace at the shelf break, extends for some 200 km (Davies and Montaggioni, 1985).

(4) Between 20° and 22°S, the shelf is an extremely wide, rimmed, high-energy platform (Fig. 3D). The shelf width ranges from 125 km in the northwest to 290 km in the southeast. The gentle, narrow continental slope passes northeastward into the Marion Plateau. Reefs are confined to the outer one-third of the shelf (Fig. 4D), but are separated from the shelf edge by several kilometers of relatively deep shoals.

(5) In the south (23°–24°S) the shelf is a narrow, unrimmed, high-energy platform (Fig. 3E). The shelf is as wide as 100 km with reefs occupying a narrow zone on the mid-shelf. The shelf edge occurs some 12–20 km east of the reefs and in about 70 m of water. No drowned barrier reefs have been found along the shelf edge, although apparently *in situ* reef materials have been reported from the slope in 175 m of water (Veeh and Veevers, 1970).

The physiographic variation throughout the modern Great Barrier Reef province includes many features that other authors have placed within an evolutionary sequence (e.g., Read, 1985). Their existence at one time suggests that process diachroneity is a fundamental factor in platform evolution.

The major characteristics of the Great Barrier Reef car-

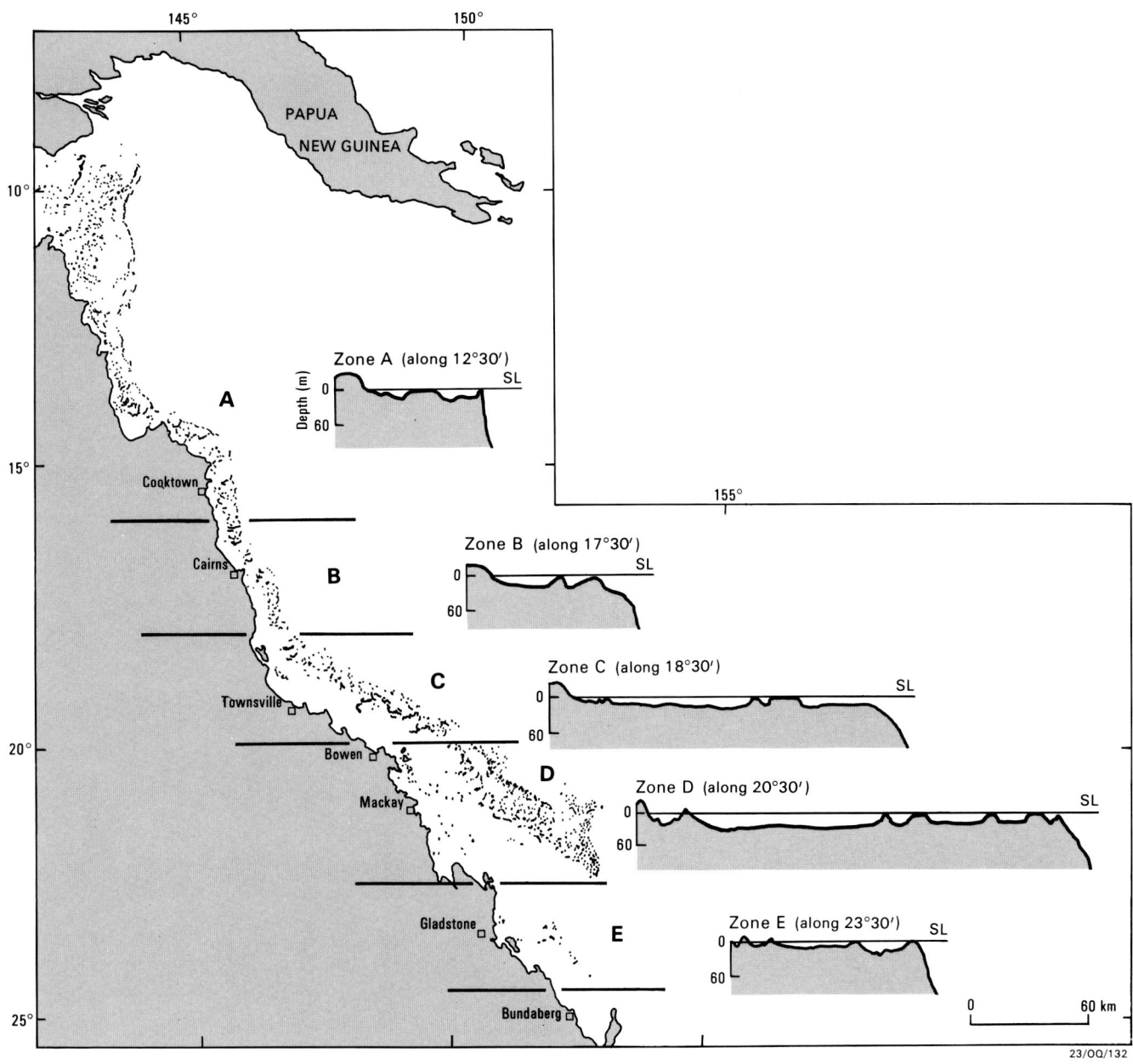

FIG. 3.—Physiographic variation throughout the Great Barrier Reef province. A–narrow rimmed platform (nomenclature after Ginsburg and James, 1974; Wilson, 1975); B–narrow, partially rimmed platform; C–wide unrimmed platform; D–extremely wide, rimmed platform; E–narrow unrimmed platform.

bonate platform have been deduced from extensive geological (e.g., Davies, 1977; Davies, 1983; Davies and Marshall, 1985) and geophysical (summarized in Symonds and others, 1983; Davies and others, 1989) investigations and are summarized here on schematic sections (Fig. 5). In the northern Great Barrier Reef and Gulf of Papua, the occurrence of subsurface reefs are well documented by seismic data and drill holes (Tanner, 1969; Tallis, 1975; Fig. 6). Miocene reefs occur in the subsurface of the Gulf of Papua (Fig. 6A–C), and Pliocene reefs and Miocene limestones containing algal rhodoliths occur at the northern end of the Great Barrier Reef (Marshall, 1983) in Anchor Cay 1. Seismic data acquired by the Bureau of Mineral Resources (Davies and others, 1989) in the northern area confirm both the presence of buried reefs and the existence of a thick reef section (Fig. 6D–F). Our seismic data indicate that a major reef structure occurs in the continental-slope sequence on the western margin of the Pandora Trough (Fig. 2) between Portlock and Boot Reefs. The data show that these modern reefs are constructed on a more extensive Miocene and Pliocene reef complex as thick as 1.5 km.

On the adjacent shelf, there are (?)Miocene-Pliocene buried reefs which may be precursors of the modern shelf edge ribbon reefs. A seismic profile across one of these features

Fig. 4.—Satellite photograph at upper left shows a large part of the northeast Australia margin. Note the shelf edge barrier reef extending for nearly 1,000 km and the huge platform reefs west of the outer barrier (location shown as 4A on Fig. 2). Satellite photograph at upper right shows the large platform reefs northeast of Cairns in the northern Great Barrier Reef (location shown as 4B on Fig. 2). Oblique air photo at lower left shows shelf edge barrier or ribbon reefs in the northern Great Barrier Reef. From left to right, note surf marking the windward margin; the reef flat; sand sheets (turquoise); and abundant patch reefs in the backreef area (location shown as 4C on Figure 1). Satellite photograph at lower right shows extremely large and complex shelf edge reefs in the Pompey complex of the south-central Great Barrier Reef (location shown as 4D on Fig. 1).

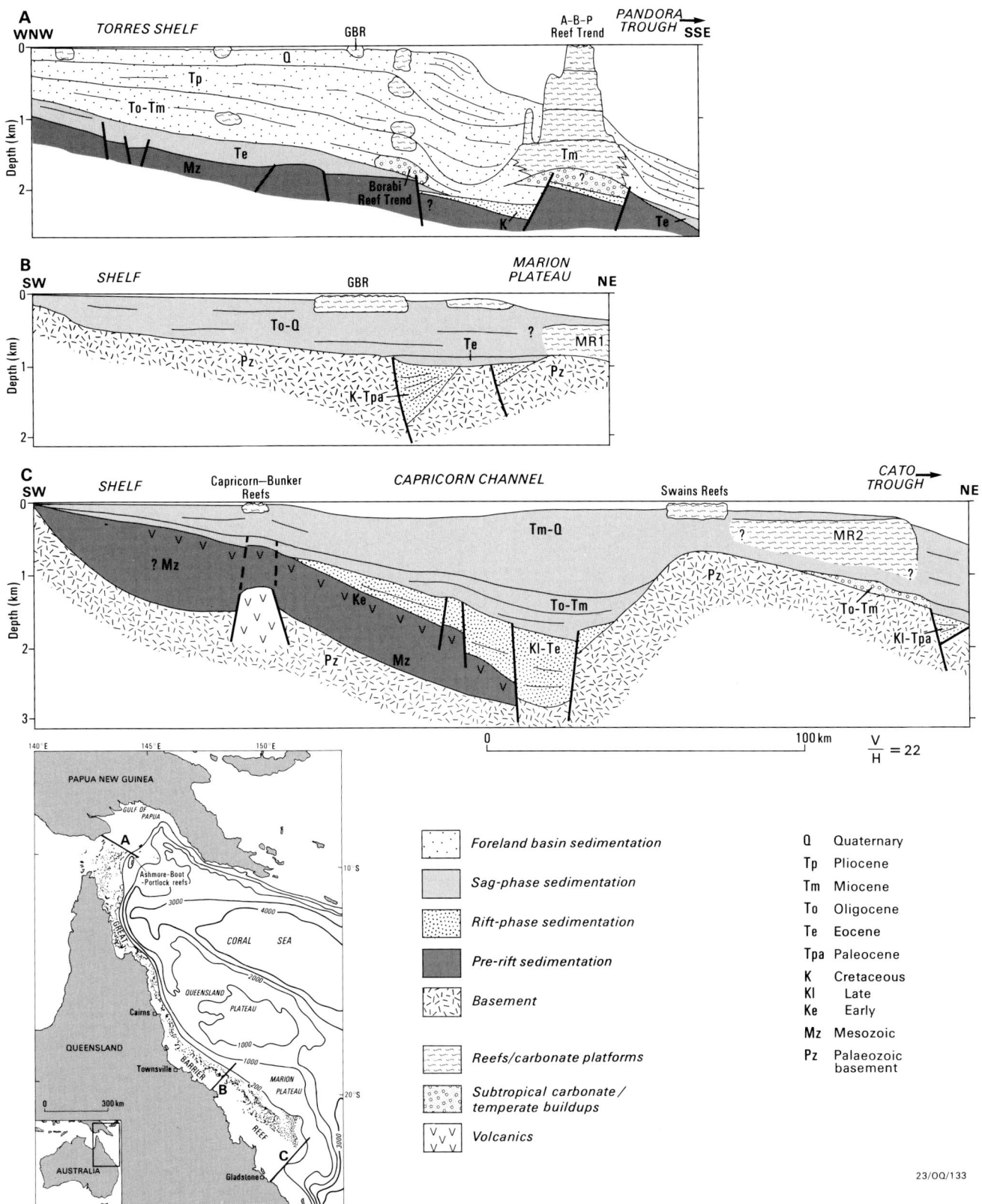

Fig. 5.—Schematic sections showing generalized structural and sedimentary geometry beneath the northern (A), central (B), and southern (C) Great Barrier Reef province. **MR1, MR2** indicate different phases of carbonate platform development on the Marion Plateau (see Fig. 9). Section locations are shown on the inset.

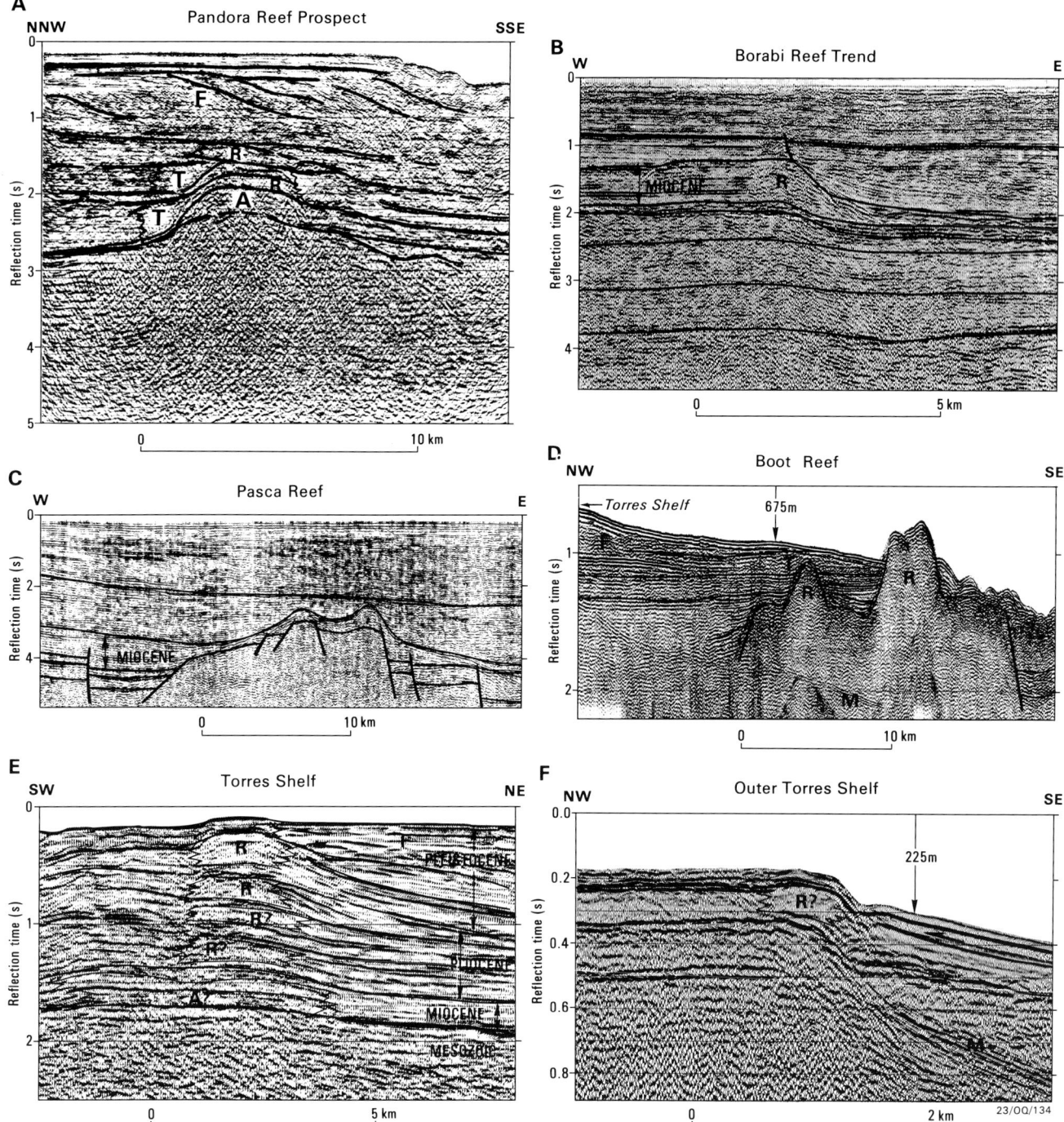

FIG. 6.—Seismic sections showing buried or partially buried reef complexes in the Gulf of Papua and northern Great Barrier Reef. Section locations are shown on Figure 1. **M** is the first water bottom multiple. (A) Airgun seismic section over a buried Miocene reef at the northern end of the Ashmore-Boot-Portlock reef complex. The early stage of carbonate buildup (perhaps a subtropical algal mound–**A**), developed on the corner of a fault block, is overlain by more areally restricted reef facies (**R**). A leeside talus facies (**T**) may also be present. The entire complex is buried by fluvioclastic sediments (**F**) derived from the west. (B) Seismic section showing the Miocene Borabi Reef trend in the Gulf of Papua. (C) Seismic section showing the Miocene Pasca Reef complex sitting on a structural high in the Gulf of Papua. (D) Sparker seismic section across the saddle between Boot and Portlock Reefs showing buried and submerged portions of reef (**R**) and a possible leeside talus facies (**T**). The reef complex has been partially buried by fluvioclastic sediments (**F**) derived from the west. (E) Airgun seismic section across the edge of the Torres shelf showing the buried northern extension of the modern shelf edge ribbon reefs. Plio-Pleistocene reefs (**R**) overlie a broader Miocene buildup, which may consist in part of a subtropical algal mound (**A**). (F) Sparker seismic profile across the outer Torres shelf, showing a 100-m-thick (?)Pleistocene reef (**R**) partially buried by terrigenous sediment.

(Fig. 6E) shows episodic reef growth throughout a 1,500-m section. A prominent forereef slope with shallow-dipping beds appears to have developed concurrently with the reef complex. The forereef slope sequence is overlain by a thick Pliocene and younger fluvio-deltaic sequence. In contrast, Quaternary buried and partially buried, relic, shelf edge reefs as thick as 100 m occur along the easterly-trending section of the outer Papuan shelf (Fig. 6F). Therefore, in the northern Great Barrier Reef and Gulf of Papua, seismic and drill hole data indicate that a reef sequence of varying thickness and age started to develop in the Miocene.

On the outer continental shelf of the central Great Barrier Reef region, the 250 to 300-m-thick reef complex (Fig. 7) is composed of a series of reef slices separated by low sea-level-generated unconformities (Davies, 1983; Symonds and others, 1983). The reef complex forms only the uppermost part of a thick outer-shelf sequence that is dominated by prograding fluvio-deltaic and onlapping slope sediments overlying a rifted basement (Symonds and others, 1983). The reef thickness and a tie to DSDP Site 209 in the Coral Sea (Fig. 2) indicate a probable Pliocene age for initiation of reef growth in this region. A borehole on Michaelmas Cay (Fig. 2) shows 100 m of (?)Plio-Pleistocene reef facies overlying siliciclastic sediments (Fig. 8B).

The boreholes on Heron Island and Wreck Reef (Fig. 8B) at the southern end of the Great Barrier Reef show that less than 150 m of reef overlies quartz sand, and that reef growth began in the Plio-Pleistocene (Lloyd, 1973; Palmieri, 1971, 1974).

The principal conclusions derived from studies of the Great Barrier Reef carbonate platform are that the reef sequence thins dramatically and the age of initial reef growth becomes younger from north to south. The Great Barrier Reef is a mixed carbonate-siliciclastic province, with reefs forming a discontinuous wedge largely enclosed within terrigenous fluvio-deltaic deposits. In some areas, however, particularly in the north, the underlying sequence is dominated by non-reefal carbonate facies.

Eastern Plateau.—

The Eastern Plateau is the most northern marginal plateau of the northeast Australian continental margin (Fig. 1). It is bounded by the Moresby, Pandora, and Bligh Troughs, the Osprey Embayment, the Moresby Canyon, and the Coral Sea Basin. The plateau has a gently convex surface with an average depth of 1,500 m and covers an area of about 31,000 km^2 at the 2,000-m isobath. Eastern Fields Reef, the only modern reef on the Eastern Plateau, is about 45 km across at its widest point and lies at the crest of the plateau near its northern margin (Fig. 1). Submerged and buried reefs extend northeast from Eastern Fields Reef beneath the Moresby Trough (Davies and others, 1989).

The Eastern Plateau is the least understood marginal plateau in the northeast Australian margin. Our knowledge of the plateau is based on reconnaissance airgun and sparker seismic data (Symonds and others, 1984; Davies and others, 1989). The major characteristics of the Eastern Plateau basement and sediment cover are summarized on a schematic section (Fig. 9A). The ages of seismic sequences are based on a tie to Anchor Cay 1 well on Torres Shelf.

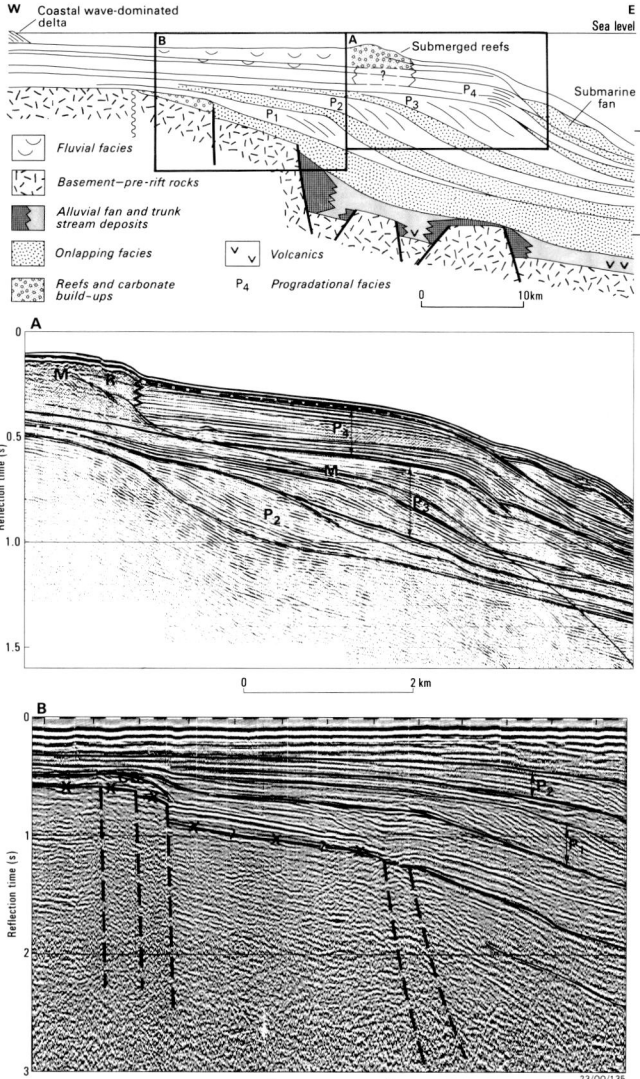

FIG. 7.—Carbonate-terrigenous facies geometry on the upper slope and outer shelf of the central Great Barrier Reef (modified after Symonds and others, 1983). Inset A is a sparker seismic section off Cairns, showing a submerged reef (**R**) and siliciclastic prograding units (**P2–P4**). **M** marks the first water bottom multiple. Inset B is an Aquapulse seismic section showing the position of outer-shelf sequences, particularly the two lower prograding units **P1** and **P2**, with respect to underlying basement (**x**) structure.

The northern and southern plateau margins are controlled by normal faults. The western margin is more complex and appears to be a product of thrusting (Fig. 9A). The Eastern Plateau is underlain by complex tilt blocks bounded by reactivated normal faults, some of which appear to have undergone both wrench and reverse movement (Davies and others, 1989). This deformation appears to have resulted from Late Oligocene and Miocene tectonism caused by the development of the New Guinea Orogen to the north (Pigram and Davies, 1987). This late phase of structuring is not represented on the other northeast Australian marginal plateaus, and it produced a Neogene topography on the Eastern Plateau quite unlike that of the other plateaus (Fig. 9A).

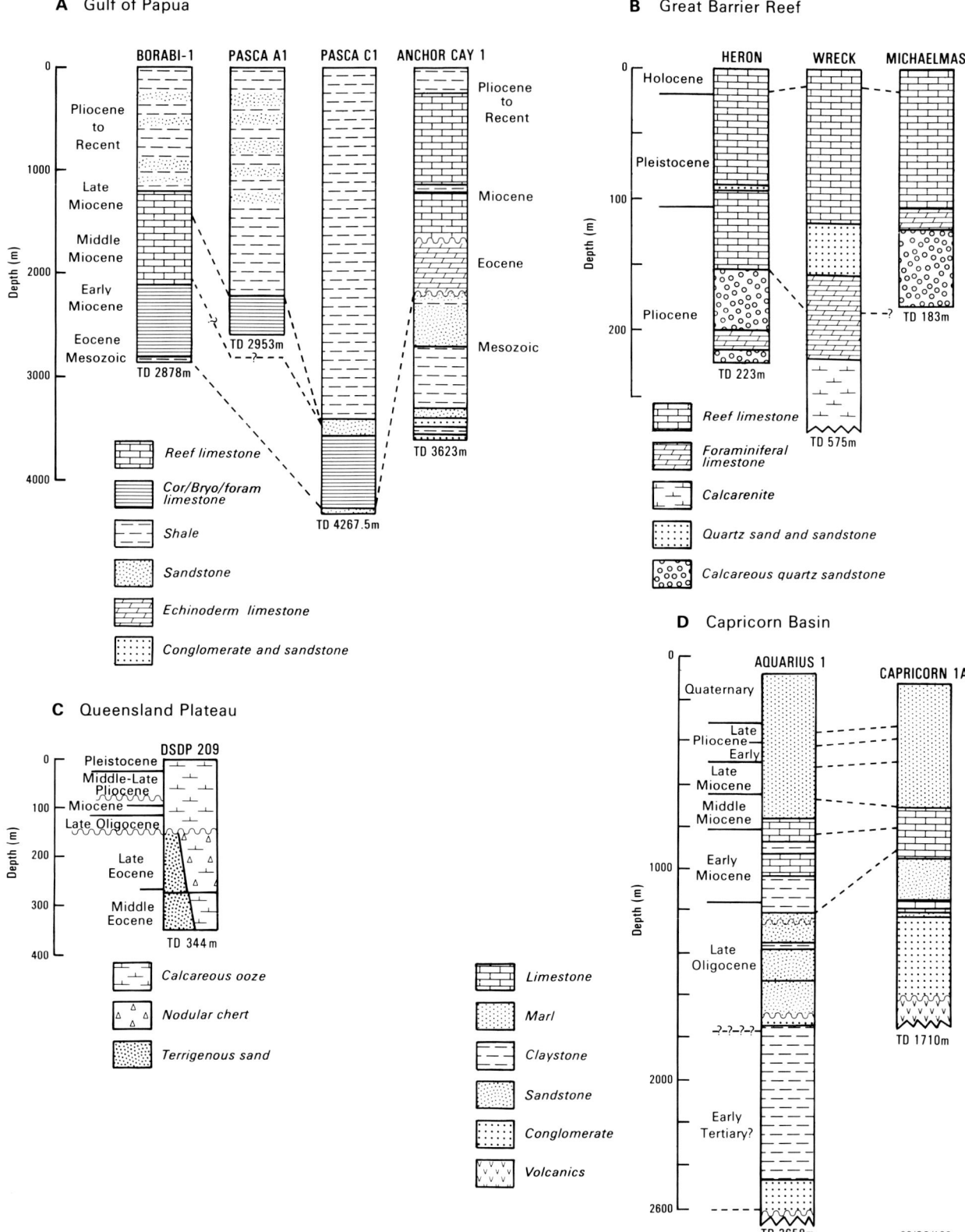

FIG. 8.—Summary lithostratigraphic logs from drill holes in the Gulf of Papua (A), Queensland Plateau (B), Great Barrier Reef (C), and Capricorn Basin (D). Locations of drill holes are shown in Figure 2.

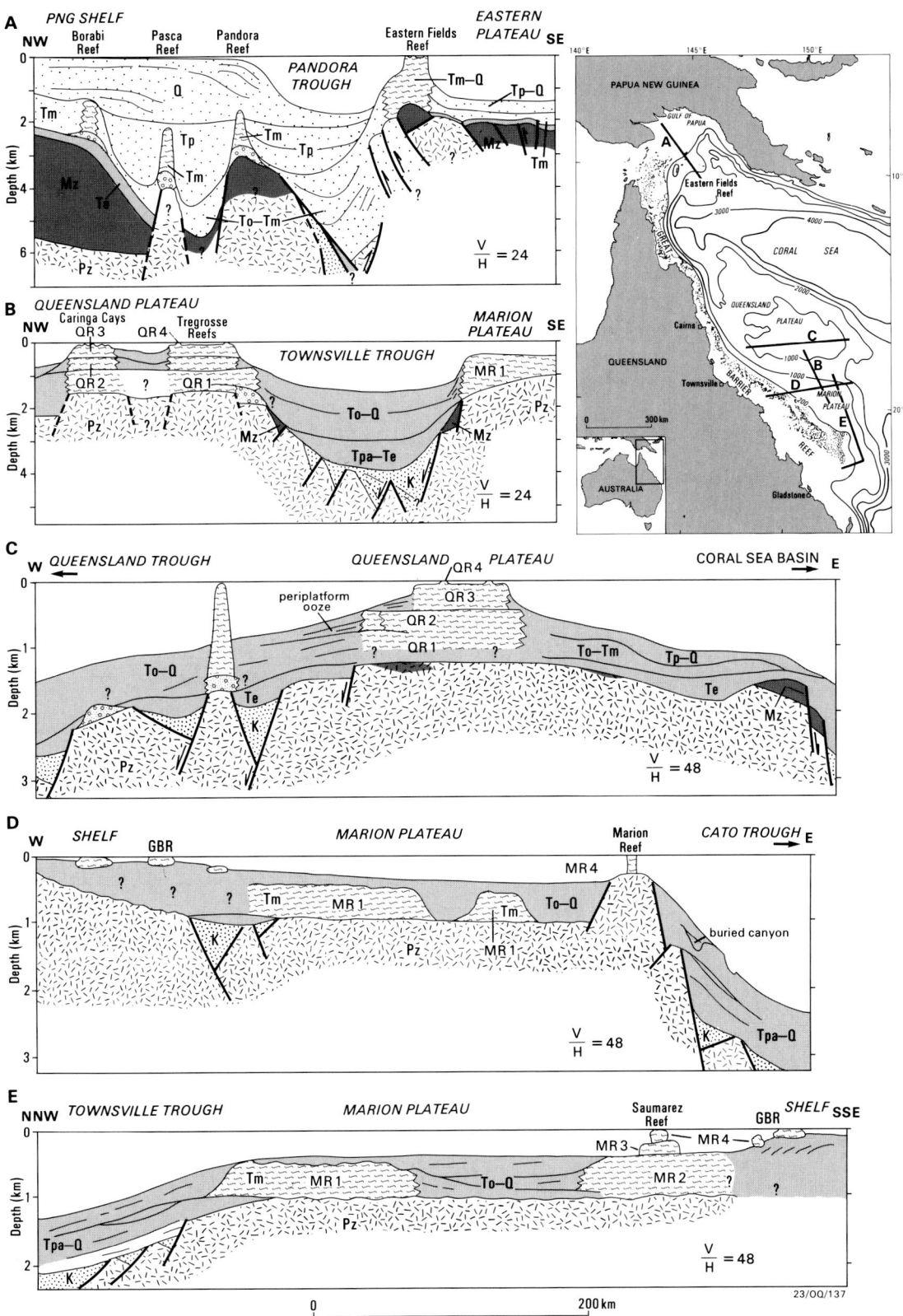

Fig. 9.—Schematic sections showing the generalized structure and sedimentary sequences on the Eastern (A), Queensland (B and C), and Marion (B, D, and E) Plateaus. **QR1** to **QR4** and **MR1** to **MR4** denote phases of carbonate platform growth on the Queensland and Marion Plateaus, respectively. Symbols and legend as for Figure 5.

Eastern Fields Reef is unusual in northeast Australia, because it has grown on a complex pedestal that appears to consist of deformed post-Oligocene sediments (Fig. 10C) and a complexly faulted basement block (Fig. 10D) (Davies and others, 1989). The elevation of this block was a result of the Late Oligocene to Miocene tectonism described earlier. The results of limited coring and dredging (Taylor, 1977) indicate that the remainder of the Eastern Plateau was covered by Miocene to Recent calcareous ooze and peri-platform detritus.

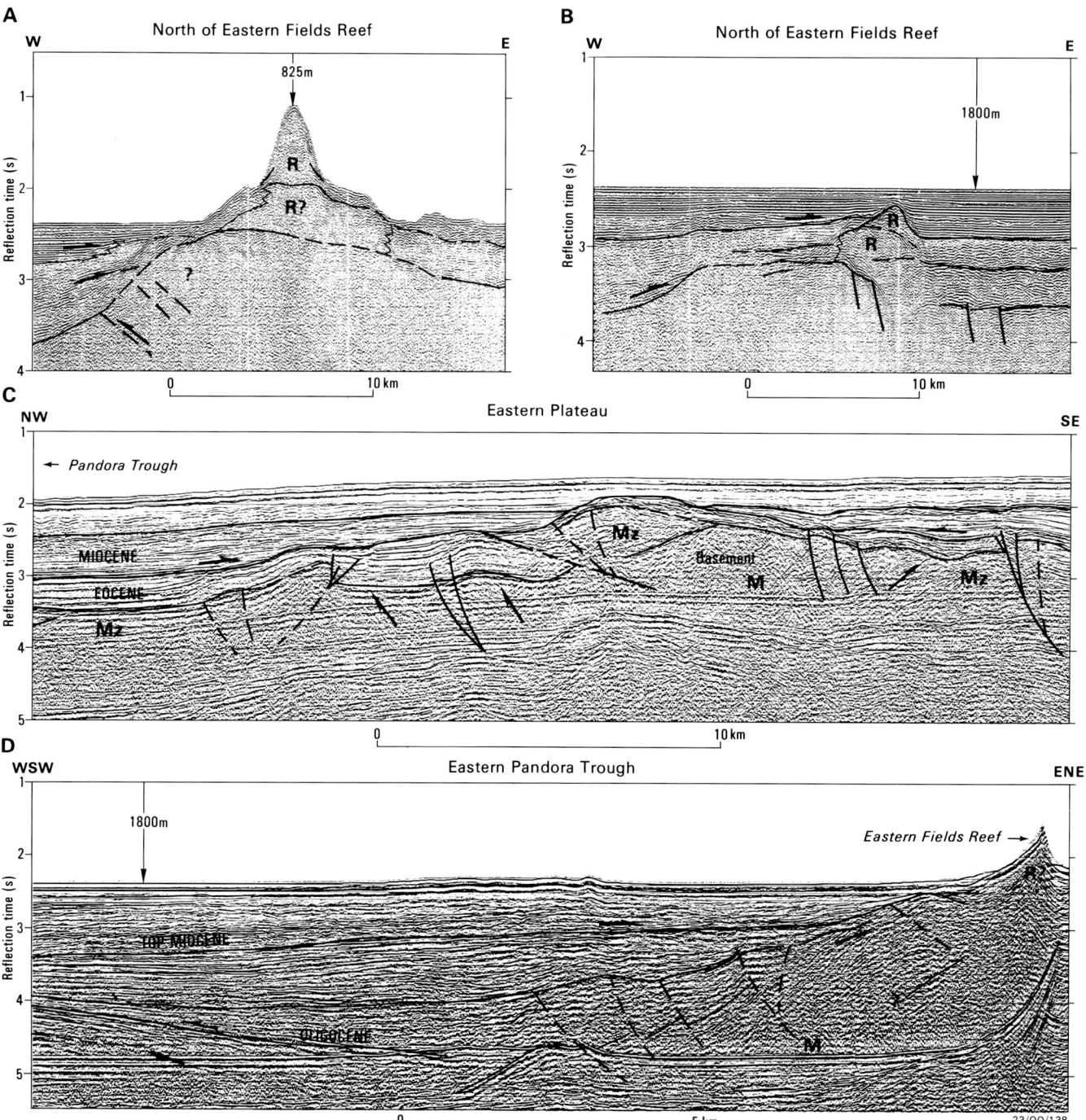

FIG. 10.—Seismic sections across the Eastern Fields Reef complex. Locations of sections are shown on Figure 1. (A) Sparker section showing the submerged northern extension of the Eastern Fields Reef complex (**R**). Note that reef development has become areally more restricted with time. (B) Sparker sections showing buried parts of the northern extension of the Eastern Fields Reef complex (**R**). (C) Airgun section showing the complex structural high, consisting of basement and Mesozoic sediments, on which part of the reef system developed. (D) Airgun section showing the deformed and uplifted post-Eocene sediments on which part of the reef system developed.

Queensland Plateau.—

The Queensland Plateau is the largest marginal plateau of the Australian continental margin. It is one of the largest features of its type in the world (Figs. 1, 2) and is approximately the same size as the Bahama Platform. It is bounded on the northeast by the Coral Sea Basin, on the west by the Queensland Trough, and on the south by the Townsville Trough. The plateau is roughly triangular in shape and extends over an area of about 165,000 km^2. Approximately half of the plateau surface lies above the 1,000-m isobath, with living reef systems at or near present sea-level forming 10 to 15% of the surface. The largest modern reef complexes are Tregrosse and Lihou Reefs, lying along the southern margin of the plateau (Fig. 1). Both these complexes are nearly 100 km long from east to west and 50 and 25 km wide, respectively, from north to south. The other major areas of modern reef growth are the Coringa, Willis, and Diana complexes, which are aligned north to south in the center of the plateau, and the large isolated pinnacles of Flinders, Holmes, Bougainville, and Osprey Reefs, which lie along the western margin of the plateau (Fig. 1). In addition, drowned reefs have been reported from at least 25 different locations (Taylor, 1977; Mutter, 1977; Davies and others, 1989). Away from reef areas, the plateau surface is generally smooth and slopes northward. A distinct terrace at approximately 450- to 500-m depths occurs between Willis and Diana reefs, and also between Tregrosse, Lihou, and Coringa reefs.

The major characteristics of the Queensland Plateau carbonate platform, as deduced from analysis of extensive airgun and sparker seismic data combined with sampling data, are summarized on schematic sections across the plateau (Fig. 9B-C). The ages of the stratigraphic sequences visible on the seismic data have been deduced by correlation with DSDP Site 209 (Burns and others, 1973; Fig. 8C), located on the northeastern margin of the plateau (Fig. 2).

Basement on the Queensland Plateau is represented by a series of fault blocks, composed of probable Paleozoic rocks, which form a basement surface that dips northeast toward the Coral Sea Basin (Mutter, 1977; Taylor, 1977). Basement beneath the western one-third of the plateau is progressively downfaulted toward the Queensland Trough (Fig. 11A). South of Tregrosse and Lihou Reefs, the basement surface slopes gently south toward the northern boundary fault of the Townsville Trough (Fig. 9B). Large parts of the basement surface were exposed and planated during the Cretaceous-Oligocene. From the Early Eocene, this surface was progressively submerged and overlain first by shallow-marine siliciclastic sediments, and then by deeper water pelagic sediments (Burns and others, 1973). A period of nondeposition or submarine erosion occurred from the Late Eocene until the Late Oligocene. The sedimentary sequence reflects constant gradual subsidence until the Late Miocene, followed by an increased subsidence rate until the present.

Although carbonate deposition may have begun earlier, the Queensland Plateau has been a carbonate-dominated province at least since the earliest Miocene. Pinchin and Hudspeth (1975) interpreted a mound structure visible on seismic data from the western margin of the Queensland Plateau as evidence of a possible Eocene barrier reef. More recent Bureau of Mineral Resources seismic data (Fig. 11B) show that this feature is a complicated mound, some 15 km in diameter and 700–800 m thick, with the steepest margin facing to the west. The mound is a composite feature, composed of both bedded and chaotic facies, and its form is at least partly the product of erosion. An Early to Middle Eocene age for this feature is suggested by onlap of the Eocene-Oligocene unconformity. Although it is possible that the structure may represent a carbonate buildup, its restricted lateral extent indicates that it is clearly not a "barrier" reef.

Along the western margin of the Queensland Plateau, steep-sided pinnacles 1–2 km across rise from depths of as much as 1,200 m to within 10 m of sea level (e.g., Fig. 11C). Dredged samples indicate that the flanks of these features are composed of reefal framework containing larger Miocene-Pliocene foraminiferids (Davies and others, 1989). Seismic data show that at least some of these pinnacles have developed on the raised corners of fault blocks.

In addition to the carbonate buildups on the plateau margins noted earlier, seismic data indicate that a thick carbonate platform sequence was deposited on the central part of the plateau (Fig. 11D). At least two phases of separate but superimposed reef and periplatform facies (QR1 and QR2; Figs. 9B–C, 11D) form the core of the carbonate platforms. Dredge samples from this complex on the southern slope of the Queensland Plateau between 1,000- and 1,300-m depths consist of Middle Miocene to Pliocene reefal material (Davies and others, 1989). The presence of shallow-water sediments at these depths confirms that there has been unusually rapid subsidence of the plateau since the Middle Miocene. The deeper water areas between reef complexes are the sites of hemipelagic sedimentation.

A terrace that occurs at 450- to 500-m depths represents the end of QR2 reef growth (Figs. 9B–C, 11D). A third, more restricted phase of reef growth (QR3) developed on this surface, with associated periplatform sedimentation in front of the reef (Fig. 11D). This reefal platform grew to sea level, and, as a result of relative sea-level rise, now forms another terrace at approximately 50-m depth.

The most recent reef complexes (QR4) developed on the 50-m terrace (Figs. 9B–C) and are even more restricted than previous phases. Descriptions of the modern coral faunas (Orme, 1977; Done, 1982) suggest that the modern reefs are oceanic equivalents of high-energy reefs present in the Great Barrier Reef. It is therefore likely that throughout their evolution, the different phases of Queensland Plateau reef development have all been products of high-energy oceanic conditions as a result of their exposed oceanic location.

Marion Plateau.—

The 77,000-km^2 area of the Marion Plateau lies directly east of the central Great Barrier Reef and is bounded along its northern margin by the Townsville Trough and along its eastern margin by the Cato Trough (Fig. 1). The present plateau surface forms a deeper water extension of the Australian continental shelf, with water depths ranging from 100 m along the western border to 500 m along the eastern

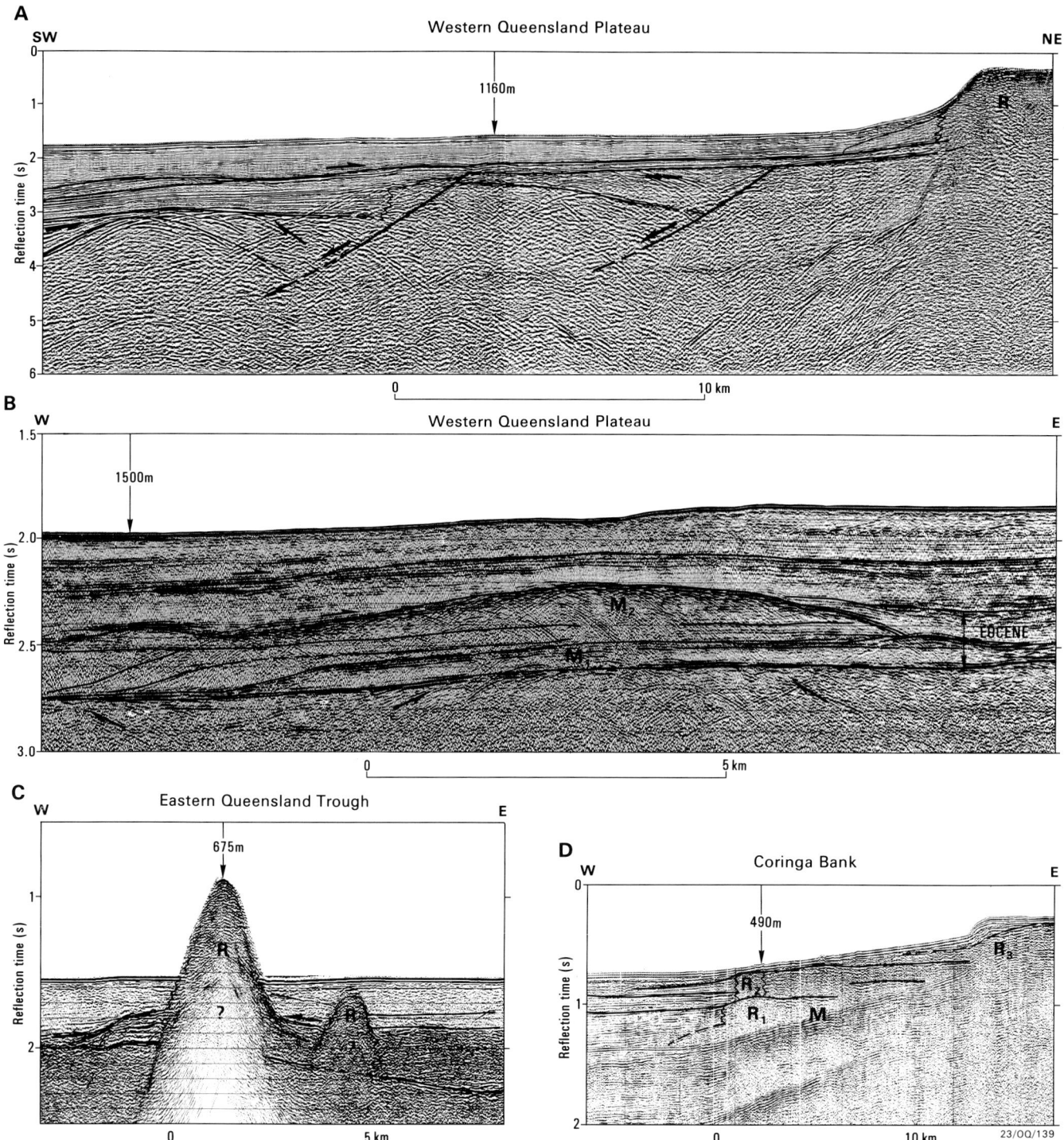

FIG. 11.—Seismic sections across the Queensland Plateau carbonate platforms. Locations of sections are shown on Figure 1. (A) Airgun section across the western Queensland Plateau showing fault-bounded tilt blocks and half-graben and an overlying carbonate bank (**R**). (B) Sparker section showing mounded features (**M1** and **M2**), possibly Eocene carbonate buildups, overlying basement on the western flank of the Queensland Plateau. Recent work indicates that this structure is probably circular and accordingly is not a barrier reef, as suggested earlier (Pinchin and Hudspeth, 1975). (C) Sparker section over a steep-sided pinnacle (**R**) rising from 1,200-m depth on the eastern side of the Queensland Trough. Miocene and Pliocene reef framework samples were dredged from the lower slopes of this pinnacle. (D) Sparker section showing three major phases of platform reef growth (**QR1** to **QR3**) at Coringa Bank.

margin. At present, reef growth is restricted to Marion Reef on the northeastern corner and Saumarez Reef at the southeastern extremity of the plateau (Fig. 1).

Little detailed subsurface structure and facies distribution information exists for the Marion Plateau (Mutter and Karner, 1980). The results of a study in progress, based on extensive airgun, watergun, and sparker seismic data combined with sampling data, are summarized on schematic sections across the plateau (Fig. 9B,D,E).

The plateau is bounded on three sides by rifts: the Cato Trough to the east; the Townsville Trough to the north; and a series of north-south-oriented, narrow half-grabens, which separate the plateau from the continent to the west (Fig. 2). During the Tertiary, siliciclastic shelf sediments prograded eastward across these half-grabens and onto the western Marion Plateau. The most northern of these half-grabens appears to join the confluence of the Townsville and Queensland Troughs. Therefore, the Marion Plateau formed a separate marginal plateau during the Early Tertiary. To the south, the Marion Plateau is separated from the Capricorn Basin by a northwest-trending basement ridge (the Swains Reef High; see Fig. 2).

The basement beneath the Marion Plateau is a planated surface, which dips gently toward the northeast. The only disruption to this surface occurs in the northeast corner of the plateau, where a basement high forms the pedestal on which Marion Reef developed. Basement beneath the plateau margins is steeply down-faulted into the troughs to the north and east. The slope sequences on the northern and eastern margins of the plateau are both onlapping and progradational. Small reef complexes overlie some of these progradational sequences along the northern margin (Fig. 12A).

The basement surface was completely transgressed during the (?)Early Miocene, resulting in development of an extensive carbonate platform (MR1; see Fig. 5). The top of this platform presently lies at 450- to 500-m depths. Shelf edge barrier reefs (Fig. 12B) and platform reefs separated by lagoons and interreef areas (Fig. 12C) can be identified over the northwestern two-thirds of the platform. Barrier reefs formed a distinct rimmed margin only along the northern edge of the plateau. The second phase of platform development (MR2) was more restricted and was confined to the southern one-third of the plateau. This phase was initiated at a level considerably below the top of the earlier phase (Fig. 9E). The top of the MR2 platform presently lies at 350–400 m below sea level. The third phase of reef growth on the Marion Plateau (MR3) is represented by small platform areas that have grown on the 350- to 400-m surface. Toward the southern Marion Plateau, part of the Great Barrier Reef overlies the third phase of carbonate platform growth. The final, very restricted, phase of growth on the Marion Plateau (MR4) is represented by Marion and Saumarez Reefs. Therefore, the successive phases of carbonate platform growth have been progressively more restricted in area (Fig. 9D–E).

At present, the top of the Marion Plateau is swept by moderately strong currents with the result that, away from the areas of modern reef growth, only thin hemipelagic sediments are accumulating in restricted areas.

FACTORS CONTROLLING THE EVOLUTION OF CARBONATE PLATFORMS IN NORTHEAST AUSTRALIA

A number of fundamental questions arise from the descriptions of the northeast Australian carbonate platforms: (1) What are the reasons for the northward-thickening of the Great Barrier Reef tropical carbonate platform? (2) Why is there such a thick platform sequence on the Queensland Plateau, with distinct levels of accumulation? (3) What are the causes of platform contraction? (4) What is the age of reef initiation in the region, and what mechanisms controlled the areal and temporal distribution of facies, e.g., what caused the stepback from the Marion Plateau to the shelf of the Great Barrier Reef? Partial answers to these questions can be obtained through analysis of the major factors that controlled platform development in the region. These factors are rifting, subsidence, plate motion, sea-level variation, and collision.

Rifting.—

Late Cretaceous extension formed the Queensland-Townsville-Cato Trough rift basin system, with another apparently less-developed system extending southeast from the Queensland Trough to the Capricorn Basin. This rift system separated the continental shelf from the Queensland and Marion Plateaus (Fig. 2). Continental breakup and seafloor spreading began in the latest Cretaceous-Paleocene and ceased in the Early Eocene. The main physical elements of the northeast Australian margin have therefore been in existence since Late Cretaceous-Paleocene time. Although the Queensland and Townsville Troughs exist as major physiographic features on the sites of the original rift troughs, small troughs along the western side of the Marion Plateau have now been infilled and are recognized only on seismic sections.

The rifting process has influenced the form of carbonate platform development off northeast Australia in both a general way, by providing large shallow-water areas suitable for platform growth, and in a very specific way, in the case of reefs that have grown on the corners of fault blocks or along major rift boundary faults.

Subsidence.—

Quantitative subsidence data have been derived from geohistory analysis (Van Hinte, 1978; Falvey and Deighton, 1982) of Anchor Cay 1, DSDP Site 209, Capricorn 1A, and Aquarius 1 (Fig. 13). The subsidence data from these wells indicate that northeast Australia has not subsided wholly as a result of uniform post-rift thermal cooling, but that subsidence pulses have occurred at different times.

The Anchor Cay 1 well at the northern end of the Great Barrier Reef (Fig. 2) contains Triassic to Middle Jurassic siliciclastic rocks unconformably overlain by Eocene to Recent carbonate-dominated sediments, with a hiatus in the Early Oligocene (Oppel, 1969; Robinson Research, 1984; see Fig. 8A). The pre-Eocene portion of the subsidence curve is not reproduced here, as it is difficult to ascertain how much section has been removed at the major unconformity (Fig. 13A). The accelerated subsidence of 50 m/m.y., which

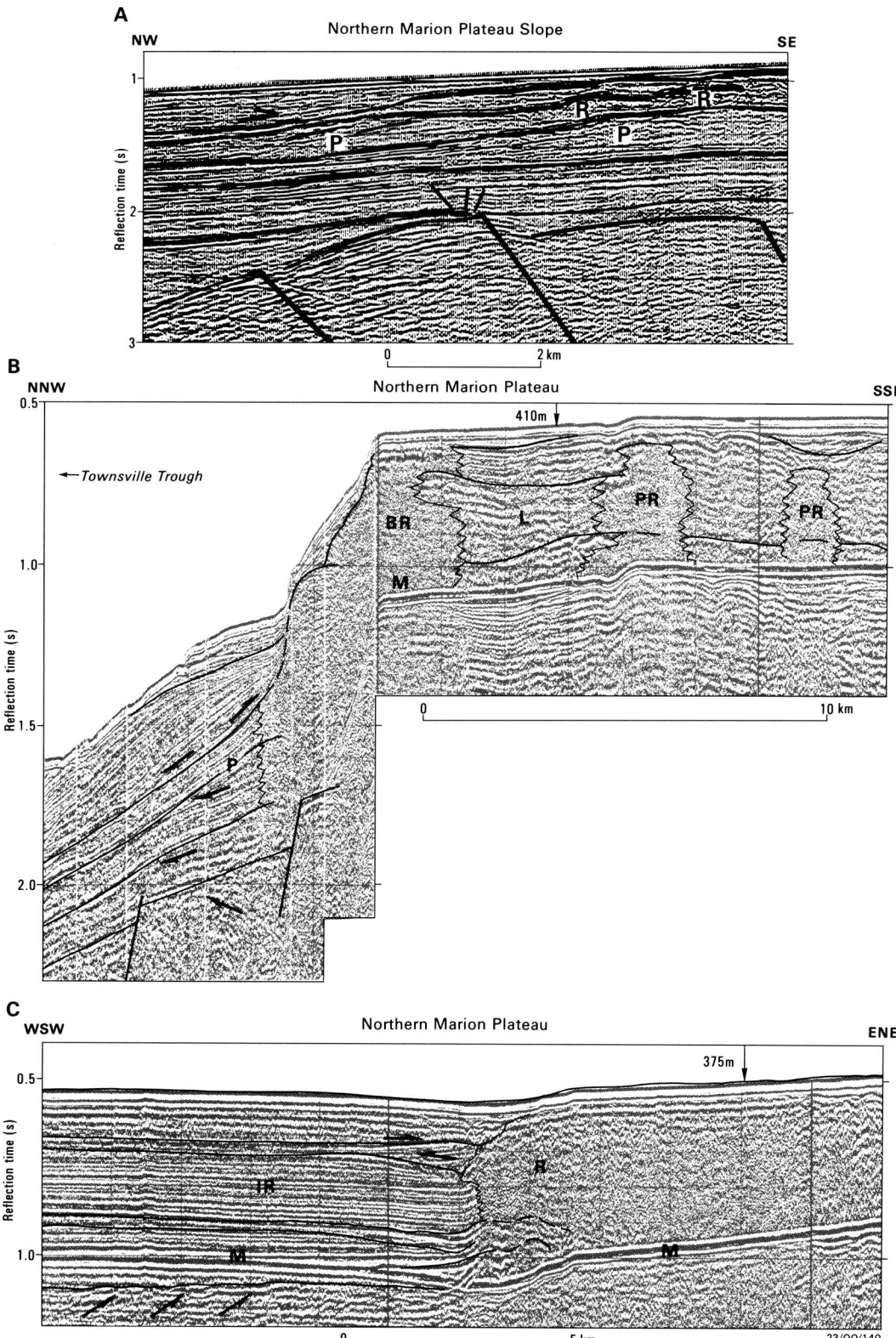

FIG. 12.—Seismic sections across the Marion Plateau carbonate platforms. Locations of sections are shown on Figure 1. (A) Airgun section across the northern margin of the Marion Plateau showing buried reefs (**R**) overlying a prograding sequence (**P**). (B) Watergun section showing the MR1 platform margin adjacent to the Townsville Trough. Note the reflection-free, thick, barrier-reef (**BR**) and patch reef (**PR**) facies, forereef periplatform facies (**P**), and back-barrier, bedded, lagoonal facies (**L**). Note that some reflectors pass into and through the reef facies and may represent low sealevel erosional surfaces. (C) Watergun section across the northeastern Marion Plateau showing development of platform reefs (**R**) and bedded, interreef sediments (**IR**).

affected this region in the Miocene (25–5 Ma), increased to 140 m/m.y. in the Pliocene (Fig. 13A).

The only source of quantitative subsidence data for this plateau is provided by DSDP Site 209, drilled in 1,428 m of water on the northeastern margin of the Queensland Plateau (Fig. 2). The well contains a Middle Eocene to Recent section, with hiatuses in the Early Oligocene and Middle Miocene (Fig. 8C). The Eocene section consists of calcareous sandstone grading upward to carbonate mudstone, with decreasing terrigenous sand content, overlain by Late Oligocene and younger calcareous ooze (Burns and others, 1973). The subsidence history of the plateau at this site was characterized by progressively increased rates of subsidence (Fig. 13B). An initial slow rate (20 m/m.y.) was succeeded by a markedly increased rate (40 m/m.y.) after the Middle Miocene (11 Ma).

Two petroleum exploration wells (Capricorn 1A and Aquarius 1) were drilled in the Capricorn Basin adjacent to the southern Marion Plateau (Fig. 2). Basement consists of Cretaceous volcanics in Capricorn 1A and indurated (?)Paleozoic shale and siltstone in Aquarius 1. In both wells, basement is overlain in turn by Paleocene to middle Oligocene basal polymictic conglomerate and arkosic redbeds, by shallow-marine glauconitic and carbonaceous sandstones, and by Miocene to Recent claystone and marl (Fig. 8D; Ericson, 1976). The geohistory curves for Aquarius 1 (Fig. 13C) and Capricorn 1A show similar subsidence patterns. A Cretaceous to middle Oligocene (88–30 Ma) slow subsidence phase (20 m/m.y.) was succeeded by increased subsidence (75 m/m.y.) until the Middle Miocene (11 Ma). Decreased subsidence followed by uplift during the Late Miocene and Early Pliocene was succeeded by a final increased subsidence pulse (75 m/m.y.) from the middle Pliocene.

Our analysis shows that the northeast Australian carbonate platforms have developed within a variety of subsidence regimes. These regimes may be categorized as simple or compound. Simple subsidence is characterized by an initial phase of rapid subsidence, followed by progressively lower subsidence rates. The facies and morphology produced by simple subsidence for a constant sea level are illustrated in Figure 14. During the rapid subsidence phase, the carbonate platform may backstep, i.e., the locus of deposition progressively migrates to maintain an approximately constant depth. During the ensuing slower subsidence phase, an aggradational or progradational carbonate platform will develop as sediment supply outstrips subsidence. Compound subsidence is characterized by alternating periods of rapid and slow subsidence. The effect of compound subsidence is to produce alternating backstepped and progradational packages.

Plate Motion and Paleoclimate-Paleoceanography.—

Hotspot (Duncan, 1981; Wellman, 1983) and magnetostratigraphic studies (Idnurm, 1985, 1986) provide a reconstruction of Indian-Australian Plate movement through the Cenozoic. These studies show that since the end of the Eocene, when northeast Australia was located between 29°S and 44°S, the region has moved almost directly northward to its present location between 9°S and 24°S. On this basis,

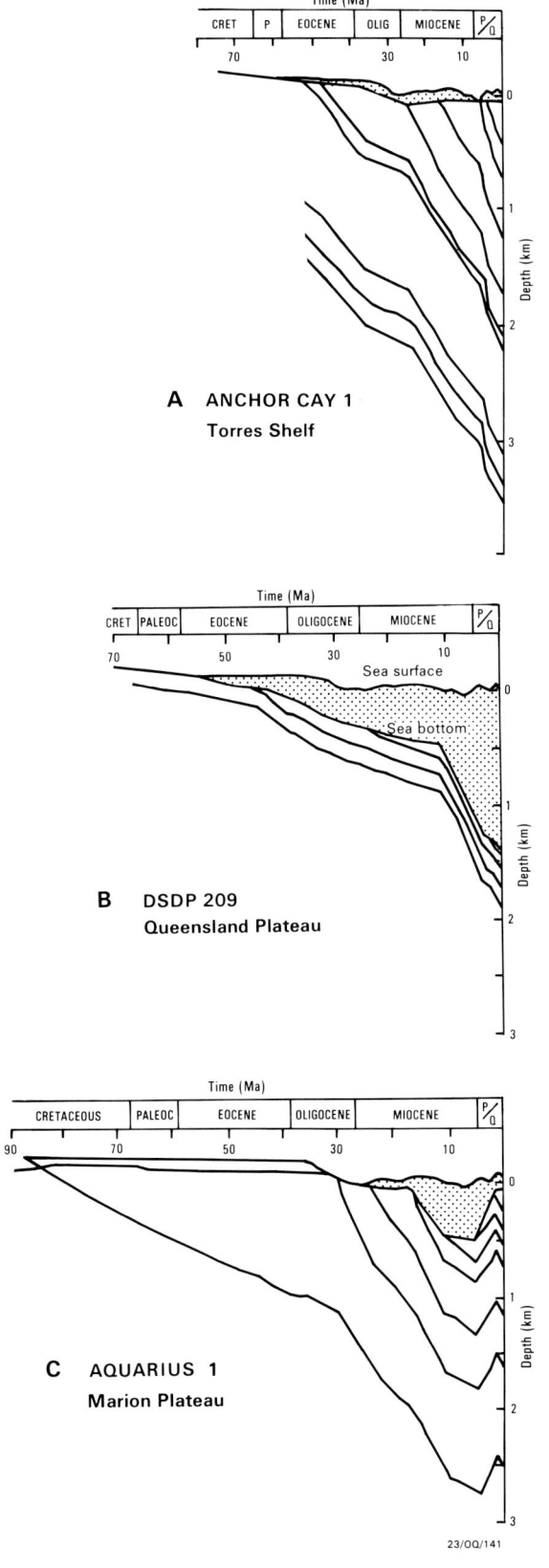

FIG. 13.—Geohistory plots for Anchor Cay 1 (A), DSDP Site 209 (B), and Aquarius 1 (C). The Capricorn 1A plot is not presented as it is essentially identical to that of Aquarius 1. Locations of holes are shown in Figure 2; summary logs are shown in Figure 8.

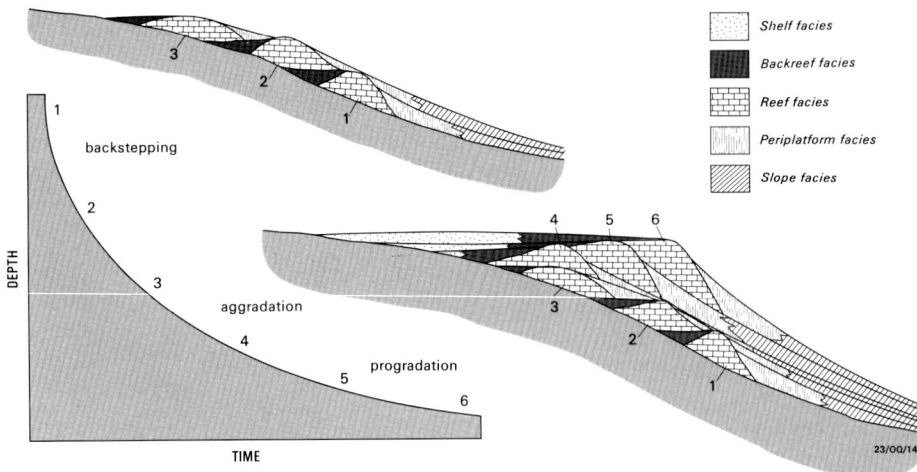

FIG. 14.—Idealized facies geometries illustrating the effects of simple subsidence on a platform margin. Reef backstepping (1–3), corresponding to the initial rapid-subsidence phase, is succeeded first by aggradation (3–4) and then by reef progradation (4–6) as the subsidence rate decreases.

the Cenozoic paleolatitudes for the northeast Australia region may be determined (Fig. 15). This latitudinal motion would have resulted in profound climatic changes along the east Australian shelf, particularly since plate movement was essentially normal to developing climatic zones. Analysis of the northward plate motion relative to 23.5°S (the Tropic of Capricorn and southern limit of the Great Barrier Reef region) indicates that, assuming present climatic conditions: the transition from temperate to tropical climatic conditions in the northern part of the Great Barrier Reef would have started about 25 Ma (Fig. 15); and the southern region has only recently entered the tropics.

The development of more complex climate distribution patterns in the southwest Pacific throughout the Cenozoic (Savin and others, 1975; Frakes, 1979; Kennett and von der Borch, 1985), however, indicates that a more detailed analysis of climatic factors is required. Since surface-water temperatures are critical to carbonate platform development, we have examined data primarily derived from geochemical and petrographic analyses of DSDP cores from the western Pacific and produced a synthesis describing Cenozoic surface-water temperature variability for the northeast Australian region (see Fig. 16). For the Paleogene, we particularly examined oxygen isotope data from DSDP Sites 167, 277, 592, and 593 (Savin and others, 1975; Shackleton and Kennett, 1975; Murphy and Kennett, 1985) and all data and syntheses relating to the development of thermal gradients (Kennett, 1977; Frakes, 1979; Murphy and Kennett, 1985; Shackleton, 1986) and oceanic circulation patterns (Kennett and others, 1975; Burns and others, 1973). For the Neogene, we placed particular emphasis on oxygen and carbon isotope data from a large number of DSDP cores (Savin and others, 1975; Loutit and others, 1983; Elmstrom and Kennett, 1985; Kennett, 1985; Kennett and von der Borch, 1985; Savin and others, 1985); variations in clay mineralogy indicating desertification (Locker and Martini, 1985; Stein and Robert, 1985); onshore palynological analyses (Kemp, 1978); and global oceanographic events (Burns and others, 1973; Mercer, 1976; Keany, 1978; Hsu and others, 1984; Weissert and others, 1984; Kennett and von der Borch, 1985). The oceanic surface-water temperature curve for northeast Australia compiled from these sources (Fig. 16) allows us to draw the following conclusions:

—Temperatures in the earliest Middle Eocene were briefly warm enough for coral reef growth. Corroboration is provided by the identification of early Middle Eocene larger foraminiferids from the northwestern margin of the Queensland Plateau (Chaproniere, 1984), indicating sea-surface temperatures of 18°–27°C (Murray, 1973).

—Temperatures from the late Middle Eocene to the middle Early Miocene were not conducive to tropical carbonate platform development. Climates at paleolatitudes of 23°–46°S were probably temperate or cool temperate, and ac-

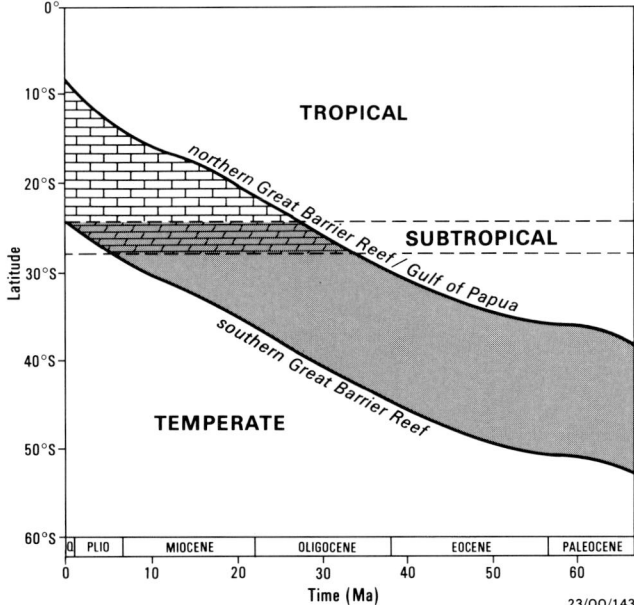

FIG. 15.—Projected latitudinal movement of the northeast Australia region throughout the Cenozoic (modified after Davies and others, 1987). The northern boundary corresponds to Anchor Cay 1 (presently at 9°30′S) and the southern boundary to Heron Island (presently at 24°S).

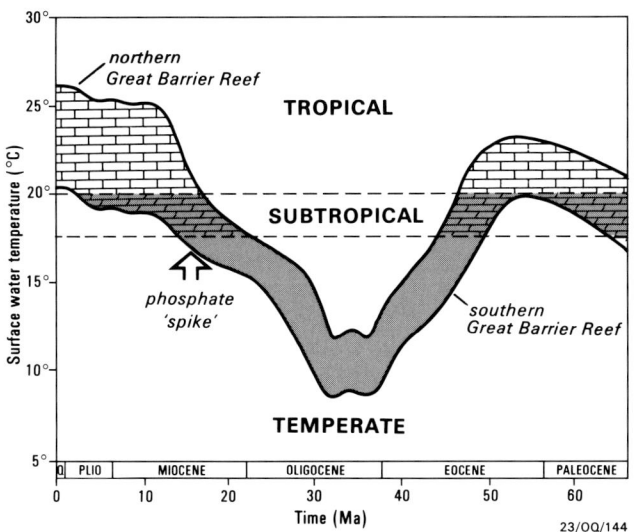

FIG. 16.—Surface-water temperature envelope for the northeast Australian region throughout the Cenozoic, showing periods when temperatures were suitable for reef growth. The Miocene "phosphate spike," which inhibited reef growth, is also shown (see Riggs, 1984).

cordingly there would have been no significant coral reef growth.
–During the Early Miocene, the northeast Australia region was bathed in surface waters marginal for supporting coral reefs, i.e., probably comparable to those off northern New South Wales and southern Queensland today. Whereas some reef growth may have been possible in the extreme north, it is most likely that prolific growth throughout much of the northern region began with the initiation of tropical climatic conditions in the early Middle Miocene.
–The Late Miocene climatic cooling would have prevented extensive reef growth in the southern part of the region, located near the subtropical-tropical climatic boundary.
–During the Pliocene, temperatures suitable for reef growth extended into the southern parts of the northeast Australian province.

The paleoclimatic and paleoceanographic data substantiate and refine the major conclusions deduced from plate-motion studies. The consequences of this interpretation are that the Great Barrier Reef tropical-shelf carbonate facies thin and become younger to the south and overlie temperate facies (Davies and others, 1987); that reefs grew first in the north, probably within the developing foreland basin; that this early reef growth was closely followed by reef growth on the Queensland and Marion Plateaus; and that reefal development occurred later in the central and southern Great Barrier Reef. Facies diachroneity must be a fundamental factor in platform evolution, where plate motion has produced movement either toward or away from the tropics.

The above conclusions can be tested by a comparison of present facies variations along the east Australian margin with the vertical facies sequence observed in cores from the Gulf of Papua. The present sediment distribution on the east Australian outer continental shelf is composed of three distinct facies (Marshall and Davies, 1978), which contain a clear climate-related signature: tropical carbonate and clastic sediments, dominated by coral and *Halimeda* debris, north of 24°S; subtropical rhodolith-encrusting foraminiferid-bryozoan facies between 24°S and 28°S, with bioherms dominated by this association occurring over large parts of the outer shelf; and temperate, branching bryozoan-foraminiferid-mollusc facies south of 28°S.

A similar facies sequence occurs vertically in the Borabi No. 1 drill hole (Fig. 8A) in the Gulf of Papua (Fig. 2). This sequence shows the development from a temperate open shelf in the Eocene and Oligocene; to a subtropical shallow outer shelf in the Early Miocene; to a tropical reef-dominated shelf in the Middle Miocene; and finally to a fluvio-clastic-dominated shelf in the Pliocene. The vertical carbonate facies variations mirror those which occur laterally on the present shelf and which are clearly climate related. Horizontal plate motion, with its attendant climatic and oceanographic effects, has therefore exerted a fundamental control on the sedimentary evolution of northeast Australia (Davies and others, 1987).

Further refinements arise from a consideration of local physical and chemical oceanographic factors. The progressive development of the east Australian current would have intensified from the Early-Middle Miocene (15–20 Ma), as the northern edge of the Australian craton began to disrupt the strong equatorial current flow (Kennett and others, 1985). Continuing northward plate motion, the elevation of New Guinea, and the closure of the east Indonesian seaway in the Late Miocene would have further restricted westerly current flow and caused diversion of warm tropical waters to the south into the northeast Australia region.

Chemical oceanographic factors will also have affected the Neogene development of carbonate platforms in northeast Australia. The late Early Miocene to early Middle Miocene apparently represented a time of increased ocean fertility commensurate with a postulated two to three orders of magnitude increase in oceanic-phosphate levels (Riggs, 1984). This "phosphate spike" (Fig. 16) is thought to have produced extensive phosphatization of continental-margin sediments throughout the world. Phosphatic facies of this approximate age occur on the northern New South Wales outer shelf (Cook and Marshall, 1981) immediately south of the northeast Australian region. This event would have had considerable effect on carbonate deposition in northeast Australia, as increased oceanic-phosphate levels inhibit the growth of coral reefs and promote a large increase in biomass production (Kinsey and Davies, 1979). This should be reflected in considerably restricted coral reef growth during this period, with the high ocean fertility resulting in high organic-carbon production and the consequent formation of petroleum source rocks in selected environments.

Sea-level Variation.—

The effects of sea-level variation on carbonate platform development have been established by detailed analysis of the nature and distribution of siliciclastic and carbonate facies on the central Great Barrier Reef shelf. High sea-level deposition on the central Great Barrier Reef shelf occurred either as progradation of prodeltaic sediments on the inner shelf, primarily concentrated on wave-dominated deltas, or as aggradation of the mid- to outer shelf as a result of reef growth and inter-reef sedimentation (Fig. 17A). Reef facies reflect both the high physical energy of the system and the transgressive/stillstand history of the Quaternary sea-level rises (Marshall and Davies, 1982, 1984; Davies, 1983; Davies and Hopley, 1983; Davies and others, 1985). The reefs are composite features composed of stacked reef facies, which grew as a consequence of successive high sea-level growth phases, separated by unconformities representing low sea-level erosion (Fig. 18). The high physical energy of the reef environment restricted reef expansion to the leeward or backreef direction.

In the inter-reef areas on the mid- to outer shelf, high sea-level platform aggradation is represented by bioherms (Davies and Marshall, 1985), biostromes, and a sediment blanket of varying thickness (<1 m to 10 m) deposited on the previously exposed shelf surface. This sediment blanket is composed of a lower, terrigenous (mud- and quartz-rich, carbonate-poor), transgressive facies, and an upper, carbonate-rich (less mud, little quartz), stillstand facies (based on studies in progress). At the present time, after 10 ka of transgression and stillstand, there has been little high sea-level progradation of coastal terrigenous facies onto the inner shelf. It seems likely that the terrigenous/carbonate facies couplet that occurs over wide areas of the mid- to outer shelf is probably representative of high sea-level sedimentation on the Great Barrier Reef platform throughout most of the Plio-Pleistocene.

Low sea-level sedimentation occurred both as aggradation of fluvial sediments on the mid- to outer shelf (Fig. 17B), and as progradational shelf edge deltas composed of terrigenous sand and sandy mud beneath the outer shelf and upper slope (Fig. 7).

Rising and high sea-level periods in the central Great Barrier Reef were therefore characterized by both reefal and inter-reef carbonate deposition, with restriction of siliciclastic deposition largely to the inner shelf. In contrast, falling and low sea-level periods were characterized by fluvial or shallow-marine siliciclastic deposition, with siliciclastic progradation on the present upper slope. The marginal plateaus are essentially isolated from terrigenous input, and accordingly their facies response to sea-level variation must have been different. Although high sea-level periods in these areas are also marked by carbonate aggradation, low sea-level periods are characterized by unconformities representing exposure of the previous reef surfaces. This response to sea-level variation can be used to interpret the evolutionary history of the Cenozoic sequences on the marginal plateaus by attributing unconformities within the carbonate platforms to low sea-level episodes and reef sequences to periods of high sea level. In the absence of a specific Cenozoic sea-level curve for northeast Australia, the global sea-level curve proposed by Haq and others (1987) can be used. On this basis, episodes of reef growth during the early Middle Miocene (QR1, MR1; see Fig. 9), The Middle to Late Miocene (QR2, MR2), the Plio-Pleistocene (QR3, MR3), and the Quaternary (QR4, MR4) are separated by unconformities representing erosion during the late Middle Miocene (1–2), late Late Miocene (2–3), and the Quaternary (3–4).

Collision.—

If, as a result of plate motion, a passive margin moves into a compressional tectonic regime, carbonate platforms developing on that margin will be profoundly affected. The effects of collision on the formation and demise of carbonate platforms may be seen in the Papuan Basin of Papua New Guinea (Fig. 2). The collision between the northern edge of the Australian craton and an island arc complex that initiated the development of the New Guinea orogen had occurred by 25 Ma (Pigram and Davies, 1987). This event affected a 150- to 200-km-wide shelf on which subtropical/

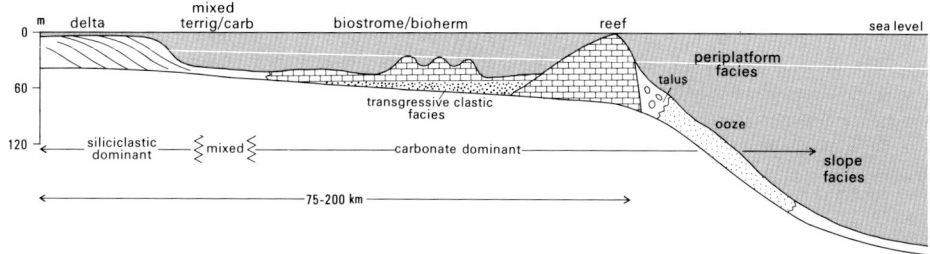

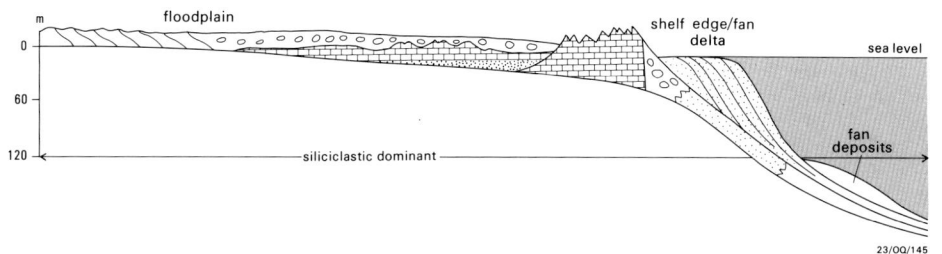

FIG. 17.—Schematic sections illustrating high (A) and low (B) sea-level control on the structural and sedimentary geometry of shelf facies in the central Great Barrier Reef province. Note the predominance of siliciclastic facies in the low sea-level situation.

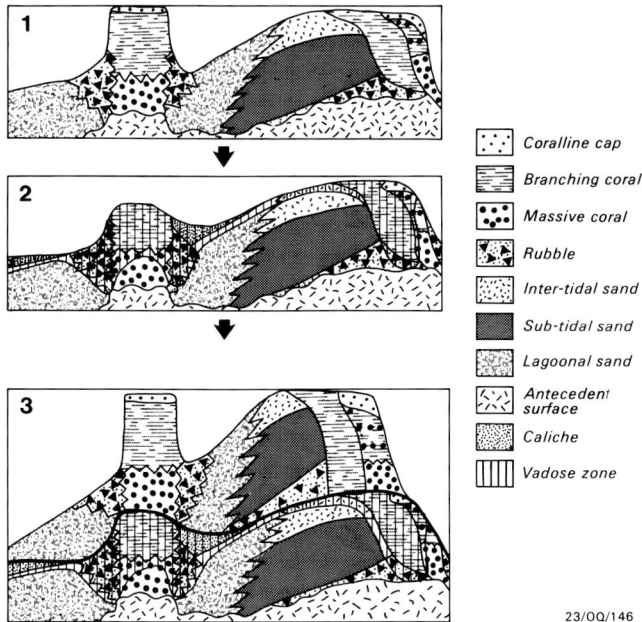

FIG. 18.—Schematic models showing high-energy reef growth in high (1 and 3) and low (2) sea-level phases. Note the leeward progradation reflecting the high-energy environment (after Davies and others, 1988).

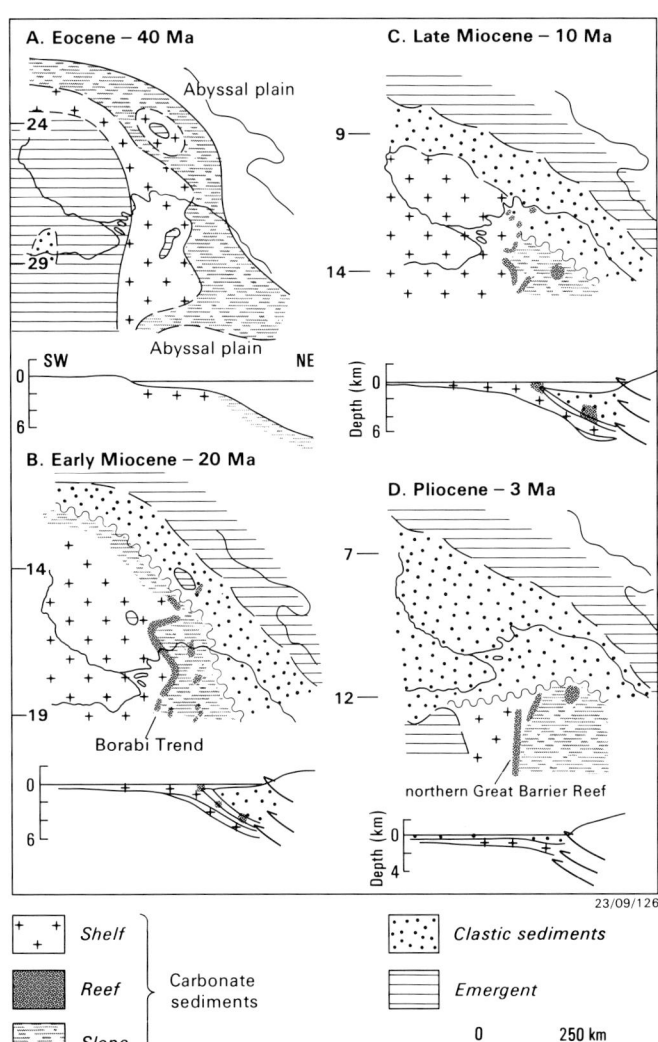

FIG. 19.—Cenozoic paleogeography of the Gulf of Papua/northern Great Barrier Reef region, showing the effects of progressive foreland basin development on carbonate platform evolution during the Eocene (A), Early Miocene (B), Late Miocene (C), and Pliocene (D). Schematic sections are oriented approximately northeast-southwest; approximate paleolatitudes are shown for each section.

temperate carbonates had accumulated during northward drift in the Paleogene (Fig. 19A). The collision caused the development of a foreland basin with a flexural wavelength of 500–700 km, with the position of the downwarp migrating south throughout the late Cenozoic as the New Guinea orogen evolved. Immediately after collision, the foreland basin consisted of a proximal deep, perhaps 100–200 km across, which was the site of thick (<5 km) clastic sedimentation; and a distal, broad, shallow epicontinental margin, some 300–500 km wide, which was ideal for carbonate deposition under the existing tropical conditions (Fig. 19B). The reef complexes that grew on this distal part of the foreland basin developed in two orientations relative to the basin axis: along the northern edge parallel to the basin axis; and along the eastern margin, on structural trends perpendicular to the basin axis (Borabi Reef Trend; Robertson Research, 1984).

With continued convergence, the Early Miocene carbonate slope facies and foredeep clastic sediments were incorporated into the foreland fold belt. Clastic sediments filled the proximal deep and started to prograde across the shelf carbonate sequence, causing marked reduction in the area of carbonate deposition. The formerly extensive reef development along the Borabi Reef Trend contracted to a series of pinnacles, and the northernmost pinnacles were sequentially buried by clastic detritus (Fig. 19C).

A Pliocene fall in sea level exposed the platform and shallow foreland basin. The clastic sediments derived from the emerging mountains to the northeast spread far to the south, covering the platform. The outpouring of clastic detritus was so great that, even during sea-level highstands, carbonate deposition was not reestablished in this area, but was restricted to the area of the present northern Great Barrier Reef (Fig. 19D). This sedimentation regime has continued to the present.

The convergence history of the northern edge of the Australian craton illustrates the effects of collision on the gross architecture of a carbonate platform and the response of reef complexes to increased subsidence arising from foreland basin development. Paradoxically, the initial effect of foreland basin development was to increase the area of carbonate platform deposition. This expansion, however, was quickly followed by contraction, demise, and burial of the carbonates by clastic detritus. The response of reef complexes was dependent on their initial orientation relative to the axis of flexure. Reef growth parallel to the collision front probably formed a discontinuous tract of thin reefs on the margin of the foredeep. As the basin axis migrated southward, the locus of reef growth migrated toward the craton (Fig. 20A). The older reefs became progressively engulfed and buried by clastic detritus derived from the developing orogen. In contrast, the location of reef growth on the Borabi Trend remained fixed along an axis oriented perpendicular to the collision zone. In this case, reef growth responded to foreland basin development and concomitant subsidence with accentuated vertical growth, rather than lateral migration (Fig. 20B). The initial barrier reef first contracted to a series of pinnacles, before being engulfed by clastic detritus. The eastern margin of the Papuan Platform and the northern end of the Great Barrier Reef are currently undergoing an evolution comparable to that of the Early to Middle Miocene reefs of the Borabi Reef Trend. Growth on the Portlock-Boot-Ashmore reef complex, east of the northern Great Barrier Reef, has now contracted to large isolated pinnacles. Farther north, reefs along this trend have been buried by more than 800 m of prograding clastic sediments (Davies and others, 1989).

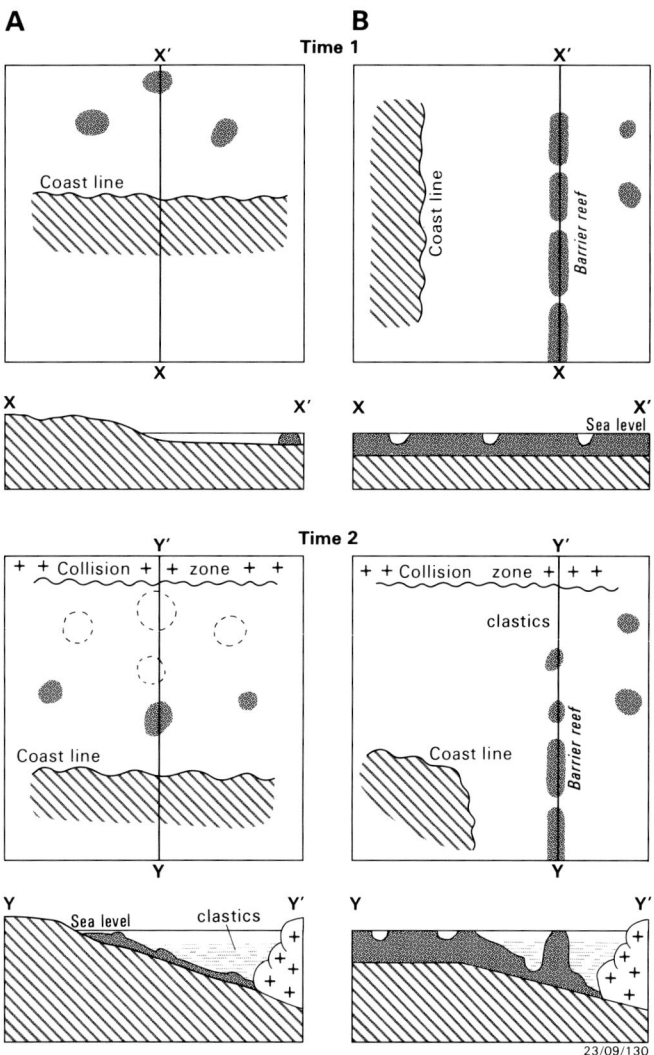

FIG. 20.—Schematic diagrams showing the differing response of reefs to foreland basin development, dependent on whether the trend of reef growth was parallel (A) or perpendicular (B) to the collision zone.

EVOLUTION OF THE NORTHEAST AUSTRALIA CARBONATE PLATFORMS–CONCLUSIONS

Carbonate platform evolution in northeast Australia was primarily controlled by five interrelated controlling factors which have acted upon the continental margin since the Cretaceous. These factors are: rifting, subsidence, plate motion (with climatic and oceanographic consequences), sea-level variation, and plate collision. These factors have interacted to give the following evolutionary sequence.

(1) Early Cretaceous northwesterly movement of the Pacific Plate may have resulted in pre-rift basin development, perhaps within an oblique wrench zone. This zone formed the site of further rifting during the Late Cretaceous and Paleocene. The western boundary fault of the rift system lay beneath the present mid- to outer shelf. This phase of extensional tectonism occurred when northeast Australia was between 28°S and 43°S.

(2) Temperate, clastic, fluvio-deltaic, and carbonate sedimentation developed along the continental margin in the Eocene and Oligocene. These sediments were deposited under a regime of sea-surface temperatures that decreased from subtropical/tropical to cool temperate, and sea levels that remained high throughout the early part of the interval, but then dropped markedly during the early and middle Oligocene. Subtropical carbonate deposition dominated toward the north, with temperate fluvio-deltaic and carbonate sedimentation along the continental margin farther south. Temperate (?)carbonate progradation may have occurred along the margins of the Queensland and Marion Plateaus.

(3) Late Oligocene to Early Miocene subsidence, coincident with a rise in sea level, led to widespread transgression over parts of all platforms. Initially, reef development was prevented by the effects of temperature and ocean chemistry, with the result that red algal bioherms developed on the outer shelf. As northward drift carried the region into the tropics, the first reefs probably formed on red algal bioherms. In the Gulf of Papua, these evolved into a barrier-reef complex (the Borabi Reef Trend; Fig. 6B) and a series of pinnacle reefs located on structural highs to the east of the barrier (e.g., Pasca Reef; Fig. 6C).

(4) In the far northern part of the Great Barrier Reef and on the Queensland and Marion Plateaus, initiation of reef growth in the early Middle Miocene was a consequence of subsidence, continued northward drift into the tropics, and a marked increase in surface-water temperatures. In the northern region, the increased subsidence resulted from development of the foreland basin, and also caused an initial expansion of shelf carbonate deposition and a later contraction of the earlier barrier reef in the Gulf of Papua to a series of pinnacles.

(5) Late Miocene subsidence counteracted the effects of the Late Miocene sea-level fall and resulted in a further, areally more restricted episode of reef growth on the Queensland and Marion Plateaus. Contraction of reef growth and progressive burial of pinnacle reefs continued in the north. Much of the central Great Barrier Reef region, although in the tropics, was either dominated by terrigenous sedimentation or exposed.

(6) Progressive foreland basin development during the

Pliocene led to burial of the northern reefs. Further subsidence produced substantial contraction of the Queensland Plateau reefs and stepback of the Miocene Marion Plateau carbonate platform to the present position of the Great Barrier Reef. The post-Pliocene evolution of the Great Barrier Reef is a function of sea-level-controlled fluvio-deltaic deposition and reef growth.

The major implications of this evolutionary history are that the Great Barrier Reef tropical carbonate platform thins to the south and overlies a temperate facies; that reefs grew first in the north, probably along the margins of the developing foreland basin; and that the platforms of the Queensland and Marion Plateaus are precursors to reef growth on the central and southern Great Barrier Reef. Although this analysis of the factors controlling carbonate platform development has accounted for much of the variability observed in the northeast Australian carbonate platforms, there remain areas where our understanding of the interactions between the controlling factors are still incomplete. We do not fully understand the factors that controlled reef contraction on the Queensland Plateau, although the interaction between subsidence and sea level is obviously important. Similarly, we do not understand why reef growth is apparently so restricted during lowstands. Although terrigenous input must be important adjacent to the continental margin, this is clearly not a significant factor on the marginal plateaus.

IMPLICATIONS FOR GLOBAL MODELS OF PASSIVE-MARGIN CARBONATE PLATFORM EVOLUTION

The continental rifting process determines the size, shape, and location of the shelf and marginal plateaus on which the carbonate platforms develop. Carbonate platform development is critically controlled by both long-term (plate motion, subsidence) and short-term (sea level, climatic, oceanographic) factors. In addition, movement of a passive-margin carbonate platform into a convergent tectonic setting will ultimately terminate platform growth.

Some recent interpretations of the development of passive-margin carbonate platforms define an evolutionary morphologic/sedimentologic sequence from pre-rift platform to platform collapse, at least in part controlled by simple subsidence scenarios (e.g., Read, 1985). On the basis of our studies in northeast Australia, we believe that these schemes represent only part of the carbonate platform evolutionary spectrum. More complete platform development models can be derived by considering the interaction of the two most significant long-term controlling processes: vertical tectonics (subsidence) and horizontal tectonics (plate motion).

The facies sequences produced as a result of both simple and compound subsidence under differing plate movement rates and directions are shown in Figure 21. The facies sequence developed in the central Great Barrier Reef region (Fig. 7) is an example of the situation illustrated in Figure 21B, where subsidence has interacted with plate movement toward the tropics to produce tropical-carbonate deposition overlying a temperate-clastic sequence. The Neogene carbonates in the Red Sea (El Haddad and others, 1984; Scott and Govean, 1985) provide an example of the development of a tropical-carbonate platform as a result of early rift subsidence (Fig. 21C).

The most complex and varied carbonate platform sequences will occur as a result of the interaction of compound subsidence and plate motion through different climatic zones over a substantial time period. The evolutionary history of the northeast Australia carbonate platforms represents a relatively complete example of such a situation. This region demonstrates that facies diachroneity is a fundamental characteristic of carbonate platform evolution, as different parts of the same platform may be at different stages in the development sequence at the same time. The extreme case occurs when different parts of the same platform are synchronously in both initiation and demise stages. It is possible for a platform to begin or end its growth anywhere with respect to climate, with the result that either the temperate or tropical stages may not develop. A carbonate platform may not progress past a particular stage, or spend a long time within one stage. For example, the southern Great Barrier Reef has only recently started to develop a tropical-carbonate platform, whereas the Papuan Basin in New Guinea has progressed very quickly into a collision setting. The Queensland Plateau may never be affected by collision unless a major reorganization of plate boundaries occurs, and will therefore remain for a very long time within a tropical-oceanographic environment under a slowing-subsidence regime. We believe that this may eventually lead to substantial shallow-water progradation comparable to that which affected the Queensland Plateau in the Miocene, and which has had such a dominant effect on the Cenozoic development of the Bahama Platform (Eberli and Ginsburg, 1987). Alternatively, the influence of a further subsidence pulse may lead to major reactivation of reef growth, and possibly even further contraction or stepback.

In contrast to the long-term influences of plate motion and subsidence, sea-level variation, climatic variation, and physical and chemical oceanographic factors all exert shorter term controls on carbonate platform development. Although the carbonate platforms of northeast Australia provide evidence of the controlling influence of all these factors, the results of eustatic sea-level variation are particularly clear. The differences between low sea-level progradational and high sea-level aggradational facies, and the stacking of reef facies as a consequence of low sea-level erosion and later high sea-level reactivation of reef growth, all attest to the importance of this factor.

Our work in northeast Australia indicates that, on a broad scale, carbonate platform development will be dependent on whether subsidence is simple or complex; whether plate motion is toward or away from the tropics; and whether movement from one climatic zone to another is slow or rapid. At a more detailed level, the particular facies sequence developed will reflect the influence of additional short-term controls. Although other areas illustrate specific aspects of platform development, the northeast Australian carbonate province provides a range of examples illustrating numerous aspects of platform evolution. This region is an ideal location to refine further our understanding of the

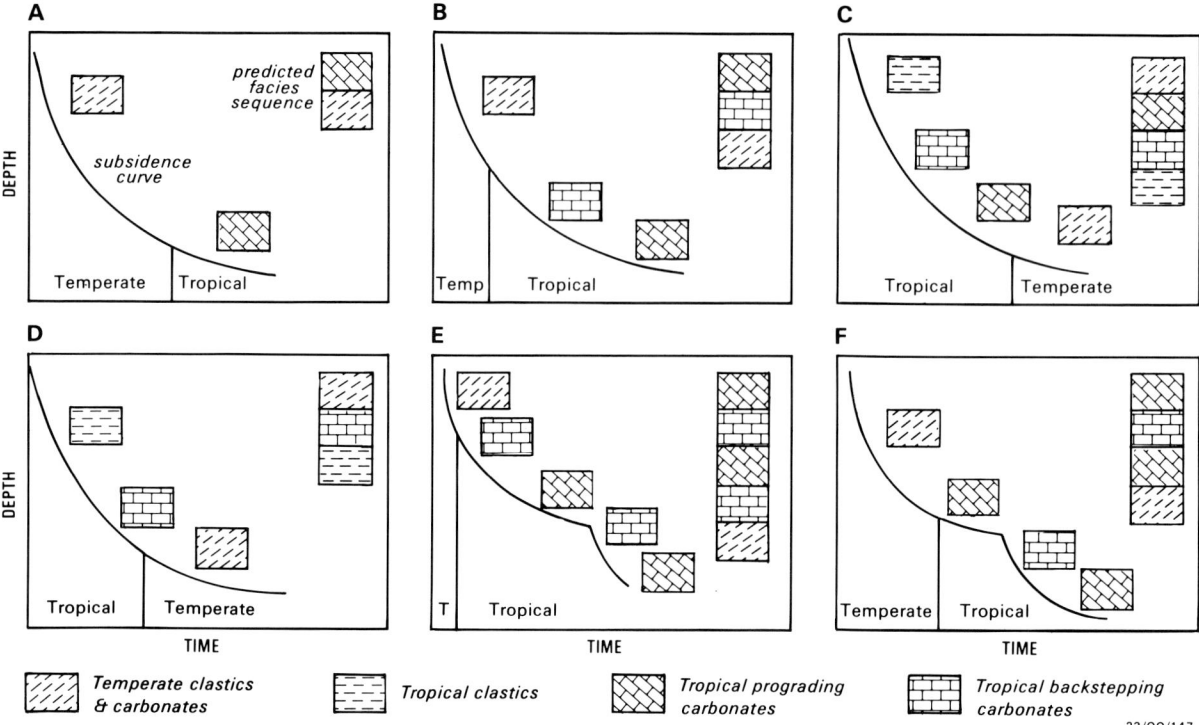

Fig. 21.—Hypothetical facies sequences produced as a result of different subsidence regimes and plate motion through different climatic zones.

factors that control platform evolution and to seek answers to the major outstanding problems.

ACKNOWLEDGMENTS

The authors thank their colleagues within the Bureau of Mineral Resources (BMR) for discussion on aspects of the work, particularly David Capon, George Chaproniere, Michael Etheridge, and John Marshall. We thank the scientific and engineering staff, who have helped us in the gathering and processing of the data base in northeast Australia, and the Master and crew of the RV Rig Seismic for their patience and professionalism during four cruises in the tropics. We also thank International Petroleum Corporation for permission to figure newly acquired (Fig. 6A) and reprocessed (Fig. 6E) seismic data, and the Geological Survey of Papua New Guinea for permission to present seismic data from a confidential report (Figs. 6B,C). Other non-BMR seismic data were provided by Geophysical Services International (Fig. 11A) and Australian Gulf Oil Company (Fig. 7B).

The perceptive and constructive reviews of an early draft of the manuscript by André Droxler (Rice University, Houston), Rick Sarg (Exxon Research and Development Corporation, Houston, Texas) and Noel James (Queens University, Kingston, Ontario) are gratefully acknowledged. Davies also wishes to thank Paul Crevello (Marathon Oil Company, Littleton, Colorado) and Rick Sarg for their encouragement, and the participants at the AAPG-SEPM Symposium on Carbonate Platforms for their stimulating contributions. This paper is published with permission of the Director, Bureau of Mineral Resources.

REFERENCES

AUSTIN, J. A., JR., SCHLAGER, WOLFGANG, PALMER, A. A., AND OTHERS, 1986, Proceedings of the Ocean Drilling Program, v. 101, Part A. Initial Reports, Bahamas: 569 p.
BURNS, R. E., ANDREWS, J. E., AND OTHERS, 1973, Initial Reports of the Deep Sea Drilling Project, v. 21: U.S. Government Printing Office, Washington, D.C., 931 p.
CHAPRONIERE, G. C. H., 1984, Oligocene and Miocene larger Foraminiferida from Australia and New Zealand: Bureau of Mineral Resources, Australia, Bulletin 188, 98 p.
COOK, P. J., AND MARSHALL, J. F., 1981, Geochemistry of iron and phosphorus-rich nodules from the east Australian continental shelf: Marine Geology, v. 41, p. 205–221.
DAVIES, P. J., 1977, Modern reef growth–Great Barrier Reef: Proceedings of the Third International Coral Reef Symposium, Miami, v. 2, p. 325–330.
———, 1983, Reef Growth, in Barnes, D. J., ed., Perspectives on Coral Reefs: Australian Institute of Marine Science, Brian Clouston, Canberra, p. 69–106.
———, AND HOPLEY, DAVID, 1983, Growth fabrics and growth rates of Holocene reefs in the Great Barrier Reef: BMR Journal of Australian Geology & Geophysics, v. 8, p. 237–251.
———, AND MARSHALL, J. F., 1985, Halimeda bioherms–Low energy reefs, northern Great Barrier Reef: Proceedings of the Fifth International Coral Reef Symposium, Tahiti, v. 5, p. 1–7.
———, ———, AND HOPLEY, DAVID, 1985, Relationships between reef growth and sea level rise in the Great Barrier Reef: Proceedings of the Fifth International Coral Reef Symposium, Tahiti, v. 3, p. 95–103.
———, AND MONTAGGIONI, LUCIEN, 1985, Reef growth and sea-level change: The environmental signature: Proceedings of the Fifth International Coral Reef Symposium, Tahiti, v. 3, p. 477–515.
———, SYMONDS, P. A., AND OTHERS, 1989, RIG SEISMIC Research Cruises 4 and 5: Northeast Australia post-cruise report: Bureau of Mineral Resources, Australia, Report 281, 101 p.
———, ———, FEARY, D. A., AND PIGRAM, C. J., 1987, Horizontal plate motion: A key allocyclic factor in the evolution of the Great Barrier Reef: Science, v. 238, p. 1697–1700.
———, ———, ———, AND ———, 1988, Facies models in ex-

ploration–The carbonate platforms of northeast Australia: The APEA Journal, v. 28, p. 123–143.

DONE, T. J., 1982, Patterns in the distributions of coral communities across the central Great Barrier Reef: Coral Reefs, v. 1, p. 95–108.

DUNCAN, R. A., 1981, Hotspots in the southern oceans–An absolute frame of reference for motion of the Gondwana continents, in Solomon, S. C., Van der Voo, R., and Chinnery, M. A., eds., Quantitative Methods of Assessing Plate Motions: Tectonophysics, v. 74, p. 29–42.

EBERLI, G. P., AND GINSBURG, R. N., 1987, Segmentation and coalescence of Cenozoic carbonate platforms, northwestern Great Bahama Bank: Geology, v. 15, p. 75–79.

EL HADDAD, A., AISSAOUI, D. M., AND SOLIMAN, M. A., 1984, Mixed carbonate-siliciclastic sedimentation on a Miocene fault-block, Gulf of Suez, Egypt: Sedimentary Geology, v. 37, p. 185–202.

ELMSTROM, K. M., AND KENNETT, J. P., 1985, Late Neogene paleoceanographic evolution of Site 590: Southwest Pacific, in Kennett, J. P., von der Borch, C. C., and others, Initial Reports of the Deep Sea Drilling Project, v. 40: U.S. Government Printing Office, Washington, D.C., p. 1361–1381.

ERICSON, E. K., 1976, The Capricorn Basin, in Leslie, R. B., Evans, H. J., and Knight, C. L., eds., Economic Geology of Australia and Papua New Guinea, v. 3, Petroleum: Australasian Institute of Mining and Metallurgy, Monograph 7, p. 464–473.

EWING, MAURICE, HAWKINS, L. V., AND LUDWIG, W. J., 1970, Crustal structure of the Coral Sea: Journal of Geophysical Research, v. 75, p. 1953–1962.

FALVEY, D. A., 1972, The nature and origin of marginal plateaux and adjacent ocean basins off northern Australia: Unpublished Ph.D. Dissertation, University of New South Wales, Sydney, Australia, 239 p.

———, AND DEIGHTON, IAN, 1982, Recent advances in burial and thermal geohistory analysis: The APEA Journal, v. 22, p. 65–81.

———, AND TAYLOR, L. W. H., 1974, Queensland Plateau and Coral Sea Basin: Structural and time-stratigraphic patterns: Bulletin of the Australian Society of Exploration Geophysicists, v. 5, p. 123–126.

FRAKES, L. A., 1979, Climates throughout geologic time: Elsevier, Amsterdam, 310 p.

GARDNER, J. F., 1970, Submarine geology of the western Coral Sea: Geological Society of America Bulletin, v. 81, p. 2599–2614.

GINSBURG, R. N., AND JAMES, N. P., 1974, Holocene carbonate sediments of continental shelves, in Burk, C. A., and Drake, C. L., eds., Continental Margins: Springer-Verlag, Berlin, 137–154.

HAQ, B. U., HARDENBOL, JAN, AND VAIL, P. R., 1987, Chronology of fluctuating sea levels since the Triassic: Science, v. 235, p. 1156–1166.

HAYES, D. E., AND RINGIS, J., 1973, Seafloor spreading in the Tasman Sea: Nature, v. 243, p. 454–458.

HOLLISTER, C. D., EWING, J. I., AND OTHERS, 1972, Initial Reports of the Deep Sea Drilling Program, v. 11: U.S. Government Printing Office, Washington, D.C., 1077 p.

HOPLEY, DAVID, 1982, The Geomorphology of the Great Barrier Reef: Quaternary Development of Coral Reefs: Wiley, New York, 453 p.

HSU, K. J., MCKENZIE, J. A., OBERHANSLI, HEDI, WEISSERT, HELMUT, AND WRIGHT, R. C., 1984, South Atlantic Cenozoic paleoceanography, in Hsu, K. J., LaBrecque, J. L., and others, Initial Reports of the Deep Sea Drilling Project, v. 73: U.S. Government Printing Office, Washington, D.C., p. 771–785.

IDNURM, MART, 1985, Late Mesozoic and Cenozoic palaeomagnetism of Australia–II. Implications for geomagnetism and true polar wander: Geophysical Journal of the Royal Astronomical Society, v. 83, p. 419–433.

———, 1986, Late Mesozoic and Cenozoic palaeomagnetism of Australia–III. Bias-corrected pole paths for Australia, Antarctica and India: Geophysical Journal of the Royal Astronomical Society, v. 86, p. 277–287.

JANSA, L. F., 1981, Mesozoic carbonate platforms and banks of the eastern North American margin: Marine Geology, v. 44, p. 97–117.

KAUFFMAN, E. G., 1984, Paleobiogeography and evolutionary response dynamic in the Cretaceous Western Interior Seaway of North America, in Westermann, G. E. G., ed., Jurassic-Cretaceous Biochronology and Paleogeography of North America: Geological Association of Canada Special Paper 27, p. 273–306.

KEANY, JOHN, 1978, Paleoclimatic trends in Early and Middle Pliocene deep-sea sediments of the Antarctic: Marine Micropaleontology, v. 3, p. 35–49.

KEMP, E. M., 1978: Tertiary climatic evolution and vegetation history in the southeast Indian Ocean region: Palaeogeography, Palaeoclimatology, Palaeoecology, v. 24, p. 169–208.

KENNETT, J. P., 1977, Cenozoic evolution of Antarctic glaciation, the circum-Antarctic Ocean, and their impact on global paleoceanography: Journal of Geophysical Research, v. 82, p. 3843–3860.

———, 1985, Miocene to early Pliocene oxygen and carbon isotope stratigraphy in the southwest Pacific, Deep Sea Drilling Project Leg 90, in Kennett, J. P., von der Borch, C. C., and others, Initial Reports of the Deep Sea Drilling Project, v. 40: U.S. Government Printing Office, Washington, D.C., p. 1383–1411.

———, HOUTZ, R. E., ANDREWS, P. B., EDWARDS, A. R., GOSTIN, V. A., HAJOS, M., HAMPTON, M., JENKINS, D. G., MARGOLIS, S. V., OVENSHINE, A. T., PERCH-NIELSON, K., 1975, Cenozoic paleoceanography in the southwest Pacific Ocean, Antarctic glaciation, and the development of the circum-Antarctic current, in Kennett, J. P., Houtz, R. E., and others, Initial Reports of the Deep Sea Drilling Project, v. 29: U.S. Government Printing Office, Washington, D.C., p. 1155–1169.

———, KELLER, G., AND SRINIVASAN, M. S., 1985, Miocene planktonic foraminiferal biogeography and paleoceanographic development of the Indo-Pacific region, in Kennett, J. P., ed., The Miocene Ocean: Paleoceanography and Biogeography: Geological Society of America Memoir 163, p. 197–236.

———, AND VON DER BORCH, C. C., 1985, Southwest Pacific Cenozoic paleoceanography, in Kennet, J. P., von der Borch, C. C., and others, Initial Reports of the Deep Sea Drilling Project, v. 40: U.S. Government Printing Office, Washington, D.C., p. 1493–1517.

KINSEY, D. W., AND DAVIES, P. J., 1979, Effects of elevated nitrogen and phosphorus on coral reef growth: Limnology and Oceanography, v. 24, p. 935–940.

LLOYD, A. R., 1973, Foraminifera of the Great Barrier Reef bores, in Jones, O. A., and Endean, R., eds., Biology and Geology of Coral Reefs, v. 1, Geology 1: Academic Press, New York, p. 347–366.

LOCKER, S., AND MARTINI, E., 1985, Phytoliths from the southwest Pacific, Site 591, in Kennett, J. P., von der Borch, C. C., and others, Initial Reports of the Deep Sea Drilling Project, v. 40: U.S. Government Printing Office, Washington, D.C., 1079–1084.

LOUTIT, T. S., KENNETT, J. P., AND SAVIN, S. M., 1983, Miocene equatorial and southwest Pacific paleoceanography from stable isotope evidence: Marine Micropaleontology, v. 8, p. 215–233.

MARSHALL, J. F., 1977, Marine geology of the Capricorn Channel area: Bureau of Mineral Resources, Australia, Bulletin 163, 81 p.

———, 1983, The Pleistocene foundations of the Great Barrier Reef, in Baker, J. T., Carter, R. M., Sammarco, P. W., and Stark, K. P., eds., Proceedings, Inaugural Great Barrier Reef Conference: James Cook University Press, Townsville, p. 123–128.

———, AND DAVIES, P. J., 1978, Skeletal carbonate variation on the continental shelf of eastern Australia: BMR Journal of Australian Geology & Geophysics, v. 3, p. 85–92.

———, AND ———, 1982, Internal structure and Holocene evolution of One Tree Reef, southern Great Barrier Reef: Coral Reefs, v. 1, p. 21–28.

———, AND ———, 1984, Last interglacial reef growth beneath modern reefs in the southern Great Barrier Reef: Nature, v. 307, p. 44–46.

MAXWELL, W. G. H., 1968, Atlas of the Great Barrier Reef: Elsevier, Amsterdam, 258 p.

MERCER, J. H., 1976, Glacial history of southernmost South America: Quaternary Research, v. 6, p. 125–166.

MURPHY, M. G., AND KENNETT, J. P., 1985, Development of latitudinal thermal gradients during the Oligocene: Oxygen-isotope evidence from the southwest Pacific, in Kennett, J. P., von der Borch, C. C., and others, Initial Reports of the Deep Sea Drilling Project, v. 40: U.S. Government Printing Office, Washington, D.C., p. 1347–1360.

MURRAY, J. W., 1973, Distribution and Ecology of Living Benthic Foraminiferids: Heinemann, London, 274 p.

MUTTER, J. C., 1977, The Queensland Plateau: Bureau of Mineral Resources, Geology and Geophysics, Bulletin 179, 55 p.

———, AND KARNER, G. D., 1980, The continental margin off northeast Australia, in Henderson, R. A., and Stephenson, P. J., eds., The Geology and Geophysics of Northeast Australia: Geological Society of Australia, Queensland Division, Brisbane, p. 47–69.

OPPEL, T. W., 1969, Tenneco-Signal Anchor Cay Number 1 Offshore Queensland, Well Completion Report: Tenneco Oil Company, Houston, 15 p.

ORME, G. R., 1977, The Coral Sea Plateau–A major reef province, in Jones, O. A., and Endean, R., eds., Biology and Geology of Coral Reefs, v. 4: Academic Press, New York, p. 267–306.

PALMIERI, V., 1971, Tertiary subsurface biostratigraphy of the Capricorn Basin: Geological Survey of Queensland, Report 52, 18 p.

———, 1974, Correlation and environmental trends of the subsurface Tertiary Capricorn Basin: Geological Survey of Queensland, Report 86, 14 p.

PIGRAM, C. J., AND DAVIES, H. L., 1987, Terranes and the accretion history of the New Guinea Orogen: BMR Journal of Australian Geology & Geophysics, v. 10, p. 193–211.

PINCHIN, J., AND HUDSPETH, J. W., 1975, The Queensland Trough: Its petroleum potential based on some recent geophysical results: The APEA Journal, v. 15, p. 21–31.

READ, J. F., 1985, Carbonate platform facies models: American Association of Petroleum Geologists Bulletin, v. 69, p. 1–21.

RIGGS, S. R., 1984, Paleoceanographic model of Neogene phosphorite deposition, U.S. Atlantic continental margin: Science, v. 223, p. 123–131.

ROBERTSON RESEARCH (AUSTRALIA) PROPRIETARY LTD, 1984, Petroleum Potential of the Papuan Basin, Papua New Guinea: Geological Survey of Papua New Guinea, Unpublished Report, 177 p.

SAVIN, S. M., ABEL, LINDA, BARRERA, ENRIQUETA, HODELL, DAVID, KENNETT, J. P., MURPHY, MARGARET, KELLER, GERTA, KILLINGLEY, JOHN, AND VINCENT, EDITH, 1985, The evolution of Miocene surface and near-surface marine temperatures: Oxygen isotopic evidence, in Kennett, J. P., ed., The Miocene Ocean: Paleoceanography and Biogeography: Geological Society of America Memoir 163, p. 49–82.

———, DOUGLAS, R. G., AND STEHLI, F. G., 1975, Tertiary marine paleotemperatures: Geological Society of America Bulletin, v. 86, p. 1499–1510.

SCHLAGER, WOLFGANG, AND GINSBURG, R. N., 1981, Bahama carbonate platforms–The deep and the past, in Cita, M. B., and Ryan, W. B. F., eds., Carbonate Platforms of the Passive-Type Continental Margins, Present and Past: Marine Geology, v. 44, p. 1–24.

SCOTT, R. W., AND GOVEAN, F. M., 1985, Early depositional history of a rift basin: Miocene in Western Sinai: Palaeogeography, Palaeoclimatology, Palaeoecology, v. 52, p. 143–158.

SHACKLETON, N. J., 1986, Paleogene stable isotope events: Palaeogeography, Palaeoclimatology, Palaeoecology, v. 57, p. 91–102.

———, AND KENNETT, J. P., 1975, Palaeotemperature history of the Cenozoic and the initiation of Antarctic glaciation: Oxygen and carbon isotope analyses in DSDP Sites 277, 279, and 281, in Kennett, J. P., Houtz, R. E., and others, Initial Reports of the Deep Sea Drilling Project, v. 29: U.S. Government Printing Office, Washington, D.C., p. 743–755.

SHAW, R. D., 1978, Sea floor spreading in the Tasman Sea: A Lord Howe Rise-eastern Australian reconstruction: Australian Society of Exploration Geophysicists Bulletin, v. 9, p. 75–81.

SHERIDAN, R. E., CROSBY, J. T., BRYAN, G. M., AND STOFFA, P. L., 1981, Stratigraphy and structure of southern Blake Plateau, northern Florida Straits, and northern Bahama Platform from multichannel seismic reflection data: American Association of Petroleum Geologists Bulletin, v. 65, p. 2571–2593.

STEIN, R., AND ROBERT, C., 1985, Siliciclastic sediments at Sites 588, 590 and 591: Neogene and Paleogene evolution in the southwest Pacific and Australian climate, in Kennett, J. P., von der Borch, C. C., and others, Initial Reports of the Deep Sea Drilling Project, v. 40: U.S. Government Printing Office, Washington, D.C., p. 1437–1455.

SYMONDS, P. A., 1983, Relation between continental shelf and margin development–Central and northern Great Barrier Reef, in Baker, J. T., Carter, R. M., Sammarco, P. W., and Stark, K. P., eds., Proceedings, Inaugural Great Barrier Reef Conference: James Cook University Press, Townsville, p. 151–157.

———, DAVIES, P. J., AND PARISI, ALFIO, 1983, Structure and stratigraphy of the central Great Barrier Reef: BMR Journal of Australian Geology & Geophysics, v. 8, p. 277–291.

———, FRITSCH, J., AND SCHLUTER, H.-U., 1984, Continental margin around the western Coral Sea Basin: Structural elements, seismic sequences and petroleum geological aspects, in Watson, S. T., ed., Transactions of the Third Circum-Pacific Energy and Mineral Resources Conference, Hawaii: American Association of Petroleum Geologists, Tulsa, p. 243–252.

TALLIS, N. C., 1975, Development of the Tertiary offshore Papuan Basin: The APEA Journal, v. 15, p. 55–60.

TANNER, J. J., 1969, The ancestral Great Barrier Reef in the Gulf of Papua: U.N. Economic Commission for Asia and the Far East, Mineral Resources Development Series, v. 41, p. 283.

TAYLOR, L. W. H., 1975, Depositional and tectonic patterns in the western Coral Sea: Bulletin of the Australian Society of Exploration Geophysicists, v. 6, p. 33–35.

———, 1977, The western Coral Sea: Sedimentation and tectonics. Unpublished Ph.D. Dissertation, University of Sydney, Australia, 193 p.

———, AND FALVEY, DAVID, 1977, Queensland Plateau and Coral Sea Basin: Stratigraphy, structure and tectonics: The APEA Journal, v. 17, p. 13–29.

VAN HINTE, J. E., 1978, Geohistory analysis–Application of Micropalaeontology in exploration geology: American Association of Petroleum Geologists Bulletin, v. 62, p. 201–222.

VEEH, H. H., AND VEEVERS, J. J., 1970, Sea level at −175 m off the Great Barrier Reef 13,600 to 17,000 years ago: Nature, v. 226, p. 536–537.

WEISSEL, J. K., AND HAYES, D. E., 1977, Evolution of the Tasman Sea reappraised: Earth and Planetary Science Letters, v. 36, p. 77–84.

———, AND WATTS, A. B. 1979, Tectonic evolution of the Coral Sea Basin: Journal of Geophysical Research, v. 84, p. 4572–4582.

WEISSERT, H. J., MCKENZIE, J. A., WRIGHT, R. C., CLARK, M., OBERHANSLI, H., AND CASEY, M., 1984, Paleoclimatic record of the Pliocene at Deep Sea Drilling Project Sites 519, 521, 522, and 523 (Central South Atlantic), in Hsu, K. J., LaBrecque, J. L., and others, Initial Reports of the Deep Sea Drilling Project, v. 73: U.S. Government Printing Office, Washington, D.C., p. 701–715.

WELLMAN, PETER, 1983, Hotspot volcanism in Australia and New Zealand: Cainozoic and mid-Mesozoic: Tectonophysics, v. 96, p. 225–243.

WILSON, J. L., 1975, Carbonate Facies in Geologic History: Springer-Verlag, Berlin, 471 p.

RECENT CARBONATE SLOPE SEDIMENTS AND SEDIMENTARY PROCESSES BORDERING A NON-RIMMED PLATFORM: SOUTHWEST FLORIDA CONTINENTAL MARGIN

GREGG R. BROOKS

Department of Marine Science, University of South Florida, St. Petersburg, Florida 33701

AND

CHARLES W. HOLMES

U.S. Geological Survey, Federal Center, Denver, Colorado 80225

ABSTRACT: The southwest Florida continental slope is part of the vast, non-rimmed west Florida carbonate platform. Bordering the southern Florida Straits, it has been the site of a thick accumulation of seaward-prograding sediments throughout at least the Late Quaternary. High-resolution seismic-reflection data and sediment cores were collected in order to determine: (1) provenance and dominant depositional processes; (2) how slope development has related to high-frequency sea-level fluctuations; and (3) how deposits compare with other carbonate slope environments.

Nine seismic sequences have been identified, each bounded by an erosional unconformity. All nine sequences consist of seaward-trending prograding clinoforms. Sediments of sequences penetrated by cores (sequences 1, 3, and 5) consist of a mixture of shallow- and deep-water biogenic carbonate sands and muds deposited rapidly (averaging greater than 2.5 m/ka) on the upper slope. Shallow-water sediments are transported to the depositional site by oceanic currents, intermittent storms, and tidal currents sweeping the southern shelf and upper slope. Based upon biostratigraphy, radiocarbon age determinations, and seismic-sequence and facies analysis, depositional patterns are interpreted to be a function of high-frequency sea-level fluctuations operating during the Late Quaternary. Most vigorous offshelf transport occurred during periods in the sea-level cycle when the shelf surface was flooded but shallow, as would occur during early transgressions and late regressions. During sea-level highstands, offshelf transport was less vigorous. During sea-level lowstands, no offshelf transport took place. Erosion of the previously deposited sequence resulted from an increase in erosive capacity of the Florida Current during glacially induced sea-level lowstands.

Four such sequences, collectively attaining a thickness of 330 m, have been deposited during approximately the last 100 ka. This would correspond to a maximum of several tens of meters on most carbonate shelves and is represented on the adjacent west Florida shelf by only a few meters of carbonate sediment accumulation.

Controls on recent margin development are dominated by high-frequency sea-level fluctuations coupled with the physical processes that act to concentrate sediments in the study area, and the lack of a shelf edge rim of reefs, enabling offshelf transport of shallow-water carbonate sediments.

INTRODUCTION

Scope.—

Carbonate platforms have existed throughout most of the geologic record (Schlager, 1981). Until recently, sedimentologic research on modern carbonate slopes has concentrated primarily on those bordering the more common rimmed platforms such as the Bahamas. Only within the past several years have slopes bordering non-rimmed platforms, such as the west Florida continental margin, begun to receive considerable attention.

The southwest Florida continental slope, which borders the southern Florida Straits (Fig. 1), has been the site of an extensive accumulation of sediments in the recent geologic past (Milligan, 1962; Holmes, 1985). Over 300 m of carbonate-rich sediments have been deposited since at least the late Quaternary, a significant accumulation considering only a few meters have accumulated on the adjacent shelf during the same time period (Stockman, Ginsburg and Shinn, 1967; Enos and Perkins, 1977). A detailed investigation of these deposits was initiated by the U.S. Geological Survey and University of South Florida in 1983. The objectives of the study were to determine: (1) sediment provenance and dominant depositional processes, and to what extent they control overall margin development; (2) the relation of these deposits and response of the slope to sea-level fluctuations; and (3) the differences and similarities of the southwest Florida slope to other carbonate slope environments.

Setting.—

The southwest Florida slope (Fig. 1) is that part of the continental margin that trends east-west and is bordered by the deep Florida Straits to the south, the Pourtales Terrace to the east, the Florida Canyon system to the west, and the west Florida continental shelf to the north. Along the shelf edge, just upslope from the study area, a series of small isolated carbonate banks occurs (Shinn and others, 1982). Two of these banks form groups of islands known as the Marquesas Keys and the Dry Tortugas (Fig. 2). Water depths in the study area range from approximately 100 m at the shelf margin (a rather arbitrary division because there is no pronounced break in slope), to over 1,000 m on the floor of the Florida Straits. The gradient is gentle, averaging less than 1° for the entire slope, but reaching approximately 3° between the 250-m and 500-m isobaths.

The physical environment is dominated by the Florida Current (Fig. 2), an eastward-flowing surface current linking the Loop Current of the Gulf of Mexico with the Gulf Stream. Surface velocities average 100 cm/sec but exceed 250 cm/sec in the core of the flow (Richardson and others, 1969). The Florida Current encounters the bottom to depths of 200 m on the adjacent Pourtales Terrace (Gomberg, 1976), where velocities of as much as 45 cm/sec have been recorded. A westward-flowing countercurrent (Fig. 3), attaining velocities of 20–25 cm/sec, has been observed between 430-m and at least 600-m depths (Stewart, 1962; Brooks and Niiler, 1975). Few data have been gathered below 600 m. Separating these two currents (between 200 m and 430 m) exists a "zone of reversal" over which the flow gradually changes direction (Stewart, 1962; Brooks and Niiler, 1975). Flow velocities are relatively sluggish throughout the entire zone, ranging between 9 and 23 cm/sec.

Superimposed on the dominantly east-west current pattern are north-south currents (Fig. 2) originating on the con-

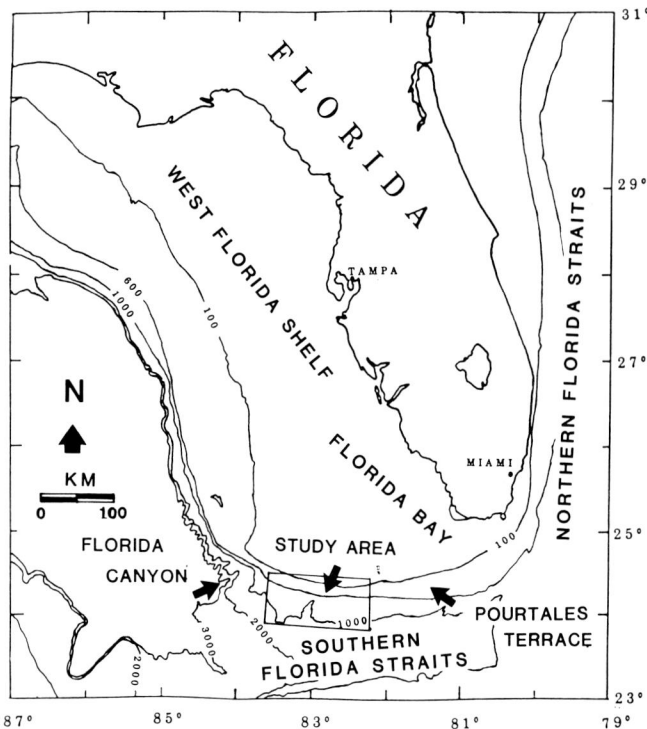

FIG. 1.—Map of southern Florida Straits showing study area.

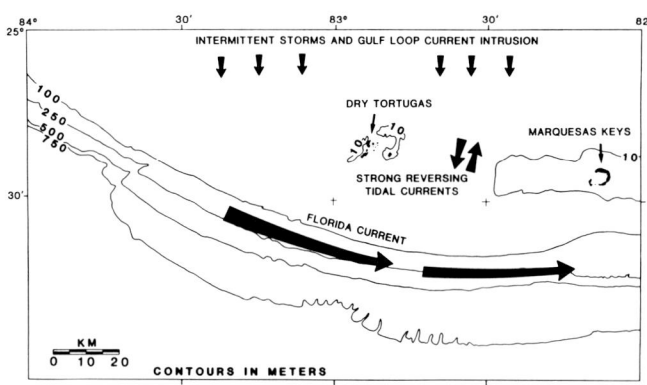

FIG. 2.—Surface currents influencing study area.

tinental shelf to the north. Strong reversing north-south tidal currents have been observed as they are funneled between the shallow carbonate banks along the outer shelf (Shinn and others, 1982). In addition, strong southerly currents develop periodically across the entire width of the shelf in response to Loop Current intrusion onto the shelf (Neurauter, 1980) and the regular passage of storm frontal systems averaging every 5 to 10 days during the winter (Niiler, 1976). Large patches of southward-trending sand waves have been identified across the entire width of the shelf south of 26°N, indicating that these currents have the capacity to transport large quantities of sediments for great distances (Neurauter, 1980; Holmes, 1984).

METHODS

High-resolution seismic-reflection data and eight gravity cores (1–3 m in length) were collected on a 13-day cruise in the study area in 1983 (Fig. 4). Over 1,500 km of seismic lines were shot with a Teledyne mini-sparker powered by a Del Norte 800J power supply, a 100-element hydrophone, and an analog recorder. Depth conversions from travel-time sections were made using seismic velocities (V_p) of 1,500 m/sec in water and 1,675 m/sec in sediments. The latter was based upon averaging a value of 1,650 m/sec calculated for Holocene south Florida shelf sediments (Enos and Perkins, 1977), and a value of 1,700 m/sec calculated for mixed carbonate and siliciclastic sands and muds on the North Carolina continental margin (S. W. Snyder and A. C. Hine, pers. commun., 1986).

Sediment cores were sampled at least three times (top, middle, and bottom) for subsequent analyses. The sampling interval was based upon core lengths and the lack of visual lithologic breaks.

Textural analysis was performed by the seive-and-pipette method (Folk, 1965). Calcium carbonate content was determined by acid leaching (Milliman, 1974). Relative percentages of low-Mg calcite, high-Mg calcite, and aragonite were determined for all bulk samples and the mud-size fraction of selected samples by X-ray diffraction (Milliman, 1974). The ratio of peak areas was used to determine percentage aragonite versus calcite (both high and low Mg). The ratio of peak heights was used to determine percentage low-Mg versus high-Mg calcite. The relative percentages of each were read from a standard curve (Milliman, 1974; Boardman, 1978) prepared for this study using pure biogenic carbonates (all species characteristic of the west Florida margin) as standards. Sediment constituent composition for the sand-size fraction was determined by visual estimation under the light microscope. Composition of the mud-size (<63 μm) fractions of selected samples was determined using the Scanning Electron Microscope (SEM). Strontium concentrations of each sample were determined by atomic absorption methods. Biostratigraphic analysis was performed on top, middle, and bottom samples from each core; fossil assemblages were used to determine the approximate age and environment of deposition of the sedi-

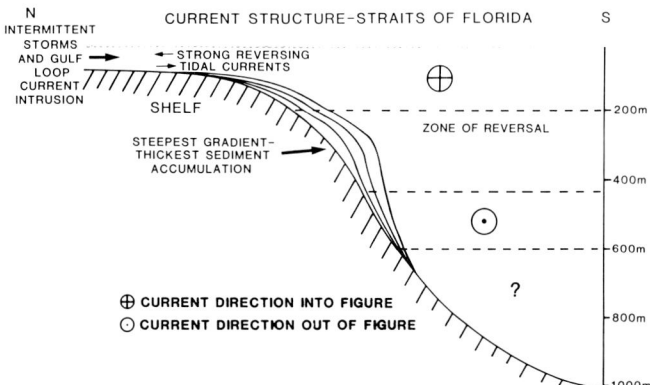

FIG. 3.—Vertical current structure in Florida Straits.

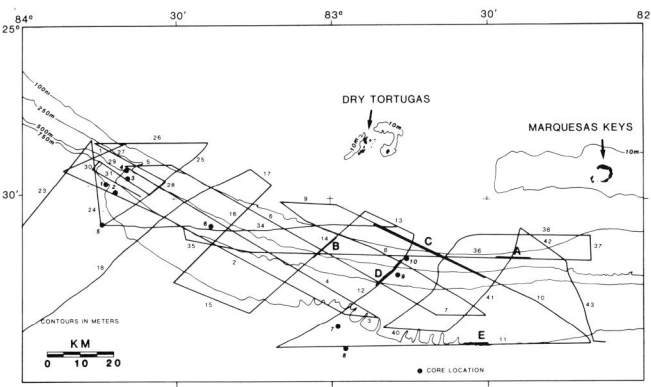

FIG. 4.—Seismic track lines and core sites in study area (A = site of Fig. 5; B = site of Fig. 6; C = site of Fig. 7; D = site of Fig. 8; E = site of Fig. 9).

ments. Bottom samples from each core were dated by radiocarbon methods.

RESULTS AND DISCUSSION

Seismic data.—

The basal reflector identified in the study area is an irregular surface (Fig. 5) that rises from a depth of approximately 700 m to crop out in the easternmost part of the study area at a depth of approximately 200 m to form the Pourtales Terrace (Fig. 1), a Miocene phosphatic limestone (Gomberg, 1976). This reflector cannot be traced very far westward because of signal attenuation caused by the increased thickness of overlying sediments, but previous studies show that it rises and crops out again on the upper slope northwest of the study area (Holmes, 1981), thereby forming a broad reentrant. The extreme irregularity of the reflecting surface and its continuity as it rises to form the Pourtales Terrace suggest that it is an erosional feature. It is doubtful that faulting is a control as no evidence of faulting, such as the vertical offset of reflectors, was identified. The erosional mechanism is unknown, but it would have to be an event of sufficient magnitude to scour out a 500-m-thick sequence of sediments.

Overlying the basal reflector is a thick sequence of post-Miocene sediments. Thickest deposits have accumulated between the 250-m and 500-m isobaths. Nine seismic sequences have been identified. Individual sequences were defined by reflector terminations, interpreted here as erosional unconformities, at both upper and lower boundaries (Figs. 5–8). Unconformable surfaces are generally restricted to upper slope (<400 m below present sea level) regions. Deeper on the slope, sequence boundaries become conformable. Unconformable sequence boundaries are not present on all seismic profiles, at which times correlative conformities are used to deliniate sequence boundaries (Figs. 5–8). Several local erosional unconformities have been identified but are too localized to be mapped. Such features may have been caused by localized mass wasting or current scour.

Each sequence consists of seaward-trending prograding clinoforms (Figs. 6–8), indicating offshelf transport of material from the shallow shelf to the north. In some sequences, and where resolution permits, there is a vertical gradation in seismic facies, which we interpret to be oblique prograding clinoforms at the base grading upward into sigmoidal prograding clinoforms (Figs. 6–8). Where no gradation exists, either oblique or sigmoidal forms dominate. Oblique prograding clinoforms imply rapid sediment accumulation in a relatively high-energy regime, whereas sigmoidal prograding clinoforms imply a slowing in sedimentation rate and a decreasing energy regime (Mitchum and others, 1977). Internal reflectors are dominantly continuous and parallel to subparallel. Hummocky to chaotic and reflection-free configurations occur locally (Fig. 7) and in some cases are interpreted to represent localized mass-wasting deposits.

Gullies have been identified in all nine sequences, primarily occurring on the lower slope between 500 m and 900 m below present sea level (Fig. 9). They range from 5 m to 20 m in relief and are generally less than 1 km across. Gullies, commonly known as the Agassiz Valleys (Minter and others, 1975), are currently being infilled by prograding sediments from the upper slope. Their origin is uncertain, but they may have been formed by submarine slope failure and gravity flows initiated on the upper slope. Numerous small submarine canyons on the upper carbonate slope north of Little Bahama Bank have been described by Mullins and others (1984) to have developed by these processes during the progradation of upper-slope periplatform oozes. Down section, gullies become more numerous and appear to head higher on the slope (Table 1), thereby suggesting that gully-forming processes have become less active throughout development of the slope. Modern burial of the gullies by prograding upper-slope sediments and direct observations of sediment fill in lower portions of some of the valleys (Minter and others, 1975) suggest that gullies are presently inactive.

Mounded configurations were identified on upper-slope sections of sequences 1, 4, and 5 (Table 1; Fig. 5). On the basis of the reflection-free or chaotic internal configurations, as well as their location along the bank edge, mounds are interpreted to be reefs. Holmes (1985) described similar bank edge mounded features in the Florida Straits as reefs attempting to migrate upslope to maintain growth during rising sea level.

All sequences have a lobate external form and appear to be radiating seaward from the shallow shelf to the north (Fig. 10). As shown in Table 1, sequences have similar characteristics, such as internal reflection configurations and erosional patterns. Thickest accumulations are confined to a relatively small portion of each sequence that represents a locus of deposition. Each succeeding depositional center is offset in an easterly direction from that of the underlying sequence (Table 1; Fig. 10). The depositional center of sequence 9, the lowermost sequence, therefore, is located in the westernmost portion of the study area immediately adjacent to the Florida Canyon, whereas the depositional center of sequence 1, the modern sequence, is located in the

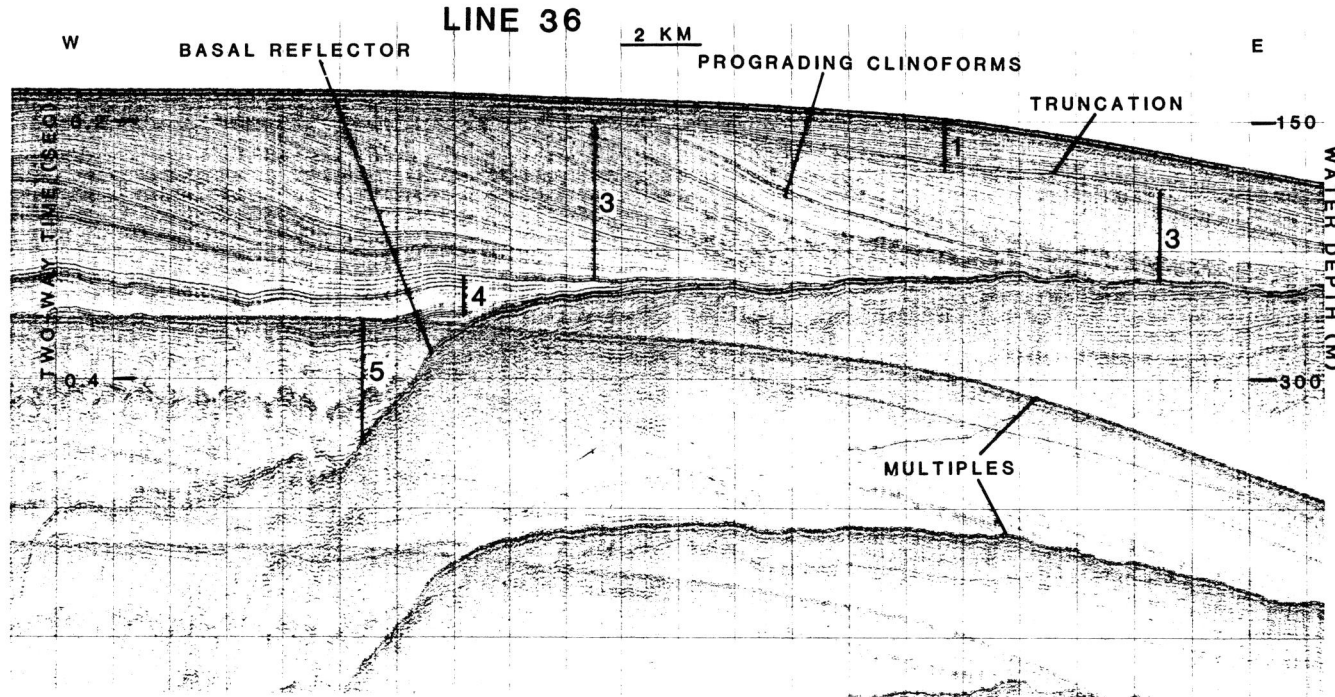

Fig. 5.—Seismic profile of line 36 showing basal reflector, seismic sequences, and prograding clinoforms (site A on Fig. 4).

easternmost portion of the study area where modern sediments are currently prograding over the Pourtales Terrace (Fig. 5).

Sediments.—

Three of the nine sequences identified were penetrated by cores. Sequence 5 was penetrated by core 5, sequence 3 was penetrated by core 6, and sequence 1, the modern sequence, was penetrated by cores 1, 4, 7, 8, 9, and 10. Since cores were all less than 3 m in length, the stratigraphic relations between sediments and seismic facies within each sequence will not be attempted. The sedimentary characteristics and relations of sequences 5, 3, and 1 will be discussed as well as the areal distribution of sediments composing sequence 1.

Core X-radiographs show sediments from most cores have a similar pattern of structures. Figure 11 shows a typical pattern of wispy cross-bedded and wavy parallel-bedded

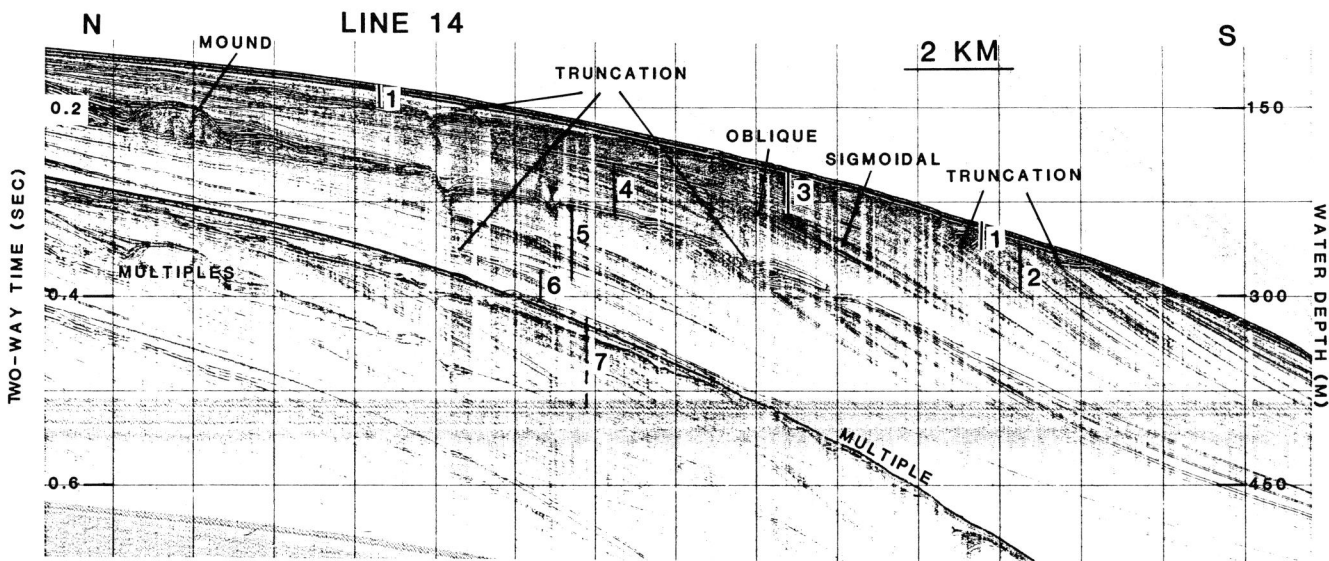

Fig. 6.—Seismic profile of line 14 showing truncated reflectors defining erosional unconformities bounding seismic sequences, and seaward-trending prograding clinoforms (site B on Fig. 4).

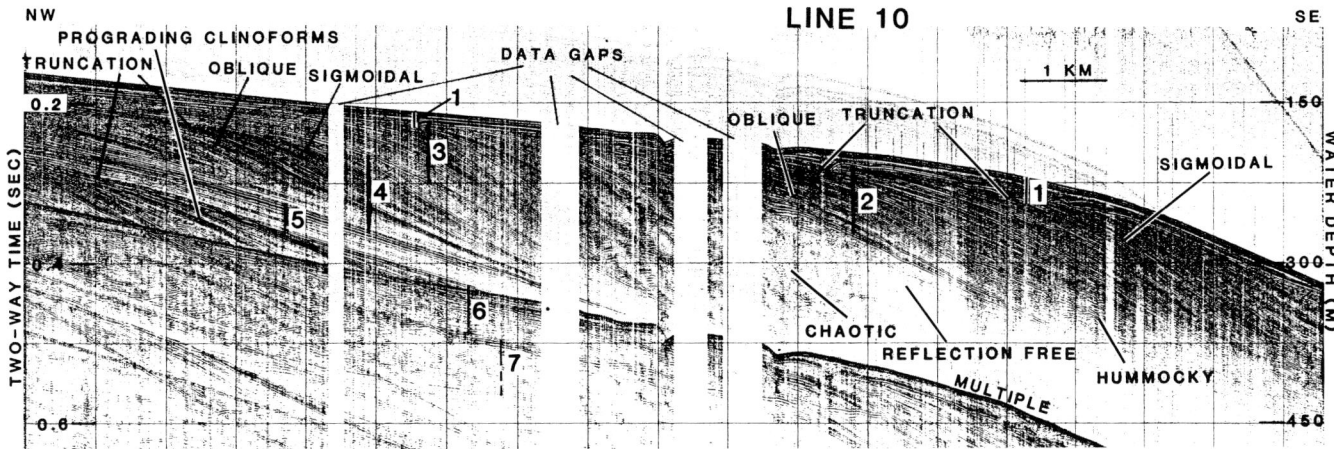

FIG. 7.—Seismic profile of line 10 showing seismic sequences bounded by erosional unconformities and seaward-trending prograding clinoforms composing seismic sequences (site C on Fig. 4).

structures concentrated in the bottom 50 cm of the core. Wispy cross-bedded and wavy parallel-bedded sediments have coarse (sand/silt) layers 2–5 cm thick intercalated with layers of finer, mud-size material. Sediments in the top 1–2 m of core consist of homogeneous to mottled muds with numerous burrows indicative of heavy bioturbation. Sediments in the middle sections of the core are transitional and therefore represent a gradation between the two structure types. Hill (1984) described wispy cross-bedded and wavy parallel-bedded structures as resulting from the same flow conditions, but accumulating at higher rates than bioturbated muds. Bouma (1973) and Walker (1984), on the other

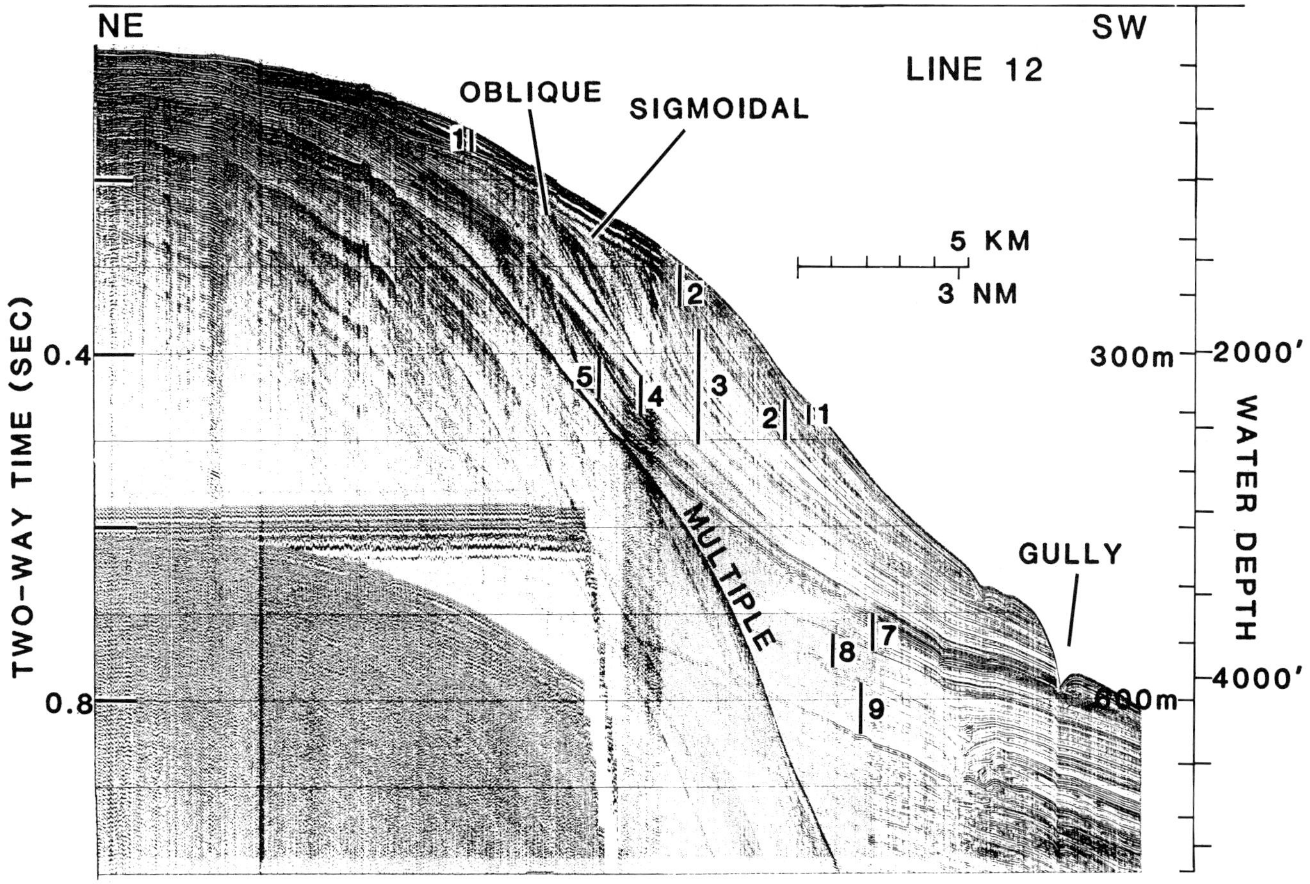

FIG. 8.—Seismic profile of line 12 showing seismic sequences bounded by erosional unconformities and upward gradation of oblique to sigmoidal prograding clinoforms within sequences (site D on Fig. 4).

TABLE 1.—MAJOR CHARACTERISTICS OF EACH SEISMIC SEQUENCE.

Seismic Sequence#	Maximum Thickness	Location of Depositiontal Center	Reflection Configurations	Outcrop Locations	Outcrop Features	Interpretations
1	50 m	Due south of Marquesas Keys overlying Pourtales Terrace	Continuous and parallel to sub-parallel; high-angle parallel-oblique prograding clinoforms occur locally	Continuous along the 150- to 200-m isobaths	70–Km-long scarp following 100-m isobath; compressional folds on upper slope, mounds (reefs?) along the 60- to 70-m isobaths	High rate of deposition in a relatively high-energy environment; burial of Pourtales Terrace continuing
2	120 m	South-southwest of Marquesas Keys	Parallel and continuous, strongly reflective in downslope segments; reflection free, hummocky, and chaotic locally; abundant in shallow easternmost section of sequence	Lower slope	Erosion in western portion of sequence; oblique prograding clinoforms and chaotic units in Pourtales Terrace area, lower slope; gullies present but appear to be less developed than those of underlying sequences	Rapid high-energy deposition in Pourtales Terrace area
3	100 m	South-southwest of Marquesas Keys	Continuous and parallel to sub-parallel; highly reflective prograding clinoforms with truncated topsets occur locally	Patchy throughout study area	Heavily eroded on western portion; chaotic units in Pourtales Terrace region; lower-slope gullies present but appear to be less developed than in underlying sequences	Rapid high-energy deposition in depositional center; mass wasting uncommon; current erosion dominant erosional process
4	60 m	South-southeast of Dry Tortugas	Sub-parallel and discontinuous, with locally hummocky areas; oblique prograding clinoforms common; downlap onto Pourtales Terrace	Broad zone southeast of Dry Tortugas	Heavily eroded on western portion; mounds (reefs?) on upper slope; gullies present but head farther downslope and are less concentrated than those of underlying sequences	Rapid high-energy deposition in depositional center; mass wasting common in shallow central portion of study area, the lower-slope gulley field, and the Florida Canyon region; beginning of burial of Pourtales Terrace
5	120 m	Due south of Dry Tortugas	Continuous and parallel to sub-parallel; chaotic locally in thickest portion	Narrow (5–10 km) band from the Florida Canyon region to the gullied lower slope	Heavily eroded on western portion and gullied lower slope; mound (reef?) on upper slope; gullies common on lower slope	Rapid high-energy deposition in depositional center; mass wasting common on lower slope and in thickest portion of sequence
6	90 m	South-southwest of Dry Tortugas	Continuous and parallel to sub-parallel; reflection free, hummocky, and chaotic in Florida Canyon and gullied lower slope areas	Narrow (5–10 km) band along lower slope	Heavily eroded on western portion and gullied lower slope	Rapid high-energy deposition in depositional center; erosion by bottom current and mass wasting common in Florida Canyon and lower slope
7	55 m	South-southwest of Dry Tortugas	Continuous and parallel to sub-parallel; chaotic locally on lower slope east of gully field	Florida Canyon region	Heavily eroded on western portion and gullied lower slope	Rapid high-energy deposition in depositional center; erosion by bottom currents and mass wasting common in Florida Canyon and lower slope
8	50 m	South-southwest of Dry Tortugas	Continuous and parallel to sub-parallel; hummocky on lower slope and reflection free in Florida Canyon area	Florida Canyon and gullied	Eroded on western portion and gullied lower slope	Rapid high-energy deposition in depositional center; erosion by bottom currents and mass wasting common in Florida Canyon and lower slope
9	160 m	Florida Canyon area	Discontinuous and parallel to sub-parallel; reflection-free units in Florida Canyon region	Florida Canyon area	Eroded on western portion; lower-slope gullies well developed	Rapid high-energy deposition in Florida Canyon region

hand, called these structures "current structure" and described them as forming in the presence of bottom currents. We interpret the upcore pattern of sedimentary structures as representing both a relative decrease in sedimentation rate and depositional energy. Fine sand and mud-size sediments in the bottom sections of core were deposited in the presence of bottom currents. Bioturbated muds identified in the top sections of core were deposited under lower flow conditions and at slower rates, thereby allowing intense bioturbation to take place.

Results of textural analysis (Table 2) show sand-size material ranges from 8 to 74 percent of the sample, with an

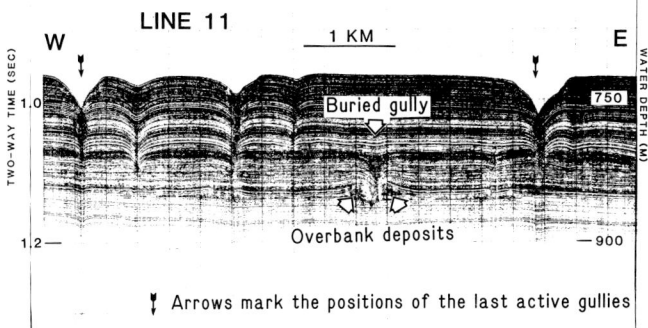

Fig. 9.—Seismic profile of line 11 showing lower slope gullies (site E on Fig. 4).

average of 27 percent. Most is in the fine to very fine sand-size range. Silt-size material composes 20 to 72 percent of the samples, with an average of 54 percent. In most samples, the silt-size fraction composes the bulk of the sediment. Clay-size material composes 6 to 40 percent of the samples, with an average of 19 percent.

Calcium carbonate content ranges from 58 to 99 percent and averages 86 percent of the samples (Table 2). The noncarbonate fraction consists dominantly of quartz with minor amounts of the kaolinite and smectite clay mineral groups. Relative percentages of dominant carbonate minerals are given in Table 2. Aragonite constitutes 34 to 61 percent of the samples with an average of 45 percent. High-Mg calcite (greater than 4 mole percent $MgCO_3$; Bathurst, 1975) composes 10 to 36 percent and averages 23 percent. Low-Mg calcite (less than 4 mole percent $MgCO_3$; Bathurst, 1975) ranges from 13 to 49 percent and averages 32 percent of all samples. Carbonate mineralogy of the mud-only fraction shows similar values to bulk samples.

Strontium concentrations may be used in distinguishing between shallow- and deep-water-derived aragonite (Boardman and others, 1986). Pelagic forms (i.e., pteropods) typically found in deep-sea sediments have consistently low (less than 2,000 ppm) values. Shallow-water forms common to the tops of carbonate platforms have higher values that range from approximately 3,000 ppm to over 10,000 ppm. Values from southwest Florida slope bulk samples range from 1,920 ppm to 4,260 ppm and average 2,370 ppm (Table 2), therefore representing a mixture of shallow-water and deep-water aragonite.

Results of sediment constituent analysis are shown in Table 3. Dominant constituents for the sand-size fraction include planktonic foraminifera, pteropods, sponge spicules, benthic molluscs, and benthic foraminifera. Echinoid fragments make a variable contribution, and ostracods, bryozoa, octacorals, coralline algae, *Halimeda*, and pelletoids contribute minor quantities. Barnacle and rock fragments,

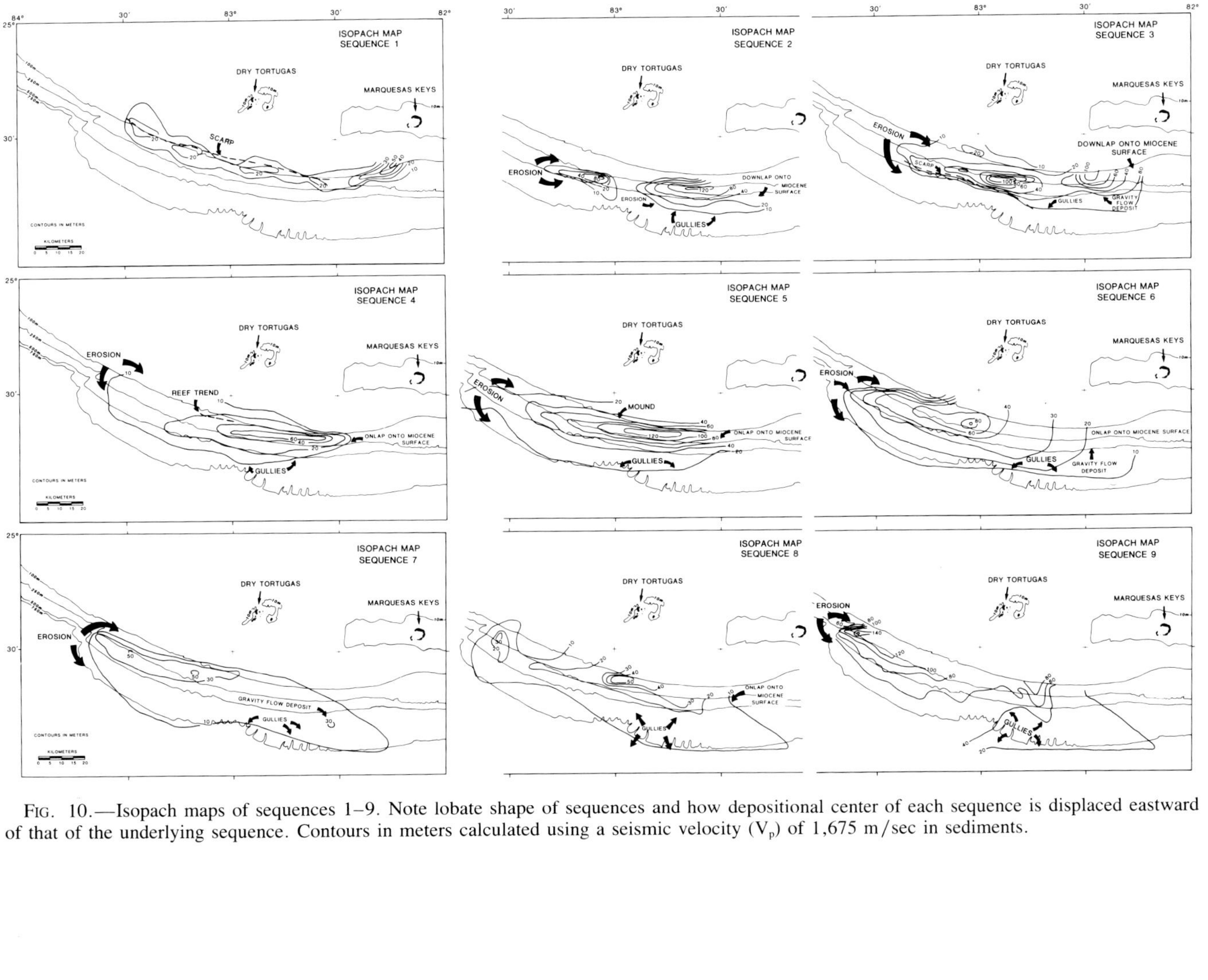

Fig. 10.—Isopach maps of sequences 1–9. Note lobate shape of sequences and how depositional center of each sequence is displaced eastward of that of the underlying sequence. Contours in meters calculated using a seismic velocity (V_p) of 1,675 m/sec in sediments.

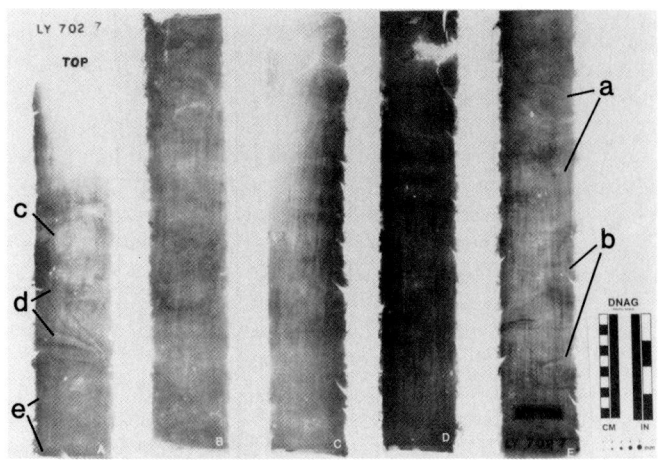

FIG. 11.—X-radiograph of core 7 showing wispy cross-bedded (a) and wavy parallel-bedded (b) structures at the base, and mottled (c), burrowed (d), and homogeneous (e) sediments in the upper section.

blackened grains, and annelid tubes are present in trace amounts. Sponge spicules, benthic molluscs, and benthic foraminifera, all present in appreciable quantities, are derived from shallow-water sources. Mollusc shells are typically fragmented, suggesting that they have undergone transport. Benthic foraminifera consist of a typical shallow-water assemblage dominated by species of *Peneroplis, Archaias, Amphistigina, Elphidium,* and *Nonionella.*

The mud-size fraction consists of aragonite needles, coccoliths, clay minerals, and unidentifiable debris, the relative amounts varying directly with water depth and distance of the sample site from the shelf margin. Clay minerals are identifiable in only a few samples. The presence of aragonite needles is indicative of input from shallow-water sources, because pteropods, the only pelagic source for aragonite, do not degrade into needles (Bathurst, 1975).

Results of sediment texture, calcium carbonate content and mineralogy, strontium concentration, and constituent composition analyses show similar values for sediments collected from sequences 1, 3, and 5, suggesting that sedimentation patterns have not differed substantially during the periods when the three sequences were deposited. In addition, sediments collected from sequence 1, the modern sequence, show a downslope decrease in grain size, calcium carbonate content, aragonite, high-Mg calcite, and strontium (Table 2), as well as sediment components, such as shallow-water molluscs and benthic foraminifera.

Radiocarbon dates from the bases of all cores, with the exception of core 5, show ages of late Pleistocene-Holocene to mid-Holocene. The base of core 5 was dated at >40 ka and, therefore, is beyond the limits for radiocarbon age dating techniques. Biostratigraphic analyses show that cores 1, 4, 8, 9, and 10 are all younger than 11 ka. Co-occurrences of the planktonic foraminifera *Globorotalia fimbriata* and *Globorotalia cultrata* indicate that sediments from these cores were deposited during Ericson zone Z (Ericson and Wollin, 1968), which is restricted to the Holocene. Core 5 is considerably older. Co-occurrences of *Globorotalia tumida flexuosa, Globorotalia ungulata,* and *Globigerina calida calida* indicate that sequence 5 (which was penetrated by core 5) was deposited between 84 and 127 ka during Ericson zone X (Ericson and Wollin, 1968). Sedi-

TABLE 2.—TEXTURE, CALCIUM CARBONATE CONTENT, CARBONATE MINERALOGIES, STRONTIUM CONTENT, SEQUENCE PENETRATED, AND DEPTH OF SEDIMENT CORE SAMPLES

Core#	Depth Down Core (cm)	%Sand	Avg	%Silt	Avg	%Clay	Avg	%CaCo$_3$	Avg	%Arag	Avg	%high-Mg CC	Avg	%low-Mg CC	Avg	[Sr] (ppm)	Avg	Sequence Penetrated
1	10–12	15		67		17		81		47		17		36		2300		
	116–118	17	13	63	68	20	19	74	81	41	43	24	25	35	32	2390	2300	1
	221–223	8		72		20		88		41		34		25		2300		
4	10–12	43		47		10		93		47		24		29		2580		
	49–51	44	45	44	44	12	11	93	93	43	44	27	29	30	27	2500	2600	1
	88–90	49		41		10		94		41		36		23		2720		
5	10–12	40		45		15		92		40		19		41		2040		
	37–39	32		48		20		91		41		19		40		1960		
	64–66	28	26	52	54	21	21	92	90	36	38	29	20	35	41	2200	2018	5
	91–93	15		61		25		94		41		10		49		1970		
	118–120	13		64		24		79		34		24		42		1920		
6	10–12	34		55		11		93		58		18		24		2520		
	49–51	14	27	75	61	11	12	93	93	48	52	25	22	27	25	2670	2607	3
	88–90	34		54		13		93		51		24		25		2630		
7	10–12	12		63		26		90		40		17		43		2320		
	62–64	7		61		32		90		42		18		40		2350		
	110–112	9	9	58	59	34	32	58	75	43	42	21	21	36	37	2400	2322	1
	160–162	7		54		40		62		47		22		31		2370		
	210–212	8		63		30		73		38		27		35		2170		
8	10–12	8		64		28		92		43		17		40		2330		
	49–51	10	10	60	59	31	31	88	89	41	42	18	18	41	39	2260	2293	1
	88–90	12		54		34		87		43		20		37		2290		
9	10–12	56		33		11		83		48		27		25		2770		
	54–56	56		33		12		77		52		23		25		2620		
	99–101	25	37	61	50	14	13	88	82	49	49	24	26	27	25	2290	2446	1
	142–144	24		62		15		89		48		30		22		2310		
	188–190	22		64		14		74		48		25		27		2240		
10	10–12	25		63		13		82		48		27		25		2550		
	92–94	55	51	33	39	12	10	86	89	51	53	29	27	20	19	2890	3233	1
	173–175	74		20		56		99		61		26		13		4260		

TABLE 3.—SEDIMENT CONSTITUENTS—SAND FRACTION

Core#	Depth Down Core (cm)	Planktonic Foraminifera	Pteropods	Sponge	Molluscs (benthic)	Benthic Foraminifera	Echinoids	Octocorals	Bryozoa	Ostracods	Barnacles	Coralline Algae	*Halimeda*	Rock Fragments	Pelletoids	Unidentified
1	10–12	A	A	C	A	C	Tr-C	Tr	—	Tr	—	Tr	—	—	Tr-r	C
	116–118	A	A	A	C	C	Tr	—	—	r	—	—	—	—	Tr-r	C
	221–223	C	C	A	C	r	—	Tr	—	—	—	—	—	—	Tr	C
4	10–12	C	C	C	C	r-C	r	Tr	—	r	—	—	—	—	—	—
	49–51	C	C	C	C	C	r	Tr	Tr	r	—	Tr	—	—	Tr	—
	88–90	C	C	r-C	C	C	r-C	r	Tr	r-C	—	—	—	—	—	—
5	10–12	A	A	A	C	C	r-C	—	—	—	—	—	—	—	—	A
	37–39	A	A	C	A	C	r	Tr	—	Tr	—	—	Tr	—	—	A
	64–66	A	A	A	C	C	r-C	—	Tr	Tr-r	—	—	—	—	—	A
	91–93	A	A	C	A	C	r-C	—	—	Tr	—	—	—	—	—	A
	118–120	C-A	C	A	A	C	r-C	—	—	—	—	—	—	—	—	A
6	10–12	A	C-A	C-A	C-A	r-C	r	Tr	Tr	Tr-r	—	—	—	—	—	—
	49–51	C-A	r-C	r-C	C-A	r-C	r	Tr	—	r	—	—	—	—	—	—
	88–90	C-A	C	C	C-A	C	r	r	Tr	r-C	—	—	Tr	—	—	—
7	10–12	A	A	A	C	r-C	r	—	—	C	—	—	—	—	C	—
	62–64	A	A	A	A	C	Tr-r	Tr	—	r	—	—	—	—	—	—
	110–112	A	A	C-A	A	r	r	—	—	r	—	—	—	—	—	r
	160–162	A	A	C	A	C	Tr	—	—	r	—	—	—	—	—	—
	210–212	A	A	A	A	C	r-C	—	—	Tr	—	—	—	—	—	r
8	10–12	C-A	C-A	r-C	C	C	r-C	Tr	—	Tr-r	—	—	—	—	—	Tr-r
	49–51	A	A	C	A	r-C	Tr	—	—	Tr	—	—	—	—	—	Tr
	88–90	A	A	C	A	C	Tr	Tr-r	—	Tr-r	—	—	—	—	—	Tr
9	10–12	A	C-A	C	A	C	r	Tr-r	Tr	r-C	—	—	—	—	—	—
	54–56	A	A	C	C-A	C	r-C	r	—	r-C	—	—	—	—	—	Tr
	99–101	A	A	C	A	A	r-C	Tr	Tr	r-C	—	Tr	—	—	—	—
	142–144	A	A	A	A	C	r	—	—	r	—	—	—	—	—	—
	188–190	A	A	A	C-A	C-A	—	—	—	r-C	—	—	—	—	—	—
10	10–12	A	A	A	C-A	C-A	r-C	—	Tr	C	—	Tr	—	—	C	—
	92–94	C-A	C-A	C	A	C-A	C	—	C	Tr-r	Tr	r	—	r-C	C	—
	173–175	R	R	C	C-A	C	C	—	r	—	—	C	Tr	C	C	—

D = Dominant (>30%)
A = Abundant (20%–30%)
C = Common (10%–20%)
r = rare (1%–10%)
TR = Trace (<1%)
— = none detected

ments from core 6, which penetrated sequence 3, and the base of core 7 contained specimens of *Globorotalia tumida flexuosa* and, therefore, were deposited somewhere between 11 and 84 ka (Ericson and Wollin, 1968).

Both radiocarbon and biostratigraphic dating techniques indicate very rapid rates of sediment accumulation. Radiocarbon dates indicate rates for Holocene sedimentation ranging from 11.0 to 41.5 cm/ka. Such values are an order of magnitude greater than normal rates for pelagic sedimentation and significantly greater than rates (6.9 to 14.3 cm/ka) reported for bank-derived input to slopes bordering the northern Bahamas (Boardman and Neumann, 1984). Biostratigraphic analysis gives even higher sedimentation rates. As mentioned previously, core 5 (which penetrated sequence 5) is dated at 84 to 127 ka. Sequences 1–4, which overlie sequence 5, collectively attain a maximum thickness of 330 m, indicating an average sedimentation rate (for approximately the last 100 ka) exceeding 2.5 m/ka.

PROVENANCE AND DEPOSITIONAL PROCESSES

Sediments on the southwest Florida slope consist dominantly of shallow- and deep-water biogenic carbonate sands and muds deposited rapidly under variable energy flow regimes. Seismic data indicate the shallow west Florida shelf to the north is the source for shallow-water sediments. Southward-trending (offshelf) prograding clinoforms indicate transport of shallow-water material to the depositional site on the slope. Sediment parameters also show the shelf to be the source. Average carbonate mineralogies for surface samples from the southwest Florida slope are shown in Table 4 along with values from the west Florida shelf, west Florida Bay (Enos and Perkins, 1977), and the west Florida slope to the north (J. Bannon, unpub. data). Values for shelf and slope samples show many similarities. The slightly higher low-Mg calcite and slightly lower high-Mg calcite values for slope samples reflect the greater input of pelagic organisms. Planktonic foraminifera and coccoliths, both common in slope sediments, consist of low-Mg calcite and their input dilutes sediment derived from shallow-water sources. The higher aragonite concentration in slope samples can be accounted for by the addition of aragonite from pteropods, or possibly input from the aragonite-rich sediments of Florida Bay. Stockman and others (1967) have shown that the rate of production of biogenic lime mud is more than enough to account for the total accumulation in

TABLE 4.—CARBONATE MINERALOGY OF SEDIMENTS FROM DIFFERENT PROVINCES OF THE WEST FLORIDA CONTINENTAL MARGIN

	S.W. Florida Slope[1]	S.W. Florida Shelf[2]	W. Florida Bay[2]	W. Florida Slope N. Part[3]
Aragonite	44%	41%	60%	11%
High-Mg Calcite	23%	30%	25%	6%
Low-Mg Calcite	33%	29%	15%	83%

[1] This study
[2] Enos and Perkins (1977)
[3] J. Bannon, unpublished data

TABLE 5.—SEDIMENT CONSTITUENTS OF SOME CARBONATE ENVIRONMENTS (SAND FRACTION)

	S.W. Florida Slope[1]	W. Florida Shelf[2]	Florida Bay[3]	Outer Shelf Banks[4]	N.W. Florida Slope[5]
Molluscs	C-A[6]	A	R-D	R	R-C
Benthic Foraminifera	C-A	C	R-C	Tr	Tr-C
Halimeda	Tr	—	C	D	—
Coral	Tr	—	—	A-D	—
Sponge	C-A	R	C	—	Tr
Pelagic Foraminifera	A	R	—	—	A-D
Pteropods	A	—	—	—	Tr-R
Quartz	—	R-A	R	—	Tr-A

[1] This study
[2] Martin, 1984
[4] Ginsburg, 1956; Enos and Perkins, 1977; Gebelein, 1977
[5] Walker, 1984
[6] D = Dominant (>40%)
A = Abundant (20%–40%)
C = Common (10%–20%)
R = Rare (1%–10%)
Tr = Trace (<1%)
— = none identified

Florida Bay. The excess must be transported elsewhere and the physical processes are consistent with transport to the southwest slope.

Strontium data show sediments on the southwest slope are a mixture of high-Sr and low-Sr varieties. High-Sr aragonite is indicative of shallow-water carbonates. Highest Sr values arise from shelf sediments rich in ooids, *Halimeda*, and coral fragments. Sediments of the southwest Florida shelf contain lesser amounts of these constituents and are dominated by the lower Sr varieties, such as molluscs and benthic foraminifera. Strontium concentrations in southwest Florida slope sediments are higher than pelagic sources but substantially lower than platform sediments rich in ooids, *Halimeda*, and coral debris (Holmes, 1973). Slope values, averaging 2,370 ppm, indicate a mixture of shelf-derived and pelagic-derived strontium.

Table 5 shows the general sediment compositions of the west Florida shelf, west Florida slope, southwest Florida slope, and Florida Bay. Shallow-water molluscs and benthic foraminifera, the dominant shallow-water components of southwest slope samples, are the principal constituents of sediments from both the west Florida shelf (Brooks, 1981; Martin, 1984) and Florida Bay (Ginsburg, 1956; Enos and Perkins, 1977). Sponge spicules, also a major constituent of southwest slope samples, may be derived from Florida Bay sediments. Gebelein (1977) described dense sponge communities in Florida Bay and found spicules account for as much as 10 percent of the sediments. Quartz and clay minerals, which compose the non-carbonate fraction in slope samples, probably also are transported from the west Florida shelf (Doyle and Sparks, 1980). Benthic foraminifera in southwest slope samples are similar to both west Florida shelf and Florida Bay forms. As previously stated, species of *Peneroplis, Archaias, Amphistigina, Elphidium* and *Nonienella* are the dominant forms in southwest slope samples. All are common shallow-water foraminifera. *Peneroplis sp.* and *Archaias sp.* are the dominant forms in Florida Bay sediments (Bathurst, 1975).

A decrease in shelf-derived material with distance from the shelf not only indicates the shelf as a source but also that most deposition of shelf material occurs on the upper slope. The perennial rain of pelagic material probably varies little, and therefore differences in relative amounts of pelagic- versus shelf-derived material are the direct result of the rate of input of shelf-derived sediments.

Considering that sediments in the study area originate from multiple sources, processes responsible for their transportation and deposition must be multiple and complex. Shallow-water sand and mud-size material is eroded and transported southward to the southwest slope by the strong southward-flowing currents that develop across the entire width of the southern shelf and Florida Bay. These processes are competent enough to erode and transport fine sand and mud-size material during normal climatic conditions. Extreme high-energy events, such as tropical storms and hurricanes, can move large quantities of coarse sand and even granule-size material. The large fields of sand waves south of 26°N (Neurauter, 1980) indicate that southward sediment transport has been active during much of the Holocene (Holmes, 1985). Processes most responsible for erosion and transport of shelf material are probably storms and Loop Current intrusion. Florida Bay, which is more protected, is primarily influenced by tides with velocities exceeding 50 cm/sec on the outer shelf during tidal exchange between Florida Bay and the Florida Straits (Shinn and others, 1982; B. Barron, pers. commun., 1984).

Shallow-water sediments were probably deposited by rapid settling of material continuously delivered to the study area by the various physical processes previously mentioned. Continuous offshelf transport is aided by the lack of a well-developed rim of shelf edge reefs, which tend to block sediment transport in some rimmed-margin settings. Once in the study area, shallow-water sediments mix with pelagic material (consisting principally of planktonic foraminifera, pteropods, and coccoliths) that is continuously settling out of the overlying water column.

The strong Florida Current, the dominant physical process influencing the study area, does not appear to play a major role as a depositional mechanism. Unlike the large carbonate sediment drifts discovered off the northwest corners of Little Bahama and Great Bahama Banks that have been attributed to contour-current origin (Mullins and others, 1980), thick accumulations in the study area show no detectable input by the Florida Current. First, sediments up current are unlike those found in the study area. The logical origin of sediments entrained by the current would be terrigenous clastics introduced by the Mississippi River and low-Mg calcite-rich slope carbonates from the west Florida upper slope where the current interacts with the margin (Doyle, 1983; Freeman-Lynde, 1983; Doyle and Holmes, 1985; Gardulski and others, 1986; Mullins and others, 1988). Sediments from both of these areas are considerably different than the shallow-water-rich carbonates found on the southwest Florida slope. In addition, seismic-reflection data show primary stratal surfaces oriented off shelf in a southerly direction, indicating that sediments are being transported from the shelf to the north and not from the west, as would be expected if the Florida Current were a major transporting agent. Also, no contourite characteristics have been found, such as thinning of stratal units down current or wedge-shaped deposits, both of which are common in Bahama sediment drifts (Mullins and others, 1980). Internal reflectors, such as wavy upper surfaces and reflection-free units, also contourite characteristics (Stow and Lovell, 1979), are found only locally. In addition, accumulation rates of sediments deposited by contour currents are consistently less than 10 cm/ka (Stow and Lovell, 1979), considerably less than rates calculated for sediments on the southwest Florida slope.

Although not a major depositional agent, the Florida Current influences upper-slope development by helping to shape the deposit. The eastward displacement of each succeeding sequence depocenter is probably a response to the Florida Current, as are the erosional unconformities concentrated on the westward margin of each sequence and the modern burial of the Pourtales Terrace (Fig. 10).

DEPOSITIONAL MODEL

The presence of nine seismic sequences, all consisting of seaward-prograding clinoforms and similar sedimentologic composition of the three sequences sampled indicating a mixture of shallow- and deep-water material, suggest that patterns of deposition and erosion on the southwest Florida continental slope are cyclic in nature and have occurred throughout at least the Late Quaternary. Seismic data, sedimentology, and age dates suggest that sediments were eroded from the west Florida shelf, transported southward, and deposited in the study area during peirods in the sea-level cycle when the shelf surface was flooded. As has been shown by Mullins and others (1980), Hine and others (1981), Boardman and Neumann (1984), Hine and Steinmetz (1984), and Mullins and others (1984), flooding of a carbonate platform surface is required in order for significant offshelf transport to occur.

Sequence 5, dated at 84 to 127 ka, was deposited during oxygen isotope stage 5 (Shackleton and Opdyke, 1973) and probably corresponds to the 5e sea-level highstand (estimated to be 6 m higher than present) of approximately 125 ka (Fig. 12). Sequence 3, dated at 11 to 84 ka, was deposited during one of the more recent sea-level highstands, possibly stage 5a at approximately 80 ka, based upon stratigraphic position of the deposit. Sequence 1 is the modern sequence, undergoing active deposition for at least the last 11 ka in response to the Holocene transgression. Although no data exist as yet, sequences 2 and 4 may represent highstand deposits associated with oxygen isotope stages 3 and 5c, respectively. Even though the precise timing of depositional sequences has not yet been established, it is important to note that four separate sequences, collectively attaining a thickness of 330 m (Table 1) and each bound by an erosional unconformity, have accumulated since the deposition of sequence 5 (dated at a maximum of 127 ka). Thus, the most recent Milankovitch 100,000-yr sea-level cycle has been characterized by extremely rapid sedimentation (at least 2.5 m/ka) punctuated by episodes of significant erosion on the southwest Florida slope.

It is interpreted that these high-frequency fluctuations in sea level are the ultimate control responsible for depositional patterns and, therefore, Late Quaternary development of the southwest Florida slope. A conceptual depositional model (Fig. 13) illustrates how cycles of deposition and erosion are controlled by sea level. During periods in the sea-level cycle when the shelf surface is flooded, carbonate sediments are produced on the shallow-shelf surface (Wilson, 1975), swept southward off the platform, and deposited rapidly on the upper slope by the previously described physical processes influencing the southern shelf and upper slope. Thickest deposits have accumulated in the form of offshelf prograding clinoforms on the upper slope between the 250-m and 500-m isobaths, within the low-energy "zone of reversal" (Fig. 3). This "zone of reversal" acts as a depositional window favoring the accumulation of sediments. Higher energy zones, both shallower and deeper on the slope, prohibit significant sediment accumulation.

The relative contribution of shallow-water material by offshelf transport is, in itself, a function of sea level. During those periods in the sea-level cycle when water depths over the shelf surface are shallow, as may occur during early transgressions and late regressions, offshelf transport is vigorous. Large quantities of shallow-water sediments are

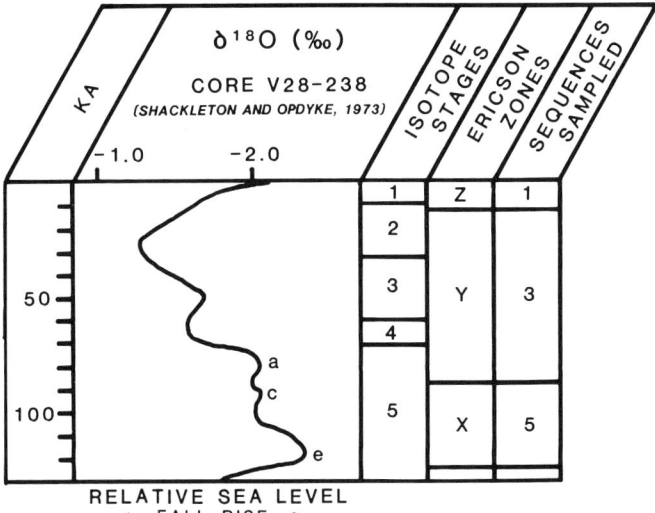

FIG. 12.—Oxygen isotope curve of core V28-238 showing ice-volume changes for approximately the last 130 ka (Shackleton and Opdyke, 1973). Included are corresponding Ericson biostratigraphic zones (Ericson and Wollin, 1968) and approximate ages of sequences sampled in this study.

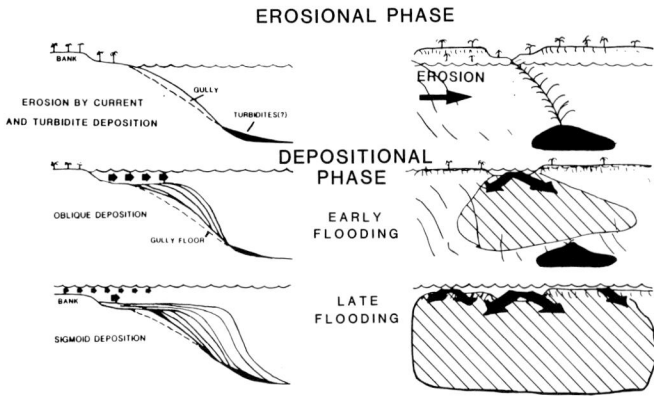

FIG. 13.—Conceptual depositional model of late Quaternary development of the southwest Florida slope showing how offshelf transport and deposition on the slope occur during sea-level highstands and erosion during sea-level lowstands. The model illustrates a single cycle among several cycles of deposition and erosion.

transported southward as physical processes operating on the shelf have more influence on the shallow bottom. Shallow-water sediments are deposited on the upper slope, mixing with pelagic sediments settling through the overlying water column. Sediments are deposited rapidly under relatively high-energy flow conditions, as shown by oblique prograding clinoforms at the bases of the sequences, and current structures in the bottoms of sediment cores. High-energy flow conditions arise from vigorous offshelf transport coupled with strong current activity.

During sea-level highstands, like the present, offshelf transport is less vigorous because much of the shelf surface lies well below wave base (approximately 10 m to 20 m under normal-energy conditions; Seibold and Berger, 1982). Except during extreme high-energy events such as tropical storms and hurricanes, when sand-size sediment can be mobilized, only fine-grained material is transported off shelf. Pelagic deposition makes an increasingly important contribution. Material is deposited at significantly lower rates under lower flow regimes, as indicated by sigmoidal prograding clinoforms and homogeneous, mottled and burrowed sedimentary structures. During sea-level lowstands, when the shelf surface is exposed, no offshelf transport takes place. Unlike terrigenous clastic margins, which have accelerated offshelf transport during these periods, carbonate margins quickly become cemented upon exposure to fresh water (Bathurst, 1975). Some shallow-water carbonate sediments are produced during this time, as the carbonate-producing zone shifts seaward down the low-angle slope with regressing sea level. Deposition of these sediments on the southwest slope, however, does not occur because the sedimentary regime switches from depositional to erosional. Erosion of the previously deposited sequence takes place in response to an increase in the erosive capacity of the eastward-flowing Florida Current, resulting from the following three factors: (1) an increase in current velocity associated with an increase in the temperature gradient between the poles and equator during glacially induced sea-level lowstands (Kennett, 1982; Brunner, 1975, 1983, 1986); (2) an increase in current velocity as the cross-sectional area of the narrow (approximately 100 km) Florida Straits decreases as sea level lowers (Brunner, 1983, 1986); and (3) a downward shift of the Florida Current with lowering sea level into the previous "zone of reversal," thereby bringing the strengthened current into contact with the previously deposited sequence.

A lowering of sea level to roughly 130 m below present (as that postulated for the previous lowstand) would drop the bottom margin of the Florida Current to approximately −330 m, assuming all other factors remained constant. This is within a reasonable range to produce the erosional unconformities identified at approximately 400 m below present sea level. Adding the effect of current strengthening by the factors mentioned above, erosive capacity to depths greater than 330 m is likely.

Heaviest erosion is concentrated on the western, or up-current portion of each sequence (Fig. 10). Eroded material is either transported out of the study area or carried to the heads of the erosional gullies where it is eventually funneled down slope in the form of gravity flows. Erosion may also tend to undermine upper-slope progradational deposits, making them unstable and susceptable to failure. Eventual failure could generate gravity flows and provide a mechanism for the initiation of lower slope gullies.

As sea level began to rise following lowstand erosional events, reefs began to form on bank edge erosional surfaces. The mound shown on Figure 5 probably represents a pinnacle reef that formed on the erosional unconformity separating sequence 4 from sequence 5. Reef growth was initiated during the sea-level rise following the erosional event responsible for the unconformity and attempted to migrate upslope, probably in order to keep pace with sea level. Eventually, reef growth was terminated, either by its inability to keep pace with sea level or by the great influx of shallow-water sediments being shed off the shelf, but the reef finally became buried by seaward-prograding sediments.

SUMMARY AND CONCLUSIONS

Thick sediment accumulations on the southwest Florida upper continental slope bordering the southern Florida Straits consist of a mixture of shallow- and deep-water biogenic carbonate sands and muds deposited rapidly (averaging greater than 2.5 m/ka) under a variable energy-flow regime. Shallow-water sediments are frequently transported off shelf by tides, storms, and oceanic currents and deposited rapidly on the upper slope. Pelagic detritus is deposited by settling through the overlying water column.

Nine seismic sequences have been identified, all consisting of seaward-trending prograding clinoforms and other similar reflection configurations. Sediment parameters of the three sequences penetrated also show similar characteristics. Similarities in seismic and sedimentologic parameters indicate a cyclic pattern of deposition and erosion. Depositional patterns are controlled by high-frequency sea-level fluctuations that have occurred throughout the Late Quaternary. During early flooding of the margin, offshelf transport is vigorous as physical processes have more influence on the shallow-shelf surface. During sea-level highstands, offshelf transport is less vigorous because much of the shelf surface lies below wave base, and sedimentation rates decline. During sea-level lowstands, no offshelf transport takes place as the shelf surface is exposed. The strengthened erosional capacity of the Florida Current erodes and reworks the previously deposited sequence.

Recent slope development is interpreted to have been controlled by high-frequency sea-level fluctuations coupled with the physical processes that act to concentrate sediments in the study area. In addition, the lack of a well-developed shelf edge reef rim allows more or less continuous offshelf transport and, therefore, continuous maintenance of upper-slope depositional units.

ACKNOWLEDGMENTS

This research was supported by the U.S. Geological Survey as part of a large ongoing program on the west Florida continental margin. We thank Fred Read and Hank Mullins for their critical reviews. We also thank Walter Dean and Rick Hildebrande for reviewing earlier versions of the manuscript. We greatfully acknowledge Charlotte Brunner (University of California, Berkeley) for aiding with biostratigraphic analyses. Atomic absorption of strontium was performed at the U.S. Geological Survey Laboratory, Corpus Christi, Texas. Radiocarbon dating was conducted by Beta Analytic, Inc., of Coral Gables, Florida.

REFERENCES

BATHURST, R. G., 1975, Carbonate sediments and their diagenesis: Elsevier, New York, 658 p.
BOARDMAN, M. R., 1978, Holocene deposition in Northwest Providence Channel, Bahamas; A geochemical approach: Unpublished Ph.D. Dissertation, University of North Carolina, Chapel Hill, 155 p.
———, AND NEUMANN, A. C., 1984, Sources of periplatform carbonates: Northwest Providence Channel, Bahamas: Journal of Sedimentary Petrology, v. 54, p. 1110–1123.
———, NEUMANN, A. C., BAKER, P. A., DULIN, L. A., KENTER, R. J., HUNTER, G. E., AND KIEFER, K. B., 1986, Banktop responses to Quaternary fluctuations in sea level recorded in periplatform sediments: Geology, v. 14, p. 28–31.
BOUMA, A. H., 1973, Leveed-channel deposits, turbidites, and contourites in deeper part of Gulf of Mexico: Gulf Coast Association of Geological Societies Transactions, v. 23, p. 368–376.
BROOKS, G. R., 1981, Recent carbonate sediments of the Florida Middle Ground reef system; Northeastern Gulf of Mexico: Unpublished M. S. Thesis, University South Florida, St. Petersburg, 137 p.
BROOKS, I. H., AND NILLER, P. P., 1975, The Florida Current at Key West: Summer, 1972: Journal of Marine Research, v. 33, p. 83–92.
BRUNNER, C. A., 1975, Evidence for intensified bottom current activity in the Straits of Florida during the last glaciation: Geological Society of America, Abstracts with Programs, p. 1012–1013.
———, 1983, Evidence for increased volume transport of the Florida Current in the Pliocene and Pleistocene: Marine Geology, v. 54, p. 223–235.
———, 1986, Deposition of a muddy sediment drift in the southern Straits of Florida during the late Quaternary: Marine Geology, v. 69, p. 235–249.
DOYLE, L. J., 1983, Shallow structure and stratigraphy of the carbonate west Florida continental slope and their implications to sedimentation and geohazards: U.S. Geological Survey Open-File Report 83-425, 19 p.
———, AND HOLMES, C. W., 1985, Shallow structure, stratigraphy and carbonate sedimentary processes of the west Florida upper continental slope: American Association of Petroleum Geologists, v. 69, p. 1133–1144.

———, AND SPARKS, T. N., 1980, Sediments of the Mississippi, Alabama and Florida (MAFLA) continental shelf: Journal of Sedimentary Petrology, v. 50, p. 905–916.
ENOS, PAUL, AND PERKINS, R. D., 1977, Quaternary sedimentation in south Florida: Geological Society of America Memoir 147, 198 p.
ERICSON, D. B., AND WOLLIN, G., 1968, Pleistocene climates and chronology in deep-sea sediments: Science, v. 162, p. 1227–1234.
FOLK, R. L., 1965, Petrology of sedimentary rocks: Hemphills, Austin, 159 p.
FREEMAN-LYNDE, R. P., 1983, Cretaceous and Tertiary samples dredged from the Florida Escarpment, eastern Gulf of Mexico: Gulf Coast Association of Geological Societies Transactions, v. 33, p. 91–99.
GARDULSKI, A. F., MULLINS, H. T., OLDFIELD, B., APPLEGATE, J., AND WISE, S. W., 1986, Carbonate mineral cycles in ramp slope sediment: Eastern Gulf of Mexico: Paleoceanography, v. 1, p. 555–565.
GEBELEIN, C. D., 1977, Dynamics of recent carbonate sedimentation and ecology, Cape Sable, Florida: E. J. Brill, Leiden, 120 p.
GINSBURG, R. N., 1956, Environmental relationships of grain size and constituent particles in some south Florida carbonate sediments: American Association of Petroleum Geologists, v. 40, p. 2384–2427.
GOMBERG, D. N., 1976, Geology of the Pourtales Terrace, Straits of Florida: Unpublished Ph.D. Dissertation, University of Miami, Coral Gables, 330 p.
HILL, P. R., 1984, Facies and sequence analysis of Nova Scotian slope muds: Turbidite vs. "hemipelagic" deposition, in Stow, D. A. V., and Piper, D. J. W., eds., Fine-Grained Sediments: Deep-Water Processes and Facies: Blackwell Scientific Publications, Boston, p. 311–318.
HINE, A. C., AND STEINMETZ, J. C., 1984, Cay Sal Bank, Bahamas–A partially drowned carbonate platform: Marine Geology, v. 59, p. 135–164.
———, WILBER, R. J., BANE, J. M., NEUMANN, A. C., AND LORENSON, K. R., 1981, Offbank transport of carbonate sands along open, leeward bank margins: Northern Bahamas: Marine Geology, v. 42, p. 327–348.
HOLMES, C. W., 1973, Distribution of selected elements in surficial marine sediments of the northern Gulf of Mexico continental shelf and slope: U.S. Geological Survey Professional Paper 814, 7 p.
———, 1981, Late Neogene and Quaternary geology of the southwestern Florida shelf and slope: U.S. Geological Survey Open-File Report 81-79, 29 p.
———, 1984, Carbonate fans in the Florida Straits: Society of Economic Paleontologists and Mineralogists, Abstracts with Programs, p. 39.
———, 1985, Accretion of the south Florida platform, late Quaternary development: American Association of Petroleum Geologists, v. 69, p. 149–160.
KENNETT, J., 1982, Marine Geology: Prentice-Hall, Englewood Cliffs, 813 p.
MARTIN, D. W., 1984, Clastic to carbonate transitions–A modern example: The west Florida continental shelf: Unpublished M. S. Thesis, University of South Florida, St. Petersburg, 89 p.
MILLIGAN, D. B., 1962, Marine geology of the Florida Straits: Unpublished M. S. Thesis, Florida State University, Tallahassee, 130 p.
MILLIMAN, J. D., 1974, Marine Carbonates: Springer-Verlag, New York, 375 p.
MINTER, L. L., KELLER, G. H., AND PYLE, T. E., 1975, Morphology and sedimentary processes in and around Tortugas and Agassiz sea valleys, southern Straits of Florida: Marine Geology, v. 18, p. 47–69.
MITCHUM, R. M., VAIL, P. R., AND SANGREE, J. B., 1977, Stratigraphic interpretation of seismic reflection patterns in depositional sequences, in Payton, C. E., ed., Seismic Stratigraphy–Applications to Hydrocarbon Exploration: American Association of Petroleum Geologists Memoir 26, p. 117–133.
MULLINS, H. T., GARDULSKI, A. F., HINCHEY, E. J., AND HINE, A. C., 1988, The modern carbonate ramp slope of central west Florida: Journal of Sedimentary Petrology, v. 58, p. 273–290.
———, HEATH, K. T., VAN BUREN, AND NEWTON, C. R., 1984, Anatomy of a modern open ocean carbonate slope: Northern Little Bahama Bank: Sedimentology, v. 31, p. 141–168.
———, NEUMANN, A. C., WILBER, R. J., HINE, A. C., AND CHINBURG, S. J., 1980, Carbonate sediment drifts in northern Straits of Florida: American Association of Petroleum Geologists, v. 64, p. 1701–1717.
NEURAUTER, T. W., 1980, Bed forms on the west Florida shelf as detected

with side scan sonor: Unpublished M. S. Thesis, University of South Florida, St. Petersburg, 120 p.

NILLER, P. P., 1976, Observations of low frequency currents on the west Florida shelf: Memoirs, Society of Royal Colloquium on Ocean Hydrodynamics Continental Shelf Dynamics, p. 331–358.

RICHARDSON, W. S., SCHMITZ, W. J., AND NIILER, P. P., 1969, The velocity structure of the Florida Current from the Straits of Florida to Cape Fear: Deep-Sea Research, v. 16, p. 225–231.

SCHLAGER, W., 1981, The paradox of drowned reefs and carbonate platforms: Geological Society of America Bulletin, v. 92, p. 197–211.

SEIBOLD E., AND BERGER, W. H., 1982, The Sea Floor: Springer-Verlag, New York, 288 p.

SHACKLETON, N. J., AND OPDYKE, N. D., 1973, Oxygen isotope and palaeomagnetic stratigraphy of equatorial Pacific core V28-238: Oxygen isotope temperatures and ice volumes on a 10^5 and 10^6 year scale: Quaternary Research, v. 3, p. 39–55.

SHINN, E. A., HOLMES, C. W., HUDON, J. H., ROBBIN, D. M., AND LIDZ, B. H., 1982, Non-oolitic, high-energy carbonate sand accumulations–The Quicksands, southwest Florida Keys (abs.): American Association of Petroleum Geologists, v. 66, p. 629–630.

STEWART, H. B., 1962, Oceanographic cruise report, U.S. Coast and Geodetic Survey Ship EXPLORER-1960: U.S. Department of Commerce, Coast and Geodetic Survey, Washington, D. C., 162 p.

STOCKMAN, K. W., GINSBURG, R. N., AND SHINN, E. A., 1967, The production of lime mud by algae in south Florida: Journal of Sedimentary Petrology, v. 37, p. 633–648.

STOW, D. A., AND LOVELL, J. P., 1979, Contourites: Their recognition in modern and ancient sediments: Earth Science Reviews, v. 14, p. 251–291.

WALKER, S. T., 1984, Sedimentary structures of the west Florida slope and Mississippi cone: Distribution and geological implications: Unpublished M. S. Thesis, University of Florida, St. Petersburg, 145 p.

WILSON, J. L., 1975, Carbonate Facies in Geologic History: Springer-Verlag, New York, 471 p.

PART III
EXAMPLES FROM THE PERMIAN BASIN CARBONATE FRINGE AND SLOPE

SLOPE SEDIMENTATION ASSOCIATED WITH A VERTICALLY BUILDING SHELF, BONE SPRING FORMATION, MESCALERO ESCARPE FIELD, SOUTHEASTERN NEW MEXICO

ARTHUR H. SALLER[1]
Cities Service Oil and Gas, Tulsa, Oklahoma 74102

JANE W. BARTON
AND
RICKY E. BARTON
OXY USA, Inc., Oklahoma City, Oklahoma 73126

ABSTRACT: Mescalero Escarpe field contains slope strata in the Bone Spring Formation that provide insight into development of the Northwestern Shelf and Delaware Basin. During Leonardian (Permian) time, the Northwestern Shelf in the vicinity of Mescalero Escarpe field grew vertically in response to rapid regional subsidence and cyclic sedimentation. Slope facies include: (1) dolomitized megabreccia, (2) dolomitized bioclast-peloid packstone, (3) laminated dolomitic mudstone, and (4) very fine-grained sandstone. The megabreccia and bioclast-peloid packstone are the reservoir facies. Stratigraphic relations established by log and seismic correlations support carbonate sedimentation in slope and basinal environments at the same time as carbonate sedimentation on the shelf. Bioclast-peloid packstones overlain by megabreccias occur at the toe of slope. Dolomitic mudstones are present higher on the slope and deeper in the basin. Very fine-grained sandstones alternate with intervals of carbonate strata on the slope. Seismic data indicate that erosion and backstepping of the carbonate shelf margin occurred at approximately the same time as deposition of the carbonate megabreccias at the toe of slope. Stable isotopic data and clast shape in the megabreccia suggest that this erosion occurred in a submarine environment. Petrographic and stable isotopic data support dolomitization of reservoir facies by sea water during or very shortly after deposition in a slope environment.

Depositional cycles on the shelf and in the basin can be divided into three main stages related to changes in relative sea level. (1) During rapid sea-level rises, the shelf was flooded, and carbonate sedimentation was initiated on the shelf. Some sediments were swept off the shelf and into the basin, forming bioclast-peloid packstones at the toe of slope and dolomitic mudstones in other slope and basinal environments. (2) As sea-level rise slowed or stopped, the vertically building shelf approached sea level. Submarine erosion of the shelf margin and debris flow transport of the coarse detritus resulted in deposition of megabreccias at the toe of slope. Deposition of carbonate mud continued in adjacent slope and basin environments. (3) Sea level remained static or dropped at the end of depositional cycles. The shelf was at sea level or emergent, and terrigenous sand was transported into the basin. To the east and west of Mescalero Escarpe field, the Northwestern Shelf margin prograded markedly during Leonardian time. Vertical building of the Northwestern Shelf margin was due, at least in part, to greater subsidence in the vicinity of Mescalero Escarpe field than in areas farther east and farther west.

INTRODUCTION

In early Permian time, the Delaware and Midland Basins underwent rapid regional subsidence. A carbonate shelf margin was established at the northern end of the Delaware and Midland Basins. In the vicinity of Mescalero Escarpe field (Fig. 1), the carbonate shelf margin grew vertically during the Leonardian before prograding to the south during the Guadalupian (Podpechan, 1959). To the east and west, the Northwestern Shelf was progradational during the Leonardian.

A critical area in shelf and basin development is the slope, which contains the shelf-to-basin transition. Stratigraphic relations between various types of slope strata in and around Mescalero Escarpe field provide much insight into the development of the Northwestern Shelf and Delaware Basin during Leonardian time. The slope is affected by processes occurring both on the shelf and in the basin. Most of the slope is below the range of eustatic sea-level falls and hence will have a more complete record of sedimentation through time than shelf strata. Portions of the slope may also receive basinal sediments and hence give information relating basinal sedimentation to shelf sedimentation. The purpose of this paper is (1) to illustrate the shelf-to-basin transition at a vertically building shelf margin, (2) to describe slope sedimentation adjacent to a vertically building shelf margin, (3) to relate those features to development of the Northwestern Shelf margin, and (4) to examine factors causing some segments of the Leonardian shelf margin to be aggradational, whereas others were progradational.

REGIONAL STRATIGRAPHIC SETTING

Leonardian carbonates of the Northwestern Shelf are divided into four parts (from top to bottom): Upper Yeso, Middle Yeso, Lower Yeso, and Abo (Silver and Todd, 1969; Fig. 2). Thin sands and shales commonly separate those shelf carbonates. The Bone Spring Formation is the slope and basinal equivalent of Leonardian shelf carbonates. The Bone Spring Formation contains four carbonate units separated by three siliciclastic sandstone units. The four carbonate units and three sandstones are (from top to bottom): First (Upper) Bone Spring limestone, First Sand, Second Bone Spring carbonate, Second Sand, Third Bone Spring carbonate, Third Sand, and Lower Bone Spring carbonate, which is colloquially known as "Wolfcamp" (Gawloski, 1987; Mazzullo and Reid, 1987; Fig. 2). The Lower Bone Spring carbonate ("Wolfcamp") and Third Sand have Early Leonard fusulinids (Mazzullo and Reid, 1987). The Second Sand and most of the Second Bone Spring carbonate have Middle Leonard fusulinids with the transition from Early to Middle Leonard fusulinids at the top of, or in the upper part of, the Third Bone Spring carbonate (Mazzullo and Reid, 1987). The First (Upper) Bone Spring carbonate and part, if not all, of the First Sand have Late Leonard fusulinids with the transition from Middle to Late Leonard fusulinids in, or at the base of, the First Sand (Mazzullo and Reid, 1987).

---

[1]Present address: UNOCAL Science and Technology, Brea, California 92621

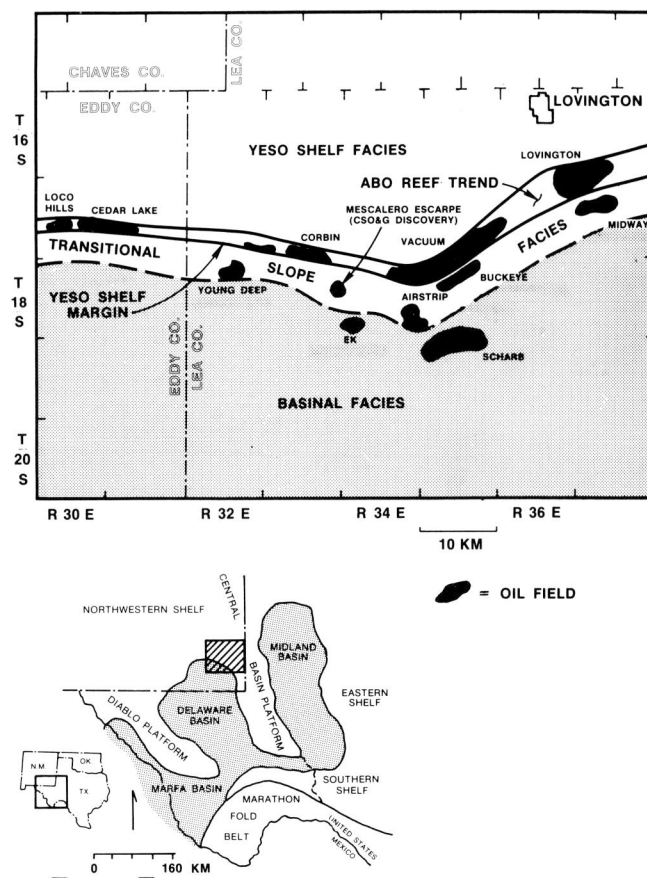

FIG. 1.—Maps showing location of study area and distribution of shelf, slope, and basinal facies in southeastern New Mexico.

Coarse shelf-derived carbonate detritus is interbedded with carbonate-rich mudstones and terrigenous sands on the slope (Wiggins and Harris, 1985; Gawloski, 1987). At the northern margin of the Delaware Basin, many oil fields including Mescalero Escarpe field produce from Leonardian slope detritus (Fig. 1; Runyan, 1965; Nottingham, 1966; Gawloski, 1987). Mescalero Escarpe field produces oil from the Second Bone Spring carbonate (Pay 2 dolomite), which is between the First and Second Bone Spring Sands in the transitional slope zone basinward of the Yeso Shelf. Slope strata and the transition from Northwestern Shelf to Delaware Basin were studied in detail at and near Mescalero Escarpe Field using cores, electric logs, and reflection seismic lines.

LITHOLOGIC FACIES ON THE SLOPE

Approximately 240 m of core in the Bone Spring Formation from six Mescalero Escarpe wells were examined (Fig. 3). Four main lithologic facies were recognized in core: (1) dolomitized megabreccia, (2) dolomitized bioclast-peloid packstone, (3) dark laminated mudstone, and (4) very fine-grained sandstone. The Pay 2 dolomite reservoir includes the dolomitized megabreccia and bioclast-peloid packstone facies.

Dolomitized megabreccia.—

The dolomitized megabreccia consists of coarse lithoclasts in a wackestone-packstone matrix (Fig. 4a). It is as much as 30 m thick. Both clast- and matrix-supported megabreccias are present. The dolomitized megabreccia is massive, lacking current and internal bedding structures. A few interbeds of dark, silty mudstone separate intervals of megabreccia. The base of the megabreccia has abrupt contacts with underyling sandstones but more gradational contacts with underlying dark laminated mudstones (Fig. 5a). The megabreccia facies is dominantly dolomite, with scattered nodular and void-filling anhydrite.

Carbonate lithoclasts are angular to subangular and range from less than 1 cm to more than 200 cm in diameter. Max-

FIG. 2.—Correlation chart for Leonardian and Wolfcampian strata on Northwestern Shelf and in Delaware Basin. Stratigraphic location of Pay 2 dolomite is shown by star (modified from Silver and Todd, 1969).

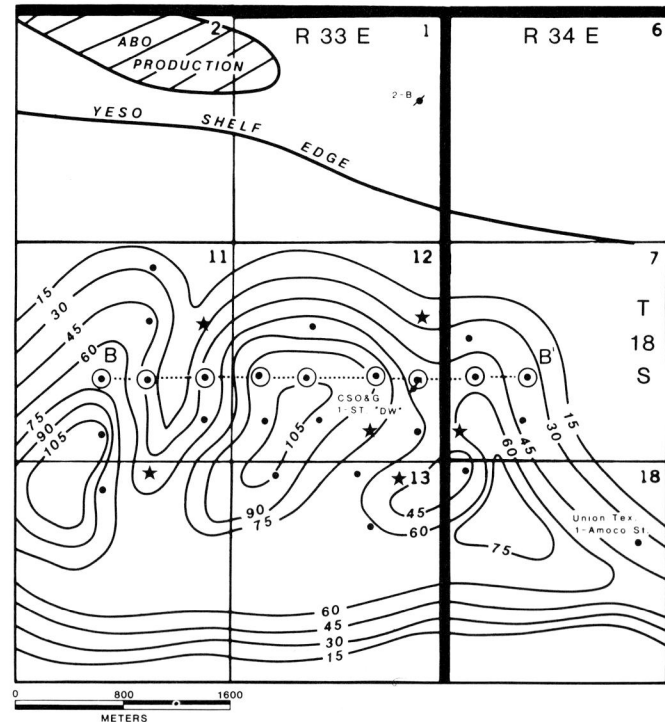

FIG. 3.—Isopach map of dolomitized slope detritus between First and Second Sands (Pay 2 dolomite) in Mescalero Escarpe field. Stars indicate cored wells. Contour interval 15 m. Section B-B' is shown in Figure 9.

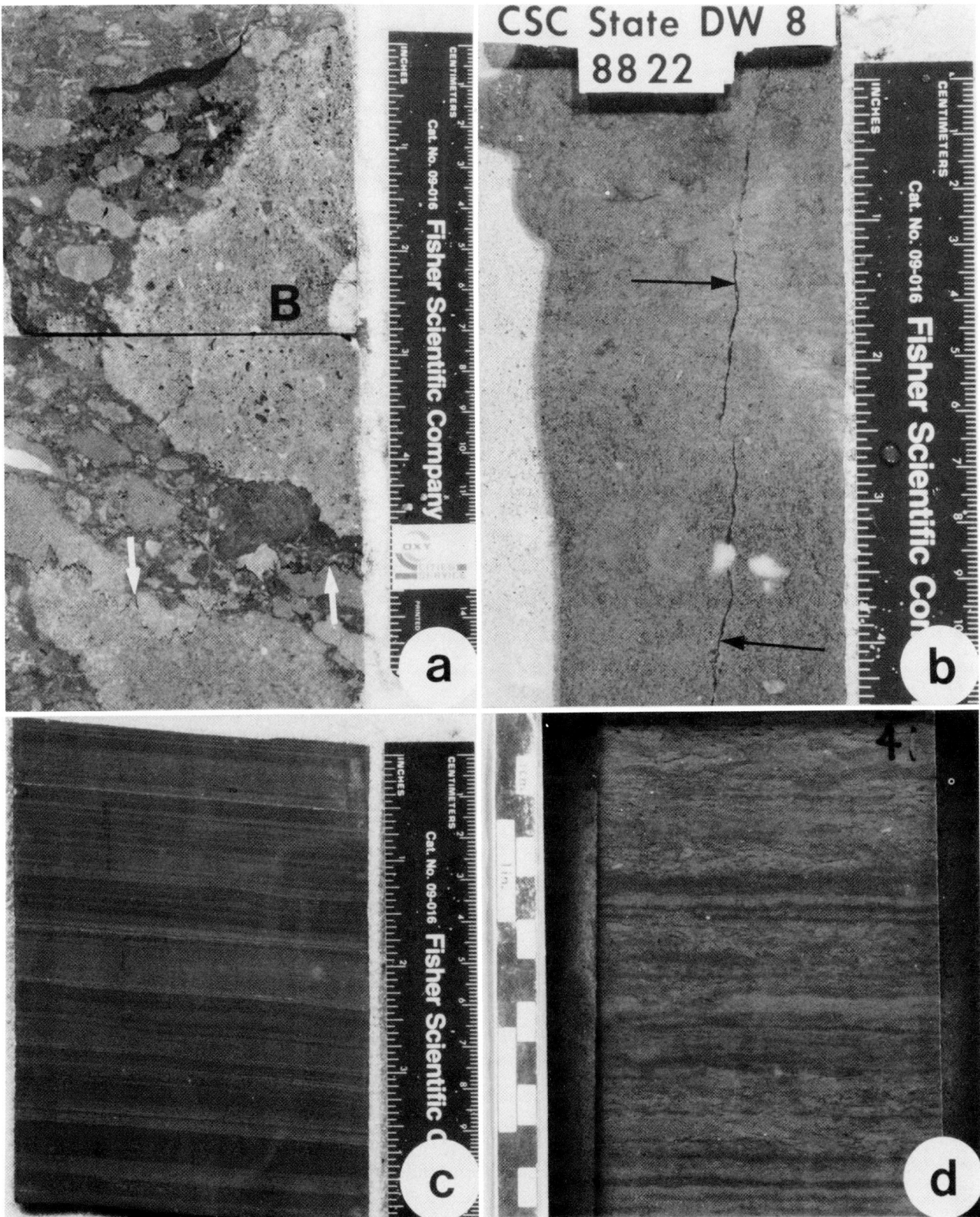

FIG. 4.—Core photographs showing the four main slope lithologies in Bone Spring Formation, Mescalero Escarpe field. (a) Dolomitized megabreccia with light-colored, irregular lithoclasts (largest one is labelled "B") surrounded by a dark, micritic matrix. Note stylolites (arrows). (b) Dolomitized bioclast-peloid packstone with vertical fracture (arrows). White anhydrite fills part of fracture immediately above lower arrow. Most of the slab is wetted with water, but leftmost part is dry and hence lighter in color. (c) Laminated dolomitic mudstone. (d) Very fine-grained sandstone.

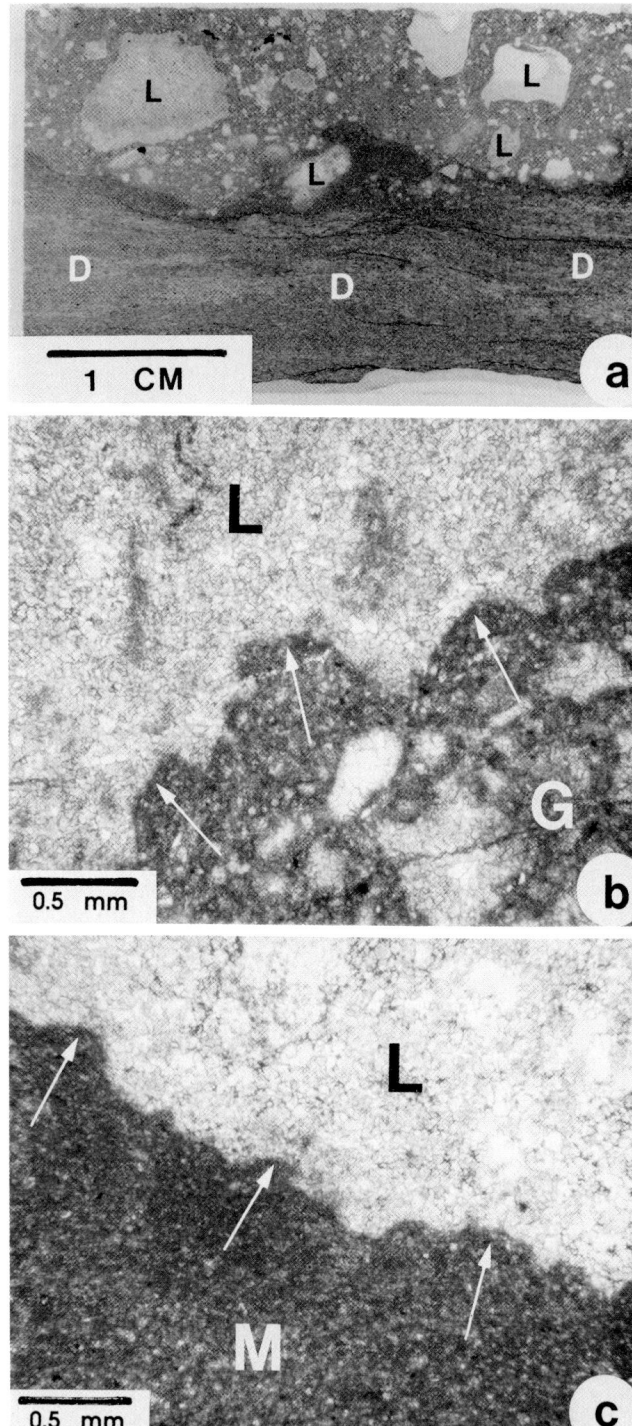

Fig. 5.—Megabreccia facies in C.S.O.G. State "DW" #3 well. (a) Photograph of thin section showing contact between dark laminated mudstone and megabreccia facies. Dark laminated mudstone (D) in lower part of photograph is overlain by lithic clasts (L) in micritic matrix. Clasts apparently project into underlying mudstone, suggesting that the mudstone was soft (unlithified) at time of megabreccia deposition (2,643 m). (b) Thin-section photomicrograph of clast boundaries. Carbonate lithoclast (L) with scalloped boundaries (arrows) in grain-rich wackestone matrix. Most grains were fossil fragments (2,639 m). (c) Thin-section photomicrograph of clast boundaries. Lithic clast (L) with scalloped boundaries (arrows) in micritic (mudstone) matrix (M) (2,641 m).

imum clast size is greatest in the most shelfward well and decreases basinward. In the uppermost 5 m of the megabreccia, maximum clast diameter decreases upward from 10 cm or more to less than 2 cm. Lithoclasts include: (1) laminated quartzose siltstones, (2) massive, fine- to medium-crystalline dolomite, (3) cross-bedded peloidal packstones and grainstones, (4) peloidal, crinoidal wackestone-packstones, (5) bioclastic packstones and grainstones, and (6) bryozoa-phylloid algae boundstones. Clasts of calcareous sandstone containing well-rounded, well-sorted, medium-grained quartz sand are also present. No clasts of spiculitic mudstone were observed. Some clasts deformed plastically around adjacent clasts, whereas others behaved rigidly and were apparently lithified at the time of deposition. Clast boundaries are stylolitic, irregular, and/or scalloped (Fig. 5b, c).

The matrix of the megabreccia varies in texture from mudstone to packstone and includes dolomitized micrite, quartz silt, peloids, and fossil fragments that have been dissolved and/or replaced by dolomite (Figs. 4a, 5b, c). Crinoids are the most abundant type of fossil fragments in the matrix, but shelly fossil fragments are also present. The matrix tends to be more micritic to the north (shelfward). Dolomitization of this unit is pervasive with both clasts and matrix being completely dolomitized. Stable carbon and oxygen isotope analyses show that megabreccia clasts and matrix have similar isotopic compositions (Fig. 6). Measured porosities in megabreccias are 0.5–9.0% (mean of 4.2%), and permeabilities are 0.01–18 md (mean of 3.3 md, excluding measurements within individual clasts). Clasts

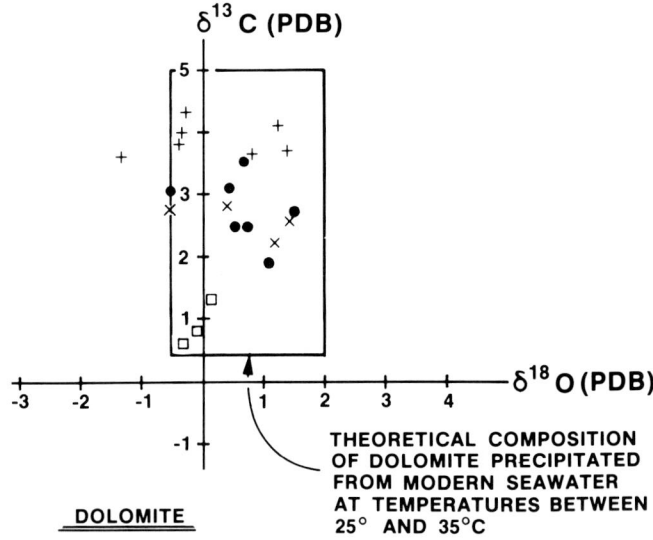

Fig. 6.—Plot of $\delta^{13}C$ versus $\delta^{18}O$ for Bone Spring dolomites in Mescalero Escarpe field. Theoretic oxygen isotope fractionation of dolomite is derived from Land (1985). Theoretic carbon isotope fractionation of dolomite is assumed to be similar to calcite. Theoretic carbon isotope composition of calcite is derived from composition of total dissolved carbon in sea water from Kroopnick and others (1977) and fractionation factor of Gonzalez and Lohmann (1985).

approximately 7 km basinward of the Yeso shelf margin (Fig. 8). Thin (1–3 cm thick), graded fossiliferous packstone to wackestone beds are rare, being present in only a few parts of the mudstone unit. Fossils in those wackestone-packstone interbeds include crinoids, brachiopods, and calcareous algae, which are often silicified. Sponge spicules are scattered throughout the laminated mudstone facies and are locally abundant (making the rock a wackestone). Except for sponge spicules and fossiliferous interbeds, this facies is unfossiliferous. Parallel laminations are common in the mudstones. Some laminations are silica rich and others are dolomite rich.

The dark laminated mudstones are organic rich. Eight samples of dark laminated dolomite mudstone from depths of 2,435 to 3,838 m (Wolfcampian and Leonardian) were analyzed from Mescalero Escarpe wells and wells along depositional strike. Total organic carbon values range from 0.8 to 4.3 weight percent with an average of 2.0 weight percent. Geochemical parameters (gas chromatograph patterns, carbon isotopes, butane-pentane ratios, aromatic hydrogen-carbon ratios, pristane-phytane-nC_{17}–nC_{18} ratios, biomarkers) indicate a fair to good correlation of organic material in the mudstone to the oil in Mescalero Escarpe field with the oil being best correlated to a mixture of samples from several different levels (W. Fallgatter, pers. commun., 1985). Results of stable carbon and oxygen isotope analyses from this facies are shown in Figure 6. The dark laminated mudstones act as an impermeable seal for the Pay 2 dolomite reservoir in an updip direction.

Very fine-grained sandstone.—

The very fine-grained sandstone lithofacies (Fig. 4d) is very poorly sorted, containing much silt and clay in addition to very fine-grained sand. Sand and silt grains are mainly angular to subangular quartz with minor feldspar. Significant amounts of clay (10–30%) and dolomite cement (10–30%) are also present in the very fine-grained sandstone. Sedimentary structures include horizontal laminations interrupted by burrows, wavy laminations with abundant bioturbation, horizontal current laminations, ripple-scale cross-

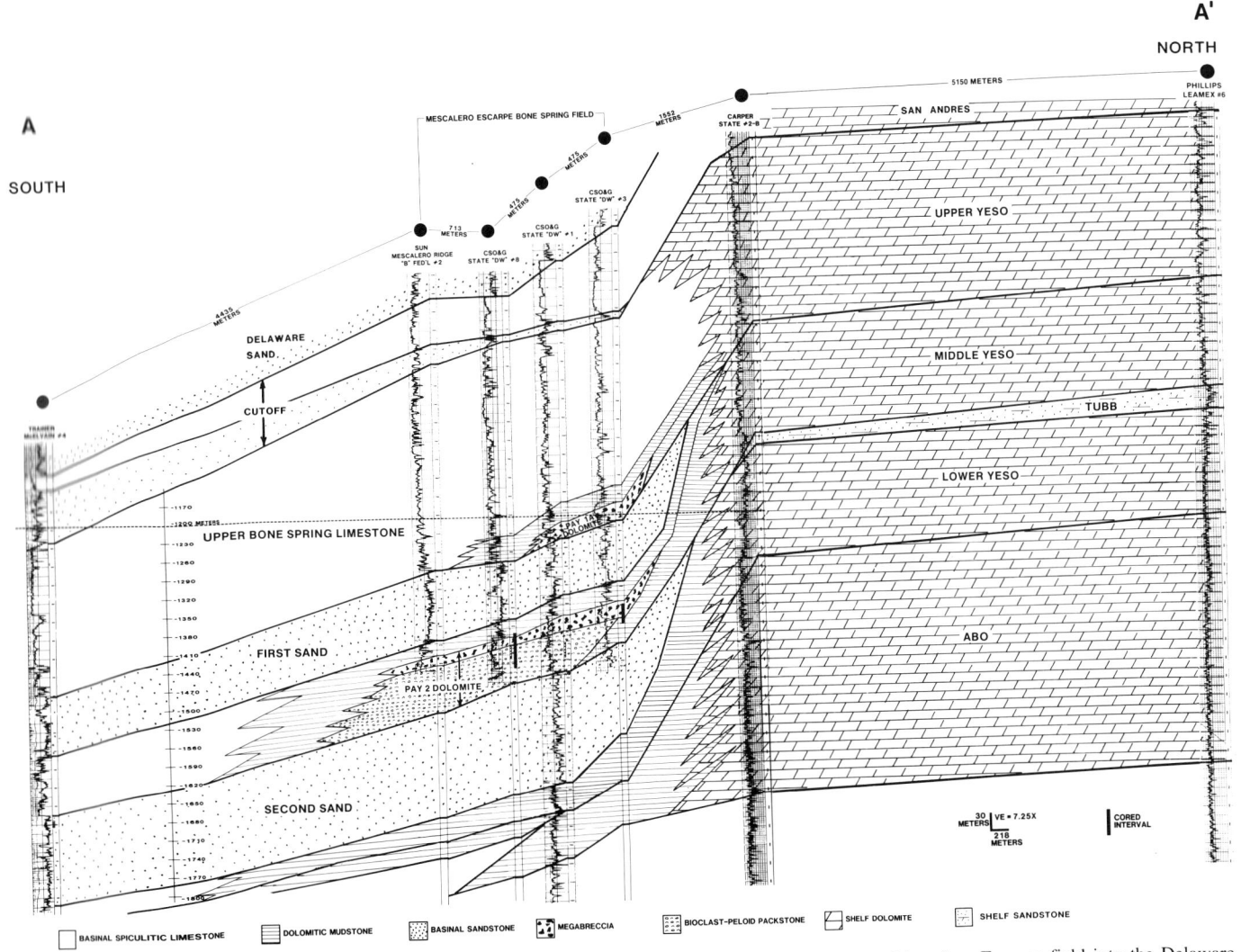

FIG. 8.—Structural cross section along depositional dip from the Yeso Shelf to the north across the Mescalero Escarpe field into the Delaware Basin. See Figure 11 for location of cross section.

with as much as 14% porosity are present; however, they are generally surrounded by a less porous and permeable matrix. The dolomitized megabreccia facies tends to be more porous and permeable in a basinward direction, which corresponds to the matrix becoming more grain rich in a basinward direction. Open fractures are rare in this facies.

Dolomitized bioclast-peloid packstone.—

The dolomitized bioclast-peloid packstone facies is a gray to light tan bioclastic-peloid packstone with subordinate amounts of interbedded wackestone and grainstone (Fig. 4b). This facies is as much as 70 m thick and is dominantly dolomite with minor anhydrite and quartz. Most dolomite replaces original calcareous material, but minor void-fill dolomite cement is also present. Anhydrite is present as replacive nodules and void-fill cement. Quartz occurs mainly as late void-fill cement. This packstone facies is much more grain rich than the matrix in the megabreccia facies. Fossil fragments are the dominant grain type in lower parts of this facies (Fig. 7a). Fossils include crinoids, brachiopods, molds of fusulinids, sponge(?) fragments, and other unidentified shelly fragments. Peloids increase in abundance upward and become dominant in upper parts of this facies. Some grains have been partially micritized (Fig. 7a).

No current structures were observed in the dolomitized bioclast-peloid packstone lithofacies. Vertical burrows, bored hardgrounds, *in situ* breccias, cracks, and fractures are scattered throughout this unit and are often concentrated below bedding surfaces (Fig. 7b, c). Thin fractures are common and often partially open (Fig. 4b). Most fractures are vertical to slightly inclined and are discontinuous with few being more than 5 cm long. In some parts of the bioclast-peloid packstone facies, intense fracturing has resulted in *in situ* breccias. Some fractures and areas around *in situ* breccia clasts are filled with micritic sediment. Often that micritic sediment appears to predate deposition of overlying strata (Fig. 7c). Stable carbon- and oxygen-isotopic analyses of the bioclast-peloid packstone facies are shown in Figure 6. Porosity in this facies ranges from 2.1 to 19.6% (average of 7.4%) with moldic, intergranular, and intercrystalline pores being dominant. Permeabilities are 0.01 md to 10 d (average of 229 md) with open fractures as much as 0.5 mm wide enhancing permeability.

Dark laminated mudstone.—

The laminated mudstone facies (Fig. 4c) is composed of microcrystalline dolomite (10–90%), quartz (10–60%; in-

cluding fine silt and sponge spicules), clay (1– minor to trace amounts of K-feldspar, plagiocl and siderite. Basinward, these mudstones are cal than dolomitic. For example, calcite is domina bonate mudstones in the Trainer McElvain #4,

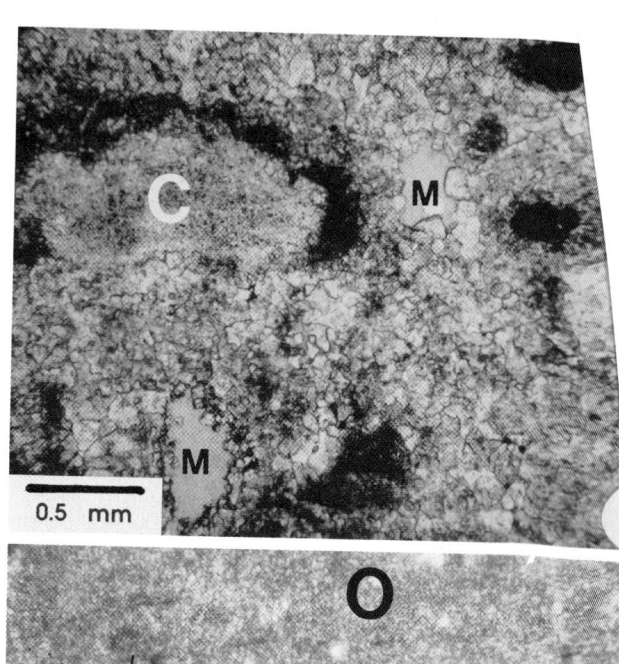

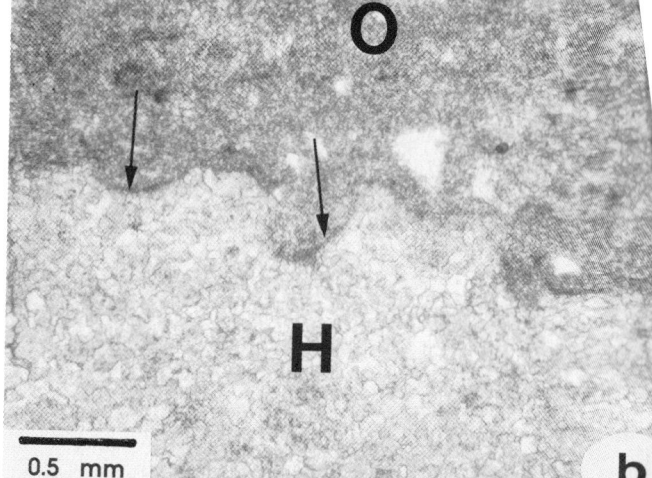

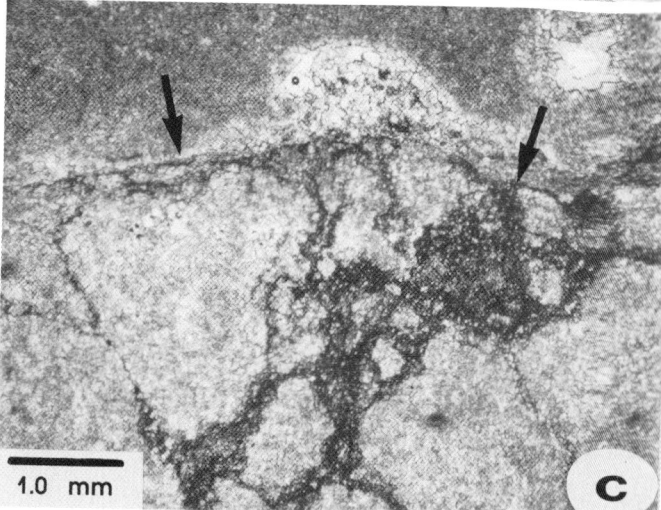

FIG. 7.—Thin-section photomicrographs of dolomitized bioclast-peloid packstone facies in the C.S.O.G. State "DW" #8 well. (a) Dolomitized bioclastic packstone with molds of bioclasts (M) that are lined with subhedral to euhedral dolomite crystals. Partially micritized crinoid fragment (C) is also present (2,720 m). (b) Bored hardground. Light-colored dolomite (H) in lower half of photograph has a scalloped (bored) upper surface (arrows), suggesting this was a syndepositional hardground that was bored and subsequently overlain by darker, finer crystalline sediment (O; 2,707 m). (c) Cracks and fractures, abundant below the bedding surface (arrows), end at the bedding surface. They formed and were infilled with micritic sediments (dark material) prior to deposition of strata above the bedding surface (2,698 m).

beds, and partial Bouma sequences. These sandstones are not reservoir rocks in Mescalero Escarpe field.

Interpretation of lithofacies.—

The dolomitized megabreccia was apparently deposited by episodic debris flows that carried shallow-marine lithoclasts into the deep basin. Textural evidence supporting deposition by debris flows includes: (1) poor sorting, (2) large, angular to irregularly shaped clasts, (3) abundant micrite, (4) lack of burrows or current structures, and (5) maximum clast size decreasing upward near the top of this unit (Fig. 4a). The megabreccia lithofacies is similar to carbonate debris flow deposits described by Cook and Mullins (1983). Wiggins and Harris (1985) noted similar features and concluded that Bone Spring detritus at Airstrip field approximately 10 km to the east was also deposited by debris flows. A lack of rounding suggests that these clasts were never subjected to abrasion on a beach or in a high-energy shallow-marine environment. The stable carbon and oxygen isotope composition of clasts are characteristic of sea water at near-surface temperatures and do not show decreased $\delta^{13}C$ or $\delta^{18}O$ values that are characteristic of freshwater diagenesis (Fig. 6; Allan and Matthews, 1977). Therefore, megabreccia clasts were probably not eroded from an island during subaerial exposure. Lithoclasts commonly have scalloped edges similar to those caused by boring sponges in modern environments (Fig. 5b, c; Futterer, 1974); therefore, marine organisms (especially sponges) probably helped erode the Yeso shelf and produce megabreccia clasts in a submarine environment. Thin interbeds of dark, silty, carbonate mudstone within the megabreccia probably represent slow rains of windblown silt and micritic sediment that accumulated between debris flow events.

The dolomitized bioclast-peloid packstone lithofacies was probably deposited in deep open-marine waters. Micritization of grains suggests derivation of those grains from a shallow shelf (Bathurst, 1975). Peloids and fossil fragments were carried off the shelf, down channels in the upper slope, and deposited at the toe of slope. Shelf-derived bioclastic material is common in modern slope environments (Cook and Mullins, 1983). Vertical burrows and bioturbation in the bioclast-peloid packstone indicate deposition in oxidized waters with near-normal-marine salinities. Hardgrounds, borings, cracks, and breccias infilled by sediments prior to deposition of overlying strata (Fig. 7c) indicate lithification penecontemporaneous with deposition. Similar early lithification is observed in modern slope environments (Mullins, 1986). Micrite filling fractures and areas around *in situ* breccia clasts suggests fracturing and brecciation in the bioclast-peloid packstone facies prior to deposition of overlying strata (Fig. 7c). Fracturing and lithification were, thus, penecontemporaneous with deposition of the bioclast-peloid packstone facies. A lack of discernable vertical offset and little vertical continuity of fractures suggest that most formed in response to local, non-tectonic, tensional stresses.

The dark laminated mudstone lithofacies of the Bone Spring Formation accumulated in slope and basinal environments when and where coarser, shelf-derived detritus was not accumulating (Harms, 1974; Wiggins and Harris, 1985). Thin, graded interbeds of fossiliferous packstone and wackestone in this lithofacies apparently represent small pulses of shallow-water detritus swept off the Yeso shelf or slope.

The very fine-grained sandstone lithofacies in the Mescalero Escarpe field was deposited in slope and basinal environments. Current laminations and Bouma sequences suggest episodic deposition of parts of this facies by turbidity currents, whereas abundant burrows within other parts of this facies suggest slower deposition.

Interpretation of dolomitization.—

Lithification of the Bone Spring Formation in and around Mescalero Escarpe field occurred penecontemporaneously with deposition and helped maintain relatively steep-slope angles. Early lithification apparently resulted from dolomitization. Therefore, understanding the processes involved in dolomitization is important to understanding the development of slope carbonates at the northern end of the Delaware Basin. Carbonate in the Bone Spring Formation in the Mescalero Escarpe field has been pervasively replaced by dolomite. The Pay 2 dolomite was deposited in relatively deep water and is surrounded by hundreds of meters of deep-marine carbonate, sandstone, and shale, making mixed-water dolomitization unlikely. Wiggins and Harris (1985) studied dolomitization at Airstrip field and concluded that (1) early replacive dolomite formed during very shallow burial with magnesium coming from marine-derived pore fluids expelled from basinal mudstones, (2) much of the dolomite precipitated during intermediate and late burial, and (3) dolomitization of megabreccia clasts occurred on the shelf.

At Mescalero Escarpe, petrographic and geochemical evidence support pervasive dolomitization in sea water shortly after deposition. Fracturing of the bioclast-peloid packstone lithofacies required lithification of the sediments. Dolomitization was probably the lithifying process. Dolomite adjacent to fossil molds has subhedral-to-euhedral terminations, indicating dolomitization during or after formation of the molds (Fig. 7a). Dolomite crystals adjacent to fractures are generally anhedral or fractured, suggesting dolomitization prior to fracturing. Since most fracturing occurred penecontemporaneously with or very shortly after deposition of the bioclast-peloid packstone (Fig. 7c), pervasive dolomitization at Mescalero Escarpe field apparently took place very shortly after deposition, while carbonate sediments were in open communication with marine water.

Stable carbon and oxygen isotope ratios of the Pay 2 dolomite support dolomitization by sea water or slightly modified sea water at approximately surface temperatures (Fig. 6). The $\delta^{13}C$ values are similar to those found in modern sea water and marine carbonate sediments (Gross, 1964; Hudson, 1977). Because carbonate is commonly dissolved and reprecipitated at higher temperatures during burial diagenesis, the $\delta^{18}O$ values of original dolomites would tend to be greater than or equal to the $\delta^{18}O$ compositions of dolomites recovered from the subsurface, which have been subjected to burial diagenesis. Therefore, the most positive $\delta^{18}O$ values (+1.5‰, PDB) should most closely resemble the composition of the original replacement dolomite. Dolomite with a $\delta^{18}O$ value of +1.5‰ could precipitate from modern sea water at approximately 26°C (Land, 1985). The micritic matrix between clasts in the megabreccia has stable isotopic compositions similar to the clasts, suggesting dolomitization of clasts and matrix at the same time or in similar waters. Dolomite in dark laminated mudstones has lower $\delta^{13}C$ values than dolomite in the megabreccias and peloid-bioclast packstones (Fig. 6). Incorporation of some organic carbon ($\delta^{13}C = -25^0/_{00}$, PDB) during dolomitization is probably responsible for dolomite in the dark mudstone having lower $\delta^{13}C$ values than the Pay 2 dolomite (Fig. 6).

Dolomitization by deep-marine waters undersaturated with respect to calcite has been described previously by Saller (1984) and Mullins and others (1985). Dissolution, which created moldic porosity, also probably occurred in deep-marine waters undersaturated with respect to calcite during dolomitization. It is not clear why Bone Spring mudstones are calcitic approximately 4–5 km basinward (to the south) of Mescalero Escarpe field, but two possible explanations are described: (1) dolomitization occurred on steeper, shallower parts of the slope, where submarine currents were able to move dolomitizing sea water through the sediments; deeper in the basin, there were less relief and probably weaker submarine currents, making it difficult to move dolomitizing water by or through the sediments; (2) carbonate mud on the slope near the shelf might have originally been more aragonite rich than more basinward muds because aragonite was largely derived from the shelf. Aragonite is often more reactive and more subject to dolomitization than calcite (especially low-magnesium calcite; Walter and Morse, 1984).

STRATIGRAPHY AND DEPOSITIONAL SYSTEMS

Mescalero Escarpe field is located at the northern margin of the Bone Spring basin and immediately south of the Abo reef trend and the Yeso shelf margin (Fig. 1). Electric logs and seismic sections have been used together to determine the stratigraphic relations of shelf, slope, and basinal Leonardian strata.

Distribution of slope lithofacies.—

Using gamma ray and neutron-density logs, the four lithologic facies were correlated into all wells in Mescalero Escarpe Field. The distribution of slope facies is shown in Figures 8 and 9. Electric-log correlations suggest 425 m of relief and a 15° slope between the Yeso shelf margin and toe of slope (not corrected for possible differential compaction; Fig. 8). The toe of slope is an inflection point where slope steepness decreases markedly. When present, the do-

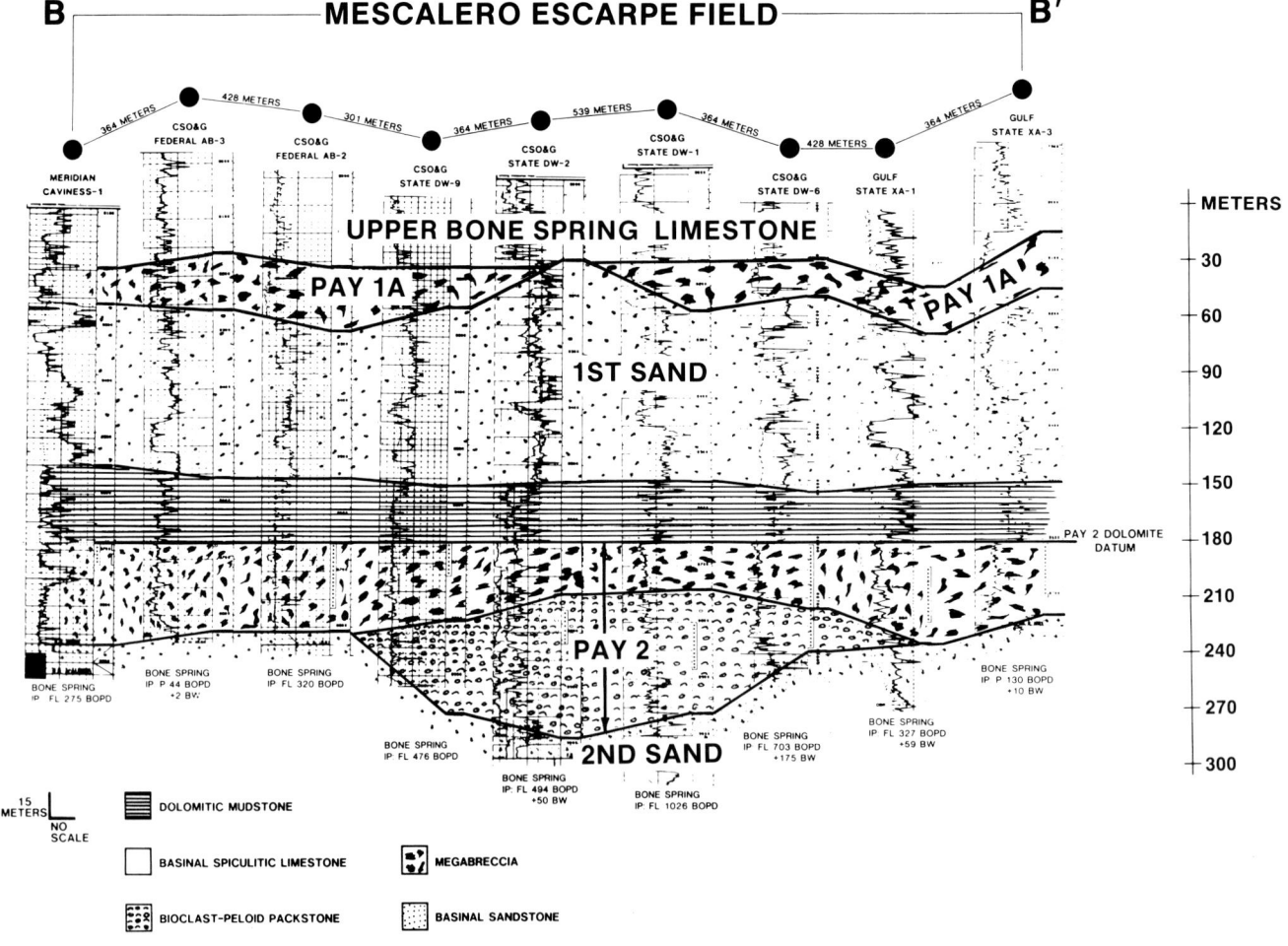

FIG. 9.—Stratigraphic cross section of slope strata roughly parallel to depositional dip. See Figure 3 for location.

with as much as 14% porosity are present; however, they are generally surrounded by a less porous and permeable matrix. The dolomitized megabreccia facies tends to be more porous and permeable in a basinward direction, which corresponds to the matrix becoming more grain rich in a basinward direction. Open fractures are rare in this facies.

Dolomitized bioclast-peloid packstone.—

The dolomitized bioclast-peloid packstone facies is a gray to light tan bioclastic-peloid packstone with subordinate amounts of interbedded wackestone and grainstone (Fig. 4b). This facies is as much as 70 m thick and is dominantly dolomite with minor anhydrite and quartz. Most dolomite replaces original calcareous material, but minor void-fill dolomite cement is also present. Anhydrite is present as replacive nodules and void-fill cement. Quartz occurs mainly as late void-fill cement. This packstone facies is much more grain rich than the matrix in the megabreccia facies. Fossil fragments are the dominant grain type in lower parts of this facies (Fig. 7a). Fossils include crinoids, brachiopods, molds of fusulinids, sponge(?) fragments, and other unidentified shelly fragments. Peloids increase in abundance upward and become dominant in upper parts of this facies. Some grains have been partially micritized (Fig. 7a).

No current structures were observed in the dolomitized bioclast-peloid packstone lithofacies. Vertical burrows, bored hardgrounds, *in situ* breccias, cracks, and fractures are scattered throughout this unit and are often concentrated below bedding surfaces (Fig. 7b, c). Thin fractures are common and often partially open (Fig. 4b). Most fractures are vertical to slightly inclined and are discontinuous with few being more than 5 cm long. In some parts of the bioclast-peloid packstone facies, intense fracturing has resulted in *in situ* breccias. Some fractures and areas around *in situ* breccia clasts are filled with micritic sediment. Often that micritic sediment appears to predate deposition of overlying strata (Fig. 7c). Stable carbon- and oxygen-isotopic analyses of the bioclast-peloid packstone facies are shown in Figure 6. Porosity in this facies ranges from 2.1 to 19.6% (average of 7.4%) with moldic, intergranular, and intercrystalline pores being dominant. Permeabilities are 0.01 md to 10 d (average of 229 md) with open fractures as much as 0.5 mm wide enhancing permeability.

Dark laminated mudstone.—

The laminated mudstone facies (Fig. 4c) is composed of microcrystalline dolomite (10–90%), quartz (10–60%; including fine silt and sponge spicules), clay (1–20%), and minor to trace amounts of K-feldspar, plagioclase, pyrite, and siderite. Basinward, these mudstones are calcitic rather than dolomitic. For example, calcite is dominant in carbonate mudstones in the Trainer McElvain #4, which is

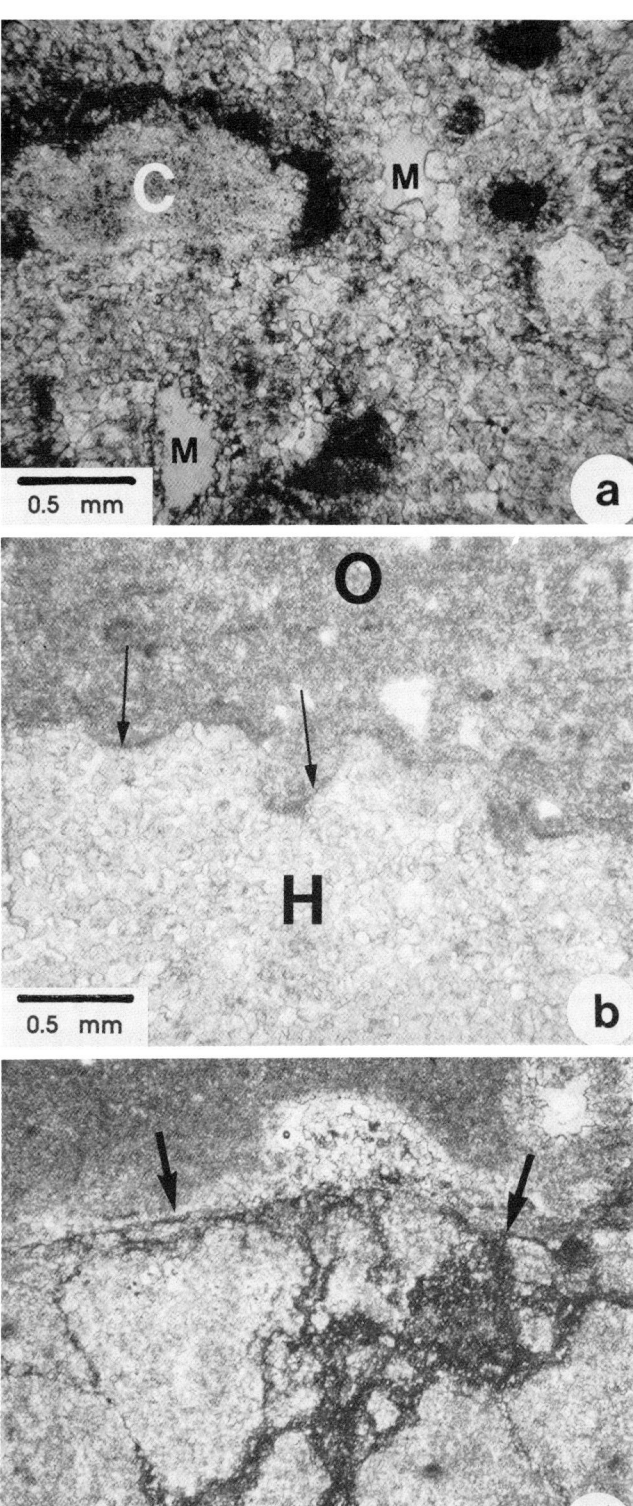

FIG. 7.—Thin-section photomicrographs of dolomitized bioclast-peloid packstone facies in the C.S.O.G. State "DW" #8 well. (a) Dolomitized bioclastic packstone with molds of bioclasts (M) that are lined with subhedral to euhedral dolomite crystals. Partially micritized crinoid fragment (C) is also present (2,720 m). (b) Bored hardground. Light-colored dolomite (H) in lower half of photograph has a scalloped (bored) upper surface (arrows), suggesting this was a syndepositional hardground that was bored and subsequently overlain by darker, finer crystalline sediment (O; 2,707 m). (c) Cracks and fractures, abundant below the bedding surface (arrows), end at the bedding surface. They formed and were infilled with micritic sediments (dark material) prior to deposition of strata above the bedding surface (2,698 m).

approximately 7 km basinward of the Yeso shelf margin (Fig. 8). Thin (1–3 cm thick), graded fossiliferous packstone to wackestone beds are rare, being present in only a few parts of the mudstone unit. Fossils in those wackestone-packstone interbeds include crinoids, brachiopods, and calcareous algae, which are often silicified. Sponge spicules are scattered throughout the laminated mudstone facies and are locally abundant (making the rock a wackestone). Except for sponge spicules and fossiliferous interbeds, this facies is unfossiliferous. Parallel laminations are common in the mudstones. Some laminations are silica rich and others are dolomite rich.

The dark laminated mudstones are organic rich. Eight samples of dark laminated dolomite mudstone from depths of 2,435 to 3,838 m (Wolfcampian and Leonardian) were analyzed from Mescalero Escarpe wells and wells along depositional strike. Total organic carbon values range from 0.8 to 4.3 weight percent with an average of 2.0 weight percent. Geochemical parameters (gas chromatograph patterns, carbon isotopes, butane-pentane ratios, aromatic hydrogen-carbon ratios, pristane-phytane-nC_{17}–nC_{18} ratios, biomarkers) indicate a fair to good correlation of organic material in the mudstone to the oil in Mescalero Escarpe field with the oil being best correlated to a mixture of samples from several different levels (W. Fallgatter, pers. commun., 1985). Results of stable carbon and oxygen isotope analyses from this facies are shown in Figure 6. The dark laminated mudstones act as an impermeable seal for the Pay 2 dolomite reservoir in an updip direction.

Very fine-grained sandstone.—

The very fine-grained sandstone lithofacies (Fig. 4d) is very poorly sorted, containing much silt and clay in addition to very fine-grained sand. Sand and silt grains are mainly angular to subangular quartz with minor feldspar. Significant amounts of clay (10–30%) and dolomite cement (10–30%) are also present in the very fine-grained sandstone. Sedimentary structures include horizontal laminations interrupted by burrows, wavy laminations with abundant bioturbation, horizontal current laminations, ripple-scale cross-

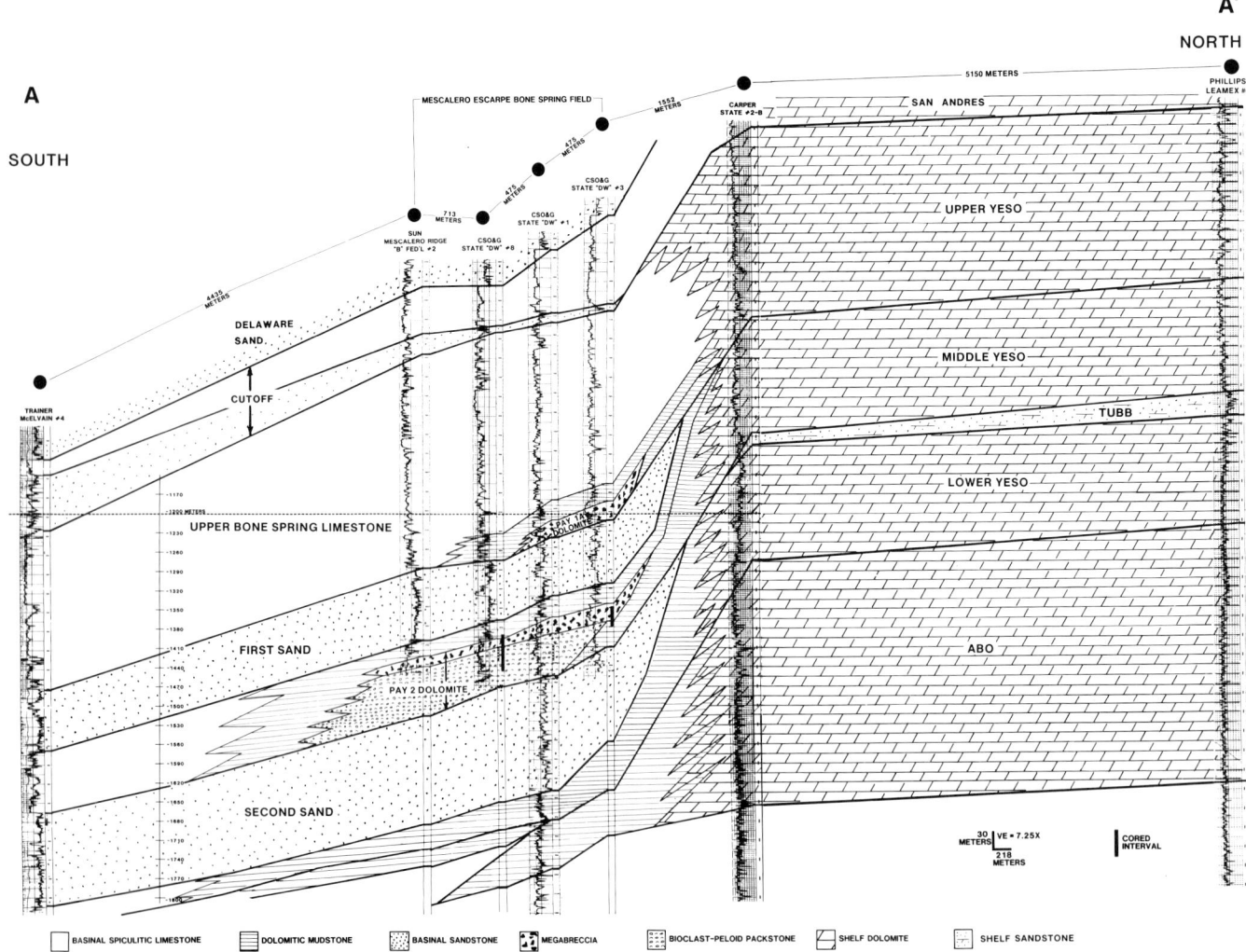

Fig. 8.—Structural cross section along depositional dip from the Yeso Shelf to the north across the Mescalero Escarpe field into the Delaware Basin. See Figure 11 for location of cross section.

beds, and partial Bouma sequences. These sandstones are not reservoir rocks in Mescalero Escarpe field.

Interpretation of lithofacies.—

The dolomitized megabreccia was apparently deposited by episodic debris flows that carried shallow-marine lithoclasts into the deep basin. Textural evidence supporting deposition by debris flows includes: (1) poor sorting, (2) large, angular to irregularly shaped clasts, (3) abundant micrite, (4) lack of burrows or current structures, and (5) maximum clast size decreasing upward near the top of this unit (Fig. 4a). The megabreccia lithofacies is similar to carbonate debris flow deposits described by Cook and Mullins (1983). Wiggins and Harris (1985) noted similar features and concluded that Bone Spring detritus at Airstrip field approximately 10 km to the east was also deposited by debris flows. A lack of rounding suggests that these clasts were never subjected to abrasion on a beach or in a high-energy shallow-marine environment. The stable carbon and oxygen isotope composition of clasts are characteristic of sea water at near-surface temperatures and do not show decreased $\delta^{13}C$ or $\delta^{18}O$ values that are characteristic of freshwater diagenesis (Fig. 6; Allan and Matthews, 1977). Therefore, megabreccia clasts were probably not eroded from an island during subaerial exposure. Lithoclasts commonly have scalloped edges similar to those caused by boring sponges in modern environments (Fig. 5b, c; Futterer, 1974); therefore, marine organisms (especially sponges) probably helped erode the Yeso shelf and produce megabreccia clasts in a submarine environment. Thin interbeds of dark, silty, carbonate mudstone within the megabreccia probably represent slow rains of windblown silt and micritic sediment that accumulated between debris flow events.

The dolomitized bioclast-peloid packstone lithofacies was probably deposited in deep open-marine waters. Micritization of grains suggests derivation of those grains from a shallow shelf (Bathurst, 1975). Peloids and fossil fragments were carried off the shelf, down channels in the upper slope, and deposited at the toe of slope. Shelf-derived bioclastic material is common in modern slope environments (Cook and Mullins, 1983). Vertical burrows and bioturbation in the bioclast-peloid packstone indicate deposition in oxidized waters with near-normal-marine salinities. Hardgrounds, borings, cracks, and breccias infilled by sediments prior to deposition of overlying strata (Fig. 7c) indicate lithification penecontemporaneous with deposition. Similar early lithification is observed in modern slope environments (Mullins, 1986). Micrite filling fractures and areas around *in situ* breccia clasts suggests fracturing and brecciation in the bioclast-peloid packstone facies prior to deposition of overlying strata (Fig. 7c). Fracturing and lithification were, thus, penecontemporaneous with deposition of the bioclast-peloid packstone facies. A lack of discernable vertical offset and little vertical continuity of fractures suggest that most formed in response to local, non-tectonic, tensional stresses.

The dark laminated mudstone lithofacies of the Bone Spring Formation accumulated in slope and basinal environments when and where coarser, shelf-derived detritus was not accumulating (Harms, 1974; Wiggins and Harris, 1985). Thin, graded interbeds of fossiliferous packstone and wackestone in this lithofacies apparently represent small pulses of shallow-water detritus swept off the Yeso shelf or slope.

The very fine-grained sandstone lithofacies in the Mescalero Escarpe field was deposited in slope and basinal environments. Current laminations and Bouma sequences suggest episodic deposition of parts of this facies by turbidity currents, whereas abundant burrows within other parts of this facies suggest slower deposition.

Interpretation of dolomitization.—

Lithification of the Bone Spring Formation in and around Mescalero Escarpe field occurred penecontemporaneously with deposition and helped maintain relatively steep-slope angles. Early lithification apparently resulted from dolomitization. Therefore, understanding the processes involved in dolomitization is important to understanding the development of slope carbonates at the northern end of the Delaware Basin. Carbonate in the Bone Spring Formation in the Mescalero Escarpe field has been pervasively replaced by dolomite. The Pay 2 dolomite was deposited in relatively deep water and is surrounded by hundreds of meters of deep-marine carbonate, sandstone, and shale, making mixed-water dolomitization unlikely. Wiggins and Harris (1985) studied dolomitization at Airstrip field and concluded that (1) early replacive dolomite formed during very shallow burial with magnesium coming from marine-derived pore fluids expelled from basinal mudstones, (2) much of the dolomite precipitated during intermediate and late burial, and (3) dolomitization of megabreccia clasts occurred on the shelf.

At Mescalero Escarpe, petrographic and geochemical evidence support pervasive dolomitization in sea water shortly after deposition. Fracturing of the bioclast-peloid packstone lithofacies required lithification of the sediments. Dolomitization was probably the lithifying process. Dolomite adjacent to fossil molds has subhedral-to-euhedral terminations, indicating dolomitization during or after formation of the molds (Fig. 7a). Dolomite crystals adjacent to fractures are generally anhedral or fractured, suggesting dolomitization prior to fracturing. Since most fracturing occurred penecontemporaneously with or very shortly after deposition of the bioclast-peloid packstone (Fig. 7c), pervasive dolomitization at Mescalero Escarpe field apparently took place very shortly after deposition, while carbonate sediments were in open communication with marine water.

Stable carbon and oxygen isotope ratios of the Pay 2 dolomite support dolomitization by sea water or slightly modified sea water at approximately surface temperatures (Fig. 6). The $\delta^{13}C$ values are similar to those found in modern sea water and marine carbonate sediments (Gross, 1964; Hudson, 1977). Because carbonate is commonly dissolved and reprecipitated at higher temperatures during burial diagenesis, the $\delta^{18}O$ values of original dolomites would tend to be greater than or equal to the $\delta^{18}O$ compositions of dolomites recovered from the subsurface, which have been subjected to burial diagenesis. Therefore, the most positive $\delta^{18}O$ values (+1.5‰, PDB) should most closely resemble the composition of the original replacement dolomite. Dolomite with a $\delta^{18}O$ value of +1.5‰ could precipitate from modern sea water at approximately 26°C (Land, 1985). The micritic matrix between clasts in the megabreccia has stable isotopic compositions similar to the clasts, suggesting dolomitization of clasts and matrix at the same time or in similar waters. Dolomite in dark laminated mudstones has lower $\delta^{13}C$ values than dolomite in the megabreccias and peloid-bioclast packstones (Fig. 6). Incorporation of some organic carbon ($\delta^{13}C = -25‰$, PDB) during dolomitization is probably responsible for dolomite in the dark mudstone having lower $\delta^{13}C$ values than the Pay 2 dolomite (Fig. 6).

Dolomitization by deep-marine waters undersaturated with respect to calcite has been described previously by Saller (1984) and Mullins and others (1985). Dissolution, which created moldic porosity, also probably occurred in deep-marine waters undersaturated with respect to calcite during dolomitization. It is not clear why Bone Spring mudstones are calcitic approximately 4–5 km basinward (to the south) of Mescalero Escarpe field, but two possible explanations are described: (1) dolomitization occurred on steeper, shallower parts of the slope, where submarine currents were able to move dolomitizing sea water through the sediments; deeper in the basin, there were less relief and probably weaker submarine currents, making it difficult to move dolomitizing water by or through the sediments; (2) carbonate mud on the slope near the shelf might have originally been more aragonite rich than more basinward muds because aragonite was largely derived from the shelf. Aragonite is often more reactive and more subject to dolomitization than calcite (especially low-magnesium calcite; Walter and Morse, 1984).

STRATIGRAPHY AND DEPOSITIONAL SYSTEMS

Mescalero Escarpe field is located at the northern margin of the Bone Spring basin and immediately south of the Abo reef trend and the Yeso shelf margin (Fig. 1). Electric logs and seismic sections have been used together to determine the stratigraphic relations of shelf, slope, and basinal Leonardian strata.

Distribution of slope lithofacies.—

Using gamma ray and neutron-density logs, the four lithologic facies were correlated into all wells in Mescalero Escarpe Field. The distribution of slope facies is shown in Figures 8 and 9. Electric-log correlations suggest 425 m of relief and a 15° slope between the Yeso shelf margin and toe of slope (not corrected for possible differential compaction; Fig. 8). The toe of slope is an inflection point where slope steepness decreases markedly. When present, the do-

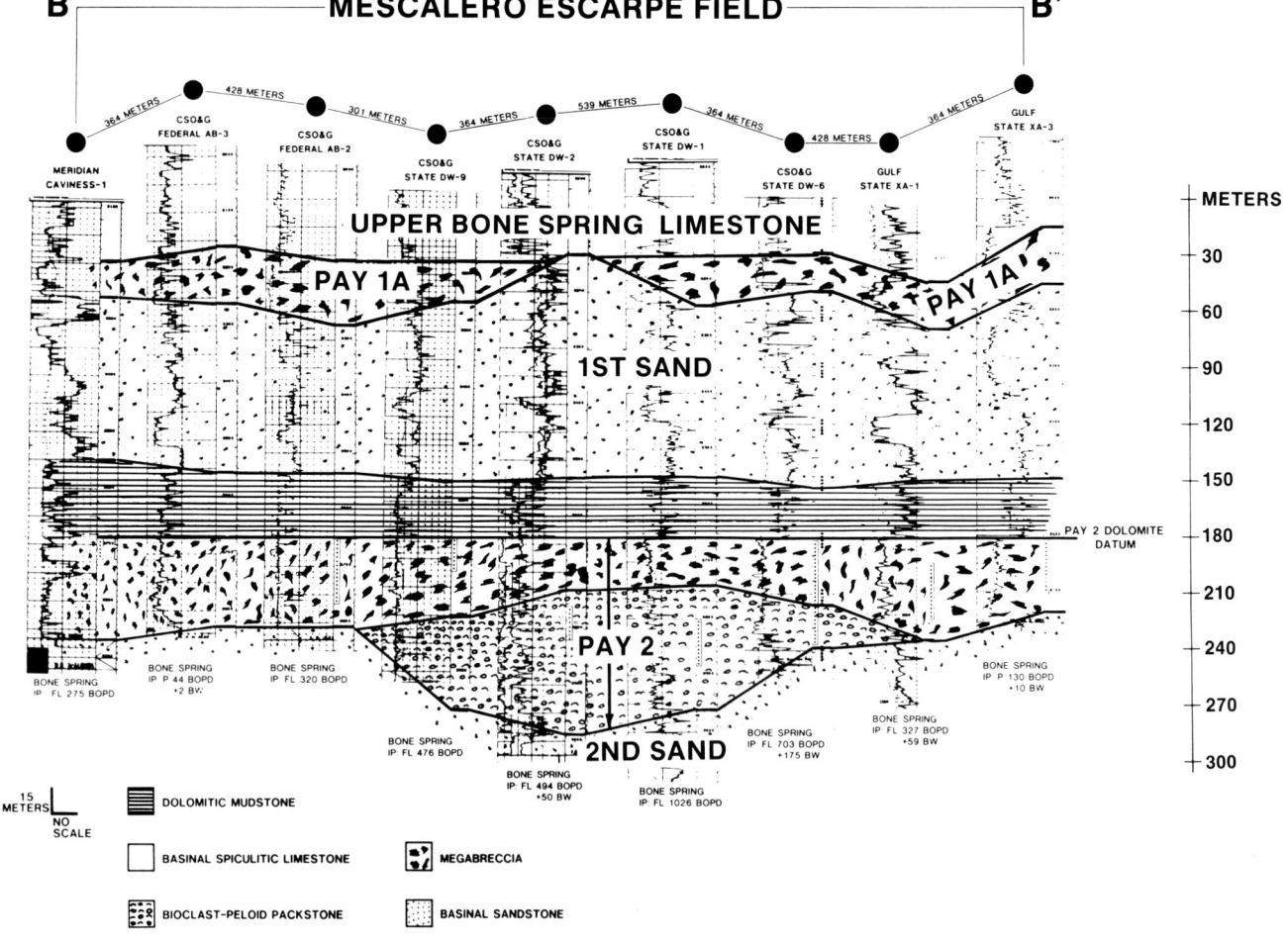

FIG. 9.—Stratigraphic cross section of slope strata roughly parallel to depositional dip. See Figure 3 for location.

lomitized bioclast-peloid packstone is overlain by the dolomitized megabreccia facies (Figs. 8, 9). The dolomitized megabreccia is present in all wells in Mescalero Escarpe field. The Pay 2 dolomite is thickest at the toe of slope in the middle of the field and thins away from the middle of the field in all directions (Figs. 3, 8, 9). Up dip (to the north), it thins and pinches out into dark carbonate mudstones (Fig. 8). Basinward, the Pay 2 dolomite thins gradually into dark laminated mudstones (Fig. 8). Well log correlations suggest deposition of carbonate mud at 2,643–2,648 m in the C.S.O.G. #3 State 'DW' well (below the megabreccia) at approximately the same time that bioclast-peloid packstones were being deposited less than 350 m basinward (to the south; Fig. 8). In more basinward portions of the field, the Pay 2 dolomite is underlain by the very fine-grained sandstone facies (Second Sand), whereas in more shelfward locations, the Pay 2 dolomite is underlain by the dark laminated mudstone facies. Depositional features at the contact of the megabreccia and underlying dark laminated mudstone suggest that the mudstone was soft at the time the lowest parts of the megabreccia were deposited (Fig. 5a). Log correlations indicate that a fairly uniform thickness of dark laminated mudstones overlies the Pay 2 dolomite (Figs. 8, 9).

Structure and isopach maps as well as cross sections along depositional strike indicate that the dolomitized bioclast-peloid packstone and megabreccia facies are concentrated in channels developed on the top of the Second Sand. Along strike, the Pay 2 dolomite pinches out against the very fine-grained sandstone facies (Fig. 9). Structural contours at the top of the Pay 2 dolomite show a relatively smooth regional structural trend (Fig. 10). Structural contours at the top of the Second Bone Spring Sand (below the Pay 2 dolomite; Fig. 11) show channels in the top of the Second Sand that are independent of the regional structure trend (Fig. 10). The location of the channels corresponds to the depocenter of the Pay 2 dolomite (Fig. 3). Dolomitized bioclastic-peloid packstones are only present in the lowest part of the channels (Fig. 9).

Seismic stratigraphy.—

Seismic lines show that the Northwestern Shelf margin grew vertically during Leonardian time (Figs. 12, 13).

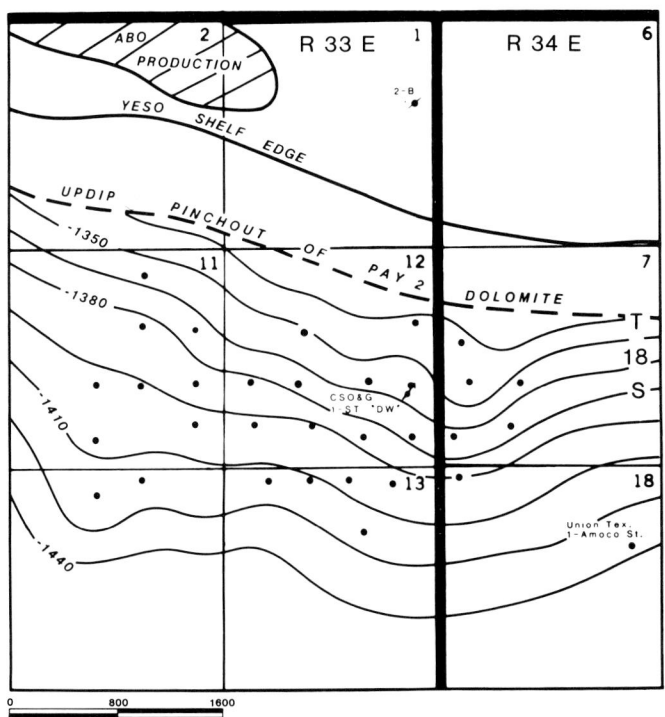

Fig. 10.—Structure contour map of top of Pay 2 dolomite in vicinity of Mescalero Escarpe field, Lea County, New Mexico.

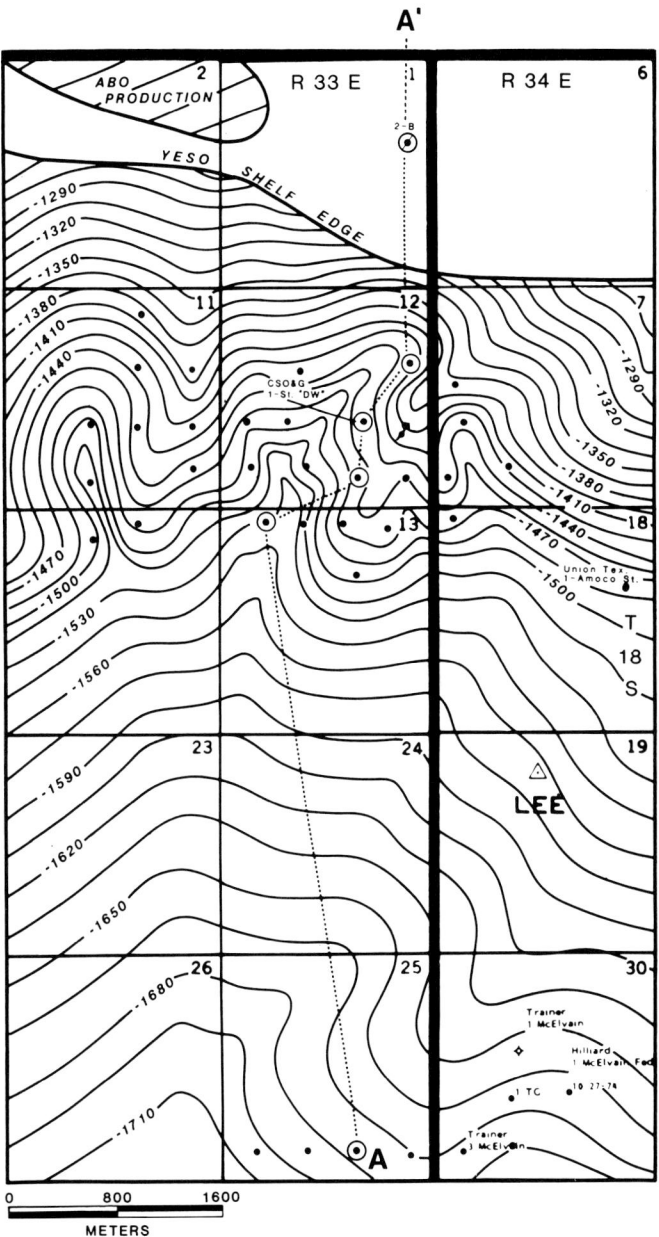

Fig. 11.—Structure contour map of top of Second Bone Spring Sand in the vicinity of Mescalero Escarpe field, Lea County, New Mexico. Section A-A' is shown in Figure 8. The northern end of section A-A' is approximately 4 km north of this map.

Seismic lines indicate 530 m of relief and a 12° slope between shelf edge and toe of slope during Middle Yeso time at the location of Mescalero Escarpe field (not corrected for possible differential compaction; Fig. 13). During periods of carbonate shelf aggradation, shelf detritus was shed into the adjacent slope environment. During Leonardian time, sedimentation in slope and basinal environments alternated between carbonate and sandstone (Fig. 8). Carbonate wedges on the slope correlate with periods of upbuilding on the shelf (Figs. 8, 12, 13). Those wedges thin and downlap basinward of the toe of slope (Figs. 8, 13). Basinal reflections (mainly sandstones) onlap those slope and toe-of-slope (carbonate) reflections (Figs. 12, 13). The Second Sand overlies a strong slope reflection at the top of the Lower Yeso. Strong positive reflections commonly occur where sandstones overlie dense slope limestones. The top of the Second Sand can be traced from Mescalero Escarpe field up slope to a position corresponding to the Tubb on the shelf (Figs. 12, 13). These two relations indicate deposition of the Second Sand after deposition of the Lower Yeso at a time that approximately corresponds to Tubb deposition (Figs. 8, 12, 13).

The synthetic seismogram from the C.S.O.G. "DW" #1 well indicates that the strong reflection immediately above the Pay 2 dolomite is at the interface between the dense laminated mudstone and the overlying First Sand (Figs. 8, 12, 13). The top of the Pay 2 dolomite is slightly below the middle of that strong reflection. Truncation (erosion) of Middle Yeso shelf reflections is observed below the strong reflection overlying Middle Yeso strata (Figs. 12, 13). The strong reflection above the Pay 2 carbonate wedge apparently correlates with the truncation of reflections at the top of the Middle Yeso, suggesting that Middle Yeso carbonates on the shelf correlate with the Pay 2 carbonate wedge on the slope (Figs. 12, 13). Truncation of shelf reflections is approximately correlative with deposition of the upper part of the Pay 2 dolomite, suggesting truncation of the shelf during deposition of the megabreccia (Figs. 12, 13). Detritus eroded during truncation of the shelf is present in the megabreccia facies and includes clasts of shallow-water carbonate.

Depositional system.—

The Bone Spring Formation generally accumulated in a deep-marine environment (Harms, 1974). Leonardian shelf margin carbonates of the Northwestern Shelf accumulated

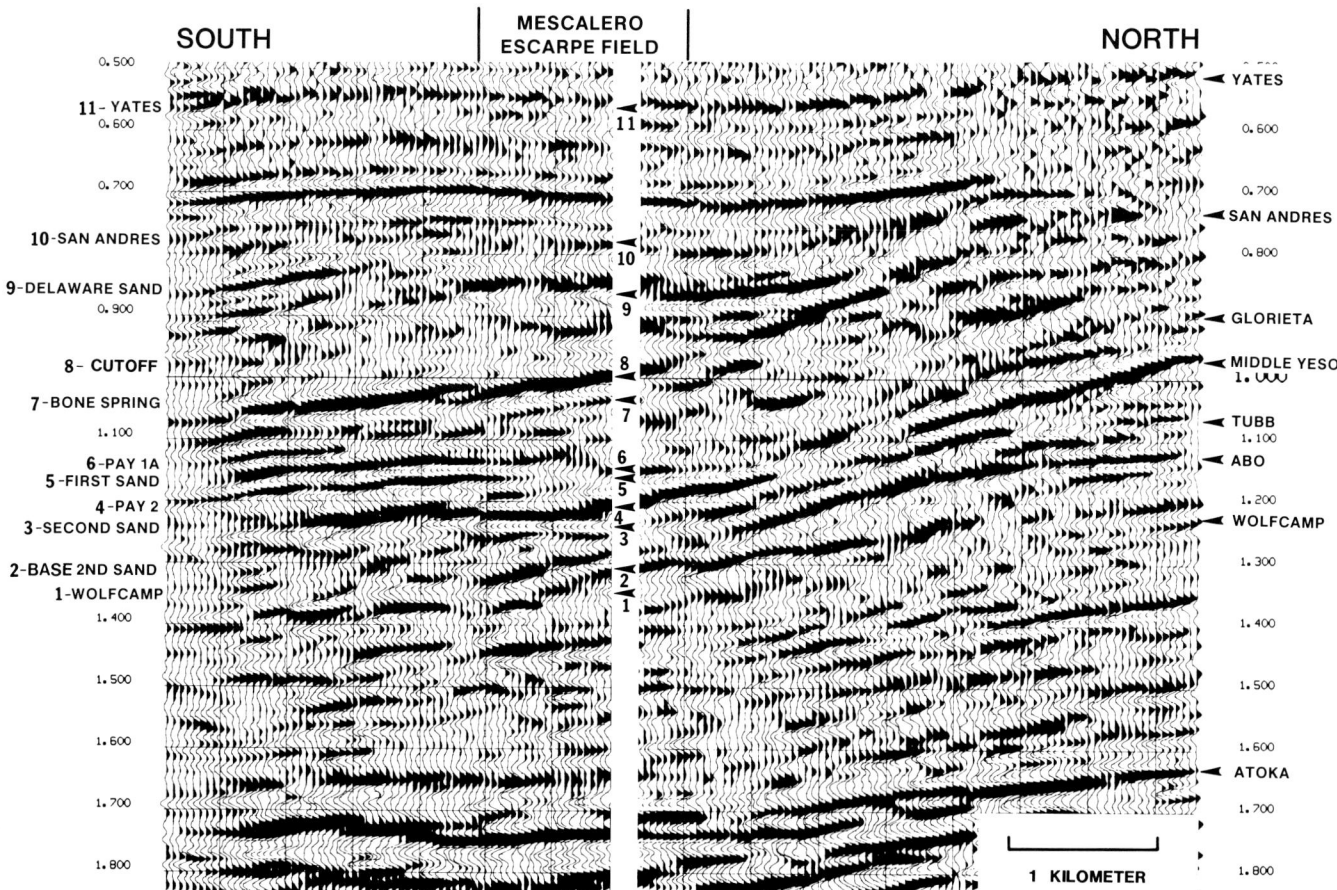

FIG. 12.—North-south (dip) seismic section across Mescalero Escarpe field. The seismic section extends from Yeso shelf margin (at right) into northern part of Delaware Basin (at left). Marks at right are tops of stratigraphic units determined by constructing synthetic seismogram from sonic log from Carper State #2-B well. White strip to left of center is at approximate location of C.S.O.G. "DW" #1 well, where a synthetic seismogram was also constructed from sonic log. Marks represent tops of various stratigraphic units described at left.

near sea level (Silver and Todd, 1969). Electric-log cross sections and seismic sections parallel to depositional dip (Figs. 8, 12, 13) show that deposition of bioclastic-peloid packstone and megabreccia facies was concentrated at the toe of slope. Isopach maps, structure-contour maps, and electric-log sections parallel to depositional strike indicate that deposition of the Pay 2 dolomite was also concentrated in or near the mouths of submarine channels (Figs. 3, 9–11). Shelf detritus in the Pay 2 dolomite was probably carried through channels on the upper slope and deposited at the toe of slope where steepness decreased markedly, while carbonate mud accumulated on the upper slope (Fig. 8). The 425–530 m of relief between Yeso shelf margin and toe of slope suggests deposition of bioclast-peloid packstone and megabreccia facies at depths of approximately 425–530 m. Wiggins and Harris (1985) describe 450–610 m of relief and a 3–6° slope on the Leonardian slope 10 km to the east at Airstrip field.

Similar carbonate depositional systems have been described in modern and ancient environments. On the northern margin of the modern Little Bahama Bank, hemipelagic muds accumulate on the steep upper slope (600–900 m; 4° slope) and coarse-grained, rubbly muds accumulate on the less steep lower slope (900–1,100 m; 1–2° slope; Mullins and others, 1984). The upper slope of the northern Little Bahama Bank is cut by numerous channels that serve as conduits for debris flows transporting rubbly muds to the lower slope. Deposition of rubbly mud transported by debris flows is greatest at the boundary between the upper and lower slope (toe of slope; Mullins and others, 1984). In contrast to coarse Bone Spring detritus, coarse grains in rubbly muds of the Bahamas are clasts eroded from the slope.

Floatstones were recovered during ODP Leg 101 in slope strata below the Northeast Providence Channel in the Bahamas (Austin and others, 1986). The floatstones contain poorly sorted, light-colored clasts of shelf and slope limestone in a darker, micritic matrix. The floatstones are Pliocene and Pleistocene, yet contain clasts ranging from the

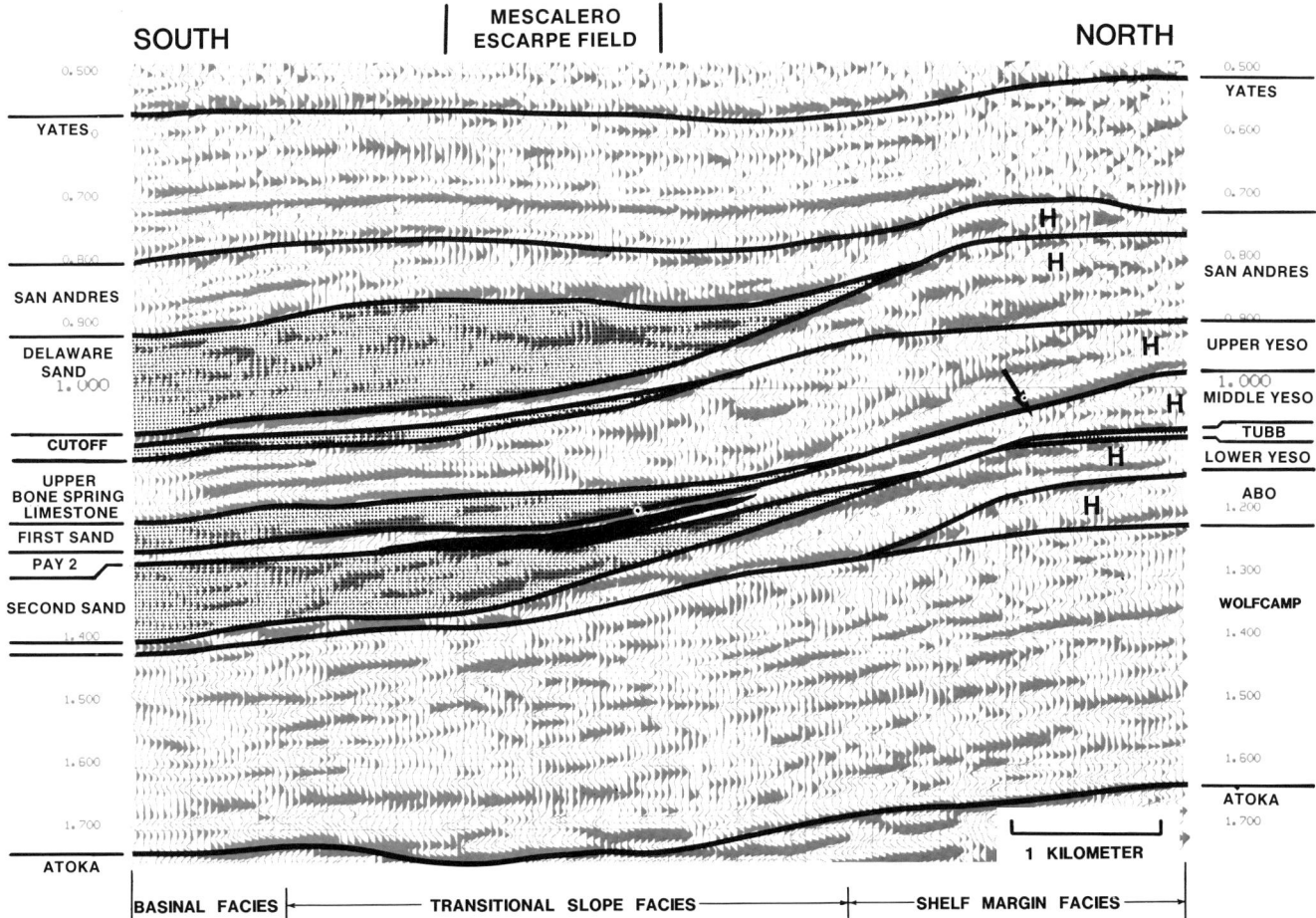

FIG. 13.—Interpretation of seismic section shown in Figure 12. Interpretations include Abo through San Andres strata and do not extend to strata above San Andres or below Abo. Labels at left and right refer to stratigraphic intervals between overlying and underlying lines, except for "Yates" and "Wolfcamp," which include other undifferentiated units below them. Heavy lines are sequence boundaries. Abo and Yeso strata on shelf are dominantly carbonate. Reflections that onlap carbonate slope strata are present in First and Second Bone Spring Sands and in Delaware Sand. Most carbonates (H; unshaded) are thought to be deposited during highstands of sea level, whereas basinal sands (shaded) are thought to be deposited during lowstands of sea level. The Tubb is interpreted as a lowstand shelf sand. The blackened area at toe of slope between First and Second Sands represents Pay 2 dolomite. Arrow shows truncation of a Middle Yeso reflection below a sequence boundary. Note erosion and backstepping shelf at top of Middle Yeso (arrow). Interpretations of San Andres, Cutoff, and Delaware Sand stratigraphy are similar to those of Sarg and Lehmann (1986).

Cretaceous to Miocene. The floatstones are interpreted as debris flow and turbidite(?) deposits with clasts eroded from Bahamian platforms and deposited in deeper slope environments (Austin and others, 1986). These floatstones are texturally similar to Bone Spring megabreccias and probably formed by similar mechanisms.

During Bone Spring time, carbonate sedimentation in the Delaware Basin alternated with siliciclastic deposition. Silver and Todd (1969) correlate intervals of carbonate deposition on Leonardian shelves with carbonate deposition in basins, and siliciclastic deposition on shelves with siliciclastic deposition in basins. Data acquired in this study support those correlations (Figs. 8, 12, 13). Alternations between carbonate and siliciclastic deposition were probably associated with fluctuations in relative sea level (Fig. 13). Presumably, shelf carbonate units accumulated when the shelf was under water, which would occur when relative sea level was high. Rates of carbonate sedimentation in Bahamian slope environments were greater during Pleistocene highstands of sea level, when the Bahamian banks were flooded, than during glacial lowstands of sea level when the banks were subaerially exposed (Droxler and Schlager, 1985; Boardman and others, 1986). Fluctuations in sea level might have been caused by advances and retreats of continental glaciers present during the Early Permian (Veevers and Powell, 1987). Therefore, carbonate deposition dominated both on the shelf and in the basin during high sea-level stands, and siliciclastic deposition occurred during low sea-level stands (Fig. 13; Silver and Todd, 1969; Gawloski, 1987). The following scenario is proposed for cyclic deposition of Leonardian shelf, slope, and basin strata in the vicinity of Mescalero Escarpe field.

(1) Rising sea level caused flooding of the Northwestern Shelf. Carbonate sediments formed on and built up the shelf. Peloids, bioclasts, and micrite were washed off the shelf. Bioclast-peloid packstones accumulated in channels at the toe of slope, and carbonate mud accumulated on steeper parts of the slope and deeper in the basin. The general decrease in bioclasts and increase in peloids upward in the bioclast-peloid packstone may reflect the Northwestern Shelf becoming shallower and more restricted during deposition of Middle Yeso carbonates (Fig. 14, T1).

(2) Most of the shelf built to sea level (due to carbonate sedimentation or sea-level fall) causing production of carbonate sediment on the shelf to decrease greatly. The shelf ceased to build upward. Only minor amounts of sediment were swept off the shelf and onto the shelf margin and slope, leaving the edge of the shelf exposed to physical and biological erosion (in marine waters). Periodically, the shelf margin became unstable, and debris flows carried lithoclasts, micrite, and minor amounts of bioclasts from the shelf into deep-marine environments, forming the megabreccia. Away from megabreccia deposition, carbonate mud accumulated on the slope and deeper in the basin (Fig. 14, T2).

(3) Sea level was near or below shelf level. Most carbonate production on the shelf ceased. Siliciclastic sand prograded across the shelf and was carried into the basin. Siliciclastic sand dominated deposition in the basin most of the time, but small pulses of shelf carbonate (lithoclasts, fossils, and micrite) were occasionally carried into the basin (Fig. 14, T3).

ACCOMMODATION RATES AND SHELF MARGIN MORPHOLOGY

The Northwestern Shelf margin in the vicinity of Mescalero Escarpe field was characterized by vertical growth during the Leonardian with slope angles of 10–15°. At other locations, Leonardian strata of the Northwestern Shelf and Northern Shelf are characterized by lower slope angles and progradation (Fig. 15). Approximately 40 km west of Mescalero Escarpe field (in the vicinity of Empire Abo field), the Northwestern Shelf margin in middle and upper Leonardian strata is characterized by progradation and slope angles of 6° or less (Sarg, 1989). Likewise, the Northern Shelf adjacent to the Midland Basin prograded markedly during

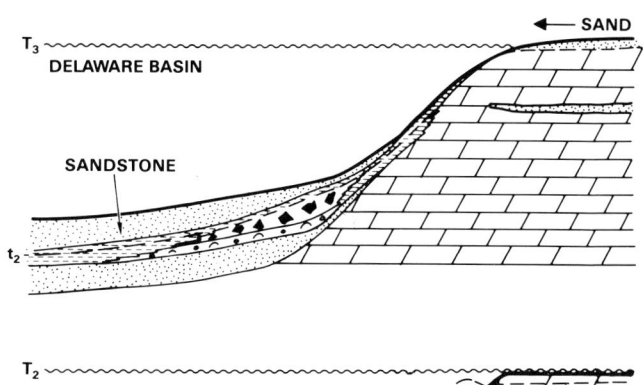

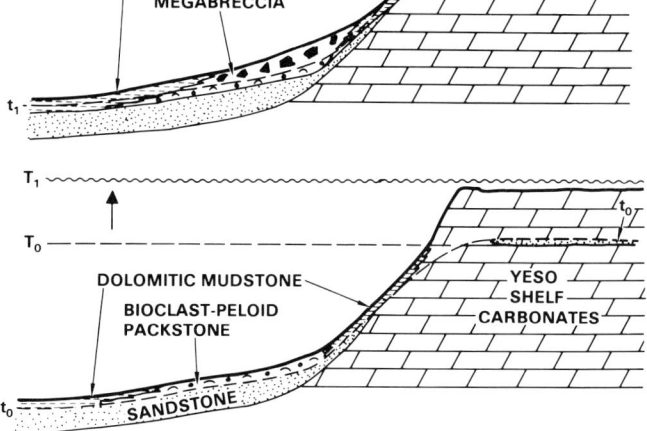

FIG. 14.—Model for deposition on slope at northern margin of the Delaware Basin during middle Leonardian time. (T_0 to T_1): rapid sea-level rise floods shelf, causing rapid vertical accretion of shelf. Peloids, bioclasts, and micrite are washed off shelf and down slope. Peloids and bioclasts are deposited at toe of slope, and carbonate mud is deposited on steeper parts of slope and in basin. (T_1 to T_2): sea-level rise slows or stops, and shelf builds to sea level. Shelf margin stops aggrading, and physical and biological processes begin to erode shelf margin. Periodically, debris flows carry eroded clasts of shelf debris into basin, depositing them as megabreccias at toe of slope. (T_2 to T_3): sea level remains constant or drops. Shelf is at sea level or exposed. Terrigenous sand is carried across shelf and deposited in basin.

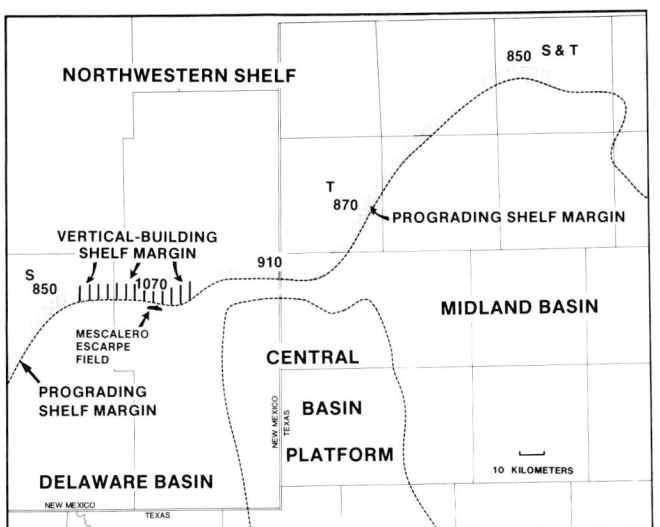

Fig. 15.—Map showing location of vertically building and prograding Leonardian shelf margins along Northwestern and Northern Shelf. Numbers indicate thickness (in meters) of Leonardian shelf margin strata at various locations. Thicknesses were always measured at the location of the Abo shelf margin. "T" and "S" superscripts refer to external sources of data ("T"–data derived from Silver and Todd, 1969; "S"–data derived from Sarg, 1989).

Leonardian time (Silver and Todd, 1969; Mazzullo and Reid, this volume). Slope angles seaward of the prograding Northern Shelf margin are 5–10° (Sarg, 1989).

Approximately 1,070 m of strata (mainly limestone and dolomite) accumulated on the Northwestern Shelf margin near Mescalero Escarpe during Leonardian time (Podpechan, 1959; Fig. 8). These limestones and dolomites were deposited in a shallow-marine environment and were probably lithified shortly after deposition. This would suggest deposition near sea level with minimal compaction. Therefore, the Northwestern Shelf margin in the vicinity of Mescalero Escarpe field experienced approximately 1,070 m of subsidence plus sea-level rise ("accommodation") during Leonardian time.

During the same interval of time, approximately 850 m of strata (dominantly carbonate) accumulated on the Northwestern Shelf margin 40 km west of Mescalero Escarpe field in the vicinity of Empire Abo field (Fig. 15; Sarg, 1989), suggesting approximately 850 m of accommodation during Leonardian time. Approximately 850 m of Leonardian, dominantly carbonate, strata accumulated on the Northern Shelf at the northern margin of the Midland Basin (Fig. 15; Silver and Todd, 1969, Fig. 10; Sarg 1989), suggesting 850 m of accommodation during Leonardian time. The Leonardian lasted approximately 10 m.y. (Harland and others, 1982). Therefore, an accommodation rate of 85 m/m.y. allowed shelf progradation into the northern Midland Basin and into the northwestern Delaware Basin (near Empire Abo field) during the Leonardian. In contrast, an accommodation rate of 107 m/m.y. at the northern margin of the Delaware Basin near Mescalero Escarpe field was too rapid to allow progradation of the shelf, but was slow enough for the carbonate shelf to keep pace by building vertically. During the Leonardian, all three shelf margin areas experienced similar "eustatic" sea-level fluctuations, so greater tectonic subsidence must have been one of the factors causing vertical building (aggradation) of the shelf margin near Mescalero Escarpe field. Other factors, such as amounts of terrigenous sediment supplied to the slope and basin, may have been important for progradation (Mazzullo and Reid, this volume). There were periods when carbonate shelf and slope accretion ceased because the shelf was either at or above sea level, causing carbonate sedimentation to cease or at least decrease greatly. Therefore, the above accommodation rates do not represent the absolute rates of carbonate sedimentation.

CONCLUSIONS

Carbonate sedimentation on the Northwestern Shelf and northern end of the Delaware Basin was apparently controlled by rapid regional subsidence and cyclic sea-level fluctuations during Leonardian time. Regional subsidence was rapid enough in the vicinity of Mescalero Escarpe field that the Northwestern Shelf grew vertically rather than prograding laterally. Cyclic sea-level fluctuations may have helped prevent lateral progradation by shutting down the "carbonate factory" during lowstands or stillstands of sea level. Depositional cycles can be divided into three main parts and related to sea-level fluctuations. (1) During rapid sea-level rises, the shelf was flooded and carbonate sedimentation was initiated on the shelf. Some sediments were swept off the shelf and into the basin, forming bioclast-peloid packstones at the toe of slope and dolomitic mudstones in other slope and basinal environments. (2) As sea-level rise slowed or stopped, the vertically building shelf approached sea level and the shelf-wide "carbonate factory" ceased or was greatly reduced. Because significant amounts of carbonate sediment were no longer being generated on the shelf and transported to the shelf margin and slope, the shelf margin and slope were subjected to physical and biological erosion while still in a submarine environment. Debris flows subsequently transported the eroded detritus to the toe of slope, while deposition of carbonate mud continued in adjacent slope and basin environments. (3) At the end of depositional cycles, sea level stayed still or dropped. The shelf was at sea level or emergent, and terrigenous sand was transported into the basin.

ACKNOWLEDGMENTS

We thank R. M. Scott and E. V. Eslinger for initiating this study. W. Fallgater supplied information on organic geochemistry. Comments and suggestions in reviews by J. F. Sarg, G. Eberli, P. M. Harris, R. A. Armin, T. L. Elliot, and G. A. Crawford greatly improved the manuscript. J. Gogas helped in preparation of the final manuscript. We thank Cities Service and UNOCAL for permission to publish this study.

REFERENCES

ALLAN, J. R., AND MATTHEWS, R. K., 1977, Carbon and oxygen isotopes as diagenetic and stratigraphic tools: Surface and subsurface data, Barbados, West Indies: Geology, v. 5, p. 16–20.

AUSTIN, J. A., JR., SCHLAGER, W., PALMER, A. A., AND OTHERS, 1986, Proceedings of ODP 101, Initial Reports, Part A: Texas A and M University, College Station, 569 p.

BATHURST, R. G. C., 1975, Carbonate sediments and their diagenesis: Elsevier, Amsterdam, 658 p.

BOARDMAN, M. R., NEUMANN, A. C., BAKER, P. A., DULIN, L. A., KENTER, R. J., HUNTER, G. E., AND KIEFER, K. B., 1986, Banktop response to Quarternary fluctuations in sea level recorded in periplatform sediments: Geology, v. 14, p. 28–31.

COOK, H. E., AND MULLINS, H. T., 1983, Basin margin environments, in Scholle, P. A., Bebout, D. G., and Moore, C. H., eds., Carbonate Depositional Environments: American Association of Petroleum Geologists Memoir 33, p. 539–617.

DROXLER, A. W., AND SCHLAGER, W., 1985, Glacial versus interglacial sedimentation rates and turbidite frequencies in the Bahamas: Geology, v. 13, p. 799–802.

FUTTERER, D. K., 1974, Significance of the boring sponge *Cliona* for the origin of fine-grained material of carbonate sediments: Journal of Sedimentary Petrology, v. 44, p. 79–84.

GAWLOSKI, T. F., 1987, Nature, distribution, and petroleum potential of Bone Spring detrital sediments along the Northwest Shelf of the Delaware Basin, in Cromwell, D., and Mazzullo, L. J., eds., The Leonardian Facies in W. Texas and S.E. New Mexico and Guidebook to the Glass Mountains: Society of Economic Paleontologists and Mineralogists, Permian Basin Section Publication No. 87–27, p. 85–105.

GONZALEZ, L. A., AND LOHMANN, K. C., 1985, Carbon and oxygen isotopic composition of Holocene reefal carbonates: Geology, v. 13, p. 811–814.

GROSS, M. G., 1964, Variations in the O^{18}/O^{16} ratios of diagenetically altered limestones in the Bermuda Islands: Journal of Geology, v. 72, p. 170–194.

HARLAND, W. B., COX, A. V., LLEWELLYN, P. G., PICKTON, C. A. G., SMITH, A. G., AND WALTERS, R., 1982, A Geological Time Scale: Cambridge University Press, Cambridge, 131 p.

HARMS, J. C., 1974, Brushy Canyon Formation, Texas: A deep-water density current deposit: Geological Society of America Bulletin, v. 85, p. 1763–1784.

HUDSON, J. D., 1977, Stable isotopes and limestone lithification: Journal of the Geological Society of London, v. 133, p. 637–660.

KROOPNICK, P. M., MARGOLIS, S. V., AND WONG, C. S., 1977, del ^{13}C variations in marine sediments as indicators of the CO_2 balance between the atmosphere and oceans, in Anderson, N. R., and Malahoff, A., eds., The Fate of Fossil Fuel CO_2 in the Oceans: Plenum Press, New York, p. 295–321.

LAND, L. S., 1985, The origin of massive dolomite: Journal of Geological Education, v. 33, p. 112–125.

MAZZULLO, L. J., AND REID, A. M., II, 1987, Stratigraphy of the Bone Spring Formation (Leonardian) and depositional setting in the Scharb Field, Lea County, New Mexico, in Cromwell, D., and Mazzullo, L. J., eds., The Leonardian Facies in W. Texas and S.E. New Mexico and Guidebook to the Glass Mountains: Society of Economic Paleontologists and Mineralogists, Permian Basin Section Publication No. 87–27, p. 107–111.

MULLINS, H. T., 1986, Carbonate depositional environments: Modern and ancient–Part 4: Periplatform carbonates: Colorado School of Mines Quarterly, v. 81, 63 p.

——, HEATH, K. C., VAN BUREN, H. M., AND NEWTON, C. R., 1984, Anatomy of modern, open-ocean carbonate slope: Northern Little Bahama Bank: Sedimentology, v. 31, p. 141–168.

——, WISE, S. W., LAND, L. S., SIEGEL, D. I., MASTERS, P. M., HINCHEY, E. J., AND PRICE, K. R., 1985, Authigenic dolomite in Bahamian peri-platform slope sediment: Geology, v. 13, p. 292–295.

NOTTINGHAM, M. W., 1966, Abo reef buildup provides five stratigraphic trap zones: World Oil, v. 162, p. 107–110.

PODPECHAN, F. W., 1959, New Empire Abo sparks rush to southeast New Mexico: Oil and Gas Journal, v. 57, p. 148–151.

RUNYAN, J. W., 1965, First New Mexico reef detritus oil pools found downdip from Abo trend: World Oil, v. 160, p. 99–106.

SALLER, A. H., 1984, Petrologic and geochemical constraints on the origin of subsurface dolomite, Enewetak Atoll: An example of dolomitization by normal seawater: Geology, v. 12, p. 217–220.

SARG, J. F., 1989, Middle-Late Permian depositional sequences, Permian Basin, West Texas-New Mexico, in Bally, A. W., ed., Atlas of Seismic Stratigraphy: American Association of Petroleum Geologist Studies in Geology #27, v. 3, p. 140–157.

——, AND LEHMANN, P. J., 1986, Lower-middle Guadalupian facies and stratigraphy San Andres/Grayburg Formations, Permian Basin, Guadalupe Mountains, New Mexico, in Moore, G. E., and Wilde, G. L., eds., Lower and Middle Guadalupian Facies, Stratigraphy, and Reservoir Geometries, San Andres/Grayburg Formations, Guadalupe Mountains, New Mexico and Texas: Society of Economic Paleontologists and Mineralogists, Permian Basin Section, Publication No. 86–25, p. 1–8.

SILVER, B. A., AND TODD, R. G., 1969, Permian cyclic strata, northern Midland and Delaware basins, West Texas and southeastern New Mexico: American Association of Petroleum Geologists Bulletin, v. 53, p. 2223–2251.

VEEVERS, J. J., AND POWELL, C. M., 1987, Late Paleozoic glacial episodes in Gondwanaland reflected in transgressive-regressive depositional sequences in Euramerica: Geological Society of America Bulletin, v. 98, p. 975–487.

WALTER, L. M., AND MORSE, J. W., 1984, Magnesian calcite stabilities: A reevaluation: Geochimica et Cosmochimica Acta, v. 48, p. 1059–1069.

WIGGINS, W. D., AND HARRIS, P. M., 1985, Burial diagenetic sequence in deep-water allochthonous dolomites, Permian Bone Spring Formation, southeast New Mexico, in Crevello, P. D., and Harris, P. M., eds., Deep water Carbonates: Buildups, Turbidites, Debris Flows and Chalks: Society of Economic Paleontologists and Mineralogists Core Workshop No. 6, p. 140–173.

EVOLUTION AND DESTRUCTION OF A CARBONATE BANK AT THE SHELF MARGIN: GRAYBURG FORMATION (PERMIAN), WESTERN ESCARPMENT, GUADALUPE MOUNTAINS, TEXAS

EVAN K. FRANSEEN
Kansas Geological Survey, Lawrence, Kansas 66046
THOMAS E. FEKETE
Arco Oil and Gas Company, Midland, Texas 79702
AND
LLOYD C. PRAY
Department of Geology and Geophysics, University of Wisconsin, Madison, Wisconsin 53706

ABSTRACT: Outcrops along the western Guadalupe Mountains, Texas, provide excellent exposures of the shelf-to-basin transition of the Grayburg Formation at the northwestern margin of the Delaware Basin. Within a distance of 4 km, the 385-m-thick Grayburg carbonate bank thins to a pinchout basinward by a combination of shelf margin erosion, depositional thinning, and facies change. The Grayburg strata at the edge of the shelf are mostly dolomite interbedded with thinner quartz sandstone-siltstone layers. They compose two shoaling-upward sections between the underlying deep-water (basin facies) Cherry Canyon Sandstone Tongue and the overlying shallow-water-to-peritidal dolomite facies of the uppermost Grayburg and overlying Queen Formations.

The lowermost one-fourth of Grayburg strata is foreslope facies exhibiting basinward-prograding strata with dips of 5° to 35°. The foreslope strata gradually decrease in dip upward into flat-bedded bank top strata. Most Grayburg strata consist of marine dolowackestones and dolopackstones with interbedded massive-to-laminated (locally cross-laminated), marine, quartz-rich sandstones-siltstones. Relief at the shelf-to-basin margin increased from 50–100 m in early Grayburg time to 200–300 m at the close of Grayburg deposition.

A spectacular basin-sloping erosion surface of listric shape truncates 115 m of uppermost Grayburg strata at a high angle (40° to 80°). From the toe of its steep slope, the erosion surface flattens, truncating an additional 70 m in the next basinward kilometer. The erosion surface may truncate an additional 10 to 100 m over the next 2 basinward kilometers. From the top of the high-angle (headwall) position, the erosion surface flattens abruptly and is traceable to the limit of outcrops 2 km farther shelfward. The shelfward trace of the surface is mostly flat, but channeling and scouring on the scale of meters to tens of meters and brecciation occur locally. This Grayburg erosion surface marks the change in depositional style of the Permian shelf margin complexes from underlying, mostly flat-bedded, clastic carbonate bank-ramp complexes to overlying reef complexes characterized by shelf edge boundstones and angle-of-repose foreslopes.

The processes and environments involved in creating the Grayburg erosion surface remain uncertain. The high-angle and basinward portions of the erosion surface are interpreted to have formed in a submarine environment. Retrograde slumping was probably a major process, and erosion by debris flows and bottom currents were likely contributing processes. Field evidence is inconclusive as to whether the shelf and uppermost shelf margin were exposed during erosion. Unequivocal karstic surfaces and soil profiles have not been recognized on the shelf. The apparent shallower water conditions at the end of Grayburg time likely caused the scouring and channeling on the shelf and may have intensified or initiated erosion at the shelf margin. The shallow-water conditions were likely caused by eustatic sea-level lowering; the erosion surface appears to correlate with a major regional sequence boundary throughout the Permian Basin area.

Very few examples of ancient submarine unconformities at shelf or platform margins have been recognized. The Grayburg erosion surface represents a spectacular example of an ancient submarine erosion surface. Features of this erosion surface may characterize other ancient submarine erosion surfaces and may aid in their recognition.

INTRODUCTION

The Guadalupe Mountain outcrops (located on the northwest edge of the Delaware Basin; Fig. 1) contain excellent exposures that reveal details of the stratal changes from the Northwestern Shelf to Delaware Basin during Permian time. The Grayburg Formation displays a spectacular shelf-to-basin margin transition along the rugged upper cliffs of the western escarpment of the Guadalupe Mountains of West Texas. The Grayburg Formation is a 385-m-thick carbonate bank that thins to a pinchout laterally into the basin within a distance of 4 km by a combination of shelf margin erosion, depositional thinning, and facies change. Approximately 185 m of middle and uppermost Grayburg strata are abruptly truncated over a distance of 1 km in a basinward direction by a basin-sloping erosion surface interpreted to be submarine in origin. Shelfward the erosion surface generally parallels shelf strata but is locally channeled and scoured. The shelfward portion of the erosion surface may have formed subaerially. The erosion surface appears to correlate with a major regional sequence boundary throughout the Permian Basin area (Sarg and Lehmann, 1986).

The Grayburg erosion surface separates the important change in depositional style of the Permian shelf margin complexes from the underlying, mostly flat-bedded, clastic-carbonate bank-ramp complexes (Victorio Peak, San Andres-Grayburg) to the overlying reef complexes characterized by shelf edge boundstones and angle-of-repose foreslopes of several hundreds of meters of relief (Goat Seep, Capitan) (Fig. 2).

This study is based on 5 months of field work in 1983, 1984, and 1985 by the two principal authors. The research resulted in two master's theses (Franseen, 1985; Fekete, 1986) that were supervised by L. C. Pray. Our work is part of a decade-long program of investigation on the Permian Strata by the University of Wisconsin carbonate research group. Earlier theses completed along the western escarpment by Crawford (1981), Harris (1982) and Kirkby (1982) have been especially useful for providing background for our work. Our studies represent the first known detailed investigations of the Grayburg and Queen Formations at the shelf margin along the western escarpment of the Guadalupe Mountains since the reconnaissance work of King (1948) and Newell and others (1953). Some of the data discussed

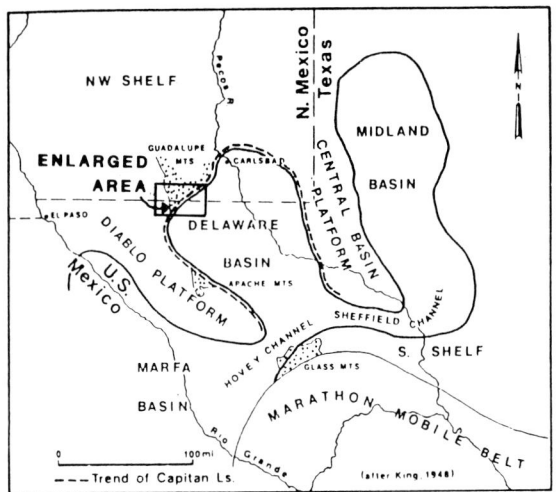
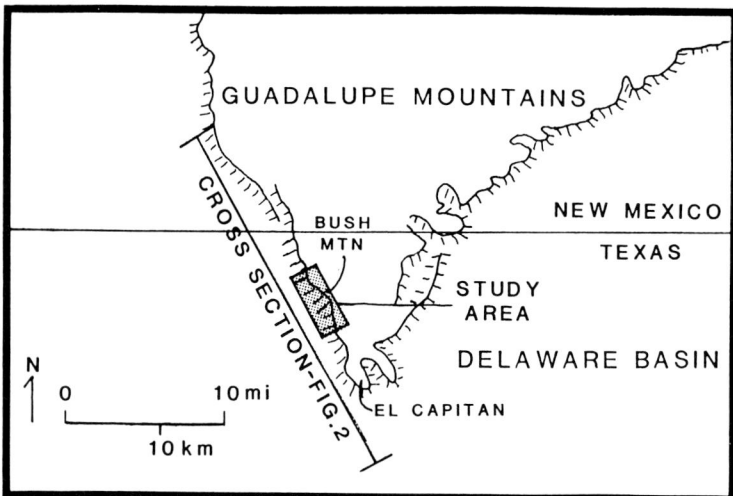

FIG. 1.—Map of the Permian Basin area showing paleogeographic setting of the Guadalupe Mountains region. Inset shows study area.

here were presented in earlier reports (Franseen, 1984; Fekete and others, 1986).

Data for this report are derived from detailed observations involving field mapping and key stratigraphic sections and from laboratory study of hand samples and thin sections. The stratigraphic sections (Fig. 3) on the western escarpment are along a line that is slightly oblique to roughly perpendicular with the inferred northeastern-southwestern shelf margin trend.

The Grayburg erosion surface is the youngest of several basin-sloping surfaces that truncate Permian shelf edge strata exposed along the western escarpment (Pray and others, 1980). At least three other erosion surfaces occur within, or separate parts of, the Victorio Peak, Bone Spring, and Cutoff Formations (Pray, 1981; Harris, 1982; Kirkby, 1982; Franseen and others, 1987). These erosion surfaces (interpreted to be of submarine origin), including the Grayburg erosion surface, form breaks that divide the Leonardian and lower Guadalupian (Permian) strata into major genetic sequences. These are interpreted as third-order cycles of Vail and others (1977) that have been correlated to Last Chance Canyon and the Algerita Escarpment in New Mexico (Sarg and Pray, 1984; Sarg and Lehmann, 1986), an estimated 16–20 km farther shelfward.

FIG. 2.—Schematic cross section (location shown on Fig. 1) of major Permian stratal complexes, their principal rock units, and erosion surfaces (wavy lines) of the Guadalupe Mountains at the Northwestern Shelf-Delaware Basin transition area. Major erosion surface truncating Grayburg Formation separates bank-ramp style of sedimentation of earlier Permian shelf margins from later reef style of shelf margin sedimentation.

The recognition of modern submarine erosion as a geologic process is a relatively new development. We believe that the portion of the Grayburg erosion surface exposed along the western escarpment of the Guadalupe Mountains represents an excellent example of an ancient submarine erosion surface, as do the earlier Permian erosion surfaces associated with the Cutoff Formation. Submarine erosion has been underestimated and is probably a common feature in other carbonate platforms in the ancient record.

STRATIGRAPHIC SETTING

The Permian strata exposed along the imposing and rugged western escarpment of the Guadalupe Mountains provide one of the finer cross sections in the world of several shelf-to-basin stratal transitions.

The stratigraphic setting of the Grayburg Formation and its associated truncating erosion surface is shown in Figure 2. The Grayburg is the uppermost shelf margin carbonate unit of the major stratigraphic package termed the "Bank-Ramp" Complex. Its overlying erosion surface (unconformity and sequence boundary) separates this major stratigraphic package from the overlying reef complex of the Goat Seep and Capitan Formations and their lateral equivalents. Designation of the "Bank-Ramp" relates to the unbound and clastic nature of the particles that compose the carbonate bank and to the interpretation of its deposition on a low-angle basin-sloping surface, in contrast to the shelf margin boundstones and the extensive high-angle foreslopes of the reef complex. The erosion surface separating the Bank-Ramp Complex from the reef complex is of particular interest as it represents the major change in sedimentologic style of the Permian shelf margin carbonates of the Guadalupe Mountains.

Earlier Permian erosion surfaces (interpreted to be submarine in origin) truncate Victorio Peak and Cutoff strata. The Victorio Peak Formation is composed of marine dolomite and limestone bank facies. The overlying Cutoff Formation is composed predominantly of mud-rich limestone-basin facies. The Cherry Canyon Sandstone Tongue is interpreted as a basinal deposit and consists of 50–120

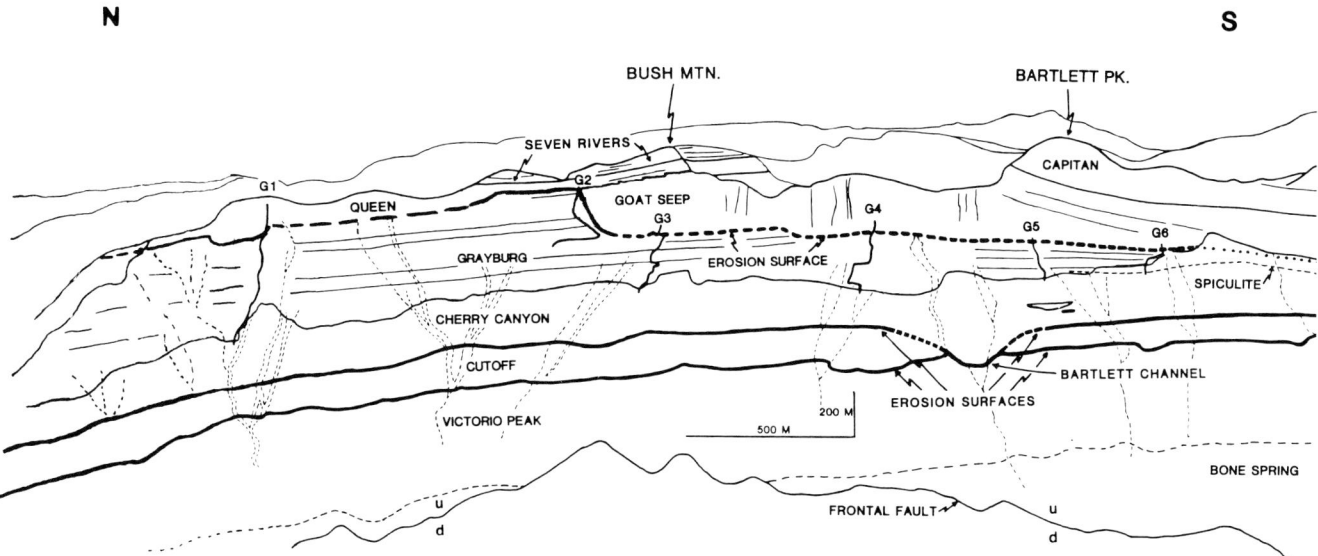

FIG. 3.—Aerial photograph and diagram of western escarpment of the Guadalupe Mountains. View to the east shows formations, erosion surfaces, and location of stratigraphic sections. Thesis area of Franseen (1985) extends north from Bush Mountain; thesis area of Fekete (1986) extends south from Bush Mountain. Vertical and horizontal scales apply at the distance of the escarpment. Photograph courtesy of Marathon Oil Company.

m of mostly quartz sandstones and siltstones. A few carbonates occur, including some dolomite conglomerates and breccias that are largely confined to local channels. The Cherry Canyon is overlain sharply by the Grayburg Formation, which is composed mainly of dolomite and some interbedded sandstones and siltstones. The Grayburg marks the return of predominantly carbonate sedimentation. The lower 75–85 m of Grayburg strata are low- to high-angle (5° to 35°) beds interpreted to be clinoform strata, here classified as toe-of-slope and foreslope deposits that appear to downlap onto a portion of the upper Cherry Canyon Sandstone Tongue (although an actual downlap surface has not been identified in the field). The strata of the upper 10–20 m of the 75- to 85-m-thick interval of clinoform deposits are termed the "transition strata" and display a decrease in dip upward from 25° to less than 5°. The 300–310 m of Grayburg strata overlying this "transition interval" in the area north of Bush Mountain (Fig. 1) are flat bedded. Under Bush Mountain, approximately 115 m of Grayburg are abruptly truncated by a high-angle (40°–80°), listric-shaped, basin-sloping erosion surface (Figs. 3, 4). Traced southward, the surface appears to flatten and truncate an additional 10 to 100 m of Grayburg strata over a distance of approximately 2 km in a basinward direction (Fig. 3).

The Goat Seep Formation immediately overlies the erosion surface and is distinguished from the flatter and thinner bedded Grayburg by its massiveness and recognizable basinward-dipping, high-angle (angle of repose) deposits (Figs. 3, 4). Until the discovery of the Grayburg erosion surface by Pray in the 1970s, the Grayburg Formation was thought to be the shelfward equivalent of the Goat Seep Formation. The contact was interpreted to be a lateral gradation from the Grayburg into the Goat Seep Dolomite (King, 1948; Newell and others, 1953).

The Queen Formation (Goat Seep equivalent) is flat bedded and overlies the Grayburg shelfward of the high-angle

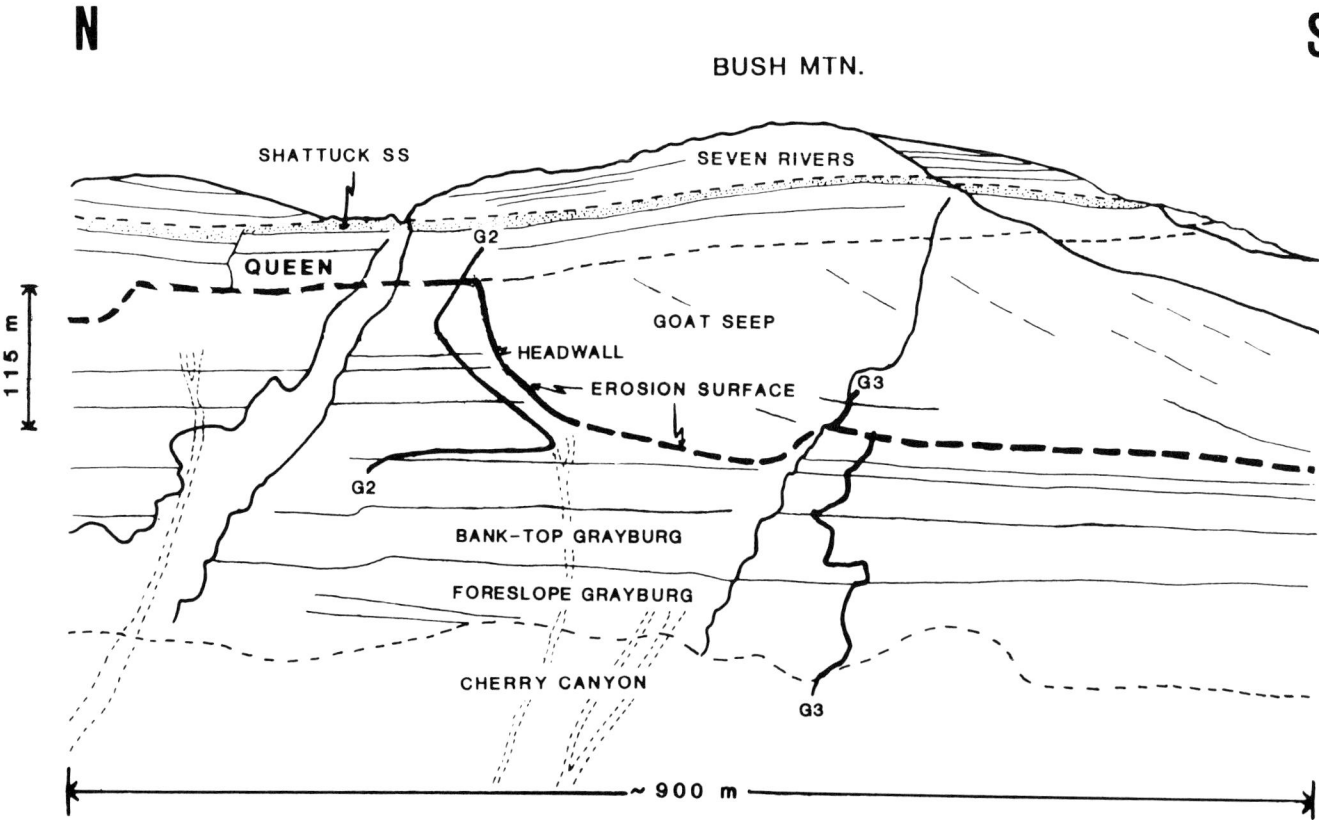

FIG. 4.—Closeup aerial photograph and diagram showing details of Grayburg erosion surface in Bush Mountain area. Note abrupt flattening of erosion surface at both top and base of headwall. Grayburg and Queen strata are flat bedded in contrast to the more massive Goat Seep: note local low-angle foreslopes in lowermost Grayburg strata. Also shown on diagram are locations of stratigraphic sections G2 and G3. Photograph by P. J. Lehmann.

portion of the erosion surface. The upper portion of the Queen Formation progrades basinward over the Goat Seep Formation in the Bush Mountain vicinity (Fig. 4).

The Queen, Seven Rivers, Yates, and Tansill Formations are backreef strata of the Goat Seep-Capitan Reef Complex (Fig. 2). The Goat Seep and Capitan are essentially reef-rimmed platforms.

GRAYBURG FORMATION STRATIGRAPHIC CORRELATIONS

Figure 5 shows the correlation of our stratigraphic sections of the Grayburg Formation measured along the western escarpment of the Guadalupe Mountains. The strata have been correlated by inspection of horizontal and oblique aerial photographs and on the basis of field observations and photographs of traceable sandy (quartz) layers that form major recessive intervals. Continuous field tracing of units is impossible due to vertical cliffs. Sedimentary structures and detailed marker beds are rare to absent, especially in the lowermost 300 m of Grayburg strata. A particularly important correlation shown in Figure 5 concerns a 3- to 10-m-thick interval of chertified dolomudstone and spiculite,

a unique lithology in the Grayburg and Cherry Canyon. This interval is massive to parallel-laminated, has even-bedding planes, fine texture, dark color on fresh surfaces, and contains very little fauna (other than siliceous sponge spicules). This unit is interpreted to represent suspension deposition in quiet, possibly poorly oxygenated water. This silicified interval directly overlies the lowest Getaway debris tongue (Fig. 5) and, although very poorly exposed, is believed to be traceable over a distance of 1.3 km northward through Cherry Canyon siliciclastics into the level of lower Grayburg foreslopes (see section G6, Fig. 5; Fekete, 1986).

The chertified dolomudstones and spiculite is correlated (tentatively) with an unusual and distinctive 3- to 4-m-thick interval of highly silicified pink (5YR8/1) dolomite located 400 m to the north in section G5 (Fig. 5). The pink dolomite occurs within the basin-sloping Grayburg foreslope facies at a level that is about 30–55 m above the Grayburg-Cherry Canyon contact. The pink dolomite is believed to be correlatable shelfward between sections G5 to G1, a distance of 3 km. If the chertified dolomudstone-spiculite overlying the Getaway tongue outcrop (basinward) and the

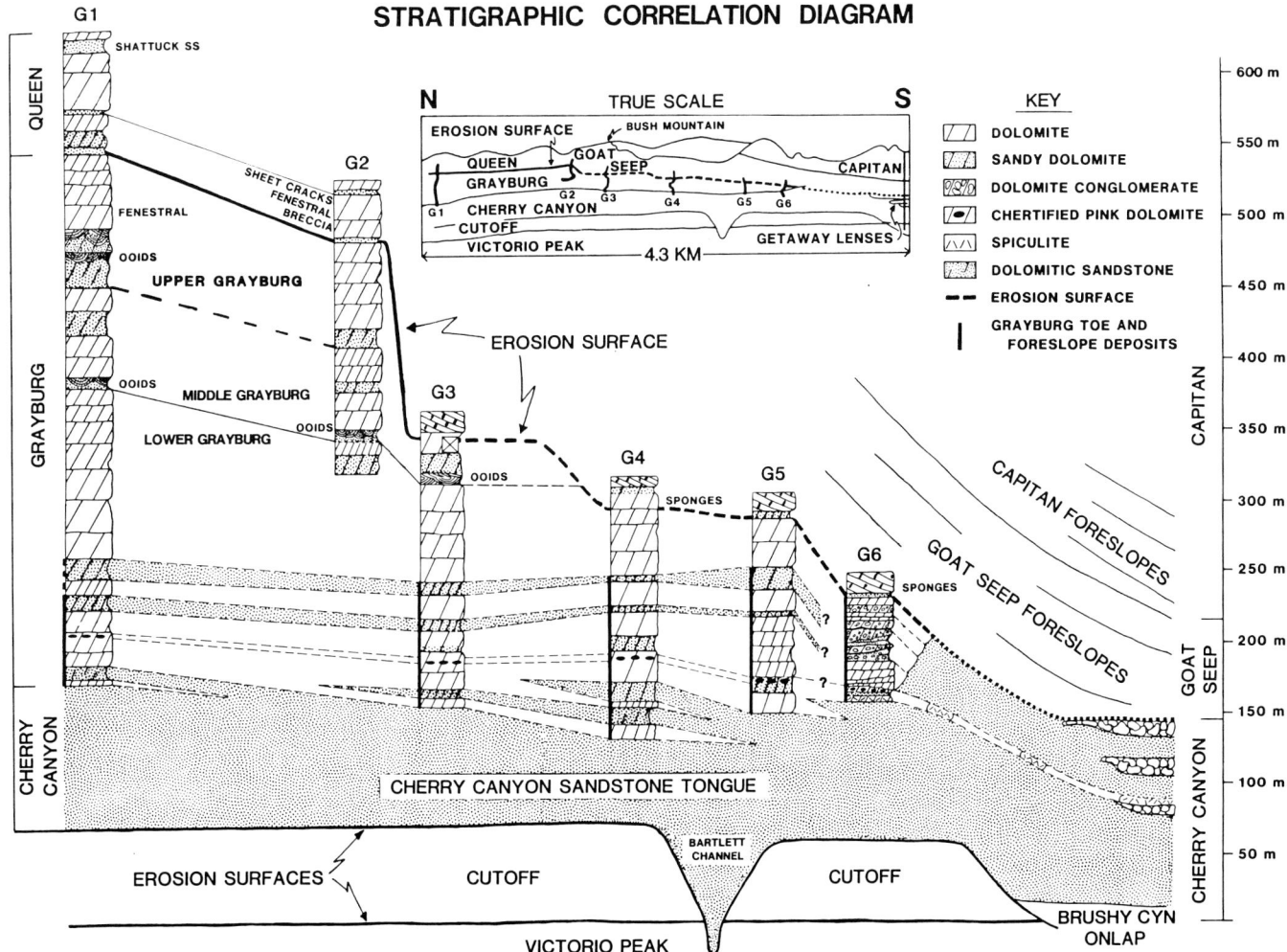

FIG. 5.—Stratigraphic sections and correlations of Grayburg strata exposed along western escarpment of Guadalupe Mountains. Note southward thinning of Grayburg strata from 385 m at section G1 to pinchout just south of section G6. Basinward thinning is largely due to erosion in area of sections G2 to G4 and depositional thinning in area of sections G5 and G6.

pink dolomite in section G1 (shelfward) are the same unit, then a major correlation exists over a 4.3-km distance that relates lower Grayburg foreslopes, Cherry Canyon siliciclastics, and Getaway debris tongues.

An oolite-rich quartz sandstone unit occurs in the Grayburg bank top facies approximately 150–200 m above the Cherry Canyon-Grayburg contact in sections G1, G2, and G3 (Fig. 5). This oolite-rich unit marks another major correlation over approximately 2.5 km, and it also marks the first of two shoaling-upward successions in the Grayburg Formation on the western escarpment. The base of this unit is the Lower to Middle Grayburg boundary of Franseen (1985).

SEDIMENTOLOGY–GRAYBURG FORMATION

Introduction

In the field, rocks have been classified according to depositional textures described by Dunham (1962). Colors are reported using the G.S.A. color index. Bed geometry is defined using the terms of Ingram (1954).

The Grayburg Formation consists mostly of very fine to finely crystalline (62–250 μm) dolomite. Strata on outcrop are light gray (N7) to medium dark gray (N4) and pinkish gray (5YR8/1). They are massive to thick and medium bedded. The dolomites form resistant cliffs (Figs. 3, 4).

Sandstones and siltstones form about 25% of the Grayburg section. These are less resistant than dolomites and form the slopes and notches along the steep cliffs (Figs. 3, 4). Some of these recessive-weathering layers provide limited north-south access within the Grayburg-Queen sequence along the western escarpment. Grayburg siliciclastics consist of interbedded quartz-rich fine- and very fine-grained sandstones and some siltstones. Siliciclastic grains are moderately well sorted and subangular to subrounded. Siliciclastic strata are as abundant in the Grayburg foreslope facies as they are in the bank top facies. These strata on outcrop are pinkish gray (5YR8/1) to very pale orange (10YR8/2) and massive to thickly laminated.

Depositional Facies

Depositional facies are recognized by the attitude of the primary bedding and are here classed simply as "foreslope facies" (with field-recognizable basinward dips) and "bank top facies" (without recognizable primary dips).

Grayburg foreslope facies.—

The Grayburg foreslope facies is located in the lowermost 75–85 m of the Grayburg Formation. This facies is here defined as those beds with a primary dip of approximately 5° to 35° (35° being the maximum dip observed; Fig. 6). Primary dips are recognized largely by the geometry of beds, but they have been confirmed by observed primary geopetal fabrics. The carbonate strata of this facies are interpreted to be of both autochthonous and allochthonous origin.

The foreslope facies consists predominantly of medium- to thick-bedded dolomites (less than 5% quartz sand) and

FIG. 6.—Photograph showing the Grayburg foreslope (GFS) to bank top (GBT) facies transition under Bush Mountain. Angle of primary dip decreases upward from base of the cliffs (foreslopes) into flat-bedded bank top facies forming upper two-thirds of cliffs in foreground. View to northeast. Basinward direction is southeast.

sandy dolomites (5–50% quartz sand). These are interbedded with some thin- to medium-bedded dolomitic sandstones and siltstones (greater than 50% quartz sand). Carbonate beds are commonly wedge shaped, with bedding thickness, and angle of primary dip decreasing southeastward toward the basin. Upper and lower contacts of beds are both gradational and sharp in this facies. Sharp contacts are planar to wavy and are commonly erosive (scoured) on the scale of centimeters to decimeters. Thin (10–15 cm) intervals of pebble conglomerate, containing randomly oriented clasts that are lithologically similar to underlying beds, locally overlie the erosive surfaces.

Pervasive dolomitization has obscured much of the original texture; however, it is possible to identify many of the grains coarser than about 1 mm. Many of the grains were fossils and most are now molds or casts or show some preserved primary features. The major texture of the foreslope facies is fusulinid dolowackestones and dolopackstones (Fig. 7A). This facies is interpreted as mostly autochthonous. Although fusulinids locally show parallel orientations, most appear to have a random orientation and are generally not in mutual grain support. Thin sections show that peloids occur throughout the Grayburg Formation and they may well be the major grain type of the dolomites. Other lithologies common to this facies are dolomitic quartz sandstones-siltstones, dolomudstones, and allodapic (allochthonous) dolomites. The sandstones-siltstones are massive to faintly parallel laminated. Some dolomites consist of graded and laminated, fining-upward skeletal-intraclast dolopackstones and dolowackestones. These overlie an erosive base and are interpreted as allodapic deposits. Other grain types that occur in the foreslope facies include brachiopods, bryozoans,

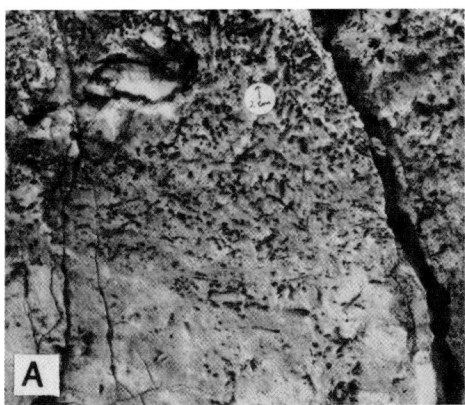

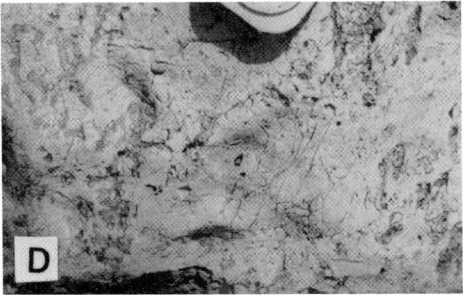

sponges, intraclasts, ooids, and unidentified skeletal fragments. Pervasive dolomitization has obscured micritic envelopes or any other structures that might support a shallow-shelf (allodapic) origin for the grains.

Tracing of partially exposed foreslope beds indicates a minimum depositional relief of 25 m. Additional observations indicate a possible shelf-to-basin relief of 75–85 m by the time that the 10- to 20-m-thick "transition interval" of the lower Grayburg formed. At this "transition interval," the Grayburg foreslopes gradually decrease in dip upward into flat-bedded (shelf or bank top) Grayburg strata (Fig. 6). The 10- to 20-m-thick "transition interval" is texturally and lithologically similar to the foreslope and bank top facies. It is only defined for the purpose of emphasizing the change from foreslope to bank top strata as gradational and not an abrupt contact.

Grayburg bank top facies.—

The Grayburg bank top facies (300–310 m-thick) forms the uppermost three-fourths of the Grayburg and overlies the "transition interval" located about 65–75 m above the Cherry Canyon. The bank top facies is here defined as those beds with primary dips of approximately 5° or less that overlie the Grayburg foreslope facies. We interpret the carbonate strata of the bank top facies to be mostly autochthonous. This interpretation is favored by the abundance of matrix, random distribution of fusulinids (unoriented), lack of primary sedimentary structures, and the relative absence of channels.

The bank top facies consists of medium- to massive-bedded dolomites and sandy dolomites. These are interbedded with minor (average 3 m-thick) intervals of thickly laminated (locally cross-laminated) to massive sandstones and siltstones (Fig. 7B). Upper and lower siliciclastic-carbonate bedding contacts are both gradational and sharp. The beds of the bank top are characteristically tabular and have flat, planar bounding surfaces where exposed. Channeling (on the scale of a few meters to possibly tens of meters) occurs locally. Most beds do not change appreciably in thickness in the study area, nor do we detect significant basinward dips (Figs. 5, 6).

Dolomitization (most crystals are 62–250 μm) has obscured much of the original texture. As in the foreslope facies, however, it is possible to identify grains that are now molds, casts, or show some preserved primary features. The major carbonate textures of the lowermost 200 m of the bank top facies are fusulinid dolowackestones and

---

FIG. 7.—Grayburg (A, B, C,) and Goat Seep (D) lithologies. (A) Fusulinid dolowackestone and dolopackstone facies showing typical, randomly oriented fusulinid molds. Lithology grades upward from dolowackestone to dolopackstone. White disc (center) is 2 cm in diameter. (B) Tabular, massive, recessive-forming siliciclastic unit characteristic of Grayburg foreslope facies and bank top facies. These units provide the limited access to otherwise sheer cliffs. (C) Peritidal-dolomite facies located in uppermost Grayburg strata in G1 section. This facies is also common in the overlying Queen strata. Shown here are laminoid fenestral fabric and wavy (algal mat?) lamination. (D) Oriented calcareous sponges, common in the Goat Seep Dolomite. Skeletal dolowackestone matrix surrounds the sponges.

dolopackstones and sandy dolopackstones (Fig. 7A). Fusulinids range in content from less than 5% to 40%. Most have a random orientation, and they are generally not in mutual grain support. Other grains occurring in this facies include peloids, brachiopods, echinoderms, detrital sponges, bryozoans, and unidentified skeletal fragments. Although difficult to recognize in the field, peloids are believed to be the predominant grain type in the bank top facies and throughout the Grayburg. Other lithologies and textures in this facies are dolomitic quartz sandstones and, to a lesser extent, dolomudstones. The sandstones are mostly massive or have faint wavy and planar laminations.

The Grayburg bank top facies are composed of two shoaling-upward trends. The first involves the lowermost 150–200 m of Grayburg strata (mostly randomly oriented fusulinid dolowackestones and dolopackstones and massive siliciclastic sandstones). These strata are capped by a cross-bedded, oolite-rich siliciclastic sandstone that is correlatable between the G1, G2, and G3 sections (Fig. 5). This correlation marks the Lower-Middle Grayburg boundary of Franseen (1985). The overlying strata change abruptly to mostly fusulinid dolowackestones to dolopackstones and massive siliciclastic sandstones.

The uppermost 85–95 m of the Grayburg bank top strata show a change in lithology from north to south. In the northern study area (shelfward; Fig. 5, G1 section), the upper Grayburg consists mostly of peritidal dolomite, sandy dolomite, dolomudstone, fine-grained quartz sandstone, and lesser amounts of fusulinid dolowackestones and dolopackstones. The peritidal dolomites are characterized by local caliche development, sheet cracks, fenestral dolomite, wavy (algal?) lamination, burrows, peloids, intraclasts, and various types of coated grains (Fig. 7C). Other grains include brachiopods, echinoderms, gastropods, red calcareous algae, bryozoans, and unidentified skeletal fragments. In contrast, the uppermost 85–95 m of Grayburg bank top strata in a more southerly (basinward) location (Fig. 5, G2 section) consists mostly of randomly oriented fusulinid dolowackestones and dolopackstones (no peritidal dolomites occur). The change in facies suggests deepening toward the basin, possibly indicating a ramp profile. The fine-grained quartz sandstone and sandy dolomite units in the uppermost Grayburg commonly show cross-lamination and cross-bedding. Most of these structures appear to indicate random or bimodal current directions.

Depositional Environment

The Grayburg Formation marks the return of predominantly carbonate sedimentation to the western escarpment area. The initial 75–85 m of Grayburg foreslope strata downlap onto and prograde over the underlying, more basinal Cherry Canyon Sandstone Tongue, thus heralding a basinward shift in the shelf margin position during early Grayburg time. A change from predominantly progradational to aggradational depositional styles for the Grayburg is indicated by the upward decrease in primary dip from the foreslope strata to the flat-bedded bank top strata. The initial Grayburg foreslope strata (75–85 m-thick) and most of the overlying 115 m of Grayburg bank top strata are interpreted to have been deposited below both normal- and storm-wave base, as indicated by the lack of sedimentary structures and winnowed deposits. These strata are interpreted to have been deposited in normal-marine water, as indicated by the stenohaline fauna.

The lowermost 150–200 m of Grayburg strata show a shoaling-upward trend, as indicated by decreasing thickness of beds upward that culminates in a cross-bedded, oolite-rich quartz sandstone unit that is correlatable between the G1, G2, and G3 sections (Lower-Middle Grayburg boundary; Fig. 5). The oolite-rich unit is interpreted to indicate shoaling of the Grayburg bank top strata into more turbulent and probably shallower, more restricted conditions than those in which the underlying strata were deposited.

Sarg and Lehmann (1986) interpret the base of the oolite-rich unit as a regionally important sequence boundary that they correlate shelfward about 16–20 km to a subaerial exposure surface in Last Chance Canyon and on the Algerita Escarpment (both in New Mexico). This correlation suggests a regional shallowing of water depth probably caused by eustatic sea-level drop.

The overlying 90 m of Grayburg bank top strata on the western escarpment change abruptly to thicker bedded, mostly fusulinid dolowackestones and dolopackstones and massive quartz sandstones, indicating a return to deposition below normal- and storm-wave base in normal-marine water. A second shoaling-upward section is indicated by strata becoming thinner bedded upward and the uppermost 90 m of Grayburg bank top strata in the G1 section (shelfward) reflecting deposition in a shallow subtidal to peritidal environment and under more turbulent conditions than the immediately underlying Grayburg strata. The correlatable 90 m of Grayburg strata in the basinward G2 section consist mostly of fusulinid dolowackestones and dolopackstones (no peritidal dolomites occur). This facies change suggests deepening toward the basin (south). Thus, the Grayburg strata are interpreted to have been deposited on a sloping ramp.

Uppermost Grayburg strata are truncated by an erosion surface that can be traced for as far as 4 km along the western escarpment. Sarg and Lehmann (1986) interpret this erosion surface as a regionally important sequence boundary (Type 1, Vail and others, 1977) that can be also identified in the subsurface in the entire Permian Basin area. The erosion surface likely formed from eustatic sea-level lowering that culminated in possible exposure of the shelfward portion of the Grayburg strata along the western escarpment of the Guadalupe Mountains (discussed later).

SEDIMENTOLOGY—QUEEN AND GOAT SEEP FORMATIONS

Following the creation of the basin-sloping erosion surface at the end of Grayburg time, massive and high-angle (angle of repose) foreslope strata of the Goat Seep Dolomite were deposited against and above the shelf margin erosion surface. Foreslope dips become more apparent to the south in the younger Goat Seep (Crawford, 1981; Figs. 3, 4). Contemporaneous deposition on the shelf and shelf margin consisted of the flat-bedded Queen Formation (Figs. 3,

4). The Queen is dominated by shallow-water and peritidal-dolomite facies characterized by local caliche development, sheet cracks, fenestral dolomite, burrows, peloids, various coated grains, echinoderms, brachiopods, fusulinids, gastropods, calcareous red algae, and unidentified skeletal fragments. Queen strata grade abruptly basinward into the Goat Seep "reef" deposits. Goat Seep strata are mostly composed of submarine-cemented, skeletally lean, fine-grained, clastic-carbonate sediment with calcareous sponges (Fig. 7D) being the dominant perserved organisms (Crawford, 1981). Locally, the uppermost Queen strata prograde basinward over Goat Seep foreslope deposits (Fig. 4) with 400 m of shelf-to-basin relief.

Shelf Margin Profile and Shelf-to-Basin Relief

Some lower Grayburg foreslopes, Cherry Canyon siliciclastics, and the lowermost Getaway debris lens appear to be overlain by a few meters of cherty dolomites (and spiculite) that are apparently correlatable over the 4.3-km distance between the G1 section and the outcrop location of the Getaway debris tongues (Fig. 5). The Grayburg strata underlying this chertified interval thin in a basinward direction over a distance of 1 km from 55 m (at the G4 section) to a pinchout (at the G6 section; Fig. 5). South (basinward) of the most basinward Grayburg outcrops in our field area (between sections G5 and G6), the chertified interval continues to drop several tens of meters in elevation as it overlies (apparently conformably) some basinal Cherry Canyon sandstones and siltstones and the lowermost Getaway carbonate-debris channel filling (Fig. 5). The chertified interval appears to be a marine drape deposit, and its shape represents an original shelf and shelf margin profile for lower Grayburg time. The total relief of this profile is 100 m over a distance of 1.9 km (from section G4 to the Getaway debris lenses, see Fig. 5), giving an average shelf margin slope of about 3°.

At the shelf margin, the profile during deposition of the lower 100 m of Grayburg was that of a flat or very gently dipping bank top (ramp) that merged basinward into foreslopes (prograding deposits with 5–35° primary dips). Tracing of partially exposed foreslope beds in the lowermost Grayburg strata indicates a minimal depositional relief of 25 m. Additional observations and measurement of relief of clinoforms on photographs indicate a possible shelf-to-toe-of-slope relief of 75–85 m at the 10- to 20-m-thick "transition interval" level, which is younger than the cherty dolomite interval.

The position and profile of the original shelf margin during deposition of the upper 100–300 m of Grayburg are unknown, because we interpret the preserved profile to represent shelfward (headward) erosion of a late Grayburg depositional shelf margin that was somewhere (tens of meters to perhaps 1 km or more) to the south of the erosion surface headwall under Bush Mountain. An estimate of the minimum shelf-to-basin relief for late Grayburg-early Goat Seep time can be derived from the relief along the erosion surface that truncates Grayburg strata and separates it from the overlying and adjacent Goat Seep Dolomite. The erosion surface has a known relief of at least 115 m at the high-angle headwall location under Bush Mountain and an additional 60–70 m of relief as traced south for 0.9 km. Additional Grayburg (10 to 100 m?) may be truncated (at a low angle) by the surface as it drops 145 m in elevation in the next 2 km to the south, but unequivocal evidence of this additional truncation has not been found in our area. Thus, the total relief of the latest Grayburg shelf margin is 330 m, of which at least 185 m is due to erosion.

Crawford (1981) estimates the shelf-to-basin relief to be a minimum of 375–400 m at the end of Goat Seep time on the basis of topographic relief between the top of the Queen Formation, which was near sea level, and the equivalent Manzanita Member of the Cherry Canyon Formation along the western escarpment of the Guadalupe Mountains.

In summary, it appears that relief at the shelf-to-basin margin increased from 50–100 m in early Grayburg time to 200–300 m at the close of Grayburg time, to 300–400 m during Goat Seep time. The increase of relief at the Grayburg shelf-to-basin margin was apparently accompanied by a decrease in water depth (from below wave base to peritidal) on the Grayburg shelf.

EROSION SURFACE

A major focus of our research has been to study an enigmatic surface that spectacularly truncates 185 m of Grayburg strata over a basinward distance of 1 km (Figs. 3, 4). The erosion surface truncates 115 m of uppermost flat-bedded Grayburg bank top strata at a high angle (up to 80°) at a position located under Bush Mountain (termed the Bush Mountain headwall) along the western escarpment of the Guadalupe Mountains. At this location, the erosion surface has a strike of about N.75°E., roughly paralleling the inferred margin of the Delaware Basin during Grayburg time. From the Bush Mountain headwall position, the erosion surface flattens both shelfward and basinward and can be traced for approximately 2 km in either direction.

The continuity of the erosion surface along its strike parallel to the shelf margin is unknown. No other outcrops exist in the study area. Grayburg and Goat Seep strata, however, are also exposed in North McKittrick Canyon of the Guadalupe Mountains, located approximately 11–13 km northeast of the western escarpment outcrops. Crawford (1981) studied outcrops of the Goat Seep and Grayburg strata in McKittrick Canyon. He did not recognize an erosion surface in that location and instead described a 100- to 200-m-wide steeply rising transition zone from the Grayburg Formation to the Goat Seep Dolomite. Crawford (1981) notes, however, that the relation between the Goat Seep and Grayburg strata is obscured by poor exposure. There are no other surface outcrops of the Grayburg-Goat Seep contact exposed in the region.

Details of the erosion surface shelfward of the Bush Mountain headwall will be discussed first, followed by the details of the Bush Mountain headwall and basinward trace of the erosion surface.

Shelfward Trace of Erosion Surface

Shelfward of the Bush Mountain headwall, the erosion surface flattens abruptly and can be traced to a position just north of the G1 section (approximately 1.5–2.0 km; Figs. 3, 8A). Farther shelfward, tracing of the erosion surface is inhibited from lack of exposure and from faulting associated with the major frontal fault of the Guadalupe Mountains. Although the erosion surface has an overall flat geometry in its shelfward trace, scouring and channeling on the scale of meters to tens of meters, and brecciation, occur locally.

At the most shelfward trace, the erosion surface is iron stained and shows scouring on the scale of meters into the underlying Grayburg dolomite. The erosion surface is overlain by fine- to coarse-grained, cross-bedded quartz sandstones of the basal Queen Formation. As traced basinward from this position, the erosion surface shows scouring and channeling on a scale of meters to tens of meters wide and deep into the underlying Grayburg dolomite (Fig. 8A). The actual surface is mostly covered in this area. Locally, the basal Queen sandstones appear to onlap the right (basin-

Fig. 8.—Features of the Grayburg erosion surface about 1.5–2.0 km shelfward of the Bush Mountain high-angle headwall. (A) View of erosion surface (dashed white line) just shelfward (north) of the G1 section (labeled at top). Surface is irregular and displays channeling on a scale of meters to possibly 10 m deep and wide. Arrow points to right (basinward) channel margin, where basal Queen sandstone appear to onlap channel margin. GB = Grayburg; QN = Queen. View to the north. (B) Trough cross-bedded siliciclastic sandstone typical of basal Queen strata in the G1 section. (C) Fine- to couarse-grained, current-rippled siliciclastic and carbonate sandstone of the basal Queen strata locally overlies the erosion surface jsut shelfward of the G1 section. Ripples indicate a shelfward transport direction.

ward) channel margin (arrow in Fig. 8A). The basal Queen strata are predominantly trough cross-bedded, mixed quartz and carbonate sandstones (Fig. 8B). Locally, the sandstones are ripple laminated (Fig. 8C) and contain coarse-grained, well-rounded quartz sand grains. Ripple laminations indicate a preserved shelfward transport direction. Exact orientation of channels is unknown, but they appear to trend obliquely to the inferred shelf margin during Grayburg time.

From the shelfward channel position, the erosion surface appears to be mostly smooth as traced basinward to the Bush Mountain headwall. The surface is mostly covered along this trace. Where it is exposed, the surface is smooth and locally iron stained. Basal Queen sandstones overlie the erosion surface as traced to Bush Mountain. Approximately 0.3 km shelfward of the Bush Mountain headwall, the erosion surface may scour 15–20 m into the underlying Grayburg dolomite (Fig. 4). This observation, however, was made from photographs and has not been confirmed in the field. Immediately shelfward of the Bush Mountain headwall, as traced for 30 m, the erosion surface is very irregular, iron stained, and locally is associated with fracturing and brecciation (G2 section breccia, Fig. 9). Locally, the brecciated areas trace laterally into bedded Grayburg strata. Details of this area are obscured by weathering of the rock, and lateral tracing farther northward is prevented by inaccessibility.

G2 section breccia.—

A breccia (Fig. 9) and the locally non-brecciated layers form the uppermost 13 m of the Grayburg 30–50 m north of the headwall at the G2 section. The breccia is characterized by both clast-support and matrix-support textures. Clasts are angular to subround, centimeters to decimeters in size and locally display in-place brecciation textures. Clast types include fusulinid, brachiopod, and peloidal dolowackestones and dolopackstones, dolomudstones, quartz sandstones-siltstones and some ooid dolograinstones. Most of the clasts are lithologically and texturally similar to underlying and adjacent Grayburg strata, which suggests they were derived from nearby rocks. The matrix (locally laminated) consists of both finely crystalline dolomite and quartz sandstone-siltstone with some minor sand-size skeletal fragments. Basal Queen sandstone overlies the breccia and locally appears to have infiltrated fractures(?) and surround Grayburg clasts in the G2 section breccia. Fragment angularity, in-place brecciation, and local occurrence of bedded material suggest little or no transport for the G2 section breccia.

Basinward Trace of Erosion Surface

At the Bush Mountain headwall, the erosion surface has a strike of about N.75°E., roughly paralleling the inferred margin of the Delaware Basin during Grayburg time. Within 100 m basinward of the high-angle headwall portion, the erosion surface flattens (less than 5°–10°) and truncates 60–70 m of additional Grayburg strata over a 0.9-km distance (Fig. 3; Fekete, 1986). Studies of aerial photographs suggest much of the additional 60–70 m of truncation occurs abruptly over a short distance rather than gradually over the

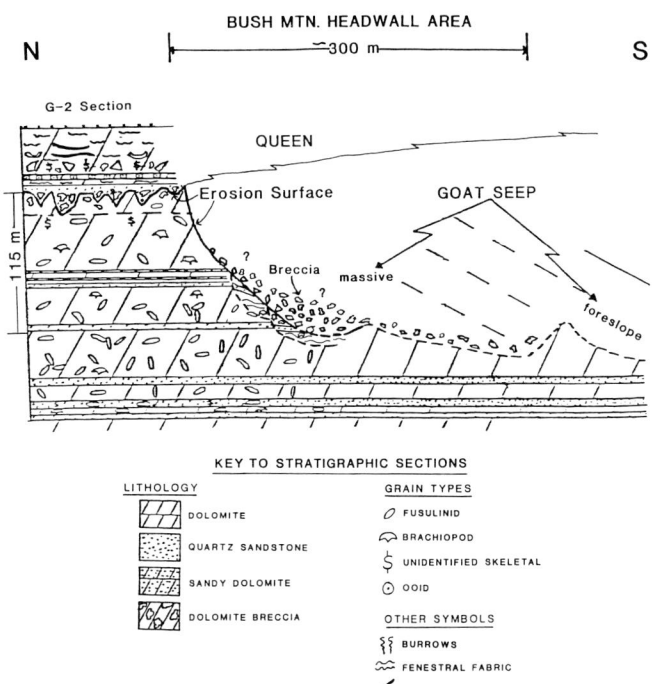

FIG. 9.—Details of Grayburg (G2 section) and erosion surface in the Bush Mountain headwall area. Note irregular shelfward trace of erosion surface and associated breccia (G2 section).

0.9 km. Figure 3 shows what appears to be a second high-angle slope or "headwall" with about 30–40 m of relief located 0.5 km south of the Bush Mountain headwall. The age relation between the two headwalls is unknown. Traced basinward for another 2 km, we believe the surface drops another 145 m in elevation (Figs. 3, 5; basinward of section G4). This could measure additional truncation, but no field evidence has been found to date of truncation along this 2-km trace. Instead, aerial photographs suggest that this basinward portion of the erosion surface parallels the bedding of the immediately underlying Grayburg, and the surface may have actually been only a surface of non-deposition. The 145-m decrease in elevation along this basinward trace of the erosion surface is apparently due to depositional thinning of the underlying Grayburg in the basinward direction.

Although the actual surface is commonly covered in the headwall area, in those places where access is possible we believe it can be located in most places south of the Bush Mountain headwall to within about 2 m. In this area, the Grayburg erosion surface truncates mostly fusulinid dolowackestones and dolopackstones, dolomudstones, and some siliciclastic units.

Details of the Bush Mountain headwall area are shown in Figure 10A. This figure also shows the trace of two apparent incipient slump surfaces in the adjacent and underlying bedded Grayburg strata. Slumping is evidenced by displacement of beds, contortion, and brecciation at the base of the displaced beds. The contorted and brecciated Grayburg beds grade upward within about 2 m into the overlying basal Goat Seep breccia (Figs. 9, 10B). Approximately 20–30 m basinward of the gradational contact, the Goat Seep breccia and underlying bedded Grayburg are in sharp contact (Fig. 10C). The details of the erosion surface in most localities are obscured by cover, however.

Basal Goat Seep breccia.—

The deposit that immediately overlies the high-angle headwall portion of the erosion surface in the Bush Mountain area is a dolomite breccia (Fig. 10B). Near the upper extent of the Bush Mountain headwall, the breccia grades upward, within 5 to 10 m, into massive, sponge-rich Goat Seep Dolomite (Fig. 7D); approximately 50–100 m farther down the "scoop," the breccia appears to be at least 20–30-m thick. The upper limit of the breccia is difficult to establish, however, due to inaccessibility on sheer cliffs and the weathered nature of the outcrop. The breccia directly overlies the erosion surface in the lower headwall area and for approximately 0.3 km south of the Bush Mountain headwall (Fig. 9). The major rock types forming the basal breccia clasts are dolomudstone, fusulinid dolowackestones and dolopackstones, other skeletal dolowackestones and dolopackstones, and some sponge-rich dolomites. All clast types, excluding the sponge-rich dolomites, resemble and presumably were derived from adjacent truncated Grayburg strata. The breccia clasts are poorly sorted and angular to subround. Most are centimeters to decimeters in size, but locally clasts as large as several meters occur. The breccia has both clast-support and matrix-support textures, with the matrix consisting mostly of very finely crystalline dolomite and quartz silt (locally sandy). The matrix is locally laminated; some lamination is wavy, contorted, and sheared.

No breccia was found in exposures located more than 0.3 km south of the Bush Mountain headwall and high-angle foreset bedding planes of the Goat Seep approach the erosion surface. Initial Goat Seep deposits overlying the erosion surface to the south consist instead of massive sponge boundstone, medium-scale trough cross-bedded sandy do-

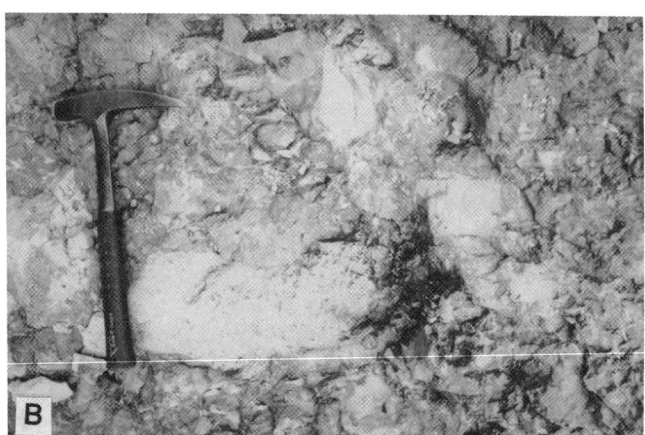

Fig. 10.—Features associated with the Grayburg erosion surface in the 115-m high-angle headwall area under Bush Mountain. (A) Flat-lying Grayburg bank top facies (GB) are truncated by the erosion surface. Massive to foreslope Goat Seep Dolomite (GS) and flat-bedded Queen strata (QN) overlie the surface. Erosion surface (arrows) strikes roughly parallel to the northeast-trending shelf margin. Locally, underlying and adjacent Grayburg beds show incipient slump surfaces (dashed white lines). Local brecciation and contorted beds occur at base of slumps. View to the north. (B) Basal Goat Seep breccia located immediately above high-angle part of erosion surface. Clast lithologies and textures resemble, and presumably were derived from, underlying and adjacent Grayburg strata. (C) View of erosion surface where it is flat, about 20–30 m basinward of the high-angle headwall. Here, erosion surface (white line) sharply truncates underlying Grayburg fusulinid dolowackestone and dolopackstone (GB). Overlying lithology is basal Goat Seep breccia (GS).

lomites and dolomitic sandstones, and fusulinid dolowackestones and dolopackstones.

Processes of Erosion

The process or processes that created the Grayburg erosion surface remain enigmatic. It is likely that the apparent shallower water conditions at the end of Grayburg time are responsible for the channeling and scouring on the shelf, and the shallower water may have intensified or initiated erosional processes at the shelf margin.

Field evidence is inconclusive as to whether or not the shelf and shelf margin were exposed during channeling, scouring, and local brecciation (G2 section) on the shelf. The process that formed the G2-section breccia remains enigmatic as well. The breccia appears to have formed essentially in place. Perhaps it formed in a submarine environment as a result of extensional forces related to gravity movements (incipient slumping) at the headwall. Alternatively, the G2 section breccia could have formed from exposure and karsting. Scouring, channeling, and iron staining in the shelfward trace of the erosion surface could indicate exposure and progradation of streams across the shelf. G2 section brecciation, scouring, and local iron staining of the erosion surface are similar to features associated with earlier Permian erosion surfaces located farther shelfward (Last Chance Canyon, Algerita Escarpment) that are interpreted as subaerial erosion surfaces (R. Sarg pers. commun., 1987).

Although a subaerial origin for the high-angle Bush Mountain headwall portion of the erosion surface cannot be disproven, we prefer to interpret its creation in a submarine environment. All overlying facies are marine lithologies, and we recognize no evidence of subaerial exposure along the sloping surface and its basinward trace. If the erosion surface were formed subaerially, a sea-level drop of nearly 300 m would be necessary to expose the more baswinward extension of the erosion surface (unless the basinward extension of the erosion surface were formed at an earlier time than its headwall portion). We know of no evidence in the mid-Permian rock record to support a sea-level drop of such major proportions. Also, the Grayburg erosion surface is morphologically similar to submarine erosion surfaces in the modern oceans. The Grayburg and older Permian basin-sloping erosion surfaces at the Delaware Basin-Northwest Shelf margin have some morphologic similarities (although the Grayburg erosion surface has a steeper headwall). These older Permian erosion surfaces (pre-Cutoff surface truncating some 200+ m of Victorio Peak and Bone Spring Formations and the post-Cutoff-pre-Brushy Canyon erosion surface truncates 200+ m of Cutoff and Victorio Peak Formations) lack any evidence for subaerial exposure (Pray, 1968, 1971; Pray and others, 1980; Harris, 1982; Kirkby, 1982, Franseen and others, 1987).

Field observations to date suggest slumping, debris flows, and bottom currents as probable mechanisms of Grayburg erosion in the Bush Mountain headwall and basinward portions of the erosion surface (Franseen, 1984, 1985; Fekete, 1986; Fekete and others, 1986; Franseen and others, 1987). We believe retrograde slumping was a major erosional process. The best evidence for this is: (1) the listric shape of the erosion surface; both the Bush Mountain headwall and a lower, more basinward headwall approximately 0.5 km from the Bush Mountain headwall may represent slump scars; and (2) the Grayburg strata immediately behind and below the trace of the erosion surface in the Bush Mountain headwall area show local brecciation and several probable incipient slump features. Some contorted beds at the base of the inferred incipient slump blocks suggest local shearing, possibly associated with mass movement.

Debris flows can be agents of erosion of bedrock over which they pass. The basal breccias have textual characteristics of debris flows, and the clasts resemble and were likely derived from the adjacent and underlying Grayburg strata. Although solution and bioerosion are viable processes of submarine erosion, we have not recognized solution features, insoluble residues or borings. The absence of breccia and debris more than 0.3 km south of the Bush Mountain headwall suggest that processes other than mass movements may have been important in forming or modifying the surface. South of the Bush Mountain headwall at the G4 section, the erosion surface is overlain by a 5- to 10-m-thick interval of sponge boundstone. This basal Goat Seep sponge boundstone is overlain by a 5- to 10-m-thick cross-bedded sandstone. If currents existed at this location in early Goat Seep time, they may have been operative at the time the erosion surface was forming. Bottom currents may have been a factor in creating or modifying the erosion surface.

DISCUSSION

The recognition of submarine erosion as a geologic process is a relatively new development. It has evolved largely as the result of turbidity current concepts and from data collected from recently developed oceanographic techniques (seismic profiling, deep-sea drilling, and the use of side-scan sonar and deep-sea submarines). These techniques reveal that deep submarine processes do move sediment, carve submarine canyons and shape slopes, and may also cause the erosion and retreat of deep submarine escarpments (Paull and Dillon, 1980; Mullins and others, 1985; Freeman-Lynde and Ryan, 1985).

Few examples of ancient unconformities that represent appreciable erosion in a submarine environment have been recognized. Some have recently been reported by Playford and others (1984) and Ward and Meyers (1987). We believe that the Grayburg erosion surface (Bush Mountain headwall and south) exposed along the western escarpment of the Guadalupe Mountains represents a spectacular example of an ancient submarine erosion surface, as do the earlier Permian erosion surfaces associated with the Cutoff Formation. Submarine erosion may have been an important and possibly common feature of other carbonate platforms in the geologic record.

Features observed along the Grayburg and earlier Permian erosion surfaces that we interpret as submarine and that may characterize other ancient submarine erosion surfaces at shelf margins and aid in their recognition are the following: (1) strike roughly parallel to the shelf margin and a basinward slope of 5° to as much as 80° on the shelf margin and a gentler slope (0°–10°) farther basinward; (2) overall or local listric (concave upward) shape to the surface; (3) sharp truncation of hundreds of meters of underlying and adjacent strata over 1 km or more laterally; (4) smooth, scoured, or locally irregular surface, commonly with evidence of lithified or partially lithified strata below the erosion surface; (5) local channeling on the scale of meters to tens and hundreds of meters may occur on the shelf, shelf margin, or in the basin; (6) local evidence of slumping, contortion, and brecciation of underlying strata; (7) marine facies directly overlying the erosion surface; and (8) local occurrence of breccias or conglomerates overlying the erosion surface.

Contemporary processes of submarine erosion, all of which are possible as major processes of erosion in the ancient record, include mass movements, bottom currents, dissolution, bioturbation, and bioerosion. Modern-day sites of major erosion in a deep submarine environment (or at which early Cenozoic submarine erosion is inferred to be important) include the west Florida carbonate platform margin (Mullins and others, 1985), the Blake Escarpment (Freeman-Lynde and Ryan, 1985; Paull and Dillon, 1980; Heezen and Hollister, 1971), the Blake-Bahama ridge (Hollister and others, 1976; Mullins and Neumann, 1979), the east flank of the Equatorial Pacific Plateau (Johnson, 1972), and the New England continental rise (Johnson and Lonsdale, 1976). The erosion surface associated with the Grayburg Formation has some morphologic features that are strikingly similar to features of the modern sites of erosion, and it is entirely possible (likely?) that the Grayburg erosion surface and modern surfaces formed from similar processes of erosion; these processes are still not fully understood. Submarine-erosion processes causing "scarp or line retreat," in contrast to those of margin-incising submarine canyons, are particularly enigmatic and deserve further investigation.

Seismic sequence-stratigraphic concepts have recently been applied to interpret and correlate Permian strata in the Guadalupe Mountains, including the strata along the western escarpment (Sarg and Lehmann, 1986). Of importance for our study is the correlation of the western escarpment Grayburg strata and bounding surfaces by Sarg and Lehmann (1986) to other regional outcrops and subsurface locations.

Sarg and Lehmann (1986) consider the base of the oolite-rich interval in the lower part of the bank top Grayburg facies (Lower-Middle Grayburg boundary) in our study area to be a sequence boundary and correlate it to an exposure surface at the top of the Upper San Andres Formation located 10–30 km farther shelfward (Last Chance Canyon and on the Algerita Escarpment in New Mexico). If this correlation is correct, the lower 150–200 m of the Grayburg Formation on the western escarpment is time equivalent to the Upper San Andres Formation. Unfortunately, there are no outcrop exposures between the western escarpment in Texas and Last Chance Canyon, and relations between the study area and the Algerita Escarpment are complicated by faulting. Identification and correlation of the Cherry Canyon Sandstone Tongue in each of the areas, however, and similarities of the Upper San Andres and Grayburg lithologies lend support to Sarg and Lehmann's correlation. Additional detailed field work along the western escarpment and at Last Chance Canyon, as well as more detailed paleontologic work, would be useful. If this sequence boundary was formed by eustatic sea-level lowering, then a detailed diagenetic and geochemical study of similarities (or differences) in trends across the sequence boundary in each of the areas may help in determining regional extent of processes responsible for creating the sequence boundary, as well as help in correlation between areas.

A comparative diagenetic study of earlier Permian subaerial-erosion surfaces located farther shelfward in Last Chance Canyon and on the Algerita Escarpment with the Grayburg erosion surface on the western escarpment may aid in determining the environment in which erosion occurred along the western escarpment.

CONCLUSIONS

(1) Outcrops on the western Guadalupe Mountains escarpment expose the critical shelf-to-basin transition of the Grayburg Formation on the northwestern margin of the Delaware Basin. The 385-m-thick Grayburg dolomite bank pinches out laterally into the basin within a distance of 4 km by a combination of bank margin erosion, depositional thinning, and facies change.

(2) The Grayburg strata are mostly dolomite with interbedded, thinner, quartz sandstone-siltstone layers. The Grayburg of the shelf is composed of two shoaling-upward sections between the underlying, more basinal Cherry Canyon Sandstone Tongue and the overlying shallow-water-to-peritidal dolomite facies of the Queen Formation. The lower one-fourth of Grayburg strata is foreslope facies consisting of basinward-prograding beds with dips of 5°–35°. Grayburg foreslope beds gradually decrease dip upward into the overlying Grayburg, composed of flat-bedded bank top facies.

(3) Grayburg strata consist mostly of pervasively dolomitized marine wackestones and packstones with interbedded, massive to laminated (locally cross-laminated), marine siliciclastic sandstones-siltstones. Most strata are interpreted to have been deposited in normal-marine water below both normal- and storm-wave base. A cross-bedded, oolite-rich quartz sandstone unit caps the first shoaling-upward section and indicates more turbulent and probably shallower, more restricted conditions than the underlying strata. Above it, a second shoaling-upward section occurs in the shelfward, uppermost Grayburg strata as shallow subtidal-to-peritidal dolomites overlie deeper water facies. Because the preserved shelf edge correlatable strata 1–2 km farther basinward contain deeper water facies, Grayburg strata are interpreted to have been deposited on a very gently sloping ramp.

(4) Shelf-to-basin relief during Grayburg time likely increased from about 50–100 m for lower Grayburg to 200–300 m at the close of Grayburg time.

(5) Approximately 185 m of middle and upper Grayburg strata are spectacularly truncated over a distance of 1 km toward the basin by a sharp basin-sloping erosion surface.

Additional truncation of ten to 100 m may occur in the next 2 km toward the basin, although field evidence to support additional truncation has not been recognized.

(6) Shelfward of the Bush Mountain headwall area, the erosion surface flattens abruptly and can be traced for approximately 2 km. The surface is mostly flat, but local scouring and channeling on the scale of meters to perhaps 10 m or more, iron staining, and brecciation occur.

(7) Processes that created the Grayburg erosion surface remain enigmatic. The Bush Mountain headwall and basinward portions of the surface are interpreted to have formed in a submarine environment. Slumping, debris flows, and bottom currents are favored for erosion in these areas. Field evidence is inconclusive as to whether the shelfward portion of the surface was formed subaerially or in a submarine environment. It is likely that the interpreted shallower water conditions at the end of Grayburg time created erosional features (scouring, channeling) on the shelf. Shallower water conditions may have intensified or initiated erosional processes at the shelf margin.

(8) The Grayburg erosion surface and the base of the oolite-rich unit at the Lower-Middle Grayburg boundary are interpreted as major sequence boundaries that are regionally correlatable in outcrop and in the subsurface (Sarg and Lehmann, 1986). These sequence boundaries likely formed from eustatic sea-level lowering.

(9) Submarine erosion is likely to have been an important and common process in the evolution of carbonate platform margins in the past. Yet, few examples of ancient submarine unconformities have been recognized. We believe that the Grayburg erosion surface represents a striking example of an ancient submarine erosion surface. Features of the Grayburg and earlier Permian erosion surfaces along the western escarpment of the Guadalupe Mountains may characterize other ancient erosion surfaces and may aid in their recognition.

ACKNOWLEDGMENTS

Financial asistance in support of the research was provided mainly by Exxon Production Research Company, Inc. Other funds were provided by the Department of Geology and Geophysics of the University of Wisconsin-Madison, the New Mexico Bureau of Mines and Mineral Resources, and the Shell Companies Foundation. All financial support is gratefully acknowledged. Much appreciated field assistance was provided by Somita Fekete, Alan Franseen, Ellen Lawson, Kevin McGinnity, Robert Monahan, Steven Roth, and Ruurdjan DeZoeten. J. F. Sarg, P. J. Lehmann, and K. Rudolph of Exxon Production Research Company, Inc., have contributed freely of their concepts and perspectives throughout the study. N. F. Hurley and J. F. Sarg critically reviewed and greatly improved an earlier version of the manuscript. The field area is located entirely within the boundaries of the Guadalupe Mountains National Park. We are grateful for the help and cooperation of the Rangers and administrative staff of the Guadalupe Mountains National Park and the Regional Headquarters of Carlsbad Caverns and Guadalupe Mountains National Parks in Carlsbad, New Mexico.

REFERENCES

CRAWFORD, G. A., 1981, Depositional History and Diagenesis of the Goat Seep Dolomite (Permian, Guadalupian), Guadalupe Mountains, West Texas-New Mexico: Unpublished Ph.D. Dissertation, University of Wisconsin, Madison, 300 p.

DUNHAM, R. J., 1962, Classification of carbonates according to depositional texture, in Ham, W. E., ed. Classification of Carbonate Rocks, a Symposium: American Association of Petroleum Geologists Memoir 1, p. 104–121.

FEKETE, T. E., 1986, The Sedimentology and Stratigraphy of the Grayburg Formation and its Associated Erosion Surface along the High Western Escarpment of the Guadalupe Mountains, Texas: Unpublished M.S. Thesis, University of Wisconsin, Madison, 174 p.

———, FRANSEEN, E. K., AND PRAY, L. C., 1986, Deposition and erosion of the Grayburg Formation (Guadalupian, Permian) at the shelf-to-basin margin, western escarpment, Guadalupe Mountains, Texas, in Moore, G. E., and Wilde, G. L., eds., Lower and Middle Guadalupian Facies, Stratigraphy and Reservoir Geometries, San Andres-Grayburg Formations, Guadalupe Mountains, New Mexico and Texas: Society of Economic Paleontologists and Mineralogists, Permian Basin Section Publication No. 86-25, p. 69–81.

FRANSEEN, E. K., 1984, The Grayburg and Queen Formations and the associated erosion surface at the shelf margin, western escarpment, Guadalupe Mountains, West Texas, in Pray, L. C., and Crawford, G. A., eds., A Field Guide to the Geology of the Permian Shelf-to-Basin Transition of the Western Escarpment, Guadalupe Mountains, Texas: Society of Economic Paleontologists and Mineralogists, Gulf Coast Section Field Conference Guidebook Preprint, p. 75–83.

———, 1985, Sedimentology of the Grayburg and Queen Formations (Guadalupian) and the Shelf Margin Erosion Surface, Western Escarpment, Guadalupe Mountains, West Texas: Unpublished M.S. Thesis, University of Wisconsin, Madison, 189 p.

———, PRAY, L. C., AND FEKETE, T. E., 1987, Mid-Permian shelf margin erosion surfaces, western escarpment, Guadalupe Mountains, Texas (abs.): American Association of Petroleum Geologists Bulletin, v. 71, p. 557.

FREEMAN-LYNDE, R. P., AND RYAN, W. B. F., 1985, Erosional modification of Bahama Escarpment: Geological Society of America Bulletin, v. 96, p. 481–494.

HARRIS, M. T., 1982, Sedimentology of the Cutoff Formation, Western Guadalupe Mountains, West Texas annd New Mexico: unpublished M.S. Thesis, University of Wisconsin, Madison, 266 p.

HEEZEN, B. C., AND HOLLISTER, C. D., 1971, The Face of the Deep: Oxford University Press, New York, 650 p.

HOLLISTER, C. D., GARDNER, W. D., LONSDALE, P. F., AND SPENCER, D. W., 1976, New evidence for northward-flowing bottom water along the Hatton sediment drift, eastern North Atlantic (abs.): EOS Transactions, American Geophysical Union, v. 57, p. 261.

INGRAM, R. L., 1954, Terminology for the thickness of stratification and parting units in sedimentary rocks: Geological Society of America Bulletin, v. 65, p. 937–938.

JOHNSON, D. A., 1972, Ocean floor erosion in the equatorial Pacific: Geological Society of America Bulletin, v. 83, p. 3121–3144.

———, AND LONSDALE, P. F., 1976, Erosion and sedimentation around Mytilus Seamount, New England continental rise: Deep-sea Research, v. 23, p. 429–440.

KING, P. B., 1948, Geology of the southern Guadalupe Mountains, Texas: U.S. Geological Survey Professional Paper 215, 183 p.

KIRKBY, K. C., 1982, Deposition, Erosion and Diagenesis of the Upper Victorio Peak Formation (Leonardian), Southern Guadalupe Mountains, West Texas: Unpublished M.S. Thesis, University of Wisconsin, Madison, 165 p.

MULLINS, H. T., GARDULSKI, A. F., AND HINE, A. C., 1985, Catastrophic collapse of the west Florida carbonate platform margin: Geology, v. 14, p. 167–170.

———, AND NEUMANN, A. C., 1979, Deep carbonate bank margin structure and sedimentation in the northern Bahamas: Society of Economic Paleontologists and Mineralogists Special Publication 27, p. 165–192.

NEWELL, N. D., RIGBY, J. K., FISCHER, A. G., WHITEMAN, A. J., HICKOX, J. E., AND BRADLEY, J. S., 1953, The Permian Reef Complex of the Guadalupe Mountains Region, Texas and New Mexico—A study in Paleoecology: Freeman and Company, San Francisco, 236 p.

PAULL, C. K., AND DILLON, W. P., 1980, Erosional origin of the Blake Escarpment: An alternative hypothesis: Geology, v. 8, p. 538–542.

PLAYFORD, P. E., KERANS, C., AND HURLEY, N. F., 1984, Platform-margin and marginal slope relationships and sedimentation in Devonian reef complexes of Canning Basin, Western Australia (abs.): American Association of Petroleum Geologists Bulletin, v. 68, p. 516–517.

PRAY, L. C., 1968, Basin-sloping submarine(?) unconformities at margins of Paleozoic banks, West Texas and Alberta: Geological Societey of America, Abstracts With Programs, p. 243.

———, 1971, Submarine slope erosion along the Permian bank margin (abs.): American Association of Petroleum Geologists Bulletin, v. 55, p. 358.

———, 1981, Submarine erosion surfaces and retreat of carbonate bank margins, (Permian), southwestern U.S.A.: Fifth Australian Geologic Convention, Perth, Abstracts 3, p. 52.

———, CRAWFORD, G. A., HARRIS, M. T., AND KIRKBY, K. C., 1980, Early Guadalupian (Permian) basin margin erosion surfaces, Guadalupe Mountains, Texas (abs): American Association of Petroleum Geologists Bulletin, v. 64, p. 768.

SARG, J. F., AND LEHMANN, P. J., 1986, Lower-Middle Guadalupian facies and stratigraphy, San Andres-Grayburg formations, Permian Basin, Guadalupe Mountains, New Mexico, in Moore, G. E., and Wilde, G. L., eds., Lower and Middle Guadalupian Facies, Stratigraphy and Reservoir Geometries, San Andres-Grayburg Formations, Guadalupe Mountains, New Mexico and Texas: Society of Economic Paleontologists and Mineralogists, Permian Basin Section Publication No. 86-25, p. 1–36.

———, AND PRAY, L. C., 1984, Unconformities and depositional sequences of seismic stratigraphic scale (Permian) of western Guadalupe Mountains, Texas-New Mexico: Society of Economic Paleontologists and Mineralogists First Annual Midyear Meeting, San Jose, Abstracts with Programs, p. 71.

VAIL, P. R., MITCHUM, R. M., JR., AND THOMPSON, S., III, 1977, Seismic stratigraphy and global changes of sea level, Part 4. Global cycles of relative changes in sea level, in Payton, C. E., ed., Seismic Stratigraphy–Applications to Hydrocarbon Exploration: American Association of Petroleum Geologists Memoir 26, p. 83–97.

WARD, W. B., AND MEYERS, W. J., 1987, Neptunian-fracture control on platform-margin geometry, Upper Devonian reef complexes, Napier Range, Canning Basin, Western Australia: Society of Economic Paleontologists and Mineralogists Fourth Annual Midyear Meeting, Austin, Abstracts with Programs, p. 89.

LOWER PERMIAN PLATFORM AND BASIN DEPOSITIONAL SYSTEMS, NORTHERN MIDLAND BASIN, TEXAS

S. J. MAZZULLO AND A. M. REID

Department of Geology, Wichita State University, Wichita, Kansas 67208; and Geological Consultant, Midland, Texas 79701

ABSTRACT: The Lower Permian (Wolfcamp-Leonard) section on the North Platform of the Midland Basin is a mosaic of lithofacies composing vertically stacked, progradational, and erosionally backstepping platform-to-basin sequences. Shelf facies include carbonates, evaporites, and siliciclastics, and contiguous basinal deposits are resedimented carbonate detritus and siliciclastics. Wolfcamp strata were deposited in a humid climatic setting and are dominantly carbonate-shale ramp and distally steepened ramp systems that pass seaward to shallow- to moderately deep-basin facies. Progradational as well as lowstand erosional phases of platform development are recognized in the section. Maximum progradation of platform facies occurred in middle early Wolfcamp time, when relatively high rates of basin subsidence were coincident with the rapid deposition of shales in the basin. The resulting shale wedge provided a foundation over which younger Wolfcamp platform depositional systems rapidly prograded. A major period of shelf deepening separates the late Wolfcamp and overlying lower Leonard sections.

The Wichita and Lower Clear Fork are rimmed-shelf systems that stacked vertically at a location basinward of the late Wolfcamp platform margin. Sedimentation and subsidence rates and shelf-to-basin depositional relief during early Leonard time represent the maxima for the Early Permian on the North Platform. Four regionally correlative megacycles are readily identifiable within shelf deposits in each of these formations, and the tops of these represent periods of sea-level lowstand and partial shelf emergence. The megacycles pass into thick, but vertically discontinuous, shelf margin reefs. The megacycles are themselves composed of innumerable subcycles that shoal upward to peritidal carbonate and/or sabkha evaporite deposits. The shelf Tubb and equivalent basinal Dean sections record a major episode of gradual, highstand-punctuated sea-level fall that terminated in complete shelf emergence. Thus began a period of alternating carbonate-evaporite and sandstone deposition on the North Platform that persisted into the late Leonardian. The subdued rimmed-shelf systems of the Middle and Upper Clear Fork prograded rapidly into the northern Midland Basin during highstands, across the sandstone wedges that were deposited during lowstands; sedimentation rates exceeded subsidence rates during this time.

The evolution of depositional systems recognized in the Lower Permian on the North Platform was affected by complex changes in several parameters, including: (1) contrasting rates of basin subsidence and sedimentation; (2) probable glacio-eustatic sea-level fluctuations; (3) the shift from dominantly carbonate to mixed carbonate-siliciclastic deposition; and (4) the evolution of reef biotic communities and extent of synsedimentary marine cementation of shelf margin deposits.

INTRODUCTION

During Early Permian time, the Permian Basin of west Texas and southeastern New Mexico (Fig. 1) was an actively subsiding depocenter in which platform deposits (carbonates, evaporites, shales) surrounded deep, dominantly siliciclastic basins. Platform-to-basin depositional sequences in subsurface Lower Permian (Wolfcamp and Leonard) strata are particularly well developed on the stable North Platform province of the northern Midland Basin (as well as on the Eastern Shelf), where cumulative sediment thicknesses exceed 1,980 m (6,500 ft). The Lower Permian section here is readily amenable to detailed subsurface geologic and seismic stratigraphic studies because of the general lack of structural complexity and the high density of well control. This stability is in marked contrast to such areas as the subsurface Central Basin Platform and outcrops on the Southern Shelf (Fig. 1), where the Lower Permian section was disturbed by syn- and post-depositional tectonism and, accordingly, is stratigraphically incomplete (i.e., Ross, 1986; Mazzullo and Reid, in prep.).

Despite the long history of geologic studies and petroleum exploration in the northern Midland Basin, details of the evolution through time of Wolfcamp and Leonard depositional systems remain poorly understood. Previously published studies on the North Platform and peripheral areas have been concerned mainly with the stratigraphy and depositional facies of the Wolfcampian and Leonardian series, and have only briefly described the sedimentology and diagenesis of hydrocarbon reservoir facies in shelf and basin carbonates in these units (Silver and Todd, 1969; Jeary, 1978; Mazzullo, 1982, 1984; Mazzullo and Reid, 1987a, b; Mazzullo and others, 1986, 1987). The depositional framework of this area is grossly similar to coeval sequences elsewhere in the Permian Basin, for example, in the northern Delaware and southern Midland basins and the Eastern Shelf (Cook, 1983; Hobson and others, 1985; Loucks and others, 1985; Wiggins and Harris, 1985; Ross, 1986; Gawloski, 1987; L. J. Mazzullo and Reid, 1987). Thus, a more thorough understanding of the stratigraphy, sedimentology, and facies evolution through time on the North Platform will provide a valuable analog for interpreting similar parameters in the Lower Permian section throughout the Permian Basin.

During the last several years, we have been involved in regional studies of the Pennsylvanian to Lower Permian section in the Permian Basin, with the specific objective of describing in detail the stratigraphic and depositional framework of the Wolfcamp and lower Leonard systems throughout the Midland Basin. This paper describes the results of these ongoing studies as they apply to our current knowledge of the stratigraphic and depositional-systems evolution of platform and basin sequences in the Lower Permian in the western portion of the North Platform of the Midland Basin (Fig. 1). The principal objective of this study is to relate the evolutionary development of depositional systems within the Lower Permian platform and basin sequence on the North Platform to such parameters as relative sea-level fluctuations, history of basin subsidence, variations in sedimentation rates, and the effects on sequence development of contrasting patterns of carbonate and siliciclastic sedimentation. Although we focus specifically on the Wolfcamp to lower Leonard (Wichita-Lower Clear Fork-Tubb-Dean) section in this paper, we also include a brief

Controls on Carbonate Platform and Basin Development, SEPM Special Publication No. 44
Copyright © 1989, The Society of Economic Paleontologists and Mineralogists, ISBN 0-918985-79-X

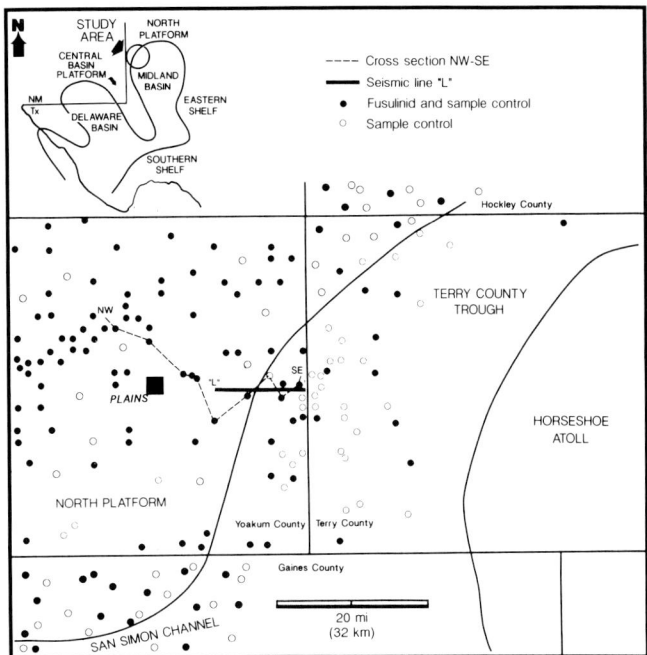

FIG. 1.—North Platform study area in the Midland Basin (inset map for reference), showing locations of wells examined for lithologic analyses and fusulinid biostratigraphic zonations. Locations of lithostratigraphic cross section (Fig. 3) and seismic line "L" (Fig. 5) indicated.

discussion of the upper Leonard (Middle and Upper Clear Fork) as it relates to the overall geologic history of the northern Midland Basin during the Early Permian.

METHODS OF STUDY

Regional aspects of the stratigraphy and depositional history of the North Platform, following the methods of Mitchum and others (1977) and Vail and Mitchum (1977), were interpreted from the examination of 25 seismic lines shot across both shelf and basin locations. A portion of one east-west line ("L") is reproduced herein (Fig. 5) as representative of the sequence stratigraphy in the study area. These interpretations were complemented with subsurface geologic studies based partly on the correlation of mechanical logs from approximately 1,900 wells in the study area (Fig. 1). Cores and well cuttings samples from approximately 300 wells (Fig. 1) were used to qualify the seismic and mechanical log correlations, and to identify and map the lithologies of shelf and basin strata in the study area (Mazzullo, 1982; Mazzullo and others, 1986, 1987a, b).

The Wolfcamp and lower Leonard sections in the study area (cumulative thickness 915–1,220 m: 3,000–4,000 ft) were further divided into component biostratigraphic and lithic units (Figs. 2, 3, 5) that have proven to be recognizable on logs and seismic lines throughout the Permian Basin (Reid and others, 1988; Mazzullo and others, in prep.). These divisions are based on the identification of fusulinid assemblage subzones in the Wolfcamp and basin sections of the lower Leonard (Reid and others, 1988), and on electric log correlations in the shelf section of the lower Leonard, where fusulinids generally are unavailable due to pervasive dolomitization. In our calculations of sedimentation rates in this paper, (Fig. 14) we have simply assumed equal durations (from the radiometric data of Harland and others, 1982) for each of the subzones recognized in the Lower Permian section (Figs. 2, 3) and divided these values into the average present thickness of each unit. Although this method may inherently be imprecise, particularly considering the erosional removal of some shelf strata, the rates discussed are nevertheless useful as comparative, relative estimates of these parameters through time. Relative basin subsidence rates (Fig. 14) were similarly derived and qualified by the known subsidence history of the Permian Basin (Adams and others, 1951; Jones, 1953; Hills, 1972, 1984).

Inferences concerning relative sea-level fluctuations are based on three lines of reasoning: (1) lithologic evidence of subaerial exposure across the study area, i.e., occurrence of colluvium on the shelf; (2) relation of the preserved stratigraphy in the study area to the complete zonation recognized in the Permian Basin (Fig. 2); and (3) facies occurrences and thickness relations in the study area relative to subsidence history (Adams and others, 1951; Jones, 1953; Hills, 1972, 1984). We agree with Burton and others (1987), however, that our suggestions on possible sea-level variations may, in fact, be the presently inseparable sum of basin subsidence and eustatic fluctuations. Inferences regarding the magnitudes and durations of relative sea-level fluctuations are based on the estimated absolute chronology considered for the Lower Permian section here, and the extent of subaerial exposure and erosion of shelf and adjoining strata as observed or interpreted on seismic lines and lithostratigraphic cross sections. Many of the sea-level fluctuations postulated for the Lower Permian in the study area are conceptually considered to have been eustatic, caused

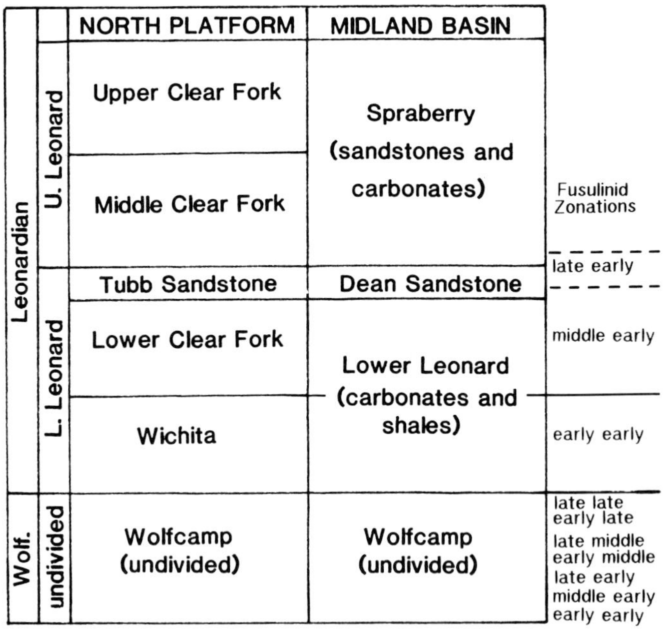

		NORTH PLATFORM	MIDLAND BASIN	Fusulinid Zonations
Leonardian	U. Leonard	Upper Clear Fork	Spraberry (sandstones and carbonates)	
		Middle Clear Fork		late early
	L. Leonard	Tubb Sandstone	Dean Sandstone	middle early
		Lower Clear Fork	Lower Leonard (carbonates and shales)	
		Wichita		early early
Wolf. undivided		Wolfcamp (undivided)	Wolfcamp (undivided)	late late early late late middle early middle late early middle early early early

FIG. 2.—Stratigraphy of Lower Permian section in study area (modified from Mazzullo, 1982), and fusulinid assemblage subzones recognized in Wolfcamp and lower Leonard.

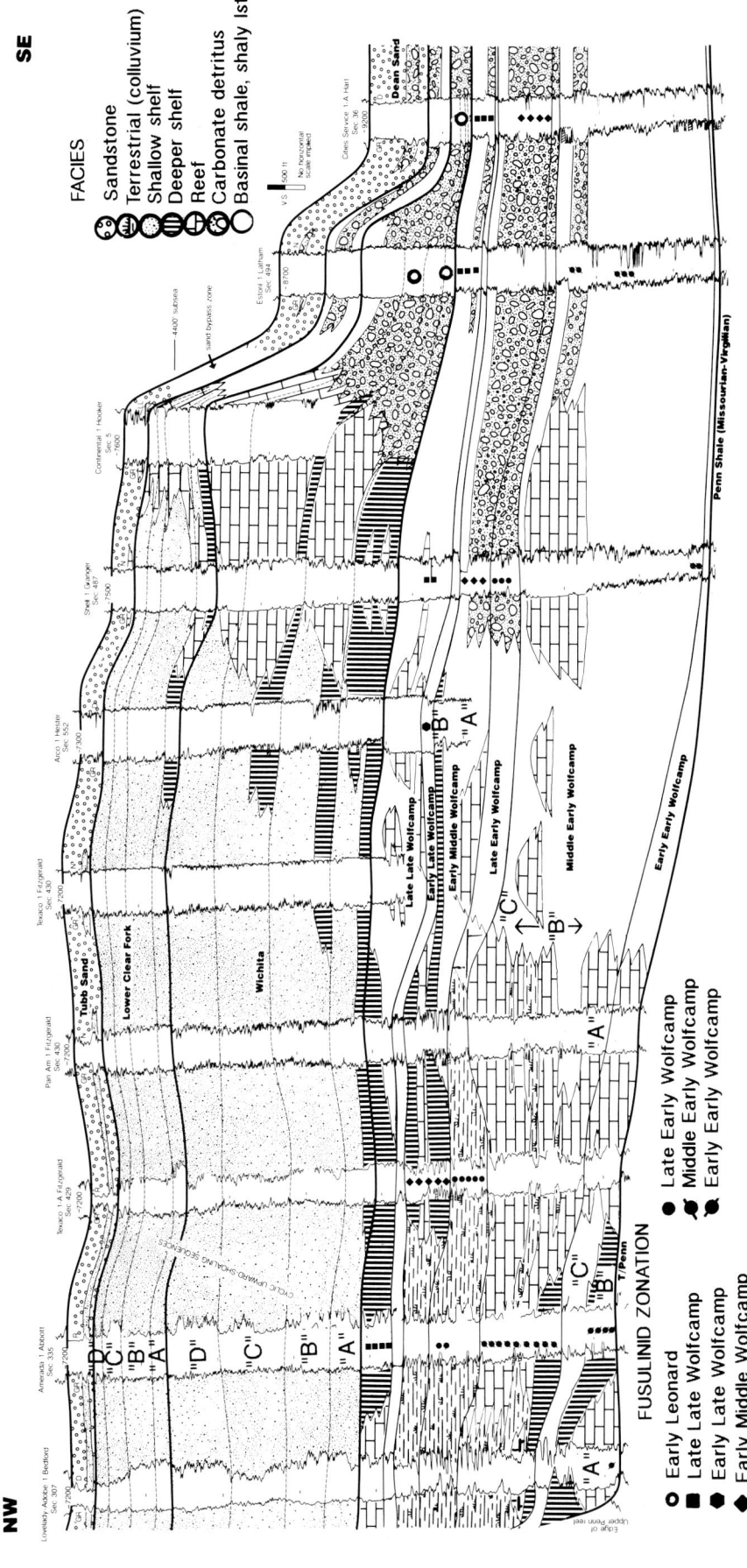

FIG. 3.—Northwest-southeast cross section illustrating formational correlations, biostratigraphic subdivisions, and facies systems in Wolfcamp to lower Leonard section; letter-designated zones referred to in text.

by glaciation in the Southern Hemisphere during the Late Paleozoic (Crowell, 1978; Veevers and Powell, 1987).

SETTING AND STRATIGRAPHY

The North Platform encompasses approximately 25,000 km^2 (9,600 mi^2) in portions of six counties along the northern margin of the Midland Basin. It is bounded by the Matador Arch and Palo Duro Basin to the north, the Central Basin Platform to the southwest, and the buried Horseshoe Atoll and Eastern Shelf province to the east (Fig. 1). The area of study in this paper includes Yoakum, Terry, southern Hockley and northern Gaines counties (Fig. 1). The North Platform was a constructional sedimentary province in Upper Pennsylvanian to Permian time and is composed of a thick section of Pennsylvanian (Atokan) to Permian (Ochoan) carbonate, evaporite, and siliciclastic rocks deposited on a folded and faulted basement of Precambrian to lower Pennsylvanian rocks. There is no evidence in the study area of tectonic activity in rocks younger than early middle Wolfcamp. Hydrocarbon fields here occur as structural and combination structural-stratigraphic traps in reservoirs of various lithology in each of the systems present in this region, the most prolific production being from Upper Permian (Guadalupian) and Silurian-Devonian dolomites.

The Lower Permian (Wolfcamp to upper Leonard) in the study area is divided into several formations and presently unnamed stratigraphic units that represent shelf (North Platform proper) and basin (Midland Basin: Terry County Trough in immediate study area) equivalents (Fig. 2). The Tubb and Dean formations here are dominantly sandstones and siltstones. Each of the other formations on the shelf consists of a complex facies mosaic of carbonates, evaporites, and siliciclastics (sandstones in the Middle and Upper Clear Fork, shales in the units beneath the Tubb). Basin sections include interbedded shales and carbonates, associated in the Middle and Upper Clear Fork (and Dean) with sandstone. Considerable confusion has existed in the past regarding the stratigraphic terminology and correlations in this area (discussed in Silver and Todd, 1969; Jeary, 1978; Mazzullo, 1982). Recent fusulinid biostratigraphic studies (Mazzullo and others, 1987; Reid and others, 1988), however, have indicated the temporal equivalence of the respective shelf strata with the undivided Wolfcamp, Wichita, and Lower Clear Fork in the basin, as shown on Figures 2 and 3. On the basis of detailed regional fusulinid biostratigraphic studies (Reid and others, 1988), we are further able to divide the Wolfcamp and lower Leonard shelf and basin sections (as well as the subjacent Pennsylvanian) into regionally recognizable fusulinid assemblage subzones (Fig. 2).

On the basis of then-existing fusulinid data, Mazzullo (1982) had placed the Wolfcampian-Leonardian series boundary on the North Platform within the lower beds of the Wichita, an assignment that subsequently was revised (Mazzullo and others, 1987) to locate it at the top of the undivided Wolfcamp stratal section (Fig. 2). Although not indicated on Figure 3, we recently have recovered middle to late early Leonardian fusulinids from immediately above and below both the Tubb and Dean sandstones in Yoakum and northern Gaines counties, thus clearly indicating the temporal equivalence of these formations (Fig. 2).

LITHOLOGIES AND DEPOSITIONAL ENVIRONMENTS

Carbonate, evaporite and siliciclastic facies are recognized in shelf deposits in the Lower Permian in the study area (Fig. 3). In the Wolfcamp and lower Leonard, shelf margin and shallow-shelf patch reef facies are mainly represented by pervasively dolomitized, phylloid algae-sponge-foraminifera-*Tubiphytes* buildups and associated biograinstones. Oolite and biograinstones alone dominate in portions of shelf margin sections at some localities (Mazzullo, 1982; Mazzullo and Reid, 1987a), but these generally are subordinate to reef facies in the study area. Similar facies are recognized in the Middle and Upper Clear Fork sections, in addition to local reef buildups dominated by crinoids, sponges, and bryozoans (Mosley, 1988). We have recognized a major change through time in the composition of shelf margin facies in the study area, from micrite-dominated, phylloid algae-foraminifera-*Tubiphytes* buildups in the Wolfcamp, to sponge-*Tubiphytes*-foraminifera buildups that are pervasively marine cemented in lower Leonard rocks (see Fig. 4; Mazzullo, 1984).

Shallow-shelf (lagoonal) facies in the Lower Permian section include heavily dolomitized micrites to packstones with a fauna of brachiopods, pelecypods, dasycladacean algae, fusulinids, foraminifera, and gastropods. Anhydritic dolomites, bedded anhydrites, and gray, green, and red shales also occur in the Leonard section. Peritidal facies in the Leonard consist of similar lithologies as well as fenestral and anhydritic, fine crystalline dolomites. Peritidal facies are not recognized in the Wolfcamp section in the study area. Unconformity-associated terrestrial (colluvial) facies in the Wolfcamp, however, include gray, green, and red shales, residual chert, and cherty limestone-pebble conglomerates. The Tubb section includes sandstones, variegated (anhydritic) shales, bedded anhydrites, and shallow-shelf and shelf margin reef carbonates (mostly dolomitized). The deeper water shelf lithologies recognized in the Lower Permian section (Fig. 3) include gray to black shales and partly dolomitized limestones and argillaceous limestones (dark mudstones to wackestones) with ostracodes, crinoids and accessory fusulinids, phylloid and dasycladacean algae, and bivalves.

Basinal facies in the Wolfcamp to lower Leonard section include black shale, gray sandstone (Dean), thick sections of carbonaceous gray shale (in the late early and early middle Wolfcamp only), and wedges of locally dolomitized, resedimented shelf carbonate detritus. This detritus includes megabreccias (carbonate clasts of shelf margin lithology in matrices of either black shale, gray shale, or carbonate micrite to grainstone), intraclastic skeletal and oolitic packstones to grainstones, and argillaceous micrites and wackestones (Fig. 4; Mazzullo and Reid, 1987a). These rocks were deposited as composite wedges and aprons in slope and base-of-slope environments seaward of the platform margins (Cook and others, 1983; Mazzullo, 1984; Mazzullo and Reid, 1987a; Mazzullo and others, 1986). Upper Leonard basinal facies are represented by gray (Spraberry)

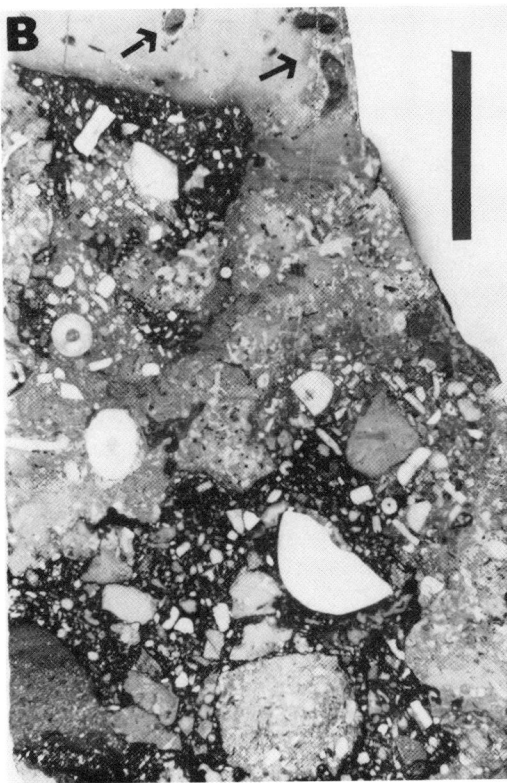

FIG. 4.—Cores of resedimented basinal-carbonate detritus. These rocks include only shelf margin lithologies and are the undolomitized representatives of late Wolfcamp and lower Leonard reef facies. (A) Late Wolfcamp detritus, closely packed intraclasts of biopackstone-grainstone ("PG") and phylloid algae-foraminifera-*Tubiphytes* wackestone ("W") in matrix of shaly carbonate-rock flour (left slab); similar detritus in black shale matrix (right slab); length of scale 6.0 cm. (B) Lower Leonard (Wichita) detritus, clasts of sponge (arrows)-*Tubiphytes*-foraminifera packstone in black shale matrix; the interparticle matrix within many of the clasts is acicular and locally, radiaxial calcite, of presumed synsedimentary marine origin (aragonite and high-Mg calcite, respectively). Length of scale 4.0 cm.

sandstones, dark shales, and argillaceous lime mudstones and, locally, dolomitized carbonate detritus (only locally intraclastic wackestones to grainstones). Dolomitized carbonates in the Wolfcamp and lower Leonard are the principal hydrocarbon reservoirs in basinal settings in the study area, whereas lime grainstones and megabreccias compose most of the fields that produce from coeval basinal rocks on the Eastern Shelf (Mazzullo and Reid, 1987a). There presently is little production in the study area from basinal carbonate facies in the upper Leonard section.

EVOLUTION OF LOWER PERMIAN DEPOSITIONAL SYSTEMS

The infilling of the northern Midland Basin due to offlapping of Walfcamp through Upper Clear Fork depositional sequences is indicated on Figure 5. This rapid period of basin filling lasted approximately 27.5 m.y. (Harland and others, 1982) and occurred subsequent to a major episode of rapid subsidence of the entire Permian Basin in the latest Pennsylvanian (Missourian-Virgilian; Adams and others, 1951). This regional drowning event is indicated by the widespread occurrence in the Permian Basin of a condensed section of dark shales (the "Penn Shale" on Figs. 3 and 5) that are the basinal equivalents of the Missourian to Virgilian shelf carbonate section to the west-northwest (Fig. 3). Reconstructed shelf-to-basin depositional relief in the study area in the latest Pennsylvanian was approximately 122–244 m (400–800 ft).

Wolfcamp

The Wolfcamp section on the North Platform is divided into seven fusulinid assemblage subzone units (Fig. 2); of these, only the late middle Wolfcamp is not recognized in the study area, either because of non-deposition or erosion (Fig. 3). The duration of the Wolfcamp was approximately 16.5 m.y. (Harland and others, 1982) and, accordingly, the estimated duration of each of the seven subzones is 2.36 m.y. For reference purposes in the following discussions, we have identified on Figure 3 letter-designated subzones in the Wolfcamp as well as in the lower Leonard section.

The seismic profile of the Wolfcamp sequence on Figure 5 is a complex of ramps and distally steepened ramps of stacked to offlapping geometry, rather than a single, progradational rimmed-shelf or ramp depositional system (terminology of Read, 1982). Each of these ramps is observed to pass eastward into a well-bedded sequence of rocks characterized seismically by thin units of relatively high and low velocity. Detailed lithologic and biostratigraphic studies in this section, however, indicate a more complex facies development than that seen by seismic stratigraphic interpretations alone.

Early early Wolfcamp.—

The basal Wolfcamp overlies the "Penn" shale and an erosionally thinned shelf carbonate section of late Pennsylvanian age (Fig. 3). The depositional relief at this time appears to have closely approximated that of the subjacent Pennsylvanian.

The edge of the early early Wolfcamp carbonate shelf of sub-unit "A" was a reef buildup that abutted the late Pennsylvanian platform margin (Fig. 3). To the west (not shown on Fig. 3), this facies passes to shallow-shelf carbonates and then a broad colluvial plain; to the east are shallow-slope to basinal facies. These stratigraphic relations suggest the establishment of a distally steepened ramp profile dur-

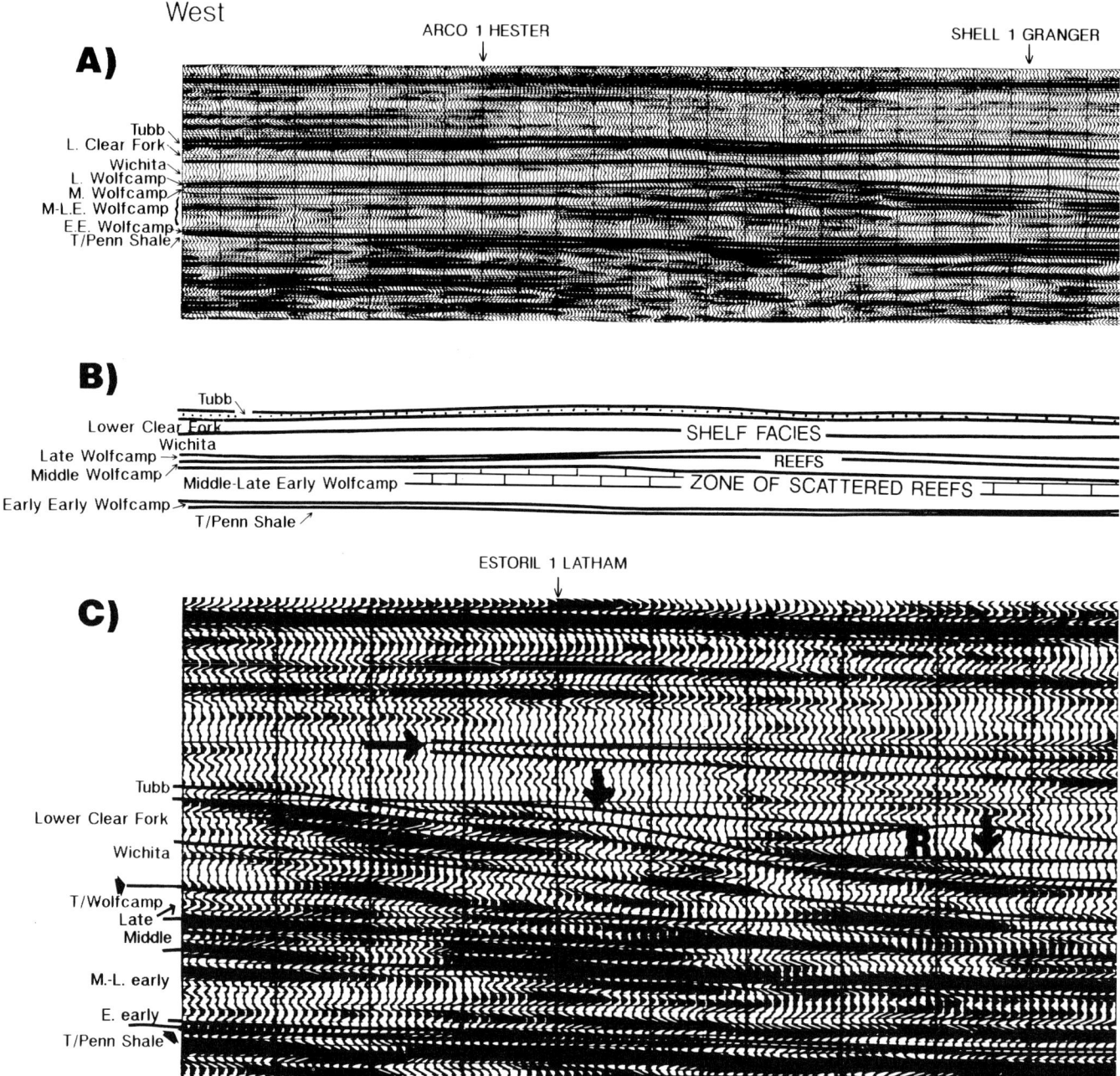

FIG. 5.—Seismic line "L" (east-west) across Yoakum County. (A and B) Regional aspect of seismic stratigraphic development and interpreted reflectors; subdivisions within the Wolfcamp by fusulinid biostratigraphy. On (A) the feathered arrows in the Wichita and Lower Clear Fork point character of depositional sequence in Middle and Upper Clear Fork section. Note basal onlapping reflectors in Middle Clear Fork directly above the sandstone by sample analyses) passing from shelf into shallow basinal setting (horizontal arrow); also note reflector interpreted as an onlap or bypass

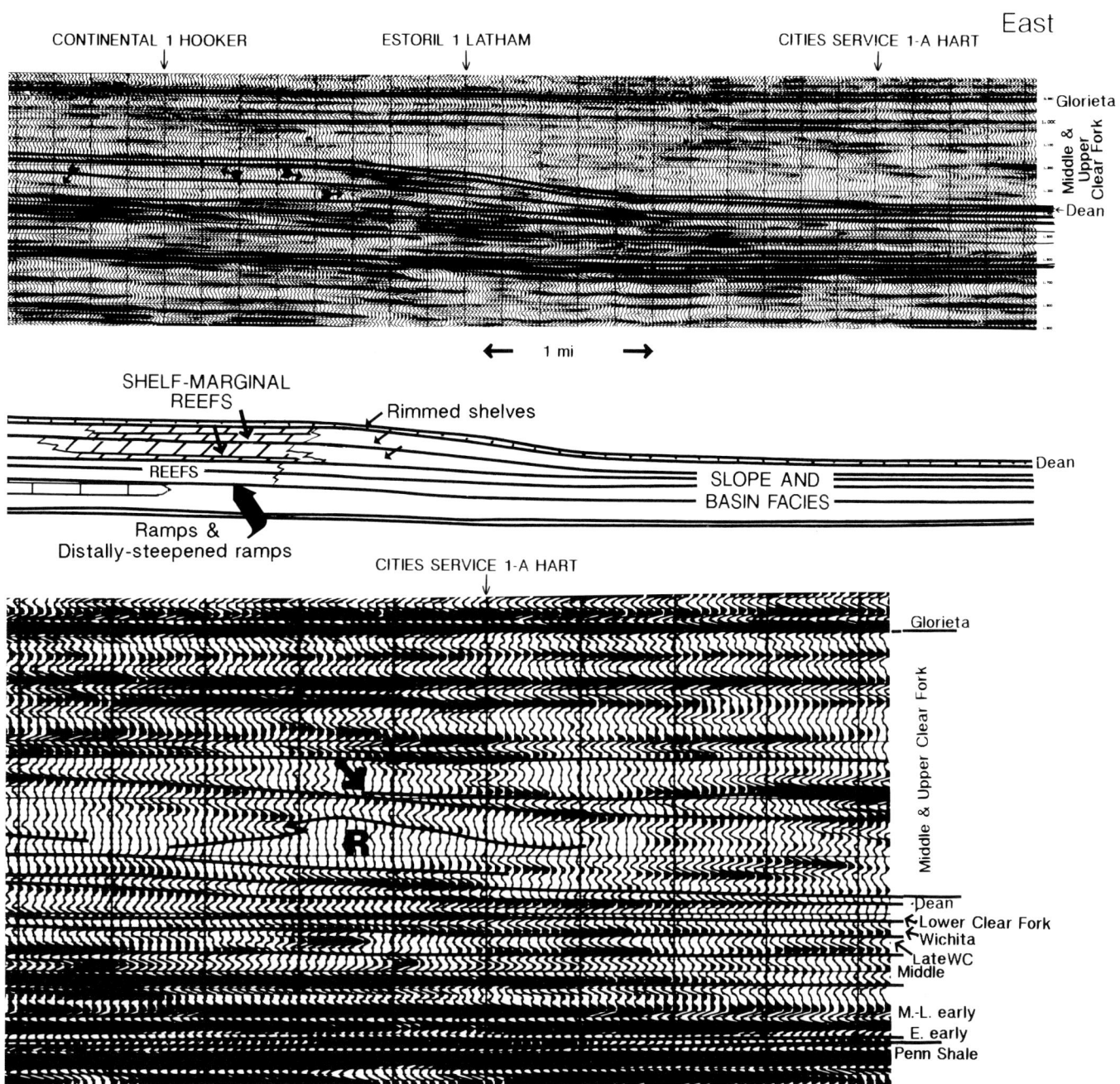

depositional systems in Wolfcamp through Tubb section. Reefs were identified seismically on the basis of the massive character and lack of seismic to reflectors that are believed to represent lowstand surfaces (tops of megacycles). (C) Enlargement of eastern part of above seismic section, illustrating Tubb and slope beds of the Dean (vertical arrows), laterally offlapping reefs ("R") in the Middle Clear Fork, and prominent lens (identified as horizon (oblique arrow) at base of this sandstone.

ing an initial period of relative lowstand. During this time, the North Platform shelf province to the west was mostly emergent. Subsequently, bypassing and the basinward shift of reef facies occurred during deposition of early early Wolfcamp sub-unit "B" (Fig. 3). The occurrence of deeper shelf facies to the lee of the reef at this time suggests that bypassing resulted from a pulse of subsidence and/or relative sea-level rise. Shelf drowning continued into sub-unit "C" time, as suggested by the establishment of a subdued distally steepened ramp (low shelf-to-basin relief) surrounded by shallow basinal facies (Figs. 3, 6). Both sub-units "B" and "C" pass westward into shallow-shelf carbonate facies and then a broad tract with a very thin section of colluvium. These relations suggest shelf emergence and probable sea-level lowstand toward the end of the early early Wolfcamp. The relatively thin section of carbonates and deeper water shales in the early early Wolfcamp suggest only minor subsidence and limited rates of sedimentation relative to the overlying Wolfcamp section (Fig. 3).

Middle early Wolfcamp.—

A thick shale section was deposited in the basin in the lower half of the middle early Wolfcamp (sub-units "A" and "B"; Figs. 3, 6), and this period of rapid basin infilling is recognized throughout the Midland Basin (Mazzullo and Reid, in prep.). Rapid sedimentation rates were coincident with an increase in basin subsidence rates (Jones, 1953;

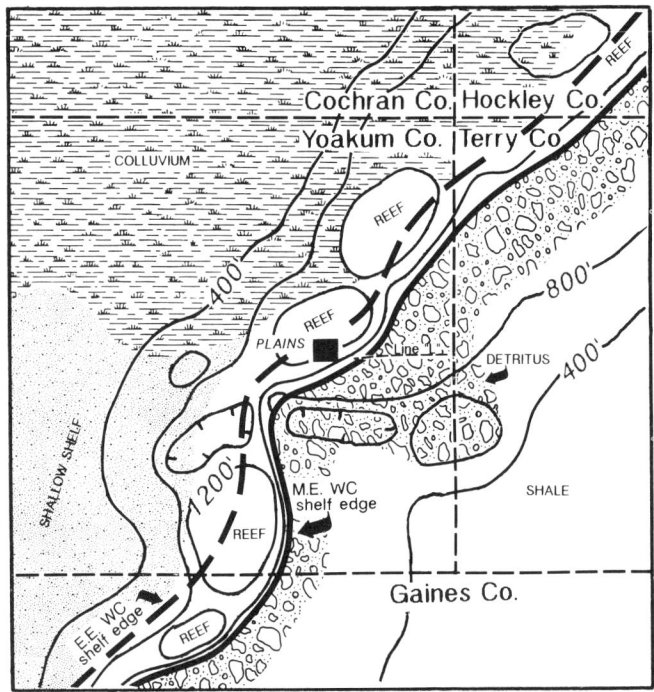

Fig. 6.—Cumulative isopach of early early and middle early Wolfcamp (CI = 30 m, 400 ft), with locations of respective shelf edges; facies shown are those at top 30 m (100 ft) of the middle early Wolfcamp. The carbonate detritus of the upper middle early Wolfcamp is underlain by a thick section of basinal shales (Fig. 3). On Figures 6–9 and 11, facies symbols are identical to those on Figure 3 except for reefs, which have no pattern. The connotation of "shale" in basinal settings on these maps refers to shales and admixed shaly limestones. Seismic line "L" (Fig. 5) shown for reference.

Hills, 1972, 1984). Carbonate ramp and distally steepened ramp systems were established in the central portion of the study area during this time, and rapidly prograded across the shale foundation and seaward of the early early Wolfcamp edge, particularly during sub-unit "B" time (Figs. 3, 5, 6). Numerous unconformities and associated colluvial deposits occur interbedded with and to the west of the carbonate facies in sub-units "A" and "B" (Fig. 3) and define a broad transition from weathered terrane to shallow-shelf and reef carbonate sedimentation across the study area (Fig. 6). These relations suggest periodically decreased rates of subsidence and likely, short-term sea-level fluctuations during deposition of these units.

The lithologic composition of slope and basin facies changed toward the close of middle early Wolfcamp time (upper sub-unit "B"), from shales to shales and resedimented carbonate detritus (Figs. 3, 6). This facies system persisted into Leonard time. This change in lithology in the Wolfcamp appears to have been coincident with the evolution from ramps to distally steepened ramps by sub-unit "B" time (Fig. 3). The middle early Wolfcamp was abruptly terminated by drowning of the outer part of the ramp beneath a thin section of dark shales. This facies passes westward to a subdued rimmed shelf (sub-unit "C") that has backstepped so that it directly overlies the shelf margin of sub-unit "A" (Figs. 3, 5). This backstepping is interpreted to have been caused by a sea-level highstand and/or renewed pulse of basin subsidence. Shelf-to-basin relief during the middle early Wolfcamp was similar to or slightly greater than that in the early early Wolfcamp, particularly during the deposition of sub-units "B" and "C".

Late early and middle Wolfcamp.—

An erosionally truncated, rimmed-shelf system may be present in the late early Wolfcamp, where the remnant of an eroded shelf margin reef passes westward into colluvial facies and eastward directly into shallow basinal shales (Fig. 3). Although shelf-to-basin relief at this time cannot be adequately determined because of erosion, it appears that it may have been slightly greater than that in the underlying Wolfcamp section. From lithofacies and thickness relations (Fig. 3), it appears that deposition of the late early Wolfcamp occurred during an initial period of sea-level highstand and/or renewed phase of basin subsidence. This was then followed by lowstand and subaerial erosion of the shelf and shelf margin reef.

Shelf and shallow basinal facies of the early middle Wolfcamp overlie colluvial deposits of the late early Wolfcamp along the outer part of the North Platform. These facies pass seaward into a prominent reef buildup in the lower beds of sub-unit "A," defining a distally steepened ramp profile (at the Texaco 1-Fitzgerald well; Fig. 3). The shelf margin reef is overlain by shallow basinal facies in the upper beds of sub-unit "A" (Fig. 3). Accordingly, the initial inundation of the platform appears to have resulted from either increased subsidence or a relative sea-level rise, which continued to the end of sub-unit "A" time. The overlying section (sub-unit "B") consists of cyclic, progradational reef and deeper shelf facies, although the reef facies had backstepped from that in sub-unit "A" (Figs. 3, 7). Estimated

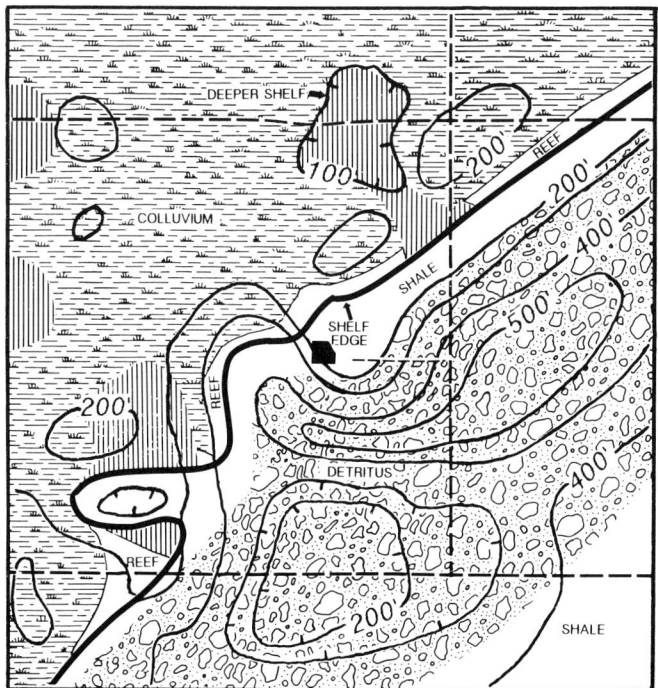

FIG. 7.—Isopach (CI = 30 m, 100 ft) of early middle Wolfcamp and generalized facies, top 60 m (200 ft) of section.

maximum depositional relief during the early middle Wolfcamp may have been on the order of 75–100 m (250–300 ft). The platform margin facies of both sub-units "A" and "B" pass westward into colluvial deposits (Figs. 3, 7). These relations suggest uniform rates of basin subsidence during sub-unit "B" time, the cyclicity seemingly related to short-term sea-level fluctuations of probable glacio-eustatic cause. During most of middle Wolfcamp time, much of the shelf province in the study area was subaerially exposed, with local, scattered areas being the sites of deposition of deeper shelf facies (Fig. 7). These depocenters likely reflect karst depressions on this and the underlying middle early and late early Wolfcamp surfaces. The prominent unconformity at the top of the early middle Wolfcamp is recognized throughout the northern Midland and Delaware Basins (i.e., Silver and Todd, 1969) as well as in several outcrop areas surrounding the Permian Basin (i.e., Ross, 1986).

Both the late early and early middle Wolfcamp slope and basin facies consist of a thick section of shales and carbonate detritus (Figs. 3, 7). The shales are distinct from those elsewhere in the Lower Permian section in the study area insofar as they are abundantly carbonaceous and light gray in color, rather than black and devoid of woody material. Furthermore, the associated carbonate detritus includes the coarsest megabreccias found in the entire Lower Permian section in the study area. Similar shale and carbonate facies can be traced as far east as Glasscock County along the Eastern Shelf (Mazzullo and others, 1986). These facies represent the detritus derived from erosion and retreat of a carbonate platform (i.e., Cook and others, 1983; Schlager and Camber, 1986) during the prolonged period of lowstands as postulated for late early to middle Wolfcamp time. In contrast to these megabreccias, the sections of relatively finer carbonate detritus (and associated black shales; Fig. 4) in other parts of the Lower Permian section in the northern Midland Basin are believed to have been deposited during highstand phases of shelf-marginal erosion, discussed later.

Late Wolfcamp.—

The early late and late late Wolfcamp in the study area represent another phase of widespread platform submergence interrupted by slight shoaling. The early late Wolfcamp consists, throughout the study area, of dark shales and argillaceous and locally detrital limestones (Fig. 3). Ensuing shallowing and the development of ramp and distally steepened ramp depositional systems characterize the late late Wolfcamp. At this time, a broad region of the platform was the site of deposition of deeper shelf facies surrounded by shallow-shelf carbonates (Figs. 3, 8). The deeper shelf depocenter that occupies the northeastern quarter of Yoakum County likely developed in response to localized increased rates of subsidence in this area. These platform facies pass eastward, into a thin section of dark shales and carbonate detritus (Fig. 4) deposited in relatively shallow (approximately 122 m, 400 ft) slope and basin settings (Figs. 3, 8). Deposition during the late Wolfcamp apparently was concurrent with low to moderate rates of basin subsidence, which maintained relative highstand. Throughout the Wolfcamp, major embayments adjoining areas of maximum carbonate accumulation along the shelf margin are present at nearly the same location in southeastern Yoakum County

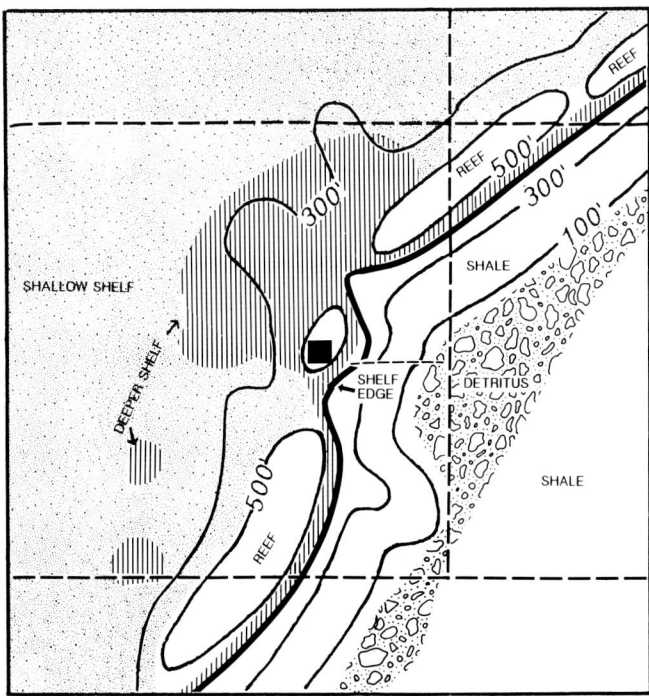

FIG. 8.—Cumulative isopach of early late and late late Wolfcamp (CI = 60 m, 200 ft), facies top 30 m (100 ft) of latter section; although shales and shaly limestones occur in the basin at the immediate top of the section, basinal detrital carbonates, which occur below that, are shown for reference.

(near the town of Plains). These embayments are coincident with areas of thick resedimented carbonate detritus in the basin (Figs. 6–8). Insofar as faults are not known to penetrate through the Wolfcamp section, these embayments probably were long-lived areas of slope weakness along which shelf margin reefs consistently failed, supplying detritus to the basin.

Discussion.—

The inferred history of deposition, subsidence and sedimentation rates, and relative sea-level fluctuations during Wolfcamp time are summarized later. Wolfcamp deposition occurred in a humid climate, in a regime of low to moderate rates of subsidence relative to that in the Early Leonard, although episodes of slightly higher as well as lower rates occurred periodically (Fig. 14). A first-order, bimodal pattern of relative sea-level highstands is readily recognized, with highstand peaks (punctuated by lowstands) in the early early to middle early and late Wolfcamp separated by a long period of general lowstand (Fig. 14). Sedimentation rates of platform and basin strata during this time include a maximum in the middle early Wolfcamp, with lower rates prior and subsequent to that time (Fig. 14). Maximum estimated sedimentation rates (0.092 m/ka, 300 ft/m.y.) are considerably less than those of modern carbonates, but are within the range of other ancient depositional systems (Wilson, 1975; Schlager, 1981).

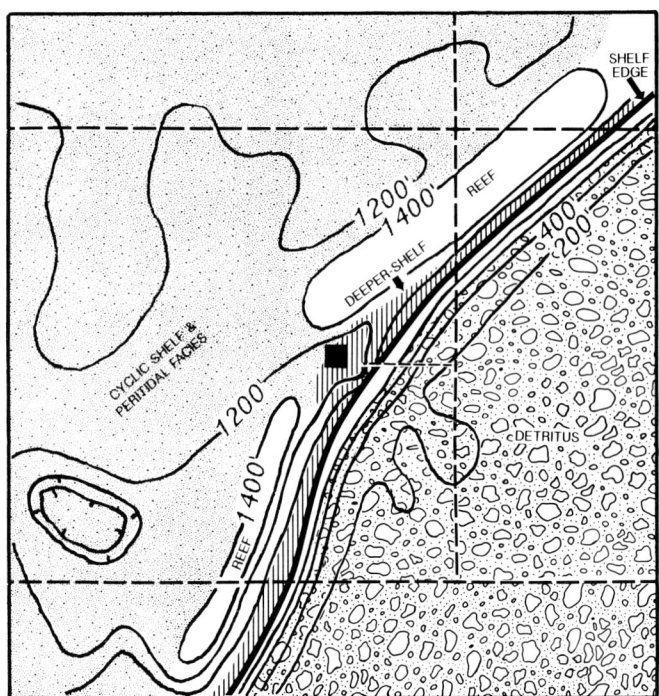

Fig. 9.—Isopach of Wichita (CI = 60 m, 200 ft) and facies in top 30 m (100 ft) of section.

Lower Leonard

The lower Leonard section on the North Platform is divided into three formations (Wichita, Lower Clear Fork, and Tubb; Fig. 2) and spanned a period of approximately 5.5 m.y. (Harland and others, 1982). Assuming equal durations, each of these units represent approximately 1.83 Ma duration. An abrupt change in depositional style, from ramp and distally steepened ramp systems in the Wolfcamp to rimmed shelves deposited in an arid climate, occurred in early Leonard time (Fig. 3). This change was concurrent with the compositional change to pervasively marine-cemented, shelf margin reef buildups (Fig. 4; Mazzullo, 1984). It is suggested that the change in depositional style to rimmed shelves was a consequence of two interrelated factors: (1) the occurrence of syndepositionally cemented reefs during this time, which resulted in the deposition of thick, wave-resistant buildups along the shelf margin; and (2) seismic stratigraphic and thickness relations (Figs. 3, 5, 9) suggest that the most rapid rate of basin subsidence in the Lower Permian was during early Leonard time. High shelf sedimentation rates coupled with rapid subsidence resulted in the rapid vertical accretion of shelf margin reefs.

Wichita and Lower Clear Fork.—

The shelf sections of these formations are each readily divided, on the basis of electric log correlations, into four megacycles (designated "A"–"D"; Fig. 3). Sample analyses indicate that each such megacycle is composed internally of innumerable shoaling-upward cycles, approximately 2.4–6.1 m (8–20 ft) thick, of shallow subtidal dolomite grading to peritidal and sabkha facies. The top units of each of the megacycles similarly are composed of such facies. These megacycles can be traced throughout the study area from electric log and sample correlations (Fig. 3). They pass into shelf margin reefs, which have stacked vertically and in slight offlapping geometry rather than in prominent offlapping fashion, but which are separated by deeper shelf facies (Fig. 3). Examination of cores of these reefs, as well as of contiguous shelf beds, indicates evidence of subaerial exposure by the occurrence of solution-collapse breccias, moldic and karstic vugular porosity and, locally, soilstone crusts (Fig. 10). Evidence of subaerial exposure is not found in rocks deposited farther basinward than uppermost slope beds. By these lines of evidence, it is suggested that the tops of each of the megacycles in the Wichita and Lower Clear Fork record sea-level lowstand phases concurrent with partial to complete shelf emergence. Some such breaks in sedimentation can also be identified seismically (Fig. 5A). This mosaic suggests that the deposition of the Wichita and Lower Clear Fork was punctuated by rapid sea-level fluctuations of relatively short duration. Such fluctuations are also indicated by the common occurrence of dual reef trends within the Wichita and Lower Clear Fork section, as readily mapped, for example, in Lower Clear Fork sub-unit "D" (Fig. 11). Although similar evidence of subaerial exposure locally is encountered in the component shoaling-upward cycles of the megacycles, they typically are not laterally continuous across the shelf (Fig. 12). By contrast then, these component cycles are believed to be mostly shoaling-upward deposits and therefore, their occurrence cannot be readily related to relative sea-level fluctuations, although cycles of lowstands and highstands almost certainly occurred during this time period.

The basal megacycle of the Wichita (subdivision "A")

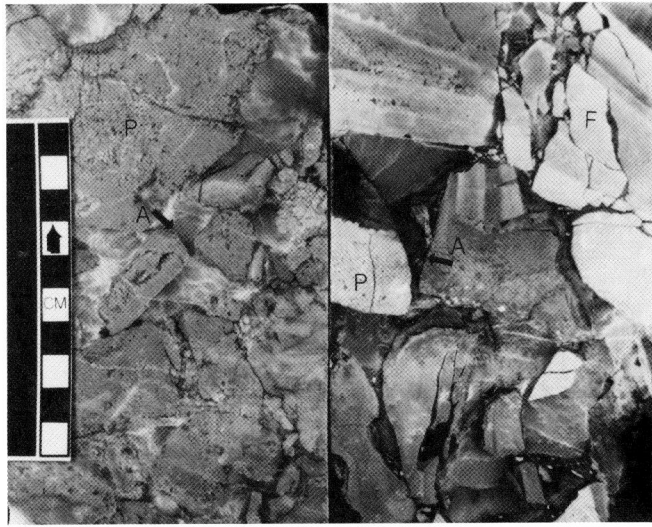

FIG. 10.—Core of dolomitized shelf margin reef facies, top of Wichita Formation, southern Hockley County, illustrating evidence of subaerial exposure. The rocks are solution-collapse breccias cemented by anhydrite ("A"). Note solution-enlarged biomoldic porosity ("P") and anhydrite-filled fractures ("F"). Scale in cm.

appears to be a distally steepened ramp that prograded over shallow slope and basin facies of the subjacent late Wolfcamp (Fig. 3). Subsequently, the remainder of the Wichita and Lower Clear Fork consists of stacked to slightly offlapping rimmed-shelf depositional systems (Fig. 3). In both units, shelf margin reefs pass landward to cyclic, shallow-shelf facies and seaward into a relatively narrow zone (average 1.0 km or 0.6 mi wide) or deeper shelf facies analogous to the upper forereef slope environments of many modern reef systems (Figs. 9, 11). The prominent shelf margin embayment observed throughout the Wolfcamp section in southeastern Yoakum County persisted into the upper Wichita, funneling detrital carbonates into the adjoining basin (Fig. 9). The slight landward shift of shelf margin reef facies in basal megacycle "A" of the Lower Clear Fork (Fig. 3) suggests a relative highstand or subsidence event at the Wichita-Lower Clear Fork boundary. By the close of Lower Clear Fork time, the platform margin had accreted at a location that is nearly directly superimposed on the Wichita shelf edge (Figs. 3, 9, 11). Locally, considerable bypassing of the Lower Clear Fork relative to the Wichita occurs in other parts of the North Platform (i.e., Hockley County; Mazzullo, 1982).

Basinal facies of the Wichita and Lower Clear Fork include a condensed section, relative to equivalent shelf deposits, of dark shales and resedimented carbonate detritus (Figs. 3, 4, 9, 11). This detritus can be traced for 16 to 24 km (10–15 mi) into the Terry County Trough, and pinches out prior to reaching the western margin of the buried Horseshoe Atoll (Mazzullo and Reid, 1987a). The intraclasts, skeletal and non-skeletal grains in the detritus in the lower Leonard section (as well as in the middle early and late Wolfcamp), are composed entirely of shelf margin lithologies (Fig. 4). The detritus is enclosed in a host of black shale that contrasts with the carbonaceous gray-shale matrix in the late early to early middle Wolfcamp lowstand detrital sections. Although parts of the lower Leonard shelf margin reef sections are known from core studies to have been subaerially exposed during periodic lowstands, there is no observable evidence in the many cores examined that would suggest that the intraclasts and grains within the detrital units were derived from shelf margin reefs during such lowstand events. Rather, it is believed that such carbonate detritus was resedimented during periods of relative sea-level highstand (Mazzullo and Reid, 1987a).

The thickest section of lower Leonard foreshelf carbonate detritus in the study area is restricted to the Wichita (Figs. 3, 9, 11). This corresponds to the development of thick shelf margin buildups and considerable (244–458 m, 800–1,500 ft; Mazzullo, 1982) shelf-to-basin relief during early Leonardian time (Fig. 3). Although Lower Clear Fork shelf-to-basin relief was similarly great, only a relatively thin section of basinal detritus is present in the contiguous basin (Fig. 3). We suggest that the highstand periods in Lower Clear Fork time were of shorter duration than those in the Wichita (Fig. 14). Accordingly, only a limited thickness of shelf margin reefs accumulated in the Lower Clear Fork (Fig. 3) and, hence, were able to shed only limited amounts of detritus into the basin. The abundance of detritus in lower Leonard basinal deposits relative to that in the middle early and late Wolfcamp (Fig. 3) is believed to be related to three factors: (1) the greater thickness of shelf margin reefs in the Wichita; (2) because of pervasive marine cementation, the reefs in the Wichita were able to build steep forereef slopes which, when they became unstable, shed considerable detritus into the basin; and (3) the accommodation potential of the basin was greater in the early Leonard because of rapid basin subsidence rates.

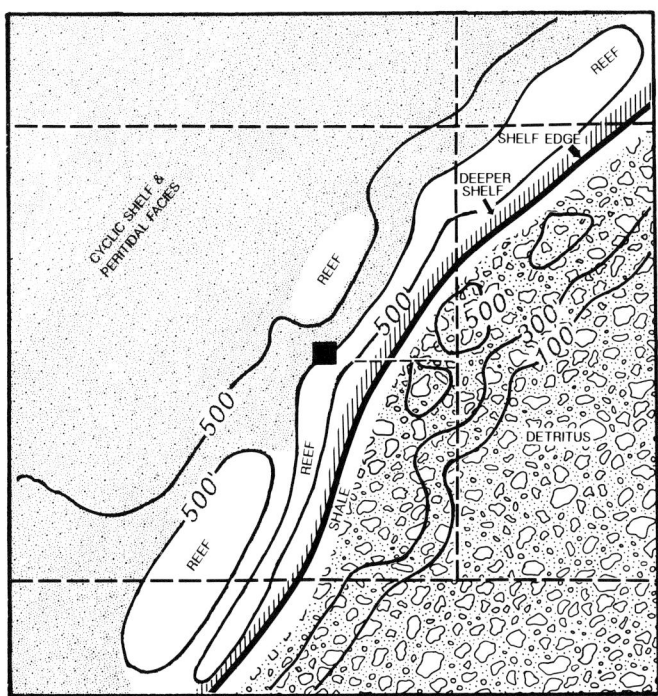

FIG. 11.—Isopach of Lower Clear Fork (CI = 60 m, 200 ft) and facies in top 30 m (100 ft) of section.

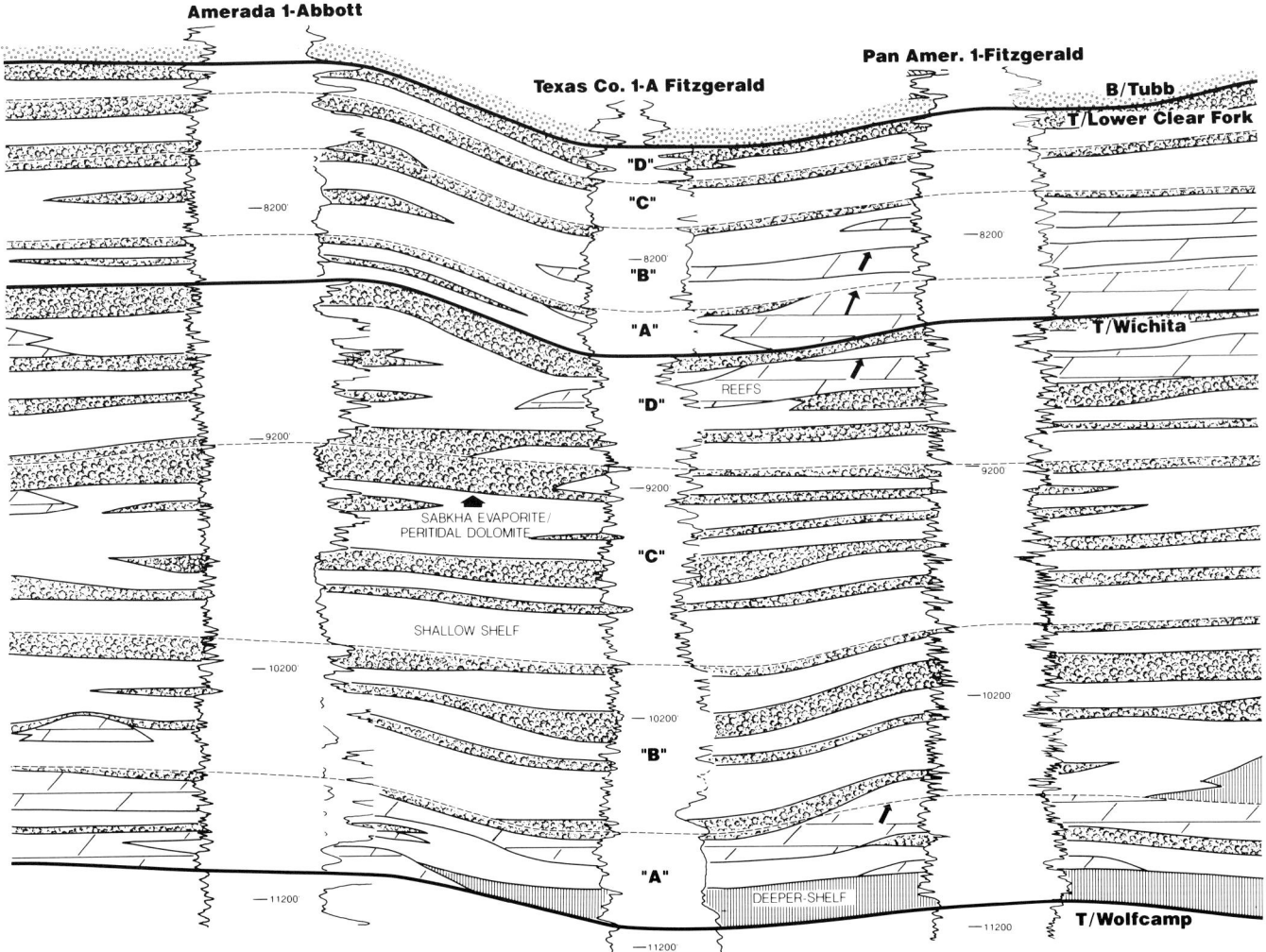

FIG. 12.—Enlargement of portion of Wichita-Lower Clear Fork section of Figure 3, showing prominent megacycles ("A"–"D") and component subcycles in shelf strata as identified by sample analyses. Note the lack of lateral continuity of most of the shoaling-upward subcycles. The tops of each of the megacycles are composed of sabkha evaporite and/or peritidal-dolomite facies. Evidence of subaerial exposure at tops of shelf patch reefs indicated by arrows.

Tubb-Dean.—

The shelf Tubb section includes sandstones, anhydritic (red and green) shales, bedded evaporites, and shallow-marine to loferitic dolomites. These facies are interpreted to have been deposited in a complex of eolian, desert fluvial, peritidal, and nearshore-marine environments in an arid climatic setting (Silver and Todd, 1969). Reef carbonates are interbedded in the Tubb section in shelf margin locations and resedimented detrital carbonates occur in the Dean (Figs. 3, 13). The contact between the Tubb and subjacent Lower Clear Fork and overlying Middle Clear Fork is sharp, as identified seismically as well as on logs and by sample analyses (Figs. 3, 5). Seismic reflectors in the basal beds of the Middle Clear Fork appear to onlap the Tubb and slope beds of the coeval Dean (Fig. 5C). These seismic and lithostratigraphic observations are consistent with the interpretation that the Tubb and Dean sections were deposited during a time of gradual relative sea-level fall, which was punctuated by cycles of relative highstand. This episode culminated at the end of Tubb time, in complete shelf emergence (also Silver and Todd, 1969). In this scenario, which is essentially similar to that during deposition of the Middle and Upper Clear Fork section (Fig. 5C), carbonates were deposited during highstand phases and shed detritus from platform margin areas into the basin; siliciclastics were deposited in the basin during intervening lowstands (Fig. 3).

Discussion.—

The deposition of the lower Leonard section occurred during a time of increased subsidence rates relative to those in the Wolfcamp (but with a decrease in subsidence rates during Tubb time), in a regime of frequent relative sea-level oscillations of apparent short duration and magnitude (Fig. 14). The maximum depositional topography in Early Permian time (maximum of 458 m, 1,500 ft) occurred during deposition of the lower Leonard section (Figs. 3, 14). Estimated average sedimentation rates of Wichita shelf strata (0.18 m/ka yrs, 600 ft/m.y.) are twice as great as those calculated for maximum sedimentation in the middle early

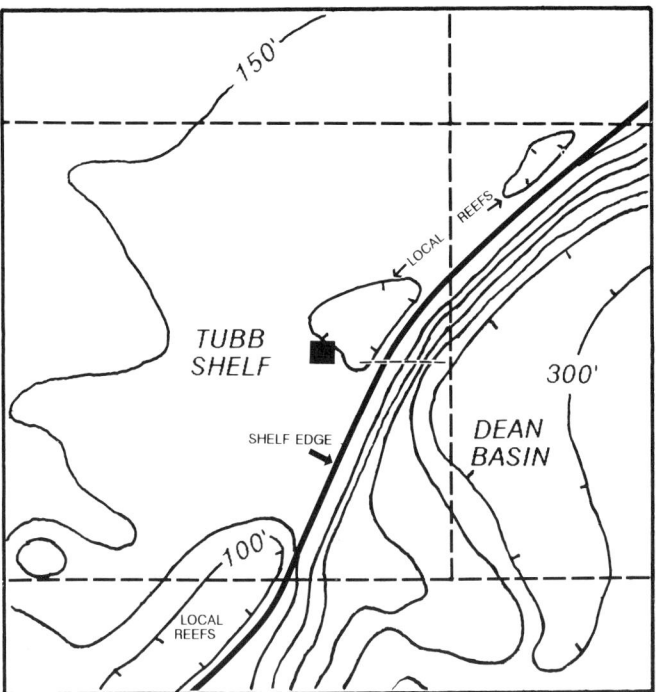

FIG. 13.—Isopach of Tubb-Dean section (CI = 15 m, 50 ft); facies predominantly sandstone in both formations, with scattered reefs along shelf margin where Tubb is thin.

Wolfcamp maximum (Fig. 14). Average sedimentation rates progressively decreased in Lower Clear Fork to Tubb-Dean time, from approximately 0.076 m/ka (250 ft/m.y.) to 0.04 m/ka (125 ft/m.y.), respectively (Fig. 14). These rates are comparable to those reported for other ancient depositional systems (Wilson, 1975; Schlager, 1981).

Upper Leonard

Following the Tubb lowstand event, the basal beds of the Middle Clear Fork transgressed the former shelf (Fig. 5C), and then a period of progradational infilling of the northern Midland Basin began. The complex lateral offlapping and downlapping character of the upper Leonard section (Fig. 5C) is believed to reflect platform and basin development in a regime of frequent sea-level fluctuations and moderate rates of basin subsidence relative to the lower Leonard sequence. The Middle and Upper Clear Fork section includes carbonates, evaporites, and sandstones on subdued rimmed shelves of moderate shelf-to-basin depositional relief, with carbonates and sandstones in the contiguous basin. The shelf sandstones are considered, by analogy to the Tubb and younger Permian units in the Permian Basin (i.e., Mazzullo and others, 1985), to be sea-level lowstand deposits.

DISCUSSION AND CONCLUSIONS

Maximum outbuilding of the North Platform into the Midland Basin occurred during Wolfcamp time when, by the middle early Wolfcamp, the shelf had prograded eastward a distance of approximately 22.5 km (14 mi) across the study area (Figs. 3, 6). This phase of platform accretion was facilitated, in large part, by the rapid deposition of a thick basinal shale foundation over which the distally steepened ramp system prograded (Fig. 3). Rates of platform carbonate sedimentation at this time were high and occurred during periods of relative sea-level highstand punctuated by only brief periods of lowstand (Fig. 14). Similar relations are known from the Eastern Shelf, where accretion of rimmed shelves and distally steepened ramps during this time locally advanced the platform basinward by as much as 32 to 48 km (20–30 mi; Mazzullo and others, 1986). Similarly, the next major episode of platform progradation in the study area occurred during Middle and Upper Clear Fork (late Leonard) time as sandstones rapidly infilled the northern Midland Basin. By the end of Upper Clear Fork time, the shelf prograded an additional 14 to 24 km (9–15 mi) into the basin. By contrast, the rimmed-shelf depositional systems of the Wichita and Lower Clear Fork, despite relatively high-sedimentation rates, generally accreted vertically along a shelf edge located along 3.2 km (2 mi) basinward of the middle early Wolfcamp edge (Figs. 5, 6, 9, 11). This limited amount of offlapping of the Wichita-Lower Clear Fork sections, in contrast to that in the Wolfcamp and Middle to Upper Clear Fork, is believed to be related to two factors: (1) drastically reduced rates of siliciclastic supply to the basin, resulting in limited development of accretionary siliciclastic wedges that served as foundations across which subsequent highstand carbonates could be deposited, and (2) high rates of basin subsidence. This second factor is discussed in more detail later.

The Wolfcamp and lower Leonard depositional systems evolved during a long-term cycle (approximate duration 22 m.y.), wherein maximum basin subsidence rates in the early Wolfcamp and late Wolfcamp to middle early Leonard were coincident with dominant relative highstands (Fig. 14). These are overprinted, however, by a second-order cyclicity of relative sea-level fluctuations of considerably shorter duration. Cycles of even shorter frequency, duration, and magnitude may, in fact, have occurred in the Wichita and Lower Clear Fork, as evidenced by the component cycles within the megacycles identified in these formations (Figs. 3, 12). The prominent accretionary ramp and distally steepened ramp systems of the early to middle early Wolfcamp were developed during a time of alternating highstands and lowstands concurrent with a major increase in basin subsidence rates (Fig. 14). The ramp systems of the middle Wolfcamp developed during short highstands punctuated by longer pereiods of widespread platform emergence and erosion (Figs. 3, 14). Those depositional systems in the late Wolfcamp were deposited in a regime of overall platform accretion followed by drowning. This drowning resulted from the coincidence of a long-term highstand and renewed basin subsidence. The above settings acted in concert with the paucity of wave-resistant reefs to preclude development of thick, laterally continuous shelf margin buildups and rimmed-shelf systems in Wolfcamp time. Although relatively steep rimmed shelves appear to have existed in the late early Wolfcamp (Fig. 3), they were mostly removed by pre-middle Wolfcamp erosion and, hence, an understanding of the conditions under which they may have formed remains uncertain.

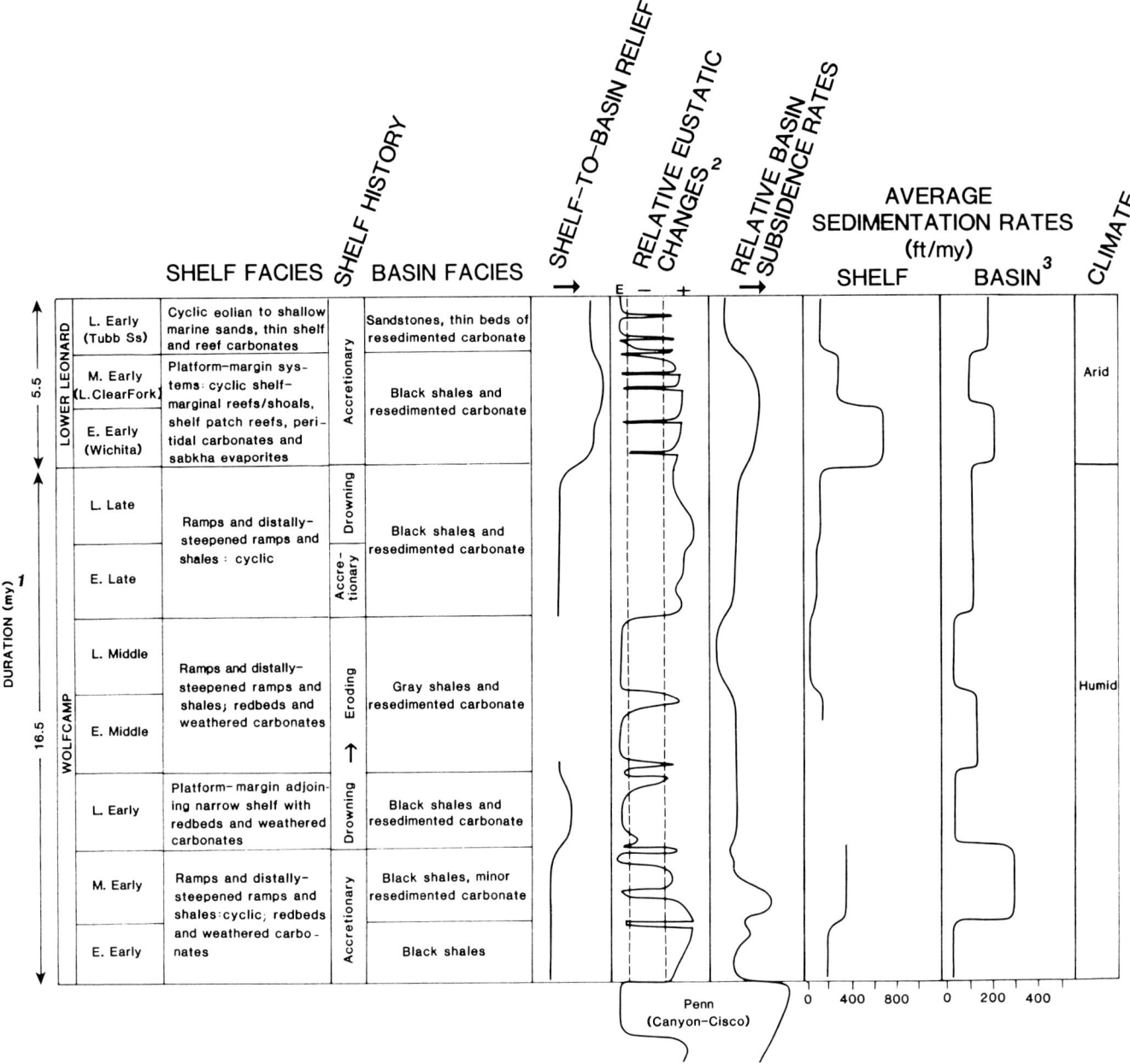

Fig. 14.—Inferred parameters of platform evolution in Wolfcamp to lower Leonard section in study area: 1—Absolute durations from Harland and others (1982); 2—Relative (eustatic) sea-level changes inferred from shelf strata; 3—Average basin sedimentation rates calculated for strata from 0–4.8 km (0–3 mi) seaward of platform margins. The relative sea-level changes in the Wichita-to-Tubb section are somewhat simplied because of space constraints: at least four lowstand cycles occur in each of the Wichita and Lower Clear Fork sections (refer to text for discussions).

Wichita and Lower Clear Fork depositional systems are similarly complex. In contrast to the Wolfcamp, a thick, rapidly deposited sequence of accretionary platform margin buildups developed during Wichita time in response to increased rates of basin subsidence and the development of wave-resistant, marine-cemented reefs. The lack of substantial basinward progradation of the platform during this time, however, appears to have resulted from its coincidence with a period of increased subsidence and overall relative highstand. This same reasoning is also inferred for the overlying Lower Clear Fork section. Such conditions also resulted in the stacking of rimmed-shelf sections along a more or less stationary shelf edge during this time (Figs. 3, 9, 11).

The limited thickness of shelf and shelf margin deposits in the Lower Clear Fork and Tubb, relative to the Wichita (Figs. 3, 9, 11, 13), may be a consequence of the deposition of the former during a period of decreased rates of subsidence and only short-duration highstands (Fig. 14), both of which combined to limit the accommodation potential of sediments on the shelf. Rates of shelf margin carbonate sedimentation may also have been substantially retarded during the initial stages of sea-level falls and subsequent rises during this time, by poisoning of shelf environments due to climate-induced hypersalinity. As a consequence of the contrasting thicknesses of shelf margin buildups in the study area, platform-derived carbonate detrital wedges here are typically thicker in the Wichita than in the Lower Clear Fork (Fig. 3).

A major sedimentologic change to alternating periods of carbonate-sandstone deposition beginning in Tubb-Dean time and heralded the rapid infilling of the Midland Basin by offlapping subdued rimmed shelves of the Middle and Upper Clear Fork sequences (Fig. 5C).

In summary, the Lower Permian (Wolfcamp to Leonard) platform-to-basin section on the North Platform is a complex wedge of carbonate, evaporite, and siliciclastic rocks representing a variety of depositional systems. A progressive evolution in depositional style is recognized in the transition from ramps and distally steepened ramps in the Wolfcamp, to steep and then subdued rimmed-shelf systems in the Wichita-Lower Clear Fork-Tubb and Middle to Upper Clear Fork, respectively. This evolution is related to changes through time, in contrasting rates of sedimentation and subsidence, relative sea-level fluctuations, the shift from dominantly carbonate to mixed carbonate-siliciclastic sedimentation, and changes in reef-building, organo-sedimentary facies. With minor variations due to local differences in subsidence and sedimentation rates, this same stratigraphy and evolution of depositional systems can be recognized over most of the Midland Basin.

ACKNOWLEDGMENTS

The concepts presented in this paper benefited from discussions with P. M. Harris, C. A. Ross, S. T. Reid, S. C. Robbins, and J. Manatt. We thank the following persons for their contributions to this study: S. C. Robbins, Paleostrat Service, Midland, Texas, for devoting long hours to the arduous task of preparing oriented fusulinid thin sections; H. E. Collins, Shell Warehouse, Midland, Texas, for the loan of many of the subsurface samples examined as part of this study; J. Manatt and C. McMillan, Permian Exploration Corporation, Roswell, New Mexico, for access to several key seismic lines across the North Platform and permission to reproduce portions of the seismic line in Figure 5; and R. Vest, Midland, Texas, for providing details of the seismic stratigraphy of the northern Midland Basin. Finally, the authors are grateful to D. K. Beach, J. F. Sarg, and A. H. Saller for their incisive reviews of the earlier version of this paper.

REFERENCES

ADAMS, J. E., FRENZEL, H. N., RHODES, M. L., AND JOHNSON, D. P., 1951, Starved Pennsylvanian Midland Basin: American Association of Petroleum Geologists Bulletin, v. 35, p. 2600–2607.

BURTON, R., KENDALL, C. G. ST. C., AND LERCHE, I., 1987, Out of our depth: On the impossibility of fathoming eustasy from the stratigraphic record: Earth-Science Reviews, v. 24, p. 237–277.

COOK, H. E., 1983, Sedimentology of some allochthonous deep-water carbonate reservoirs, Lower Permian, West Texas: Carbonate debris sheets, aprons, or submarine fans? (abs.): American Association of Petroleum Geologists Bulletin, v. 63, p. 442.

———, HINE, A. C., AND MULLINS, H. T., 1983, Platform Margin and Deep Water Carbonates: Society of Economic Paleontologists and Mineralogists, Lecture Notes, Short Course No. 12, 573 p.

CROWELL, J. C., 1978, Gondwanan glaciation, cyclothems, continental positioning, and climatic change: American Journal of Science, v. 278, p. 1345–1372.

GAWLOSKI, T. F., 1987, Nature, distribution, and petroleum potential of Bone Spring detrital sediments along the Northwest Shelf of the Delaware Basin, in Cromwell, D., and Mazzullo, L., eds., The Leonardian Facies in W. Texas and S.E. New Mexico and Guidebook to the Glass Mountains, West Texas: Society of Economic Paleontologists and Mineralogists, Permian Basin Section, Publication No. 87-27, p. 85–105.

HARLAND, W. B., COX, A. V., LLEWELLYN, P. G., PICKTON, C. A. G., SMITH, A. G., AND WALTERS, R., 1982, A Geologic Time Scale: Cambridge University Press, Cambridge, England, 131 p.

HILLS, J. M., 1972, Late Paleozoic sedimentation in West Texas Permian basin: American Association of Petroleum Geologists Bulletin, v. 56, p. 2303–2322.

———, 1984, Sedimentation, tectonism, and hydrocarbon generation in Delaware basin, West Texas and southeastern New Mexico: American Association of Petroleum Geologists Bulletin, v. 68, p. 250–267.

HOBSON, J. P., CALDWELL, C. D., AND TOOMEY, D. F., 1985, Early Permian deep-water allochthonous limestone facies and reservoir, West Texas: American Association of Petroleum Geologists Bulletin, v. 69, p. 2130–2147.

JEARY, G. L., 1978, Leonardian strata in the northern Midland basin of west Texas, in Energy Quest for the Southwest: West Texas Geological Society Publication 78-69SWS, p. 30–47.

JONES, T. S., 1953, Stratigraphy of the Permian Basin of West Texas: West Texas Geological Society, 63 p.

LOUCKS, R. G., BROWN, A. A., ACHAUER, C. W., AND BUDD, D. A., 1985, Carbonate gravity-flow sedimentation on low-angle slopes off the Wolfcampian Northwest Shelf of the Delaware Basin, in Crevello, P. D., and Harris, P. M., eds., Deep-Water Carbonates: Buildups, Turbidites, Debris Flows and Chalks: Society of Economic Paleontologists and Mineralogists Core Workshop No. 6, p. 56–92.

MAZZULLO, L. J., AND REID, A. M., 1987, Stratigraphy of the Bone Spring Formation (Leonardian) and depositional setting in the Scharb field, Lea County, New Mexico, in Cromwell, D., and Mazzullo, L., eds., The Leonardian Facies in W. Texas and S. E. New Mexico and Guidebook to the Glass Mountains, West Texas: Societey of Economic Paleontologists and Mineralogists, Permian Basin Section, Publication No. 87-27, p. 107–111.

MAZZULLO, S. J., 1982, Stratigraphy and depositional mosaics of Lower Clear Fork and Wichita Groups (Permian), northern Midland Basin, Texas: American Association of Petroleum Geologists Bulletin, v. 66, p. 210–227.

———, 1984, Foreshelf carbonate facies mosaics, Permian, Midland Basin, Texas: Society of Economic Paleontologists and Mineralogists, First Annual Midyear Meeting, San Jose, Abstracts with Programs, p. 52.

———, MAZZULLO, J., AND HARRIS, P. M., 1985, Significance of eolian quartzose sheet sands on emergent carbonate shelves: Permian of West Texas-New Mexico (abs.): American Association of Petroleum Geologists Bulletin, v. 69, p. 284.

———, AND REID, A. M., 1987a, Basinal Lower Permian facies, Permian Basin: Part II. Depositional setting and reservoir facies of Wolfcampian-Lower Leonardian basinal carbonates: West Texas Geological Society Bulletin, v. 26, p. 5–10.

———, AND ———, 1987b, Contrasting evolutionary patterns of Lower Permian shelf and basinal facies, Midland Basin, Texas (abs.): American Association of Petroleum Geologists Bulletin, v. 71, p. 590–591.

———, ———, AND MAZZULLO, L. J., 1987, Basinal Lower Permian facies, Permian Basin: Part I. Stratigraphy of the Wolfcampian-Leonardian boundary: West Texas Geological Society Bulletin, v. 26, p. 5–9.

———, ———, ———, AND REID, S. T., 1986, Evolution of Permian carbonate shelf and foreshelf detrital systems, Midland Basin, Texas (abs.): American Association of Petroleum Geologists Bulletin, v. 70, p. 347.

MITCHUM, R. M., VAIL, P. R., AND THOMPSON, S., 1977, Seismic stratigraphy and global changes of sea level, Part 2. The depositional sequence as a basic unit for stratigraphic analysis, *in* Payton, C. E., ed., Seismic Stratigraphy–Applications to Hydrocarbon Exploration: American Association of Petroleum Geologists Memoir 26, p. 53–62.

MOSLEY, B. R., 1988, Deposition and diagenesis of an Upper Clear Fork reef trend, Palm Sunday field, Hockley County, Texas (abs.): American Association of Petroleum Geologists Bulletin, v. 72, p. 102.

READ, J. F., 1982, Carbonate platforms of passive (extensional) continental margins: Types, characteristics and evolution: Tectonophysics, v. 81, p. 195–212.

REID, A. M., REID, S. T., MAZZULLO, S. J., ROBBINS, S. T., 1988, Revised fusulinid biostratigraphic zonation and depositional sequence correlation, subsurface Permian basin (abs.): American Association of Petroleum Geologists Bulletin, v. 72, p. 102.

ROSS, C. A., 1986, Paleozoic evolution of southern margin of Permian Basin: Geological Society of America Bulletin, v. 97, p. 536–554.

SCHLAGER, W., 1981, The paradox of drowned reefs and carbonate platforms: Geological Society of America Bulletin, v. 92, p. 197–211.

———, AND CAMBER, O., 1986, Submarine slope angles, drowning unconformities, and self-erosion of limestone escarpments: Geology, v. 14, p. 762–765.

SILVER, B. A., AND TODD, R. G., 1969, Permian cyclic strata, northern Midland and Delaware Basins, West Texas and southeastern New Mexico: American Association of Petroleum Geologists Bulletin, v. 53, p. 2223–2251.

VAIL, P. R., AND MITCHUM, R. M., 1977, Seismic stratigraphy and global changes of sea level, Part 1. Overview, *in* Payton, C. E., ed., Seismic Stratigraphy–Applications to Hydrocarbon Exploration: American Association of Petroleum Geologists Memoir 26, p. 51–52.

VEEVERS, J. J., AND POWELL, C. MCA., 1987, Late Paleozoic glacial episodes in Gondwanaland reflected in transgressive-regressive depositional sequences in Euramerica: Geological Society of America Bulletin, v. 98, p. 475–487.

WIGGINS, W. D., AND HARRIS, P. M., 1985, Burial diagenetic sequence in deep-water allochthonous dolomites, Permian Bone Spring Formation, southeast New Mexico, *in* Crevello, P. D., and Harris, P. M., eds., Deep-Water Carbonates: Buildups, Turbidites, Debris-Flows and Chalks: Society of Economic Paleontologists and Mineralogists Core Workshop No. 6, p. 140–173.

WILSON, J. L., 1975, Carbonate Facies in Geologic History: Springer-Verlag, New York, 470 p.

PART IV
PINNACLE REEFS—ISOLATED OFFSHORE BANKS ON PASSIVE MARGINS

EUSTATIC CONTROLS ON THE STRATIGRAPHY AND GEOMETRY OF THE LATEMAR BUILDUP (MIDDLE TRIASSIC), THE DOLOMITES OF NORTHERN ITALY

ROBERT K. GOLDHAMMER[1]
AND
MARK T. HARRIS[1]
Department of Earth and Planetary Sciences, The Johns Hopkins University, Baltimore, Maryland 21218

ABSTRACT: Superimposed short-term and long-term eustatic sea-level fluctuations directly controlled Latemar platform stratigraphy and influenced the deeper water facies and overall buildup geometry. Deposition of deeper water sediments (foreslope and toe of slope) was linked to alternating submergence (highstand shedding) and subaerial exposure (lowstand lithification) of the platform top and thus recorded a periplatform signal of the eustatic fluctuations.

The Latemar consists of a platform core (3-4 km wide, 700 m thick) with a narrow margin (tens of meters wide), flanked by foreslope (30°–35° dips), toe of slope, and basinal deposits. The platform sequence is interpreted to record a long-term (about 10 Ma) third-order eustatic sea-level oscillation with an amplitude of about 60 m. The lower 250 m of the platform section marks the initial third-order rise (subtidal carbonates), and the upper 450 m marks the subsequent highstand and is characterized by meter-scale cyclic carbonates. These cycles record platform submergence and exposure interpreted to be caused by short-term (10^4–10^5 yr) Milankovitch eustatic oscillations superimposed on the long-term trend.

The short-term platform submergence and exposure conditions result in alternating styles of foreslope deposition. During highstands, platform-derived sands bypass the foreslope, accumulating as toe-of-slope graded beds and basinal turbidites. During lowstands, sand supply ceases, producing basin hardgrounds. Marginal boundstones supplied clasts to the foreslope breccias during both highstands and lowstands, with only minor amounts of platform-derived sands (highstands) and lithified clasts (lowstands).

Since the platform margin/foreslope contact is nearly vertical, a progressively increasing volume of foreslope breccia was needed to maintain the depositional geometry. This coincided with increasing amounts of exposure in the platform section related to the long-term sea-level change, suggesting that increasing clast production (at sea-level highstand and subsequent fall) was the basic control on the depositional geometry.

INTRODUCTION

The controls on platform stratigraphy and buildup geometry are difficult to quantify, because facies patterns are linked to complex combinations of eustasy and tectonics (Kendall and Schlager, 1981; Schlager, 1981). The Middle Triassic Latemar buildup is an unusual example in which the effects of both short-term and long-term sea-level fluctuations are recognizable and unambiguous. We believe that eustasy is the dominant influence on both platform and deeper water facies and buildup geometry. The recognition of short-term cyclic events in the platform establishes a time scale, which allows direct calculation of the duration and rates of sea-level change.

Throughout this paper, we will refer to stratigraphic cycles of different orders of temporal magnitude. Extending the terminology of Vail and others (1977) and Miall (1984), we recognize and will address three different orders of eustatic sea-level fluctuations. These are: third-order cycles (1–10 m.y. duration), fourth-order cycles (100 ka duration), and fifth-order cycles (20 ka duration).

GEOLOGIC SETTING

The Dolomites are located in northern Italy on the southern margin of the Alpine chain within the major autochthonous structural element of the little-deformed Southern Alps. During the Permo-Triassic, the Southern Alps region was part of a coherent continental block (consisting of Eurasia, Africa, and the Americas; Bernoulli and Lemoine, 1980), which underwent rifting (Bechstadt and others, 1978;

---

[1]Present address: Exxon Production Research Co., P.O. Box 2189, Houston, Texas 77252-2189

Winterer and Bosellini, 1985). In the Dolomites region, Middle Triassic (Late Anisian/Ladinian) rifting was marked by sinistral, strike-slip tectonics (Doglioni 1984a, b), with both extensional and compressional structures, followed by volcanism and magmatic intrusions (Fig. 1). Thick, upper Anisian-Ladinian carbonate buildups (700–1,300 m thick) grew from structurally induced topographic highs, as vertical aggradation matched the combined effects of regional subsidence and eustatic sea-level changes (Bosellini and Rossi, 1974; Bosellini, 1984). Deep-water starved basins developed adjacent to the buildups, which prograded and partly infilled basinal areas (Fig. 1). The differential topography was short-lived, because volcanics and volcanogenic clastics partially filled the depositional basins, onlapping and in some cases burying pre-existing carbonate buildups (Bosellini and Rossi, 1974). In Early to Middle Carnian time (late Middle Triassic), shallow-marine carbonate platforms (as thick as 1 km) nucleated on submerged topographic highs of the Ladinian buildups and prograded out into remaining basinal areas (Bosellini and others, 1977; Bosellini, 1984).

Upper Anisian-Ladinian paleogeography of the Dolomites has been reviewed by numerous authors (Bosellini and Rossi, 1974; Gaetani and others, 1981; Bosellini, 1984). The Latemar buildup was an isolated, atoll-like entity based on the occurrence of basinal strata (Livinallongo Formation) and foreslope deposits (Marmolada Limestone), which essentially surrounded the platform (Fig. 2). Another isolated coeval buildup occurs immediately southwest of the Latemar (Fig. 1), which we informally refer to as the Southern Latemar buildup. It is positioned structurally on the southern side of the main syndepositional strike-slip fault, which crosscuts the region (Doglioni, 1984a, b), and ex-

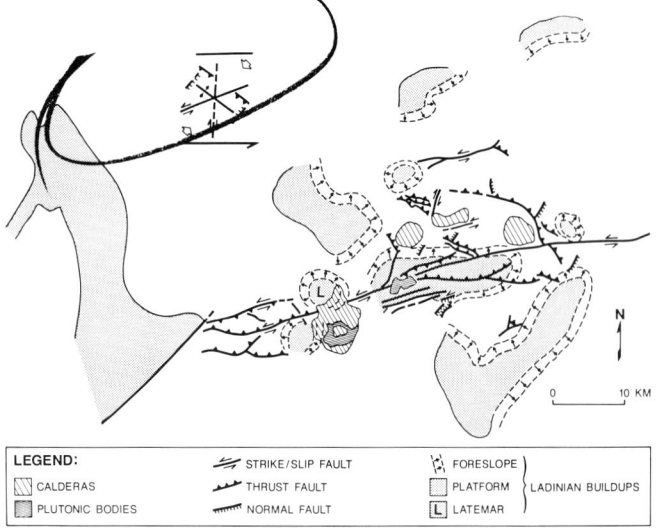

FIG. 1.—Middle Triassic paleogeography of the central Dolomites superimposed on the paleotectonic map of Doglioni (1984a). Southern Latemar buildup is the platform immediately to the south-southwest of the Latemar.

perienced a different rate of tectonic subsidence compared to the Latemar (Goldhammer, 1987).

LATEMAR BUILDUP

The stratigraphy and sedimentology of the Upper Anisian-Ladinian buildups of the Dolomites have received attention in recent years by various authors (Cros and Lagny, 1969; Bosellini and Rossi, 1974; Cros, 1974; Gaetani and others, 1981; Fois and Gaetani, 1981; Fois, 1982; Blendinger, 1986). The Latemar buildup is an ideal case study because of its preservation as limestone, relatively small size, the lack of structural complexity, and spectacular exposure.

The Latemar buildup (Figs. 2, 3) consists of a central platform core (4 km wide) rimmed by a narrow (tens of meters wide) margin rim that passes abruptly into steeply dipping (30°–35°) foreslope clinoforms. At the toe of slope, the clinoforms flatten abruptly and merge into nearly flat-lying basinal strata. This profile was maintained throughout the growth of the buildup, even as the relief increased from perhaps a few tens of meters to about 500 m. At the end of carbonate deposition, the foreslope was a kilometer or more in width. The platform margin is nearly vertical throughout most of the Ladinian buildup (Fig. 4). The extent of the foreslope indicates, however, that the uppermost Ladinian margin prograded rapidly, similar to other Ladinian margins (Catinaccio, Southern Latemar).

Platform Facies

The horizontally bedded platform interior of the Latemar was divided into two stratigraphic intervals by Gaetani and others (1981): (1) the Upper Anisian "Lower Edifice" (250–300 m thick), and (2) the Ladinian Latemar Limestone (450 m thick) with a thick tepee belt in the middle part. Our work indicates that this stratigraphy can be extended in a layer-cake arrangement across the entire platform (Fig. 5). From bottom to top, four vertically stacked facies are recognized: (1) a lower platform facies (250–300 m thick); (2) a lower cyclic facies (90 m thick); (3) a tepee facies (120 m thick); and (4) an upper cyclic facies (210+ m thick). The lower platform facies is equivalent to the Upper Anisian "Lower Edifice" of Gaetani and others (1981), whereas

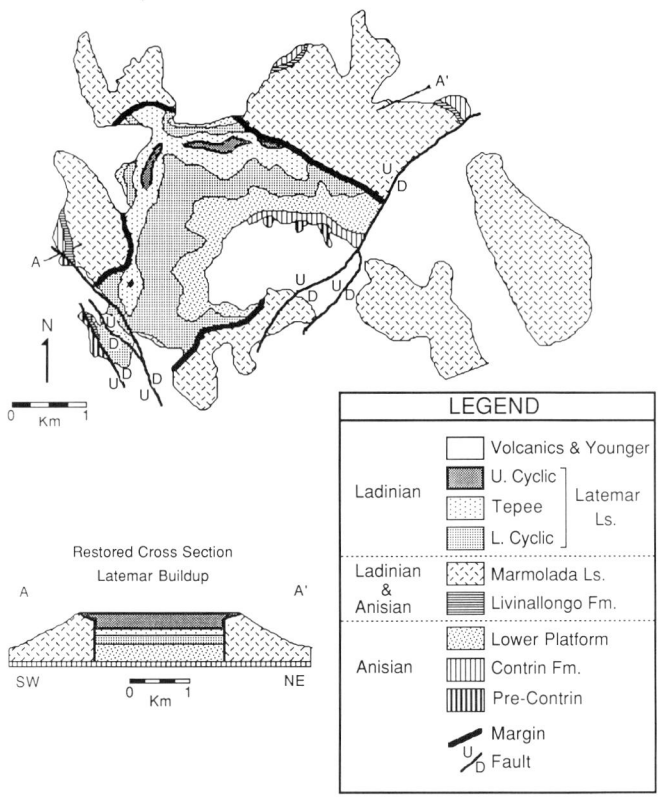

FIG. 2.—Generalized geologic map of the Latemar massif and restored cross section. Note the narrow margin zone that separates the platform interior from the surrounding foreslopes.

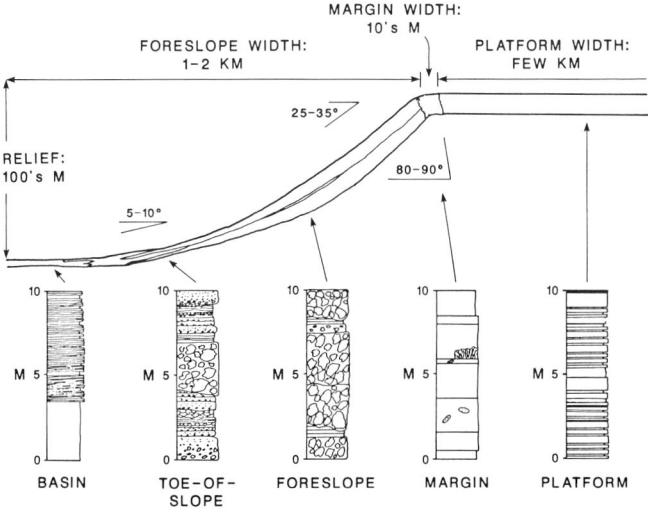

FIG. 3.—Depositional facies and geometry of the Latemar buildup.

FIG. 4.—Margin geometry at the northeast edge of the platform, viewed along the margin strike. Flat-lying platform facies (to the left) pass through a vertical margin zone into steeply dipping foreslope clinoforms on the right. The projected progradational geometry at the end of buildup growth is based upon regional thickness considerations and the occurrence of uppermost foreslope facies along the top of the preserved foreslopes.

the remaining three overlying facies compose the Ladinian Latemar Limestone.

Lower platform facies.—

The lower platform facies (Fig. 6) is characterized by flat-lying, very thin- to thick-bedded, dark grey limestones and lighter dolostones composed essentially of skeletal lithoclastic packstone-grainstone intercalated with submarine-cemented sheet cracks (1–50 cm thick). Subaerial exposure surfaces are uncommon in the lower half of the lower platform facies, becoming more abundant in the upper half with one present every 5 to 15 m. Lower platform facies coarse-grained skeletal lithoclastic sands contain a restricted fauna, dominated by dasycladacean (*Diplopora*) and blue-green algae, with lesser foraminifera and ostracod debris. Non-skeletal components are primarily reworked, early cemented lithoclasts and peloids. Burrowed and bioturbated textures are common, and depositional lithologies lack primary, mechanically generated sedimentary structures.

Syndepositional early diagenetic structures and associated cements are the most distinctive attribute of the lower platform facies and include: hardgrounds and associated macroborings, microborings and encrusting organisms; cement-filled sheet cracks; submarine antiform structures (tepees); neptunian fractures; and a variety of early marine-cement types (Goldhammer, 1987). Hardground surfaces range from single surfaces to stacked (10–20 in 1 m) or superimposed surfaces. Hardgrounds may be overlain by grainstone, but typically are overlain by laterally continuous, parallel bedding, sheet cracks filled with marine cements. Where sheet crack density is very high (as much as one every 10 cm) sheet cracks can be traced laterally into antiformal submarine tepee structures, which consist of uparched, disrupted submarine-cemented grainstone layers. Lower platform facies tepees lack any evidence of having formed in association with subaerial exposure. Early marine cements compose 30–40 percent of the volume of lower platform facies lithologies and include: isopachous fibrous calcite lining pores, spherulitic-botryoidal calcite occluding interparticle pores and filling sheet cracks, and radiaxial fi-

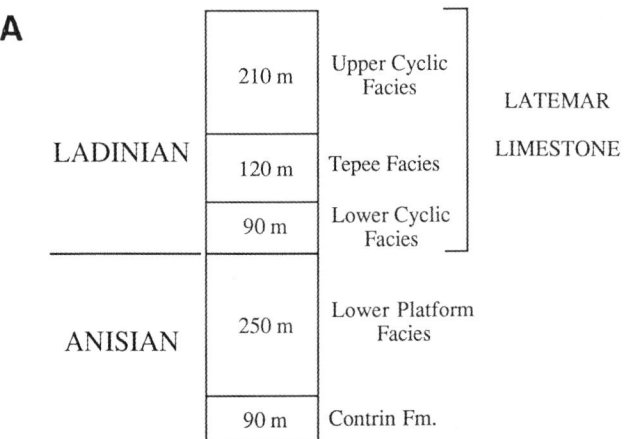

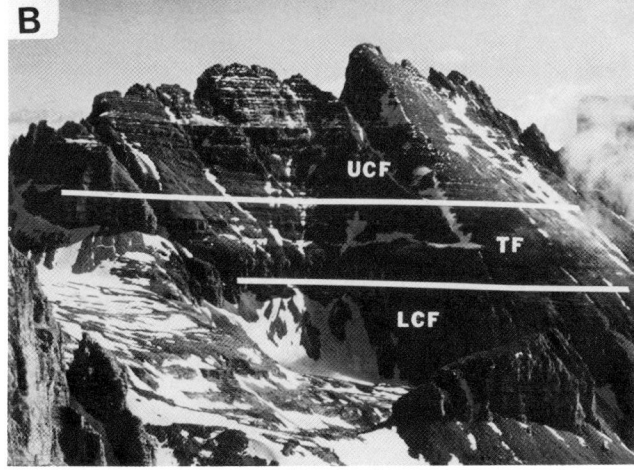

FIG. 5.—(A) Platform stratigraphy of the Latemar buildup. (B) South face of the Latemar Towers (platform interior) with the vertically stacked facies of the Latemar Limestone: LCF = lower cyclic facies; TF = tepee facies; UCF = upper cyclic facies; lower platform facies not shown.

Fig. 6.—Lower platform facies. (A) Stratigraphic contact (arrow) between the lower platform facies and the overlying lower cyclic facies. Scale bar equals 5 m. (B) Thin-bedded, layered aspect of lower platform facies resulting from intercalated grainstones-packstones (A) and cement-filled sheet cracks (B). Scale is 6 cm across.

brous calcite. These cement types incorporate and are interlayered with marine fossiliferous sediment and commonly are bored and organically encrusted, testifying to their syndepositional, early marine origin.

The lower platform facies is interpreted as primarily subtidal in origin, representing a shallow-bank lagoon. Bedding in the lower platform facies stems mainly from the intercalation of grainstone-packstone layers, hardground surfaces, and cemented sheet cracks. The depositional conditions of the lower platform facies consisted of monotonous subtidal sedimentation and submarine diagenesis. Submarine sedimentary hiatuses and syndepositional marine diagenesis are very common, whereas subaerial depositional breaks are not, pointing at generally submergent conditions and a dominantly subtidal origin.

Lower and upper cyclic facies.—

Both the lower cyclic facies and upper cyclic facies consist of flat-lying carbonate cycles (Fig. 8A) composed of limestone-dolomite couplets ranging in thickness from 0.1 to 5 m and averaging just less than 1 m thick (Fig. 7; Goldhammer and others, 1987). Cycles extend laterally across the platform and consist of two parts, a subtidal limestone composed of bioturbated skeletal peloidal packstone-grainstone (Fig. 8B), and a thin (5–15 cm), dolomite-rich, subaerial-exposure cap (Fig. 8C). Internally, the subtidal limestone grades upward from a burrowed, fine-grained, restricted skeletal peloidal wackestone at the base to coarse-grained, skeletal lithoclastic grainstones that are typically capped by oncolitic gravels. In contrast to the subtidal deposits of the underlying lower platform facies, the subtidal portion of the cycles lack evidence for significant syndepositional early marine diagenesis. The thin cycle caps contain overwhelming evidence for subaerial exposure, such as meniscus and pendant cements, solution vugs, and caliche fabrics (Goldhammer and others, 1987). The dolomitic caps are simply the underlying, pre-existing subtidal-limestone lithology modified to varying degrees by subaerial vadose diagenesis. The thickness, mineralogy, and petrographic features of the cycle caps are similar to the thin supratidal dolomitic crusts of modern carbonate tidal flats (for example, Hardie, 1977).

Each limestone-dolomite couplet represents subtidal-carbonate sediment that was abruptly subaerially exposed. Eustatic sea-level oscillations are the likely mechanism for forming these cycles, because progradation would produce cycle caps with both intertidal and supratidal deposits.

Tepee facies.—

The tepee facies separates the evenly bedded sequences of the lower cyclic facies from the upper cyclic facies (Fig. 7) and is characterized by 35 discrete tepee zones (Fig. 9), which add a different component to the normal cyclic pattern. The tepee zones are separated by cycles that are analogous to those that characterize both the lower and upper cyclic facies (Fig. 7).

The tepees of the tepee facies are non-tectonic antiform structures displaying symmetrical to asymmetrical chevron-like cross sections that have flat bases and irregular tops. Tepee horizons range in thickness from less than 1 m to about 15 m and consist of buckled, pre-existing limestone-dolomite cycles (identical to those occurring within both cyclic facies; Hardie and others, 1986; Goldhammer, 1987) separated by centimeter-scale sheet cracks, fractures, and shelter pores filled with a variety of exotic calcite cements, laminated internal sediments, and pisoids. Tepee crests have a relief of as much as 3 m and exhibit truncated tops. Overlying beds thin laterally as they onlap tepee flanks, indicating a syndepositional origin for these tepees (Assereto and Kendall, 1977). Typically, the number of cycles incorporated within the tepee zones varies from one to four couplets, each cycle being on the order of 0.10–0.40 m thick (Fig. 7).

Tepee zones are interpreted as the diagenetic product of massive expansive cementation associated with prolonged subaerial exposure under semiarid to arid conditions. Tepees of the tepee facies are associated with sediments that were subaerially exposed (i.e., vadose caps of cycles), contain fitted pisoids and vadose cements (Assereto and Kendall, 1977), and may be capped by thin, red (iron oxide-stained) soil zones. All argue for an origin primarily related to subaerial exposure, in contrast to submarine tepees of the underlying lower platform facies. Details of tepee genesis and rates of tepee formation can be found in Goldhammer (1987).

Interpretation of Short-Term Eustasy

The lower and upper cyclic facies.—

With the exception of the lower platform facies, the interior platform record of the Latemar buildup is characterized by a continuous succession of meter-scale limestone-dolomite cycles (600 cycles in 420 m within the lower cyclic facies, tepee facies, and upper cyclic facies; average 0.70 m/cycle). Despite the overprint of tepee formation, the cycles persist through the tepee facies (Fig. 7). Based upon the lack of true peritidal (i.e., progradational) deposits between the subtidal portion and the overprinted diagenetic cap, the individual cycles are interpreted as fifth-order stratigraphic cycles representing the response to short-term eustatic sea-level oscillations (Hardie and others, 1986; Goldhammer and others, 1987). Two factors contributed to the lack of progradational tidal-flat subfacies within the cycles. The Latemar was an isolated buildup of small size (4-km platform diameter) and was not landlocked. Thus, no nucleus existed from which tidal flats could accrete and prograde. Instead, most of the carbonate mud generated *in situ* on the platform was winnowed during storms and redeposited in the adjoining basins.

The Latemar meter-scale (fifth-order) cycles are normally arranged into asymmetric fourth-order megacycles, which typically contain five thinning-upward cycles per megacycle (Fig. 10). The grouping of cycles into asymmetric megacycles and the repetition of megacycles vertically are confirmed by both graphical time-space and statistical relative-time series analyses (Fig. 11; Goldhammer and others, 1987). The graphical time-space analysis follows the method of Fischer (1964). On this "Fischer diagram" (Goldhammer and others, 1987), the horizontal axis represents relative time and individual fifth-order cycles are plotted as triangles evenly spaced along the axis, on the assumption that each cycle is of equivalent duration. The vertical axis is a measured section (here, from the upper cyclic facies) with arrows indicating boundaries between megacycles and the numbers referring to the number of fifth-order cycles per megacycle. The straight line that connects

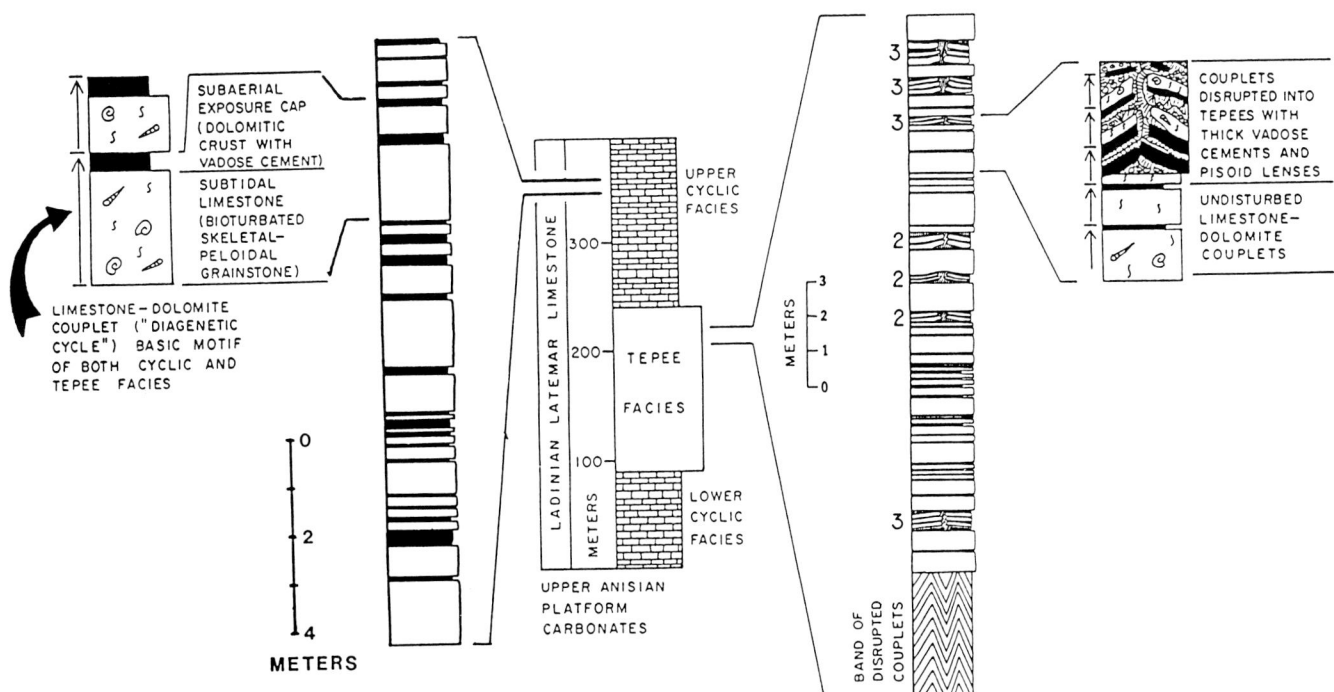

FIG. 7.—Representative measured sections of the upper cyclic and tepee facies of the Latemar Limestone. Note how meter-scale cycles are arranged into thinning-upward megacycles. Numbers to left of tepee zones indicate the number of cycles incorporated within each tepee zone.

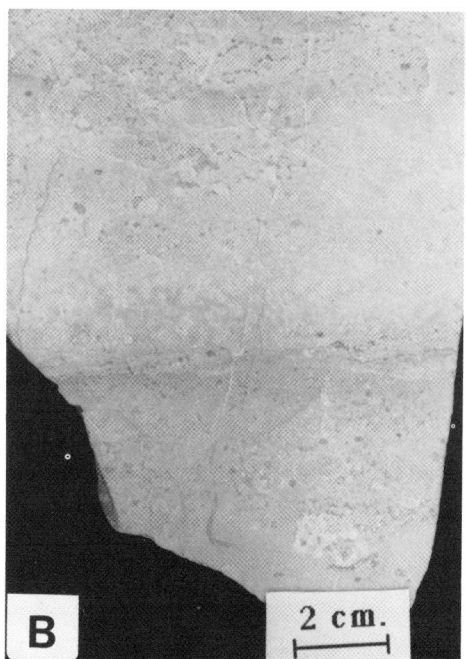
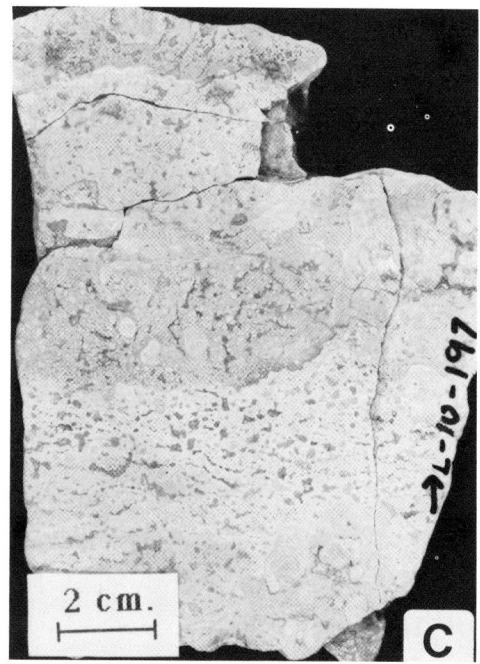

FIG. 8.—Cyclic facies. (A) Flat-lying cyclical carbonates of the upper cyclic facies. Thin cycle caps weather as recessive bands. Scale bar equals 2 m. (B) Subtidal member of a cycle consisting of burrowed skeletal lithoclastic packstone-grainstone; polished slab. (C) Subaerial-exposure cap of a cycle, here a coated grain packstone with a variety of caliche fabrics and fenestrae related to subaerial vadose diagenesis; polished slab.

the base of the stratigraphic section to time zero is the mean subsidence vector, depicting the mean subsidence. Individual cycles are plotted above this vector. Any slope (positive or negative) to a string of cycles reflects deviations from mean subsidence and, hence, changes in relative sea level. Treated in this manner, groupings of five cycles plot as concave-downward "waves," illustrating packaging of five fifth-order cycles per asymmetric, fourth-order megacycle. Such a rhythmic pattern strongly indicates an underlying periodicity and suggests that the cycles involved in a megacycle rhythm were of equal duration (Goldhammer and others, 1987).

The Pleistocene sea-level record serves as an excellent analog for further understanding the Latemar cyclicity. Pleistocene sea-level curves constructed from data coral reef terraces (Fig. 11) and deep-sea oxygen isotope data (Broecker and Van Donk, 1970; Shackleton and Opdyke, 1973) reveal an approximately 20-ka (fifth-order) symmetrical pulse superimposed upon an asymmetric approximately 100-ka (fourth-order) oscillation. The asymmetry of the fourth-or-

Fig. 9.—Tepee facies. (A) Thin (small arrows) and thick (large arrows) tepee zones interrupting the evenly bedded, flat-lying meter-scale cycles; scale bar equals 10 m. (B) Three tepee zones separated by flat-lying, undisturbed cycles. Tepee crests are marked by arrows. Scale bar equals 1 m.

der oscillation results from the rate of sea-level rise being greater than the rate of fall. Pleistocene sea-level fluctuations were driven by a glacio-eustatic mechanism directly related to Milankovitch climatic cycles (Mesolella and others, 1969; Hays and others, 1976; Berger, 1977; Imbrie and others, 1984). The fourth-order asymmetrical pulse is linked to the orbital eccentricity cycles that average approximately 100 ka (95 and 103 ka). The fifth-order symmetrical oscillation coincides with the periodicities of the precessional cycles that average approximately 20 ka (19 and 23 ka).

Thus, the Pleistocene sea-level record indicates that a 5:1 ratio of superimposed orders of sea-level oscillations with

Fig. 10.—One platform megacycle consisting of five individual cycles, each composed of a subtidal layer and a thin, recessive, exposure cap; upper cyclic facies.

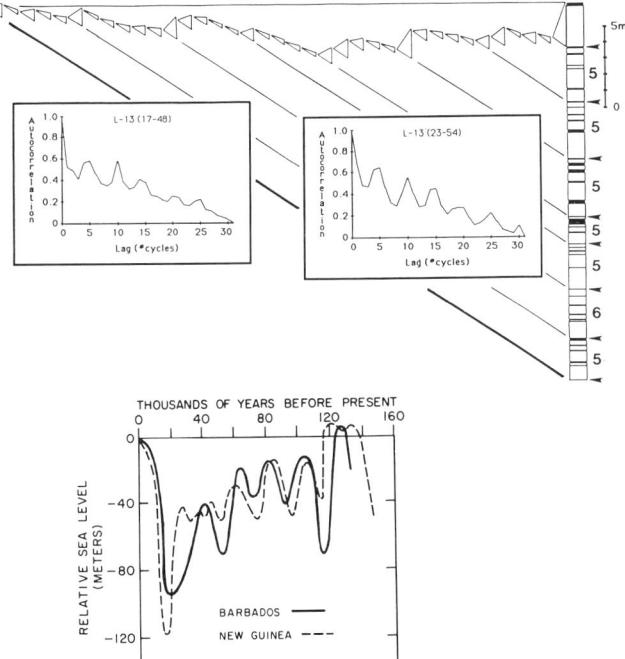

Fig. 11.—Graphical time-space (large plot) and relative-time series analyses (center insets) for 37 consecutive meter-scale cycles from the upper cyclic facies, with Pleistocene sea-level curves (lower insert). Graphical time-space analysis described in text. Relative-time series analysis: autocorrelation graphs refer to two different, but overlapping, sets of 32 consecutive cycles within the measured section; repeated autocorrelation maxima at a lag equal to five confirms packaging of five cycles per megacycle. Pleistocene sea-level curves (Barbados curve from Steinen and others, 1973; New Guinea curve from Aharon, 1984) illustrate eustasy with five symmetrical 20-ka cycles superimposed on an asymmetric 100-ka megacycle.

an asymmetry in the higher order fluctuation are characteristic of glacio-eustasy driven by Milankovitch climatic cycles. These are exactly the characteristics of the Latemar cycles and megacycles as revealed by graphical time-space analyses and relative-time series (Fig. 11). In addition, whereas the Triassic age dates are imprecise (Hardie and others, 1986), the average duration of the Latemar cycles is between 10 ka and 20 ka, similar to the Pleistocene fluctuations. We conclude that superimposed fifth- (approximately 20 ka) and fourth-order (approximately 100 ka) eustatic sea-level oscillations dictated the development of individual meter-scale stratigraphic cycles and their grouping into megacycles (Fig. 12, cyclic facies), which is the dominant theme of the Latemar platform record. The amplitudes of these short-term oscillations were probably small, however, because large-scale glaciation is not known during the Triassic. Computer simulations of Latemar cycles (Goldhammer and others, 1987) provide estimated rise/fall rates of 2–10 cm/ka for the fourth-order (100 ka) pulse, and 5–25 cm/ka for the fifth-order (20 ka) oscillation.

The lower platform and tepee facies.—

If Milankovitch-related glacio-eustatic cycles were operating throughout the entire accumulation of the buildup, then we must consider the significance of both the lower platform facies and tepee facies, which do not contain an uninterrupted succession of meter-scale cycles arranged in megacycles. In examining the relation between cyclic events in time (rhythms) and the resulting stratigraphic record, one needs to be aware of two possibilities (Sander, 1936): (1) each temporal rhythm of a periodic cycle-producing mechanism is recorded as a stratigraphic entity or cycle, or (2) despite rhythmic periodicity of the cycle-producing mechanism (e.g., glacio-eustasy), the corresponding stratigraphic record may not preserve every potential cycle-producing pulse as a stratigraphic cycle. The first situation refers to both cyclic facies, where the interaction of short-term eustasy and relative rate of subsidence yielded a rather complete stratigraphic record. The second situation is proposed to explain the variations from the normal cyclic pattern observed in the lower platform facies and tepee facies, where the interaction of short-term eustasy and accommodation potential (i.e., relative rate of subsidence, equal to third-order eustasy plus subsidence) resulted in "missed beats" of deposition, resulting in an incomplete record (Fig. 12).

The lower platform facies lacks repetitive meter-scale fifth-order cycles arranged into fourth-order megacycles; rather, it is marked by subaerial-exposure caps spaced approximately 1 every 10 m. Based on its thickness and age constraints, the relative-subsidence rate during lower platform facies accumulation was greater than that of the lower cyclic facies, tepee facies, or upper cyclic facies (Goldhammer, 1987), because the third-order rise (see later discussion) was superimposed upon subsidence. This increase in accommodation potential resulted in thick subtidal packages stemming from several "missed beats" of fifth-order deposition, during which time amalgamated subtidal fifth-order cycles (*without* subaerial-exposure caps) were generated. Most of the 20-ka symmetrical sea-level rhythms oscillated above

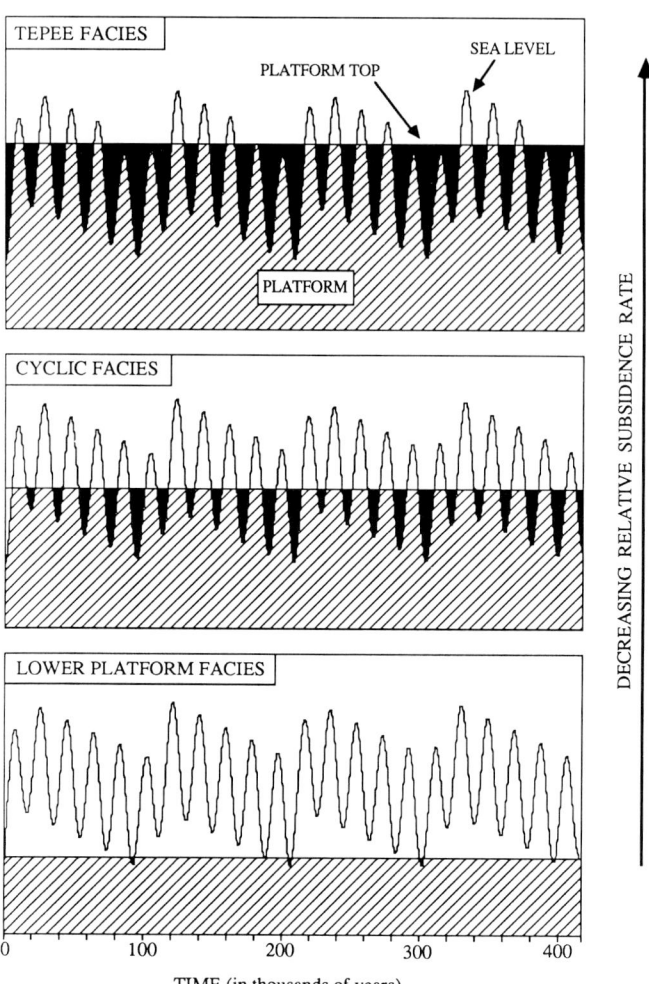

FIG. 12.—Conceptual model for short-term eustasy. Rapid rates of relative subsidence (lower cyclic facies) result in "missed beats" as sea-level oscillations occur above the platform top, resulting in long periods of submergence (white) with only brief exposure events (black), where sea-level curve falls below platform top. At intermediate relative-subsidence rates (cyclic facies), the alternations of submergence and exposure are recorded as individual cycles and megacycles, reflecting the fifth- and fourth-order eustatic changes. At slow relative-subsidence rates (tepee facies), "missed beats" are due to oscillations occurring below the platform top.

the platform top (Fig. 12), thus no subaerial-exposure caps formed to delineate fifth-order cycles. In this manner, lower platform facies deposits were subjected to long, uninterrupted periods of submergence.

The presence of fully developed megacycles between the tepee zones in the tepee facies indicates that the same composite, eustatic sea-level rhythm operated during deposition of the tepee facies. Despite the overprint of tepee formation, an accurate counting of cycles within tepee zones reveals that the normal composite cycle rhythm can still be read (Fig. 7). Tepee zones with disrupted thin cycles occur as tops to megacycles marked by thicker basal cycles. Megacycles with tepee tops may contain five cycles, or they may be incomplete, containing only three or four cycles. We interpret tepee-capped megacycles as "condensed" megacycles that do not necessarily record all depositional pulses of the composite short-term eustatic curve (Fig. 12; Hardie and others, 1986).

Based on the thickness of the tepee facies and its age constraints, the accommodation potential during tepee facies accumulation was less than that of the lower platform facies, lower cyclic facies, or upper cyclic facies, because the third-order fall was superimposed upon subsidence. This decrease in accommodation potential resulted in the formation of condensed megacycles generated by missing one or more of the fifth-order pulses during the falling phase of the asymmetric fourth-order sea-level oscillation (Fig. 12). Extended periods of subaerial exposure ensured that pre-existing thin cycles would be susceptible to tepee formation and associated subaerial diagenesis.

Flank Facies

The flank facies extend from the platform margin to the basin and relate directly to the depositional profile (Fig. 3). Both the margin and foreslope portions of the Latemar buildup are included in the Marmolada Limestone (Gaetani and others, 1981), which surrounds the platform interior (Fig. 2). Previous studies of the flank facies of Ladinian buildups in the Dolomites (Gaetani and others, 1981; Fois and Gaetani, 1981; Biddle, 1981; Fois, 1982) concentrated on description of the biota, especially from the platform margin. Because many buildups are dolomitized, the petrography of redeposited boulders of margin facies was examined in most cases. The platform margin of the Latemar buildup is well preserved and in place, however, and indicates that Ladinian biota could form significant platform edge reefs.

Margin facies.—

The Ladinian margin consists of two basic rock types: peloidal skeletal grainstones/packstones and algal boundstones. The grainstones/packstones are similar to the platform sands; peloids and micrite clots ("structure grumeleuse") are the most abundant grains. The biota is much more diverse however: brachiopods, ostracods, gastropods, bivalves, echinoderms, foraminifera, dasycladaceans, and tufts of blue-green algae are all common. Various genera of microproblematica also occur in the margin grainstones. The algal boundstones (Fig. 13A) are dominated by porostromate and spongiostromate algae, *Tubiphytes*, and ubiquitous submarine cement. Sphinctozoan and inozoan sponges (largely recrystallized) are common, but hexacorals are rare. Cements include isopachous fibrous, botryoidal, and radiaxial fibrous calcites interlayered with fine internal sediment. The boundstone framework contains numerous borings and patches filled by both cements and peloid-skeletal grainstones.

The margin is massive to thickly bedded and ranges from 10 m to about 30 m in width. The basinward half of this zone is grainstone, commonly with the microproblematica *Bacinella* and small micritic thrombolite clots. The remainder is boundstone with subordinate grainstone, except at the transition to the platform facies, which is marked by grainstone. The subaerial exposure horizons that mark the platform section are lacking, although exposure horizons at the platform edge are developed locally on boundstone beds. The margin lithologies interfinger with the foreslope and platform deposits on a fine scale (few meters), but the transitional zone remains nearly vertical.

In map view (Fig. 2), the margin forms a nearly continuous rim around the central platform section. In one locality equivalent to a tepee zone, the boundstone is missing due to erosion. Whereas the margin did not form a raised platform rim (as evidenced by the lack of exposure surfaces), it was a true ecologic barrier that sharply separated the low-diversity biota of the platform interior from those of the outer margin and foreslope grainstones.

Foreslope facies.—

The foreslope deposits (Fig. 13B) are predominantly breccia and megabreccia sheets and lenses (several meters thick) that lack a fine matrix but contain abundant cements. Clasts are almost exclusively margin-derived (Fig. 13C) boundstones with only minor amounts of lithified platform-derived grainstones in the tepee zone equivalents (including some red-soil clasts). The maximum clast size increases down the foreslope from approximately 10 cm adjacent to the margin to several meters (megabreccias) at the base of the slope. The breccias have erosional bases, indicating that scouring alternated with deposition. In places, grainstones overlie the breccias and are truncated by the scour surfaces below the succeeding breccia. These scoured surfaces form the obvious clinoform surfaces, which extend from the margin to the toe of slope. The megabreccia facies pinches out abruptly at the toe of slope, where the depositional slope decreases rapidly from 30°–35° to about 5°–10°.

Toe-of-slope and basin facies.—

At the toe of slope, the megabreccias are replaced by graded beds (<1 m thick) of peloidal skeletal grainstone (Fig. 13D). These sands clearly originated in the platform and margin settings as they contain the entire spectrum of shallow-water biota. The grainstones fine upward into wackestone or mudstone, and some beds are capped by a nodular limestone. The nodular fabric decreases in intensity downward from the top of the bed and in places preserves the primary fabric.

The toes of the breccia bodies thin and pinch out over a distance of a few tens of meters basinward from the decrease in depositional dip. Generally, the foreslope prograded over the toe of slope as the platform relief increased. On a scale of several meters, the breccia lenses alternate with grainstone intervals. Based on the overall thickness of the toe-of-slope facies, these alternations roughly correspond to platform megacycles.

Laterally, the graded grainstones merge basinward into thin (<10 cm), graded, carbonate grainstones, interpreted as turbidites (Fig. 13E; Bosellini and Rossi, 1974; Bosellini and Ferri, 1980).

Depositional processes along the platform-to-basin transition.—

The facies along the platform-to-basin transition reflect the different depositional settings and processes along this

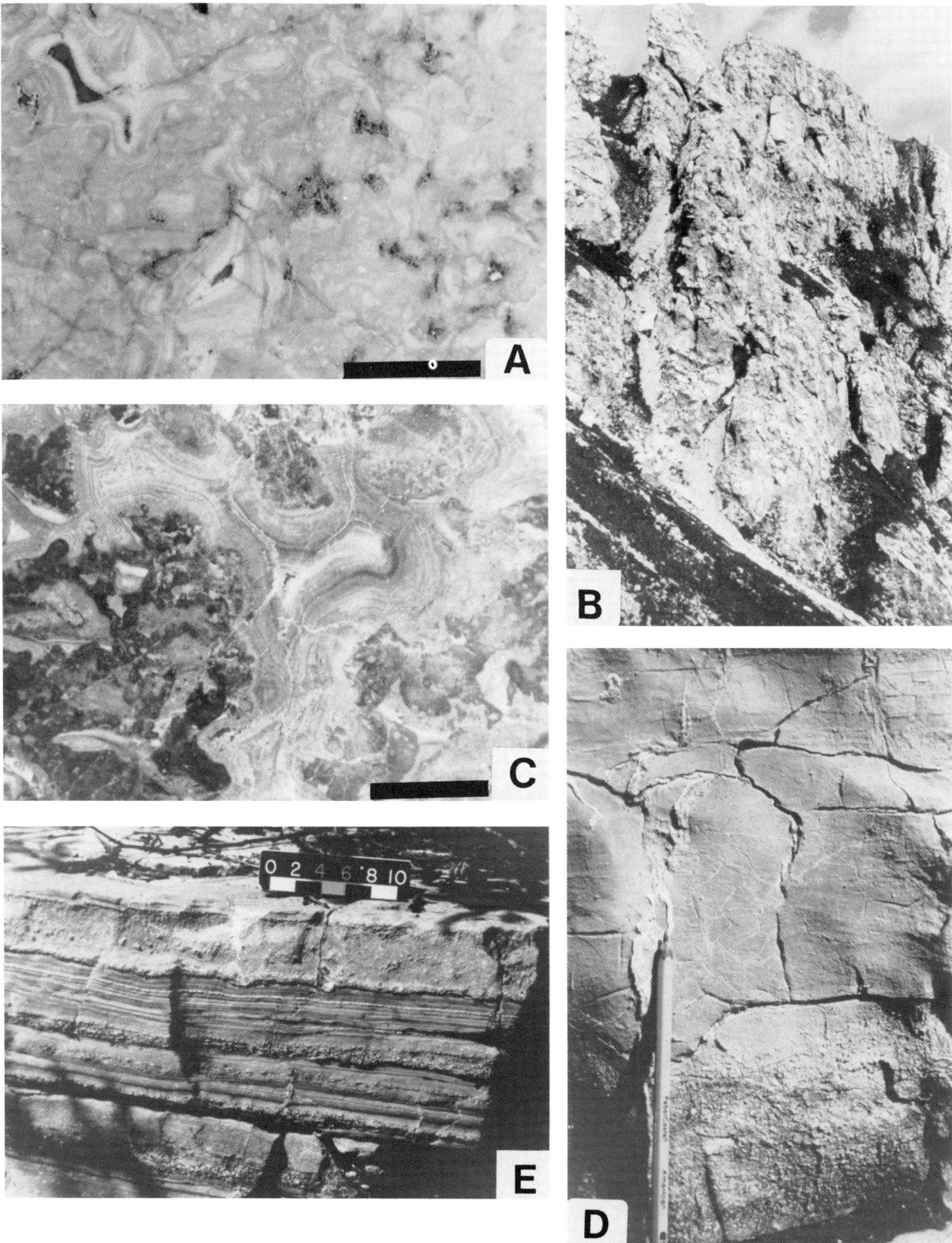

FIG. 13.—Flank facies. (A) Algae-*Tubiphytes* boundstone with abundant cement and recrystallized biotics (sponge or coral?) from the margin facies. Scale bar equals 2 cm; polished slab. (B) Steeply dipping foreslope clinoforms; approximately 100 m of vertical section. (C) Foreslope breccia composed of boundstone clasts and isopachous cement. Note the lack of matrix. Scale bar equals 2 cm; polished slab. (D) Graded bed of grainstone from the toe of slope. (E) Basinal deposits composed of carbonate mudstone and turbidites; scale in cm.

profile. These processes can be related to the exposure events recorded in the platform stratigraphy.

The marginal boundstones and grainstones do not record the meter-scale exposure horizons found in the platform deposits. Apparently, the magnitude of the sea-level changes was inadequate to expose the margin of the buildup, suggesting the changes were limited to a few meters at most. The margin boundstones continued virtually uninterrupted throughout the history of the buildup. Erosion of the boundstones probably increased during sea-level falls due to increased wave effects. During formation of tepee zones, erosion locally removed the margin and cut into platform facies. The marginal boundstones and grainstones were stabilized by submarine cementation and organic processes.

The foreslope breccia beds are large-scale talus piles, based upon their steep dip, lenticular bedding, lack of matrix material, increase of clast size toward the toes, and abrupt pinchout at the toe of slope. These talus piles consist almost entirely of margin-derived material, with only a minor amount of lithified platform material. Markedly absent are any debris flow deposits, probably due to the lack of mud in the well-winnowed platform section. Some thin grainstones occur, but their eroded tops indicate that erosion along the foreslope scoured much of this finer material. The large clinoforms are scour surfaces, formed by erosion prior to deposition of breccia beds.

At the toe of slope, two depositional regimes interfinger with each other. The megabreccia toes of the talus debris pinch out abruptly into a grainstone section consisting of shallow-water (platform and margin) sands that bypassed the steep foreslopes and were deposited as graded beds as the flows decelerated at the edge of the basin floor. The nodular limestone caps on some of the beds record periods of non-deposition and the formation of submarine hardgrounds, due to interruption in the sand supply. Basinward, the flows that deposited the graded beds changed facies into carbonate turbidites.

Highstand versus Lowstand Deposits: Eustatic Control on Buildup Deposition

The depositional pattern is interpretable in terms of a highstand versus lowstand model (Fig. 14), similar to that inferred for the Bahama Platform by some workers (Boardman and Neumann, 1984; Boardman and others, 1986; Droxler and Schlager, 1985). During sea-level highstands, both the platform and the margin were submerged. Shallow-water subtidal sands accumulated in the platform interior, which was surrounded by a reef margin. Redeposited, early marine cemented boundstones built the foreslope talus slopes, but redeposited shallow-water sands largely bypassed the foreslope to form the toe of slope grainstones and basinal turbidites.

Sea-level lowstands resulted in a different set of processes. Subaerial exposure of the platform resulted in formation of the exposure surfaces. The margin boundstones continued to supply the bulk of the foreslope clasts, probably at an increased rate due to increased erosion. Some lithified platform clasts were incorporated into the talus slopes, however. The supply of redeposited shallow-water

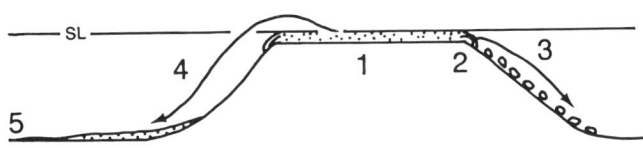

HIGHSTAND MODEL

1. GRAINSTONES
2. ALGAL BOUNDSTONES
3. MEGABRECCIAS OF ALGAL BOUNDSTONES
4. GRADED GRAINSTONES
5. CARBONATE TURBIDITES & MUD

LOWSTAND MODEL

1. EXPOSURE & CEMENTATION OF GRAINSTONES
2. ALGAL BOUNDSTONES
3. MEGABRECCIAS OF ALGAL BOUNDSTONES
   & CEMENTED GRAINSTONES
4. STARVED BASIN

FIG. 14.—Highstand versus lowstand model for platform margin, slope, and basin sedimentation in response to short-term eustatic sea-level changes. Actual facies distributions are symmetrical, and no significant leeward-windward effects are recognized.

sands ceased, resulting in the formation of the nodular hardgrounds atop the grainstones at the toe of slope and the interruption of basinal turbidites.

The platform megacycle rhythm reflects a regular shift in the duration of exposure versus submergence. The resulting changes in the supply of highstand grainstones and the erosion of the margin boundstones are recorded in the toe of slope interfingering of breccia and redeposited grainstone.

Whereas the platform cycles clearly record the short-term sea-level fluctuations, the flank facies are more heterogeneous. The depositional processes and resulting facies along the platform-to-basin transition, however, also relate directly to the sea-level fluctuations recorded in the platform stratigraphy.

THIRD-ORDER EUSTATIC CONTROL ON BUILDUP STRATIGRAPHY AND GEOMETRY

Vertical Platform Succession and Long-Term Sea Level

The vertical facies succession of the Latemar platform (from lower platform facies to lower cyclic facies to tepee facies) depicts a third-order (10 m.y.) depositional sequence developed atop the inherited Lower Anisian topographic banks. The upper cyclic facies does not necessarily fit into the Latemar third-order sequence, but may represent the beginning of another sequence. The Latemar sequence records various changes in stratigraphic and lithologic character that reflect continuously decreasing accommodation potential.

The lower platform facies lacks repetitive fifth-order cycles, but subtidal carbonate beds are punctuated by subaerial-exposure caps that increase in frequency toward the top of the facies. Within the overlying facies, fifth-order cycles decrease in thickness from the lower cyclic facies (average 1.24 m/cycle) to the tepee facies (average 0.40 m/cycle). This progression from a non-cyclic interval to cyclic intervals marked by decreasing cycle thickness upward indicates that the accommodation space (net subsidence plus eustasy) for each depositional pulse was progressively reduced.

The distribution of grain types, sedimentary structures, and early diagenetic features varies within the vertical sequence. For example, the lower platform facies is dominated by early marine, syndepositional diagenetic features (hardgrounds, sheet cracks, marine cements) that are conspicuously absent in the overlying lower cyclic facies and tepee facies, which are full of early vadose diagenetic features (caliche fabrics, vadose cements). In the tepee facies, tepees are evidence of increased subaerial exposure in the uppermost facies within the sequence. The progressive increase in subaerial-exposure features clearly records a shift from generally submergent conditions during lower platform facies deposition to conditions whereby subtidal deposition was frequently interrupted by intervals of subaerial exposure (lower cyclic facies and tepee facies), reflecting continuously decreasing accommodation potential.

To understand the vertical variation in subaerial exposure better, we compared sedimentation rates against platform aggradation rates. Using a range of reasonable sedimentation rates and the total time available to accumulate each facies (see later discussion), the potential thickness of each facies was calculated by multiplying subtidal sedimentation rate (net accumulation rate of 7.5–15 cm/ka; Goldhammer, 1987) by the duration of each facies. Holocene sediment production rates provide a constraint on sedimentation rates. For example, carbonate shelf lagoon production rates range from about 2.5 cm/ka (Stockman and others, 1967) to about 20 to 30 cm/ka (Neumann and Land, 1975). Sedimentation rates used in our calculations may appear low, but it is necessary to take into account the large volume of sediment redeposited over the platform edge. Hence, our sedimentation rates are net platform accumulation rates; true sediment production was probably at least twice this rate. This value is the maximum thickness each facies could have achieved if sedimentation was continuous. Sedimentation was not continuous, however, principally due to subaerial exposure (although there were submerged periods of slow deposition or non-deposition). The ratio of actual to potential (or maximum) thickness yields the percentage of time during which deposition took place, roughly equal to submergence. In this way, a relative estimate of the percentage of time the platform was submerged versus subaerially exposed in each facies was calculated and plotted at the midpoint (in thickness) of each facies (Fig. 15). The plot reveals progressively increasing amounts of subaerial exposure, from the lower platform facies through the lower cyclic facies to the tepee facies. The trend reverses in the overlying upper cyclic facies, as expected.

By utilizing the fifth-order, meter-scale cycles as stratigraphic time units of fixed duration (20 ka), it is possible to derive more quantitative information regarding the third-order sequence. An accommodation potential (or relative subsidence) curve was constructed (Fig. 16A) with the temporal duration of each facies derived by counting the number of fifth-order cycles per facies, assuming 20 ka per cycle. This diagram shows total accommodation versus time, including both tectonic subsidence and eustatic components. It substantiates the notion that the Latemar third-order sequence (lower platform facies to tepee facies) depicts progressively decreasing accumulation rates. Potential error in this analysis stems from the "missed beats" of fifth-order eustasy recorded in the lower platform and tepee facies; thus, estimated time is a minimum. Within the tepee facies, fourth-order megacycles with tepee caps were assigned 100 ka, even if less than five fifth-order cycles were present. For the lower platform facies, the duration was constrained by age dates and computer simulations of the entire Latemar buildup (Goldhammer, 1987).

Assuming that subsidence (tectonic subsidence plus isostatic loading and compaction) was constant over the life span of the Latemar buildup, a continuous "Fischer diagram" for the entire Latemar buildup can be constructed

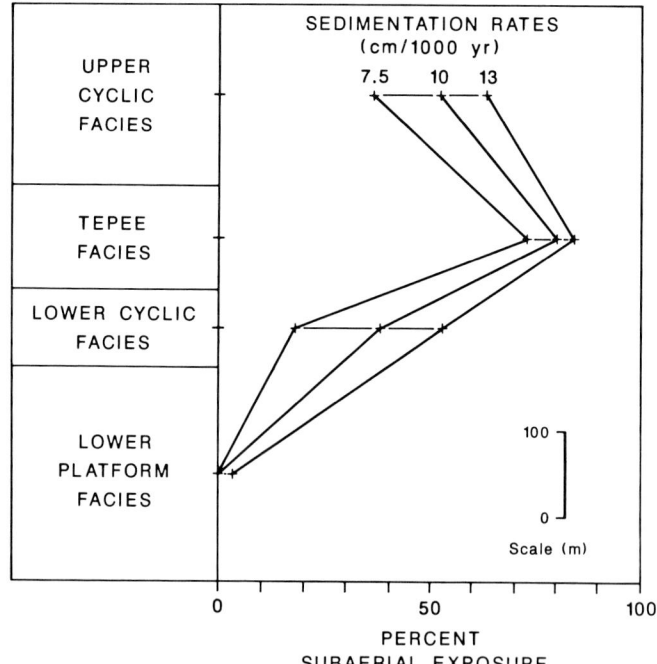

FIG. 15.—Plot of Latemar platform facies versus percentage of time each facies was subaerially exposed, showing increasing amount of subaerial exposure from the lower platform facies to the tepee facies. Percent subaerial exposure for each facies is plotted at the stratigraphic midpoint for each facies.

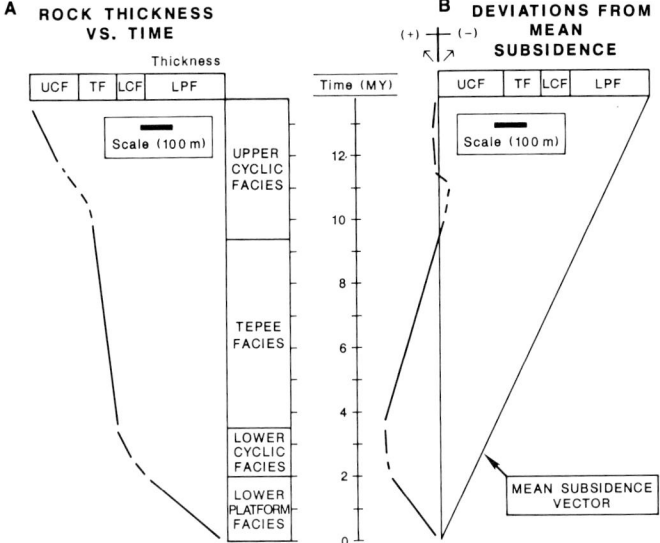

FIG. 16.—(A) Accommodation potential curve for Latemar platform facies, plotting cumulative thickness (horizontal axis, increasing to left) against time (vertical axis). Duration of each facies derived by counting cycles and assigning 20 ka/cycle. (B) Graphical time-space analysis for Latemar platform facies. Vertical axis is again time; horizontal axis is sediment thickness. Mean subsidence vector assumes constant background subsidence and represents average rate of sediment accumulation. Deviations from the mean subsidence rate plot as shifts away from the time axis and are interpreted as third-order eustasy after correction for isostatic loading and compaction. Maximum positive deviation (at 3 m.y.) is 170 m, which corresponds to a 60-m eustatic sea-level rise.

via graphical time-space analysis (Fig. 16B; Goldhammer, 1987). This diagram contains the trace of the long-term "Fischer diagram" (each individual cycle is not plotted). It illustrates deviations from constant subsidence (mean subsidence equals approximately 5 cm/ka) plotted in time against successive Latemar facies. Positive deviations (shifts to the left) depict increased subsidence, whereas negative deviations (to the right) depict decreased subsidence relative to mean subsidence. The trend, which forms an asymmetric cycle of positive (rapid accumulation) to negative (slow accumulation) deviations for the lower three Latemar facies, could be interpreted as follows: (1) totally eustatic in origin–deviations from the mean are due to variations of eustatic sea level superimposed on continuous platform subsidence; (2) totally tectonic in origin–deviations from the mean delineate the true path of platform subsidence; or (3) mixed eustatic and tectonic origin–deviation trend results from both eustatic and tectonic effects.

Comparing the platform facies stratigraphy of the Latemar and Southern Latemar buildups (Fig. 1) suggests that the third-order trend is eustatic in origin. The Southern Latemar platform section is thicker (over 1 km), probably because of the location on a different tectonic block, but the same four-fold vertical stratigraphy occurs, thickened proportionately. The similarity in vertical sequence favors the eustatic interpretation, allowing the variations in accumulation rate (third-order deviation from mean subsidence) to be used as a measure of eustatic sea level. After removing the contribution of isostatic loading and compaction from the deviation trend, the eustatic component yields a maximum positive deviation of about 60 m (compared to the maximum excess accumulation value of 170 m; Fig. 16B).

If the deviation curve is due to eustasy, the shape and slope record a eustatic rise and fall of sea level (60-m amplitude, 10-m.y. period) during the evolution of the lower platform facies-lower cyclic facies-tepee facies third-order sequence, followed by a turnaround and successive smaller rise during upper cyclic facies deposition. Other independent estimates of the Upper Anisian-Ladinian eustasy support our interpretation. The magnitude of the rise is close to the Upper Anisian-Ladinian third-order rise of about 50 m, shown in the recent Exxon cycle chart (Haq and others, 1987). The global sea-level curve of Hallam (1984) documents a major eustatic rise (amplitude 90 m) beginning in the Anisian, reaching a maximum in the Ladinian. Ager (1981) argues that the most important event in the European Triassic is a widely recognized late Middle Triassic transgression.

The rate of sea-level rise and fall is on the order of 1–3 cm/ka, which is consistent with rates of third-order eustatic sea-level fluctuations documented elsewhere (Haq and others, 1987; Kendall and Schlager, 1981). Hine and Steinmetz (1984) estimate third-order eustatic curves to have amplitudes of 50–75 m, and rates of 2–10 cm/ka. Hancock and Kauffman (1979) calculated rates of 1–9 cm/ka for third-order sea-level cycles in the Late Cretaceous.

A eustatic model for the development of the Latemar vertical facies packaging under conditions of constant subsidence includes the following (Fig. 17):

(1) Lower platform facies–A prolonged, relatively rapid third-order sea-level rise (superimposed on subsidence) maximized accommodation potential and maintained subtidal conditions over the entire platform as the high-relief buildup accreted atop the low-relief Middle Anisian bank. During this phase of subtidal carbonate deposition, subaerial exposure occurred infrequently and lengthy periods of marine submergence promoted abundant syndepositional, early marine diagenesis. "Missed beats" of deposition (stemming from fifth-order eustatic pulses failing to expose the platform) resulted in the formation of thick, amalgamated subtidal beds.

(2) Lower cyclic facies–The rate of third-order sea-level rise progressively decreased, resulting in reduced accommodation potential and meter-scale cycles with subaerial-exposure caps developed. This progressive slowing of sea-level rise resulted in sequentially thinner cycles upward. The interaction of third-order eustasy with high-frequency (fourth-order and fifth-order) eustasy and subsidence was favorable for generating complete megacycles composed of five meter-scale cycles.

(3) Tepee facies–The third-order sea-level fall commenced, but due to continuous subsidence, the net effect was close to stillstand and accommodation potential was minimized. Numerous thin cycles (thinner than those of the lower cyclic facies) developed along with "condensed" megacycles with tepee tops, representing long periods of subaerial exposure. The entire tepee facies is considerably condensed, recording approximately 6 Ma of geologic time in 120 m of section.

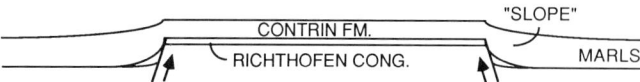
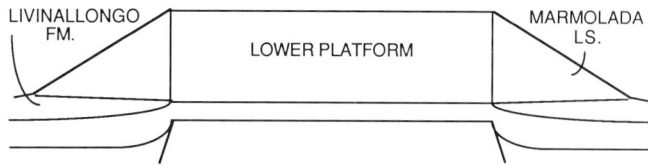
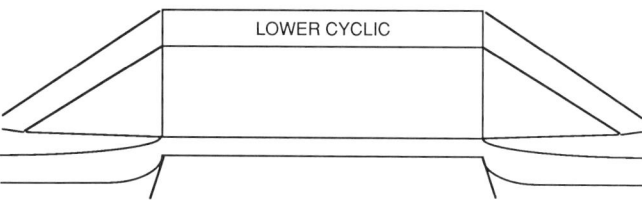
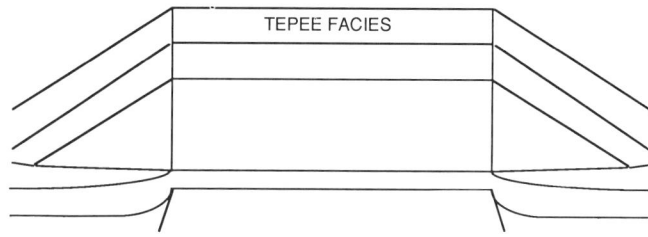

Fig. 17.—Diagrammatic view of the response of the Latemar buildup to the third-order sea-level oscillation. Originating on a Middle Triassic topographic high, the lower platform facies accumulated during the sea-level rise (60 m in 2 m.y.) that drowned the basin. During the stillstand (duration of 1.5 m.y.), the lower cyclic facies formed. The slowly accumulating tepee facies records the long (6 m.y.) fall.

(4) Upper cyclic facies–Following the long slow fall in sea level, a renewed third-order rise occurred with the rate progressively increasing, producing thicker cycles and complete megacycles without tepees.

Margin Geometry

The geometry of the margin-to-foreslope transition is vertical from the lower platform facies through the base of the upper cyclic facies (Fig. 4). Higher levels are not preserved, but a prograding geometry is inferred based upon the extent of the preserved foreslopes, the occurrence of uppermost foreslope facies along the top of the preserved clinoforms, and comparison with other Ladinian buildups in the Dolomites (Bosellini, 1984). In the Latemar buildup, the depositional surface maintains a constant dip, while the depositional relief steadily increases to over 500 m. No escarpment is recognized along the margin; rather, the facies interfinger with each other abruptly. To maintain such a geometry, an increasing volume of slope material is required (Schlager, 1981). In other words, the geometry of the Latemar margin requires a progressive increase in the ratio of foreslope volume to platform volume (Fig. 18A)

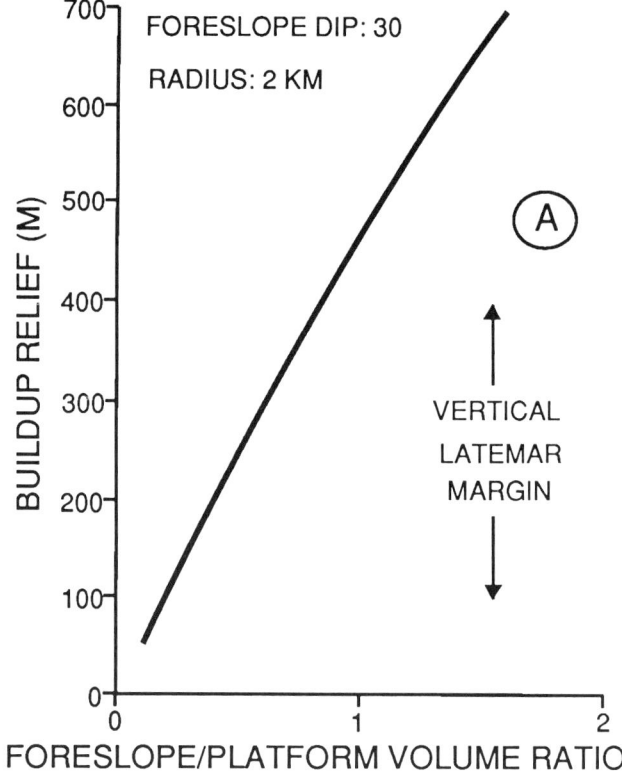
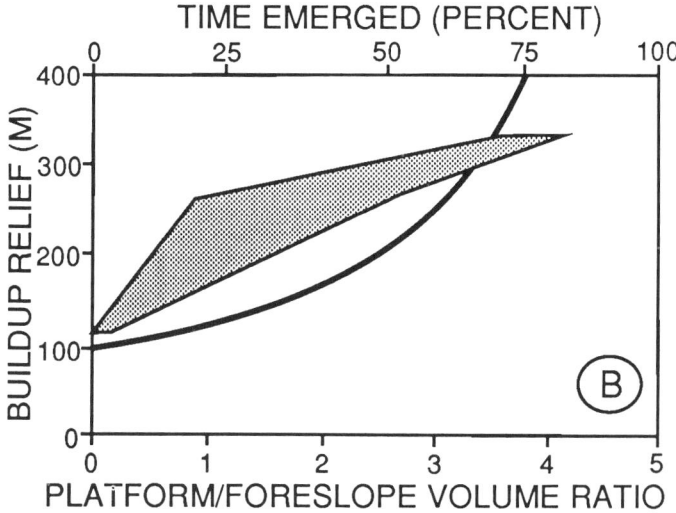

Fig. 18.—(A) Calculated ratio of foreslope volume to platform volume based on the geometry of a Latemar-size platform as the relief increases from 50 to 700 m. The estimated range of the syndepositional relief during the third-order sea-level cycle is shown. (B) For the range of syndepositional relief during the third-order sea-level cycle, the buildup relief is plotted against the ratio of platform volume to foreslope volume (heavy line; lower horizontal axis) as determined from the geometry of the Latemar. In addition, the percent of time the platform is exposed (darkened field, upper horizontal axis) is replotted from Figure 15 against the buildup relief.

during the long-term relative sea-level rise/fall cycle recorded in the platform stratigraphy.

The platform facies equivalent to the vertical geometry (lower platform facies, lower cyclic facies, and tepee facies) record a progressive increase in the duration of platform exposure events. Based on the lack of exposure horizons in the margin facies, the boundstones were available to be reworked into the talus piles without interruption. Thus, the talus slopes continued to be built during exposure events. With the increase in platform exposure during the third-order sea-level cycle, progressively larger amounts of boundstone debris were deposited without any vertical accretion of the platform section.

The agreement between the stratigraphic data and the geometric constraints is illustrated by consideration of the variations in duration of platform exposure. In this depositional system, the rate of platform aggradation varied (Fig. 16A) due to platform exposure events (Fig. 15), while foreslope volume was added more continuously. Thus, the ratio of platform volume to foreslope volume is a rough indication of the duration of the platform submergence. Comparing this geometric relation to the platform exposure pattern derived from the platform stratigraphy (Fig. 18B), the similarity is apparent. Although the exact equivalence is uncertain, both approaches indicate similar changes in the extent of platform exposure related to the third-order sea-level change. This relation suggests that long-term sea-level changes can directly affect the volume of redeposited carbonates and the resulting facies geometry.

In summary, the geometric volume constraints of the Latemar geometry can be satisfied by the interaction of the third-order sea-level change with the depositional processes controlled by the short-term oscillations. During the third-order sea-level fall, the supply of breccia increased relative to the rate of platform aggration. Thus, the eustatic sea-level fluctuations controlled the geometry of the buildup margin by dictating the volume of material available to build the foreslope talus pile.

CONCLUSIONS

(1) The Latemar buildup is an isolated carbonate buildup with a central platform core rimmed by a marginal reef and surrounded by steeply dipping foreslopes on all sides.

(2) The platform stratigraphy consists of four vertically stacked facies units. The lowermost (lower platform facies) consists of subtidal grainstones. The upper three facies consist of meter-scale cycles, each consisting of a thick subtidal grainstone capped by a thin subaerial-exposure unit. The middle unit between the two cyclic facies is marked by numerous tepee zones formed by prolonged exposure events.

(3) The cycles are packaged into megacycles typically consisting of five individual cycles. The cycles and megacycles are interpreted as the product of Milankovitch fifth- (10^4 yr) and fourth-order (10^5 yr) sea-level changes.

(4) The vertical stacking of platform facies units defines a third-order (10^7 yr) sea-level change with a magnitude of 60 m. The rate of the third-order sea-level change is approximately 1–3 cm/ka.

(5) The flank facies can be related directly to the depositional profile. The margin is an ecologic barrier and consists of algal boundstone and grainstone. The foreslope is dominated by talus breccias derived from the margin boundstones. At the toe of slope, graded grainstone beds are redeposited shallow-water sands, some with nodular limestone caps, which grade laterally into basinal turbidites.

(6) The short-term sea-level changes (documented in the platform record) directly control depositional processes along the platform-to-basin profile.

(7) The vertical geometry of the buildup margin relates to the third-order sea-level change recorded in the overall platform stratigraphy. The basic geometry requires a progressive increase in the ratio of foreslope volume to platform volume. The variation in platform aggradation rate (during the third-order cycle), coupled with continual addition of foreslope talus, supplied the increasing volume of foreslope talus needed to maintain the depositional profile.

ACKNOWLEDGMENTS

We thank our advisor Dr. Lawrence Hardie (Johns Hopkins University) for his direction and advice during this study. Drs. Alfonso Bosellini and Carlo Doglioni of the University of Ferrara (Italy) introduced us to the geology of the central Dolomites, and their help is gratefully acknowledged. Discussions with Paul Dunn considerably helped us clarify our ideas. We thank J. F. Sarg, J. F. Read, and J. L. Wilson for their suggestions, which improved this manuscript. Our field work was funded by Exxon Production Research Company, Marathon Oil Company, National Science Foundation grants to Dr. Hardie, the Bulk Fund of the Johns Hopkins University, and Grants-in-Aid of Research from Sigma Xi, the Scientific Research Society.

REFERENCES

AGER, D. V., 1981, Major marine cycles in the Mesozoic: Journal of the Geologic Society of London, v. 138, p. 159–166.

AHARON, P., 1984, Implications of the coral-reef record from New Guinea concerning the astronomical theory of ice ages, in Berger, A., Imbrie, J., Hayes, J., Kukla, G., and Saltzman, B., eds., Milankovitch and Climate: Reidel Publishing Company, Boston, p. 379–389.

ASSERETO, R. L. A. M., AND KENDALL, C. G. ST. C., 1977, Nature, origin and classification of peritidal tepee structures and related breccias: Sedimentology, v. 24, p. 153–210.

BECHSTADT, T., BRANDNER R., MOSTLER, H., AND SCHMIDT, K., 1978, Aborted rifting in the Triassic of the Eastern and Southern Alps: Neues Jahrbuch fur Geologie und Palaontologie Abhandlungen, v. 156, p. 157–178.

BERGER, A., 1977, Long-term variation of the earth's orbital elements: Celestial Mechanics, v. 15, p. 53–74.

BERNOULLI, D., AND LEMOINE, M., 1980, Birth and early evolution of the Tethys: The overall situation: Memoire du Bureau de Recherches Geologiques et Minieres, No. 115, p. 168–179.

BIDDLE, K. T., 1981, The basinal Cipit boulders: Indicators of Middle to Upper Triassic buildup margins, Dolomite Alps, Italy: Rivista Italiana di Paleontologia e Stratigrafia, v. 86, p. 779–794.

BLENDINGER, W., 1986, Isolated stationary carbonate platforms: The Middle Triassic (Ladinian) of the Marmolada area, Dolomites, Italy: Sedimentology, v. 33, p. 159–183.

BOARDMAN, M. R., AND NEUMANN, A. C., 1984, Sources of periplatform carbonates: Northwest Providence Channel, Bahamas: Journal of Sedimentary Petrology, v. 54, p. 1110–1123.

———, ———, BAKER, P. A., DULIN, L. A., KENTER, R. J., HUNTER, G. E., AND KIEFER, K. B., 1986, Banktop responses to Quaternary fluctuations in sea level recorded in periplatform sediments: Geology, v. 14, p. 28–31.

BOSELLINI, A., 1984, Progradation geometries of carbonate platforms: Examples from the Triassic of the Dolomites, northern Italy: Sedimentology, v. 31, p. 1–24.

———, CASTELLARIN, A., ROSSI, P. L., SIMBOLI, G., AND SOMMAVILLA, E., 1977, Schema sedimentologico per il Trias medio della Val di Fasse ed aree circostanti (Doloimiti centrali): Giornale di Geologia (Bologona), v. 42, p. 83–108.

———, AND FERRI, R., 1980, La Formazione di Livinallongo (Buchenstein) nella Valle di S. Lucano (Ladinico Inferiore, Dolomite Bellunesi): Annali dell'Universita di Ferrara, nouva serie, Sez. IX, v. 6, p. 63–89.

———, AND ROSSI, D., 1974, Triassic carbonate buildups of the Dolomites, northern Italy, *in* Laporte, L. F., ed., Reefs in Time and Space: Society of Economic Paleontologists and Mineralogists Special Publication 18, p. 209–233.

BROECKER, W. S., AND VAN DONK, J., 1970, Insolation changes, ice volumes, and the O^{18} record of deep-sea cores: Reviews of Geophysics and Space Physics, v. 8, p. 169–197.

CROS, P., 1974, Evolution sedimentologique et paleostructurale de quelques plates-forms carbonatees biogenes (Trias des Dolomites italiennes): Sciences de la Terre, v. 19, p. 299–379.

———, AND LAGNY, P., 1969, Paleokarsts dans le Trias moyen et superieur des Dolomites et des Alpes Carniques occidentales, importance stratigraphique et paleogeographique: Sciences de la Terre, v. 14, p. 139–195.

DOGLIONI, C., 1984a, Tettonica Triassica transpressiva nelle Dolomiti: Giornale di Geologia (Bologna), v. 46, p. 47–60.

———, 1984b, Triassic diapiric structures in the Central Dolomites: Ecologae Geologicae Helvetiae, v. 77, p. 261–285.

DROXLER, A. W., AND SCHLAGER, W., 1985, Glacial versus interglacial sedimentation rates and turbidite frequency in the Bahamas: Geology, v. 13, p. 799–802.

FISCHER, A. G., 1964, The Lofer Cyclothema of the Alpine Triassic: Kansas Geological Survey Bulletin 169, p. 107–149.

FOIS, E., 1982, The Sass da Putia carbonate buildup (western Dolomites): Biofacies succession and margin development during the Ladinian: Rivista Italiana di Paleontologia e Stratigrafia, v. 87, p. 565–598.

———, AND GAETANI, M., 1981, The northern margin of the Civetta buildup, evolution during the Ladinian and the Carnian: Rivista Italiana di Paleontologia e Stratigrafia, v. 86, p. 469–542.

GAETANI, M., FOIS, E., JADOUL, F., AND NICORA, A., 1981, Nature and evolution of Middle Triassic carbonate buildups in the Dolomites (Italy): Marine Geology, v. 44, p. 25–57.

GOLDHAMMER, R., 1987, Platform carbonate cycles, Middle Triassic of northern Italy: The interplay of local tectonics and global eustasy: Unpublished Ph.D. Dissertation, Johns Hopkins University, Baltimore, 468 p.

———, DUNN, P. A., AND HARDIE, L. A., 1987, High-frequency glacioeustatic oscillations with Milankovitch characteristics recorded in northern Italy: American Journal of Science, v. 287, p. 853–892.

HALLAM, A., 1984, Pre-Quaternary sea-level changes: Annual Review of Earth and Planetary Sciences, v. 12, p. 205–243.

HANCOCK, J. M., AND KAUFFMAN, E. G., 1979, The great transgressions of the Late Cretaceous: Journal of the Geologic Society of London, v. 136, p. 175–186.

HAQ, B. U., HARDENBOL, J., AND VAIL, P. R., 1987, Chronology of fluctuating sea levels since the Triassic (250 million years ago to present): Science, v. 235, p. 1156–1167.

HARDIE, L. A., 1977, Sedimentation on the modern carbonate tidal flats of northwest Andros Island, Bahamas: Johns Hopkins University Press, Baltimore, 202 p.

———, BOSELLINI, A., AND GOLDHAMMER, R., 1986, Repeated subaerial exposure of subtidal carbonate platforms, Triassic, northern Italy: Evidence for high frequency sea level oscillations an a 10^4 year scale: Paleoceanography, v. 1, p. 447–457.

HAYS, J. D., IMBRIE, J., AND SHACKLETON, N. J., 1976, Variation in the earth's orbit: Pacemaker of the ice ages: Science, v. 194, p. 1121–1132.

HINE, A. C., AND STEINMETZ, 1984, Cay Sal Bank, Bahamas–A partially drowned carbonate platform: Marine Geology, v. 59, p. 135–164.

IMBRIE, J., HAYS, J. D., MARINSON, D. G., MCINTYRE, A., MIX, C. C., MORLEY, J. J., PISIAS, N. G., PRELL, W. L., AND SHACKLETON, N. J., 1984, The orbital theory of Pleistocene climate: Support from a revised chronology of the marine $\delta^{18}O$ record, *in* Berger, A., Imbrie, J., Hayes, J., Kukla, G., and Saltzman, B., eds. Milankovitch and Climate: Reidel Publishing Company, Boston, p. 269–306.

KENDALL, C. G., ST. C., AND SCHLAGER, W., 1981, Carbonates and relative changes in sea level: Marine Geology, v. 44, p. 181–212.

MESOLELLA, K. J., MATTHEWS, R. K., BROECKER, W. S., AND THURBER, D. L., 1969, The astronomical theory of climatic change: Barbados data: Journal of Geology, v. 77, p. 250–274.

MIALL, A. D., 1984, Principles of sedimentary basin analysis: Springer-Verlag, New York, 490 p.

NEUMANN, A. C., AND LAND, L. S., 1975, Lime mud deposition and calcareous algae in the Bight of Abaco, Bahamas: A budget: Journal of Sedimentary Petrology, v. 45, p. 763–786.

SANDER, B., 1936, Beitrage zur Kenntnis der Anlagerungsgefuge: Mineralogische und Petrographische Mitteilungen, v. 48, p. 27–139.

SCHLAGER, W., 1981, The paradox of drowned reefs and carbonate platforms: Geological Society of America Bulletin, v. 92, p. 197–211.

SHACKLETON, N. J., AND OPDYKE, N. D., 1973, Oxygen isotope and paleomagnetic stratigraphy of equatorial Pacific core V28-238: Oxygen isotope temperatures and ice volumes on a 10^5 and 10^6 year scale: Quaternary Research, v. 3, p. 39–55.

STEINEN, R. P., HARRISON, R. S., AND MATTHEWS, R. K., 1973, Eustatic low stand of sea level between 125,000 and 105,000 B. P.: Evidence from the subsurface of Barbados, West Indies: Geological Society of America Bulletin, v. 84, p. 63–70.

STOCKMAN, K. W., GINSBURG, R. N., AND SHINN, E. A., 1967, The production of lime mud by algae in south Florida: Journal of Sedimentary Petrology, v. 37, p. 633–648.

VAIL, P. R., MITCHUM, R. M., AND THOMPSON, S., 1977, Seismic stratigraphy and global changes of sea level, Part 4. Global cycles of relative changes of sea level, *in* Payton, C. E., ed., Seismic Stratigraphy–Applications to Hydrocarbon Exploration: American Association of Petroleum Geologists Memoir 26, p. 83–97.

WINTERER, E. L., AND BOSELLINI, A., 1981, Subsidence and sedimentation on a Jurassic passive continental margin, Southern Alps, Italy: American Association of Petroleum Geologists Bulletin, v. 65, p. 394–421.

CENOZOIC PROGRADATION OF NORTHWESTERN GREAT BAHAMA BANK, A RECORD OF LATERAL PLATFORM GROWTH AND SEA-LEVEL FLUCTUATIONS

GREGOR P. EBERLI
Geological Institute ETH, Sonneggstrasse 5, CH-8092, Zürich, Switzerland

AND

ROBERT N. GINSBURG
Comparative Sedimentology Laboratory, University of Miami, Fisher Island Station, Miami Beach, Florida 33139

ABSTRACT: Seismic profiles across the top of northwestern Great Bahama Bank reveal that the modern bank is formed by the coalescence of three smaller platforms. This coalescence resulted from progradation of the bank margin during the Cenozoic. Since the Late Cretaceous, vertical aggradation on the banks has been approximately 1,500 m, whereas the leeward bank margin has migrated as much as 25 km, indicating that lateral growth can dominate the growth direction of a healthy platform with the capability of transporting offbank excess sediment. In addition, asymmetric progradation of the leeward margin indicates the important role of physical energy on the direction of platform expansion. A phase of filling preceded the progradation and led to shallowing of the intraplatform seaways and reduction of their slope height and declivity. Progradation began when slopes were approximately 500 m high and/or had angles less than 5°.

The pattern and direction of progradation is governed by prevailing currents and changes in sea level. Progradation occurred in pulses that are recognized in the seismic lines as discrete sequences of sigmoidal clinoforms. Each prograding sequence is interpreted as the record of a relative rise and highstand of sea level, when sediment production on the bank top was high and cross-bank currents moved excess sediment offbank. Repeated fluctuation of sea level resulted in a series of horizontally stacked sigmoidal sequences that form two distinct bundles of sequences.

Correlation with the Great Isaac exploration well suggests that progradation began in the late Oligocene in the Straits of Andros, a buried intraplatform seaway, and in the late Miocene in the Straits of Florida. If these age assignments are correct, the two bundles of sequences coincide with the youngest second-order cycles TB2 and TB3 of the sequence stratigraphy curve of Haq and others (1987). The number of sequences within the bundles nearly matches the number of third-order cycles in the curve of Haq and others (1987). Furthermore, the shape of the onlap curve in the Bahamas is similar to that of the curve determined in siliciclastic terrains. Assuming that the correlation between the Bahamas curve and the global curve is correct, we propose specific ages for the sequence boundaries and a demonstration of the potential of platform carbonates to provide a legible record of the relative sea-level fluctuations of a second- and third-order magnitude.

INTRODUCTION

Progradation is known from shelves throughout geologic time (Wilson, 1975). For example, in the Devonian Canning Basin in Australia, 200-m-thick, steep forereef deposits can be mapped horizontally for some kilometers (Playford and Lowry, 1966), or in the Permian Reef Complex in Texas, massive carbonates 300 to 400 m thick have prograded basinward for 20 km across the shelf (Van Siclen, 1958; Silver and Todd, 1969; Sarg and Lehmann, 1986). In contrast, the growth pattern of large isolated carbonate platforms was thought to be predominantly vertical with sedimentation keeping pace with subsidence (Wilson, 1975). Prograding margins along isolated platforms are reported only from the Triassic platforms of the Southern Alps (Bosellini, 1984), an indication that isolated platforms prograde in a way similar to that of attached platforms (shelves; Van Siclen, 1958; Playford and Lowry, 1966; Meissner, 1972).

In the Bahamas, results from several investigations produced evidence that these isolated platforms also expand laterally, despite the fact that their steep and high slopes are more reminiscent of vertically growing atolls than prograding shelves (Newell, 1955; Paulus, 1972; Beach and Ginsburg, 1980). First indications of lateral growth came from single-channel seismic profiles along the platform edge that documented a sequence approximately 10 to 15 m thick of prograding Holocene sediment, the result of offbank transport on the leeward side of the platform (Hine and Neumann, 1977; Mullins and Neumann, 1979; Hine and others, 1981). High-resolution seismic-reflection profiles off the corner of both the Great and Little Bahama Banks revealed the accretion of large (as thick as 600 m) hemiconical sediment drifts on the lee side of the present oceanic circulation pattern (Mullins and others, 1980). More recently, the results from ODP Leg 101 indicate the retreat and re-expansion of both the Little and Great Bahama Banks for some tens of kilometers since the mid-Cretaceous (Austin, and others, 1986).

Multichannel seismic profiles across the top of northwestern Great Bahama Bank have documented progradation so extensive that, during the Cenozoic, it led to the coalescence of initially smaller platforms (Eberli and Ginsburg, 1987). The seismic profiles also offer the special opportunity to examine geometry and extent of the progradation, as well as to determine the relative importance of aggradation versus progradation, in order to characterize the dynamics of the platform growth and its controlling factors. In addition, by comparing the Bahamian sequences with those of the sequence stratigraphy developed from siliciclastic sediments (Haq and others, 1987), we are able to date the pulses of progradation and to suggest that platform carbonates may have a complete and legible record of small-scale sea-level fluctuations.

METHODS

The data set consists of approximately 700 km of mostly unmigrated, multichannel seismic profiles shot by Western Geophysical (Fig. 1). In order to obtain best results, parameters were adjusted several times. Ten to 12 airguns with variable volumes (720-2280 in.^{3}) were used; shotpoint interval was 25 m and occasionally 50 m. A "drag yo-yo" cable allowed the geophones to be lowered between each

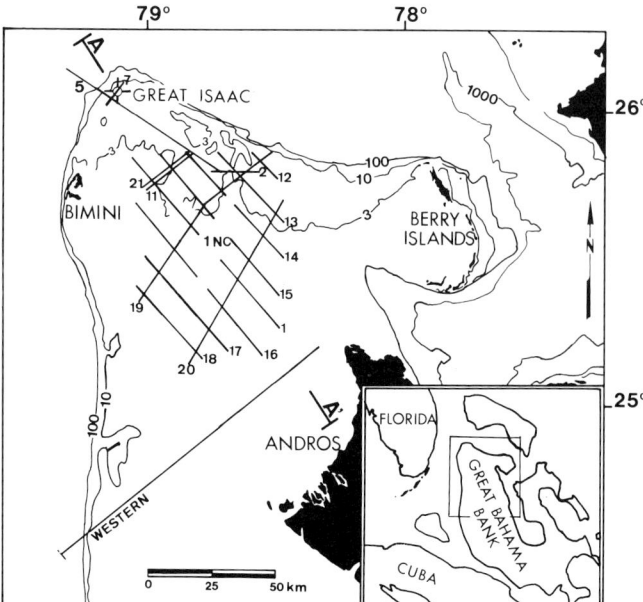

FIG. 1.—Location map of the multichannel seismic profiles. Grid of profiles is connected with exploratory well Great Isaac-1 at the northwest corner of Great Bahama Bank. A-A' = cross section shown in Figure 2.

shot point on the sediment surface, as well as constant movement of the boat which was crucial for navigation. The processing sampling interval of the incoming signal was 2 ms; a F-K filter and a time-variant filter were used and deconvolution applied before stacking. For this study, only the top 1.1 s (two-way travel time) of the grid of profiles and the top 1.7 s of the cross-bank profile (WESTERN, Fig. 1) south of the grid were accessible. The grid of profiles is connected with the Great Isaac-1 exploration well (Fig. 1), for which there are biostratigraphic age determinations (see Schlager and others, 1988) and a profile of velocities necessary to calculate depths.

INTERNAL STRUCTURE OF NORTHWESTERN GREAT BAHAMA BANK

The internal structure of northwestern Great Bahama Bank seen in seismic profiles shows that the present bank developed in three stages by the combination of segmentation and coalescence of smaller platforms (Eberli and Ginsburg, 1987). First, a segmentation event in the mid-Cretaceous produced two separate banks, Andros Bank to the east and Bimini Bank to the west, separated by a fault-bounded, north-south-trending seaway, 25 km wide and 1,500 m deep, termed the Straits of Andros (Fig. 2). Platform growth started above what is interpreted as a fault-controlled topography (Eberli and Ginsburg, 1987). In mid-Tertiary time, growth was interrupted by segmentation of the Bimini Bank, by which the Bimini Embayment formed. The segmentation was probably the combined result of large-scale folding and differential sedimentation (Eberli and Ginsburg, 1987). The Bimini Embayment, –a blind-ended seaway– was also oriented North-South and had a maximum width of approximately 10 km and an initial depth of 450 m (Fig. 2). During the second phase of growth, the intraplatform seaways were closed; the western margin of the Bimini Bank and the Bimini Embayment were filled by prograding sequences and the Straits of Andros by sub-horizontally layered deposits. The final phase is seen as a thin sequence of horizontally layered deposits covering all preceding sequences.

VERTICAL AND LATERAL GROWTH

Vertical growth of a platform is controlled by the rate of subsidence and the stand of sea level (Kendall and Schlager, 1981). Lateral growth or progradation of platforms requires production of excess sediment and energy to transport these sediments over the platform edge. From the seismic profiles across northwestern Great Bahama Bank, it is estimated that approximately 40 percent of the present platform is built by sediments transported offbank and deposited in intraplatform seaways. The large amount of offbank sediments resulted in extensive lateral growth and finally closure of the seaways.

Lateral growth, which is the combined result of infilling of the seaways and progradation of the platform edge, outpaces vertical growth by more than an order of magnitude. Migration of the position of the platform edge displays best the amount of the two growth vectors of the platforms (Fig. 3). Since formation of the platform margins in the mid-Cretaceous (Austin and others, 1986; Eberli and Ginsburg, 1987), the platform edge has grown upward approximately 1,500 m, whereas the horizontal component is approximately 15 km in the Straits of Andros and more than 25 km in the Straits of Florida (Fig. 3). Horizontal movement of the platform edge, however, occurs only at the western leeward side of the platforms, where prograding sequences have built the margins out into the adjacent seaways. Along these leeward margins, lateral growth outpaces vertical growth between 10 and 17 times, considering the whole time interval from the mid-Cretaceous to Recent (Figs. 3, 4). If only the time interval of progradation is taken into account, the ratio is much higher, i.e., approximately 50:1, and on the low-angle slope into the Straits of Florida, the ratio is as high a 80:1 (Fig. 3).

The large amount of lateral growth documents the impressive growth potential of a "healthy" platform on which currents transport excess sediment offbank. In the northern Bahamian region, easterly trade winds produce a net east-to-west energy flux (Hine and Neumann, 1977; Hine and others, 1981), which is considered to be responsible for the extreme asymmetric platform expansion with a near-vertical growing windward margin and a prograding leeward margin (Figs. 3, 5).

INFILLING OF THE INTRAPLATFORM SEAWAYS

The intraplatform deposits accumulated in two major stages: aggradation in the depressions, followed by successive pulses of progradation, both of which proceeded from east to west (Eberli and Ginsburg, 1988).

The aggrading phase transformed a relatively steep, west-facing, leeward margin into a low-angle acretionary slope that became the base of the subsequent prograding sequences. First, the initially asymmetric, fault-bounded

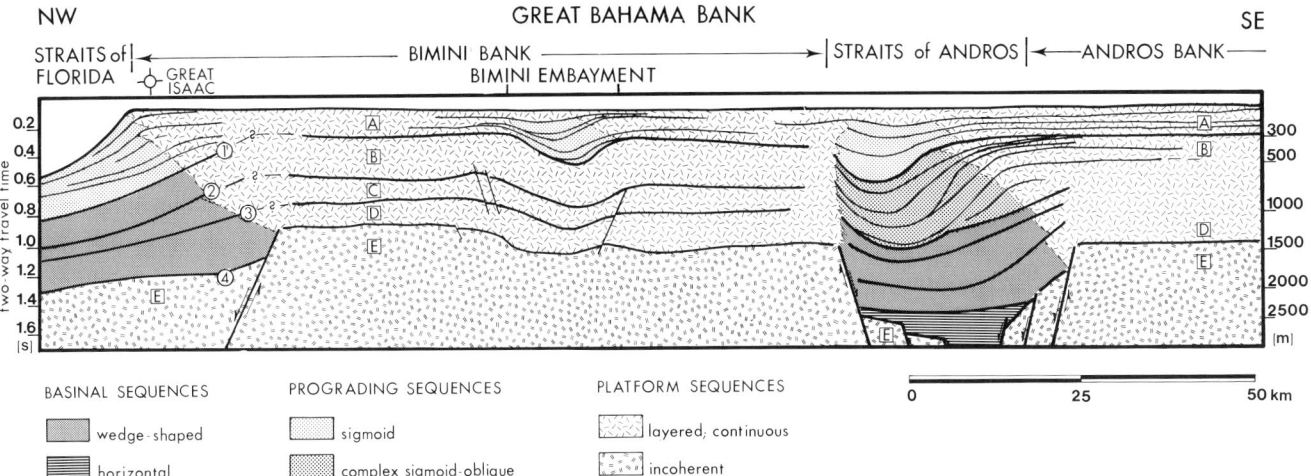

FIG. 2.—Schematic cross section over northwestern Great Bahama Bank displaying intraplatform seaways and nuclear platforms. Circled numbers = reflections identified and correlated with biostratigraphic data from the Great Isaac-1 well by Schlager and others (1988). Capital letters = megasequences defined by Eberli and Ginsburg (1987). For location, see Figure 1.

depressions were leveled by a prism of continuous reflections forming a sub-horizontal basin floor (Figs. 3, 4; Eberli and Ginsburg, 1988). In the Straits of Andros, a second wedge with mostly incoherent reflections overlies the first wedge and ends with a slope of approximately 10° on the east side of the Straits (Fig. 4). In the next overlying unit, the slope is less inclined, and it is this slope that forms the base of the prograding sequences (Fig. 4). A similar style of basinal aggradation is seen along the western margin of the Bimini Bank (Fig. 3). As in the Straits of Andros, the slope angle of the aggrading slope decreases and finally is overlain by prograding sequences. The jump correlation from

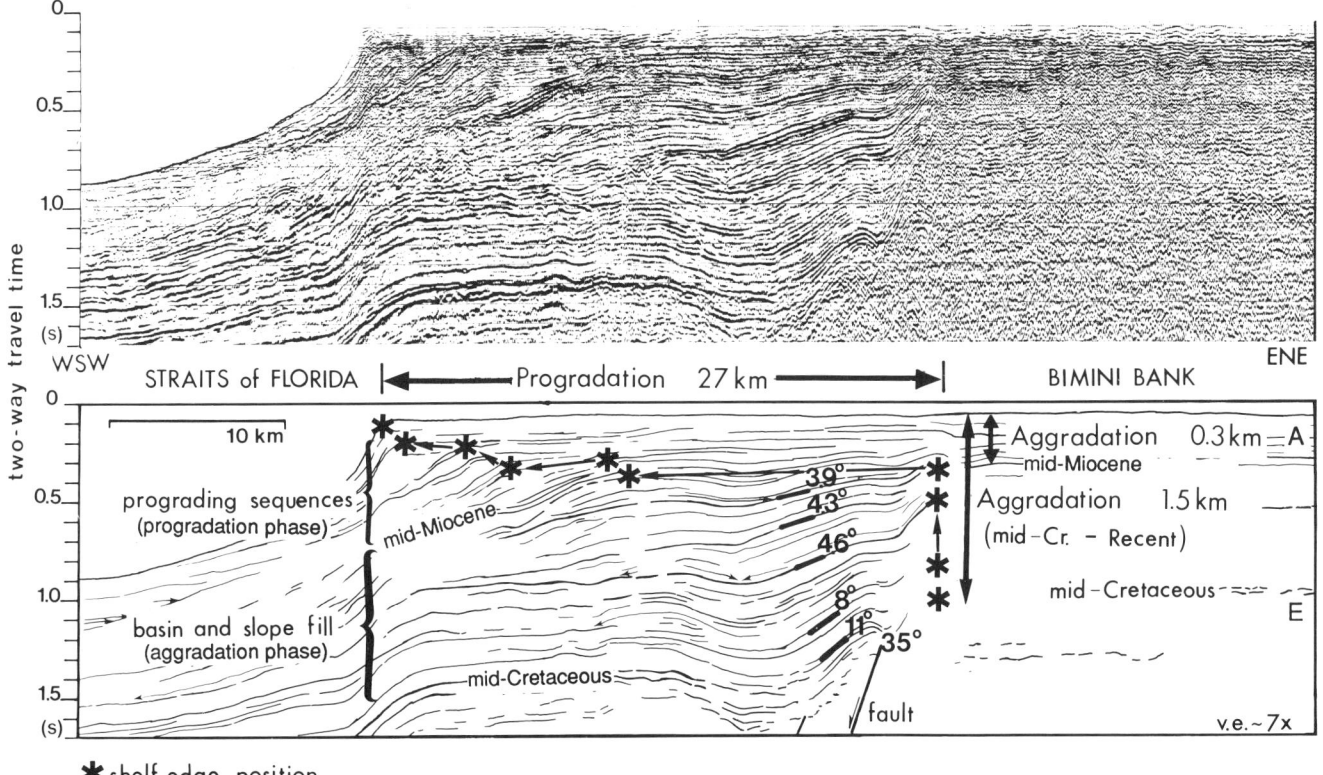

FIG. 3.—West margin of Bimini Bank documenting extensive lateral platform growth. Evolution of the slope shows two stages: a basinal aggradation phase with basin-and-slope fill sequences, followed by a progradation phase. Note decreasing slope angle during infilling phase (mid-Cretaceous to mid-Miocene). During progradation phase, shelf edge (stars) prograded 27 km to the west, while aggradation on the Bimini Bank was approximately 0.3 km. Total aggradation on the bank since the mid-Cretaceous was approximately 1.5 km. West end of WESTERN line (Fig. 1).

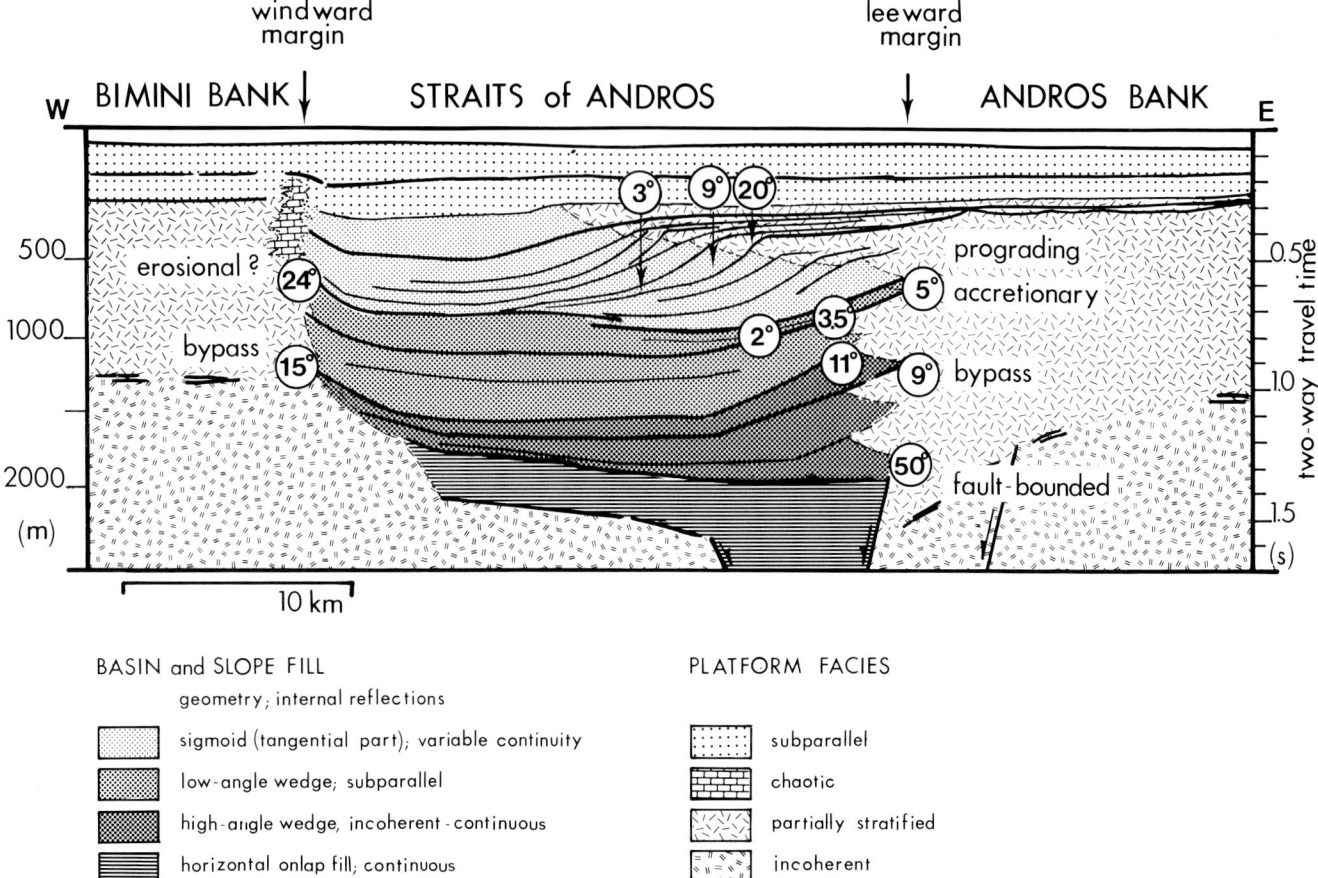

Fig. 4.—Schematic cross section through Straits of Andros. Preceding progradation phase, sediment wedges fill the seaway. On the leeward margin, decreasing slope angles indicate a slope evolution from a fault-bounded to a bypass to an accretionary slope and finally to a prograding slope.

the western margin of the Bimini Bank into the Straits of Andros indicates, however, that progradation occurred at different times in the two straits (see below). In the Straits of Andros, progradation started earlier and had nearly closed the seaway when it began at the western margin of Bimini Bank (Fig. 2; see also Fig. 10). Nevertheless, in both straits, the depositional geometry before progradation displays a similar evolution. The slope of the leeward margin evolved from a steep, fault-bounded margin into a bypass and later into a low-angle accretionary slope, which then formed the base of the prograding sequences (Fig. 3, 4).

THE PROGRADATION

The progradation that follows the phase of infilling is the more important phase in the growth of the platform, because it leads to closure of the two seaways and expansion of the bank to its present size. The east-to-west progradation (Fig. 5) occurs in pulses, which are recognized in the seismic record as discrete sequences. Each sequence is generally sigmoidal and has approximately 450 m of shelf-to-basin relief. Individual sequence boundaries can be defined by downlap and onlap of the reflections on the underlying sequence and occasionally by toplap on the upper boundary. Progradation occurs in three sites: (1) a composite of eight successive sequences in the Straits of Andros that is approximately 500 m thick and 15 km wide; (2) a similar composite in the Bimini Embayment that is 475 m high and 10 km wide; and (3) at the western margin of the Bimini Bank, where a wedge of eight prograding sequences advanced the platform edge more than 25 km into the Straits of Florida (Fig. 5).

Conditions Necessary for Progradation

Progradation depends on overproduction of sediment, transporting currents, and the proper seafloor profile. In the Straits of Andros and the western margin of the Bimini Bank, prograding sequences overlie a low-angle slope built during the aggradation phase. In the Straits of Andros, prograding sequences develop only after the seaway was filled to about 500 m. In the Bimini Embayment that was never deeper than 475 m, there is no basinal aggradation and the filling is dominated by progradation. In these two depressions progradation began when the relief was below 500 m and, as long as the western margin of the Bimini Bank, when the slope angle was less than 5°. The hypothesis that the seafloor profile is a controlling factor for progradation is sup-

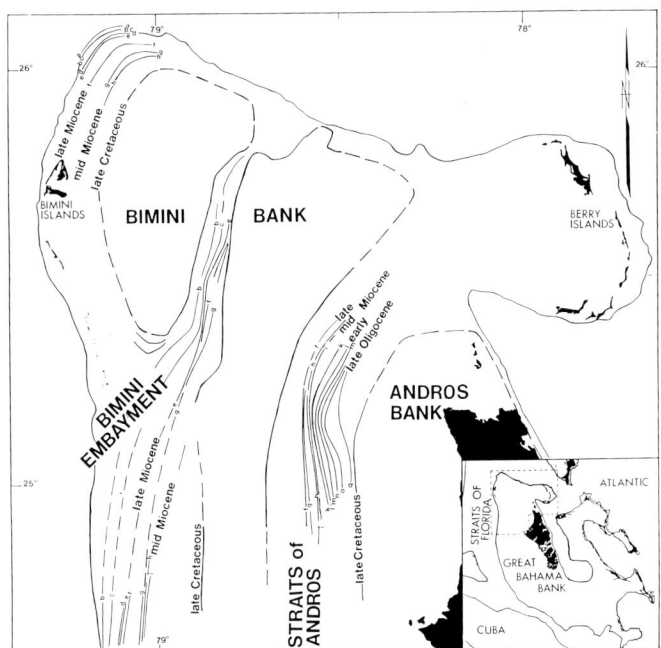

FIG. 5.—Plain view of the prograding margins of northwestern Great Bahama Bank since the mid-Cretaceous. Progressive westward shift of the margins results in a narrowing and shallowing of the seaways and ultimately their closure. The east-west direction of progradation coincides with prevailing wind direction in the area and is therefore thought to be the result of preferred offbank sediment transport on the western leeward margins (Hine and Neumann, 1977).

ported by the fact that progradation is of different ages in the Straits of Florida and the Straits of Andros. The necessary conditions for progradation were achieved earlier in the Straits of Andros.

The Types of Prograding Sequences

Two major types of sigmoids are distinguished on the basis of their slopes and their internal reflections.

Simple sigmoids.—

Sequences of this type are composed of sigmoids characterized by simple internal reflection pattern with a horizontal or very low-angle topset and a sigmoid front from 1° and 6°. In the Bimini Embayment, the internal reflections are continuous, especially in the sigmoid front parallel to the sequence boundary (Fig. 6), but toward the topset the reflections occasionally fade out into an incoherent zone.

Sequences with simple sigmoids of this type are best developed in the Bimini Embayment, where they retain their thickness and geometry laterally for some 50 km, i.e., for most of the recorded length of the embayment. Simple prograding sigmoids are also recognized in the western margin of the Bimini Bank, but in comparison to those in the embayment, they have a somewhat lower foreset angle and often an incoherent sigmoid edge.

Complex oblique sigmoids.—

This type of sequence is characterized by complicated internal reflection pattern with steep reflections (up to 20°)

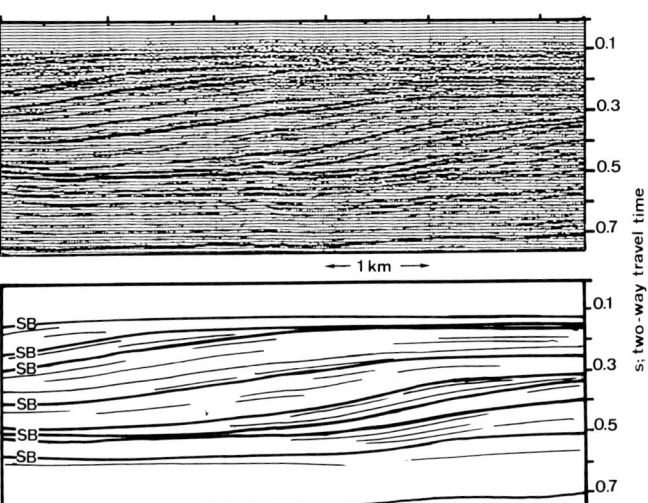

FIG. 6.—Prograding sequence geometry, type 1: simple sigmoids characterized by low-angle foreset and sub-parallel internal reflections. Line 1-N-C, Bimini Embayment.

that fade out into a reflection-free area at the edge of the sigmoid that interrupts equally both the sequence boundary and the internal reflections (Fig. 7). In the back of the reflection-free area, a series of high-amplitude reflections parallels the upper boundary of the sequence. In the bottomset, a high-amplitude reflection parallel to the top of the underlying sequence occasionally forms the basal unit of the sequence. The basal unit acts as a downlap surface for the reflections running basinward from the reflection-free area (Fig. 7). In most of these complex sequences, an onlap can be recognized in the foreset of the underlying sequence climbing progressively higher and ultimately forming the aggrading part of the new sequence.

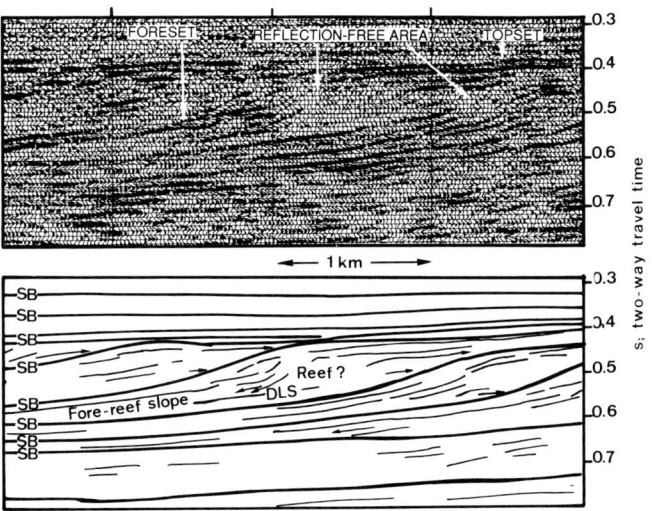

FIG. 7.—Prograding sequence geometry, type 2: complex sigmoid-oblique sequences characterized by a reflection-free area at the sigmoid edge, steep high-amplitude reflections seaward, and flat continuous reflections shelfward in the topset. WESTERN line, Straits of Andros.

Seismic Facies

The geometry and the internal reflections of the sequences suggest that the sediment deposited in the sequences is mainly derived from the platforms. This interpretation is in accordance with the results from the Great Isaac-1 exploratory well and the DSDP and ODP drill holes in the region (Hollister and others, 1972; Meyerhoff and Hatten, 1974; Austin and others, 1986). The inclined reflections are assumed to be a mixture and alternation of periplatform ooze and downslope deposits from the platform margin, slopes, and possibly the platform interior (Mullins and others, 1984; Austin and others, 1986). The chaotic to reflection-free seismic signal at the edge of the complex sigmoid is indicative of a reefal buildup (Bubb and Hattlelid, 1977). Along modern leeward margins, however, reef growth is limited, as many of the early Holocene reefs are inundated by sediment transported offbank once the adjacent platform was completely flooded (Hine and Neumann, 1977; Hine annd Mullins, 1983). The reflection-free area may therefore not be a massive reef complex, but either a rim of skeletal sediments or a series of reefs overlain by carbonate sand. The steep reflection segments basinward of the reflection-free area may represent the forereef area with a mixture of talus blocks, reefal debris, and pelagic material. The well-developed sub-horizontal reflections in the topset might consist of lagoonal sediments.

Development of the two different types of sigmoids seems to be dependent upon the occurrence of a skeletal buildup that, in turn, is related to energy flux (Hine and others, 1981). In the Bimini Embayment, continuous reflections in the sigmoidal sequences suggest that the prograding margin did not have extensive reefs, but instead there was a continuous slope of platform-derived sediments extending into the basin. These sigmoid sequences with a simple internal geometry may indicate a lower energy environment (Mitchum and others, 1977), which is in concert with their more protected position within the bank. The Straits of Andros, an open seaway with oceanic circulation, were probably a higher energy environment.

Influence of Sea Level on Development of the Sequence

Two features within the sequences indicate their relation to the relative position of sea level. The progressive onlap over the underlying sequence is considered to be the result of a relative rise of sea level, and the progressive downlap onto the basal unit is believed to record progradation during a highstand of sea level. Therefore, it is proposed that each sequence was formed mainly during a period of sea-level rise and highstand, and the boundaries between the sequences record a lowstand of sea level with little or no progradation. The interpretation that progradation occurs during a relative sea-level highstand is supported by the presence of nascent sigmoidal sequences in the Holocene bank margin (Hine and Neumann, 1977; Palmer, 1979) and the increased sediment production during sea-level highstands in the carbonate environment (Kendall and Schlager, 1981; Mullins, 1983; Droxler and Schlager, 1985).

A proposed scenario for the development of sigmoidal sequences on leeward margins is show in Figure 8. Figure

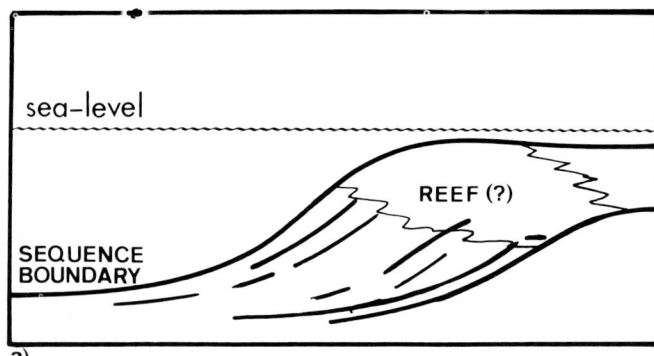

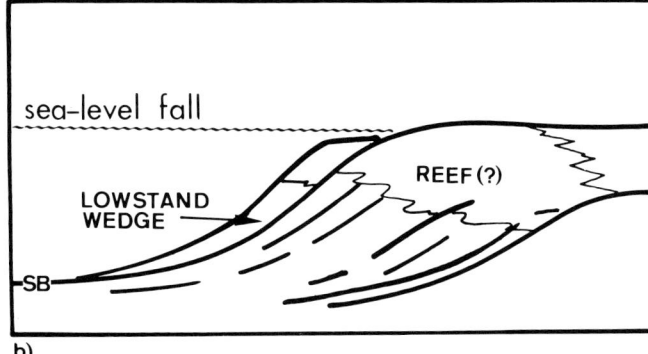

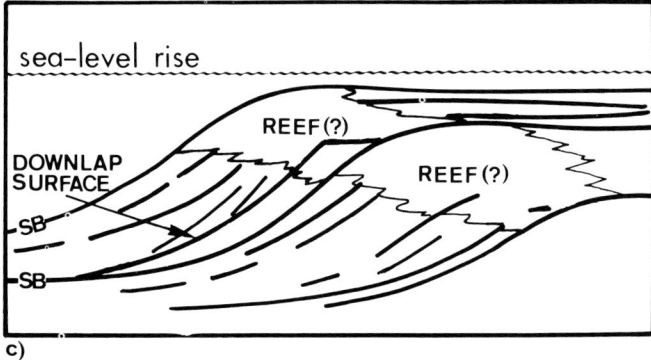

Fig. 8.—Schematic evolution of the individual sequences and the proposed relation to fluctuations in sea level. (a) A reef-rimmed(?) margin is established during a highstand of sea level. (b) A fall of sea level shuts off sediment production on the bank and reduces growth of the marginal buildup. As a result, sediment input into adjacent deep-water area is minimal and only a thin lowstand wedge accumulates. (c) A rise of sea level reestablishes sediment production on the platform, and the resultant offbank transport of sediment progressively covers the evolving reef.

8a depicts a reef or shoal-fringed bank margin that has built to sea level. Figure 8b illustrates the result of a fall in sea level that shuts off sediment production on the bank and reduces reef growth to that of a fringing-reef situation. We believe that this period of reduced progradation is represented in some of the sequences by the thin basal unit with a single reflection. Subsequently, a rise of sea level provides room for the buildup to grow upward and, when the banktop is flooded, for producing excess sediment. When offbank transport is resumed, most of the reefs on the leeward margin are buried by carbonate sand (Hine and Neumann, 1977). During this rise, shorter term, lower amplitude sea-level fluctuations of a higher order are visualized to be the mechanism by which the reefs thrive basinward on the prograding margin. A short-term sea-level fall may be sufficient to reduce offbank sediment transport for the reestablishment of reef growth on the new platform edge before the reefs are buried again during the next low-amplitude rise. In the profiles, the horizontal reflections over the reflection-free area record the smothering of the reefs and their basinward shift. The short-term cyclity producing the shift is not clear in the multichannel data, but the interpretation is supported by drill hole data long the western margin of the Great Bahama Bank, where a sequence of reefal debris intercalated with carbonate sand was discovered (Beach and Ginsburg, 1980).

SEQUENCE STRATIGRAPHY

If, as proposed above, a relative rise of sea level leads to the development of individual sigmoids and a fall of sea level separates successive sigmoids, then the succession can be used to generate onlap/offlap curves for the northwestern Great Bahama Bank. These curves, for which the biostratigraphic data from the Great Isaac well provide some age brackets, can be compared with the existing global curve derived from the siliciclastic environment (Vail and others, 1977; Haq and others, 1987). The correlation is used to test the application of the carbonate environment for refined sequence stratigraphy and, if successful, to date precisely the seismically defined sequences.

Age Correlations

Using standard seismic stratigraphic techniques, four megasequences were defined within the Bimini Bank (Eberli and Ginsburg, 1987). The term megasequence is used here to describe a depositional unit that, on the platform, appears as one seismic sequence, but in the basinal area is split into a bundle of sigmoidal sequences. The prograding sequences of the Bimini Embayment and the western margin of the Bimini Bank are within the uppermost megasequence A (Fig. 2). In the Straits of Andros, the bundle of prograding sequences lies under the A/B boundary and was probably deposited during the aggradation of megasequence B (Fig. 2).

Ages of the megasequences can be estimated by correlating them with dated horizons in the only exploratory well in the area, Great Isaac-1. Schlager and others (1988) examined cuttings from the top 2,100 m of the well, identified four biostratigraphic boundaries within the Cretaceous and Tertiary section, and correlated them to the seismic lines (Fig. 9b). There are some uncertainties in using these horizons to date the megasequences defined by Eberli and Ginsburg (1987), because the location of the Great Isaac well is outside the former Bimini Bank, and the seismically incoherent edge of the bank makes it necessary to jump correlate over approximately 12 km.

Important for dating the prograding sequences are the ages of megasequences A and B and their respective boundaries. The youngest age reported in the well is middle Miocene (*Lepidocyclina, sp., Orbulina universa*) at a depth of 800 m, just above "reflector 1" of Schlager and others (1988). Schlager and others (1988) carried this reflector to the dip line (5A–N; Fig. 9a), where they identified the overlying reflector (1') as a sequence boundary. This sequence boundary probably coincides with the A/B boundary of Eberli and Ginsburg (1987) and is believed to be the boundary between the Middle and Upper Miocene section (Fig. 9). Schlager and others (1988) place the next older age boundary (Oligocene/Miocene) at approximately 1,200 m and correlate it with "reflector 2" at the top of a series of high-amplitude reflections. A similar transition of the seismic-reflection characteristics suggests that "Reflector 2" coincides with the B/C boundary.

Unfortunately, the key profile of this study (WESTERN, Fig. 1) is not connected to the grid of profiles, and a correlation with the grid is based on jump correlation (Fig. 9d). Within the WESTERN line itself, correlation between the west margin of the Bimini Bank and the Straits of Andros can be achieved by a simple foldover, which produces a good fit of the reflections (Fig. 10). Eight prograding sequences, of which the last one forms the modern platform edge, are recognized above the A/B megasequence boundary. Below the boundary, again eight sequences form another prograding system. The base of the older system is not dated, but the reflector marking the top of the Oligocene ("reflector 2", B/C boundary) lies at the top of the first (or second?) sequence of the system. The system therefore includes part of megasequence C and started in the Oligocene. We speculate that the base of the system is of Mid-Oligocene age, because the onlap pattern suggests that progradation began after a sea-level fall, which might coincide with the major sea-level fall in the mid-Oligocene recorded on the continental margins (Vail and others, 1977; Miller and others, 1985) and in the mid-Pacific (Schlanger and Premoli Silva, 1986).

If these age assignments and the correlation from the western margin into the Straits of Andros are correct, a continuous record of progradation from the mid-Oligocene to the Recent is displayed in the WESTERN line (Fig. 10).

Sequence Analysis

The three horizons dated by correlation with those established by biostratigraphy of Great Isaac-1 well and the seismic data (Schlager and others, 1988) provide a time frame for the sequence analysis. Within the 30 m.y. from the beginning of progradation at the mid-Oligocene to the present, the mid-Miocene horizon is the major unconformity separating the 16 sequences in pre- and post-mid-Miocene sequences (Fig. 11 a, b).

The pre-mid-Miocene sequences (i–q).—

Prograding sequences deposited before the end of the mid-Miocene, i.e., below the A/B boundary, are found only in the Straits of Andros where progradation probably started in the late Oligocene. In this time period, eight sequences advanced the eastern margin for more than 10 km to the west. In the profiles, the three lowest sequences are of simple sigmoidal geometry and are followed by complex sigmoid-oblique sequences (Fig. 11b). The first of these com-

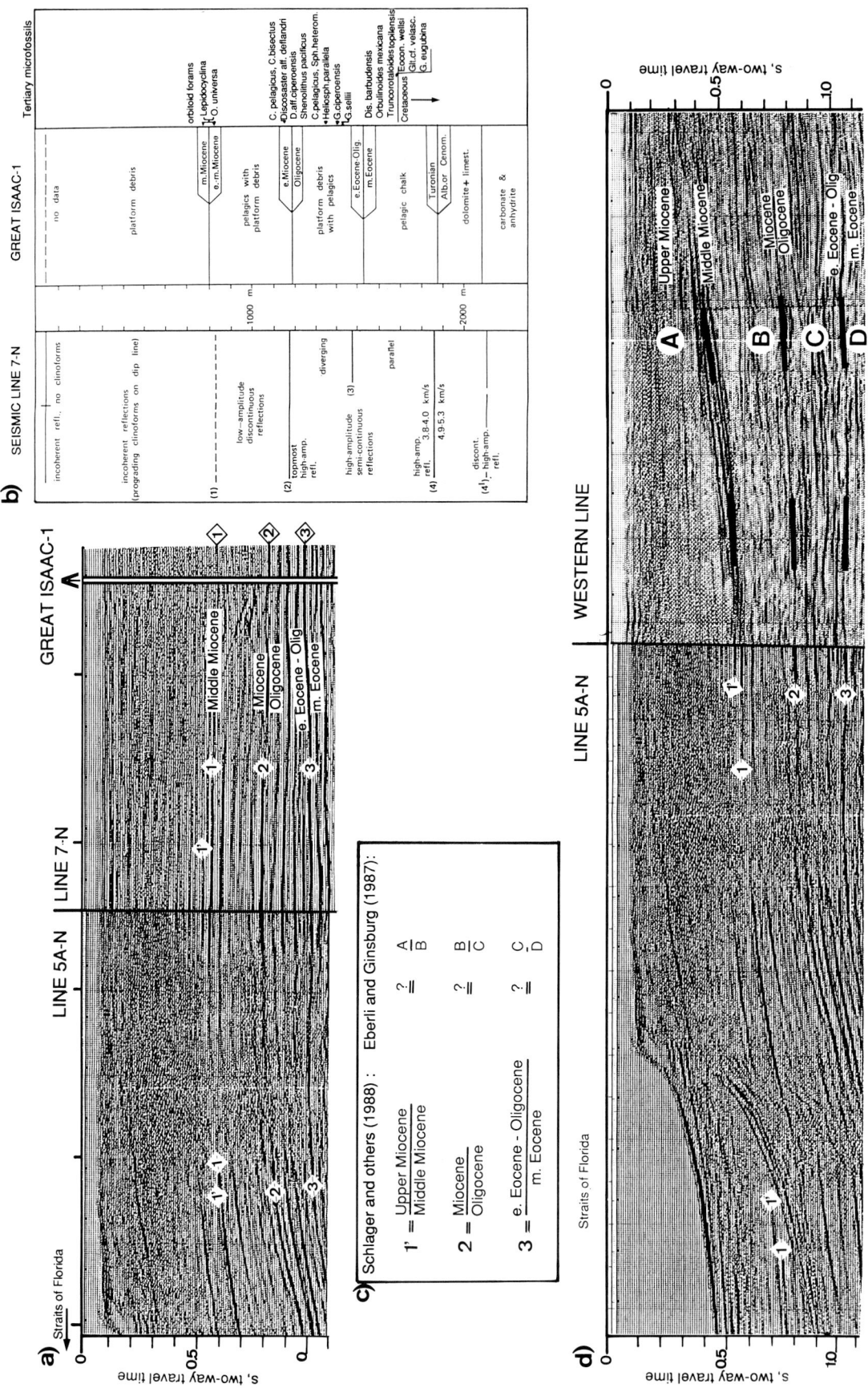

FIG. 9.—Correlation of the Great Isaac well data with the seismic lines. (a) Correlation between line 7-N and line 5A-N crossing bank margin southwest of Great Isaac-1. Numbers indicate reflectors identified on line 7-N by Schlager and others (1988). (b) Comparison of seismic reflection character, lithology, and biostratigraphy at Great Isaac-1 well; modified after Schlager and others (1988). Depth of seismic reflectors was calculated with Dix average and internal velocities (Schlager and others, 1988). (c) Possible correlation of sequence boundaries defined by Schlager and others (1988) and Eberli and Ginsburg (1987). (d) Jump correlation between line 5A-N, which runs near the Great Isaac-1 well into the Straits of Florida, and part of the WESTERN line crossing the platform margin farther south. For location of profiles, see Figure 1.

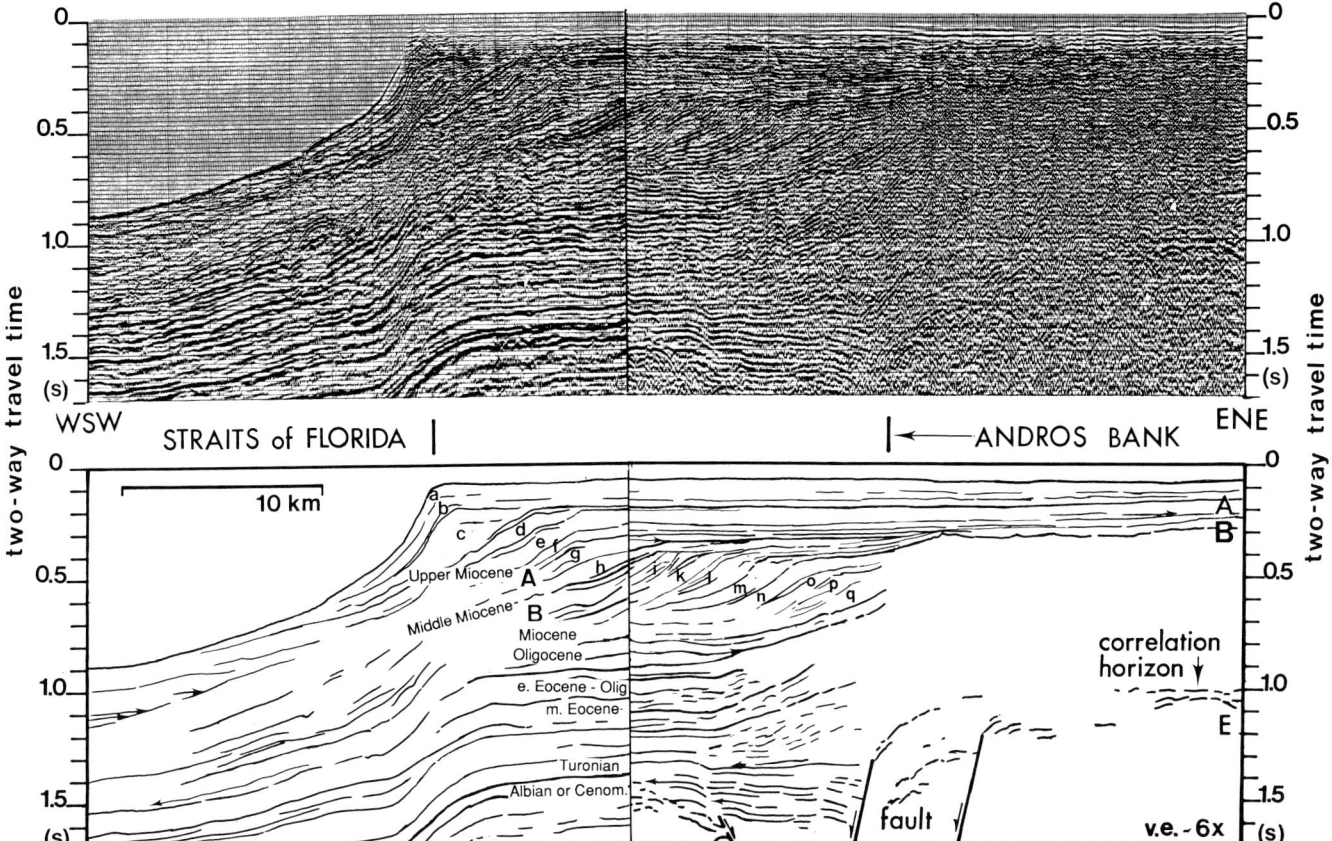

FIG. 10.—Jump correlation by foldover within WESTERN line from the west margin of the Bimini Bank into the Straits of Andros. High-amplitude reflection horizon of presumably mid-Miocene age is the prominent correlation horizon. Correlation is supported by the good match of many sequence boundaries. The two profile sections combined result in a complete record of progradation from the mid-Oligocene to Recent.

plex sequences overlies the Andros Bank and creates a new bank top, which is seen as a prominent reflector in the seismic line (Fig. 11b). The new bank top is draped by the thin shelf members of the next two sigmoids. The two youngest sequences of the system, characterized by toplap, do not reach the new surface.

The post-mid-Miocene sequences (a–h).—

Prograding sequences overlying a strong reflector, which is believed to be the top of the mid-Miocene, are seen in all three seaways. They are best developed along the western margin of the Bimini Bank (Fig. 11b) and within the Bimini Embayment. In the Straits of Andros, only the four oldest sequences (e–h) of the system are recognized. Within the Bimini Embayment, seven sequences can be distinguished. They are characterized by simple sigmoid geometry with progressively lower slope angles as the embayment gets filled (Fig. 6). Eight sequences have built the leeward margin of the Bimini Bank to its present position at the western edge of the Great Bahama Bank. The onlap pattern of the younger system is characterized by a alternation of sequences with an aggrading shelf member and sequences that only onlap the preceding sequence. For example, the first sequence (h) onlaps the mid-Miocene surface, whereas the shelf member of the following sequence (g) overlies the platforms. Sequence e displays the thickest shelf member of all sequences of this post-mid-Miocene system, indicating at least 130 m of platform aggradation. The newly formed platform top is onlapped by a thin and restricted sequence (d) and then overlain by the three youngest sequences (a, b, c; Fig. 11b).

The Onlap/Offlap Curve

The onlap/offlap pattern of the sequences is displayed in a chart of relative onlap (Fig. 12) constructed as described by Vail and others (1977). Constructing an onlap chart on a detached carbonate platform such as the Great Bahama Bank is a problem, however. There are times when the entire platform is flooded and, because in this setting no coastline exists, no onlap is seen. During such times, aggradation occurs on the platform, resulting in a new surface. The sequences with a shelf member indicating such aggradation are marked with an "A" in Figure 12.

Comparison on the Onlap/Offlap Curve from the Bahamas with the Global Chart of Coastal Onlap

From the correlation of seismic horizons with dated horizons in the Great Isaac-1 well, we consider that all the prograding sequences were deposited between the mid-Oligocene and the Recent. On the global sequence stratigraphy chart, this time interval is the supercycle set TEJAS B, consisting of three supercycles (second-order cycles); Haq and others, 1987). The second-order cycles as well as third-order cycles therein appear to be present in our seismic data.

The prograding sequences in the Bahamas profiles are divided into two distinct prograding systems (Fig. 10) by a prominent horizon, which is believed to be the top of the mid-Miocene. The bundle of sequences above the horizon would be time equivalent with the youngest second-order cycle (TB3) of Haq and others (1987; Fig. 12). In the Straits of Andros, the prograding sequences that lie below the horizon are probably of mid-Oligocene to mid-Miocene age, thus spanning the same time interval as the two supercycles TB1 and TB2. The older system starts with three simple sigmoidal sequences, followed by a bundle of complex oblique sequences (Fig. 11a). We speculate that the three basal sequences represent cycle TB1, because the Oligocene/Miocene boundary probably lies within the first three sequences, and the continuous landward onlap of the three sequences is consistent with the onlap pattern of TB1 in the global chart (Fig. 12). If this interpretation is correct, the bundle of complex sigmoid-oblique sequences would coincide with the second-order cycle TB2.

The number and onlap pattern of the sequences within the second-order cycles of the global chart suggest that the individual sequences represent third-order cycles. Within both bundles of sequences, which we correlate with TB2 and TB3, the number of sequences nearly matches the number of third-order cycles of Haq and others (1987; Fig. 12). In addition to the correspondence in number of cycles, the similar onlap pattern of the sequences compared with those of the global chart strongly supports the interpretation that the individual sequences are correlative with the third-order cycles (Fig. 12). For example, the supercycle TB2, characterized by progressive landward shift, followed by more basinal restricted onlap (Fig. 12), is seen in the Bahamas profile in sequences with aggradation on the platform (sequence l, m, n; Fig. 11b), followed by the two sequences (i, k) that remain in the seaways and only onlap the platform (Fig. 11b). Similarly, the proposed progressive landward shift at the base of the younger cycle TB3 is observed in our profiles in sequence h through e (Fig. 11b).

On the basis of comparison with the global curve of coastal onlap, we assign tentative ages for the sequence boundaries within the prograding systems of northwestern Great Bahama Bank as shown in Figure 11a and b.

Comparison of the Prograding Sequences with Eustatic Sea-Level Fluctuations

Comparison of the Great Bahama Bank onlap curve with the eustatic sea-level curve of Haq and others (1987) suggests a close relation between development of the prograding sequences and sea-level fluctuations. As indicated by the close match of the number of sea-level fluctuations with the number of prograding sequences, it seems that each short-term sea-level fluctuation produced one prograding sequence (Fig. 12). For example, from the mid-Miocene to the Recent, nine sea-level fluctuations are proposed on the eustatic sea-level curve, whereas eight sequences are seen in our seismic profiles over the time interval (Fig. 12).

In times of long-term sea-level highstands, the sequences are more complete and better developed. For example, the three fluctuations during the long-term sea-level highstand at the Early/Middle Miocene boundary probably produced the three well-defined "highstand" sequences l,m,n, whereas the sequences (o, p, q) deposited during the Late Oligocene long-term lowstand are poorly developed (Fig. 11a). In addition, the internal reflection pattern varies; "highstand" sequences tend to have steep sigmoids and thick shelf members, whereas "lowstand" sequences show toplap terminations and have no or only thin shelf members. This depositional response to the sea-level stands seems to be consistent in the carbonate environment (Mitchum and Uliana, 1985).

Comparison of the extent of basinward migration of the platform edge in "highstand" sequence and in "lowstand" sequences shows that progradation is not only a function of sediment production but also of space available for deposition. It is generally believed (Lynts and others, 1973; Neumann and Land, 1975; Kendall and Schlager, 1981; Boardman and Neumann, 1984) that sediment production is highest in times of relative sea-level highstands, a hypothesis supported by the well-developed aggrading sequences. Nevertheless, in thick aggrading sequences, the amount of progradation is less than in thin onlapping sequences (Fig. 13), indicating that amount of progradation and sediment production is not a one-to-one relation and the space available for deposition has to be taken into account. The thick shelf member of aggrading sequences indicates that there is space on the platform for deposition of sediment. In contrast, when a sea-level rise barely floods the platform top, there is less space and almost all excess sediment is transported offbank, resulting in extensive platform edge migration. Comparison with the eustatic curves suggests that this situation is achieved more frequently during long-term sea-level lowstands, whereas times of major aggradation correspond to times of long-term sea-level rise.

DISCUSSION

Seismic profiles across the top of northwestern Great Bahama Bank document that an impressive 40 percent of the modern platform volume consists of deposits in intraplatform seaways. We assume that most of the sediment is produced on the platform, but it is not possible to estimate the relative contribution from the margin and interior. There is, however, an indication that sediment production from the marginal area is high enough to build prograding sequences without contribution from the platform interior. There are onlapping sequences, especially during long-term sea-level lowstands, that were deposited when a large part of the adjacent platform was exposed (Fig. 11). Therefore, if most of the sediment in the prograding sequences was derived from the platform margin of some few kilometers, production rate was much higher there than generally recognized.

In the Bahamas, the intraplatform seaways are first filled to approximately 500 m before prograding sequences narrow and finally close them. These prograding carbonate sequences are very similar to those found in the siliciclastic environment (Mitchum and others, 1977). There are vari-

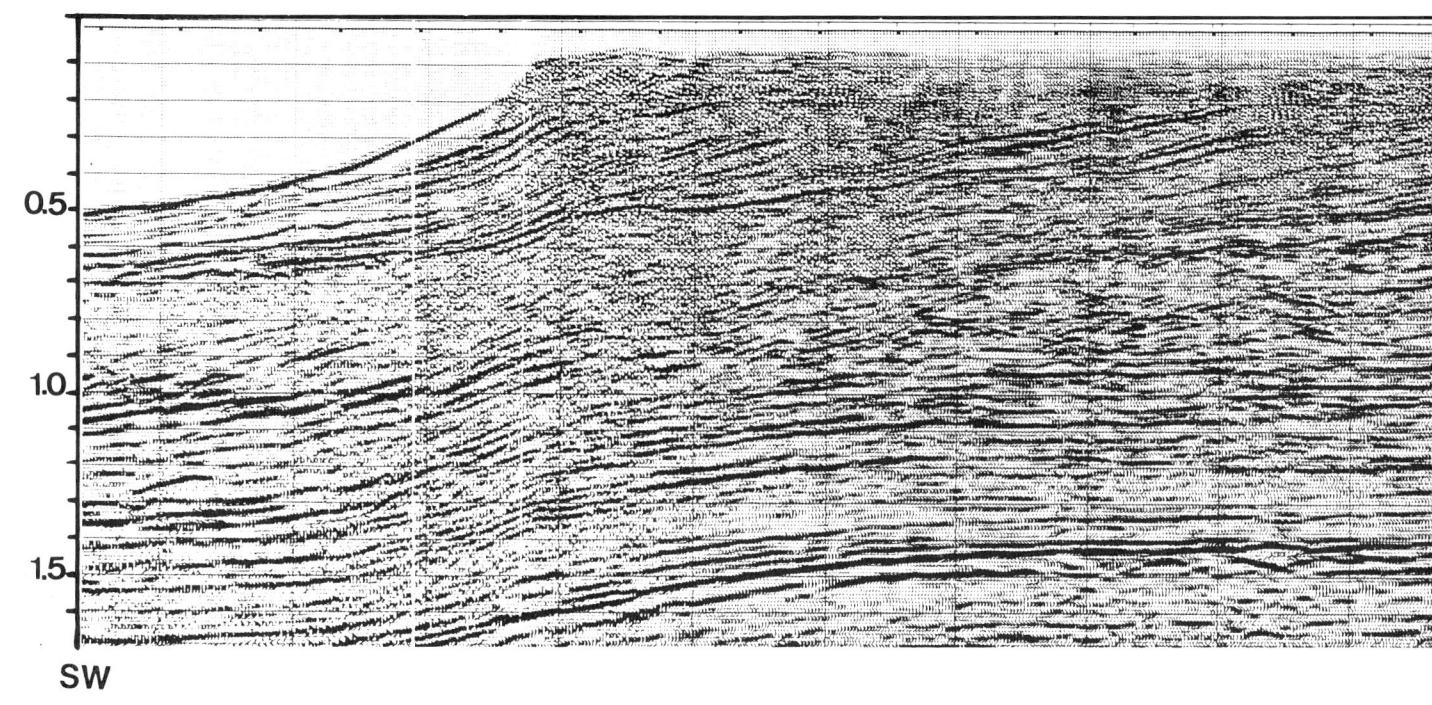

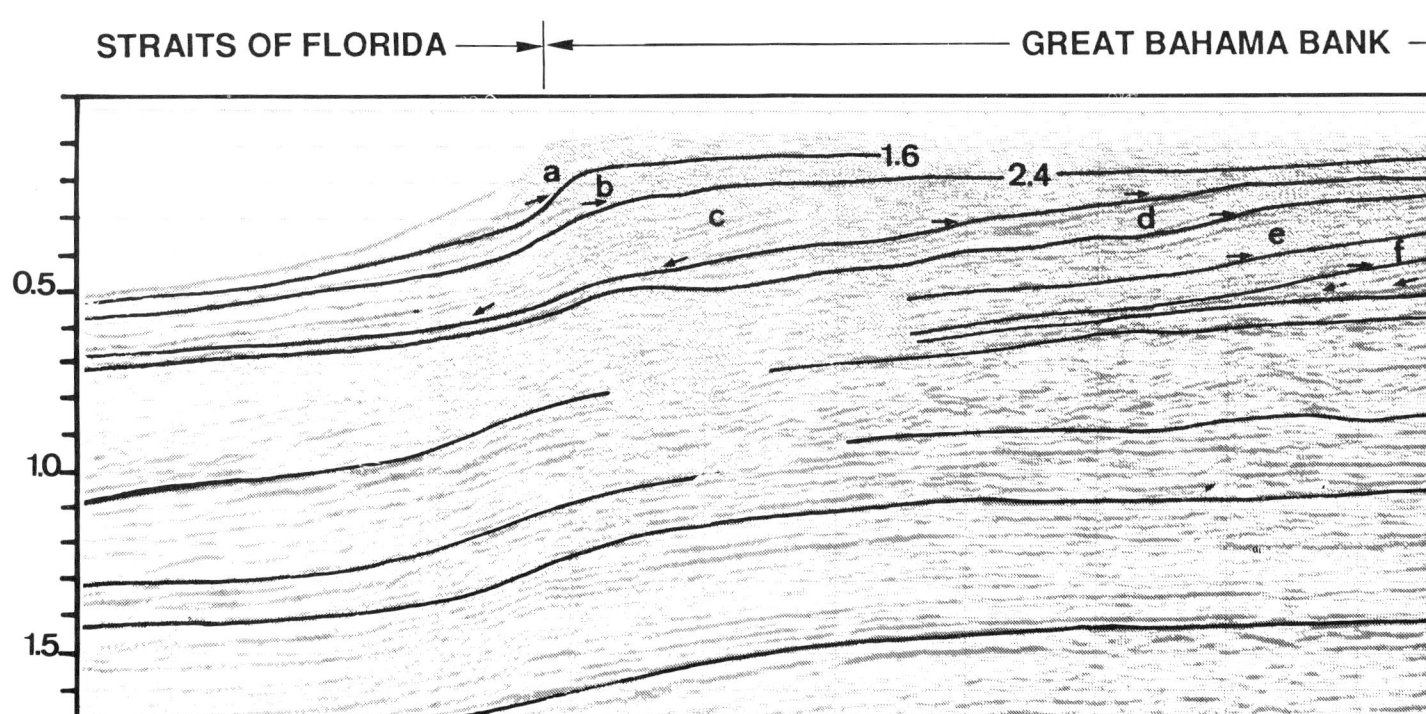

FIG. 11.—(b) Part of WESTERN line across west margin of the Great Bahama Bank, displaying prograding sequences from the late Miocene to Recent. In the interpreted section, the sequence boundaries are identified and labeled (a–h). The numbers give tentative ages to the sequence boundaries, as indicated by comparison of the onlap with the global chart of coastal onlap (see Fig. 12).

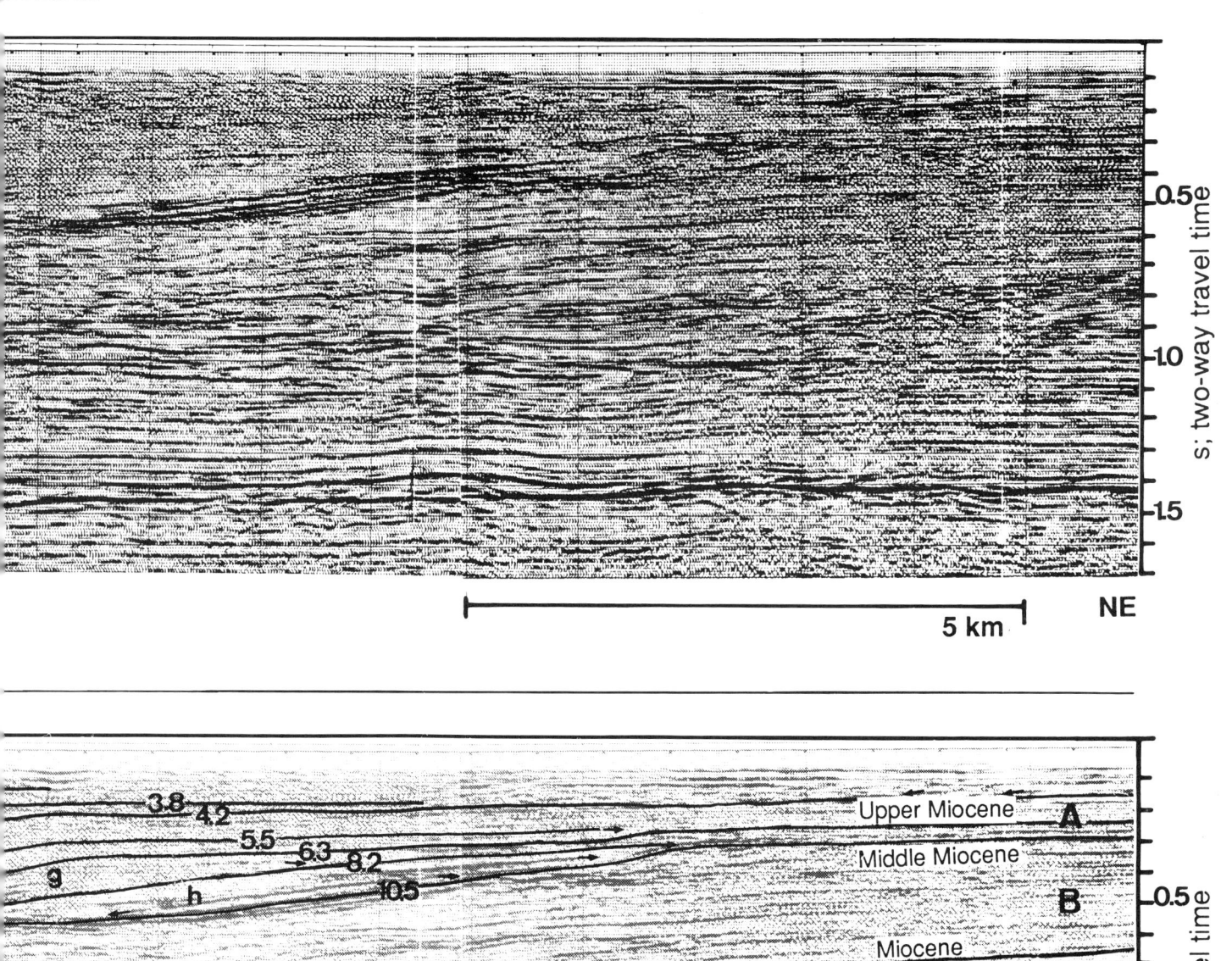

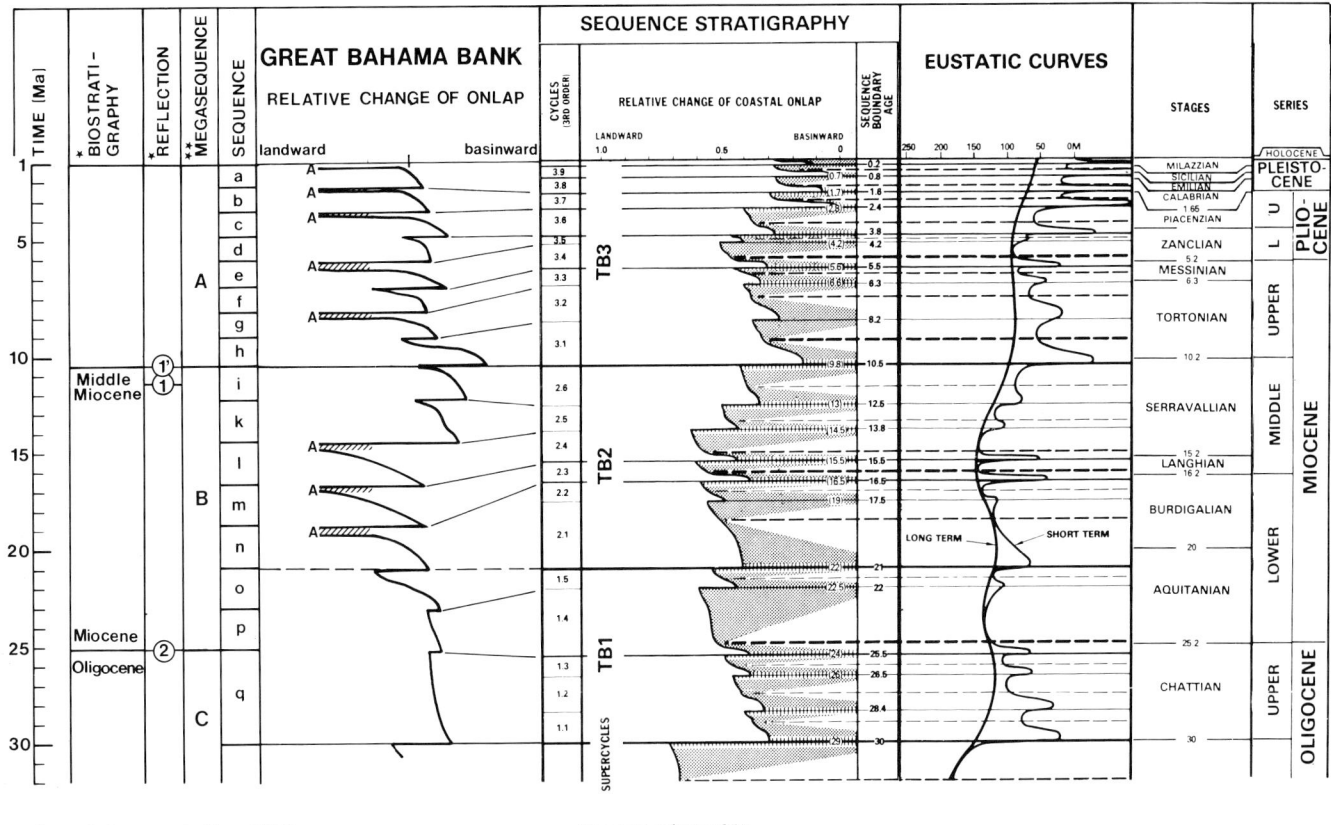

FIG. 12.—Great Bahama Bank onlap curve and correlation with the global chart of coastal onlap and eustatic curves of Haq and others (1987). Biostratigraphic data from the Great Isaac-1 well and correlation with seismic reflectors provide a time frame for dividing the 16 prograding sequences. Note the close match of numbers of prograding sequences with third-order cycles within TB2 and TB3, which strongly suggests that the prograding pulses are a record of these cycles and the result of a single third-order cycle of sea-level fall and rise.

ations, however, which might be used to distinguish the two. Especially indicative of a carbonate setting are the angles of the foresets. In carbonate sequences, the slope angle can be as much as 6°, and in complex sigmoid sequences, reflections with angles of as much as 20° are observed, whereas in the siliciclastic environment the depositional angle is usually less than 1° (Mitchum and others, 1977; Schlager and Ginsburg, 1981; Schlager and Camber, 1986).

Numerous studies have documented the high potential of platform carbonates to record sea-level fluctuations of different orders (Kendall and Schlager, 1981; Read and others, 1986; Sarg and Lehmann, 1986; Grotzinger, 1986). Stimulated by these results, we used the onlap/offlap pattern of the prograding sequences for a sequence stratigraphy analysis and compared it with the curve developed in the siliciclastic environment (Vail and others, 1977; Haq and others, 1987). A number of studies, have suggested, however, that the carbonate environment responds differently to sea-level fluctuations than the siliciclacstic environment. Sediment is exported at different stands of sea level into the deeper water in the two environments with the consequence that basinal fill occurs at different times (Kier and Pilkey, 1971; Mullins, 1983; Boardmann and Neumann, 1984; Droxler and Schlager, 1985; Austin and others, 1986). For this sequence analysis, we considered the onlap/offlap pattern of the prograding sequences along the platform edge rather than the basinal unconformities. The analysis is based on the assumption that the progressive onlapping reflections were produced during a relative rise and highstand of sea level. The global curve of coastal onlap monitors the progressive encroachment of the coastal deposits and is similarly the record of a relative rise and highstand of the sea level (Vail and others, 1977; Haq and others, 1987). Therefore, we think the two curves are the result of the same events and can be correlated.

The major uncertainties in our sequence stratigraphy are the age assignments. They are based on the scarce biostratigraphic data from the cuttings of the Great Isaac-1 well (Schlager and others, 1988). Two reflections, the mid-Miocene and the top of the Oligocene, can be dated biostratigraphically. They are then jump correlated to reflections in our key line, and the uncertainty of the correlation has to be kept in mind. A calibration of the lithology and the ages of the sequences will be necessary to test our interpretation that the pulses of progradation coincide with third-order sea-level fluctuations.

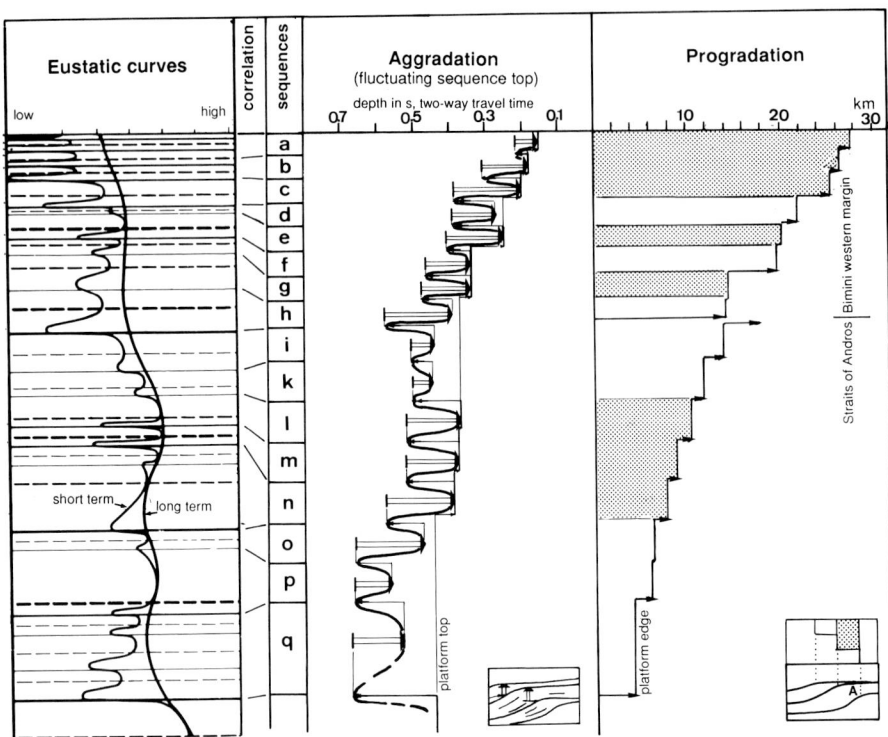

FIG. 13.—Comparison of the extent of aggradation, progradation, and eustatic curves of Haq and others (1987). Note that, except for the youngest sequences, aggradation occurred during long-term sea-level rises. In the right column, extent of progradation in aggrading (shaded) and in onlapping sequences is displayed. Note that extent of progradation is slightly higher in onlapping sequences, namely 17.6 km compared to 12.6 in aggrading sequences.

CONCLUSIONS

In northwestern Great Bahama Bank, a combination of infilling and prograding sequences is responsible for welding together smaller platforms, thus creating the large modern Great Bahama Bank. Platform growth is controlled by several factors: (1) tectonism, which created the initial topography in the mid-Cretaceous and interrupted growth in the mid-Miocene (Eberli and Ginsburg, 1987); (2) seaway morphology that determines the filling history (3) regional prevailing currents that strongly influence the direction of growth; and (4) sea-level fluctuations, which result in variable rates of sediment production and produce pulsed platform progradation.

The recognition of extensive lateral growth of the northwestern Great Bahama Bank documents the dynamic quality of a "healthy" platform. Overproduction of 40 percent of excess sediment resulted in a lateral growth that outpaces vertical aggradation on the platform more than tenfold. The initiation of platform edge migration seems to be dependent upon seaway morphology, as progradation starts only when the seaways have shoaled to approximately 500 m and/or the slopes have decreased to about 5°. The lateral growth displays an extreme asymmetry with nearly vertically growing eastern margins and prograding western leeward margins. The asymmetry indicates the important role of the direction of the energy flux –in the northwestern Bahamas, the prevailing easterly trade winds (Hine and Neumann, 1977)– for the direction of platform growth.

Sea-level fluctuations produce pulses of progradation. The prograding sequences are arranged into two major systems that are time equivalent with the youngest supercycle set TEJAS B of Haq and others (1987). The number of sequences matches nearly completely with the number of third-order cycles on the global chart (Haq and others, 1987). Sequence analysis suggests that individual sequences are the result of third-order sea-level fluctuations, and bundles of sequences represent second-order cycles.

The intraplatform seaways are major contributors to the volume of the modern platform as they were filled and incorporated into the platform. We speculate that other similar hidden seaways, over which shallow-water conditions became established after filling, will be found in the rock record of ancient epicontinental platforms.

ACKNOWLEDGMENTS

The results presented in this paper were obtained during Eberli's tenure at the Comparative Sedimentology Laboratory. During this time, we received financial support and scientific input from various people and organizations. We thank the Swiss National Science Foundation, the Industrial Associates of the Comparative Sedimentology Laboratory, and the Exxon Production Research Company for financial support; Texaco Inc. for releasing the grid of profiles; Western Geophysical for the cross-bank profile; and especially Wolfgang Schlager and his co-workers for providing us with important, unpublished results. Much appreciated are comments and discussions with Peter Vail, Stephen Greenlee, and Alfonso Bosellini, which in combination with the perceptive questions and suggestions of our colleagues and students at the Comparative Sedimentology Laboratory, encouraged the research and improved the interpretation.

REFERENCES

AUSTIN, J. A. JR., SSCHLAGER, W., AND OTHERS, 1986, Proceedings, Preliminary Report, Ocean Drilling Program, Leg 101, Part A: Texas A and M University, College Station, p. 1–569.

BEACH, D. K., AND GINSBURG, R. N., 1980, Facies succession, Pliocene-Pleistocene carbonates, northwestern Great Bahama Bank: American Association of Petroleum Geologists Bulletin, v. 64, p. 1634–1642.

BOARDMAN, M. R., AND NEUMANN, A. C., 1984, Sources of periplatform carbonates: Northwest Providence Channel, Bahamas: Journal of Sedimentary Petrology, v. 54, p. 1110–1123.

BOSELLINI, A., 1984, Progradation geometries of carbonate platforms: Examples from the Triassic of the Dolomites, northern Italy: Sedimentology, v. 31, p. 1–24.

BUBB, J. N., AND HATLELID, W. G., 1977, Seismic stratigraphy and global changes of sea level, Part 10: Seismic recognition of carbonate build-ups, in Payton, C. E., ed., Seismic Stratigraphy–Application to Hydrocarbon Exploration: American Association of Petroleum Geologists Memoir 26, p. 185–204.

DROXLER, A., AND SCHLAGER, W., 1985, Glacial versus interglacial sedimentation rates and turbidite frequency in the Bahamas: Geology, v. 13, p. 799–802.

EBERLI, G. P., AND GINSBURG R. N., 1987, Segmentation and coalescence of Cenozoic carbonate platforms in northwestern Great Bahama Bank: Geology, v. 15, p. 75–79.

———, AND ———, 1988, Aggrading and prograding infill of buried Cenozoic seaways, Northwestern Great Bahama Bank, in Bally, A. W., ed., Atlas of Seismic Stratigraphy: American Association of Petroleum Geologists, Studies in Geology Series No. 27, v. 2, p. 97–103.

GROTZINGER, J. P., 1986, Cyclicity and paleoenvironmental dynamics, Rocknest platform, northwest Canada: Geological Society of America Bulletin, v. 97, p. 1208–1231.

HAQ, B. U., HARDENBOL, J., AND VAIL, P. R., 1987, Chronology of fluctuating sea levels since the Triassic: Science, v. 235, p. 1156–1167.

HINE, A. C. AND MULLINS, H. T., 1983, Modern carbonate shelf-slope breaks, in Stanley, D. J., and Moore, G. T., eds., The Shelf Break: Critical Interface on Continental Margins: Society of Economic Paleontologists and Mineralogists Special Publication 33, p. 169–188.

———, AND NEUMANN, A. C., 1977, Shallow carbonate-bank-margin growth and structure, Little Bahama Bank, Bahamas: American Association of Petroleum Geologists Bulletin, v. 61, p. 376–406.

———, WILBER, R. J., BANE, J. M., NEUMANN, A. C., AND LORENSON, K. R., 1981, Offbank transport of carbonate sands along open, leeward bank margins: Northern Bahamas: Marine Geology, v. 42, p. 327–348.

HOLLISTER, C. D., EWING, J. I., AND EIGHT OTHERS, 1972, Initial Reports of the Deep Sea Drilling Project 11: U.S. Government Printing Office, Washington, D.C., 1077 p.

KENDALL, C. G. ST. C., AND SCHLAGER, W., 1981, Carbonates and relative changes in sea level: Marine Geology, v. 44, p. 181–212.

KIER, J. S., AND PILKEY, O. H., 1971, The influence of sea-level changes on sediment carbonate mineralogy, Tongue of the Ocean, Bahamas: Marine Geology, v. 11, p. 189–200.

LYNTS, G. W., JUDD, J. B., AND STEHMAN, C. F., 1973, Late Pleistocene history of Tongue of the Ocean, Bahamas: Geological Society of America Bulletin, v. 84, p. 2665–2684.

MEISSNER, F. F., 1972, Cyclic sedimentation in the Middle Permian strata of the Permian basin, West Texas and New Mexico, in Elam, J. C., and Chuber, S., eds., Cyclic Sedimentation in the Permian Basin, second edition: West Texas Geological Society, p. 203–232.

MEYERHOFF, A. A., AND HATTEN, C. W., 1974, Bahamas salient of North America: Tectonic framework, stratigraphy, and petroleum potential: American Association of Petroleum Geologists Bulletin, v. 58, p. 1201–1239.

MILLER, K. G., MOUNTAIN, G. S., AND TUCHOLKE, B. E., 1985, Oligocene glacio-eustacy and erosion on the margins of the Atlantic: Geology, v. 13, p. 10–13.

MITCHUM, R. M., JR., AND ULIANA, M. A., 1985, Seismic stratigraphy of carbonate depositional sequences, Upper Jurassic-Lower Cretaceous, Neuquen Basin, Argentina, in Berg, O. R., and Woolverton, D. G., eds., Seismic Stratigraphy II: A Integrated Approach to Hydrocarbon Exploration: American Association of Petroleum Geologists Memoir 39, p. 255–274.

———, ———, AND SANGREE, J. B., 1977, Seismic stratigraphy and global changes of sea level, Part 6: Stratigraphic interpretation of seismic reflection patterns in depositional sequences, in Payton, C. E., ed., Seismic Stratigraphy–Applications to Hydrocarbon Exploration: American Association of Petroleum Geologists Memoir 26, p. 117–133.

MULLINS, H. T., 1983, Comment on "Eustatic control of turbidites and winnowed turbidites": Geology, v. 11, p. 57–58.

———, HEATH, K. C., VAN BUREN, H. M., AND NEWTON, C. R., 1984, Anatomy of a modern open-ocean carbonate slope: Northern Little Bahama Bank: Sedimentology, v. 31, p. 141–168.

———, AND NEUMANN, A. C., 1979, Deep carbonate bank margin structure and sedimentation in the northern Bahamas, in Doyle, L. J., and Pilkey, O. H., eds., Geology of Continental Slopes: Society of Economic Paleontologists and Mineralogists Special Publication 27, p. 165–192.

———, WILBER, R. J., HINE, A. C., AND CHINBURG, S. J., 1980, Carbonate sediment drifts in the northern Straits of Florida: American Association of Petroleum Geologists Bulletin, v. 64, p. 1701–1717.

NEUMAN, A. C., AND LAND, L. S., 1975, Lime mud depositionn and calcareous algae in the Bight of Abaco, Bahamas: A budget: Journal of Sedimentary Petrology, v. 45, p. 763–786.

NEWELL, N. D., 1955, Bahamian platforms, in Poldervaart, A., ed., Crust of the Earth: Geological Society of America Special Publication 62, p. 303–316

PALMER, M. S., 1979, Holocene facies geometry of the leeward bank margin, Tongue of the Ocean, Bahamas: Upublished M.S. Thesis, University of Miami, Coral Gables, 199.

PAULUS, F. J., 1972, The geology of Site 98 and the Bahama platform, in Hollister, C. H., Ewing, J. I., and eight others, Initial Reports of the Deep Sea Drilling Project, v. 11: U.S. Government Printing Office, Washington, D.C., p. 877–897.

PLAYFORD, P. E., AND LOWRY, D. C., 1966, Devonian reef complexes of the Canning basin, Western Australia: Geological Survey of Western Australia Bulletin, v. 118, 50 p.

READ, J. F., GROTZINGER, J. P., BOVA, J. A., AND KOERSCHNER, W. F., 1986, Models for generation of carbonate cycles: Geology, v. 14, p. 109–144.

SARG, J. F., AND LEHMANN, P. J., 1986, Lower/Middle Guadalupian facies and stratigraphy San Andres/Grayburg Formations, Permian Basin, Guadalupe Mountains, New Mexico, in, Moore, G. E., and Wilde, G. L., eds., Lower and Middle Guadalupian Facies, Stratigraphy and Reservoir Geometries, San Andres, Grayburg Formations, Guadalupe Mountains: Society of Economic Paleontologists and Mineralogists, Permian Basin Section, Special Publication 86-25, p. 1–35.

SCHLAGER, W., BOURGEOIS, F., MACKENZIE, G., AND SMIT, J., 1988, Boreholes, Great Isaac-1, ODP 626 and the history of the Florida Straits, in Austin, J. A., Jr., Schlager, W., and 22 others, Proceedings, Scientific Results, Ocean Drilling Program Leg 101, p. 425–438.

———, AND CAMBER, O., 1986, Submarine slope angles, drowning unconformities, and shelf-erosion of limestone escarpments: Geology, v. 14, p. 762–765.

———, AND GINSBURG, R. N., 1981, Bahamas carbonate platforms – The deep and the past: Marine Geology, v. 44, p. 1–24.

SCHLANGER, S. O., AND PREMOLI SILVA, I., 1986, Oligocene sea-level falls recorded in mid-Pacific atoll and archipelagic apron settings: Geology, v. 14, p. 392–395.

SILVER, B. A., AND TODD, R. G., 1969, Permian cyclic strata, northern Midland and Delaware basins, West Texas and southeastern New Mexico: American Association of Petroleum Geologists Bulletin, v. 53, p. 2223–2251.

VAIL, P. R., MITCHUM, R. M., JR., AND THOMPSON, S., III., 1977, Seismic stratigraphy and global changes of sea level, Part 3: Relative changes of sea level from the coastal onlap, in Payton, C. E., ed., Seismic Stratigraphy–Applications to Hydrocarbon Exploration: American Association of Petroleum Geologists Memoir 26, p. 63–82.

VAN SICLEN, D. C., 1958, Depositional topography–Examples and theory: American Association of Petroleum Geologists, v. 42, p. 1897–1913.

WILSON, J. L., 1975, Carbonate Facies in Geological History: Springer-Verlag, New York, 471 p.

PLATFORM EVOLUTION AND SEQUENCE STRATIGRAPHY OF THE NATUNA PLATFORM, SOUTH CHINA SEA

KURT W. RUDOLPH AND PATRICK J. LEHMANN
Exxon Production Research Company, P.O. Box 2189, Houston, Texas 77001

ABSTRACT: By integrating seismic, well-log, and core data into a sequence framework, we are able to recognize seven complete depositional sequences in the Miocene Terumbu Formation carbonates of the Natuna Platform, South China Sea. Each sequence consists of a lowstand-systems tract, a transgressive-systems tract and condensed section, and a highstand-systems tract.

Terumbu carbonates display a downward shift of reservoir facies in the lowstand-systems tract, deepen upward (retrograde) in the transgressive-systems tract, and shoal upward (prograde) in the highstand-systems tract. At each sequence boundary, there is erosional truncation of the platform margin and upper slope and exposure of the platform crest.

The highest porosity occurs in grain-prone shoal-water carbonates of the late highstand-systems tract on the platform crest. Porosity also occurs downdip from the platform crest in the onlapping lowstand-systems tract. Sequence stratigraphy, seismic facies, and seismic-modeling analysis are used to map and predict reservoir distribution on the Natuna Platform.

Increased subsidence from the Middle Miocene onward caused the retreat of the Natuna Platform. Retreat occurred in an asymmetric fashion with more retreat on the west, or low-productivity, shelfward side of the platform. Platform retreat occurred incrementally, during deposition of transgressive-systems tracts and the condensed sections. The large eustatic sea-level rise in the early Pliocene, combined with continued rapid subsidence, drowned the platform and ended carbonate sedimentation.

INTRODUCTION

The Natuna L-Structure is a large Miocene carbonate platform complex located on the western margin of the East Natuna-Sarawak Basin (Fig. 1). In 1973, Agip, the national oil company of Italy, drilled the AL-1X well on the crest of the L-Structure ("Natuna Platform" of this paper) and encountered approximately 1,600 m of porous, gas-productive carbonates. Analysis of the gas indicates it is 27 percent methane, 72 percent CO_2, and 0.5 percent H_2S, the remaining 0.5 percent being other hydrocarbon gases.

Esso Exploration successfully bid for the block containing the Natuna Platform in 1979 after Agip relinquished its concession. Later that year, Esso shot 2,574 km (1,600 mi) of 66-fold CDP seismic data over the block on a 1.6×1.6-km or a 1.6×0.8-km grid. These records constitute the principal seismic data base for this investigation. The four wells that have subsequently been drilled indicate that there is great lateral variation in stratigraphy and reservoir quality.

The oldest sedimentary rocks in the area are Oligocene to Lower Miocene clastics of the Arang and Gabus Formations. The deltaic Arang Formation is considered the most likely source of gas in the L-Structure.

The overlying Terumbu Formation consists of Middle and Upper Miocene carbonates, which compose the reservoir on the Natuna Platform. Similar Miocene carbonate platforms in the Luconia Province (Fig. 1) to the southeast are gas-productive (Epting, 1980). The Terumbu is thickest on the Natuna Platform crest and the coeval shelf to the west (Fig. 2). Natuna Platform carbonate deposition is nucleated on an antecedent basement-involved structural high, and carbonate sediments thin depositionally away from the platform crest.

The Terumbu Formation is sealed by the Pliocene-Pleistocene Muda Shale, a stacked series of prograding clastic wedges. The lowest portion of the Muda, which drapes the Natuna Platform, consists of prodelta to basinal shales and was deposited in about 900 m of water, based on paleontologic and seismic interpretation.

The AP-1X well, drilled on the Terumbu shelf margin 25 km west of the Natuna Platform, penetrated Cretaceous granodiorite below the Arang Formation. A similar Cretaceous granitic basement crops out on Borneo, Natuna Island, and the Anambas Islands (Hamilton, 1979). From geohistory analysis and normal faulting of the basement, a rifting episode appears to have occurred in late Oligocene to Early Miocene time in the area (Fig. 3). Rift-style tectonism, enhanced by differential carbonate deposition, created Terumbu paleobathymetry. The basement at Natuna is probably a thinned and extended continental borderland, since oceanic crust underlies much of the South China Sea to the north (Taylor and Hayes, 1983). The Natuna Platform may represent a foundered microcontinental fragment or basement warp. Hamilton (1979) has described analogous continental fragments with capping Tertiary reef complexes 320 km to the northeast.

LITHOFACIES AND POROSITY

Four lithofacies have been identified from 293 m of core at Natuna: (1) coral-red algae boundstone (grainstone matrix): reef, average porosity = 20 to 25 percent (Fig. 4A); (2) coral-red algae-echinoderm packstone/grainstone: platform margin sands, reef/reef flat, forereef debris; average porosity = 15 to 25 percent (Fig. 4B, C); (3) red algae-echinoderm-mollusc packstone: open platform, platform interior; average porosity = 10 to 25 percent; and (4) planktonic foraminifer-ostracod wackestone: low-energy outer platform; average porosity = 2 to 10 percent (Fig. 4D).

Despite extensive diagenetic modification, porosity distribution of the Terumbu is facies controlled. The best porosity occurs in the coral-red algae boundstone and coral-red algae-echinoderm packstone/grainstone lithofacies. Major pore types include moldic (dominant type), interparticle, solution-enlarged interparticle, intraparticle, and micropore.

There is good agreement among porosity, seismic facies, and cored lithofacies (Table 1). Low-amplitude, shingled, mounded, and parallel seismic facies correlate with the porous coral-red algae boundstone and coral-red algae-echinoderm packstone/grainstone lithofacies. High-amplitude

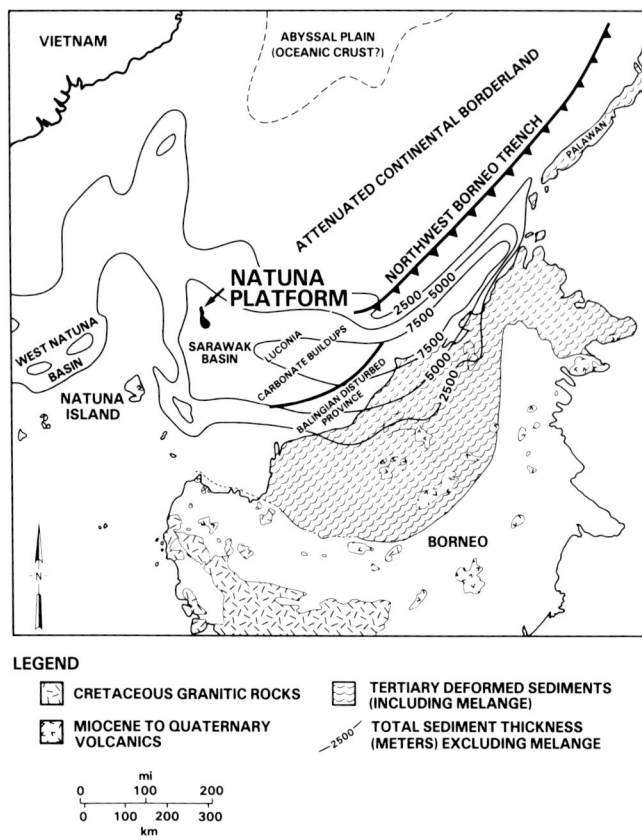

FIG. 1.—Regional tectonic map of Sarawak-East Natuna Basin (adapted from Hamilton, 1979, and Epting, 1980). The Natuna Platform occurs on attenuated continental crust on the west flank of the Sarawak Basin. Portions of this attenuated continental borderland have been subducted beneath the Northwest Borneo Trench, at least into the Pleistocene. An accreted sedimentary prism of Tertiary deformed sediments and melange lies to the southeast of the subduction zone.

sigmoidal and onlapping seismic facies correlate with the non-porous planktonic foraminifer-ostracod wackestone facies.

SEQUENCE STRATIGRAPHY

Seven complete sequences[1] with their component systems tracts[2] have been recognized within the Middle and Upper Miocene at Natuna (Figs. 5, 6). Seismic-sequence analysis and well control show that the sequences display similar generalized stratigraphic patterns (Figs. 7, 8). Each depositional sequence consists of a lowstand-systems tract, a transgressive-systems tract and condensed section (seismic downlap surface), and a highstand-systems tract. At Natuna, the base of each sequence is characterized by the erosional truncation of the underlying highstand strata. This basal boundary is expressed as erosionally truncated seismic reflections, especially along the margin and upper slope of the platform. Meteoric leaching associated with each unconformity is interpreted as causing karstification and/or porosity enhancement in the underlying highstand-systems tract. Porosity is best developed beneath each sequence boundary in the highstand-systems tract (Fig. 8).

In drilling the L-4X well, cavernous porosity was encountered at both the 15.5- and 13.8-Ma sequence boundaries. The drill stem dropped 3.7 m at the 15.5-Ma unconformity and 1.2 m at the 13.8-Ma unconformity (Fig. 8). From drilling mud losses, the volume of each megapore is judged to exceed 32 m^3 (J. A. May, pers. commun., 1983). Subaerial exposure, karstification, cavern formation, and meteoric diagenesis are probably related to the large falls in global sea level that occurred when these unconformities formed.

Lowstand-systems tract.—

The basal sequence boundary is seismically onlapped by the lowstand-systems tract along the platform slope and updip to the seismically interpreted platform margin (Figs. 5, 9, 10). Reflections within the underlying highstand-systems tracts are truncated at the sequence boundary on the platform margin and slope (Fig. 9). The lowstand is absent on the platform top. The lowstand-systems tracts in the early to middle Tortonian and Messinian sequences are characterized by a downdip displacement of reservoir lithofacies to the flanks of the platform. Porous limestones of interpreted lowstands are encased in relatively non-porous limestone of the highstand- and transgressive-systems tracts (Fig. 4C). As an example, the lower Messinian lowstand-systems tract penetrated in L-2X well contains coarse, abraded coralgal debris with core porosity of as much as 23 percent and depositional dips of 10° to 16° from dipmeter information (May and Eyles, 1985). Deposition of allochthonous carbonate sediments downslope from erosion of the platform margin is enhanced during sea-level lowstands. On the basis of occurrence of high-relief mounded seismic-reflection geometries (carbonate buildups), at least some of the Terumbu lowstand carbonates were deposited autochthonously (Figs. 11, 12).

An example of lowstand-systems tracts at Natuna is the lowstand deposited during the 10.5-Ma lowstand (upper Serravallian to lower Tortonian; Fig. 6). The 10.5-Ma lowstand is characterized by high-relief seismic mounds that form a belt around the entire platform (Figs. 11, 12, 13). Individual mounds have so much as 300 m of estimated depositional relief.

The large mounds exhibit two distinct map patterns, with elongate mounds parallel to the platform margin on the western side and oriented perpendicular to the margin on the northern and eastern sides of the platform. Mounds on the western side built relief vertically (Fig. 11) and have a parallel reflection geometry along strike. Mounds on the eastern side of the platform display oblique seismic-reflection geometry in dip section (Fig. 11) and are mounded along strike (Fig. 12). The high-relief mounded facies is interpreted as a fringing-reef complex deposited on the underlying platform slope.

---

[1]Sequence: a conformable succession of genetically related depositional systems bounded by interregional unconformities (e.g., a significant hiatus occurring at global falls of sea level) or their correlative conformities (sequence boundaries; Mitchum and others, 1977).

[2]Systems tract: a linkage of contemporaneous depositional systems (Brown and Fisher, 1977).

Variation in mound geometries is probably caused by varying energy and carbonate-productivity levels. The mounds on the western side were deposited largely as aggradational features in a protected setting on the shelfward side of the platform. The mounds on the northern and eastern sides were deposited as basinward-prograding mounds in a high-energy, high-productivity setting on the open-ocean side of the platform.

Although the 10.5-Ma lowstand has not been drilled, analogous lowstands have been penetrated by the L-2X and L-5X wells (Fig. 8). In these wells, lowstands contain rel-

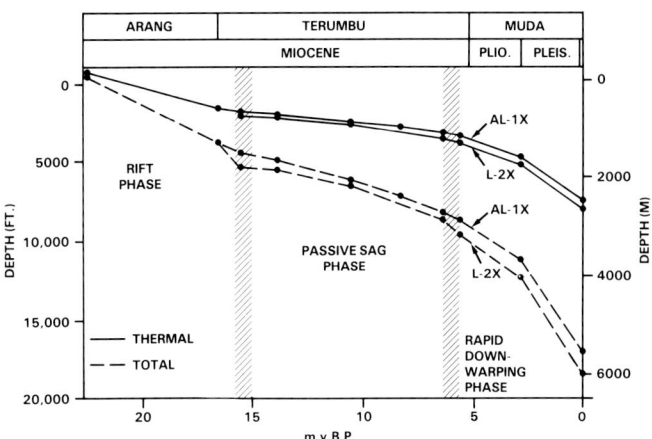

FIG. 3.—Geohistory analysis of Natuna Platform.

atively porous limestone encased in less porous limestone. Although seismically thin, these lowstands also show evidence of a mounded seismic facies and are interpreted as a basinward shift of reservoir lithofacies.

Transgressive-systems tract.—

Overlying the lowstand-systems tract is the transgressive-systems tract, which displays a decrease in porosity upward in most wells. The transgressive-systems tract is capped by a condensed section, a thin, starved section that corresponds to a seismic downlap surface. In the L-2X well, the lower Tortonian condensed section consists of planktonic foraminifer-ostracod wackestone, representing relatively low-energy deposition (Fig. 4D). The transgressive-systems tract is thinnest at the platform top; it thickens basinward.

The transgressive-systems tract is interpreted as having been deposited during a rapid rise in sea level as the platform was reflooded. This deepening provides accommodation both off the platform and on the platform top, resulting in deposition of retrograding carbonate facies (Fig. 8). Typically, this is expressed as a decrease in porosity upward. The platform retreated during deposition of the transgressive-systems tract. Condensed sections are deposited during the most rapid rise in sea level, during which time sedimentation is starved and deep-water sediments may be deposited relatively high on the platform.

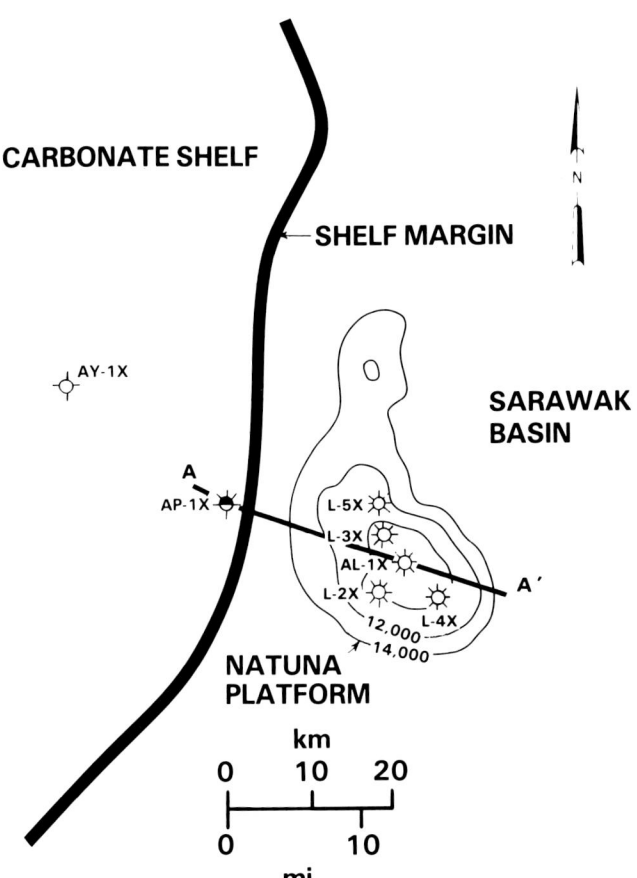

FIG. 2.—(A) Terumbu subsea depth structure in feet, Natuna Platform (from Eyles and May, 1982).

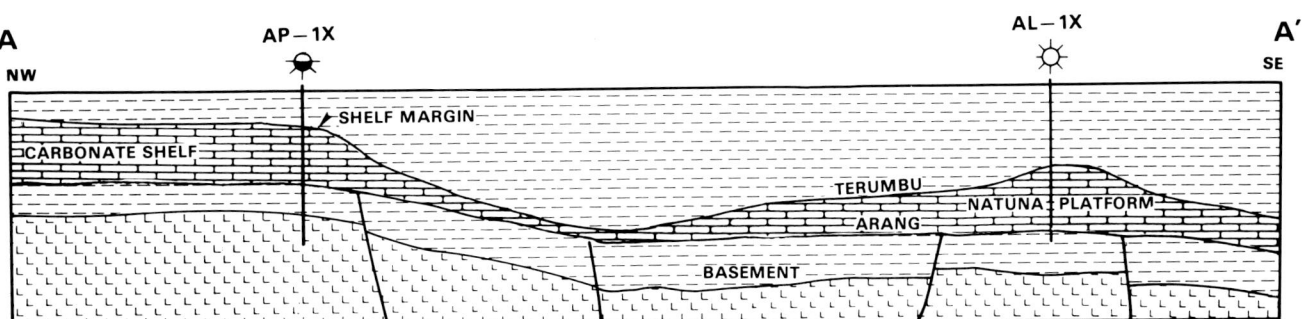

FIG. 2.—(B) Geologic cross section showing coeval shelfal and Natuna platform carbonates (vertical exaggeration = 1.5).

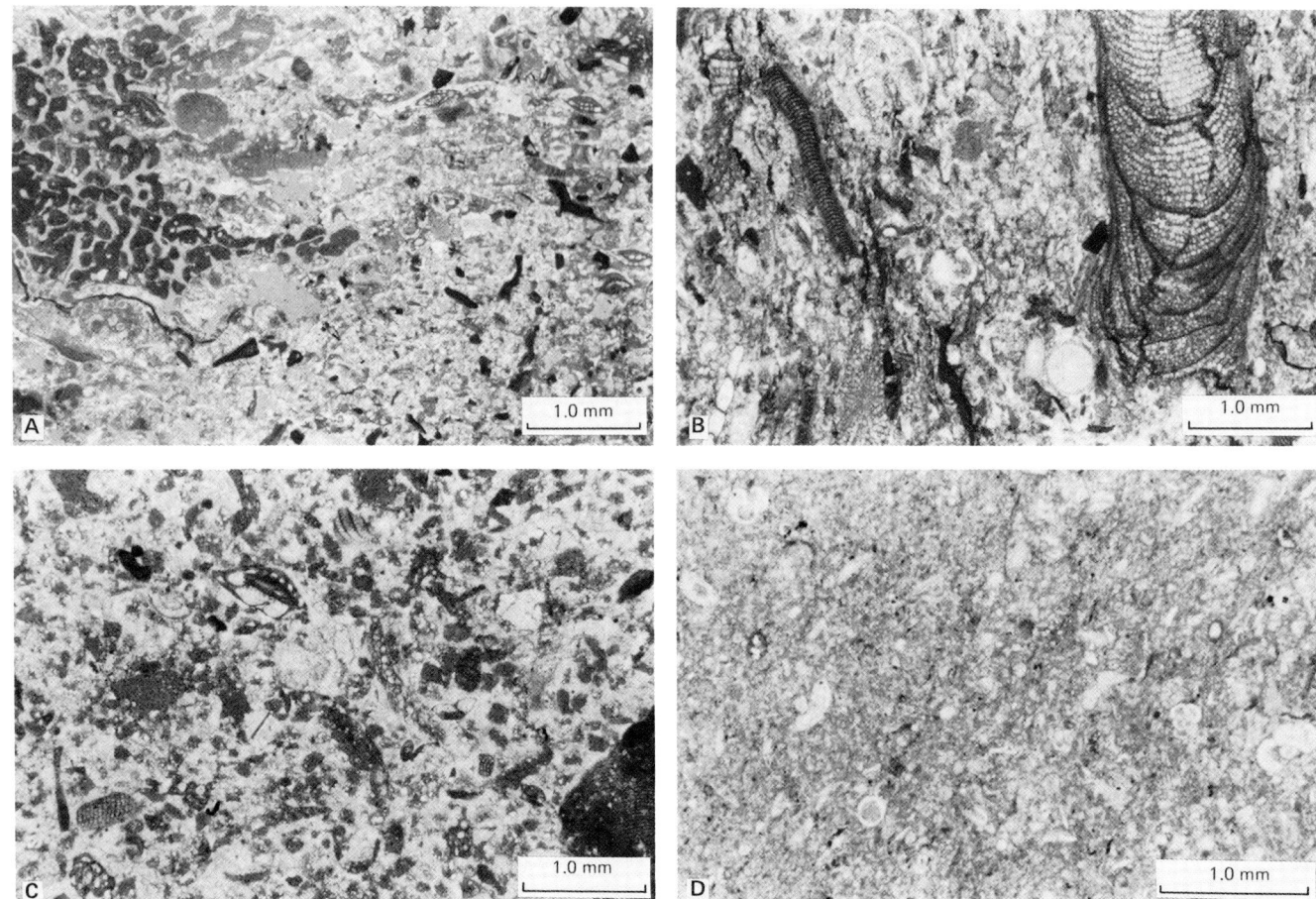

FIG. 4.—Thin section photomicrographs of Natuna Platform lithofacies, plane-polarized light. (A) Coral-red algae boundstone lithofacies in a grainstone matrix (highstand-systems tract). A large moldic (mud-filled) coral encrusted by red algae is present to the left. Light gray areas are moldic and vug porosity. Porosity = 24 percent and permeability = 40 md (from core). AL-1X, 2,764 m. (B) Coral-red algae-echinoderm packstone/grainstone lithofacies (highstand-systems tract). Porosity = 12 percent and permeability = 2 md. L-2X, 3,801 m. (C) Coral-red algae-echinoderm grainstone lithofacies (lowstand-systems tract). Moldic porosity is shown as light gray areas. Porosity = 20 percent and permeability = 8 md. L-2X, 3,091 m. (D) Planktonic foraminifer-ostracod wackestone from condensed section from near the top of the transgressive-systems tract. Porosity = 3 percent and permeability = 0.01 md. L-2X, 3,425 m.

TABLE 1.—CORE LITHOFACIES, SEISMIC FACIES, AND POROSITY

Systems Tracts	AL-1X Well	L-2X Well	L-3X Well
Messinian highstand	**Lithofacies 2** Average porosity = 26% Mounded, high-ampliude, peak-trough seismic facies	**Lithofacies 3** Average porosity = 12% Mounded, high-amplitude, trough-peak seismic facies	**Lithofacies 1 and 2** Average porosity = 25% Thin, high-amplitude, peak-trough seismic facies
Middle Tortonian highstand	**Lithofacies 2** Average porosity = 23% Low-amplitude shingled seismic facies	**Lithofacies 4** Average porosity = 3% Onlapping parallel seismic facies	**Lithofacies 1 and 2** Average porosity = 24% Low-amplitude mounded seismic facies
Early Tortonian transgressive	**Lithofacies 2** Average porosity = 17% Low-amplitude parallel seismic facies	**Lithofacies 4** Average porosity = 7% High-amplitude sigmoidal/wavy seismic facies	**Not cored** Average porosity = 20% Low-amplitude parallel seismic facies

Lithofacies 1. Coral-red algae boundstone
Lithofacies 2. Coral-red algae-echinoderm packstone/grainstone
Lithofacies 3. Red algae-echinoderm-mollusc packstone
Lithofacies 4. Planktonic foraminifera-ostracod wackestone

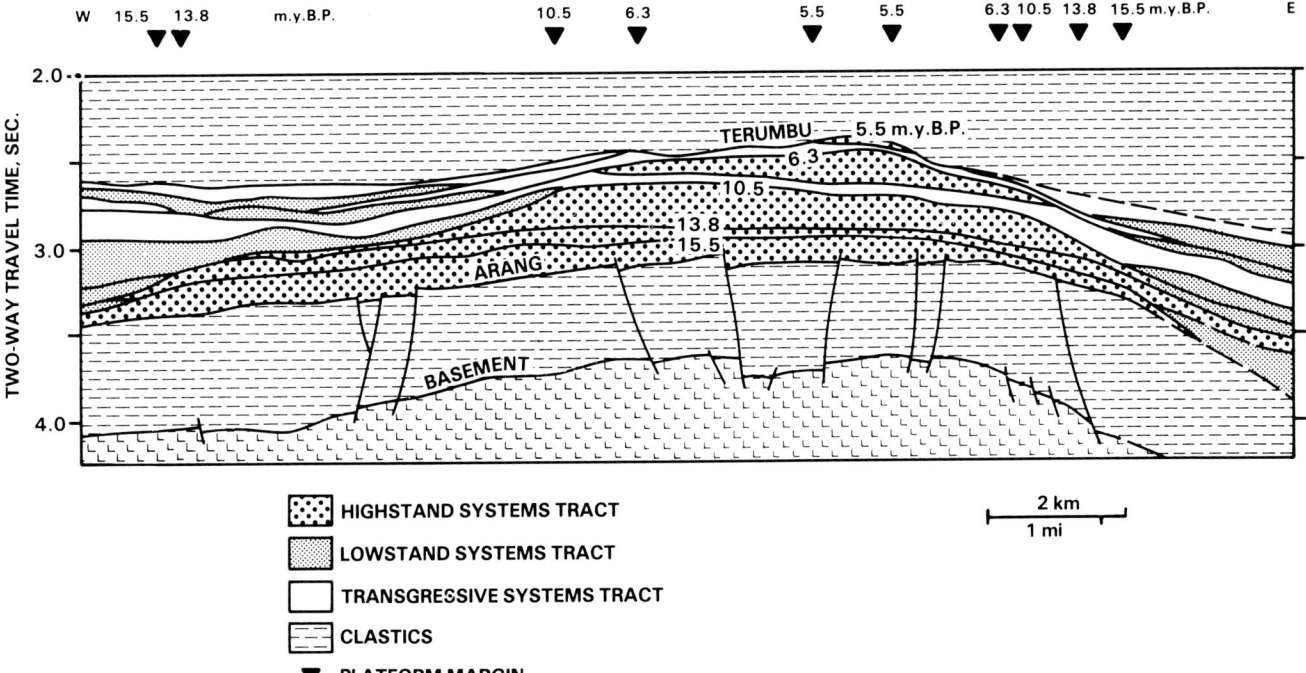

FIG. 5.—Interpretation of seismic line A, showing distribution of lowstand-, transgressive-, and highstand-systems tracts. The carbonate platform has nucleated on a faulted basement high. Platform margins (triangles) have retreated more on western, shelfward side than on eastern side.

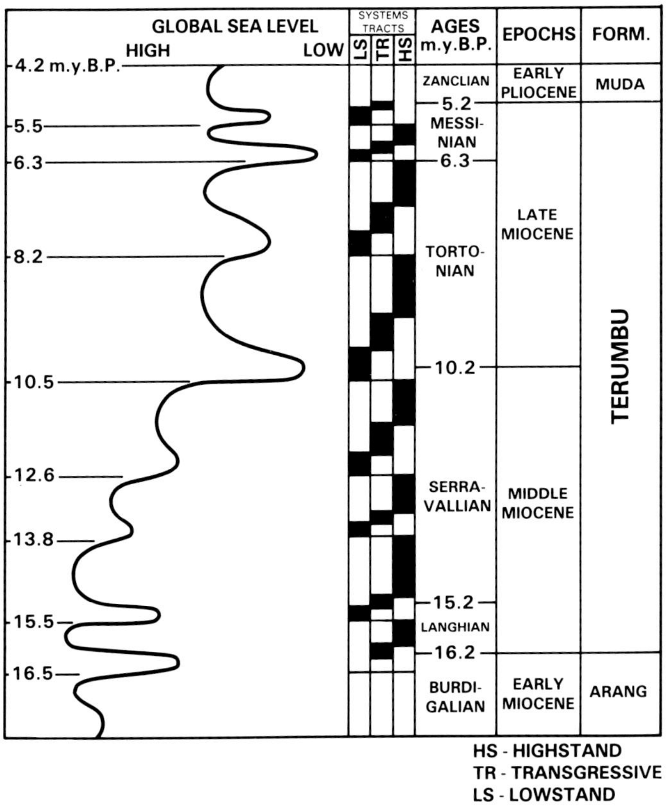

FIG. 6.—Natuna sequence stratigraphy showing relation of global sea level, systems tracts, and lithostratigraphy for the Early Miocene to Early Pliocene. The sea-level curve is from Haq and others (1987).

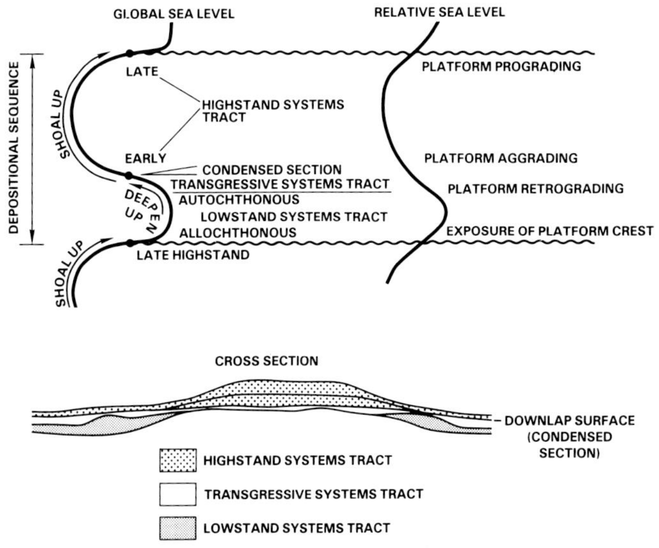

FIG. 7.—Idealized Natuna eustatic cycle.

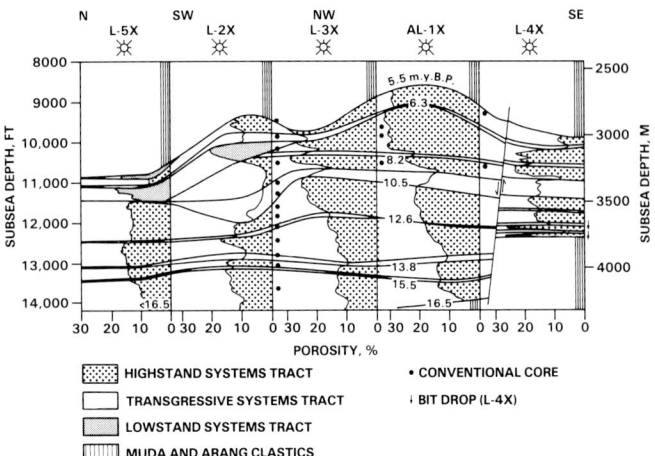

Fig. 8.—Natuna sequence stratigraphy, with well correlations tied to seismic stratigraphic interpretation. On platform crest (AL-1X), Terumbu consists of alternating thick shoaling-upward highstands and thin deepening-upward transgressive deposits. Condensed sections mark tops of transgressive deposits. On platform flanks (upper portion of L-5X), Terumbu consists of porous lowstands encased in thin, less porous highstand and transgressive deposits. Logs represent calibrated and smoothed porosity (petrophysical analysis by J. A. May).

Highstand-systems tracts.—

Highstand-systems tracts are thickest over the highest parts of the Natuna Platform, thinning below seismic resolution into the basin (Fig. 5). Most highstand-systems tracts show an upward increase in porosity (Fig. 8). This increase upward is an expression of the overall shoaling upward (progradation) of lithofacies within the highstand. Shallow, agitated conditions conducive to carbonate production prevail on the crest of the platform when sea level is high, resulting in thick accumulations. On the platform crest, highstand-systems tracts contain porous coral-red algae boundstone, coral-red algae echinoderm packstone/grainstone and red algae-echinoderm-mollusc packstone lithofacies (Fig. 4A, B). Carbonte deposition was able to keep up with the slowly rising sea level, filling all available accommodation in platform areas. Platform-to-basin relief is increased during highstands because of differential sedimentation.

An example of highstand-systems tracts is the highstand deposited before the 10.5-Ma sequence boundary (upper Serravallian; Fig. 6). This unit was deposited immediately before the 10.5-Ma lowstand described earlier.

The 10.5-Ma sequence boundary is characterized by erosional truncation of reflections at the seismically defined platform margin and upper slope (Figs. 10, 11, 14). On the platform crest, a low-amplitude oblique seismic facies occurs along the eastern or open-ocean side of the platform. This facies is interpreted to correspond to strongly prograding platform carbonates that are associated with high-carbonate productivity.

A high-amplitude parallel seismic facies is developed on the western side of the platform, where organic productivity and energy conditions are thought to be lower. On the basis of well information, the high-amplitude events within this facies are reflections from less porous interbeds that are not present in other higher energy platform facies.

Although subtle, the lower portion of the highstand-systems tract is characterized by parallel seismic-reflection geometry and is more oblique in the upper portion (Fig. 9). This is interpreted as an evolution from aggradational to progradational deposition. The highstand-systems tracts are divided into an early and a late portion. The early highstand-systems tract is characterized by aggradational sedimentation, since more accommodation is available during the early portion of the highstand. The late highstand-systems tract is characterized by more strongly progradational sedimentation. Accommodation on the platform top is reduced during the later portion of the highstand cycle, when global sea level is at a stillstand or is falling slowly.

PLATFORM EVOLUTION

Platform retreat.—

The Natuna Platform became more areally restricted with time (Figs. 5, 15). The oldest highstand carbonate-systems tracts were deposited on a large platform separated from the shelf to the northwest by a gentle saddle. The younger highstands, which were increasingly areally restricted, formed an isolated pinnacle buildup surrounded by deep water on all sides. Retreat of the platform occurred discretely during deposition of the transgressive-systems tracts, time of rapid sea-level rise.

The oldest carbonate highstand-system tracts at Natuna–Langhian (15.5 Ma) to early Serravallian (13.8 Ma)–were deposited on a widespread platform separated from the shelf by a shallow sag. We estimate from seismic interpretation that there is 300 m of relief from platform crest to toe of slope. These highstands show only modest lateral variations in seismic facies and log porosity. Seismic facies are either oblique or wavy/parallel on the platform. Calibrated log porosity varies from 10 to 14 percent. Deposition was relatively uniform in an open-platform setting.

As carbonate deposition continued, the foundering Natuna Platform became more areally restricted, and platform-to-basin relief increased to about 900 m. Younger carbonate highstands–late Tortonian (6.3 Ma) and Messinian (5.5 Ma)–display the greatest lateral variations in seismic facies and lithofacies. With increased relief, the platform differentiated into a more varied pinnacle-reef-facies mosaic surrounded by deep water on all sides. Log porosity varies from 6 to 26 percent.

Retreat of the Natuna Platform occurred in an asymmetric fashion (Fig. 15). Platform margins have retreated the farthest on the west side, about 8 km between the early Serravallian (13.8 Ma) and late Tortonian (6.3 Ma) highstands. Retreat on the east side was much less, averaging about 1.5 km between the early Serravallian and late Tortonian highstand margins.

The eastern platform margin contains the best developed oblique seismic facies, evidence of greater progradation on the eastern, or open-basin, side (Fig. 10). As a result, the Natuna Platform displays less retreat along this margin. The asymmetric pattern of platform evolution and seismic facies may be the result of higher organic productivity on the open-basin side. Controlling factors may include prevailing wind

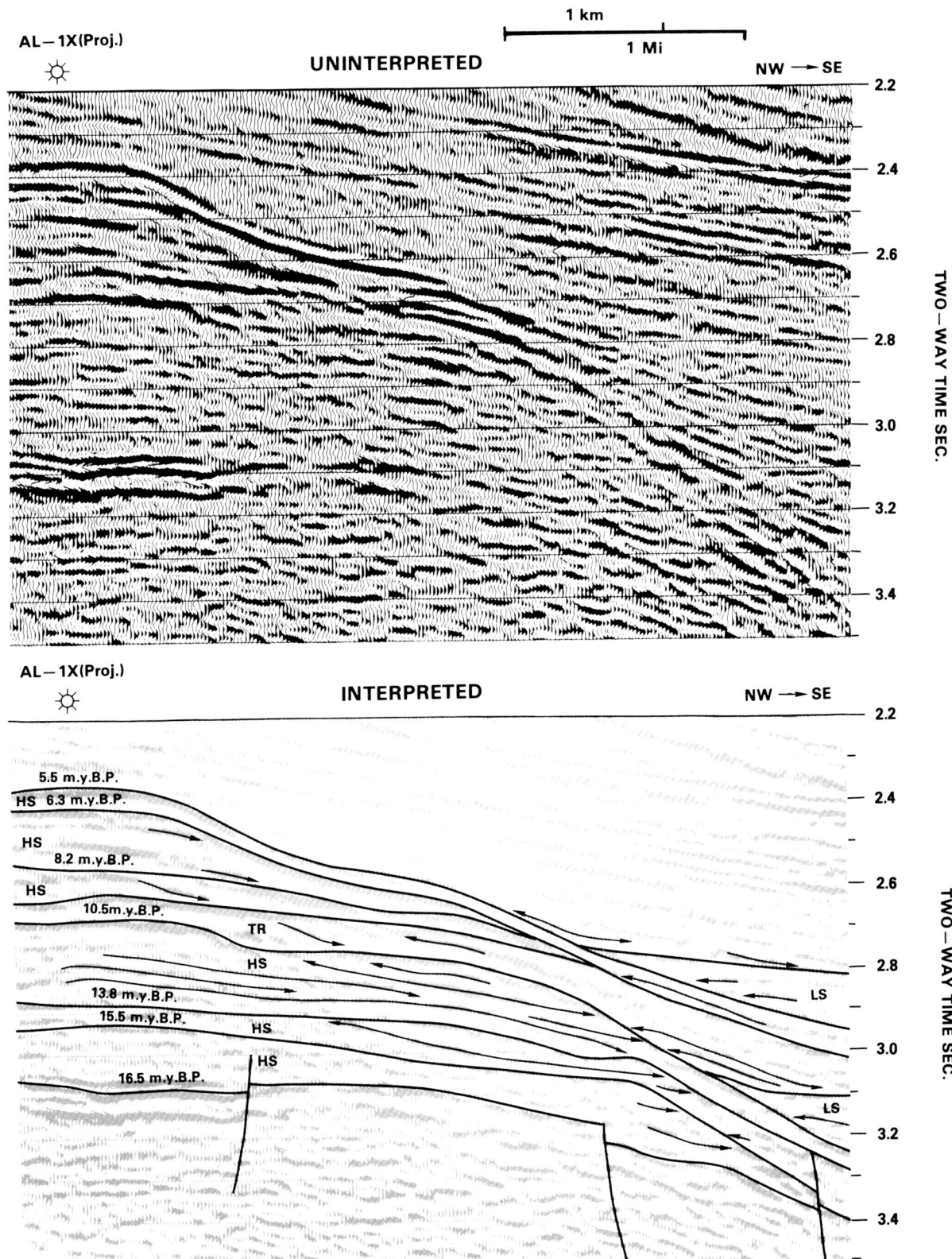

FIG. 9.—Seismic line A (segment) showing erosional truncation of seismic reflections at interpreted sequence boundaries at platform margin and slope position.

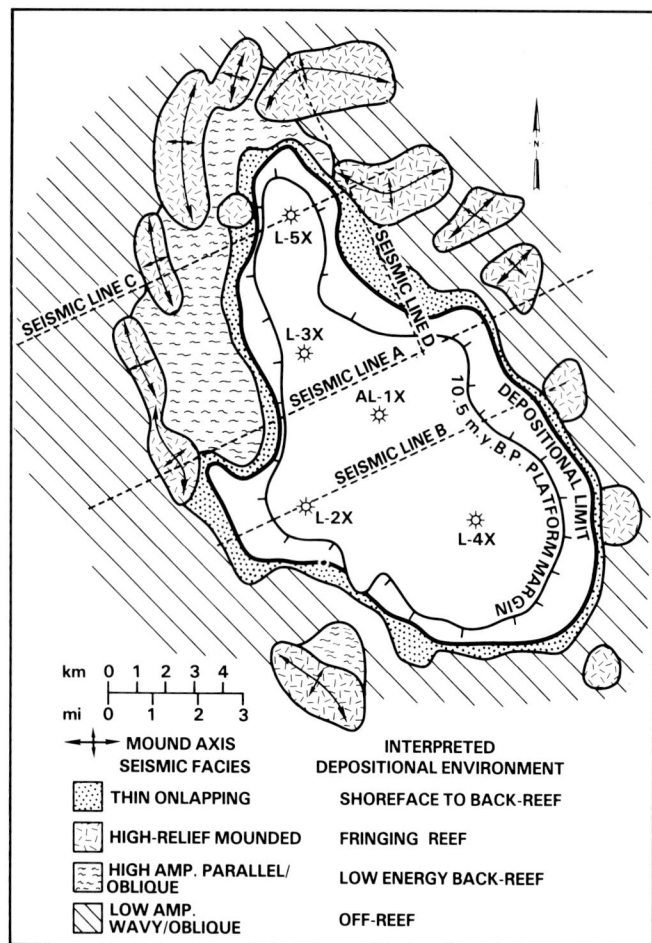

FIG. 13.—Seismic facies map, upper Serravallian to lower Tortonian lowstand-systems tract.

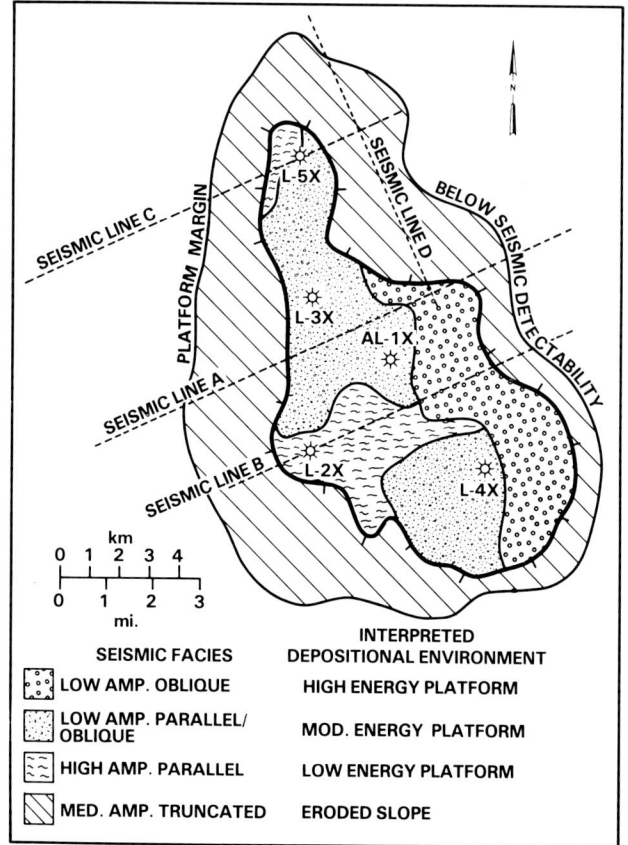

FIG. 14.—Seismic facies map, upper Serravallian highstand-systems tract.

direction, upwelling of nutrient-rich basinal waters, and hydrodynamic energy level.

Subsidence history and platform drowning.—

Evolution of the Natuna Platform is partially controlled by the subsidence history. Three phases of subsidence can be recognized from geohistory analysis (Fig. 3). Rifting occurred in the early Miocene and is evidenced by normal faulting and rapid subsidence (rate ≈ 250 m/m.y.) during Arang deposition. Crustal extension in the Natuna area is related to the opening of the South China Sea Basin (Taylor and Hayes, 1983). The middle Miocene is characterized by slower subsidence during post-rift passive sag or thermal cooling (rate from 15.5 to 10.5 Ma ≈ 110 m/m.y.). Subsidence increased exponentially in the late Miocene, Pliocene, and Pleistocene (rate from 5.5 Ma to present ≈ 450 m/m.y.). The mechanism for this rapid downwarping phase is not well understood, but it may include both flexural loading and bending of the crust associated with subduction along the Northwest Borneo Trench (Fig. 1).

The increased subsidence rate from post-Middle Miocene onward (rate from 10.5 to 5.5 Ma ≈ 140 m/m.y.) caused the retrograding of the Natuna Platform. As the platform foundered more and more rapidly, it became increasingly difficult for carbonate production to keep up. The large global sea-level rise (approximately 60 m) in the Early Pliocene

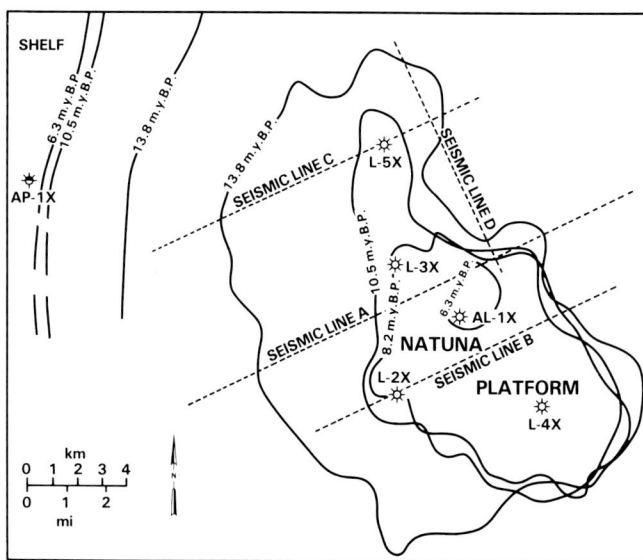

FIG. 15.—Platform margins of highstand-system tracts, Natuna Platform. Greater retreat has occurred on western than on eastern side.

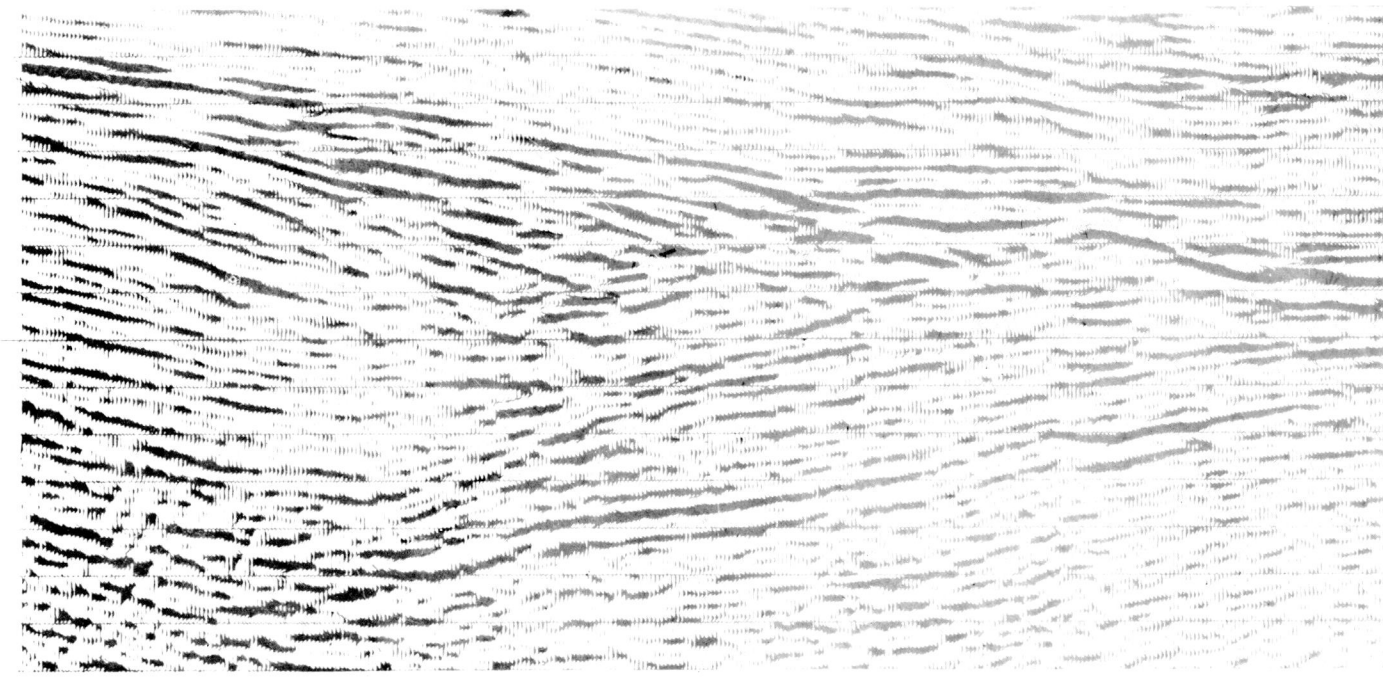

NW → SE
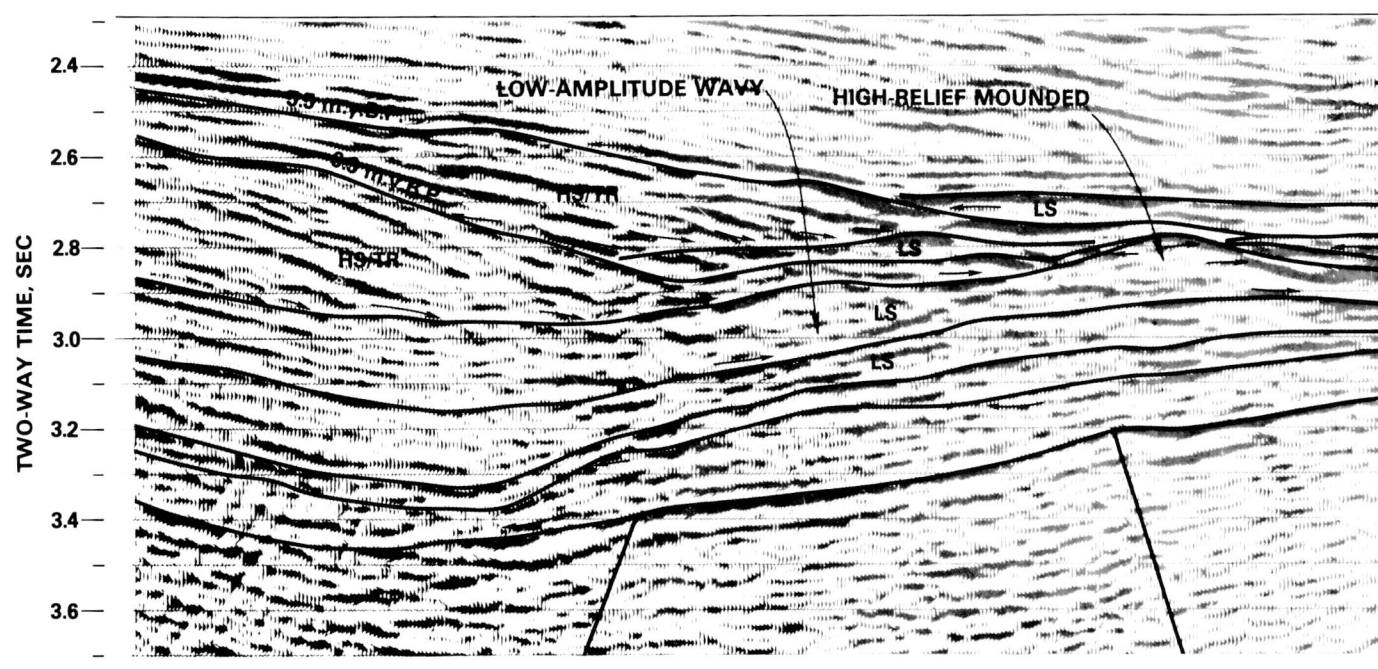

Fig. 11.—Seismic line C. Upper Serravallian to lower Tortonian (above 10.5 Ma) lowstand-systems tract onlaps underlying platform margin and is characterized by a concentric belt of high-relief seismic mounds. These mounds are strike-oriented and aggradational on west side and dip-oriented and progradational on east side. See Figures 13, 14, and 15 for location of line.

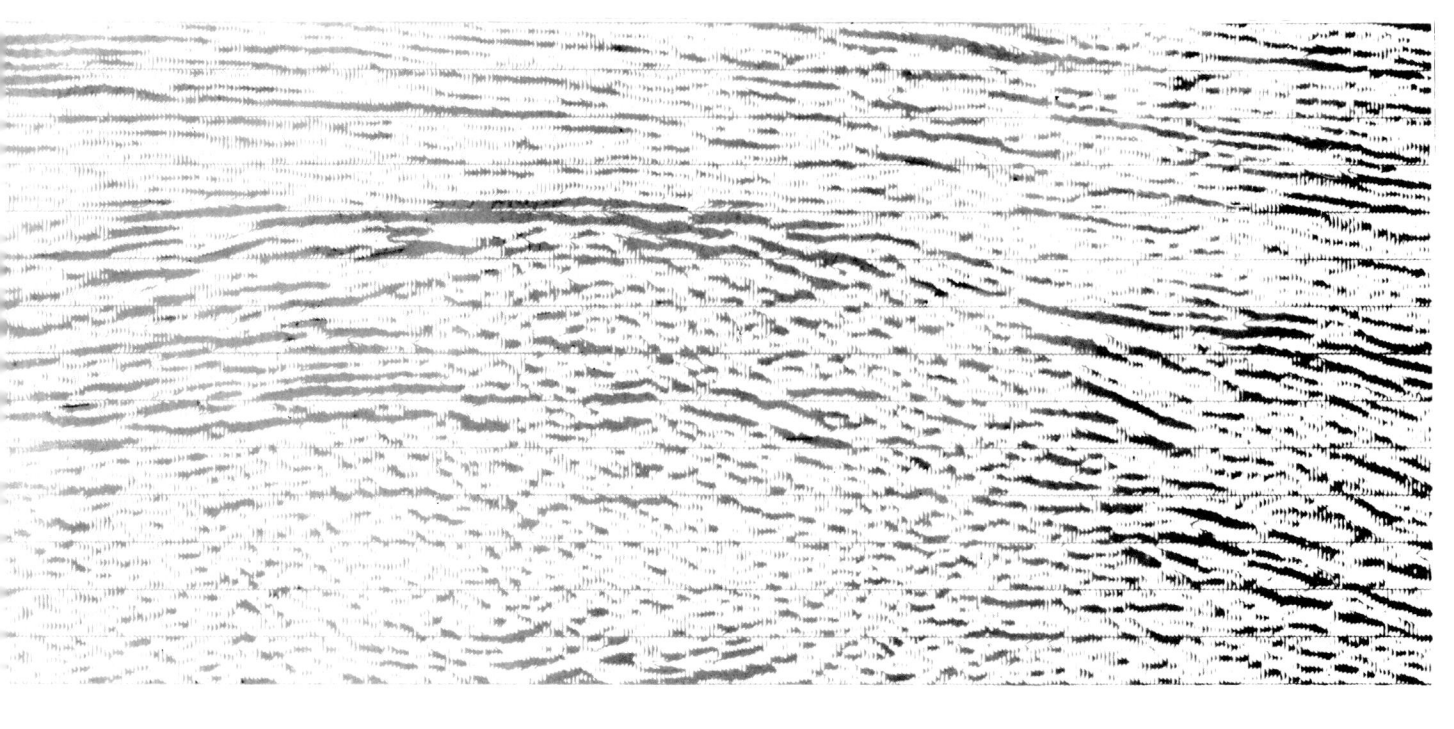

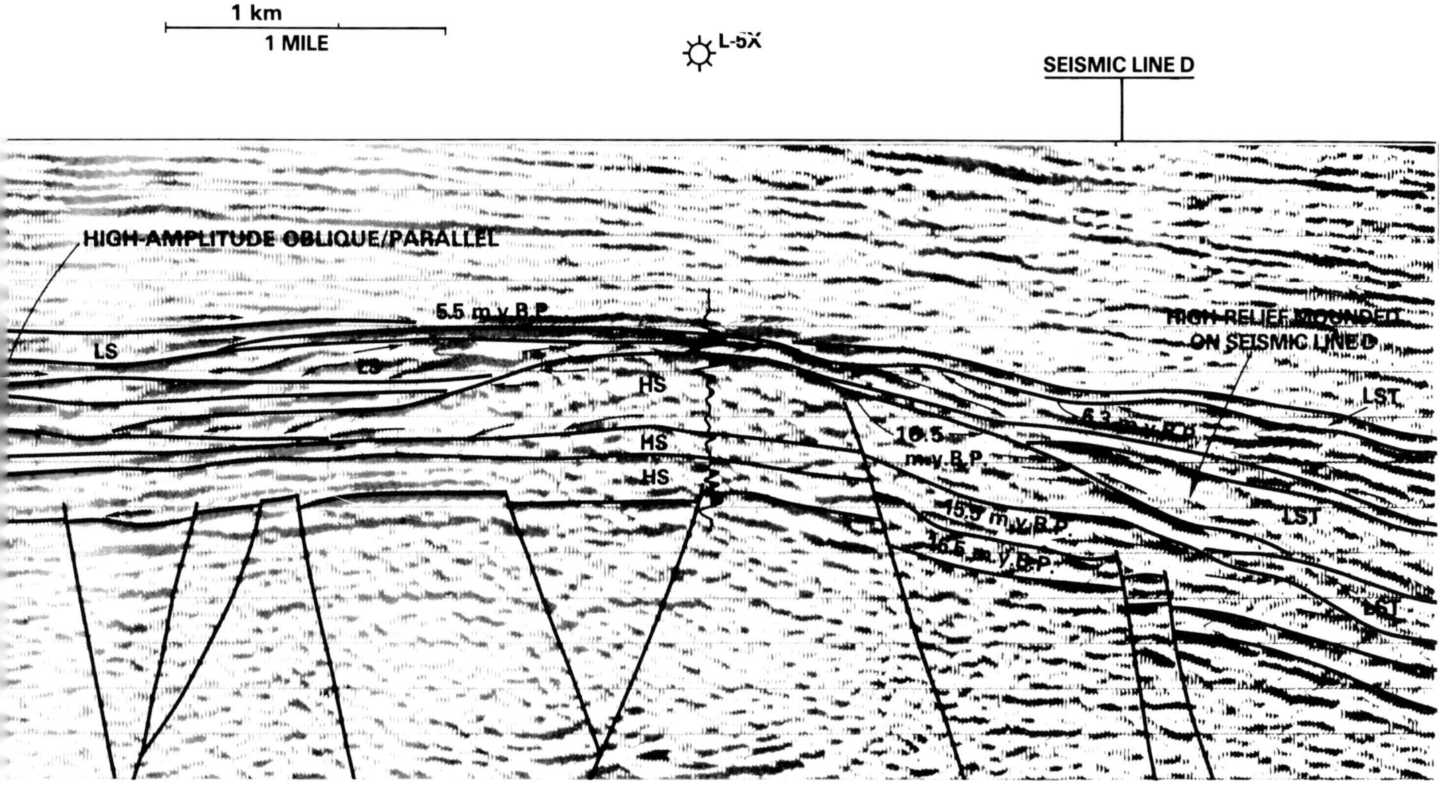

SYSTEMS TRACTS
HS - HIGHSTAND
TR - TRANSGRESSIVE
LS - LOWSTAND

associated with the 4.2 – Ma transgressive/highstand cycle drowned the platform and ended carbonate production (Fig. 6).

CONCLUSIONS

The evolution of the Natuna Platform is controlled by eustasy, subsidence, and differential carbonate productivity. Eustasy imposes a cyclicity that is expressed through seven complete depositional sequences.

Constituent systems tracts are: (1) *Lowstand*–erosional modification of platform margin, exposure of platform crest and resultant dissolution, downward shift of reservoir lithofacies; (2) *Transgressive*–deepening upward of lithofacies, reflooding of platform crest, retreat of platform margin or retrogradation; and (3) *Highstand*–shoaling upward of lithofacies, progradation of platform margin.

Increasing subsidence in the Late Miocene caused the progressive retreat of the Natuna Platform. Retreat occurs incrementally, during deposition of the transgressive-systems tract.

The Natuna Platform is characterized by differential carbonate productivity, with greater productivity on the open-basin (east) side of the platform. As a result, platform retreat occurs in an asymmetric fashion, with less retreat and better reservoir development on the open-basin side of the platform.

During the course of this study, two pre-drill predictions of porosity in the L-4X and L-5X wells were made by using the platform evolution model, seismic-facies analysis, and seismic-modeling analysis (Table 2).

ACKNOWLEDGMENTS

The writers thank Exxon Production Research, Exxon Company International, and Pertamina for permission to publish this paper. J. F. Read, T. Simo, and J. F. Sarg reviewed the manuscript and made many helpful suggestions.

REFERENCES

BROWN, L. F., AND FISHER, W. L., 1977, Seismic-stratigraphic interpretation of depositional systems: Examples from Brazil rift and pull-apart basins, *in* Payton, C. E., ed., Seismic Stratigraphy–Applications to Hydrocarbon Exploration: American Association of Petroleum Geologists Memoir 26, p. 213–248.

EPTING, M., 1980, Sedimentology of Miocene carbonate buildups, central Luconia, offshore Sarawak: Geological Society of Malaysia Bulletin, v. 12, p. 17–30.

EYLES, D. R., AND MAY, J. A., 1982, Exploration of the L-Structure, Natuna D-Alpha Block, Offshore Indonesia: CCOP-ASCOP Seminar on Hydrocarbon Occurrence in Carbonate Rocks, Surabaya, Indonesia, 15 p.

HAMILTON, W., 1979, Tectonics of the Indonesian region: U.S. Geological Survey Professional Paper 1078, 345 p.

HAQ, B. U., HARDENBOL, J., AND VAIL, P. R., 1987, Chronology of fluctuating sea levels since the Triassic: Science, v. 235, p. 1156–1167.

MAY, J. A., AND EYLES, D. R., 1985, Well log and seismic character of Tertiary Terumbu carbonate, South China Sea, Indonesia: American Association of Petroleum Geologists Bulletin, v. 69, p. 1339–1358.

MITCHUM, R. M., JR., VAIL, P. R., AND SANGREE, J. B., 1977, Seismic stratigraphic interpretation of seismic reflection patterns in depositional sequences, *in* Payton, C. E., ed., Seismic Stratigraphy–Applications to Hydrocarbon Exploration: American Association of Petroleum Geologists Memoir 26, p. 117–133.

TAYLOR, B., AND HAYES, D. E., 1983, Origin and history of the South China Sea Basin, *in* Hayes, D. E., ed., The Tectonic and Geologic Evolution fo Southeast Asian Seas and Islands, Part 2: Geophysics Monograph 27, Lamont-Doherty Geological Observatory, Palisades, New York, p. 23–57.

TABLE 2.—RESERVOIR PREDICTIONS–AVERAGE POROSITY PERCENT

Systems Tract	L-4X Predicted	L-4X Logged	L-5X Predicted	L-5X Logged	Range in all wells
Messinian highstand	25	8	10	6	6–26
E. Messinian lowstand	*	*	15	14	14–20
L. Tortonian early highstand	24	24	*	*	24–26
M. tortonian highstand	24	25	*	*	3–25
E. Tortonian transgressive	17	17	5	3	7–20
M.-L. Serravallian highstand	20	19	15	13	12–19
E. Serravallian highstand	13	12	13	13	12–14
Langhian highstand	13	13	13	10	10–13

*Very thin or absent

PART V
BANK DEVELOPMENT IN A FORELAND BASIN AND PELAGIC SEDIMENTATION IN AN ACTIVE—MARGIN BASIN

UPPER CRETACEOUS PLATFORM-TO-BASIN DEPOSITIONAL–SEQUENCE DEVELOPMENT, TREMP BASIN, SOUTH-CENTRAL PYRENEES, SPAIN

ANTONIO (TONI) SIMO

Department of Geology and Geophysics, University of Wisconsin, 1215 West Dayton Street, Madison, Wisconsin 53706

ABSTRACT: Upper Cretaceous depositional-sequence architecture of the Tremp Basin in the Pyrenees reflects the effects of both subsidence and eustasy. Overall geometry is controlled by thermal subsidence, and depositional-sequence geometry, depositional facies, and sequence boundaries are controlled by relative sea-level oscillations as well as subsidence.

Five depositional sequences, of seismic stratigraphic scale, are recognized. (A) The three oldest sequences show moderate extent of the bayline coastal onlap, backstepping of each successively younger carbonate platform, and shelfward basin encroachment, which increased the water depth in the basin. (B) The fourth sequence is the thickest sequence and displays a dramatic backstepping of the carbonate platform with coastal onlap occurring over 60 km shelfward of the previous sequences. The fourth sequence is divided into a lower siliciclastic-rich subsequence and an upper carbonate-rich subsequence. Depositional sequences one through four can be divided into a volumetrically small transgressive lower unit and a volumetrically large regressive upper unit. (C) The fifth depositional sequence shows an overall regressive cycle that filled the basin. Two subsequences are differentiated, each made of siliciclastic nearshore, shelf, and basin facies.

Sequence boundaries, over most of the study area, are concordant surfaces where a sharp transition from shallower to deeper water facies occurs. These conformable surfaces can be traced (1) shelfward into subaerial karst surfaces, or channelized surfaces filled with sandstone, and (2) basinward into submarine erosional truncations over which chaotic sediments onlap.

Tectonism is recorded in the geometry and sequence boundaries of the Upper Cretaceous sediments of the Tremp Basin. Thermal subsidence during Cenomanian-Coniacian time created depositional space in the basin, but uplift and slow-subsidence rates reduced bayline onlap on the shelf. Subsidence increased during late Santonian time, creating new depositional space and a widespread coastal onlap. Uplift concurrent with thrusting during Maastrichtian time induced erosion around the thrust area. Post-thrust crustal relaxation created new depositional space for accumulation of the fifth sequence. The formation of the sequence boundaries was influenced by listric-normal faulting, uplift, and periods of rapid subsidence.

Relative sea-level changes of as much as 150 m are recorded in the Upper Cretaceous of the Tremp Basin. Comparison of the sediment thickness with the predicted space created by thermal subsidence requires relative sea-level oscillations to create the additional depositional space needed for sediment accumulation. The geometry of each depositional sequence, with a thin transgressive unit and a thick regressive unit, suggests an initial rapid sea-level rise followed by a stillstand with subsequent sea-level drop toward the end of the sequence. The relative lowering of sea-level exposed the inner shelf, and in some cases channels were cut on the platform and filled with sandstones. However sea-level drops do not appear to have exposed the entire shelf and shelf edge. These sea-level drops may have triggered erosion in the basin by increasing the current activity and instability of sediments.

The depositional sequences and sequence boundaries recognized in the Tremp Basin are time equivalent to other sequences and sequence boundaries in other Cretaceous basins. These similarities between different basins support a "global" mechanism controlling depositional sequences during the Upper Cretaceous, with distinctive tectonic factors in each basin (or a segment of a basin) controlling each depositional-sequence geometry and facies pattern and depositional-sequence succession.

INTRODUCTION

Upper Cretaceous sediments of the south-central part of the Pyrenees were deposited during spreading of the European and Iberian plates and during later subduction of the Iberian plate beneath the European plate. As a result of spreading, the Pyrenean basin opened and a cyclic succession of four depositional sequences consisting of backstepping carbonate platforms occurred. A fifth sequence, composed of siliciclastic sediments, formed as a result of continent-continent collision.

This article describes the geometries and facies of the Upper Cretaceous depositional sequences, which reflect the effects of subsidence (thermal cooling, sediment loading, uplift) and eustasy, in relation to the evolution of the Iberian continental margin. The sequence boundaries and genetic packages were defined from detailed field mapping and observations of the Upper Cretaceous of the south-central Pyrenees (Simó, 1985). One geologic map (at 1:25,000 scale) of the study area was generated during approximately one year of field work. During the mapping, the stratigraphy was reviewed and published stratigraphic sections were correlated (Mish, 1934; Rosell, 1967, 1970; Souquet, 1967; Mey and others, 1968; Nagtegaal, 1972; Garrido-Mejias, 1973; Gallemi and others, 1983).

Cretaceous sediments show only moderate structural deformation and are exceptionally well exposed. Stratigraphic sections and geometry descriptions are based on physical and aerial photography correlation at a scale comparable to seismic-reflection profiling. Concepts of carbonate platform interpretation (Wilson, 1975; James and Mountjoy, 1983; Read, 1982, 1985), depositional-sequences description (Vail and others, 1977), and subsidence analysis (Steckler and Watts, 1978) are used in this paper to describe and interpret sedimentology and margin evolution.

REGIONAL SETTING

The study area (Fig. 1) is the Tremp Basin and is located in the Montsec thrust sheet (Williams, 1985) of the south-central Pyrenees Upper Thrust Sheet unit (Muñoz, 1985). The Montsec thrust sheet (Fig. 2), a part of the Iberian continental margin, moved southward during the continent-continent collision of the Iberian and European plates during the Eocene. Restored cross sections suggest a minimum shortening of 43.5 km (55 percent) in the south-central Pyrenees (Williams, 1985) and 20 km (60 percent) in the southeastern Pyrenees (Muñoz, 1985). Internally, the Montsec thrust sheet shows only moderate tectonic deformation, with the exception of some synsedimentary thrusting (Boixols thrust).

The Mesozoic Pyrenean basin opened by a 35°-counter-

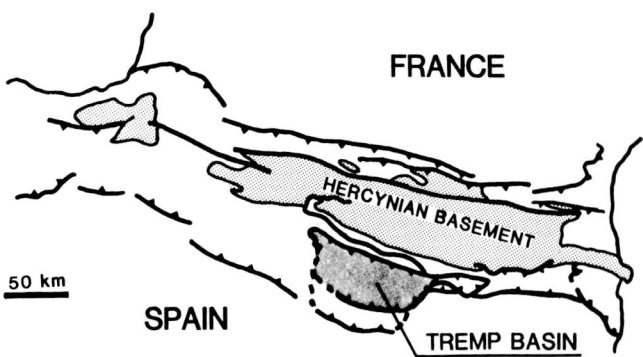

FIG. 1.—Structure map of the Pyrenees and location of the Tremp Basin.

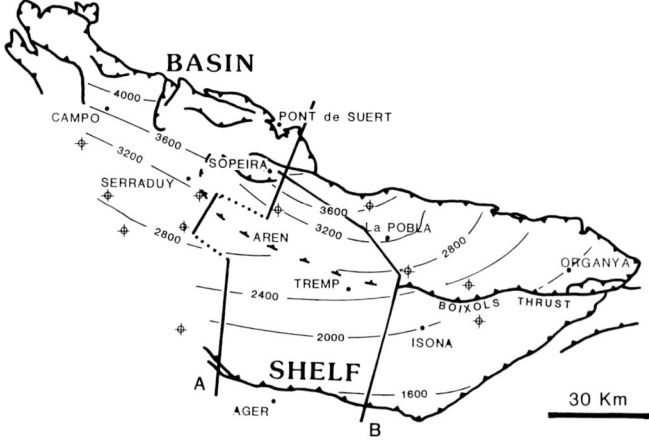

FIG. 3.—Isopach map (in meters) of the Upper Cretaceous rocks of the Tremp Basin. This map does not show the sediment thickness eroded away during synsedimentary thrusting; see Simó (1985) for a more detailed isopach map. Note basin is to the north and shelf to the south. Note location of oil wells (⊕) and cross sections A (Fig. 2) and B (Fig. 5).

clockwise rotation of the Iberian plate with respect to the European plate (Masson and Miles, 1984). The collision and formation of the Pyrenees resulted from a 120-km displacement of the Iberian plate from southeast to northwest during Cenozoic times (Grimaud and others, 1982). Souquet (1984), and Puigdefabregas and Souquet (1986) give field evidence for these movements, and Banda and Wickham (1986) review the evolution of the Pyrenees.

Upper Cretaceous paleogeography of the Tremp Basin shows an expanded and subsiding basin in the north-northwest and a more stable platform in the south. Upper Cretaceous sediments thicken toward the north-northwest (Fig. 3). Cretaceous sediments are continuously exposed in the south, east, and north, but are covered in the central and western parts of the Tremp Basin by Tertiary sediments (Fig. 4). Sedimentation during the opening of the basin was mostly of a marine-carbonate nature. With the onset of plate collision, sedimentation changed to siliciclastic sandstones and non-marine shales. South-verging synsedimentary thrusts created changes in the Tremp Basin, including subaerial erosion in the northeast, non-marine sedimentation in the south and east, and marine sedimentation in the northwest. The Upper Cretaceous Pyrenean basin ended with formation of the Pyrenees and development of Pyrenean foreland basins in the north and south (Fig. 1). The thin (900 m) Tertiary sediment cover in the center of the Tremp Basin suggests that the Tremp Basin has undergone only shallow burial since the Maastrichtian.

The Mesozoic stratigraphy of the southern Pyrenees was mapped and reported in Mish (1934), Rosell (1967, 1970), Souquet (1967), Mey and others (1968), and Garrido-Mejias (1973), among others. Detailed work of the Upper Cretaceous stratigraphy (Pons, 1977; Gallemi and others, 1983) and interpretation of the depositional environments (Van Hoorn, 1970; Rosell and others, 1972; Nagtegaal, 1972) resolved some of the stratigraphic uncertainties. The above works, however, resulted in different stratigraphic nomenclature at each locality, and thus, correlation between sections was confusing. Simó (1985, 1986) simplified the stratigraphic nomenclature and defined the Upper Cretaceous depositional sequences of the Tremp Basin.

An extensive paleontologic literature has been published on the Upper Cretaceous of the Tremp Basin. Some workers (Bilotte and Souquet, 1972; Liebau, 1973; Fondecave, 1975; Bilotte, 1978; Gomez-Garrido, 1981) refer to local sections, whereas others (Pons, 1977; Caus and others, 1981) either synthesize or publish their own planktonic foraminifer and rudist data. The most valuable paleontologic and biostratigraphic studies are those of Martinez (1982) and Gomez-Garrido (1987), which were carried out independently of the sequence-stratigraphic analysis. Both studies complement the ammonite (Martinez, 1982) and the planktonic foraminiferal (Gomez-Garrido, 1987) biostratigraphy. I correlate the faunal assemblages of Martinez and Gomez-Garrido to the depositional sequence framework (Fig. 7) and those together to the time scale of Haq and others (1987). This time scale is used for convenience in relating the depositional record of the Upper Cretaceous of the Tremp Basin to the eustatic history and global depositional cycles defined by Haq and others (1987).

DEPOSITIONAL FACIES

The Upper Cretaceous of the Tremp Basin is divided into two depositional facies tracts (such as in Wilson, 1975): a carbonate-dominated and a siliciclastic-dominated tract. The

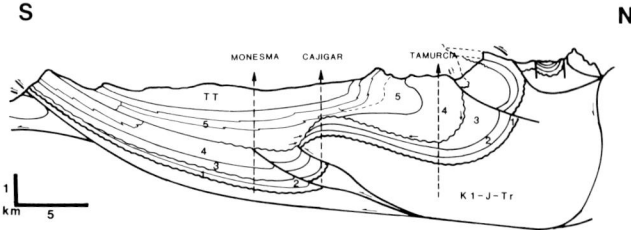

FIG. 2.—Structural cross section through the Tremp Basin (Simó, 1985). See location on Figure 3. The numbers here correspond to the Upper Cretaceous depositional-sequence numbers (Figs. 4, 5, and 7). K1-J-Tr = Lower Cretaceous, Jurassic, Triassic, and TT = Tertiary.

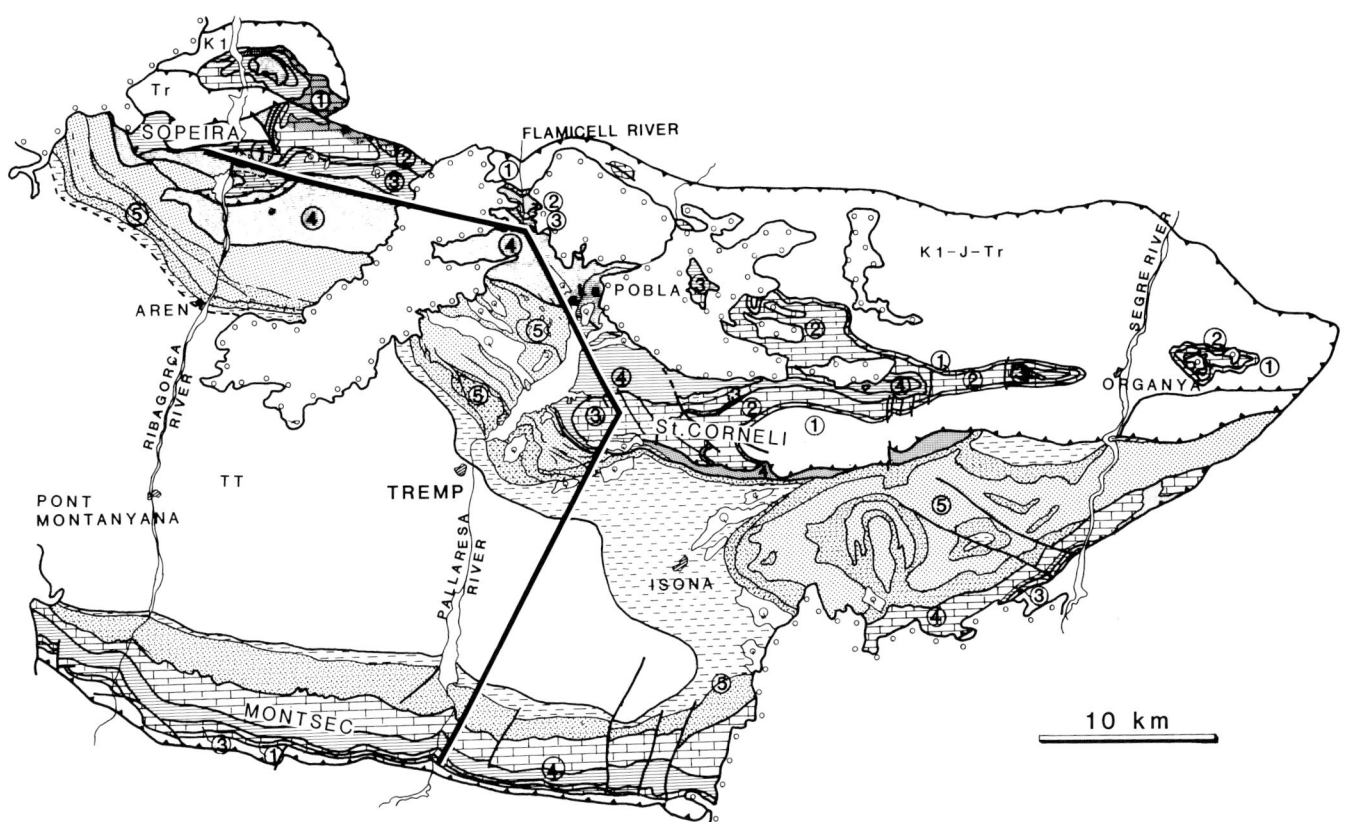

Fig. 4.—Geologic map of Upper Cretaceous rocks of the Tremp Basin east of Sopeira. Numbers and patterns are the same as in Figure 5.

four youngest sequences (Santa Fe, Congost, Sant Corneli, and Vallcarca) are carbonate dominated, and an ideal facies tract reveals six depositional facies (Fig. 6), which are summarized here.

Carbonate Platform Facies.—

The carbonate platform facies represents the most shallow-water environment. The platform facies is aggradational (Sant Corneli sequence), progradational (Congost and Vallcarca sequences), or a combination of both (Santa Fe sequence). Carbonate platforms are characterized by two facies: (1) lagoonal (subtidal burrowed lime mudstones to packstone, with occasional rudist buildups), and (2) shoal-reef complex or bank margin facies (fine to very coarse packstones and grainstones, and coral-rudist framestone).

Carbonate Shelf Facies.—

The shelf facies differs from the platform facies in that it lacks bank margin facies, contains a retrogradational succession, and is lens shaped. The carbonate shelf deposits occur as a thinning-upward succession of shallow-water, low-energy facies that pinch out in both a landward and a basinward direction. The carbonate shelf facies is dominated by fine-grained sediments, and the major lithologies are (A) well-bedded mudstones and packstones (Pinyana limestone), (B) cross-bedded calcarenites and black shales (Collada Calcarenites), and (C) rudist-coral buildups and nodular marls (Bastus Limestones).

Upper-Slope Facies.—

The upper-slope facies occurs adjacent to bank margin facies and extends basinward for a few kilometers. The upper-slope facies is composed of fine-grained sediments, skeletal fragments, and clasts derived from the platform. Most of upper-slope facies deposits are allochthonous with the dominant transport mechanism being offbank dispersion and downslope mass movement. The upper-slope deposits are either massive, coarsening-upward packages composed of calcarenites and blocks (Santa Fe sequence), coarsening- and thickening-upward, fine-grained calcarenites (Congost and Sant Corneli sequences), or sheetlike breccia beds within mudstones (Vallcarca sequence).

Outer-Shelf Facies.—

The outer-shelf facies occurs as a relatively "deep" shelf, largely below fair-weather wave base, between the upper-slope and lower-slope facies, and with a depositional basinward dip of a few (1°–3°) degrees. The sediments are mostly autochthonous. The sediment rate, compared to that of the other facies, is low. Shallow-water-derived sediments bypass the outer shelf. The outer-shelf facies consists of nodular limestones, burrowed marls and shales, with isolated coral and equinoids mixed with ammonites and planktonic foraminifera. Glauconite is abundant locally (Sant Corneli and Vallcarca sequences). Slump scars and thin-bedded calcarenite beds are rare. Basinward, the outer-shelf facies consists of laminated claystone (Vallcarca sequence).

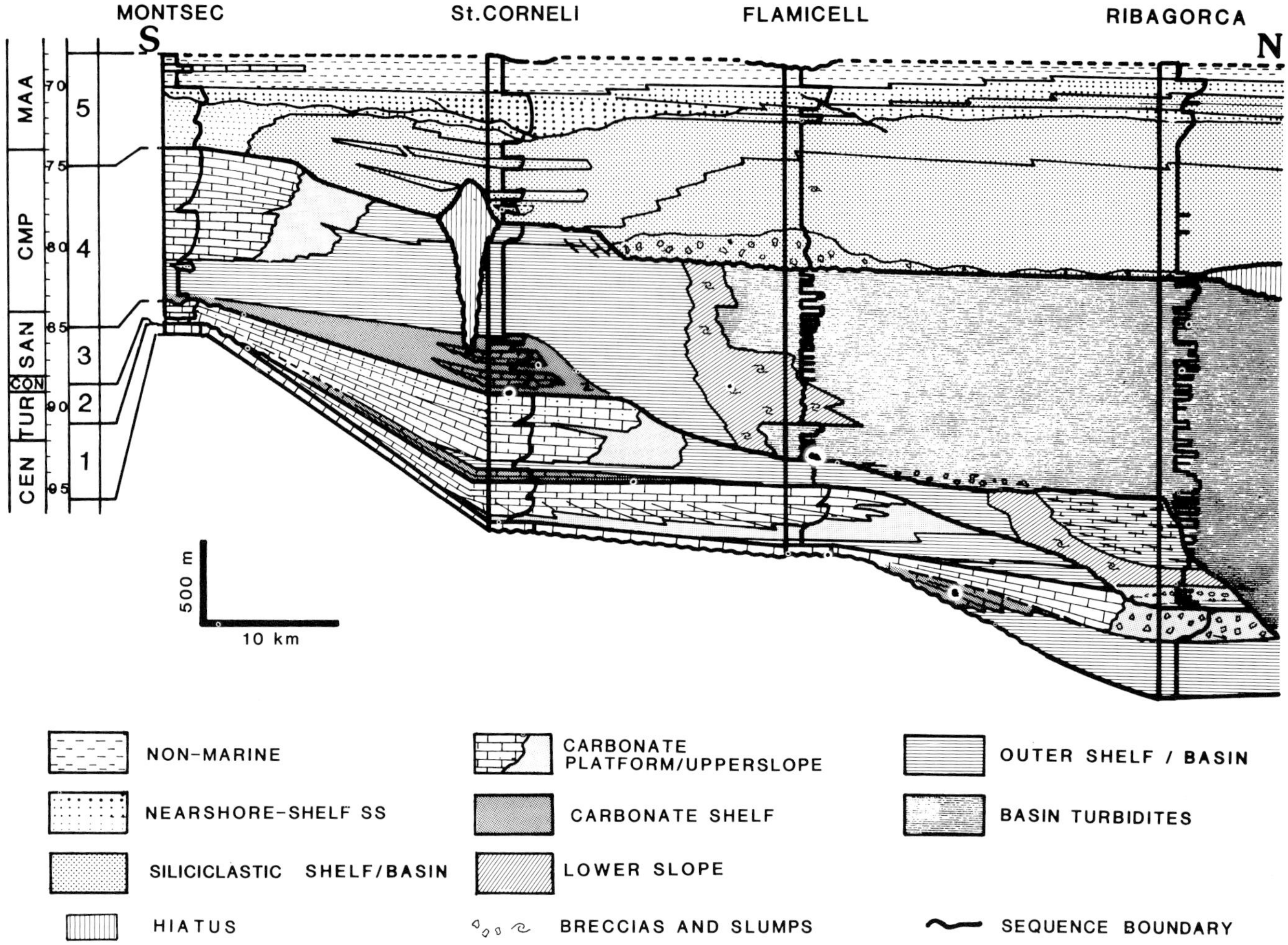

FIG. 5.—Stratigraphic framework, stratigraphic sections, and depositional sequences of the Upper Cretaceous of the Tremp Basin. Stratigraphic section location is shown in Figure 3 (section B) and 4. Depositional sequences and numbers as in text and Figure 7.

Lower-Slope Facies.—

The lower-slope facies occurs between the outer-shelf and basin facies and is adjacent to a submarine escarpment. The sediments are allochthonous, having been derived from outer-shelf and more proximal lower-slope facies, and from the erosion of older depositional sequences. This facies is not recognized in the Santa Fe and Congost sequences. The lower-slope facies consists of thin, well-bedded, and burrowed lime-siltstone containing many slump scars (Sant Corneli sequence), or massive shale with conglomerates (Vallcarca sequence). The dominant transport mechanism is downslope mass movement.

Basin Facies.—

The carbonate basin facies consists of glauconitic marls and shales (Santa Fe sequence), and black, thin-bedded, calcareous mudstones with radiolarians and sponge spicules (Sant Corneli sequence). The Vallcarca sequence basin facies (formally called the Mascarell Turbidites) is siliciclastic dominated and consists of normal graded, coarse- to fine-grained sandstone (quartz and some skeletal fragments) and claystones.

The carbonate-dominated facies tract distribution is different for each of the four major sequences (see Fig. 8). The platform profile changes from the lower and oldest Santa Fe sequence to the upper and youngest Vallcarca sequence. Platform backstepping and shelfward basin encroachment

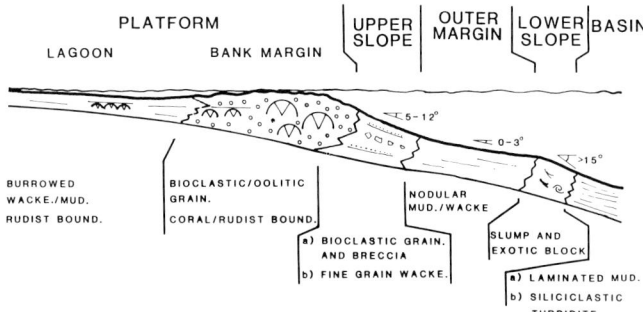

FIG. 6.—Diagram of morphologic profile of the Tremp Basin Upper Cretaceous carbonate platform and schematic representation of facies tracts.

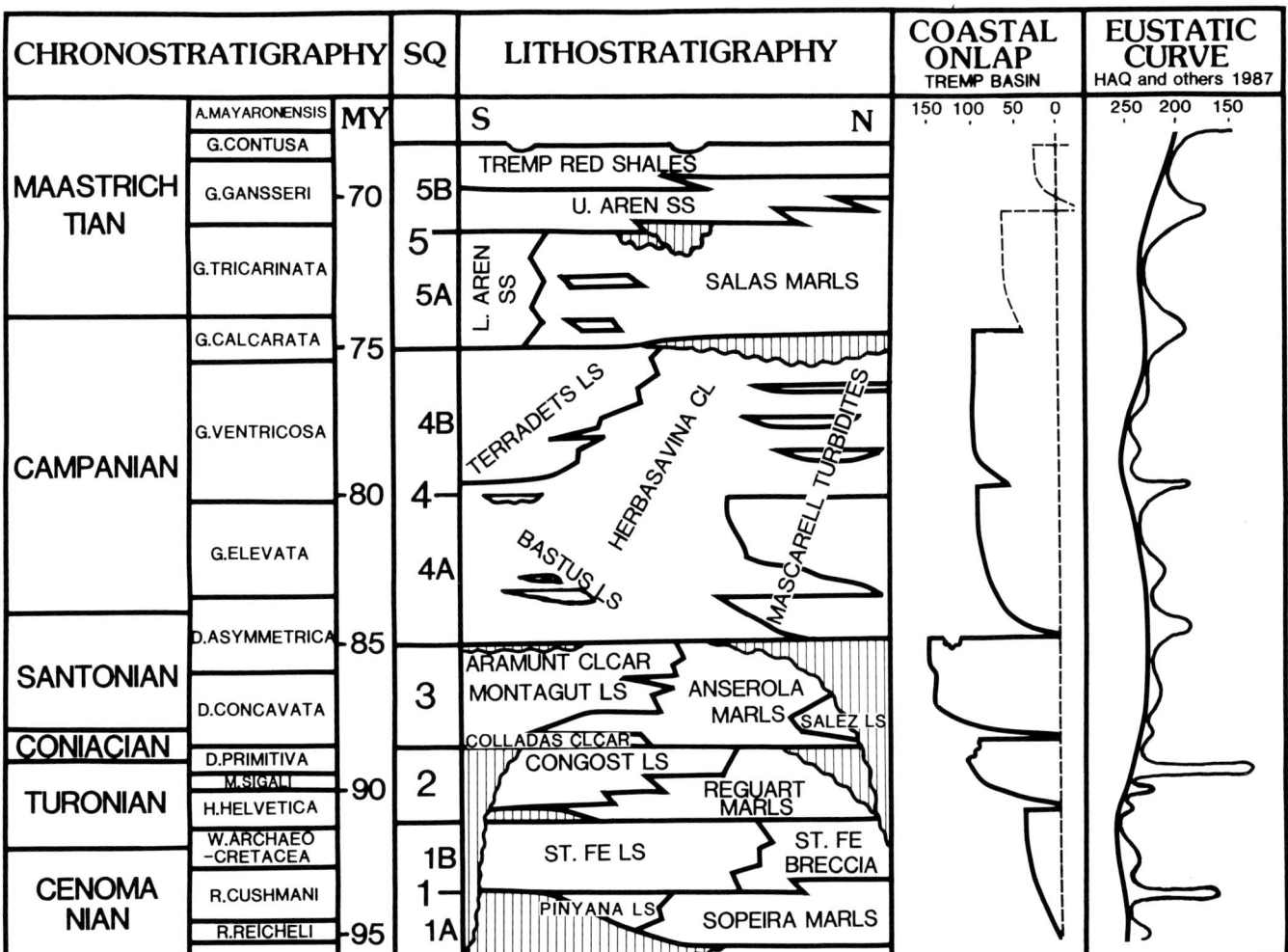

FIG. 7.—Chronostratigraphic framework, depositional sequences, and stratigraphic units of the Upper Cretaceous Tremp Basin. Note the Tremp Basin relative sea-level curve, as compared with the global eustatic curve of Haq and others (1987). Chronostratigraphy is from Haq and others (1987). SS = sandstone, LS = limestone, CL = clay, CLCAR = calcarenite.

created basin margin differentiation into an outer shelf and a lower slope. The carbonate shelf facies is not incorporated into the facies tract diagram and is discussed in the evolution of each depositional sequence.

The fifth and youngest Aren sequence is siliciclastic dominated with three recognized depositional facies.

Non-Marine Facies.—

Formally called the Tremp Red Shales, the non-marine facies consists of coastal-swamp deposits (as much as 90 m thick, gray to black, fine-grained carbonaceous shales, and sandy marls), lacustrine limestones (as much as 4 m thick and a few kilometers of lateral extension), fluvial channel fill (conglomerates and sandstones), and red shales.

Siliciclastic Nearshore Facies.—

The nearshore facies contains coarsening- and shallowing-upward packages of tidal, wave-barrier, estuarine-channel, and aeolian-dune deposits. Multidirectional cross-bedded quartz-bearing sandstones are the dominant lithology with minor occurrences of quartz-rich calcarenites. Grain sizes range from very fine to pebbly sandstone; local conglomerate layers occur at the top of the unit.

Siliciclastic Shelf and Basin Facies.—

The siliciclastic shelf facies is composed of cross-bedded sandstones and shale. Three separate stages of sand bar development occur, each as much as 40 m thick with hundreds of meters of lateral extent. Quartz and marine skeletal grains form the fine-to-coarse sand fraction, and the proportion of bioclasts increases seaward. The shales are massive with horizons of intense burrowing. Basinward, these shales are more laminated and have darker gray colors. The basin facies, formally called the Salas Marls, consists of dark gray clays with thin-bedded, quartz-bearing calcarenites and red carbonate nodules.

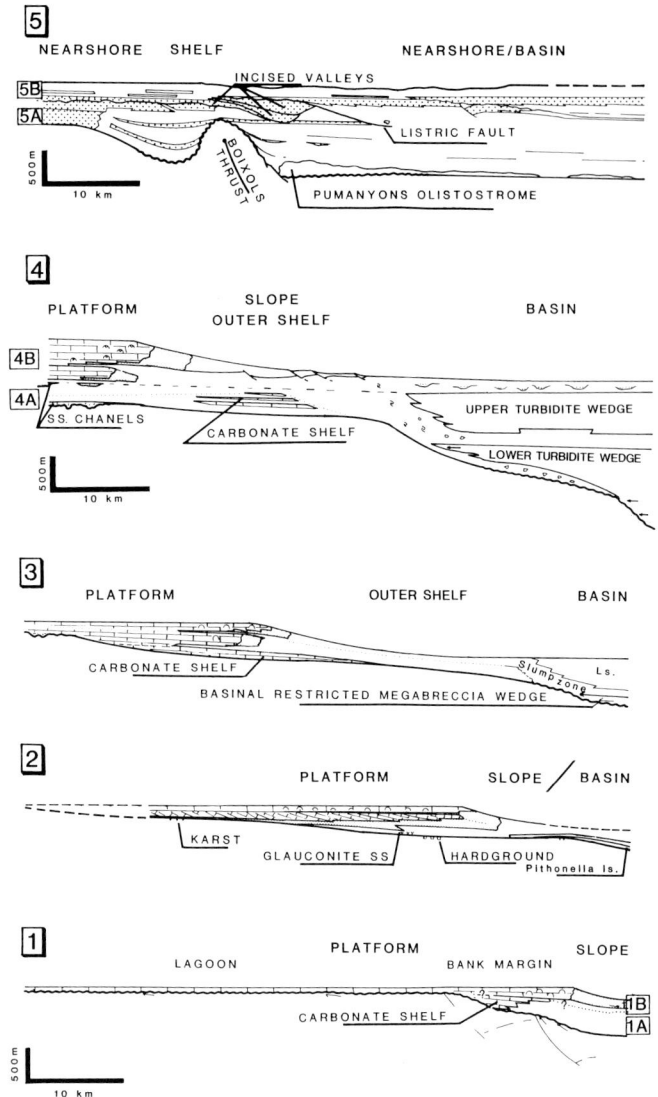

FIG. 8.—Enlargements of Upper Cretaceous depositional sequences along section in Figure 4. Facies descriptions and geometries presented in text. Sequence numbers (cf. Fig. 5) are shown in the upper left corner.

UPPER CRETACEOUS DEPOSITIONAL SEQUENCES

Introduction

Cretaceous transgressive-regressive cycles of global distribution and with a frequency of 5.5 ± 1 m.y. have been recognized world-wide (see summary in Schlanger, 1986). The Upper Cretaceous sediments of the Tremp Basin show a sequential character defined by transgressive-regressive cycles, each bound by surfaces of erosion, non-deposition, or their correlative conformities (Simó, 1986). Cycles of this type have been referred to as depositional sequences and the bounding surfaces as sequence boundaries by Vail and others (1977, 1984). Because this concept is useful in field geology and seismic interpretation, the terminology of Vail and others (1984) is followed in this paper.

Five depositional sequences can be recognized in the Upper Cretaceous rocks of the Tremp Basin (Figs. 7, 8). The oldest three sequences are composed of carbonate platform, slope, and basin facies. These three sequences show: (1) very little coastal onlap; (2) a backstepping of each successively younger carbonate platform; (3) a shelfward basin encroachment, which increased the water depth in the basin; and (4) basin margin differentiation into upper-slope, outer-shelf, and lower-slope facies (Fig. 8).

The fourth and thickest sequence is divided into two subsequences: (1) a lower siliciclastic-rich subsequence with outer-shelf, lower-slope, and basin facies, a thick turbiditic succession fills the basin, and (2) an upper carbonate-rich subsequence with a thick carbonate platform succession, upper-slope, outer-shelf, and basin facies. The fourth sequence shallow-water sediments backstepped considerably compared to lower sequences with coastal onlap occurring across 60 km.

The fifth depositional sequence (the youngest) shows an overall regressive succession that filled the basin with siliciclastic nearshore, shelf, and basin facies.

Depositional sequences one through four can be divided into a volumetrically minor transgressive lower unit and a volumetrically thicker regressive upper unit. The transgressive units are represented by carbonate shelf facies or by a centimeter-thick lag on top of a sequence boundary. The regressive units are volumetrically the most important units of the depositional sequences and show a shallowing-upward character. They are represented by carbonate platform, slope, and basin facies.

In the Tremp Basin, the sequence boundaries are identified by concordant surfaces, where a sharp transition from shallower to deeper facies occurs. The conformity surfaces can be traced (1) shelfward into karst surfaces, or surfaces scoured by channels, and (2) basinward into submarine erosional truncations that occur within relatively deep-water facies without evidence of shallow-water facies. Basinal restricted wedges that onlap basinal erosion surfaces resulted from the same conditions that generated the erosional truncation.

The following summary describes the five sequences and focuses on the stages of development and evolution of the carbonate platforms and basin evolution throughout the entire Upper Cretaceous history of the Pyrenees.

Description

Santa Fe sequence (sequence #1).—

This sequence is bounded below by the lower Cenomanian and above by the middle Turonian sequence boundaries and lasted approximately 4.5 m.y. The lower Cenomanian sequence boundary (planktonic foraminifer zone Rotalipora brotzeni 95.5 Ma; Fig. 9) occurs as a regional erosion surface (subaerial and submarine). At this time, the basin depocenter shifted from east to west (Rosell, 1967; Van Hoorn, 1970) as a result of listric-normal-fault development during regional extensional tectonism. Part of the Lower Cretaceous was subaerially exposed during this time, and the unit was partially truncated before deposition of Upper Cretaceous marine sediments.

Congost sequence (sequence #2).—

The Congost sequence is bounded at its base by the middle Turonian sequence boundary and above by the late Coniacian sequence boundary. The Congost sequence is mid-Turonian to late-Coniacian age (Pons, 1977; Gomez-Garrido, 1981; Martinez, 1982) and lasted about 2.5 m.y. The middle Turonian sequence boundary (planktonic foraminifer zone Helvetogtobotruncana helvetica 91 Ma) occurs as a local preserved karst surface above the platform facies of the Santa Fe sequence. Above the karst, glauconitic sand or a hardground occurs. Pelagic mudstone (condensed section) drapes the entire sequence boundary.

The Congost sequence is composed of two parts: (1) a lower onlap wedge and (2) an upper carbonate platform, slope, and basin facies wedge. The lower wedge is thicker in the basin, where it consists of laminated pelagic limemudstone (Pithonella Limestone) with some synsedimentary folds that thin shelfward from 30 m to a few centimeters.

The upper-wedge carbonate slope/basin facies thickens from 40 to 200 m basinward, and the carbonate platform facies retains a constant thickness (170–200 m thick) throughout the 45- to 50-km-long outcrop. In the basin, the thickness of the upper wedge is reduced to 50–100 m because of both depositional thinning and erosion. In the inner shelf the Congost sequence is mostly eroded away (Fig. 8). Carbonate platform facies can be divided into two parts: a lower part containing large-scale prograding clinoforms and upper slope/basin facies (Fig. 10), and an upper part containing flat-bedded coral-rudist reefs. The contact between the two parts of the carbonate platform is sharp. Basinward, both parts can be traced as two coarsening- and thickening-upward sections, within an overall shoaling-upward succession. The carbonate slope/basin facies is composed of well-bedded mudstones and wackestones with very few shallow-water fauna fragments. The basin facies is black ribbon limestone and shale preserved as a thin section 40–60 m thick.

FIG. 9.—Angular unconformity (Lower Cenomanian sequence boundary) between the Lower Cretaceous (L.K.) and the Upper Cretaceous Santa Fe Limestone (S.Fe Ls.). Note the continuity of the Santa Fe and Congost (CO.Ls) Limestones. Dashed line shows bedding. Thickness of the Santa Fe Limestone in the photograph is about 55 m. Location south flank, Santa Fe syncline.

The Santa Fe sequence is represented by the Pinyana and Santa Fe Limestones, Santa Fe breccia, and Sopeira Marls (Fig. 7) and was deposited from lower mid-Cenomanian to lower Turonian (Martinez, 1982; Bilotte and Souquet, 1972; Bilotte, 1978).

The Santa Fe sequence fills relief created by listric-normal faulting associated with the lower Cenomanian sequence boundary (Fig. 8). Listric faulting controlled not only changes in sequence thickness but also the location of depositional facies. The thicker and more basinal section is localized on the downthrown side of the fault, where a continuous marine section from Lower to Upper Cretaceous occurs. Santa Fe sequence sediments lap outward against the Lower Cretaceous to Jurassic rocks in the upthrown block, which was subaerially exposed to the southeast.

The Santa Fe sequence (Figs. 5, 8) is divided into two subsequences (Santa Fe 1-A and 1-B) on the basis of facies and geometries. Santa Fe 1-A, of about 1.5-m.y. duration, is confined to the downthrown block, where carbonate shelf facies (Pinyana limestone) stack upward and prograde over basinal facies (Sopeira Marls). The carbonate shelf facies has a brown-yellow color and is a low-energy mud-rich facies. The basinal facies is bounded above by a 2-m-thick glauconitic and bioclastic-rich condensed section.

Santa Fe 1-B, of about 3-m.y. duration, is characterized by carbonate platform and upper-slope facies. A 150-m-thick, white, high-energy, carbonate bank margin facies rapidly developed over the low-energy shelf facies of Santa Fe 1-A. A coarsening- and thickening-upward upper-slope facies (Santa Fe breccia) downlaps over the condensed basinal section of Santa Fe 1-A (Fig. 4 of Simó, 1986). A lagoonal facies was deposited behind the bank margin on the upthrown block. The lagoonal facies is homogeneous in thickness and lithology (Fig. 8) and consists of a lower packstone-grainstone to wackestone milliolid-rich unit and an upper bioturbated mudstone to wackestone unit.

FIG. 10.—Congost Limestone clinoforms prograding toward left of picture (basinward) and downlapping over Santa Fe Limestone (S.Fe Ls.). L.K. is Lower Cretaceous. Thickness of the Congost Limestone in the photograph is about 110 m. South flank, Santa Fe syncline.

Sant Corneli sequence (sequence #3).—

The Sant Corneli sequence is bounded below and above by the late Coniacian and late Santonian sequence boundaries and lasted about 3.5 m.y. The late Coniacian sequence boundary (planktonic foraminifer zone Dicarinella primitiva 87.5 Ma) is characterized by subaerial exposure and local scouring on the Congost platform and submarine erosion in the basin. Carbonate megabreccias onlap the basin unconformity, and shelf quartz calcarenites partially cover the platform. Deposition of this sequence took place from late Coniacian to late Santonian (Martinez, 1982; Gomez-Garrido, 1987).

Different units can be recognized within the Sant Corneli sequence (Figs. 5, 8): (1) a basinal restricted wedge of megabreccia beds derived from the erosion of the underlying Santa Fe margin and on sedimentary onlap above the late Coniacian erosional sequence boundary; this unit was first included into the Congost sequence (Simó and others 1985; Simó, 1986), but planktonic foraminiferal dating (Gomez-Garrido, 1987) indicates a Sant Corneli sequence age; (2) a transgressive carbonate shelf facies (Collada Calcarenites), characterized by cyclic black fetid shales and cross-bedded quartz-rich calcarenites; this facies has a lens shape with a maximum thickness of 180 m and thins to a pinchout in both a landward and a basinward direction; a marine hardground occurs on the top of this unit; (3) a carbonate platform, upper-slope, outer-shelf, lower-slope, and basin facies. Two platforms can be recognized: (a) a lower aggradational platform with mud-rich wackestone-packstone limestone containing patchy coral-rudist buildups, and (b) an upper progradational platform with quartz-rich calcarenites and well-developed rudist buildups in the platform/slope transition. Lower and upper platforms are separated by a siliciclastic shelly unit with some rudists and corals. The lower platform upper-slope facies is recognized by fine-grained carbonate sediments derived from the platform and by depositional thinning from 200 to 80 m basinward. The upper-platform upper-slope facies is a combination of coarse-grained carbonate and fine-grained siliciclastic sediments derived from the platform. The outer-shelf facies occurs as a 180-m-thick unit (Anserola Marls) composed of glauconitic nodular marls and shales. The outer-shelf sequence thins upward, siliciclastic shale increases upward, and slump scars become more frequent basinward. Deposition of the lower-slope facies resulted from both the relief originated on the late Coniacian sequence boundary and an increase of depositional space from backstepping of the carbonate platforms. The lower-slope facies is composed of highly burrowed, thin-bedded wackestone with very abundant slumps and slump scars. The basin facies is composed of a 500- to 1,500-m-thick succession of unburrowed, even-bedded, gray to black lime mudstone and black shale; small slumps occur locally.

Vallcarca sequence (sequence #4).—

The Vallcarca sequence is bounded below by the late Santonian sequence boundary and above by the late Campanian sequence boundary and lasted about 10 m.y. The late Santonian (planktonic foraminifer zone Dicarinella asymetrica 85 Ma) sequence boundary is characterized in the basin by erosion and deep channeling (as much as 800 m deep, Fig. 11B) and on the platform by channeling. This is the thickest (100–1,500 m) and most extensive sequence of the Upper Cretaceous.

The Vallcarca sequence (Fig. 8) can be divided into two subsequences (Vallcarca 4-A and 4-B). The Vallcarca subsequence 4-A is composed of three sedimentary wedges: (1) a lower wedge characterized by a siliciclastic basin facies, which onlaps the late Santonian unconformity and cuts deeply into the Sant Corneli, Congost, and Santa Fe sequences (Fig. 11B); this coarsening- and thickening-upward turbiditic section corresponds shelfward to localized quartz sandstone-filled channels that were cut into the Sant Corneli sequence; (2) a middle wedge characterized by a carbonate shelf facies (Colladas Limestones) that overlies the conformable part of the late Santonian sequence boundary and thins both landward and basinward; and (3) an upper wedge characterized by outer-shelf facies (Herbasavina

FIG. 11.—Vallcarca sequence. (A) Aramunt calcarenites (Ar. calca.) overlain by outer-shelf Herbasavina clays (He. Cl.; thickness in the photograph is about 150 m). Sandstone-filled channels (Ss. channels) cut into the Herbasavina clays. The channels are overlain by the Terradets Limestones (Te. Ls.). Montsec area. (B) Erosional truncation (Upper Santonian sequence boundary) between the Mascarell Turbidites (turbidites, Vallcarca sequence) and the Santa Fe breccia (S.Fe, Santa Fe sequence). This unconformity cuts out depositional sequences 2 and 3 in this area. Lower Cretaceous (L.K.) crops out in the background. Width of the photograph in the foreground is about 250 m. Sopeira area.

Clays) and siliciclastic-basin facies composed of a coarsening- and thickening-upward turbiditic section. Farther shelfward, this upper wedge correlates with time-equivalent nearshore sandstones (Caus and others, 1982).

The Vallcarca subsequence 4-B basal boundary is characterized on the shelf by channels cut directly into the outer-shelf shale facies of subsequence 4-A (Fig. 11A) and filled with quartz sandstone of shallower shelf origin. These sandstone channels are not recognized basinward; instead, the section is mostly composed of outer-shelf claystones. The Vallcarca subsequence 4-B is composed of carbonate platform, upper-slope, outer-shelf, lower-slope, and basin facies. The carbonate platform facies (Figs. 8, 11A) contains two coarsening- and shallowing-upward bank margin facies successions. The upper-slope facies is dominated by fine-grained deposits with debris sheets and chaotic masses derived from the platform. The outer-shelf facies consists of claystone and well-bedded white limestones, and near the upper slope, is affected by a large number of growth faults (Simó, 1985, 1986). The basin facies is composed of small-scale thickening- and coarsening-upward turbidite beds filling discrete channels; the turbidites are composed of skeletal fragments and rounded pebble-cobble-size clasts (quartzarenite, quartz) encrusted by corals and rudists.

Aren sequence (sequence #5).—

The Aren sequence is bounded below by the late Campanian sequence boundary and above by the late Maastrichtian sequence boundary and lasted about 7 m.y. The late Campanian sequence boundary (planktonic foraminifer zone *Globotruncanita calcarata* 75 Ma) is characterized by erosional truncation (Fig. 12) and by emplacement of a basin-restricted olistostrome (Pumanyons Olistostrome; Simó and others, 1985). The late Maastrichtian sequence boundary (? 68 Ma) is characterized by erosion and an amalgamation of fluvial channels. This sequence boundary occurs within non-marine deposits, and its formation is concurrent with the southward movement of the axis of the Tremp Basin.

The Aren sequence represents the Upper Cretaceous basin fill resulting from a general northwestward progradation of nearshore to offshore/basin siliciclastic sandstones and shale deposits. This sequence is represented in outcrop by the lower and upper Aren Sandstone, Salas Marls, and Tremp Red Shales (Fig. 7) of uppermost Campanian to late mid-Maastrichtian age (Liebau, 1973; Martinez, 1982; Gomez-Garrido, 1987).

Deposition of the Aren sequence occurs concurrently with thrusting, which is evidenced by thickness reduction and facies changes in the area around the Boixols thrust (Figs. 3, 4) and by uplift (Simó, 1985, 1986). Along the Boixols thrust, the sediment thickness is reduced from 1,000 m to 30 m in 6 km (Fig. 4) as a result of the siliciclastic-shelf facies onlap against the uplifted hangingwall ramp (an enlargement of this area has been published in Simó, 1986, fig. 11). Also, the Tremp Red Shales rest unconformably above an erosion surface that cuts into sequences 3, 2, and 1 (Simó, 1985).

The Aren sequence can be divided into two subsequences: (1) a lower Aren subsequence 5-A with nearshore, shelf, and basin facies, and (2) an upper subsequence 5-B with non-marine and nearshore/basin facies. The subsequences 5-A and 5-B are separated by a karst surface, by a quartz-pebble conglomerate, and by incised valleys (with as much as 350 m of incision; Fig. 13).

Subsequence 5-A shows two beach series containing offshore limestones at the base, high-energy nearshore quartz sandstones and a karst surface on top, with well-developed dissolution pockets filled with green shale and/or coal (Simó and others, 1985). Basinward, the nearshore facies changes to shelf facies. The shelf facies shows three coarsening- and shallowing-upward sequences, with a lower, thick (150 m) shale unit and a thinner (60–40 m), upper, sand bar unit. Paleocurrent direction from cross-bedding indicates a strong west-northwest direction of transport. The basin facies consists of dark massive clays, with minor sandstone-filled channels, thin-bedded sandstones, and thin-bedded nodular limestones.

The Aren subsequence 5-B lower boundary is characterized by a quartz-pebble conglomerate, which fills a karst topography, and by three incised valleys. The incised valleys are stacked and cut into each other, and cut into the karst surface and shelf/slope facies. The valley fill consists of coarse quartz-feldespathic sandstone with large-scale cross-bedding. Paleocurrent directions are toward the southwest. The incised valleys are truncated by a listric-normal fault. A succession of nearshore sandstone and basin shale deposits occurs over the listric fault and prograded toward the northwest. This succession is well exposed west of Tremp and has been described by Ghibaudo and others (1974). Non-marine red shales and nearshore facies were deposited in the south and east of the Tremp Basin.

SUBSIDENCE

Measured field sections have been used to calculate the amount of subsidence caused by sediment load and tectonic-basement subsidence in the inner-shelf, mid-shelf, and basin positions. The subsidence curves in this article were generated at Exxon Production Research Company by J. F. Sarg. The interacting variables used in the computer program (e.g., correction for compaction, density of sediment, water, and mantle) are described in Hardenbol and others (1981), although the program has since been updated. The program assumes an Airy-type crust and uses the backstripping technique described in Steckler and Watts (1978).

Figure 14 shows the total subsidence curve and the predicted thermal-subsidence curve of three stratigraphic sections (see location on Figs. 4 and 5). Data points on the curves are upsection lithological changes, and the thickness has been corrected for compaction and estimated paleo-water depth. The long-term eustatic change has not been removed from the thermo-tectonic curve, and the error incorporated into the predicted thermal-subsidence curve does not exceed 50 m.

Thermal-subsidence curves (Fig. 14) show: (1) a convex-upward shape (sector A, Fig. 14B) from Cenomanian (93.5 Ma) to middle Santonian (85 Ma), and a concave-upward shape (sector B, Fig. 14B) from middle Santonian to Maastrichtian (71 Ma); (2) two uplift events in the inner-shelf and basin sections (late Coniacian and early Maastrichtian);

Fig. 12.—(A) Shelf erosional unconformity (Late Campanian sequence boundary) between the Aren sequence (Aren Ss.) and the Vallcarca sequence (Bastus rudist reefs, Herbasavina clay, and Terradets outer-shelf facies, see Fig. 7). South flank, Sant Corneli anticline.

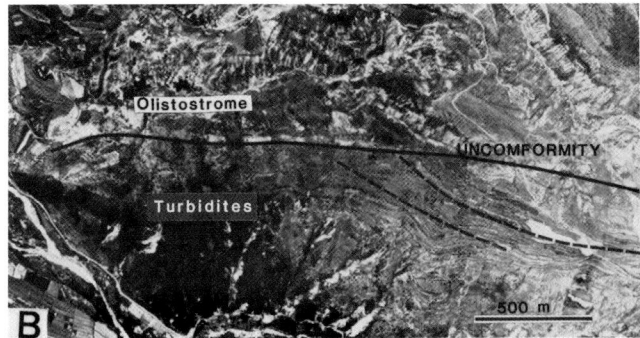

Fig. 12.—(B) Basin erosional unconformity (Late Campanian sequence boundary) between the Aren sequence (Pumanyons Olistostrome) and the Vallcarca sequence (turbidites, see Fig. 7). La Pobla area.

and (3) higher subsidence rates in the basin than on the shelf with shelfward movement of the axis of maximum subsidence through time. The first part of the curve (sector A) contrasts with the general concave-upward shape of a post-rift subsidence curve. This convex-upward shape is interpreted to be a result of basinal tilt and sediment load because of the preferable accumulation of sediments basinward (Figs. 5, 8). The second sector (sector B) is characterized by a rapid subsidence followed by a progressive decay of the subsidence rate with time. The rapid-subsidence event is interpreted to be a result of an initial thrusting and tectonic load of the crust basinward (east-northeast) of the Tremp Basin. Souquet and others (1977) described thrusting of Campanian age.

The uplift events are localized to specific parts of the basin. The late Coniacian uplift event is recognized in the Montsec section, where truncation of an estimated 30 m of section occurs. The early Maastrichtian uplift is well evidenced in the Boixols thrust hangingwall ramp (Figs. 4, 12) by 600 m of erosion (Sant Corneli section).

The predicted thermal subsidence rates for sector A and sector B are: (Montsec section) 6 m/m.y. and 20.6 m/m.y., respectively, in the inner shelf; 11 m/m.y. and 51 m/m.y., respectively, in the mid-shelf; and 19 m/m.y. and 70 m/m.y., respectively, in the basin.

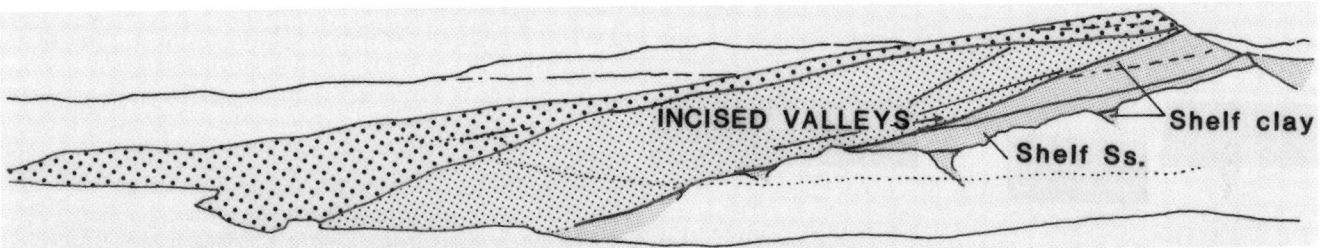

FIG. 13.—Incised valleys of the Aren sequence. The valleys divide the Aren sequence into two subsequences. The lower incised valley cuts into the shelf facies of sequence 5-A and is filled by turbidites; the upper valley is filled by estuarine-fluvial sandstones. Width of the panorama is about 2 km. Location north of Tremp.

The thermal-subsidence curves match well with the coastal-onlap curve and the backstepping geometry of the Upper Cretaceous depositional sequences. The more moderate coastal onlap of the three oldest sequences can be explained by slow-subsidence rates and erosion on the inner shelf (sector A). The major shelfward shift in coastal onlap, of about 60 km, at the base of the fourth depositional sequence correlates to the onset of subsidence (sector B). The backstepping of each successively younger carbonate platform (sequence 1 through 4) and shelfward encroachment of the basin are explained by (1) the contrast of subsidence rate on the shelf versus that in the basin (three times higher in the basin than on the shelf), and (2) by the shelfward increase of subsidence rate with time (Fig. 14).

RELATIVE SEA-LEVEL CHANGES

One purpose of this paper is to compare the Upper Cretaceous sequences of the Tremp Basin to sequences in other Cretaceous basins and the Tremp Basin sea-level curve to the global eustatic curve of Haq and others (1987). In order to compare the global eustatic curve with the Tremp Basin relative sea-level curve, the biostratigraphy of the study area (Martinez, 1982; Gomez-Garrido, 1987) has been compared to the chronostratigraphy chart of Haq and others (1987) (Fig. 7).

The coastal onlap curve of Figure 7 was calculated following the methodology described by Greenlee and others (1988). The curve was obtained subtracting the predicted thermal subsidence and estimated paleowater depth from the sediment thickness (thickness corrected for compaction from the first coastal-onlap horizon of each sequence to the top of the sequence). The predicted thermal subsidence was plotted to reflect periods of lineal subsidence and periods of uplift or rapid subsidence (Fig. 14B). Subsidence rates for the uppermost and youngest sequence are estimated (dashed line, Fig. 7). The paleowater depth was considered to be 0 m at the top of sequences 1 through 3 and 5, and about 35 m at the top of sequence 4. Errors in calculations may result from stratigraphic and chronostratigraphic data, decompaction and subsidence analysis techniques, miscalculated amount of erosion, and assuming constant rates of sedimentation. Calculations and modeling show that tectonic subsidence creates space for potential sediment accumulation, but does not account for the total accumulated sediment thickness. The additional space needed (coastal onlap, Fig. 7) is thus attributed to sea-level movements. Magnitudes of the relative sea-level rise in the Tremp Basin (6–10 cm/ka) are compatible with Cretaceous eustatic movements (see Schlanger, 1986).

Depositional sequences, sequence boundaries, and relative sea-level fluctuations recognized in the Tremp Basin in general are similar to those of Haq and others (1987). The major differences are: (1) the sequence boundaries 94 Ma and 90 Ma (type 1) of Haq and others (1987) have not being recognized in the Tremp Basin, and the timing of

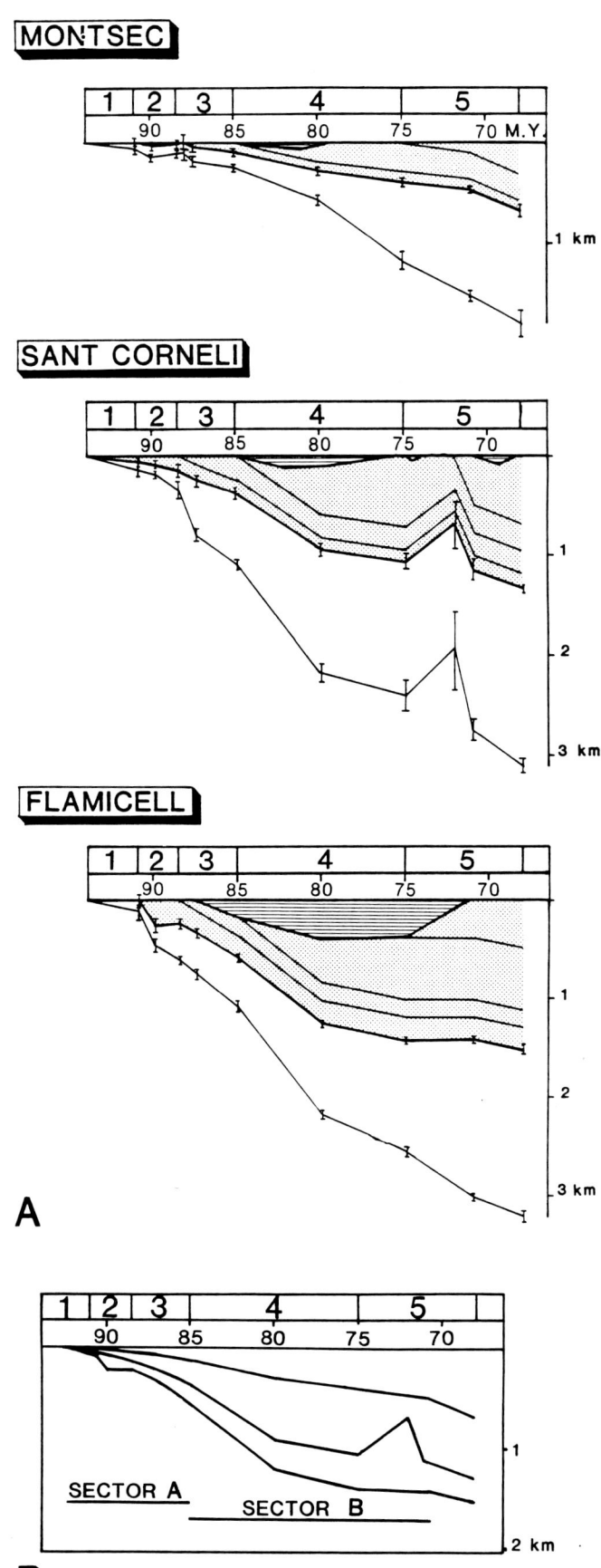

these two sequence boundaries is roughly equivalent with high-carbonate platform progradation rates in the Tremp Basin; (2) the Tremp Basin 91 Ma- and 88.5 Ma-sequence boundaries are minor sequence boundaries in Haq and others (1987); and (3) the envelope of the Tremp Basin sea-level curve indicates a maximum highstand during the lower Santonian, instead of the lower Turonian of Haq and others (1987).

The Tremp Basin sequences can be grouped into three supersequences, which are roughly analogous to the three Upper Cretaceous supercycles of Haq and others (1987). Sequences 1, 2, and 3 can be clustered into one supersequence, and sequences 4 and 5 are each a supersequence. The supersequence boundary ages of the Tremp Basin differ from the supercycle ages of Haq and others (1987; Fig. 7). Other workers have recognized sequences in other Cretaceous basins correlative with and similar to the Tremp Basin sequence boundaries. These include regression peaks documented by Juignet (1980) in France, Flexer and others (1986) in Israel, and Weimer (1984) in the United States Western Interior seaway. Equivalent depositional hiatuses have been identified by Poag and Schlee (1984) in the United States Atlantic Continental Margin. The similarities in these different locations support a "global" mechanism for controlling depositional-sequence formation during the Upper Cretaceous. The differences indicate that distinctive features of each basin (or a segment of a basin) controlled depositional-sequence geometry and depositional-sequence succession.

SUMMARY AND INTERPRETATION

The Upper Cretaceous depositional sequences of the Tremp Basin are seismic stratigraphic in scale and are a result of a combination of "global rhythms" and tectonics. The three oldest sequences are carbonate rich, and each sequence exhibits: (1) moderate coastal onlap, with each successively younger carbonate platform backstepping the former, and (2) shelfward encroachment of the basin, resulting in submarine-slope erosion and increasing water depth in the basin. The fourth sequence is the thickest and laterally most extensive with a major shelfward shift of the carbonate platform. The fourth sequence is divided into a siliciclastic-rich lower subsequence and an upper carbonate-rich subsequence. The fifth depositional sequence is predominantly siliciclastic rich and shows an overall regression in response to regional thrusting, uplift, and the late Cretaceous sea-level stillstand.

The geometries of the Upper Cretaceous depositional sequences were influenced by both tectonics and eustacy. High

FIG. 14.—Subsidence diagrams of three selected stratigraphic sections (A; see Fig. 5 for location) and a summary of predicted thermal-subsidence rates discussed in the text. (B). Large numbers (upper row) correspond to depositional-sequence numbers and smaller numbers (lower row) are millions of years. Water depth is shown with horizontal ruling. Thermal subsidence is designated by shading. Curves in the shaded area are thermal-subsidence curves for each sequence boundary (lower Cenomanian and middle Turonian sequence boundaries plotted together). Total subsidence is the lowest curve in each diagram (A).

thermal-subsidence rates in the basin induced backstepping of the carbonate platform, and slow subsidence on the shelf reduced bayline coastal onlap. An increase of thermal subsidence, both in the basin and on the shelf, induced a major landward shift in coastal onlap and created space for the deposition of a thick succession of sediments. Uplift concurrent with thrusting mostly affected the hangingwall ramp, and relaxation after the thrusting created new depositional space.

Relative sea-level changes are recorded in the depositional sequences and the sequence boundaries of the Upper Cretaceous of the Tremp Basin. Comparison of the sediment thickness with the predicted space created by the thermal-subsidence rate requires relative sea-level oscillations of as much as 150 m to create the additional depositional space needed for sediment accumulation. Sequence boundaries show subaerial erosion on the inner shelf, non-deposition or erosion on the outer shelf and at the shelf edge, and submarine erosion in the basin. Sequence boundaries developed as a response to relative sea-level changes and tectonics. Eustatic lowering of sea level exposed the inner shelf, but did not appear to have exposed the shelf edge. The sea-level drops may have triggered erosion in the basin by increasing current activity and instability of sediments. Tectonics played a role in the formation of the sequence boundaries, including enhancement of the boundaries expression (listric-normal faulting and uplift), or diminishment of the boundaries expression during times of rapid-subsidence rates.

In summary, the climate and sediment accumulation rates of the Upper Cretaceous of the Tremp Basin are comparable to those of modern carbonate environments. Subsidence and relative sea-level oscillation controlled the depositional-sequence architecture. Thermal subsidence regulated the large-scale geometry. The combined effects of tectonics and eustatic sea-level oscillations controlled the depositional-sequence geometries, distribution of depositional facies, and sequence boundary formation.

The Tremp Basin Cretaceous outcrops contain excellent exposures that reveal depositional sequences and sequence boundaries similar to those described from different Cretaceous basins. Although the Tremp Basin relative sea-level curve is similar overall with the eustatic curve of Haq and others (1987), some marked differences occur, and these differences are attributable to local (Tremp Basin) tectonics. Studies of individual basins, such as presented here for the Pyrenees, may be useful in modifying published global eustatic sea-level curves.

ACKNOWLEDGMENTS

Financial support during the field mapping was provided by the "Servei Geologic de Catalunya (Generalitat de Catalunya)." I thank C. Puigdefabregas (Servei Geologic de Catalunya) and J. F. Sarg (Exxon Production Research Company) for their interest and technical support. The manuscript was significantly improved by the reviews of E. K. Franseen (University of Wisconsin), D. Kerr (University of Wisconsin), J. F. Sarg (Exxon Production Research Company), N. F. Hurley (Marathon Oil Company), and J. L. Wilson. This paper was written during a Fulbright-MEC (Ministerio de Educación y Ciencia) fellowship under the supervision of L. C. Pray, at the Department of Geology and Geophysics, University of Wisconsin, Madison.

REFERENCES

BANDA, E., AND WICKHAM, S. M., 1986, (eds.), The geological evolution of the Pyrenees: Tectonophysics, v. 129, 380 p.

BILOTTE, M., 1978, Proposition pour une biozonation des séries épicontinentales du Cénomanien des Pyrénées: Géologie Méditerranenne, v. 5, p. 39–46.

———, AND SOUQUET, P., 1972, Les biozones des Foraminifères benthiques du Cénomanien Pyrénéen: Comptes Rendus, Académie Science, Paris, v. 274, p. 3352–3355.

CAUS, E., CORNELLA, A., GALLEMI, J., GILI, E., MARTINEZ, R., AND PONS, J. M., 1981, Field guide: Excursions to Coniacian-Maastrichtian of south central Pyrenees: Publicaciones de Geologia de la Universidad Autonoma de Barcelona Spain, v. 13, 170 p.

———, ———, AND GOMEZ-GARRIDO, A., 1982, Evolución de la cuenca sedimentaria del Cretácico superior surpirenaico entre los rios Segre y Noguera Ribagorzana (NE España): Cuadernos de Geología Ibéricos, v. 8, p. 965–977.

FLEXER, A., ROSENFELD, A., LIPSON-BENITAH, S., AND HONIGSTEIN, A., 1986, Relative sea-level changes during the Cretaceous in Israel: American Association of Petroleum Geologists Bulletin, v. 70, p. 1685–1699.

FONDECAVE, M. J., 1975, Essai de biozonation par les foraminifères pélagiques du Sénonien sud-pyrénéen; Description d'une nouvelle espèce "Hedebergella aubertae n. sp.": Géologie Méditerranéenne, v. 2, p. 5–10.

GALLEMI, J., MARTINEZ, R., AND PONS, J. M., 1983, Coniacian-Maastrichtian of the Tremp area (south central Pyrenees): Newsletter Stratigraphy, v. 12, p. 1–17.

GARRIDO-MEJIAS, A., 1973, Estudio geológico y relación entre tectónica y sedimentación del Secundario y Terciario de la vertiente meridional pirenaica en su zona central (provincia de Huesca y Lérida): Unpublished Ph.D. Dissertation, University of Granada, Spain, 395 p.

GHIBAUDO, G., MUTTI, E., AND ROSELL, J., 1974, Le spiagge fossili delle Arenarie di Aren (Cretacico Superiore) tra Isona e il Rio Noguera Ribagorcana (Pirenei centro-meridionali, province di Lerida e Huesca, Spagna): Memoire della Società Geologica Italiana, v. 13, p. 497–537.

GOMEZ-GARRIDO, A., 1981, Foraminíferos planctónicos de la Formación Reguard (Turoniense) en el Valle del Flamicell (Prepirineo de Lerida): Publicaciones de Geología, Universidad Autónoma de Barcelona, Spain, No. 16, 48 p.

———, 1987, Foraminíferos planctónicos del Cretácico superior surpirenaico: Unpublished Ph.D. Dissertation, Universidad Autónoma de Barcelona, Spain, 184 p.

GREENLEE, S. M., SCHROEDER, F. W., AND VAIL, P. R., 1988, Seismic stratigraphic and geohistory analysis of Tertiary strata from the continental shelf off New Jersey; Calculation of eustatic fluctuations from stratigraphic data, in Sheridan, R. E., and Grow, J. A., eds., The Atlantic Continental Margin, U.S.: Geological Society of America, The Geology of North America, v. I-2, p. 437–444.

GRIMAUD, S., BOILLOT, G., COLLETE, B. J., MAUFFRET, A., MILES, P. R., AND ROBERTS, D. G., 1982, Western extension of the Iberia-European plate boundary during early Cenozoic (Pyrenean) convergence: A new model: Marine Geology, v. 45, p. 63–77.

HAQ, B. U., HARDENBOL, J., AND VAIL, P. R., 1987, Chronology of fluctuating sea levels since the Triassic: Science, v. 235, p. 1156–1167.

HARDENBOL, J., VAIL, P. R., AND FERRER, J., 1981, Interpreting paleoenvironments, subsidence history and sea-level changes of passive margins from seismic and biostratigraphy: Proceedings, 26th International Geological Congress, Paris, Oceanologica Acta, p. 33–44.

JAMES, N. P., AND MOUNTJOY, E. W., 1983, Shelf-slope break in fossil carbonate platform: An overview, in Stanley, D. J., and Moore, G. T., eds., The Shelf Break: Critical Interface on Continental Margins: Society of Economic Paleontologists and Mineralogists Special Publication 33, p. 187–206.

JUIGNET, P., 1980, Transgressions-régressions, variations eustatiques et influences tectoniques de l'Aptien au Maastrichtien dans le Bassin de Paris Occidental et sur la bordure du Massif Armoricain: Cretaceous Research, v. 1, p. 341–357.

LIEBAU, A., 1973, El Maastrichtiense lagunar (Garumniense) de Isona: Coloquio Europeo de Micropaleontología, España, p. 87–111.

MARTINEZ, R., 1982, Ammonoideos Cretácicos del Prepirineo de la Provincia de Lleida: Unpublished Ph.D. Dissertation, Universidad Autónoma de Bellaterra, Publicaciones de Geología, Universidad Autónoma de Barcelona, Spain, v. 17, 198 p.

MASSON, D. G., AND MILES, P. R., 1984, Mesozoic seafloor spreading between Iberia, Europe, and North America: Marine Geology, v. 56, p. 279–287.

MEY, P. H. W., NAGTEGAAL, P. J. C., ROBERTI, K. J., AND HARTEVELT, J. J. A., 1968, Lithostratigraphic subdivision of post-Hercynian deposits in the south-central Pyrenees, Spain: Leidse Geologische Mededelingen, v. 41, p. 221–228.

MISCH, P., 1934, Der Bau der Mittleren Sudpyrenean: Adhandlungen der Gesellschaft der Wissenschaftenzu Göttingen Mathematish Physikalischen Klasse, Berlin, v. 13, Folge chapter 3, Mediterranebeite, p. 1597–1764.

MUNOZ, J. A., 1985, Estructura Alpina i Herciniana a la vora sud de la zona axial del Pirineu oriental: Unpublished Ph.D. Dissertation, University of Barcelona, Spain, 305 p.

NAGTEGAAL, P. J. C., 1972, Depositional history and clay minerals of the Upper Cretaceous Basin in the south-central Pyrenees, Spain: Leidse Geologische Mededelingen, v. 47, p. 251–275.

POAG, C. W., AND SCHLEE, J. S., 1984, Depositional sequences and stratigraphic gaps on submerged United States Atlantic Margin, in Schlee, J. S., ed., Interregional Unconformities and Hydrocarbon Accumulation: American Association of Petroleum Geologists Memoir 36, p. 165–182.

PONS, J. M., 1977, Estudio estratigráfico y paleontológico de los yacimientos de rudístidos del Cretácico superior del Prepirineo de la Provincia de Lérida: Unpublished Ph.D. Dissertation, Universidad Autónoma de Barcelona, Publicaciones de Geología, Universidad Autónoma de Barcelona, Spain, v. 3, 105 p.

PUIGDEFABREGAS, C., AND SOUQUET, P., 1986, Tecto-sedimentary cycles and depositional sequences of the Mesozoic and Tertiary from the Pyrenees: Tectonophysics, v. 129. p. 173–203.

READ, J. F., 1982, Carbonate platforms of passive (extensional) continental margins: Types, characteristics and evolution: Tectonophysics, v. 81, p. 195–212.

———, J. F., 1985, Carbonate platform facies models: American Association of Petroleum Geologists Bulletin, v. 69, p. 1–21.

ROSELL, J., 1967, Estudio geológico del sector del Pireneo comprendido entre los ríos Segre y Noguera Ribagorzana (provincia de Lérida): Pirineos, v. 21, p. 9–214.

———, J., 1970, Mapa geológico de España 1:50,000, Tremp: Instituto Geologico y Minero de España, Mapa Nacional 252.

———, OBRADOR, A., AND PONS, J. M., 1972, Significación sedimentológica y paleogeográfica del nivel arcilloso con corales del Senoniense superior de los alrededores de Pobla de Segur (provincia de Lérida): Acta Geológica Hispánica, v. 7, p. 7–11.

SCHLANGER, S. O., 1986, High frequency sea-level fluctuations in Cretaceous time; An emerging geophysical problem, in Hsu, K. J., ed., Mesozoic and Cenozoic Oceans: America Geophysical Union, Geodynamic Series, v. 15, p. 61–74.

SIMO, A., 1985, Secuencias Deposicionales del Cretácico superior de la Unidad del Montsec Pirineo Central: Unpublished Ph.D. Dissertation, University of Barcelona, Spain, 325 p.

———, 1986, Carbonate platform depositional sequences, Upper Cretaceous, south-central Pyrenees (Spain): Tectonophysics, v. 129, p. 205–231.

———, PUIGDEFABREGAS, C., AND GILI, E., 1985, Transition from shelf to basin on an active slope, Upper Cretaceous Tremp area, southern Pyrenees, in Mila, M. D., and Rosell, J., eds., International Association of Sedimentologists, Sixth European Regional Meeting: Excursion Guidebook, p. 63–108.

SOUQUET, P., 1967, Le Crétacé Supérieur sud-Pyrénéen en Catalogne, Aragon et Navarre: Thèse Doctorale Sciences Naturelles, Univeresité de Toulouse, France, 529 p.

———, 1984, Les cyclès majeurs du Crétacé de la paléomarge ibérique dans les Pyrénées: Strata, Université de Toulouse III, France, v. 1, p. 47–70.

———, PEYBERNES, B., BILOTTE, M., AND DEBROAS, E. J., 1977, Nouvelle esquisse structurale des Pyrénées: Laboratoire Géologie, Université Paul Sabatier, Toulouse, France, p. 1–16.

STECKLER, M. S., AND WATTS, A. B., 1978, Subsidence of the Atlantic continental margin off New York: Earth and Planetary Science Letters, v. 41, p. 1–13.

VAIL, P. R., MITCHUM, R. M., AND THOMPSON, S., 1977, Seismic stratigraphy and global changes of sea-level. Part 4. Global cycles and relative changes of sea-level, in Payton, C. E., ed., Seismic Stratigraphy–Application of Hydrocarbon Exploration: American Association of Petroleum Geologists Memoir 26, p. 83–87.

———, HARDENBOL, J., AND TODD, R. G., 1984, Jurassic unconformities; Chronostratigraphy and sea-level changes from seismic stratigraphy and biostratigraphy: American Association of Petroleum Geologists Memoir 36, p. 129–144.

VAN HOORN, B., 1970, Sedimentology and paleogeography of an Upper Cretaceous turbidite basin in the south-central Pyrenees, Spain: Leidse Geologische Mededelingen, v. 45, p. 73–154.

WEIMER, R. J., 1984, Relation of unconformities to tectonics and sea-level changes, Cretaceous of Western Interior, U.S.A., in Schlee, J. S., ed., Interregional Unconformities and Hydrocarbon Accumulation: American Association of Petroleum Geologists Memoir 36, p. 7–36.

WILLIAMS, G. D., 1985, Thrust tectonics in the south-central Pyrenees: Journal of Structural Geology, v. 7, p. 11–17.

WILSON, J. L., 1975, Carbonate Facies in Geologic History: Springer-Verlag, New York, 471 p.

SYNSEDIMENTARY TECTONICS IN THE LATE CRETACEOUS-EARLY TERTIARY PELAGIC BASIN OF THE NORTHERN APENNINES, ITALY

ALESSANDRO MONTANARI
Department of Geology and Geophysics, University of California, Berkeley, California 94720
LUNG S. CHAN
Department of Geology, University of Wisconsin, Eau Claire, Wisconsin 54701
AND
WALTER ALVAREZ
Department of Geology and Geophysics, University of California, Berkeley, California 94720

ABSTRACT: The sequence of Upper Cretaceous-Lower Tertiary pelagic limestones in the Umbria-Marches Apennines of Italy has recorded, with remarkable continuity, the geologic history of an epeiric sea on the eastern continental margin of the Ligurian Ocean, during a time of widespread tectonism in the western Tethys domain. Sedimentary facies and paleocurrent analyses indicate that intrabasinal depocenters and structural highs have formed in response to extensional tectonic movements that started to affect the central part of the paleobasin in early Turonian. The topography of the paleobasin was probably controlled by a complex pattern of buried faulted blocks formed during the Jurassic passive-margin phase of the western Tethys, and then reactivated in the Turonian after a prolonged time (Aptian to Cenomanian) of tectonic quiescence. Calcareous turbidites, essentially made of remobilized pelagic mud, were generated on the newly formed intrabasinal slopes and deposited in the adjacent depocenters. Conspicuous sedimentary events, such as maxima in turbiditic deposition and soft-sediment slumps in these intrabasinal depocenters, are attributed to major syndepositional earthquakes of regional extent. A detailed event stratigraphy based on these sedimentary features indicates that the level of syndepositional tectonic activity reached a peak in late Maastrichtian-early Paleocene time and rapidly diminished in the Eocene.

INTRODUCTION

The eastern part of the northern Apennines (Umbria and Marches regions) consists of northeast-verging folds and thrusts emplaced during the Late Tertiary (Fig. 1). The sedimentary sequence in this area records with great continuity the complex tectonic evolution of an Early Jurassic-Oligocene deep-water basin, developed on thinned continental crust, from an extensional regime to a compressive regime that culminated with the Late Tertiary orogenesis (Bortolotti and others, 1970; Centamore and others, 1980). The change of tectonic regime is reflected, in the sedimentary record, by a lithologic change from a Lower Jurassic-Paleocene carbonate sequence to a synorogenic and post-orogenic siliciclastic sequence deposited in the Neogene and Quaternary. The orogenic phase is mainly represented by the Marnoso Arenacea flysch, the bulk of which was derived from the erosion of the Alps and deposited in northwest-southeast-elongated foredeep basins. Deep-water, biomicritic, and mostly pelagic limestones deposited on a subsiding Late Triassic-Early Jurassic carbonate platform compose the carbonate sequence underlying the Marnoso-Arenacea (Fig. 2). Coltorti and Bosellini (1980) attributed the subsidence to the Early Jurassic opening of the Ligurian Ocean, a branch of the western Tethys Ocean adjacent to the northern Apennines epeiric sea. These authors have also compared the breakup of the carbonate platform, represented by the Calcare Massiccio limestone, to the subsidence of modern passive margins (the Red Sea). From the beginning of the subsidence in the Early Jurassic to the Tertiary orogenesis, the northern Apennines continental crust experienced tectonic stretching and subsidence (D'Argenio and Alvarez, 1980). Four major phases compose the tectonic evolution of the Umbria-Marche paleobasin prior to the onset of the synorogenic siliciclastic deposition (Fig. 2): (1) Lower Jurassic: rapid subsidence characterized by block faulting and formation of a horst-and-graben topography (Crescenti and others, 1969; Colacicchi and others, 1970; Centamore and others, 1971; Coltorti and Bosellini, 1980; Bice and Stewart, 1985). (2) Middle Jurassic to Neocomian: regional subsidence and deep-water carbonate sedimentation, which filled in the irregular pre-existing topography through episodic slumping, winnowing, and turbiditic deposition (Cardellini, 1982; Lowrie and Alvarez, 1984); (3) Aptian to Cenomanian: tectonic quiescence characterized by rhythmic pelagic deposition in a remarkably flat basin (Schwarzacher and Fischer, 1982; De Boer, 1983; Herbert and Fischer, 1985; Montanari, 1985); and (4) Turonian to Oligocene: resumption of tectonic activity and reactivation of buried Jurassic normal faults (Carbone and others, 1971; Montanari, 1985); this newly resumed synsedimentary tectonism caused the formation of intrabasinal depocenters and structural highs, winnowing and mass-remobilization of unconsolidated pelagic sediment to form turbidites and slumps, and localized fault-controlled hydrothermal activity (Montanari, 1979; Baldanza and others, 1982; Alvarez and Lowrie, 1984; Chan and others, 1985; Montanari, in prep.).

This paper documents the sedimentary manifestations of the fourth tectonic phase, focusing attention on distal sequences not directly influenced by the sedimentary and tectonic activity of the margins of the basin. Excellent biostratigraphic and magnetostratigraphic information on several sequences of continuously exposed, mostly pelagic, limestones in the northern Umbria and Marches regions (Crescenti, 1969; Baumann, 1970; Premoli Silva and others, 1974, 1981; Premoli Silva and others, 1976; Roggenthen and Napoleone, 1977; Lowrie and Alvarez, 1977, 1980; Alvarez and Lowrie, 1978, 1984; Napoleone, 1980; Wonders, 1980; Lowrie and others, 1982; Chan and Montanari, 1984), allow a precise correlation of major tectonic events recorded at a regional scale by slumps and maxima of turbidity deposition. Finally, we attempt to frame the Late Cretaceous-Paleogene tectonic evolution of the Umbria-Marches basin in the broader picture of the plate tectonics of the western Tethys, which was characterized by the inception of plate subduction in the Ligurian Ocean (Treves, 1984) and by

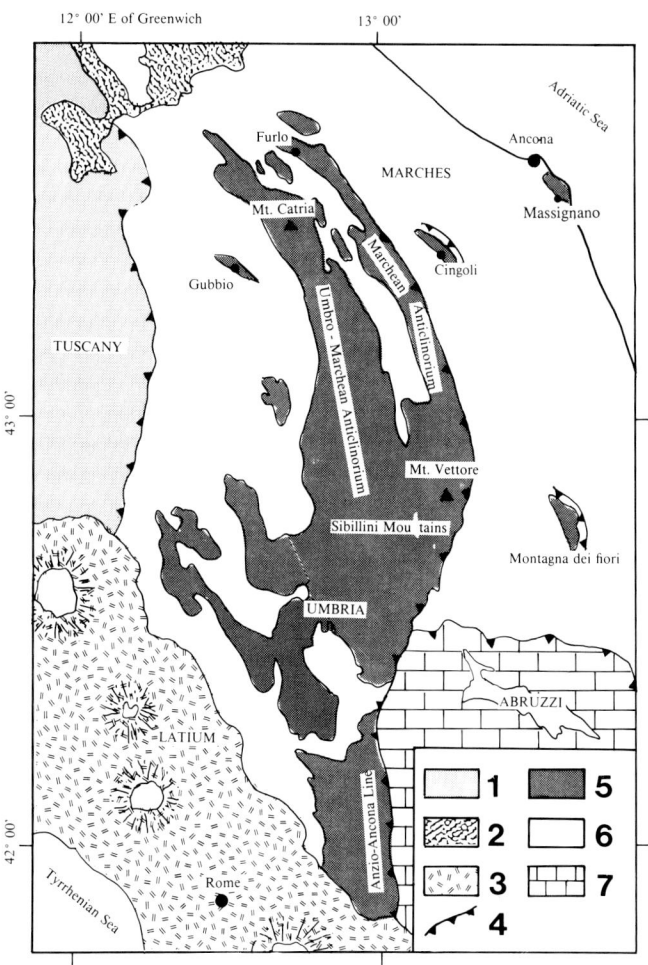

FIG. 1.—Simplified geologic map of the northern Apennines. (1) Tuscan Nappe; (2) Val Marecchia chaotic complex (3) Latium volcanic province; (4) major thrust; (5) Lower Jurassic to Oligocene deep-water carbonate rocks; (6) Upper Tertiary syn- and post-orogenic siliciclastic rocks; (7) Upper Triassic to Oligocene carbonate platform limestones.

the onset of orogenic deformations in the central and southern Alps (i.e., Trumpy, 1973).

GEOLOGIC SETTING

The Umbria-Marches Apennines are an unmetamorphosed fold-and-thrust foreland belt bordered on the north by the overlying tectonic slides of the Val Marecchia chaotic complex, on the east by the Adriatic sea, on the south by the Anzio-Ancona tectonic line and the Latium volcanic province, and on the west by the front of the Tuscan Nappe (Fig. 1). During the Mesozoic and Early Tertiary, the Ancona-Anzio Line represented a major facies boundary delimiting the Latium-Abruzzi carbonate platform to the east from the subsiding Umbria-Marches pelagic basin to the west.

The Ancona-Anzio Line represented a facies boundary already in the Late Triassic. The Umbria-Marches area was the site of deposition of evaporites, represented by the Burano anhydrites, whereas the Latium-Abruzzi region was mainly characterized by platform dolomites and dolomitic limestones. A thick carbonate platform represented by the Calcare Massiccio Formation developed in the Early Jurassic on top of the Burano anhydrites of the Umbria-Marches region, and similar facies developed in the adjacent Latium-Abruzzi region. At this point, the Ancona-Anzio line became a prominent paleogeographic feature following the differentiated subsidence between the two regions (Cantelli and others, 1982). While the Umbria-Marches region broke apart into faulted blocks (horst-and-graben topography) leading to the deposition of seamount and basinal, mosly pelagic, carbonate sequences of highly differentiated thicknesses and facies, the subsiding Latium-Abruzzi platform remained fairly uniform until the Miocene, hosting a variety of shallow-water carbonate facies. The Ancona-Anzio Line developed into an abrupt platform scarp from which large blocks of carbonate platform sequences detached and slumped down into the adjacent deep-water environment to form the well-known megabreccia found in the southern Umbrian basinal facies (Castellarin and others, 1982).

The regional differentiation through time of Triassic to Paleogene carbonate platform and transitional facies in the Latium-Abruzzi region is fairly well known and documented (Colacicchi and Praturlon, 1965 a, b; Colacicchi, 1966, 1967; Cantelli and others, 1978; Consiglio Nazionale Ricerche, 1986), although a reliable palinspastic reconstruction is still precarious due to the structural complexity of this area, which is mainly characterized by conspicuous thrusting (Accordi, 1966; Angelucci and Praturlon, 1968).

In the Miocene, at the beginning of the Apennine orogenesis, the western margin of the Latium-Abruzzi platform represented by the Ancona-Anzio Line was tectonically activated as a dextral transcurrent fault evolving into a northeast-verging thrust in the Early Pliocene (Castellarin

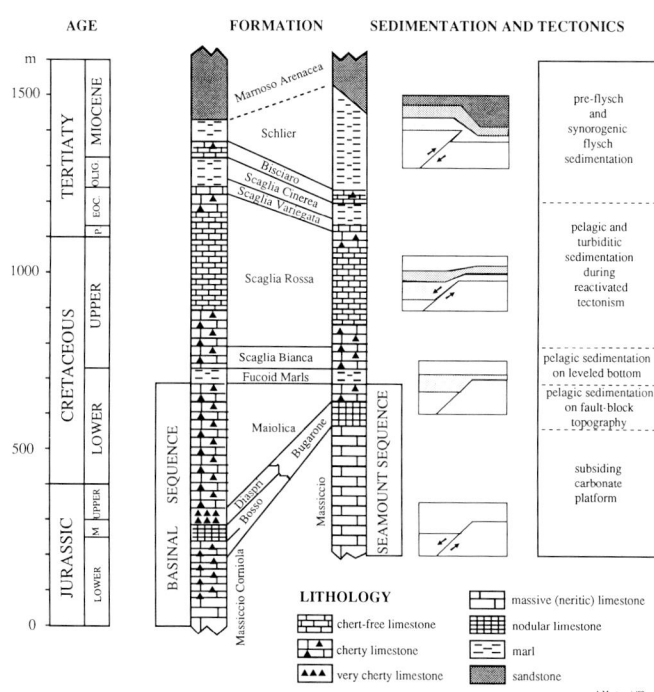

FIG. 2.—Lithostratigraphy of the Umbria-Marches sedimentary sequence.

and others, 1982). Emerging thrusts and frequent recumbent folds characterize the Sibillini Mountains (Fig. 1), which are located immediately west of the Ancona-Anzio Line and are the most elevated (maximum elevation, Mt. Vettore, 2,474 m) and structurally complex area in the Umbria-Marches region.

The Upper Cretaceous-Lower Tertiary sequence is represented by the Scaglia Rossa Formation, which is composed typically of pink, biomicritic limestones intercalated with calcarenitic turbidites and synsedimentary slumps. This sequence, although useful for the paleogeographic and sedimentologic definition of the relation between the carbonate platform and the adjacent deep-water basin (Baldanza and others, 1982; Alvarez and others, 1985; Colacicchi and Baldanza, 1986), does not allow a precise stratigraphic reconstruction of major tectonic events in a continuous time sequence. Most of the exposures are, in fact, incomplete and discontinuous due to common erosional gaps, synsedimentary slumps, and intense tectonic deformation.

On the other hand, the northern part of the Umbria-Marches Apennines (see Fig. 1) exhibits milder deformation and lower relief (maximum elevation, Monte Catria, 1,701 m). Two broad anticlinoria expose, in this area, Lower Jurassic to Paleogene carbonate rocks and are separated by synclinoria filled with Upper Tertiary siliciclastic sequences. The fold axes, along with back-limb normal faults and front-limb thrusts, are generally oriented northwest-southeast and are frequently offset by northeast-southwest transcurrent faults (Lavecchia, 1981). Isolated anticlines emerge outside and between these two major anticlinoria (e.g., Gubbio, Acqualagna, Sassoferrato, Cingoli). Complete and nearly continuous exposures of Scaglia Rossa are frequent, whereas synsedimentary slumps and turbidites are a minor and localized component of this pelagic unit. This situation allows a detailed basin analysis of the pelagic Scaglia Rossa Formation, which, with its well-defined biostratigraphy and magnetostratigraphy, records the major tectonic events occurring in this part of Tethys Ocean during Late Cretaceous and Early Tertiary time.

STRATIGRAPHY

The term "Scaglia" refers to a biomicritic, pelagic limestone with a scaly aspect (scaglia = scale or flake) in the sense that it is characterized by thin tabular bedding. Units termed Scaglia are extensively exposed in Italy, from Sicily to the southern Alps. In the Umbria-Marche Apennines, the Scaglia consists of a 550-m-thick sequence of pelagic, more or less marly limestones extending from the Cenomanian to the Oligocene. The Scaglia sequence is divided into several lithostratigraphic units according to lithologic characteristics of practical use for geologic mapping. Figure 3 shows a complete picture of the lithostratigraphy, foraminiferal biostratigraphy, and magnetostratigraphy of the Scaglia sequence. The biostratigraphy and magnetostratigraphy rely on previous studies carried out on the classic sections near Gubbio (eastern Umbria) by Premoli Silva (1977), Lowrie and Alvarez (1977), Alvarez and others (1977), Wonders (1980), Lowrie and others (1982), and Napoleone and others (1983). The present study elaborates in detail the lithostratigraphic subdivision of the Scaglia sequence. Although this subdivision is not ubiquitous over the entire Apennines, lithologic markers, such as first and last appearance of chert, can be used as boundaries between lithostratigraphic units and provide approximate time correlation between Scaglia sections as well (Chan and others, 1985). A brief description of each lithologic unit of this sequence is reported below.

Scaglia Bianca

We divide this unit into two members: (1) a lower Nodular Chert Member (lower Cenomanian, ~40 m thick), and (2) an upper Bedded Chert Member (upper Cenomanian, 22 m thick). The Nodular Chert Member consists of a rhythmic sequence of white, gray, and pink biomicritic limestones containing abundant nodular chert. This unit rests on the variegated, chert-free Fucoid Marls of Albian-Aptian age. The Bedded Chert Member is characterized by a rhythmic alternation of white, biomicritic limestones containing nodular chert, and black bedded cherts composed of reworked radiolarian skeletons and foraminiferal tests (Montanari, 1985). The top of the Scaglia Bianca is marked by the Bonarelli level, a 1-m-thick regional marker bed made of thin layers of radiolarites and bituminous shales (Arthur and Premoli Silva, 1982; Van Graas, 1983). The Scaglia Bianca, as well as the underlying Fucoid Marls, are remarkably consistent in thickness and facies throughout the region and represent, therefore, an essentially flat-bottomed and tectonically quiet paleobasin.

Scaglia Rossa

We divide the Scaglia Rossa Formation into four members on the basis of presence or absence of chert and marly layers. From the youngest to the oldest, these members are defined as follows (age and thickness based on the Gubbio sequence): the R4 member is 32 m thick and consists of white and pink marls and limestones containing nodular chert (upper Ypresian to lower Lutetian); the R3 member is 50 m thick and is made of pink and red, chert-free limestones and marls. The top is defined by the first appearance of nodular chert and the bottom by the Cretaceous-Tertiary (K-T) boundary marked by the last appearance of globotruncanids (Danian to lower Ypresian); the R2 member is about 130 m thick and consists of pink and white, chert-free limestones. In areas characterized by proximal facies (e.g., Sibillini Mountains, Montagna dei Fiori, Monte Cònero; see Fig. 4), nodules of chert are found within calcareous turbidites and cannot be used as a reliable lithologic criterion to distinguish this member from the underlying R1 member. In numerous sections in the region, including Gubbio, the lower part of the member consists of a red marly interval 8 to 10 m thick, here referred to as R2m submember (upper Campanian to Maastrichtian); the R1 member is 110 m thick and is composed of pink and white limestones containing nodular chert. The lower part of the member is characterized by a variegated (white, green, and red) interval 3 to 10 m thick. The Bonarelli level marks the lower boundary of this member, whereas the last appearance of nodular chert marks the upper boundary. Two laminated black chert layers separated by 1 m of white cherty limestones located

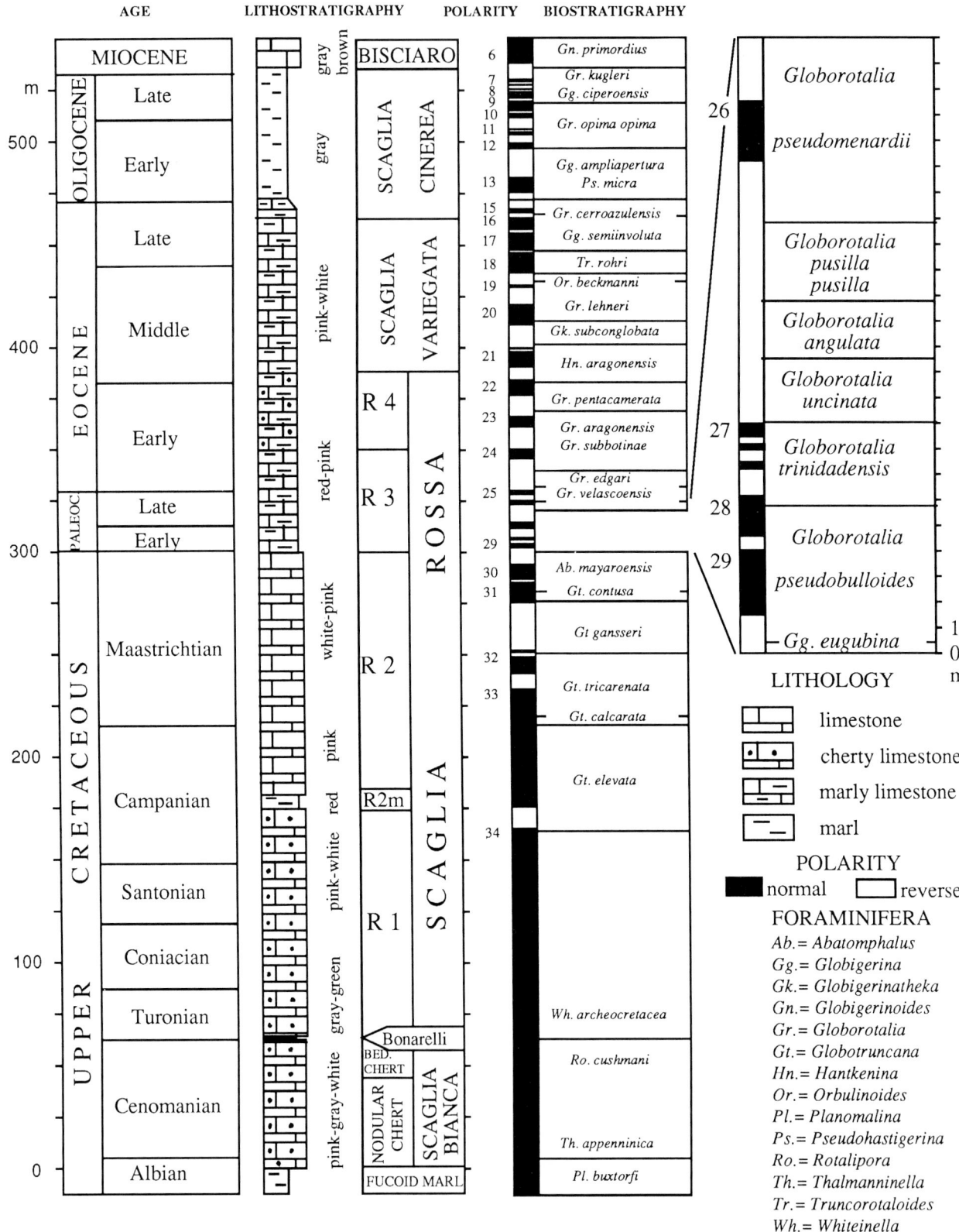

FIG. 3.—Stratigraphy of the Scaglia sequence at Gubbio. The Cretaceous portion of the sequence is derived from the Bottaccione section (Alvarez and others, 1977) and the Tertiary portion from the Contessa sections (Lowrie and others, 1982).

about 8 m above the Bonarelli (Montanari, 1979) mark the top of the basal variegated interval of the R1 member in sections exposed along the Marchean anticlinorium (Turonian to lower Campanian).

Scaglia Variegata

This unit consists of white and pink, chert-free limestones and marls ranging from the lower Lutetian to the upper Priabonian. In the Contessa sequence, near Gubbio (Lowrie and others, 1982), the Scaglia Variegata is 80 m thick. The last appearance of nodular chert marks the lower boundary of this unit, whereas the last occurrence of red marly limestone marks the gradational upper boundary.

Scaglia Cinerea

This formation is a 90-m-thick sequence of uniform, gray, marly limestones spanning epochs from the upper Priabonian to the lowermost Miocene. Calcarenitic turbidites are locally present in the southern Umbria region and are related to the nearby Latium-Abruzzi carbonate platform (Baumann, 1970). Over the rest of the region, the lack of turbidites and synsedimentary slumps suggests a stable depositional environment. The first layer of calcareous sandstone marks the boundary with the overlying Bisciaro Formation. Above some calcareous and glauconitic sandstones found in the lower part of the unit, the Bisciaro consists of an alternation of gray limestones and altered volcanic ashes (Borsetti and others, 1984). Several levels containing datable volcanic biotite have been found in the Scaglia Cinerea (Montanari and others, 1985) and attributed to the volcanic activity that accompanied the orogenesis in the Alps. In short, the Scaglia Cinerea represents the end of a long period of extensional tectonism and continuous pelagic sedimentation and precedes the onset of the synorogenic deposition of the overlying siliciclastic sequence.

SEDIMENTARY FACIES OF THE SCAGLIA ROSSA

The Scaglia Rossa Formation exhibits, in essence, three distinct sedimentary facies (Fig. 4): (1) a proximal turbiditic facies characterized by platform-derived calcareous turbidites interbedded with biomicritic pelagic limestones; (2) a distal turbiditic facies characterized by fine to very fine calcarenitic turbidites trapped in intrabasinal depocenters and interbedded with biomicritic pelagic limestones; and (3) a turbidite-free facies consisting of pelagic limestones and marls.

Soft-sediment slumps occur in each of these facies, indicating that syndepositional tectonic movements were not restricted to marginal areas but were affecting local areas throughout the paleobasin (Alvarez and Lowrie, 1984; Alvarez and others, 1985; Chan and others, 1985). The proximal turbiditic facies is found in the southern part of the region and in the isolated anticline of Monte Cònero, near the city of Ancona. Detrital beds range from pebbly mudstones and calcirudites to fine calcarenites and are composed fragments of bivalves, calcareous algae, large benthic foraminifera, and clasts of pre-existing rocks derived from an adjacent carbonate platform. Baldanza and others (1982) and Colacicchi and Baldanza (1986) have thoroughly described this facies.

The turbidite-free facies of the Scaglia Rossa is characterized by pink biomicritic limestones and marls. The texture of these pelagites consists of a suspension of planktonic foraminiferal tests in a coccolith matrix that contains 5 to 35 percent wind-blown silt and clay (Arthur and Fisher, 1977). The sedimentology and geochemistry of the turbidite-free Scaglia Rossa have been studied by Arthur (1976) in the complete and nearly continuous section of Gubbio. Arthur and Fisher (1977, p. 370) pointed out that, "Within the Campanian through Paleocene portion of the Scaglia Rossa [at Gubbio], original laminations are virtually absent, and though some burrow-mottling is present, limestones and marls are basically homogeneous; bioturbation eliminated the details of depositional history." In other words, although the homogeneous Scaglia Rossa of Gubbio has recorded with great continuity the geochemical and biological history of this Late Cretaceous-Early Tertiary pelagic basin, it lacks clear evidence of turbiditic deposition, which would give information on synsedimentary tectonic activity. On the other hand, the proximal facies does indicate intense

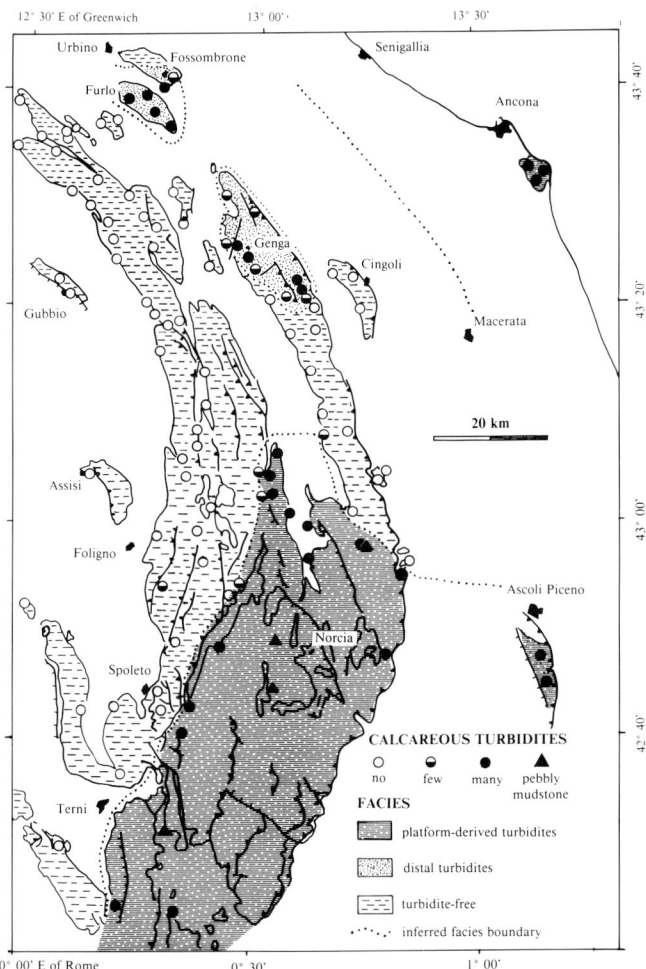

FIG. 4.—Distribution of Maastrichtian-Paleocene sedimentary facies in the Umbria-Marches anticlinoria.

syndepositional tectonism (Baldanza and others, 1982) but lacks the stratigraphic continuity required for a precise reconstruction of the tectonic and topographic evolution of the paleobasin. For these reasons, we concentrate our attention on the distal turbiditic facies, which is extensively exposed in numerous sections along the Marchean anticlinorium. The magnetostratigraphy and biostratigraphy of the Furlo and Poggio San Vicino sequences, which exhibit this facies, have been described in detail by Alvarez and Lowrie (1984) and Chan and others (1985). In essence, this particular facies constitutes a link between the quiet turbidite-free facies of Gubbio and the chaotic and incomplete proximal facies of the southern part of the region, and permits a detailed reconstruction of major tectonic events that controlled the sedimentation and topographic evolution of the paleobasin.

Distal Turbidites

The calcarenitic turbidites that distinguish the Scaglia Rossa facies exposed along the eastern anticlinorium in the Marches region constitute, in most of the cases, pure white, tabular layers that contrast with the enclosing pink limestones. In some rare cases, however, white turbidites grade upward into pink micrite, and the change in color is often marked by a bedding-parallel pressure-dissolution stylolite. These distal turbidites range in thickness from 1 to 120 cm with a mean thickness of 10–20 cm. In most cases their bottom surface is flat and marked by *Panolites* ichnofossils and, more rarely, by *Helminthoidea* and *Paleodictyon* (Fig. 5). Flute cases are less common and exhibit a variety of shapes, shown in Figure 6. In the Furlo area, the lowermost turbidites are found in the upper 27 m of the R1 member, where many of them contain nodules of structureless chert (Fig. 7A), or immediately underlie a bed of homogeneous whitish chert. Parallel lamination is the most common internal structure in the lowermost turbidites, whereas the B-C Bouma sequence is more frequent in the upper part of the R1 member. In the middle part of the R2 member and in the lowermost part of the R3 member, calcarenitic turbidites are more frequent and characterized by Bouma intervals B, B-C, and C (Fig. 7). The Bouma interval A is rather rare, and grading in the turbiditic beds is not evident in most of the cases. The C interval is represented by cross-laminae and, more rarely, by convolution and flame structures. The Bouma D interval is also very rare, probably due to the intense bioturbation that has obliterated the primary structures in the finer sediment. In the upper part of the R3 member and in the R4 member, calcarenitic turbidites are rare, thin, and very fine. In these members, internal turbiditic structures are seldom visible and are predominantly represented by parallel lamination.

The thickest turbidite in the Scaglia Rossa of the Furlo area is found in the lower R3 member, within the lower part of the *Globorotalia pseudomenardii* Zone. Because of its conspicuous thickness, ranging from 65 to 120 cm over an area of about 30 km^2 (Fig. 11), this particular turbidite has been called the MegaT (Montanari, 1979) and provides an unmistakable local marker bed (Fig. 8). The MegaT, like other turbidites of the distal facies, is composed of abundant planktonic foraminiferal tests (in this case of Cretaceous and Paleocene species), more or less recrystallized intraclasts of micritic mud, and fragments of shallow-water organisms, such as calcareous encrusting algae, benthic foraminifera, calcite prisms of bivalve shells, and possibly echinoderms. Some of the clasts seem to be coated by encrusting algae, indicating that they indeed originated at shallow depth. Fragments of large benthic foraminifera (i.e., *Orbitoides Spp.*), corals, and rudists, which are abundant in proximal turbidites at Monte Cònero and in the southern part of the region (Crescenti and others, 1969; Colacicchi and Baldanza, 1986), however, have not been found in the MegaT. The source of this turbidite, as well as of other turbidites in the Furlo sequence, could have been, therefore, an intrabasinal shallow plateau. Another possibility is that the shallow-water material contained in distal turbidites was derived from the remobilization of proximal turbiditic sequences from central parts of the basin and not necessarily from the margins of the basin. Remobilization may have been triggered by earthquakes, which would have liquified pre-existing unconsolidated sediments resting on forming intrabasinal slopes and controlled by active buried faults. It must be pointed out, however, that the areas with a distal turbiditic facies (i.e., Furlo, Genga-Poggio San Vicino) are surrounded by turbidite-free facies (Fig. 4), and feeding channels connecting proximal areas with these distal intrabasinal depocenters have not yet been identified.

FIG. 5.—Ichnofossils on the bottom surface of distal turbidites of the Scaglia Rossa at Furlo: (1) *Planolites sp.*; (b) *Helminthoidea labirintica*; (c) *Paleodictyon (minutissimus?)*.

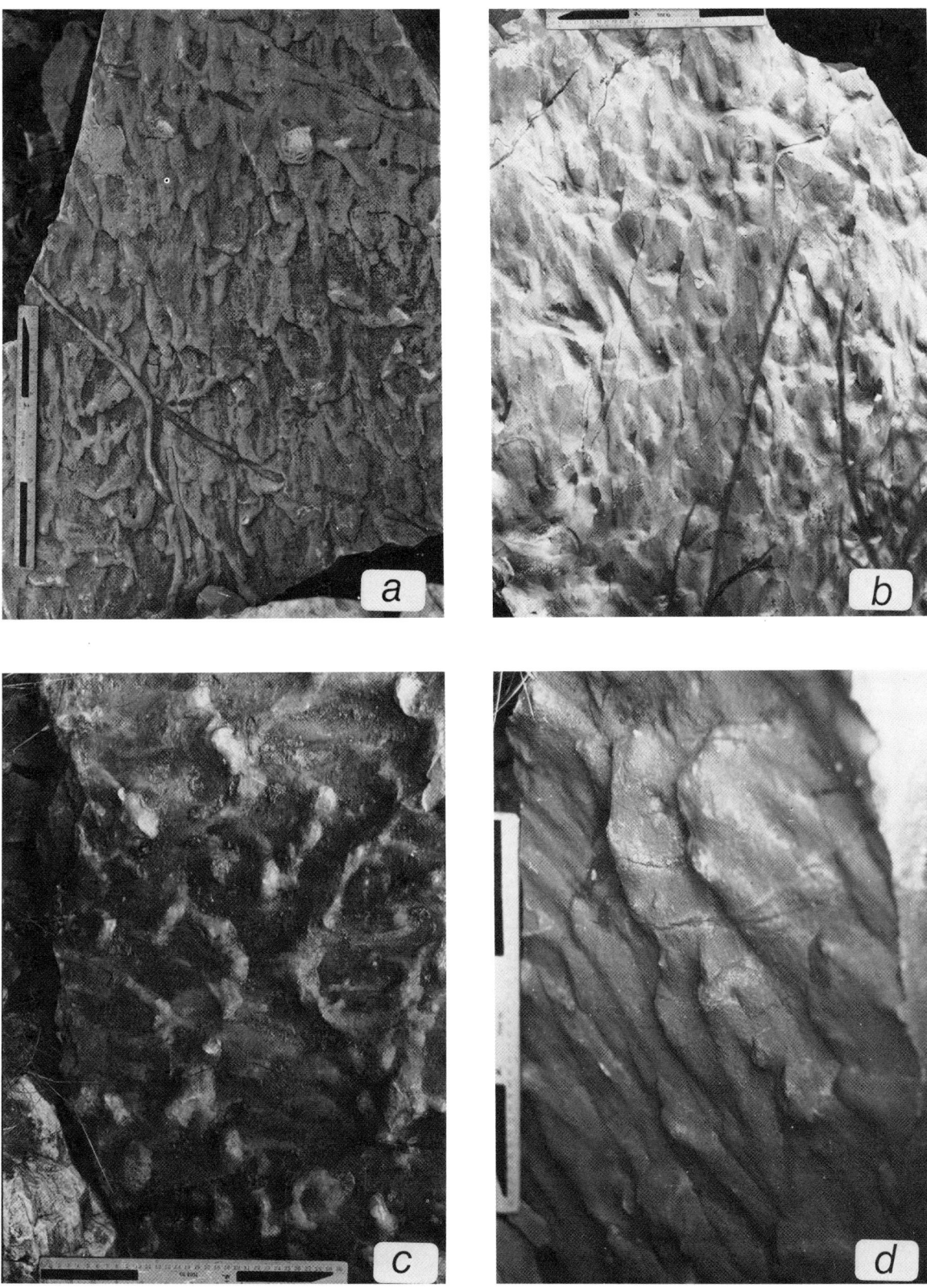

FIG. 6.—Flute casts on the bottom surface of distal turbidites at Furlo: (a) tongue-shaped flute casts; (b) tongue- and fan-shaped flute casts; (c) incomplete twisted flute casts; (d) bulb-shaped flute casts.

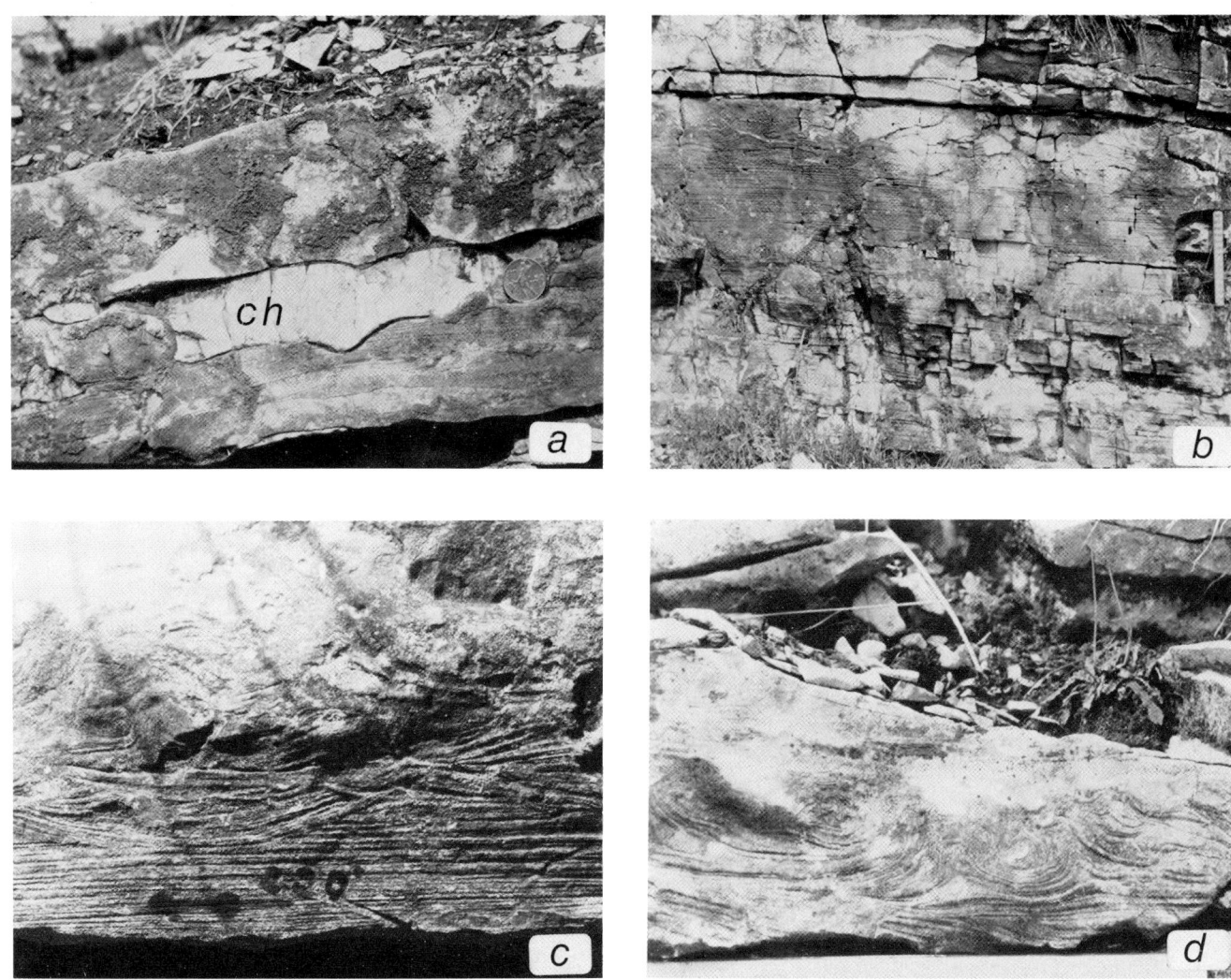

Fig. 7.—Internal structures in distal turbidites of the Scaglia Rossa at Furlo: (a) calcarenitic turbidite in the upper part of the R1 member containing light colored nodular chert (CH) above the Tb Bouma interval; the coin is about 2.5 cm in diameter; (b) calcarenitic turbidite exhibiting a thick Tb and a relatively thin convoluted Tc; (c) typical calcarenitic turbidite (about 10 cm thick) of the Furlo facies exhibiting a basal Tb and a Tc with ripple cross-lamination; (d) calcarenitic turbidite entirely composed of the interval Tc with convolution and flame structures.

More commonly, the turbidites in the distal facies of the Marchean anticlinorium are essentially made of reworked planktonic foraminiferal tests, scarce interstitial micritic matrix, intraclasts of micritic mud, prisms of inoceramids, and rare deep-water benthic foraminifera. Because of this texture, we prefer to refer to them as foraminiferal turbidites, for this term describes their composition and depositional mechanism better than the more familiar petrographic terms of packstone and grainstone (Dunham, 1962). The lack of shallow-water detritus in the foraminiferal turbidites of the Scaglia Rossa undoubtedly indicates that these bioclastic turbidites originated in a deep-water pelagic environment isolated from the carbonate platform. We stress that syndepositional seismic events were the most probable trigger for these turbidites, because storms or other violent events that could trigger turbidites in proximal (shallow) environments are not likely to affect deep and distal areas of the paleobasin. As a consequence of syndepositional tectonic movements inferred by a number of authors (Carbone and others, 1971; Montanari, 1979; Baldanza and others, 1982; Alvarez and Lowrie, 1984; Chan and others, 1985; Colacicchi and Baldanza, 1986), earthquakes would have triggered the foraminiferal turbidites by liquidifying unconsolidated pelagic sediment on unstable, sufficiently steep slopes located in the middle part of the paleobasin. The source sediment of these turbidites was probably a normal pelagic mud composed of 10 ± 5 percent foraminiferal tests suspended in a coccolith matrix. This is suggested by the fact that condensed sequences representing the probable source for foraminiferal turbidites (i.e., Fossombrone, see later discussion) are not anomalously enriched in foraminiferal tests.

Possible evidence of a synsedimentary earthquake that occurred in the central part of the basin is represented by a 10-cm-thick foraminiferal turbidite found at the top of a soft-sediment slump (Chan and others, 1985) in the lower part of the *Globorotalia pseudomenardii* Zone. A textural analysis of the Poggio San Vicino turbidite suggests, despite its minor thickness, a large submarine slump probably related to a major (regional?) seismic event. Unlike most of the distal turbidites observed in the present study, this turbidite is characterized by the Bouma intervals A and B (Fig. 9a). The bottom is constituted by a sole of intraclasts of porcellanaceous limestone as large as 1 cm in diameter derived from the *Globigerina eugubina* Zone and the *Abathomphalus mayorensis* Zone. The Bouma A interval is essentially composed of large upper Maastrichtian globotruncanids and small intraclasts of micritic matrix (Fig. 9b), whereas the B interval is made of small lower Paleocene globigerinids and globorotalids Fig. 9c). In the classic section of the Contessa Highway, near Gubbio (see Fig. 3), the sequence between the Cretaceous-Tertiary boundary and the bottom of the *Globorotalia pseudomenardii* Zone is about 15 m thick (Lowrie and others, 1982). In consideration of this, and also taking into account the half of the turbidite is made of reworked Maastrichtian sediment, the total thickness of unconsolidated sediment that was brought into suspension to form this turbidite must have been at least 30 m.

The importance of sediment reworking in the distal part of the Scaglia Rossa basin has been recognized by Alvarez and Lowrie (1984), who demonstrated that a high influx of turbidity currents into the Furlo basin corresponded to relatively high-sedimentation rates in the turbidite-free section of Gubbio. As a result, the upper Campanian-lower Paleocene turbidite-free sequence at Gubbio (175 m) is twice as thick as the turbiditic sequence in the Furlo basin. This has been explained by invoking hydraulic sorting during turbiditic transport and turbidite scouring of the muddy sea floor. In addition to these mechanisms of sediment reworking, we want to stress that the foraminiferal turbidites *per se* represent liquefaction and remobilization of large amounts of pelagic sediment, as the example described earlier suggests.

The relation between sedimentation rate, turbidite activity, and foraminiferal abundance in pelagic sediments is shown in Figure 10. The correlation, based on magnetic stratigraphy, shows that the characteristic maximum in sedimentation rate in the interval between the top of polarity Zone 32N and the bottom of polarity Zone 31N (*Globotruncana gansseri* and *Globotruncana contusa* Zones) in the turbidite-free sequence of Gubbio corresponds to a sudden increase in turbidite deposition in the Furlo sequence. Furthermore, the intensity of turbidite activity throughout the interval between the top of polarity Zone 32N and the middle of polarity Zone 30R correlates very well with the abundance of planktonic foraminiferal tests contained in the pink pelagic limestones of the same sequence. In the turbidite-free section of Acqualagna, located just 4 km southwest of Furlo (Chan and others, 1985), the abundance of foraminiferal tests in pelagic limestones shows one distinctive maximum across the boundary between polarity Zones 31N and 30R, which correlates with maxima in turbidite activity and foraminiferal abundance in the Furlo sequence. These observations strongly suggest that winnowing of the micritic fraction from pre-existing, unconsolidated pelagic sediments related to fluctuation in turbidite activity. Although it is still not understood why the abundance of foraminiferal tests in the Gubbio section is essentially constant through time, it seems clear that the sedimentation rate in this apparently quiet pelagic sequence is tied to turbiditic deposition well represented in nearby coeval sequences, and probably controlled by the regional tectonic activity.

Paleocurrents, Slumps, and Paleotopography

It has not yet been possible to make a precise palinspastic restoration of the Scaglia Rossa basin for the unavailability of accurate seismic profiles. The amount of shortening and the actual number of juxtaposed thrusted sheets in any given place of the northeast-verging Umbria-Marches thrust-and-fold foreland belt are still unknown. Nevertheless, the facies distribution along a northwest-southeast direction should be paleogeographically representative, and the paleocurrent and slump structures should be useful for a cautious inter-

FIG. 8.—The MegaT of the Furlo sequence (90 to 100 cm thick); (a) Furlo Via Flaminia section; (b) Torricella section.

FIG. 9.—Foraminiferal turbidite from the lower *Globorotalia pseudomenardii* Zone at Poggio San Vicino: (a) polished slab showing the basal conglomerate made of intraclasts of pelagic limestones reworked with the Maastrichtian *Abatomphalus mayorensis* Zone and the Danian *Globigerina eugubina* Zone. Unlike most turbidites of the distal facies of the Scaglia Rossa, this one at Poggio San Vicino exhibits the Bouma sequence Ta-b; (b) thin section photomicrograph of lower portion (Ta) of turbidite shown in Figure 9a; note concentration of relatively large tests of Upper Cretaceous globotruncanids; (c) thin section photomicrograph of upper portion (Tb) of same turbidite: note small tests of Lower Tertiary globigerinids and globorotalids sorted by the low regime of the turbidity current.

pretation of paleotopographic features that were present in the Scaglia Rossa basin. Accordingly, we have analyzed paleocurrents and slumps in the upper Campanian-lower Paleocene sequence of the Scaglia Rossa in the northern part of the Umbria-Marches region, with special attention to the areas of the Furlo anticline and the middle part of the Marchean anticlinorium.

The Furlo anticline is a large, coherent, simple fold in which original paleogeographic relations are unaffected (within the anticline) by Tertiary thrusting or strike-slip faulting. The paleocurrent directions of distal turbidites are preferentially directed toward the northwest in the southern part of the anticline, and toward the southwest in the northern part (Fig. 11). A conspicuous soft-sediment slump is found in the southern part of the anticline over an area of about 15 km² (Fig. 11) and involves upper Maastrichtian and lower Paleocene rocks. Near the village of Torricella, in the southernmost tip of the anticline (Alvarez and others, 1985), this slump shows a clear detachment surface about 10 m below the K-T boundary and underlies undisturbed pelagic limestones essentially composed of reworked pelagic foraminifera belonging to the indistinct *Globorotalia pseudobulloides-Globorotalia trinidadensis* Zone (Fig. 12). The verges of slump folds indicate a sliding direction toward the north-northwest, which is in agreement with the preferentially northerly directions of the paleocurrents in this part of the anticline (Fig. 11). The paleocurrent directions obtained from the MegaT mimic fairly well the general paleocurrent trend, which is characterized, in this area, by a sharp shifting of the paleocurrents in the downslope direction from northwest to southwest. The remarkable lateral continuity of the MegaT indicates also that the sea floor has remained essentially flat after the *G. pseudobulloides-G. trinidadensis* slump. Such a slump was probably caused by an earthquake, which would have liquefied the water-saturated pelagic mud just below the sediment-water interface and plastically deformed the more compacted sediments at greater depths below the sea floor. Liquefied pelagic sediment would have been brought into suspension to form a detrital mat of sorted foraminiferal tests atop in-

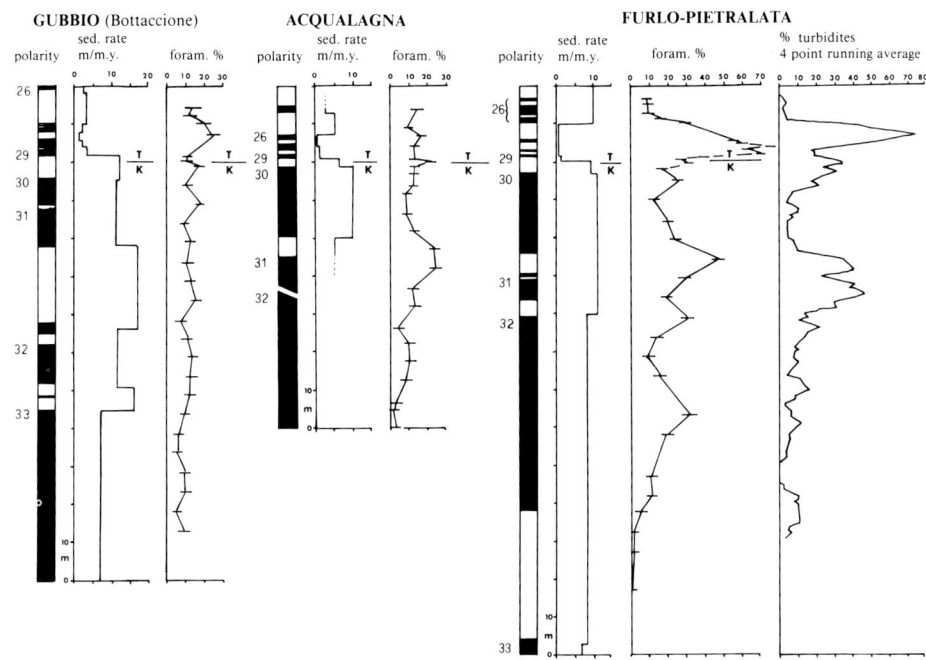

FIG. 10.—Relation between sedimentation rate, abundance of planktonic foraminiferal tests (by volume) relative to micritic matrix in pelagic limestones, and abundance of calcarenitic turbidites in the Gubbio, Acqualagna, and Furlo sequences. Relative abundance of foraminiferal tests was obtained by point counting thin sections. Error bars represent 2α.

tensely deformed pelagic sediments. In the Acqualagna anticline, just 4 km west of Furlo, the Scaglia Rossa lacks calcarenitic turbidites (Chan and others, 1985). This seems to indicate that the Acqualagna area was a structural high and was probably functioning as a sill deviating, possibly toward the north, the flow of turbidites and suspended micritic mud arriving from the northeast. About 7 km northeast of Furlo, in the Pian del Ponte quarry near the village of San Lazzaro, the partial exposure of the R2 member exhibits a facies very similar to that of Furlo. Two flute casts from different turbidites indicate a paleocurrent direction toward 210° (Fig. 11). In the Fossombrone section 3 km east of Pian del Ponte, the R2 member shows a few thin turbidites and is reduced in thickness with respect to the Furlo sections by a factor of two. The presence of sedimentary hiatuses in the lowermost portion of the R3 member led Chan and others (1985) to interpret the Fossombone area as an elevated margin of the Furlo basin from which pelagic muds were remobilized to form southwest-flowing turbidity currents.

A situation similar to that of the Furlo basin is found in the central part of the Marchean anticlinorium. Campanian to lower Paleocene turbiditic facies in outcrops near Genga and Poggio San Vicino, surrounded by coeval facies with very few or no turbidites, support the inference that the central part of the Marchean anticlinorium also represented an intrabasinal depocenter. This depression was separated from the Furlo basin by a structural high, probably located near Pergola (Fig. 13). Unfortunately, the lack of Scaglia Rossa outcrops in the area around Pergola, about 4 km southeast of Torricella, does not allow a direct correlation of this part of the paleobasin with the Furlo basin. Preferential paleocurrent directions seem to vary from place to place in this part of the basin (Fig. 13). In the Poggio San Vicino area, turbidity flows are dispersed between the southwest and north, with a predominant direction toward west-northwest. At Vallemania, near Genga, paleocurrents in the lower Paleocene sequence are preferentially directed toward the north-northeast. North from there, in the vicinity

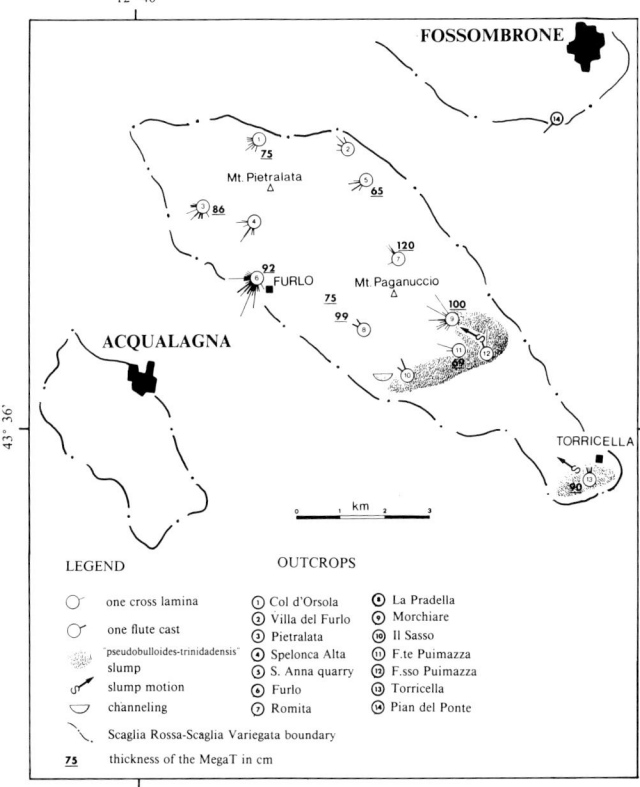

FIG. 11.—Paleocurrent and slump directions of the Scaglia Rossa in the Furlo area.

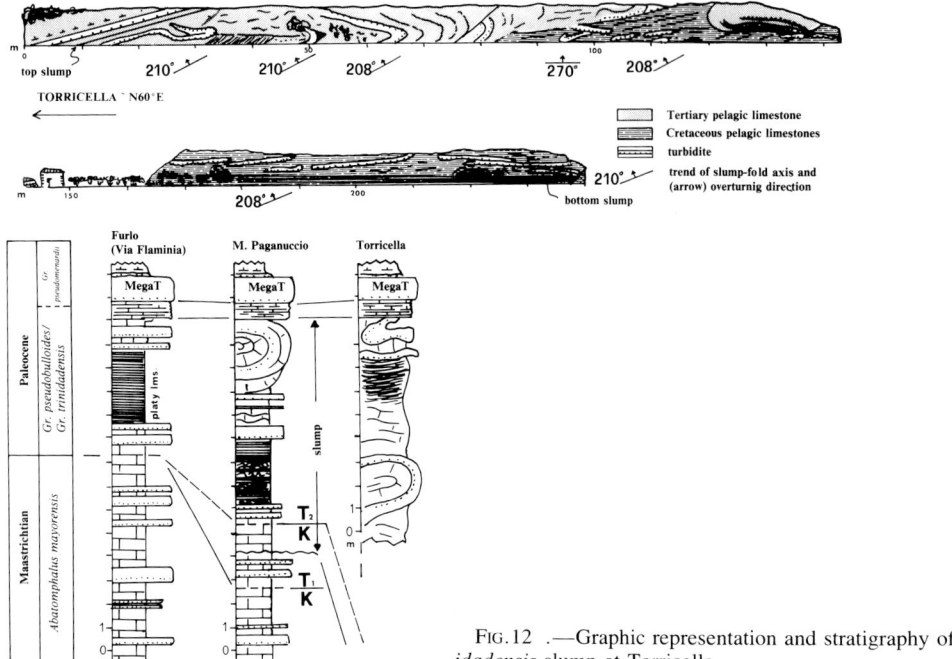

FIG. 12.—Graphic representation and stratigraphy of the *Globorotalia pseudobulloides-G. trinidadensis* slump at Torricella.

of Arcevia, paleocurrent indicators seem to return to a preferential west-northwest direction. A synsedimentary slump of large proportions is found at Vallemania and Pierosara, near Genga (Fig. 14). Conspicuous slump folds indicate, in both localities, a consistent overturning direction toward the east. The slump is overlain by undisturbed layers of limestone essentially made of reworked planktonic foraminifera belonging to the *G. pseudobulloides-G. trinidadensis* Zone, and containing some rare tests of Cretaceous globotruncanids. The substantial difference between the inferred slump motion (east) and the mean paleocurrent direction (north-northeast) suggests that in this locality the paleobasin was characterized by a valley with an axis slightly plunging toward the north-northeast and bordered to the west by a north-trending rise. On the other hand, at Arcevia about 7 km north of Genga, a lower Paleocene synsedimentary slump seems to indicate a westward-dipping slope (Chan and others, 1985), and paleocurrent directions at Prosano (~3 km southeast of Arcevia) also indicate a paleoslope toward the west-northwest. Such a complicated pattern of paleocurrents and intrabasinal slopes seems to reflect the structural complexity of this area. Similarly, in the Furlo area relatively simple paleocurrents and slump patterns seem to reflect the simplicity of the local structural setting.

Structural Control

The inferred irregular topography of the Scaglia Rossa basin requires differential tectonic movements and faulting, which would have caused some parts of the basin to subside with respect to others. Direct evidence of synsedimentary faulting, however, such as high-angle unconformities, onlap of Scaglia Rossa strata on older units, or visible paleofaults within the formation, is lacking. The Tertiary orogenesis may account for the obliteration of normal paleofaults which, very likely, would have been reactivated after the inversion of the tectonic regime (Koopman, 1983; Montanari and others, 1983). This seems to be the case for the major basinward-dipping faults that bordered the Latium-Abruzzi carbonate platform. After reversal of the tectonic

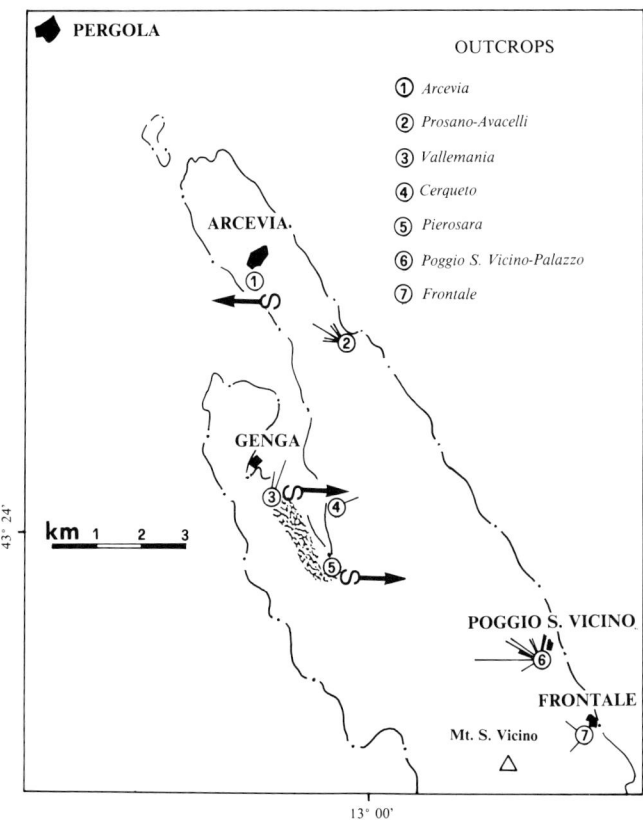

FIG. 13.—Paleocurrent and slump directions in the Scaglia Rossa of the Genga area (central part of the Marchean anticlinorium).

regime in the Late Tertiary, the western margin of the platform evolved into the Ancona-Anzio Line thrust and the northern margin into the Gran Sasso thrust (Castellarin and others, 1982). This would have caused the obliteration of facies transitional to the Scaglia Rossa basin (Baldanza and others, 1982). Nevertheless, syndepositional faulting has been indirectly inferred from paleomagnetic declination anomalies between the sections of Gubbio and Moria (Chan, 1984). In addition, unusual minerals, such as domeykite, barite, digenite, native copper, and celestite, found in several foraminiferal turbidites, were attributed to syndepositional hydrothermal activity related to syndepositional normal faults at the margins of structural highs (Montanari, in prep.).

Direct evidence for synsedimentary reactivation of buried normal faults during Scaglia Rossa time is provided by Galdenzi (1986), who describes neptunian dikes filled with Campanian biomictic sediment in a Jurassic megabreccia facies exposed in the Vernino gorge near Genga. The dikes cut through the sedimentary sequence for at least 500 m and are found exactly where a normal fault, active in the Early Jurassic, displaced a structural high from an adjacent basinal area.

The intraformational detritus composing the foraminiferal turbidites indicates, however, that in the distal part of the paleobasin, no rocks older than the Scaglia Rossa were ever exposed on the sea floor. Therefore, rather than a mosaic of faulted blocks with the high relief typical of extensional basins, we propose that the topography of the Scaglia Rossa paleobasin consisted of gently sloping intrabasinal highs and depocenters generated by slight flexures of the sea floor. Differential vertical movements may have been provided by the reactivation of a complex setting of buried Jurassic faults. The Jurassic horst-and-graben topography in the northern Apennines is known to have been very complicated, and tentative paleogeographic reconstructions have not established a regular pattern of faults (Coltorti and Bosellini, 1980). This may be due to early modifications of the shape of the horst by submarine erosion (Coltorti, 1979) and subsequent reactivation of faulted blocks during the orogenesis (Montanari and others, 1983), which would have obliterated the evidence of original structural trends. Nevertheless, shape modification of the tip of these horsts should not have affected the actual role that deep faults in a thick and rigid basement (the Calcare Massiccio limestone) must have had in the structural development of the region. In well-known cases of non-reactivated extensional basins, such as the Bay of Biscay (Montadert and others, 1980), the Bass Strait in southeastern Australia (Etheridge and others, 1985), and the United Kingdom Continental Shelf (Gibbs, 1984), the structural setting is dominated by complex patterns of planar- and listric-normal faults intersecting transfer faults at various angles. The complexity of these patterns is such that different structural styles may be found in separated areas of the same basin (Gibbs, 1984). According to Gibbs (1984), the differentiation of structural styles and the for-

Fig. 14.—Slump-deformed turbidites (prominent layers) and pelagites overlain by undeformed limestones belonging to the *Globorotalia pseudobulloides-G. trinidadensis* Zone at (a) Pierosara and (b) Vallemania near Genga (see Fig. 13 for location).

mation of intrabasinal depocenters and structural highs are derived from the sinuous shape of the major detachment surface that develops at depth in extending and subsiding basins. Such a detachment surface was inferred from detailed structural analysis based on well data and seismic profiles across the Central Graben in the North Sea, which is described by Gibbs (1984). The model of the Central Graben structural cross section proposed by Gibbs (1984) is here used for comparison to interpret the topographic and structural development of the Scaglia Rossa basin on a transect from the eastern margin (Monte Cònero) to the distal basin of Gubbio (Fig. 15). In the case of the Central Graben, the eastern area is characterized by a listric fan departing from the shallower plateau of the major detachment surface. A horst is found in the central part of the basin at the edge of a steepening of the major detachment surface. To the west of this intrabasinal horst, the so-called Central Graben high, a structural depocenter is controlled by a complex setting of listric-normal faults that root on the deep ramp of the major detachment surface. The Auck Ridge constitutes the west border of this intrabasinal depocenter and consists of a horst geometrically similar to the Central Graben high. The geometric arrangement of the principal and antithetic listric-normal faults along this transect of the North Sea is such that it would generate differential vertical movements in the case of further extension of the crust. The area with the listric fan and the intrabasinal depocenter would subside, whereas the Central Graben high and the Auck Ridge would remain "stable" and relatively elevated. In the northern Apennines, the resumption of extensional tectonic movements in the Cenomanian-early Turonian, at a time when the paleobasin was essentially flat and homogeneous (Bonarelli level), would have caused the proximal area and the Furlo-Genga area to subside to form two distinct basins bordered by structural highs (Fig. 15c). The model of Gibbs is not only useful to explain differential subsidence in the Scaglia Rossa basin, but is also consistent with the variability of deformational styles in the modern Umbria-Marches fold-and-thrust belt. The structures within each anticline were very likely controlled by the more or less complicated setting of the original mosaic of faulted blocks, whereas the location of the northwest-southeast-trending anticlinoria may be controlled by both the shape of the Jurassic detachment surface and the presence of stable structural highs.

EVENT STRATIGRAPHY AND TECTONICS

Late Cretaceous-Eocene Tectonic Events in the Umbria-Marches Apennines

Combined magnetostratigraphic and biostratigraphic analyses of several Upper Cretaceous and Lower Tertiary sections throughout the northern part of Umbria-Marches Apennines (Chan and others, 1985) has allowed us to date and correlate seven major sedimentary events, including submarine slumps and maxima of turbidite deposition, which appear to be synchronous in widely separated areas of the paleobasin (Fig. 16). We interpret these sedimentary events as manifestations of tectonic movements that were responsible for the topographic and structural evolution of the paleobasin. The topographic and sedimentary evolution of the Late Cretaceous-Paleogene pelagic basin of the northern Apennines is described here from the oldest to the youngest event.

Turonian.—

The uniformity throughout the region of mid-Cretaceous sedimentary sequences, represented by the pelagic limestones and marls of the Fucoid Marls and Scaglia Bianca Formations, suggests that a flat-bottomed paleobasin and a period of tectonic quiescence lasted about 25 Ma from the early Aptian to the late Cenomanian (Montanari, 1985; Fig. 17A). Some detrital limestones at the bottom of the Fucoid Marls are found in areas close to the Latium-Abruzzi carbonate platform (Crescenti and others, 1969). Turbiditic calcarenites associated with black cherts in the Bedded Chert Member of the Scaglia Bianca (Montanari, 1985) have been attributed to the resumption of tectonic movements in the late Cenomanian, first documented on the carbonate platform sequences by Carbone and others (1971). At the time

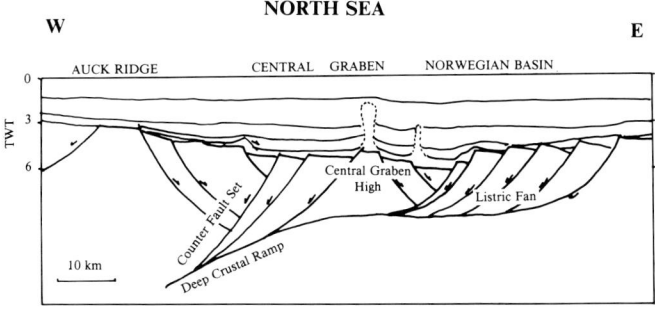

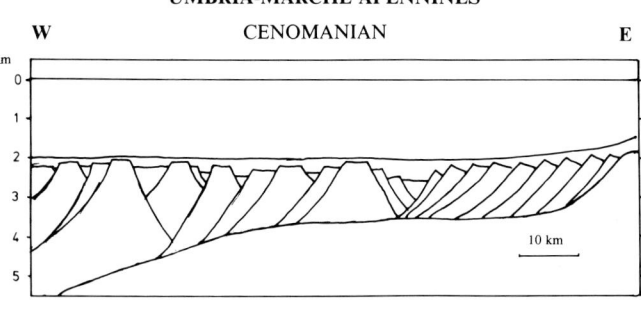

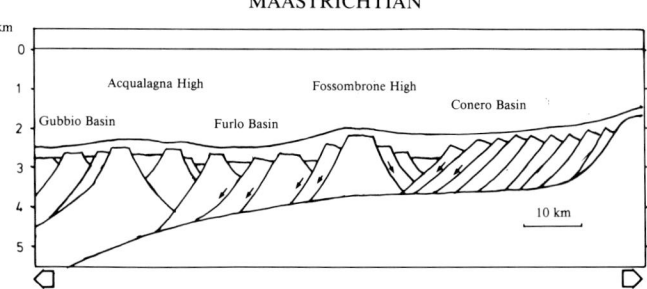

Fig. 15.—Comparison between the setting of buried normal faults across the passive margin of the North Sea (interpretative model after Gibbs, 1984) and that of the Late Cretaceous-Early Tertiary subsiding pelagic basin of the northern Apennines (see text for explanation).

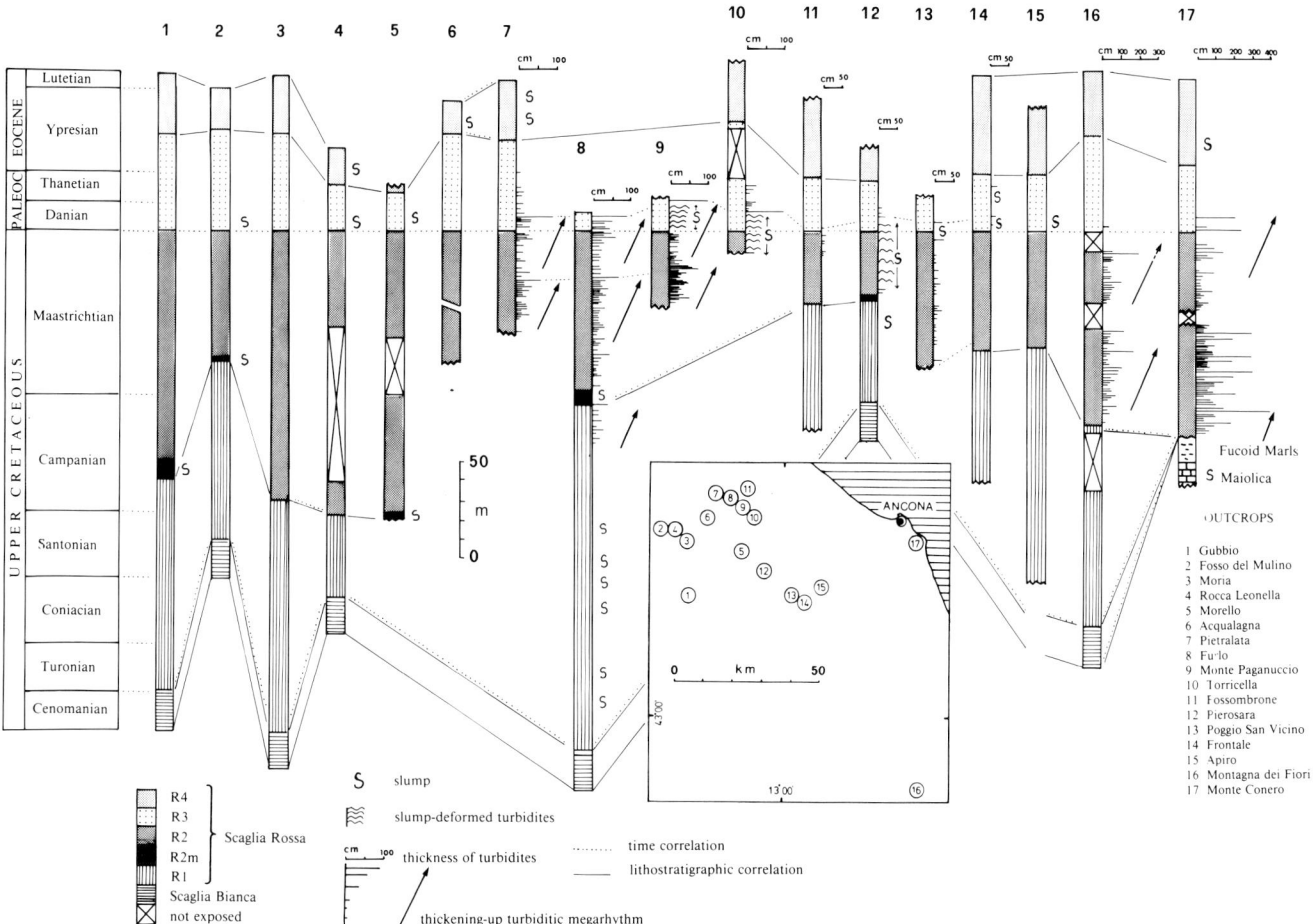

FIG. 16.—Lithostratigraphic correlation among 17 representative sections of Scaglia Rossa in the Umbria-Marches Apennines.

of deposition of the Bonarelli level (Cenomanian-Turonian boundary), the topography was still essentially flat, but isopachs of the Bonarelli in the northern part of the region seem to prefigure the morphology of the Scaglia Rossa paleobasin, which was to evolve during the upper Cretaceous and Lower Paleocene (Fig. 17D-E). This is here being interpreted as an effect of the incipient deformation of the sea floor (Montanari, 1985), probably caused by the reactivation, under an extensional tectonic regime, of buried Jurassic normal faults (see next section). Lower Turonian syndepositional slumps are present in the Furlo section in close proximity to the edge of a Jurassic fault-block horst (Alvarez and Lowrie, 1984; Alvarez and others, 1985) and at Avacelli, near Arcevia (Montanari, 1979). In both cases overturned anticlinal folds indicate slump movement toward the north-northeast (Montanari, 1979). These slumps are the oldest post-Aptian syndepositional mass movement recorded in the Umbria-Marches pelagic sequence and indicate topographic irregularities in the distal part of the paleobasin. Substantial lateral variations of thickness of the variegated interval at the bottom of the R1 member also indicate that the homogeneous and flat Cenomanian paleobasin became topographically irregular during the Turonian.

Santonian.—

The Furlo sequence contains four juxtaposed syndepositional slide slabs overlain by undisturbed middle Santonian pelagic limestones (Alvarez and Lowrie, 1984; Fig. 17B). This sequence of slumps is followed by the first appearance of distal turbidites, many of which contain chert (e.g., Fig. 7A). Coniacian and Santonian syndepositional slumps are present also in the Genga area, whereas turbidites containing cherts in the upper part of the R1 member are present in the Poggio San Vicino area. These sedimentary features indicate that intrabasinal depocenters in the northern part of Umbria-Marches region became well defined in the late Santonian when they started to collect turbidites generated by the remobilization of unconsolidated pelagic sediment.

Campanian.—

The tectonic event in the lower Campanian is represented, in the Furlo sequence, by a thickening-upward megarhythm of turbidites in the upper portion of the R1 member (Figs. 10, 16, 17C). This mega-rhythm culminates with a 40-cm-thick turbidite and is overlain by a synsedimentary slump located within the R2m marly interval. Despite the similarity of sedimentary facies with the Furlo sequence,

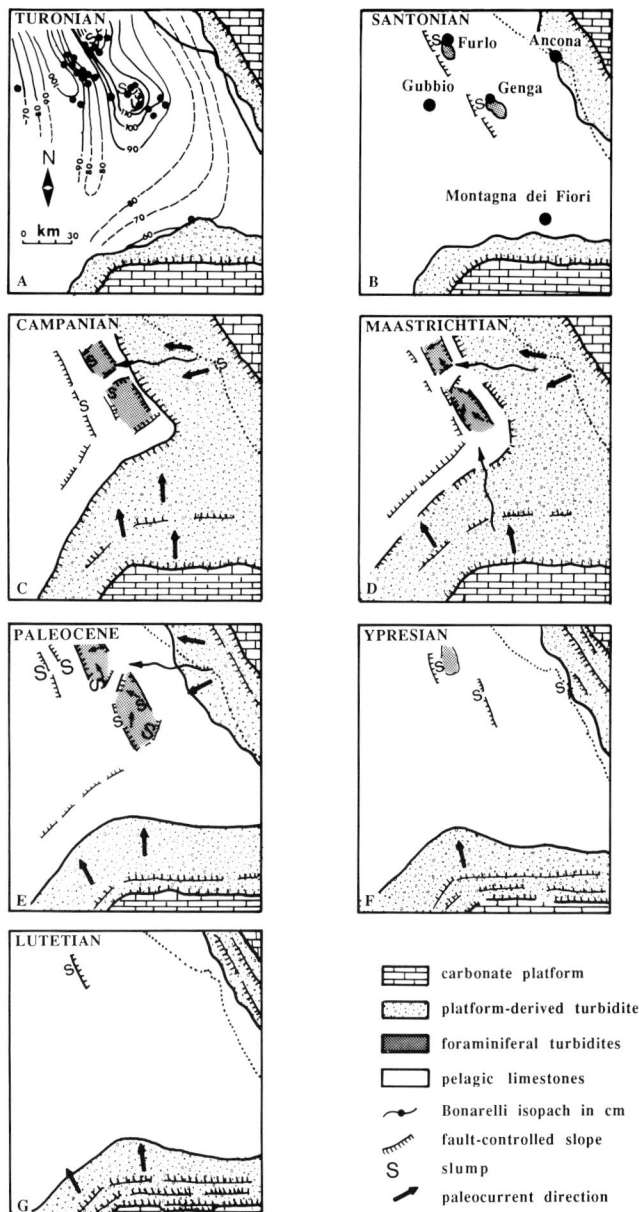

FIG. 17.—Sedimentary evolution of the Scaglia Rossa paleobasin in the Umbria-Marches Apennines.

the R2m submember in the sections of Vallemania and Pierosara, near Genga, lacks evidence of synsedimentary slumping. Synsedimentary slump deformations are present in the turbidite-free sections of Gubbio, Morello, and Fosso del Mulino, indicating that slump events in R2m were scattered throughout the region. At Monte Cònero, lower Campanian turbidites are paracomformably resting on the Fucoid Marls of probable Aptian age. This considerable hiatus, first discovered by Crescenti (1969) and confirmed by Chan (1984) after magnetostratigraphic analysis, was probably produced by a large slump that removed the sedimentary sequence that spans the Albian through Santonian (e.g., part of the Fucoid Marls, the Scaglia Bianca, and the R1 member of the Scaglia Rossa). In the Due Sorelle section (Montanari, 1979), two 40-cm nodular limestone layers at the level of the hiatus contain reworked foraminifera dispersed in the micritic matrix and angular fragments of variegated cherts identical to those found in the underlying Fucoid Marls Formation. This suggests that the scar of the slump was located not far upslope from this site and exposed Aptian rocks. The Scaglia Rossa at Monte Cònero shows a proximal turbiditic facies, whereas the pre-Scaglia units (e.g., the Fucoid Marls and the Maiolica) exhibit deep-water, essentially pelagic, facies (Crescenti, 1969; Cardellini, 1982). Sometime in the period between the Aptian and the late Santonian, therefore, the margin of the eastern carbonate platform was tectonically reactivated and the area of Monte Cònero became the site of intense turbiditic sedimentation. Because of the hiatus, however, we do not know precisely when the tectonism started to affect this platform margin.

In the southern part of the region, calcarenitic turbidites derived from the Latium-Abruzzi carbonate platform are present in the lower Campanian of the Scaglia Rossa at Montagna dei Fiori (Fig. 16). In this sequence, the R1 member is free of slumps and turbidites, suggesting that the Turonian-Santonian period was tectonically tranquil in the southern part of the region. These observations agree well with the inference by Baldanza and others (1982) that a sudden subsidence of the platform margin in the Campanian caused the beginning of massive turbiditic sedimentation and slumping in proximal areas that, until that time, were characterized by calm, essentially pelagic sedimentation.

The R2m marly interval is absent in all the other studied sections, as are turbidites and synsedimentary slumps. In the topographic and sedimentary interpretation sketched in Figure 17C, it appears that the Gubbio area and the area around Furlo and Genga were tectonically active and were probably subsiding with respect to adjacent, relatively elevated areas such as Frontale and Rocca Leonella, both with a R1 member of reduced thickness. In summary, the early Campanian tectonic event was responsible for drastic topographic and sedimentologic changes in both proximal and distal areas of the paleobasin.

Maastrichtian.—

In the middle Maastrichtian, at the time corresponding to magnetic polarity Zone 31R and the *Globotruncana gansseri* Zone, the Scaglia Rossa paleobasin experienced the maximum of turbiditic sedimentation in both proximal and distal facies (Fig. 17D). In the Furlo area, the *G. gansseri* event is represented by the culmination of a thickening-upward turbiditic mega-rhythm (Figs. 10, 16). Relatively numerous turbidites are also found in the Poggio San Vicino section and near Genga. Platform-derived calcareous turbidites also peak in abundance at Monte Cònero and Montagna dei Fiori (Fig. 16), whereas slumps associated with high turbidity influx are found in the southern part of the basin (Baldanza and others, 1982). The thickening-upward sequence of distal turbidites, including foraminiferal turbidites at Furlo, and the lack of slumping in the northern part of the paleobasin suggest that the middle Maastrichtian tectonic phase consisted of a progressive sequence of seismic shocks able to liquefy and remobilize pelagic sediments without further altering the pre-existing basinal topography.

Paleocene.—

The synsedimentary tectonism that accompanied the evolution of the Scaglia Rossa basin culminated in the Paleocene (Fig. 17E). According to Baldanza and others (1982), in the southern part of the basin the high influx of turbidites and slumps is the expression of extensional tectonism, which caused a rapid subsidence of the Latium-Abruzzi carbonate platform and retreat of the platform margin. In the southern part of the paleobasin, however, Paleocene sequences are intensely disturbed by synsedimentary slumps, massive detrital deposition, and sediment reworking, which reduce the stratigraphic resolution for definition of major tectonic and sedimentary events.

In the Furlo area, two major sedimentary events occur in the Paleocene: a large slump in the upper Danian, the "*pseudobulloides-trinidadensis*" slump of Torricella (Figs. 11, 16), and the deposition of the MegaT in the Thanetian. A conspicuous slump in the Danian occurs also in the Genga area and is followed by several turbidity events. The Thanetian event is recorded by slumps and turbidites also at Arcevia, Frontale, and Poggio San Vicino (Chan and others, 1985). In the proximal sequence of Monte Cònero, two strongly channelized calciruditic turbidites as much as 4 m thick are found, respectively, in the Danian and in the Thanetian. Slumps of the same age occur in many other sites of the paleobasin in turbidite-free sequences (Fig. 17E), confirming the regional extent of the synsedimentary tectonism. At Fossombrone, on the eastern border of the Furlo depocenter, several hiatuses are present in the lowermost Tertiary and are attributed to sediment removal by intense bottom winnowing, turbidite generation, and perhaps slumping (Luterbacher and Premoli Silva, 1964; Chan and others, 1985; Montanari, in prep.).

The level of tectonic activity seems to have decreased drastically after these two major events. In the northern part of the basin, sporadic and very thin turbidites are present in the Furlo area, near Genga, and at Frontale. At Monte Cònero, turbidites in the R3 member are also very sporadic, suggesting a major retreat of the platform margin (Calesini, 1981; Stagnozzi, 1981).

Ypresian.—

Synsedimentary tectonic activity caused minor sediment sliding in the areas of Furlo, Arcevia, Genga, and Poggio San Vicino (Fig. 17F). Deformed intervals within the R4 member were observed also at Fosso del Mulino, Rocca Leonella, and at Monte Cònero. Foramininiferal turbidites are sporadic and very thin in the northern part of the Furlo depocenter (e.g., Pieralata section) and virtually absent in other areas of the distal paleobasin and at Monte Cònero. Few turbidites are found in the southern part of the paleobasin and are related to sediment mass movement on the retreating platform margin (Baldanza and others, 1982). This situation indicates that at the end of deposition of the Scaglia Rossa, the topographic irregularities of the sea floor were nearly leveled and tectonic activity was slowly fading.

Lutetian.—

The regional homogeneity of facies of the Scaglia Variegata Formation indicates that the paleobasin was again essentially flat and that tectonic movements were unimportant in the middle Eocene (Fig. 17G). The last seismic shock of the Late Cretaceous-Early Tertiary tectonic cycle in the Umbria-Marches Apennines is represented by a slump in the lower part of the Scaglia Variegata exposed in several sections in the Furlo anticline. This suggests that some topographic reliefs still existed around this intrabasinal depocenter, but the absence of turbidites indicates that relief was minor, perhaps barely perceptible.

Calcareous turbidites in the Scaglia Variegata and overlying Scaglia Cinerea are found only in the southern part of the paleobasin (Baumann, 1970; Monaco and others, 1987), where they are related to the proximity of the carbonate platform margin.

Late Cretaceous-Early Tertiary Plate Tectonics in the Western Tethys: The Response of the Scaglia Rossa Basin

The late Cretaceous resumption of tectonic movements in the Umbria-Marches Apennines occurred at the same time as a major plate-tectonic event in the western Tethys realm: the initiation of convergence between the European plate and the African plate. In this process, the Ligurian Ocean, which rifted in the Early Jurassic in response to the opening of the North Atlantic Ocean (Bernoulli and Lemoine, 1980; see also Fig. 18B), started to be consumed along the orogenic fronts of the Alps and the northern Apennines. The tectonic history of the convergence and collision between Africa and Europe along the Alpine front is extremely complex. The so-called paleo-Alpine orogenesis, which pre-dated the major uplift in the mid-Tertiary (meso-Alpine orogenic phase), started probably in the early Aptian and was punctuated by several tectonic phases (Trumpy, 1973).

Plate convergence in the Apenninic Ligurian Ocean probably started in the early Late Cretaceous (Treves, 1984). Clear evidence of subduction of oceanic crust under the European continent, however, dates from the late Campanian (Treves, 1984). The inception of subduction was followed by the evolution of an accretionary wedge and the deposition of flysch units in east-migrating foredeep basins. One of the most important aspects of the orogenesis in this part of Tethys is that, since the earliest time of plate convergence, the Apennines domain was distinct from the Alpine domain. The subduction planes dipped toward opposite directions in the respective domains: the Ligurian Ocean was subducting toward the south or southeast under the African plate along the Alpine orogenic front, whereas the same oceanic crust was subducting toward the west or northwest under the European plate along the Apenninic front. Bosellini (1981) documents the existence of a major Jurassic-Early Cretaceous crustal fracture zone in the central part of the Ligurian Ocean. According to Bosellini (1981), during Late Cretaceous-Paleogene time, this fracture evolved into a sinistral wrench fault, the so-called Emilia Fault, which would have separated the Alpine domain to the north from the Apennines to the south. A schematic picture of the paleogeographic and tectonic situation of the western Tethys in the late Cretaceous is shown in Figure 18A. The interesting aspect of this picture, despite the enormous uncertainties that still exist in such a paleogeographic reconstruction, is that the extending and subsiding Umbria-Marches basin was practically surrounded by compressional domains. The question to be answered now is whether the syndepositional tectonic events in the Scaglia Rossa basin are related to the paleo-Alpine phase or to the early subduction in the Apennines, or both. Chan and Montanari (1984) have noticed a fairly good correlation between major Scaglia Rossa events and paleo-Alpine orogenic phases (Fig. 19). In most parts of the Alps, folding and metamorphism were occurring, especially in the latest Cretaceous. Cretaceous tectonic phases, however, did not continue in to the Paleocene, which represents, in the Alpine domain, a time of relative tectonic quiescence that persisted until the Eocene (Trumpy, 1973). Similarly, the Umbria-Marches basin experienced a maximum of syndepositional tectonic activity at the transition between the Cretaceous and Tertiary. The peak of synsedimentary tectonism was then followed by a drastic decrease of activity in the mid-Paleocene.

On the other hand, a major tectonic event in the Apennine subduction complex occurred in the Campanian and is represented by the accretion of the Vara supergroup and consequent eastward shifting of the trench (Treves, 1984). This seems to correlate well with the time of initiation of turbiditic sedimentation in both proximal and distal areas of the Scaglia Rossa basin. The subduction process in the Apennines domain is punctuated by other accretion events, which lasted until the Paleocene (Treves, 1984; see also Fig. 19). Continental crust started to underthrust in the Eocene, whereas arc magmatism seems to have begun only in the Oligocene. The extensional regime in the Umbria-Marches region may have been generated by a time of "lithospheric pull" during the subduction of oceanic crust, which would have caused the Umbria-Marches continental crust to stretch and subside after reactivation of buried Jurassic normal faults. The sudden decrease of activity in the Eocene may be related to the initiation of subduction of the thinned continental crust. In conclusion, it seems reasonable to relate the extensional synsedimentary tectonism in the Umbria Marches paleobasin to the early phases of the Apennines orogenesis, whereas the apparent coincidence of paleo-Alpine tectonic phases with major Scaglia Rossa events suggests that there might have been a relation between the two orogeneses. Differences in the timing of uplift, structural complexity, and tectonic verges between the Alps and the northern Apennines are due to the original plate geometry and paleogeographic setting. Nevertheless, a single regional tectonic phase related to the reciprocal motion between Africa and Europe and the evolution of the Atlantic Ocean was likely the unique cause of Late Cretaceous and Early Tertiary tectonism in the western Tethys, in the Alps as well as in the northern Apennines.

CONCLUSIONS

A detailed basin analysis of the Scaglia Rossa sequence in the northeastern Apennines has provided evidence relevant to: (1) deep-water carbonate sedimentation, (2) the relation between the topography and structural geology of an epeiric basin during syndepositional extensional tectonism, and (3) the timing of major tectonic events during an important preorogenic phase occurring in the domain of the western Tethys.

(1) In the deep Scaglia Rossa basin, accumulation rates of pelagic sediment were controlled by syndepositional tectonic movements that triggered turbidites in distal areas of the paleobasin after liquefaction of unconsolidated pelagic muds. Turbidites, mostly composed of planktonic foraminifera and other sand-sized grains derived from remobilized intraformational pelagic mud, were deposited in tectonically controlled intrabasinal depocenters. The micritic mud suspended in the turbidity events was probably re-dispersed in distal areas of the paleobasin in the form of hemipelagic sediment and then homogenized with pre-existing pelagic sediment by the intense bioturbation. Relatively high con-

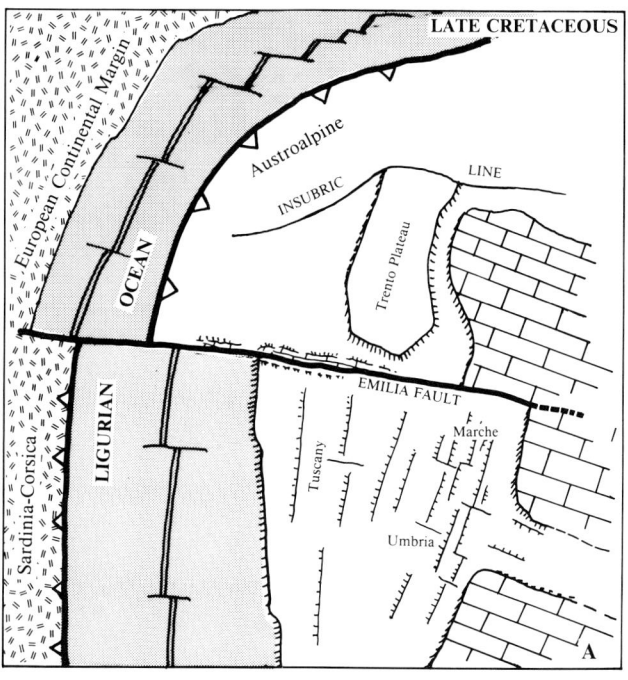

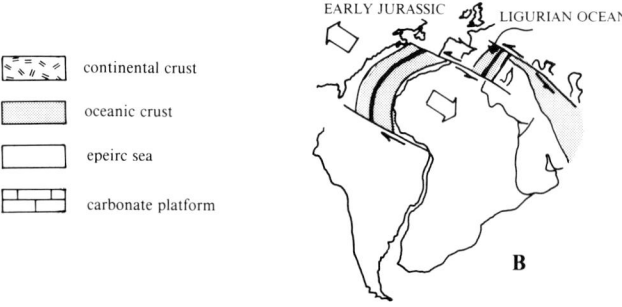

FIG. 18.—(a) Schematic representation of the tectonic setting in the Late Cretaceous Ligurian Ocean and adjacent epeiric seas (modified after Bosellini, 1981); (b) general tectonic setting of the opening of the western Tethys Ocean in the Early Jurassic.

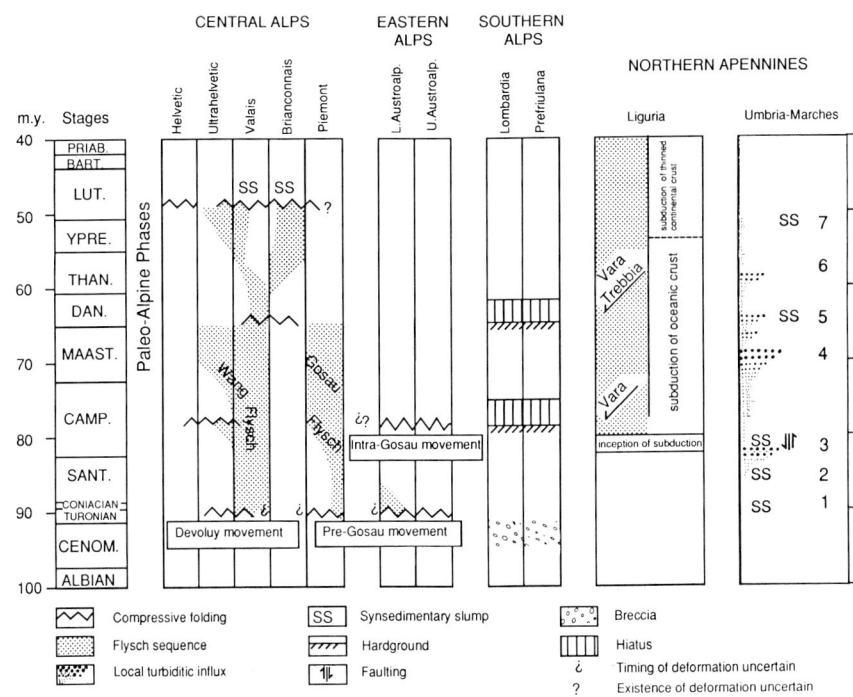

FIG. 19.—Sequence of seven tectonically controlled sedimentary events in the Umbria-Marches basin occurring during the compressive tectonic phases in the northwestern Apennines and in the Alpine domain.

centrations of planktonic foraminiferal tests were deposited in turbidite-free sequences (e.g., Acqualagna) downcurrent and near sites of turbiditic deposition (e.g., Furlo) in times of maximum turbidite activity, suggesting that bottom currents triggered by seismic events have winnowed the fine coccolith matrix of the pelagic sea floor, leaving the observed concentrations of foraminifera. On the other hand, in areas where the turbidites were probably generated (e.g., Fossombrone), the sequence is reduced in thickness and contains several sedimentary hiatuses, indicating remobilization of normal pelagic sediment.

(2) The topography of the paleobasin was characterized by intrabasinal depocenters and structural highs produced by gentle flexures of the sea floor. These topographic irregularities were generated by the reactivation of buried Jurassic normal faults during an extensional tectonic regime. The fault pattern, as in modern extensional basins, was probably characterized by listric- and planar-normal faults intersected by transfer faults. The comparison with the structural situation in the Central Graben of the North Sea suggests that differential vertical movements in the Scaglia Rossa basin were controlled by a sinuous detachment surface that was also responsible for substantial differentiation of the structural style in different parts of the paleobasin.

(3) The tectonic activity that controlled the sedimentary and topographic development of the Scaglia Rossa basin was probably a distant effect of the regional plate-tectonic situation in that part of Tethys Ocean. In the Late Cretaceous important deformational events were occurring in the Alps at the same time as the inception of subduction in the western part of the northern Apennines. Subsidence of the Scaglia Rossa basin, however, was likely a direct effect of the oceanic subduction along the Apennines orogenic front, which would have caused further stretching of a pre-faulted and thinned continental crust prior to the continent-continent collision between the Adriatic (African) plate and the Corso-Sardinian (European) plate.

ACKNOWLEDGMENTS

This research was supported by National Science Foundation Grant EAR-83-18660 and a grant from Chevron Overseas Petroleum, Inc. We thank Leonardo Polonara and Assessorato al Territorio of the Regione Marche for the logistic support in the field, and Richard Hay, Charlotte Brunner, David Bice, Mark Heckman, and Ric Fergeson for reviewing the manuscript.

REFERENCES

ACCORDI, B., 1966, La componente traslativa nella tettonica dell'Appennino calcereo laziale-abruzzese: Geologica Romana, v. 5, p. 351–406.
ALVAREZ, W., COLACICCHI, R., AND MONTANARI, A., 1985, Synsedimentary slides and bedding formation in Apennines pelagic limestones: Journal of Sedimentary Petrology, v. 55, p. 720–734.
———, ARTHUR, M. A., FISCHER, A. G., LOWRIE, W., NAPOLEONE, G., PREMOLI SILVA, I., AND ROGGENTHEN, W. M., 1977, Upper Cretaceous-Paleocene magnetic stratigraphy at Gubbio, Italy. V. Type section for the Late Cretaceous-Paleocene geomagnetic reversal time scale: Geological Society of America Bulletin, v. 88, p. 383–389.
———, AND LOWRIE, W., 1978, Upper Cretaceous paleomagnetic stratigraphy at Moria (Umbrian Apennines, Italy): Verification of the Gubbio section: Geophysical Journal of the Royal Astrophysical Society, v. 55, p. 1–17.
———, 1984, Magnetic stratigraphy applied to synsedimentary slumps, turbidites, and basin analysis: The Scaglia limestone at Furlo (Italy): Geological Society of America Bulletin, v. 95, p. 324–336.
ANGELUCCI, A., AND PRATURLON, A., 1968, Raddoppio tettonico della serie mesocenozoica nelle Gole del Celano (Appennino Centrale): Geologica Romana, v. 7, p. 431–446.
ARTHUR, M. A., 1976, Sedimentology of the Gubbio sequence and its

bearing on paleomagnetism: Memorie della Società Geologica Italiana, v. 15, p. 9–20.

———, AND FISCHER, A. G., 1977, Upper Cretaceous Paleocene magnetics stratigraphy at Gubbio, Italy. I. Lithostratigraphy and sedimentology: Geological Society of America Bulletin, V. 88, p. 367–371.

———, AND PREMOLI SILVA, I., 1982, Development of widespread organic carbon-rich strata in the Mediterranean Tethys in Schlanger, S. O., and Cita, M. B., (eds.,) Nature and Origin of Cretaceous Carbon-Rich Facies: Academic Press, London, p. 6–54.

BALDANZA, A., COLACICCHI, R., AND PARISI, G., 1982, Controllo tettonico sulla deposizione dei livelli detritici della scaglia cretacico-paleocenica (Umbria orientale): Renaliconti della Società Geologica Italiana, v. 5, p. 11–14.

BAUMANN, P., 1970, Micropaläontologische und stratigraphische Untersuchung der oberozänen-oligozänen Scaglia im Zentralen Apennin (Italien): Eclogae Geologicae Helvetiae, v. 63, p. 1133–1211.

BERNOULLI, D., AND LEMOINE, M., 1980, Birth and early evolution of the Tethys: The overall situation: 26th International Geological Congress, Paris, Colloquium C5, p. 167–179.

BICE, D. M., AND STEWART, K. G., 1985, Ancient erosional grooves on exhumed bypass margins of carbonate platforms: Examples from the Apennines: Geology, v. 13, p. 565–568.

BORSETTI, A. M., CATI, F., MEZZETTI, R., SAVELLI, C., AND TONI, G., 1984, Le intercalazioni vulcano-clastiche nei sedimenti oligo-miocenici dell'Appennino settentrionale e centrale: Giornale di Geologia, v. 45, p. 159–198.

BORTOLOTTI, V., PASSERINI, P., SAGRI, M., AND SESTINI, G., 1970, Development of the northern Apennines geosyncline: Sedimentary Geology, v. 4, p. 341–344.

BOSELLINI, A., 1981, The Emilia Fault: A Jurassic fracture zone that evolved into Cretaceous-Paleogene sinistral wrench fault: Bollettino della Società Geologica Italiana, v. 100, p. 161–169.

CALESINI, N., 1981, Rilevamento e studio sedimentologico della "Scaglia Rossa" del Monte Cònero (parte sud): Università degli Studi di Urbino, Istituto di Geologia, Laurea Thesis, p. 1–116.

CANTELLI, C., CASTELLARIN, A., COLACICCHI, R., AND PRATURLON, A., 1982, La scarpata tettonica mesozoica lungo il settore nord della "Linea Ancona-Anzio": Memorie della Società Geologica Italiana, v. 24, p. 149–153.

———, AND PRATURLON, A., 1978, Tettonismo giurassico lungo la "Ancona-Anzio" nel settore Monte Terminillo-Antrodoco: Geologica Romana, v. 17, p. 85–97.

CARBONE, F., PRATURLON, A., AND SIRNA, G., 1971, The Cenomanian shelf edge facies of Rocca di Cave (Prenestini Mts., Latium): Geologica Romana, v. 10, p. 131–198.

CARDELLINI, S., 1982, La formazione della "Maiolica" nelle Marche settentrionali: I. Rilevamento formazionale e biostratigrafia delle aree campione; II. Correlazioni regionali: Università degli Studi di Urbino, Istituto di Geologia, Laurea Thesis, p. 1–231.

CASTELLARIN, A., COLACICCHI, R., AND PRATURLON, A., 1982, Fasi distensive, trascorrenze e sovrascorrimenti lungo la "Linea Ancona-Anzio" dal Lias medio al Pliocene: Geologica Romana, v. 17, p. 161–189.

CENTAMORE, E., CHIOCCHINI, M., DEIANA, G., MICARELLI, A., AND PIERUCCINI, U., 1971, Contributo alla conoscenza del Giurassico nell'Appennino umbro-marchigiano: Studi Geologici Camerti, v. 1, p. 1–89.

———, CHIOCCHINI, M., JACOBACCI, A., MANFREDINI, M., AND MANGANELLI, V., 1980, The evolution of the Umbrian-Marchean Basin in the Apennine section of the Alpine orogenic belt (central Italy), in Cogne, J., and Slansky, M., Memoires Bureau de la Recherche Geologique et de Minieres, No. 108, p. 298–305.

CHAN, L. S., 1984, Paleomagnetic determination of the mechanism of rotation in the Umbria-Marches Apennines: Unpublished Ph.D. Dissertation, University of California, Berkeley, 250 p.

———, AND MONTANARI, A., 1984, Paleomagnetism and tectonic studies of the Scaglia Rossa pelagic limestones (Upper Cretaceous-Eocene) from the Umbria-Marches Apennines, Italy: EOS, American Geophysical Union, v. 65, p. 866.

———, AND ALVAREZ, W., 1985, Magnetic stratigraphy of the Scaglia Rossa: Implications for syndepositional tectonics of the Umbria-Marches basin, Italy: Rivista Italiana di Paleontologia e Stratigrafia, v. 91, p. 219–258.

Consiglio Nazionale Ricerche, 1986, Progetto Finalizzato di Geodinamica 4: Ricostruzioni paleogeografiche, paleotettoniche e applicazione ai giacimenti minerari; Lithofacies map of the Latium-Abruzzi and neighboring area, 1:250,000: Accordi, G., and Carbone, F., eds., Stabilmento Solomone, Roma.

COLACICCHI, R., 1966, Le caratteristiche della facies abruzzese alla luce delle moderne indagini geologiche: Memorie della Società Geologica Italiana, v. 5, p. 1–18.

———, 1967, Geologica della Marsica orientale: Geologica Romana, v. 6, p. 189–316.

———, AND BALDANZA, A., 1986, Carbonate turbidites in a Mesozoic pelagic basin: (Scaglia Formation, Apennines). Comparison with siliciclastic depositional models: Sedimentary Geology, v. 48, p. 81–105.

———, PASSERI, L., AND PIALLI, G., 1970, Nuovi dati sul Giurese umbro-marchigiano ed ipotesi per un suo inquadramento regionale: Memorie della Società Geologica Italiana, v. 9, p. 839–874.

———, AND PRATURLON, A., 1965a, Il problema delle facies nel Giurese della Marsica nord-orientale: Bollettino della Società Geologica Italiana, v. 84, p. 55–66.

———, AND ———, 1965b, Stratigraphical and paleogeographical investigations on the Mesozoic shelf-edge facies in eastern Marsica: Geologica Romana, v. 4, p. 89–118.

COLTORTI, M., 1979, Geologia della regione di M. Petroso-M. Murano (Appennino Marchigiano): Annali dell'Università di Ferrara, Scienze Geologiche e Paleontologiche, v. 7, p. 1–36.

———, AND BOSELLINI, A., 1980, Sedimentazione e tettonica nel Giurassico della dorsale Marchigiana: Studi Geologici Camerti, v. 6, p. 189–316.

CRESCENTI, U., 1969, Stratigrafia della serie calcarea dal Lias al Miocene nella regione marchigiano-abruzzese (Parte I. Descrizione delle serie stratigrafiche): Memorie della Società Geologica Italiana, v. 8, p. 155–204.

———, CROSTELLA, A., DONZELLI, G., AND RAFFI, G., 1969, Stratigrafia della serie calcarea dal Lias al Miocene nella regione Marchigiano-Abruzzese (Parte II. Lithostratigrafia, biostratigrafia, paleogeografia): Memorie della Società Geologica Italiana, v. 8, p. 343–420.

D'ARGENIO, B., AND ALVAREZ, W., 1980, Stratigraphy evidence for crustal thickness changes on the southern Tethyan margin during the Alpine cycle: Geological Society of America Bulletin, v. 91, p. 681–689.

DE BOER, P. L., 1983, Aspects of middle Cretaceous pelagic sedimentation in southern Europe: Geologica Ultraiectina, No. 31, p. 1–112.

DUNHAM, R. J., 1962, Classification of carbonate rocks according to depositional texture, in Ham, W. E., ed., Classification of Carbonate Rocks: American Association of Petroleum Geologists Memoir 1, p. 108–121.

ETHERIDGE, M. A., BRANSON, J. L., AND STEWART-SMITH, P. G., 1985, Extensional basin-forming structures in Bass Strait and their importance for hydrocarbon exploration: Australian Petroleum Exploration Association Journal, v. 25, p. 344–361.

GALDENZI, S., 1986, Megabrecce giurassiche nella dorsale marchigiana e loro implicazioni paleotettoniche: Bollettino della Società Geologica Italiana, v. 105, p. 371–382.

GIBBS, A. D., 1984, Structural evolution of extensional basin margins: Journal of the Geological Society of London, v. 141, p. 609–620.

HERBERT, T. D., AND FISCHER, A. G., 1985, High-resolution history of carbonate and silica accumulation rates in a mid-Cretaceous pelagic core: Geological Society of America, Abstracts with Programs, v. 17, p. 608.

KOOPMAN, A., 1983, Detachment tectonics in the central Apennines: Geologica Ultraictina, No. 30, p. 1–155.

LAVECCHIA, G., 1981, Appunti per uno schema strutturale dell'Appennino Umbro-Marchigiano. 3. Lo stile deformativo: Bollettino della Società Geologica Italiana, v. 100, p. 271–278.

LOWRIE, W., AND ALVAREZ, W., 1977, Upper Cretaceous magnetic stratigraphy, in Upper Cretaceous-Paleocene Magnetic Stratigraphy at Gubbio, Italy: Geological Society of America Bulletin, v. 88, p. 367–371.

———, AND ———, 1981, One hundred million years of geomagnetic polarity history: Geology, v. 9, p. 392–397.

———, AND ———, 1984, Lower Cretaceous magnetic stratigraphy in Umbrian pelagic limestones: Earth and Planetary Science Letters, v. 71, p. 315–328.

———, NAPOLEONE, G., PERCH-NIELSEN, K., PREMOLI SILVA, I., AND

TOUMARKINE, M., 1982, Paleogene magnetic stratigraphy in Umbrian pelagic carbonate rocks: The Contessa sections, Gubbio: Geological Society of America Bulletin, v. 93, p. 414–132.

LUTERBACHER, H. P., AND PREMOLI SILVA, I., 1964, Stratigrafia del limite Cretacico-Terziario nell'Appennino centrale: Rivista Italiana di Paleontologia e Stratigrafia, v. 70, p. 67–128.

MONACO, P., NOCCHI, M., AND PARISI, G., 1987, Analisis stratigrafica e sedimentologica di alcune sequenze pelagiche dell'Umbria sud-orientale dall'Eocene inferiore all'Oligocene inferiore: Bollettino della Società Geologica Italiana, v. 106, p. 71–91.

MONTANARI, A., 1979, Lineamenti sedimentologici della "Scaglia Bianca" e della "Scaglia Rossa" nelle Marche settentrionali: Università degli Studi di Urbino, Istituto di Geologia, Laurea Thesis, p. 1–289.

―――, 1985, Cenomanian anoxic foreslope inferred from turbiditic cherts in the pelagic basin of the northern Apennines, Italy: Geological Society of America, Abstracts with Programs, v. 17, p. 667.

―――, DRAKE, R., BICE, D. M., ALVAREZ, W., CURTIS, H. G., TURBIN, B. D., DEPAOLO, D. J., 1985, Radiometric time scale for the upper Eocene and Oligocene based on K/Ar and Rb/Sr dating of volcanic biotites from the pelagic sequence of Gubbio, Italy: Geology, v. 13, p. 596–599.

―――, STEWART, K., BICE, D., AND ALVAREZ, W., 1983, Apennine foldbelt tectonics. 3. Reactivation of fault-block mosaics: EOS, American Geophysical Union, v. 64, p. 861.

MONTADERT, L., CHARPEL, O., ROBERTS, D., GUENNOC, P., AND SIBUET, J., 1980, Northeastern Atlantic passive margins: Rifting and subsidence processes, in Talwani, M., Hay, W. W., and Ryan, W. B. F., eds., Deep Drilling Results in the Atlantic Ocean: Continental Margins and Paleoenvironment: American Geophysical Union, Washington, D.C., p. 153–186.

NAPOLEONE, G., 1980, Lo sviluppo della stratigrafia magnetica nella sezione tipo di Gubbio e sua incidenza nella ricerca paleomagnetica: Memorie della Società Geologica Italiana, v. 21, p. 289–299.

―――, PREMOLI SILVA, I., HELLER, F., CHELI, P., COREZZI, S., AND FISCHER, A. G., 1983, Eocene magnetic stratigraphy at Gubbrio, Italy, and its implications for Paleogene geochronology: Geological Society of America Bulletin, v. 94, p. 181–191.

PREMOLI SILVA, I., 1977, Upper Cretaceous-Paleocene magnetic stratigraphy at Gubbio, Italy. II. Biostratigraphy: Geological Society of America Bulletin, v. 88, p. 371–374.

―――, NAPOLEONE, G., AND FISCHER, A. G., 1974, Risultati preliminari sulla stratigrafia della Scaglia cretaceo-paleocenica della sezione di Gubbio (Appennino centrale): Bollettino della Società Geologica Italiana, v. 93, p. 647–659.

―――, ―――, AND ―――, 1980, La sezione magnetostratigrafica di Gubbio: Indagini nella storia del Cretacico-Paleogene: Memorie della Società Geologica Italiana, v. 21, p. 301–311.

―――, PAGGI, L., AND MONECHI, S., 1976, Cretaceous through Paleocene biostratigraphy of the pelagic sequence at Gubbio, Italy: Memorie della Società Geologica Italiana, v. 2, p. 21–32.

ROGGENTHEN, W. M., AND NAPOLEONE, G., 1977, Upper Maastrichtian-Lower Paleocene magnetic stratigraphy, in Upper Cretaceous-Lower Paleocene magnetic stratigraphy at Gubbio, Italy: Geological Society of America Bulletin, v. 88, p. 374–377.

SCHWARZACHER, W., AND FISCHER, A. G., 1982, Limestone bedding and perturbations of the earth's orbit, in Einsele, G., and Seilacher, A., eds., Cyclic and Event Stratification: Springer-Verlag, Berlin, p. 72–95.

STAGNOZZI, C., 1981, Rilevamento e studio sedimentologico della "Scaglia Rossa" del Monte Cònero (parte nord): Università degli studi di Urbino, Istituto di Geologia, Laurea Thesis, p. 1–139.

TREVES, B., 1984, Orogenic belts as accretionary prisms: The example of the Northern Apennines: Ofioliti, v. 9, p. 577–618.

TRUMPY, R., 1973, The timing of orogenic events in the Central Alps, in De Jong, K. A., and Scholten, R., eds., Gravity and Tectonics: John Wiley and Sons, New York, 229–252.

VAN GRASS, G., VIETS, T. C., DE LEEUW, J. W., AND SHEMK, P. A., 1983, A study of the soluble and insoluble organic matter from the Livello Bonarelli, a Cretaceous black shale in the central Apennines, Italy: Geochimica et Cosmochimica Acta, v. 47, p. 1051–1059.

WONDERS, A. A. H., 1980, Middle and Late Cretaceous planktonic foraminifera of the western Mediterranean area: Utrecht Micropaleontological Bulletin, v. 24, p. 1–157.

SUBJECT INDEX

SUBJECT INDEX

A

Accretionary margin, 86, 88, 90
Aggradation, 53, 63, 65, 70, 75, 90, 91, 93, 110, 175, 182, 213, 214, 223, 228, 229, 231, 252, 275, 284, 287, 296, 305, 317, 334, 337, 340, 343, 348, 350, 367
Algal buildup, 44, 129, 141, 153, 154
Algal-cryptalgal laminites, or microbial laminites, 48, 50, 83, 102, 110, 152
Allocyclic deposition, 141, 156
Almalgamated conglomerate, 129, 131
Ammonitico Rosso, 7, 8, 9, 119
Apennines, 8, 379–381, 391, 392, 395–397
Appalachian orogen and miogeocline, 39, 50, 51, 52, 53, 120, 121, 123, 124, 142, 147, 148, 151, 213
Archaeocyathids (ans), 107–109, 112, 117, 127, 149
Archean, 80, 81, 92, 94, 101
Australian continental margin, 233, 234, 245
Autocyclic deposition, 5, 141, 156

B

Backstepping, 5, 59, 63, 65, 66, 70, 71, 75, 93, 117, 132, 149, 155, 188, 192, 193, 195, 199, 249, 305, 312, 358, 360, 365, 370, 375–377
Bahamas, 156, 236, 245, 259, 261, 285, 286, 302, 333, 339
Banks (platform), tops and shoals of banks-platforms, 32, 187, 214, 290, 295, 300, 302, 334, 347
Basin analysis, 110, 281, 317, 365, 394, 397
Basinal facies, 4, 101, 107, 117, 189, 192, 222, 279, 284, 285, 290, 308, 312, 315, 368, 371–373, 379
Bauxite, 8, 82, 83
Bioherms and lithistid sponge type bioherm, 87, 88, 89, 152, 153, 156, 159, 161, 211, 252, 254
Blina oil field, 195, 197, 198
Bone Springs formation, 275–277, 281–283, 286
Breccias (megabreccia), 84, 116, 117, 119, 160, 162, 181, 209, 279, 299, 300, 301, 331, 332, 337
Buildups, 27, 30, 31, 32, 86, 135, 150, 169, 182, 331
Buildups, carbonate (general), 27, 323, 337
Buildups, geometry, 323
Buildups, mudmounds (Waulsortian), 169, 203, 207, 208
By-pass margins on platforms, 86, 90

C

Calcarenites-grainstones (see grainstone and calcarenites)
Calcirudite, 171, 204
Cambrian, 39, 49, 56, 107, 114, 119–121, 155, 162
    Cambro-Ordovician boundary, 107, 129, 136, 142, 147, 151
Canada, 51, 63
Canadian Rocky Mountains, 39, 49, 51, 52, 56, 57
Canning basin, 187, 199, 200
Carbonate-growth potential and carbonate production, 28, 32, 69, 70, 75, 93, 132
Carboniferous (Pennsylvanian-Mississippian), 211, 307, 309
"Catch up" buildup growth, 69
Cenozoic, 233, 249, 302, 339, 345, 366
Chert, 80, 81, 85, 90, 114, 136, 205, 208, 293, 297, 308, 381
Climate controls on sedimentation, 9, 120, 123, 162, 181, 233, 250, 255, 256, 314, 330, 377
Computer simulation and modelling, 27, 29, 30, 44, 63, 147, 158–160
Corals (see reefs)
Cordilleran orogen, 39
Cretaceous, 5, 6, 22, 142, 198, 213, 214, 233, 236, 340, 345, 350, 365, 366, 370, 371, 375, 379, 382, 395, 397
    Mid-Cretaceous unconformity, 17, 18, 354
Cyclicity-cycles, 5, 84, 93, 112, 114, 129, 138, 140, 149, 154, 156, 195, 269, 270, 305, 312, 314, 317, 323, 326–328, 337, 345, 361, 370
    Fifth order cycles, 85, 327, 329–331, 334, 335
    Fourth order eustatic cycles (parasequences), 63, 69, 74, 94, 139, 141, 323, 328–330, 331, 334, 335, 337
    Meter scale sequences or cycles, 85, 127, 139, 141, 147, 150, 327, 329–331, 334, 335, 337
    Milankovich cycles (see Milankovich)
    Peritidal cycles, 128, 141, 161, 297
    Second order cycles, 139, 141, 155, 348, 354
    Shoaling or shallowing upward cycles, 40, 63, 94, 112, 131, 141, 153, 171, 195, 199, 296, 305, 313, 316, 352, 358
    Third order cycles, 48, 90, 93, 94, 139, 141, 147, 150, 153, 154, 156, 157, 159, 290, 323, 330, 331, 333, 335, 337, 348, 349, 354

D

Debris flow aprons, deposits, 16, 84, 114, 116, 117, 119, 136, 187, 189, 191, 200, 209, 281, 285–287, 290, 297, 301, 303, 314, 315, 373
Delaware basin, 275, 276, 281, 284, 286, 287, 289, 290, 297, 299, 301, 305, 312
Delithification of rock types and delithification thickness, 47, 48, 375
Depocenters, 147, 154, 155, 157, 159, 391, 402, 407
Depositional phases or sequences or facies, 365, 366, 368, 370, 376
Devonian, 59, 148, 187, 199, 205, 339
Devonian "Great Barrier Reef of Australia", 187
Dololaminites, 127, 131, 282
Dolomites and dolomitization, 6, 81, 82, 87, 97, 114, 116, 128, 129, 131, 149, 152, 153, 158, 171, 187, 199, 208, 276, 278, 281, 283, 293–295, 297, 300, 302, 306, 308, 326, 327, 331, 380
Drowned banks or platforms (see platforms)
Drowning processes, 7, 63, 70, 84, 88, 150, 312
Drowning unconformity (see unconformity)
The Dolomites, 6, 323, 324, 331, 336

E

Earthquakes, 394, 396
Epiphyton, 108, 109, 154
Eroded (bare platform top) or erosion, 18, 19, 191, 193, 289, 290, 293, 297, 299, 300, 302, 303, 312, 353, 358
Eustacy and eustatic fluctuations, 27, 39, 75, 93, 94, 109, 120, 139, 141, 155, 178, 188, 192, 196, 200, 223, 229, 233, 255, 289, 296, 306, 323, 326, 327, 330, 333, 335, 348, 350, 353, 361, 365, 373, 377
Eustatic cycles, 3, 34, 50, 56, 139, 140, 197, 199, 335, 357, 375
Evaporites, 22, 81, 83, 94, 97, 110, 120, 132, 163, 305, 308, 316, 319, 380

F

Fans of calcite-aragonite druse, 95, 96
Faulting, (see synsedimentary tectonics and also block faulting tectonics)
Fenestral fabric, 127, 171, 175, 296, 297
Fischer plots or diagrams, 57, 157, 158–160, 327, 334, 335
Flexural bulge and bending, 32, 34, 58, 83, 87, 92, 136, 142, 155, 253, 360
Florida platform and Florida current, 259, 260, 268–271, 302
Foreland basins and platforms, 84, 93, 133, 233, 253, 254
Fourth order cycles (see cycles)
Frasnian-Famennian, mass extinction, 194–201

G

Geohistory curves, 197, 226, 249, 386
"Give up" cessation of buildup growths, 69, 73, 213, 215
Glacial and cool climate controls, 9, 140, 162, 270, 286, 308, 313, 329
Grainstone and calcarenites, 48, 50, 83, 86, 89, 116, 125, 127, 133, 152, 156, 161, 171, 203, 204, 208, 215, 217, 230, 279, 309, 325, 326, 331, 333, 337, 353, 367, 396
Grand Cycles, 40, 49, 50, 56, 90, 127, 129, 131, 139, 140, 141, 152, 157, 161, 181
Grayburg Formation, 289–303
Great Bahama Bank, 268, 339, 340, 341, 345, 347, 350, 352, 353
Great Barrier Reef of Australia, 169, 233, 236–239, 242, 245, 247, 250–252, 254, 255
Great Basin, 39, 43, 50, 51, 57, 59
Guadalupe Mountains, 289–291, 296–298, 303

H

Hardgrounds, 7, 16, 22, 63, 65, 71, 74, 126, 152, 160, 175, 279, 325, 333
Hawks Bay event, 120, 127, 128, 138, 162
Hinge lines, 147, 148, 155, 203, 210
House Embayment, Utah, 181

I

Iapetus ocean, 121, 123, 138
Interplatform basin, 27, 32
Intraplatform seaway or intrashelf basin, or intracratonic basins, 22, 27, 32, 86, 91, 141, 147, 151, 152, 169, 172, 177, 181, 182, 340, 348, 350
Intertidal sediments (see peritidal sediments)
Intraformational limestone conglomerate, 136, 391
Isostacy, 8, 27, 29, 32, 34, 44
Italy, 323, 389

J

Judy Creek, Alberta, 63, 66, 71, 76
Jurassic and Liassic, 5, 7, 11, 19, 21, 22, 119, 142, 198, 213, 214, 233, 236, 371, 379, 380, 391, 395

K

Karst, paleokarst, and karst plane, 8, 11, 19, 82, 93, 94, 117, 120, 127, 131, 133, 134, 136, 160, 173, 178, 179, 181, 290, 313, 314, 353, 370, 371, 373, 375
"Keep up" buildup growth, 69, 213, 215, 229

L

Lagoonal sedimentation and cyclicity, 65, 67, 71, 85, 91, 177, 308, 326, 367, 371
Lag time in cycles or lag surfaces, 22, 70, 71, 157, 159, 160, 230, 370
Laminites (general), 98, 109, 112, 114, 116, 119, 141, 152–154, 171, 206, 208, 276, 279, 280, 282, 294, 299, 367, 381
Latemar buildup, 323–325, 327, 328, 330, 334–337
Lead zinc, MVT deposits, 117, 187, 201
  Sphalerite mineralization, 119, 131
Leeward margin, 5, 65, 74, 340, 343, 344, 347
Leonardian (Permian) facies, 275, 283, 284, 286, 287, 305, 306, 308, 309, 313, 315, 316
Limestones, 116, 125, 127, 149, 204, 327
  Black, 116, 119
  Conglomeratic, 131, 132, 136, 141, 156, 161, 308
  Hemipelagic, 131, 133, 136, 404
  Nodular, 40, 48, 65, 117, 119, 127, 129, 133, 149, 204, 206, 333, 337
  Non-cyclic, 152
  Parted, 48, 131, 141
  Ribbon, 125, 131–133, 136, 152–154, 160, 161, 171, 175, 371
Lithistid sponge bioherm, 133, 136
Lowstand wedge, 153, 198, 255, 305, 358
Luconia, central, 27

M

Massiccio Limestone, 379, 380, 391
Mediterranean, 7, 9, 11
Megabreccia, 5, 6, 90, 114, 153, 187, 276, 278, 281, 283, 285, 286, 308, 313, 331, 333, 372, 380, 391
Megaconglomerate, 133, 134, 140
Midland basin, north platform, 275, 276, 286, 287, 305, 306, 308, 312, 313, 317, 318
Milankovitch cycles or sea level fluctuation, 5, 65, 94, 150, 158–162, 196, 329, 330, 337
Miocene reefs, 237, 240, 244, 249, 251, 252, 254, 255, 345
Mound, 108, 132, 140, 141, 261, 270, 353
  Calathid, 169, 171, 172, 175, 182,
  Coral, 131
  Mudmound complex, 149, 155, 171, 172, 175, 217
  Thrombolite, 131, 134

N

Natuna platform, South China Sea, 353, 355, 358, 360, 370, 371
Neptunian dykes, 9, 114, 117, 119, 120, 189, 191, 192, 195, 200, 325, 391
Northwestern Shelf-Delaware basin, 275, 283, 286
Nutrient stress, 68, 74, 100, 229

O

Offlap, 124, 127, 139, 142, 309, 311
Olistolith, 136
Oncolite, 110, 125, 136, 171, 175, 177, 215
Oolite-oncolite grainstone, 4, 5, 6, 7, 9, 19, 20, 48, 81, 87, 88, 91, 97, 109, 110, 112, 116, 119, 120, 123, 125–130, 140–142, 149, 152, 154, 171, 177, 182, 268, 296, 299, 302, 308
Ordovician, 107, 120, 121, 133, 155, 167
Oversteepened platform margins (see platform margin)
Oxygen minimum or anoxic zone, 8, 131, 132, 169

P

Paleobathymetric High, 205, 209, 211
Paleo-caves, paleokarst (see karst)
Paleolatitude-paleoclimate, 249, 250
Parasequences (see cyclicity-Fourth Order)
Passive continental margin, 8, 22, 39, 40, 55, 57, 92, 123, 138, 147, 148, 155, 233, 252, 255, 379
Pelagic sediments, 7, 9, 17, 18, 95, 270, 371, 379, 381, 383, 387, 389, 396
Peripheral bulge (see flexural bulge)
Periplatform talus or sediment, 40, 90, 93, 110, 150, 162
Peritidal carbonate environment and sediments, 4, 40, 101, 110, 127, 129, 131–133, 135, 139–142, 150, 151, 159, 171, 177–181, 296, 297, 305, 308, 314, 316
Permian, 198, 289, 290, 301–303, 305, 306, 308, 312, 313, 316, 339
Phosphatic pebble conglomerates, ooids, and sandstones, 19, 127, 132, 251
Pinnacle reefs and banks, 32, 34, 88, 89, 193, 230, 245, 253, 254
Pisolites, 89, 125, 327
Platforms (carbonate), 3, 85, 89, 110, 116, 138, 142, 178, 179, 203, 213, 239, 247, 255, 317, 324, 340, 353, 363, 367, 371, 380
  Carbonate rimmed and non-rimmed shelves, 101, 107, 117, 129, 135, 140, 143, 150–152, 155, 156, 159, 182, 188, 199, 229, 236, 237, 259, 293, 305, 309, 314, 317, 319
  Collapse margin, 133, 188, 191, 195
  Drowned platform (foundered platform), 15, 70, 117, 123, 133, 136, 138, 141, 143, 155, 187, 188, 192, 193, 195, 200, 213, 230, 309, 319, 360
  Evolution (or platform development in general), 4, 107, 191, 195, 203, 231, 233, 318, 358
  Marginal rim, 20, 22, 100, 119, 129, 229, 344, 353, 359, 367
  Nullara cycle, 191, 193, 195, 196, 198, 199, 200
  Pillara cycle, 191–193, 196, 198–201
  Pinnacle reef evolution (see also pinnacle reefs), 188, 192, 195, 213, 223
  Facies, 169, 324, 325, 331, 367
  Growth potential, 19, 39, 138
  Initiation, 4, 9
  Margin and facies, 171, 177, 181, 188, 191, 213, 217, 223, 259, 325, 336
  Platform to basin relief, 150, 297, 302, 337, 358
  Top of platform shoals, 70, 156, 171, 175, 177
PreCambrian, 79, 80, 94, 143, 147, 187, 192
Prograding of sediments, 5, 15, 18, 63, 65, 66, 70, 75, 87, 91, 93, 94, 128, 129, 133, 140, 141, 160, 168, 177, 210, 213, 214, 217, 223, 227, 231, 241, 247, 254, 259, 261, 263, 267, 270, 286, 287, 296, 301, 305, 312, 315, 317, 318, 323, 324, 339, 340, 342–345, 347, 348, 350, 352, 358, 367, 371, 373, 376
Prograding of oncolite complex, 168
Proterozoic, 39, 80, 81, 83, 85, 94, 120, 138, 142, 155
Proto-Atlantic (see also Iapetus), 123
Pyrenees, 365, 370, 376, 387

Q

Quaternary platform development, 259
Quaternary sea level fluctuations, 270, 328, 329

R

Radiolarite, 7, 378
Ramp model, 65, 75, 85, 86, 100, 101, 108, 119, 147, 149, 150, 154,

155, 175, 203, 209–211, 289, 290, 296, 309, 313–315, 317, 319, 373
Rates of sedimentation or accumulation, 63, 66, 93, 109, 121, 139, 142, 227, 229, 267, 334, 387
Redbeds, 4, 149, 150, 155, 228
Reef, 98, 99, 132, 169, 194, 200, 208, 211, 236, 245, 250, 270, 308, 312, 344, 353
  Barrier, 86, 101, 254
  Coral, 132, 169, 194, 200
  Fringing, 344, 358
  Margin, 65, 67, 192, 211, 333, 344
  Talus, 66, 85
Renalcis, 108, 109, 131, 132, 154
Rhythmites (laminites and bedding rhythms), 85, 89, 94, 381
Rifting, 4, 8, 11, 32, 39, 44, 81, 83, 93, 121, 124, 139, 147, 148, 155, 233, 236, 247, 254, 353, 360
Rub al Khali basin, 28

S

Sandstone (quartz), 127, 150, 153, 157, 158, 177–181, 275, 280, 283, 294, 296, 299, 310, 365, 366, 369, 373, 383
  By-passing, 228
  Cross-bedded, 109, 125, 127, 280, 296, 298, 299, 302, 369
  Eolian, 169, 316, 369
  Fluvial, 108, 125, 316
  Marine pure quartzose, 39, 81, 153, 167, 290, 294, 302
Sardinia, 107, 119
Sauk Sequence, 138, 139
Sea level curves, 27, 28, 32, 63, 69, 74, 197, 200, 223, 226, 345, 349, 353, 357, 366, 369, 375, 376
Sea level fluctuations (relative sea level rise and fall), 15, 27, 35, 116, 131, 139, 140, 147, 175, 178, 192, 199, 213, 227, 230, 233, 251, 254, 270, 286, 287, 305, 306, 314, 316, 319, 323, 327, 333, 335, 344, 345, 350, 353, 365, 375, 377
  Eustatic fluctuations (see also eustatic cycles), 27, 39, 55, 63, 109, 120, 123, 140, 142, 155, 157, 161, 175, 188, 275, 289, 295, 305, 335
  Cycles of sea level change (see also cyclicity), 150, 153, 269, 270
  High frequency fluctuations, 5, 28, 259, 269, 270
  High stands of sea level, 5, 23, 142, 254, 259, 270, 271, 285, 286, 312, 314, 315, 317, 333, 344, 348, 349, 353, 358, 376
  Low stands of sea level, 5, 15, 19, 83, 229, 252, 259, 270, 285–287, 296, 305, 313–315, 333, 344, 348, 353, 371
  Rapid rises of sea level, 4, 5, 7, 27, 29, 63, 65, 66, 72, 74, 76, 140, 175, 187, 192, 193, 195, 196, 199, 229, 230, 287, 312, 355
Sea water composition-changes, 94
Sediment production, 132, 309, 317, 348, 361
Seismic stratigraphy, 22, 23, 187, 198–200, 215, 218, 220, 231, 236, 240, 259, 261, 263, 283, 302, 310, 344
Seismic unconformity (see unconformity)
Sequence boundary, 22, 23, 57, 261, 285, 289, 290, 296, 303, 344, 346, 353, 359, 365, 370, 371
Sequence stratigraphy, 15, 57, 262, 345, 352, 353
Shale, 150, 161, 205, 379
  Black, graptolitic black shale, 7, 136, 167, 169, 171, 308, 309, 317, 368
  Red, 131, 132, 150, 308, 369, 373
Shelf margins, 119, 147, 149, 171, 203, 210, 211, 247, 275, 285, 286, 296, 297, 305, 312, 314, 315, 359
Siliciclastic burial of platforms, 17, 21, 23, 136, 139, 142, 193, 230, 233, 253–255, 270
Siliciclastic influence on carbonates, 58, 68, 100, 108, 124, 191, 213, 223, 227, 231, 241, 247, 286, 302, 308, 316, 319, 354, 379, 393
Slide scars on platform margins, 6, 19, 23, 195
Slope angle of platform margins (increasing with development), 5, 16, 19, 188, 195, 216, 286, 362
Slope apron, 6, 114, 131, 275, 308, 331
Slope to basin facies (foreslope), 160, 171, 189, 190, 217, 222, 241, 259, 275, 282, 289, 294, 324, 331, 367, 371, 372
Slopes, carbonate sediments, 15, 114, 116, 140, 156, 175, 177, 191, 259, 291, 284, 368
Slump folds, slumping, 109, 112, 114, 116, 117, 119, 131, 136, 175, 289, 299, 301, 303, 367, 379–381, 383, 387, 388, 390, 393, 394

Southern Alps, 323, 339
Spain, 121, 365, 366
St. George unconformity, 133, 136
St. Lawrence platform, 123
Starved sedimentation or starved basin, 7, 18, 20, 150, 323, 333, 355
Storm events, 148, 152, 154, 156, 181, 182, 259, 270
Stromatolites, 40, 79, 80–85, 87–89, 96–98, 110, 112, 127, 128, 131, 194, 195
Stromatoporoids, 64, 65, 129, 194, 196, 200, 214
Subaerial exposure, 8, 21, 22, 27, 63, 65, 93, 94, 125, 131, 133, 141, 187, 193, 227, 301, 306, 312, 315, 325, 326, 330, 331, 333–335, 353, 365, 366, 372, 377
Submarine cementation, 15, 188, 200, 217, 297, 305, 308, 314, 315, 318, 325, 333
Submarine erosion, 290, 297, 301–303
Subsidence, 28, 139, 142, 192, 196, 229, 233, 247, 249, 254, 256, 306, 309, 328, 353, 360, 361, 365, 373
  Rates of subsidence, 55, 56, 109, 120, 121, 147, 148, 150, 155, 158, 159, 213, 223, 286, 287, 306, 309, 313, 316, 317, 330, 340, 357, 375, 376
  Regional subsidence, 155, 178, 275, 379
  Jerky or sporadic subsidence, 155, 156
  Tectonic subsidence, 28, 41, 55, 57,
  Thermal subsidence (cooling), 39, 47, 53, 56, 155, 229, 365, 373–375, 377
Swan Hills platform, 64
Synorogenic sediment, 133, 136, 393
Synsedimentary tectonics, 123, 142, 365, 373, 383, 384, 386, 389, 390, 396

T

Talus, 66, 85, 89, 154, 333, 337, 344
Tectonics, 108, 116, 117, 120, 123, 141, 142, 178, 182, 199, 203, 209, 233, 305, 308, 323, 350, 357, 365, 376, 377, 379, 390
  Block-faulting, 209, 210, 241, 245, 246, 353, 357, 379, 380, 391
  Control on sedimentation, 5, 69, 92, 108, 144, 155, 354
  Extensional, 181, 379, 383, 392, 393, 396, 397
  Retreat of platforms (see backstepping)
  Subsidence, 63, 69, 93, 121, 139, 147, 148, 157, 175, 229, 287, 324, 334, 374
Tepees, 326, 327, 329, 330, 333–337
Tethys, 3, 4, 7, 9, 119, 121, 379, 395, 397
Thrombolites (see also mounds), 40, 53, 130, 131, 132, 152, 154, 217
Tidal flat caps, 141, 161
Tidal flat facies (see also peritidal), 67, 85–87, 90, 96, 101, 114, 116, 125, 128, 156
Tidal flat island, 141
Tippecanoe sequence, 140
Trento platform, 8, 9, 11
Triassic, 5, 6, 119, 198, 323, 330, 335, 339, 380
Turbidite, 5, 15, 19, 20, 126–130, 150, 151, 189, 191, 192, 323, 332, 333, 337, 372–374, 379, 381, 384, 386–389, 393–396
Turbidity, 16, 68, 82, 161, 301, 399

U

Unconformity, 6–8, 40, 49, 69, 140, 142, 148, 150, 153, 155, 158, 193, 204, 209, 255, 259, 261, 270, 290, 301, 303, 313, 374
  Drowning, 15, 17, 19, 21, 22, 309
  Lowstand, 17, 21, 23, 241
  Seismic, 15, 23
  Unconformity bounded sequences, 129, 263

V

Volcanics, 9, 15, 81, 82, 124, 148, 151, 323, 383

W

Whiterockian Series, 133, 142, 167, 175, 177, 178
Windward margin of platform, 5, 65, 74, 340
Wolfcampian Series, 305, 306, 308–310, 312–315, 318

Y

Yeso Formation, 275, 276, 280–282, 284–286

Z

Zoophycos, 205, 207, 208